Theory
and Practice
of Aircraft
Performance

AEROSPACE SERIES LIST

Theory and Practice of Aircraft Performance

AJOY KUMAR KUNDU

Queen's University Belfast
Belfast
UK

MARK A. PRICE

Queen's University Belfast
Belfast
UK

DAVID RIORDAN

Bombardier
Belfast
UK

This edition first published 2016
© 2016 John Wiley & Sons, Ltd

Registered Office

John Wiley & Sons, Ltd, The Atrium, Southern Gate, Chichester, West Sussex, PO19 8SQ, United Kingdom

For details of our global editorial offices, for customer services and for information about how to apply for permission to reuse the copyright material in this book please see our website at www.wiley.com.

Library of Congress Cataloging-in-Publication data applied for

ISBN: 9781119074175

A catalogue record for this book is available from the British Library.

Cover image: Learjet 45 (Image provided courtesy of Bombardier Inc.)

Set in 9.5/11.5pt Times by SPi Global, Pondicherry, India

1   2016

Contents

2

Aerodynamic and Aircraft Design Considerations　37

3

Air Data Measuring Instruments, Systems and Parameters   109

4
Equations of Motion for a Flat Stationary Earth   153

5
Aircraft Load   169

6

Stability Considerations Affecting Aircraft Performance 185

7

Aircraft Power Plant and Integration 209

8

Aircraft Power Plant Performance 247

9

Aircraft Drag   **275**

10

Fundamentals of Mission Profile, Drag Polar and Aeroplane Grid   *339*

11
Takeoff and Landing 379

12
Climb and Descent Performance 419

13

Cruise Performance and Endurance   463

14

Aircraft Mission Profile   491

15

Manoeuvre Performance   509

16
Aircraft Sizing and Engine Matching   533

17
Operating Costs   559

18

Miscellaneous Considerations   579

Appendices

Preface

This book is about estimating and appreciating the performance of conventional fixed winged aircraft of given designs. It is primarily meant for undergraduates, taking from introductory to advanced intensive courses on aircraft performance. It will also be useful for those in industry as training courses. Practising engineers will also find it helpful, especially for retraining and for those wishing to broaden their knowledge beyond their main area of specialization. We have left out treating VTOL/STOL and helicopters in their entirety – these are subjects by themselves, which require voluminous extensive treatment. We have also left out tilted rotor/vector thrust aircraft, UAVs and high-altitude aircraft performance analyses.

Today's engineers must have strong analytical and applied abilities to convert ideas into profitable products. We hope that this book will serve this cause by combining analytical methods and engineering practices, and adapting them to aircraft performance. New engineers are expected to contribute to the system almost immediately, with minimal supervision, and to do it "right-first-time." The methodology adopted herein is in line with what is practised in industry; the simplifications adopted for classroom use are supported by explanations so that an appreciation of industrial expectations will not be lost. The singular aim of this book is to prepare the reader, as far as possible, for industry-standard engineering practices: to enable new graduates to join industry seamlessly and to become productive as soon as possible. Technology can be purchased, but progress must be earned. We hope to prepare readers to contribute to progress.

The readers are assumed to have exposure to undergraduate engineering mathematics, aerodynamics and mechanics coursework. It is also assumed the reader will have some aircraft design experience. While the book is not on aircraft design, those design topics affecting aircraft performance are included here.

The presentation begins with the derivation of aircraft performance equations, followed by industry standard worked examples. Supporting materials are provided in the Appendices. Examples from engineering practice and "experience" are included. We would be grateful to receive suggestions and criticisms from readers; please contact the publisher or email to a. kundu@qub.ac.uk or M.Price@qub.ac.uk with any relevant information.

There are many excellent books treating the subject matter at various levels; all offer valuable exposure to aircraft performance. There are those that approach the topic classically, treating close-form analyses of exact equations obtained through assumptions. The examples do not represent real aircraft, but are close enough and powerful enough to represent capabilities and characteristics of the aircraft quickly and extensively. At the other extreme, there are simpler books appropriate to an undergraduate curriculum, exposing the barest essentials of aircraft performance. Our goal is to produce a new textbook reflecting some of the advances and presenting relatively detailed analyses as used in industry, tailored to an academic curriculum.

This book can be used in preparing an aircraft performance manual, which is nowadays computerized. Therefore, some operational aspects of commercial transport aircraft are included.

We thank Professor Michael Niu, Professor Jan Roskam (DARcorp), Professor Egbert Torenbeek, Dr Bill Gunston, the late Dr John McMasters (Boeing Aircraft Company), and the late Dr L. Pazmany. Richard Ferrier, Yevgeny Pashnin and Pablo Quispe Avellaneda allowed us to use their figures.

We are indebted to Jane's *All the World Aircraft Manual*, Flightglobal, BAE Systems, Europa Aircraft Company, Airbus, NASA, MIT, Boeing, Defense Advanced Research Projects Agency, Hamilton Standards, Virginia Tech Aerospace and Ocean Engineering, and General Atomics for allowing us to use their figures free of cost. All names are duly credited in their figures.

We offer our sincere thanks to Ms Anne Hunt, Associate Commissioning Editor, Mechanical Engineering, of John Wiley & Sons Ltd, the publisher of this book. Her clear, efficient and prompt support proved vital to reaching our goal. Dr. Samantha Jones's editing made substantial improvement to our book - our heartfelt thanks to her. Also, we thank Mr. Radjan Lourde Selvanadin of SPi Global for managing so efficiently and courteously the publication process.

There are many excellent web sites in the public domain. We gratefully acknowledge benefiting from them. We apologize if we have inadvertently infringed on any proprietary diagrams for educational purposes.

Ajoy K. Kundu
Mark A. Price
David Riordan

Series Preface

The field of aerospace is multidisciplinary and wide-ranging, covering a large variety of products, disciplines and domains, not merely in engineering but in many related supporting activities. These combine to enable the aerospace industry to produce innovative and technologically advanced vehicles. The wealth of knowledge and experience that has been gained by expert practitioners in the various aerospace fields needs to be passed onto others working in the industry, including those just entering from university.

The Aerospace Series aims to be a practical, topical and relevant series of books aimed at people working in the aerospace industry, including engineering professionals and operators, engineers in academia, and allied professions such as commercial and legal executives. The range of topics is intended to be wide-ranging, covering design and development, manufacture, operation and support of aircraft, as well as topics such as infrastructure operations and current advances in research and technology.

Aircraft performance concerns the prediction of how well an aircraft functions throughout its operation, and provides most of the key considerations for aircraft design, leading to the configurations and geometries that we see in today's modern aircraft. The topic is inherently multidisciplinary, requiring not only an understanding of a wide range of individual disciplines such as aerodynamics, flight mechanics, power plant, loads, and so on, but also an appreciation of how these fields interact with each other.

This book, *Theory and Practice of Aircraft Performance*, is a welcome addition to the Wiley Aerospace Series and complements a number of other titles. It tackles the subject from an industrial viewpoint, but is written in such a way as to be very suitable for the curriculum of undergraduate aero engineering courses. Following a comprehensive introductory section covering some fundamentals and background, a complete range of relevant topics is examined, including: aircraft design considerations, loads, stability, power plant, aerodynamics, operational performance and costing. Each chapter contains the appropriate detailed analysis combined with plenty of industrial-standard worked studies and classroom examples.

Peter Belobaba
Jonathan Cooper
Alan Seabridge

Road Map of the Book

Organization

In a step-by-step manner, I have developed a road map to learning industry standard aircraft performance methodology that can be followed in classrooms. Except for Chapter 1, the book is written in formal third-person grammatical usage. The chapters are arranged quite linearly, and there is not much choice in tailoring a course. While the course material progresses linearly, the following diagram depicts how the topics are interlinked.

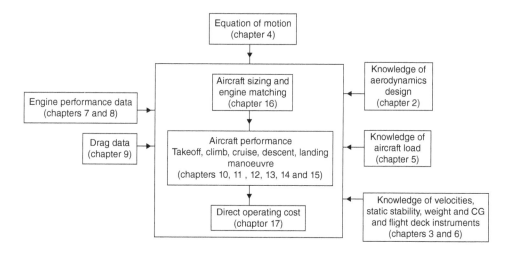

Chapter 1 introduces some background material to prepare readers on the scope of aeronautical engineering. It gives broad coverage of some historical perspective, future trends, role of marketing, project phases from conception to completion, role of airworthiness requirements, and some miscellaneous topics. The main purpose of this chapter is to motivate readers to explore and learn about aircraft.

Chapter 2 covers atmosphere, aircraft aerodynamics and design considerations that influence aircraft performance and must be known to the engineers. This is the only chapter that could be browsed, as these topics are normally covered separately in academies. Chapter 3 introduces the definitions of various kinds of aircraft velocities, related topics on static stability, and some related flight deck instruments. The equations of motion for a flat stationary Earth are derived in Chapter 4, and classroom work starts from here. Next, the aircraft load limits are introduced in Chapter 5, as these define the aircraft performance envelope. Stability considerations are dealt with in Chapter 6; understanding aircraft stability is essential to performance analyses. Aircraft performance computations cannot start without engine performance data, and they are dealt with in Chapters 7 and 8. Aircraft drag data is evaluated in Chapter 9.

Chapters 10 to 15 present the core treatment of aircraft performance. Aircraft sizing and engine matching are done by aircraft performance engineers, and dealt with in Chapter 16 in a formal manner after completing studies on performance. The methodology is treated uniquely in close conformation with industry practices, and is an indispensable part of analyses at the conceptual design phase of a project, as it finalizes aircraft configuration and demonstrates compliance with customer specifications. The procedure offers a "satisfying" solution to generating a family of aircraft variants. This approach is widely practised by all major aircraft manufacturing organizations. The chapter also permits parametric sensitivity studies, which will eventually prove the key to success through balancing comfort with cost in a fiercely competitive market. Readers are encouraged to study aircraft design considerations along with performance analyses. It is to be stressed all the time that aircraft safety is never compromised. Each chapter starts with an overview, a summary of what is learnt, and the classroom work content.

Direct operating cost estimates are looked at in Chapter 17. For commercial aircraft the economic factors are the most important considerations, and for military aircraft it is the performance. Safety and reliability is never sacrificed. The importance of developing the configurations in a family concept is emphasized. The variants can emerge at a low cost by retaining component commonality and covering a wider market area: one might say, lightheartedly, "Design one and get the second at half the development cost". Finally, Chapter 18 concludes by discussing some miscellaneous topics related to aircraft performance.

Appendices A and B give the conversion factors and the ISA day atmospheric tables for both the SI and Imperial systems. Appendix C covers some important formulae. Appendix D gives a basic review on matrices and determinants required to study equations of motion and some important equations. Appendix E gives the problem sets for the pertinent chapters. It is recommended that these should be worked out. Appendix F gives a case study of the class Airbus 320 aircraft. Finally, Appendix G considers some interesting aerofoil in wide usage. This book is meant to reflect what aircraft performance engineers do in industry and airline operation.

Jane's Aircraft Manual [1] is an indispensable book for the vital statistics of aircraft geometries, design data and relevant performance details. This yearly publication has served many generations of aeronautical engineers around the world for more than half a century. The data from *Jane's Aircraft Manual* can be used to compare classroom work on similar types of aircraft. Flightglobal.com is another good source to study cutaway diagrams of aircraft and engines. Products from different origins do show similarities, and this is picked up as a strong statistical pattern, which can help to give an idea of what is to be expected in a new design (Section 18.9). Readers should prepare their own statistics for the type of aircraft under study. Other useful publications are [2] to [5]. I would recommend the readers to look at the Virginia Tech web site on Aircraft Design Bibliographies [6]. Their compilation of aircraft design information sources is comprehensive. Related web sites also give useful data.

Many categories of aircraft have been designed. I have chosen the important ones that will broadly cover a wide range of classroom exercises; these will provide adequate exposure for the students. The associated examples in the book would be those of four cases: (i) a turbofan powered Learjet 45 class business jet (Bizjet); (ii) a turboprop powered propeller-driven Tucano class military trainer aircraft (TPT); (iii) a military advanced jet trainer (AJT) in the class of BAe Hawk that has a close support role variant; and (iv) performance computation of a high subsonic jet in the class of Boeing 737/Airbus 320 aircraft is given in Appendix E. These are the types recommended as the most suitable for classroom projects. Classroom methodology should be in harmony with industrial practices, otherwise the gap between academy and industry might widen.

Case studies are indispensable in the course of learning in the classroom. Example exercises must bear high fidelity with the real ones – I take some satisfaction in providing real world examples modified to classroom usage to maintain commercial confidence. These are not from any academic projects, but follow the designs of the real ones worked out by myself. At this point, I highlight that the results are not those from the industry, but have been compared with the performance data available. Industry is not liable for what I present here.

The book gives full coverage of worked-out examples. Instructors have the flexibility to generate problem assignment sets at the level of class requirements. I strongly recommend the adoption of manual computation, leaving the repetitive aspects to spreadsheets to be developed by the students as part of their learning process. This is essential if students are to develop a feel for numbers and to learn the labour content of a design (it is expensive to make mid-course changes). Also, extensive theoretical treatment is embedded for research workers to extend their analytical work. It is common nowadays to provide CDs as companion software. I have elected not to follow this practice because the supplied software to handle repetitive tasks constrains students to interact more deeply with the governing equations, which is an important part of the learning experience.

If students elect to use off-the-shelf software, then let it be the reputable ones. However, these are more meaningful after the subject is well understood, that is, after completing the course with manual computation. This will lead to an appreciation of how realistic the computer output is and how to make changes in input to improve results. It is better to postpone the usage of aircraft performance software until one joins industry. In academia, a student can use computational fluid dynamics (CFD) and finite element method (FEM) analyses as computer-aided engineering (CAE). Today's students are proficient with computers and can generate their own programs.

Suggested Structure for the Coursework

The author suggests a typical pathway for one term of 36 hours of classroom contact hours – 24 hours of lectures and 12 hours of tutorials. [A note of caution: what is done in about 36 hours of classroom lectures takes about 36 weeks in industry.]

		Classroom lecture (contact) hours
1.	Coverage of Chapter 1 and some topics from Chapter 2 as selected by the instructor (other topics are to be sandwiched as and when required).	1
2.	Chapter 3 – to cover velocities, stability criteria, etc.	1.5
3.	Equation of motion – Chapter 4.	2
4.	Aircraft load – Chapter 5.	1
5.	Air stability, weights and CG – Chapter 9.	0.5
6.	Engine performance (establish engine data) – Chapters 7 and 8.	2.5
7.	Drag estimation – Chapter 9.	2
8.	Fundamentals of aircraft performance analyses, payload range – Chapter 10.	2.5
9.	Field performance (takeoff and landing) – Chapter 11.	2

(Continued)

(*Continued*)

		Classroom lecture (contact) hours
10.	Climb and descent performance – Chapter 12.	2
11.	Cruise performance – Chapter 13.	2
12.	Aircraft mission analyses – Chapter 14.	2
13.	Manoeuvre – Chapter 15.	1
14.	Aircraft sizing and engine matching – Chapter 16.	1.5
15.	Aircraft operating costs– Chapter 17.	1
16.	Miscellaneous (optional) – Chapter 18.	0.5
	Total	25 lecture hours

Of course, instructors are free to plan their course as it suits them.

Ajoy K. Kundu

Acknowledgements

By A.K. Kundu

Throughout my career, which began in the 1960s and continues in the twenty-first century, I have had the good fortune of witnessing many aerospace achievements, especially putting mankind on the Moon. A third of my career has been spent in academia and two-thirds in industry. I owe a lot to many.

I thank my teachers, heads of establishments/supervisors, colleagues, students, shop-floor workers, and all those who taught and supported me during my career. I remember the following (in no particular order) who have influenced me – the list is compact for the sake of brevity, and there are many more individuals to whom I owe my thanks.

Teachers/Academic Supervisors/Instructors:
The late Professor Triguna Sen of Jadavpur University.
The late Professor Holt Ashley and Professor Samuel McIntosh of Stanford University.
Professor Arthur Messiter and Professor Martin Sichel of University of Michigan.
Professor James Palmer of Cranfield University.
The late Squadron Leader Ron Campbell, RAF, of Cranfield University (Chief Flying Instructor).

Heads of Establishments/Supervisors:
The late Dr Vikram Sarabhai, Indian Space Research Organisation.
James Fletcher, Short Brothers and Harland, Belfast.
Robin Edwards, Canadair Limited, Montreal.
Kenneth Hoefs, Head of the New Airplane Project group of Boeing Company, Renton, USA.
Wing Commander Baljit Kapur, Chairman of Hindustan Aeronautics Limited (HAL), Bangalore.
The late Mr Raj Mahindra, MD (D&D), Hindustan Aeronautics Limited (HAL), Bangalore.
Tom Johnston, Director and Chief Engineer, Bombardier Aerospace-Shorts (BAS), Belfast.
Dr Tom Cummings, Chief Aerodynamicist, Bombardier Aerospace-Shorts (BAS), Belfast.

I am grateful to Boeing Company, Hindustan Aeronautics Ltd (*HAL*) and Bombardier Aerospace-Short (*BAS*) and proud to be associated with them. I learnt a lot from them. I started my aeronautical career with BAS (then Short Brothers and Harland Ltd), and after a long break rejoined and then retired from the company, the first aerospace company to celebrate its centenary. Many of my examples are based on my work in those companies.

I am indebted to my long-time friend and ex-colleague at Boeing Company, Mr Stephen Snyder, a registered Professional Engineer and now independent aviation consultant. I offer my thanks to Anthony Hays of Aircraft Design and Consulting, San Clemente, for his help in professional matters. Suggestions by Professor (Emeritus) Bernard Etkin of the University of Toronto and Professor Dieter Scholz of Hamburg University are gratefully appreciated with thanks. I thank my present and former colleagues (Rev. Dr John Watterson and Dr Theresa Robinson) and former students (Dr Mark Bell and Christina Fanthorpe) for their help.

I am thankful to have my former colleagues Colin Elliott, Vice President of Engineering and Business Development, James Tweedie and Lesley Carson, both Senior Engineers, BAS, help me bring out an industry-standard book.

I offer my thanks to Cambridge University Press for allowing me to use some of the materials of my earlier book entitled *Aircraft Design* (ISBN 978-0-521-88516-4), assisted by the support of their Senior Editor Mr Peter Gordon.

My grandfather, the late Dr Kunja Behari Kundu, my father the late Dr Kamakhya Prosad Kundu, and my cousin-brother the late Dr Gora Chand Kundu are long gone, but they kept me inspired and motivated to remain studious. My wife Gouri's tireless support saved me from being a hunter-gatherer, keeping me comfortable while sparing the time to write this book. Would mere thanks be enough for them?

I had my aeronautical education in the UK and in the USA; I worked in India, the UK and North America. In today's world of cooperative ventures among countries, especially in the defence sector, the methodologies adopted in this book should apply. I dedicate this book to all those organizations (listed under acknowledgements) where I learnt a lot, and that is what I have included in this book. These organizations gave me the best education, their best jobs and their best homes.

I am very fortunate to be able to join with my long-standing colleagues Professor Mark Price, Pro-Vice-Chancellor for the Faculty of Engineering and Physical Sciences at Queen's University Belfast. Formerly, he was the Head of School of Mechanical and Aerospace Engineering. (Queen's University, Belfast) and Mr David Riordan, Senior Engineering Advisor, Nacelles Design and Powerplant Integration (Bombardier Aerospace, Belfast) as co-authors. I have gained a lot from these two brilliant young friends, beyond being just colleagues, and naturally sought their contributions to get this old man supported.

By Mark Price

Before taking up an academic position in the late 1990s, I spent a number of years in industry applying what knowledge I had gained to a variety of design and analysis problems. Often this was with difficulty, and often with that feeling of weakness than any engineer feels in the formative stages of their career, in the depth and extent of their knowledge. However, the challenges that come with bringing a working product to life help to mature an engineer, forcing them to understand the limits of their ability, and the theory on which they have based their decisions. It drives them to strive for better. And it was this desire for improvement that brought me into academia, where I encountered Ajoy, someone with the same motivation and commitment to always do things better, except he had vast experience and so much more knowledge. I have learned much from him, in both detail and approach, and it has been fun, challenging our way of thinking.

It is therefore a real honour for me to join with him in his publication ventures, but moreover I am delighted to do so. The concept of this book is to provide that bridge in knowledge between the undergraduate curriculum and the complex world within which a professional engineer exists. This is something which is lacking in most textbooks, and something which we all recognize we can do better. In this book Ajoy brings this combination of an intelligent approach with practical examples, and real scenarios, to the student. It both challenges and excites, creating a learning experience that will accelerate the formation of an engineer by embedding them from the start within real-world applications. I have learned much by being involved with the book, and I hope that all of you who read it will similarly learn, and that this book will be your stepping stone to developing your engineering knowledge to the highest standards.

I am thankful to Queens University, Belfast, for providing an environment supportive of educational development, and in particular the noble aim to provide graduates valuable to industry. Together with Ajoy, my contribution in this effort has been to shape the book to offer course material in line with industrial standard treatise.

I have many people to thank who have supported me in my career and my life thus far. My mentors, Mr Sam Sterling, Professor Raghu Raghunathan, and Professor Cecil Armstrong provided much in the way of guidance, wise words and sharp wit, in addition to standing as exemplars of their profession, providing excellence in education and research. My colleague Dr Adrian Murphy has worked alongside me from the start and has shared the many risks we took in developing new ideas to bring to a sceptical world. We have learned much together as we trod the path of mistakes and blind alleys. I thank the outstanding team of academics and support staff in Mechanical and Aerospace Engineering who make work such an enjoyable part of my life. I cannot thank enough my family, my wife Denise and my daughter Rachel, who have patience beyond their calling in allowing me space and time to fulfil my dreams. And lastly, my late father, Matt, instilled in me the virtue of delivering to the customer what they actually need, and hence my enthusiasm for this book, to fulfil a need for industry and the graduates they need.

By David Riordan

I count it a privilege to have been invited by Ajoy to contribute to this textbook. The enthusiasm of Ajoy in writing the book has been impressive, and at the same time inspiring to recognize that his motivation has solely been to share an accumulated lifetime's worth of knowledge and experience with the younger generation who may choose aerospace as a career, and subsequently contribute to the further development of aircraft design.

I first met Ajoy when, in the late 1980s as a relatively young engineer, I was assigned to the Aerodynamics department at Short Brothers PLC, Belfast. Even then, Ajoy was recognized as one who would not only provide technical direction and advice when required to help progress the task, but who would also help to ensure you understood the basic fundamentals of aerodynamics and the relevance of the assigned task to the development of new aircraft designs. Reflecting on those times of new aircraft design concepts for regional jet aircraft design, it is amazing to realise what was achieved at a time when computer-based analysis tools were not as prolific as they are today.

Each chapter of this textbook has necessitated many long hours of effort and research for Ajoy. The content reflects Ajoy's broad exposure to the many specialist disciplines required to integrate a successful new aircraft design. For me, the explanations of aircraft aerodynamic drag are the best to appear in contemporary textbooks. With this new book, Ajoy's contribution to both academic and industrial learning is admirable. I trust that those reading the book will both benefit therefrom and at the same time appreciate the abilities and diligence of its principal author, who I have found to be a true gentleman: a colleague and friend.

I am thankful to my employer, Bombardier Aerostructures and Engineering Services, Belfast (Short Brothers PLC), who has afforded me the opportunity to work with many different aerospace companies and engineering professionals from all over the world, since I started with the company in September 1978. No other career could have been so enjoyable or as rewarding.

The patient support of my wife, Hazel, over the years, and also of my two sons, Matthew and Jack, has been appreciated. They have each taught me to realise that, no matter what enjoyment aerospace engineering might bring, nothing surpasses the pleasure of having a supporting family with which to share life's experiences.

Nomenclature

Symbols

A	area
A_1	intake highlight area
APR	augmented power rating
A_{th}	throat area
A_W	wetted area
AR	aspect ratio
A_W	wetted area
a	speed of sound, acceleration
$\bar{a}$	average acceleration at $0.7\,V_2$
ac	aerodynamic centre
b	span
C_R, C_B	root chord
C_D	drag coefficient
C_{Di}	induced drag coefficient
C_{Dp}	parasite drag coefficient
C_{Dpmin}	minimum parasite drag coefficient
C_{Dw}	wave drag coefficient
C_v	specific heat at constant volume
C_F	overall skin friction coefficient, force coefficient
C_f	local skin friction coefficient, coefficient of friction
C_L	lift coefficient
C_l	sectional lift coefficient, rolling moment coefficient
C_{Li}	integrated design lift coefficient
$C_{L\alpha}$	lift curve slope
$C_{L\beta}$	side-slip curve slope
C_M	pitching moment coefficient
C_n	yawing moment coefficient
C_p	pressure coefficient, power coefficient, specific heat at constant pressure
C_T	thrust coefficient
C_{HT}	horizontal tail volume coefficient
C_{VT}	vertical tail volume coefficient
C_{xxxx}	cost with subscript identifying parts assembly
C'_{xxxx}	cost heading for the type
CC	combustion chamber

CG	centre of gravity
c	chord
c_{root}	root chord
c_{tip}	tip chord
cp	centre of pressure
D	drag, diameter
D_{skin}	skin friction drag
D_{press}	pressure drag
d	diameter
E	modulus of elasticity
e	Oswald's factor
F	force
f	flat plate equivalent of drag, wing span
f_c	ratio of speed of sound (altitude to sea level)
F_{ca}	aft fuselage closure angle
F_{cf}	front fuselage closure angle
F_B	body axis
F_I	inertial axis
F_W	wind axis
F_{xxx}	component mass fraction, subscript identifies the item
F/m_a	specific thrust
FR	fineness ratio
g	acceleration due to gravity
H	height
h	vertical distance, height
J	advance ratio
k	constant, sometimes with subscript for each application
L	length, lift
L_{FB}	nacelle fore-body length
L_N	nacelle length
L_{VT}	vertical tail-arm
L_{HT}	horizontal tail-arm
L	length
M	mass, moment
M_f	fuel mass
M_i	component group mass, subscript identifies the item
M_{xxx}	component item mass, subscript identifies the item
m	mass
$\dot{m}_a$	air mass flow rate
$\dot{m}_f$	fuel mass flow rate
$\dot{m}_p$	primary (hot) air mass flow rate (turbofan)
$\dot{m}_s$	secondary (cold) air mass flow rate (turbofan)
N	number of blades, normal force
N_e	number of engines
n	revolutions per minute, load factor
nm	nautical miles
P, p	static pressure
p'	angular velocity about Y-axis
p_e	exit plane static pressure
p_∞	atmospheric (ambient) pressure

P_t, p_t	total pressure
Q	heat energy of the system
q	dynamic head, heat energy per unit mass
q'	angular velocity about Z-axis
R	gas constant, reaction
Re	Reynolds number
Re_{crit}	critical Reynolds number
r	radius, angular velocity
r'	angular velocity about X-axis
S	area, most of the time with subscript identifying the component
S_H	horizontal tail reference area
S_n	maximum cross-sectional area
S_w	wing reference area
S_V	vertical tail reference area
sfc	specific fuel consumption
T	temperature, thrust, time
T_C	non-dimensional thrust
T_F	non-dimensional force (for torque)
T_{SLS}	sea-level static thrust at takeoff rating
T/W	thrust loading
t/c	thickness to chord ratio
tf	turbofan
U_g	vertical gust velocity
U_∞	freestream velocity
u	local velocity along X-axis
V	freestream velocity
V_A	aircraft stall speed at limit load
V_B	aircraft speed at upward gust
V_C	aircraft maximum design speed
V_D	aircraft maximum dive speed
V_S	aircraft stall speed
V_e	exit plane velocity (turbofan)
V_{ep}	primary (hot) exit plane velocity (turbofan)
V_{es}	secondary (cold) exit plane velocity (turbofan)
W	weight, width
W_A	useful work done on aircraft
W_E	mechanical work produced by engine
W/S_w	wing loading
x	distance along X-axis
y	distance along Y-axis
z	vertical distance

Greek Symbols

α	angle of attack
β	CG angle with vertical at main wheel, blade pitch angle, side-slip angle
Γ	dihedral angle, circulation
γ	ratio of specific heat, fuselage clearance angle
Δ	increment measure
δ	boundary layer thickness
ε	downwash angle

η_t	thermal efficiency
η_p	propulsive efficiency
η_o	overall efficiency
Λ	wing sweep, subscript indicates at the chord line
λ	taper ratio
μ	friction coefficient, wing mass
ρ	density
θ	elevation angle, flight path angle, fuselage upsweep angle
π	constant $= 3.14$
σ	atmospheric density ratio
τ	thickness parameter
υ	velocity
ϕ	roll angle, bank angle
ψ	azimuth angle, yaw angle
ω	angular velocity

Subscripts

[In many cases the subscripts are spelled out and not listed here.]

a	aft
ave	average
ep	primary exit plane
es	secondary exit plane
f	front, fuselage
f_b	blockage factor for drag
f_h	drag factor for nacelle profile drag (propeller driven)
fus	fuselage
HT	horizontal tail
M	middle
N, nac	nacelle
o	freestream condition
p	primary (hot) flow
s	stall, secondary (cold) flow
t, tot	total
w	wing
VT	vertical tail
∞	freestream condition

Abbreviations

AB	afterburner
ACAS	advanced close air support
ACM	air combat manoeuvre
ACT	active control technology
ADC	air data computer
AEA	Association of European Airlines
AEW	airborne early warning
AF	activity factor
AFM	aircraft flight-track monitoring, aircraft flight manual
AGARD	Advisory Group for Aerospace Research and Development
AGL	average ground level

AHM	aircraft health monitoring
AIAA	American Institute for Aeronautics and Astronautics
AJT	advanced jet trainer
ALD	actual landing distance
AMPR	aeronautical manufacturer's planning report
AO	angle off
AOA	angle of attack
AOB	angle of bank
APM	aircraft performance monitoring/management
APR	augmented power rating
APU	auxiliary power unit
AR	aspect ratio
ARINC	Aircraft Radio Inc.
ASD	accelerate-stop distance
ASDA	accelerate-stop distance available
ASI	airspeed indicator
AST	Air Staff Target
ATA	Aircraft Transport Association
ATC	air traffic control
ATF	advanced tactical fighter
AVGAS	aviation gasoline (petrol)
AVTUR	aviation turbine fuel (kerosene)
BAS	Bombardier Aerospace–Shorts
BFL	balanced field length
BFM	basic fighter manoeuvre
BHP	brake horse power
BOM	bill of material
BPR	bypass ratio
BRM	brake release mass
BVR	beyond visual range
BWB	blended-wing body
CAA	Civil Aviation Authority (UK)
CAD	computer-aided design
CAE	computer-aided engineering
CAM	computer-aided manufacture
CAS	close air support, calibrated air speed, control augmentation system
CAT	clear air turbulence
cc/c	flap chord to aerofoil chord ratio
CCV	control configured vehicle
CFC	chlorofluorocarbon
CFD	computational fluid dynamics
CFL	critical field length
CFR	Code of Federal Regulation
CG	centre of gravity
COC	cash operating cost
CS	Certification Standard
CTOL	conventional takeoff and landing
CV	control volume
CWY	clearway
DCPR	defence contractor's planning report

DBT	design built team
DF	drift angle
DFFS	design for Six Sigma
DFM/A	design for manufacture/assembly
DLM	design landing mass
DOC	direct operating cost
DOM	dry operational mass
DOT	Department of Transport
EAS	equivalent airspeed
EASA	European Aviation Safety Agency
ECS	environmental control system
ED	engine display
EF	endurance factor
EFIS	electronic flight information system
EGT	exhaust gas temperature
EI	emissions index
EPA	Environmental Protection Agency
EPNL	effective perceived noise level
EPR	exhaust pressure ratio, engine pressure ratio
ESHP	equivalent SHP (shaft horse power)
ETOPS	extended twin-engine operations
ETP	equal time point
EW	electronic warfare
FAA	Federal Aviation Administration
FADEC	full authority digital engine control
FAR	Federal Aviation Regulation
FBL	fly-by-light
FBW	fly-by-wire
FCOM	flight crew operating manual
FE	field elevation
FEM	finite element method
FF	fuel flow
FI	Fatigue Index
FL	flight level
FOC	fixed operating cost
FPS	foot, pound, second
FR	fineness ratio
FS	factor of safety
FTM	flight test manual
HMD	helmet-mounted display
HOTAS	hands-on throttle and stick
HP	horse power, high pressure
HSC	high-speed cruise
HST	hypersonic transport
H-tail	horizontal tail
HUD	head-up display
IA	indicated altitude
IATA	International Air Transport Association
ICAO	International Civil Aviation Organization
IGE	in-ground effect

ILS	instrument landing system
INCOSE	International Council of Systems Engineering
IOC	indirect operating cost
IPPD	integrated product and process development
ISA	International Standard Atmosphere
JAA	Joint Aviation Authorities
JAR	Joint Airworthiness Requirements
JPT	jet pipe temperature
JUCAS	Joint Unmanned Combat Air System
KE	kinetic energy
KEAS	knots equivalent air speed
LCA	light combat aircraft
LCC	life cycle cost
LCD	liquid crystal display
LCN	load classification number
LCR	lip contraction ratio
L/D	lift-to-drag (ratio)
LDA	landing distance available
LE	leading edge
LF	load factor
LFL	landing field length
LO	low observable
LOH	liquid hydrogen
LP	low pressure
LPO	long period oscillation
LRC	long-range cruise
MAC	mean aerodynamic chord
MDA	multidisciplinary analysis
MDO	multidisciplinary optimization
MEM (W)	manufacturer's empty mass (weight)
MFD	multifunctional display
MFR	mass flow rate
MGC	mean geometric chord
MLM	maximum landing mass
MoD	Ministry of Defence
MOGAS	motor gasoline (petrol)
MPM	manufacturer process management
MRM (W)	maximum ramp mass (weight)
MTBF	mean time before failure
MTM	maximum taxi mass
MTOM (W)	maximum takeoff mass (weight)
MTTR	mean time to repair
NACA	National Advisory Committee for Aeronautics
nam	nautical air miles
NASA	National Aeronautics and Space Administration
NBAA	National Business Aviation Association
NCS	normal climb speed
ND	navigational display
nm	nautical miles
NP	neutral point

NRC	non-recurring cost
NTC	normal training configuration
OAT	outside air temperature
OEM (W)	operating empty mass (weight)
OEMF	operating empty mass fraction
OGE	out-of-ground effect
OPR	overall pressure ratio
PCU	power control unit
PE	potential energy
PFD	primary flight display
PHA	positive high angle of attack
PIA	positive intermediate angle of attack
PLA	positive low angle of attack
PLM	product life cycle management
PNL	perceived noise level
PNR	point of no-return
PPR	product, process and resource
psfc	power specific fuel consumption
RAeS	Royal Aeronautical Society
RAF	Royal Air Force
RAE	Royal Aircraft Establishment
RAT	ram air turbine
RC	recurring cost
RCS	radar cross-section
RF	range factor
RFP	Request for Proposal
RJ	regional jet
RLD	required landing distance
RPM	revolutions per minute, revenue passenger miles
RPS	revolutions per second
RPV	remotely piloted vehicle
RSA	runway safety area
RTOW	regulatory takeoff weight
SAS	stability augmentation system
SAT	static air temperature
SD	system display
SE	specific energy
SEP	specific excess power
sfc	specific fuel consumption
SHP	shaft horse power
SI	Système International
SPL	sound pressure level
SPO	short period oscillation
SR	specific range
SST	supersonic transport
STOL	short takeoff and landing
SWY	stopway
TA	true altitude
TAF	total activity factor
TAS	true airspeed

TAT	total air temperature
TBO	time between overhauls
TE	trailing edge
TET	turbine entry temperature
TGT	turbine guide-vane temperature
TOC	total operating cost
TOD	takeoff distance
TODA	takeoff distance available
TOFL	takeoff field length
TOR	takeoff run
TORA	takeoff run available
TPT	turboprop trainer
TQM	total quality management
TR	thrust reverser
tsfc	thrust specific fuel consumption
TTOM	typical takeoff mass (military)
T&E	training and evaluation
UAV	unmanned air vehicle
UBFL	unbalanced field length
UCA	unmanned combat aircraft
UHBPR	ultra-high bypass ratio
UHC	unburned hydrocarbon
ULD	unit load device
VSI	vertical speed indicator
V-tail	vertical tail
VTOL	vertical takeoff and landing
WAT	weight, altitude and temperature
WS	wind speed
ZFM (W)	zero fuel mass (weight)

CHAPTER 1

Introduction

1.1   Overview

This book begins with a brief historical introduction, surveying our aeronautical legacy to motivate readers by describing the remarkable progress we have made from mythical conceptions of flight to high-performance aircraft with capabilities unimagined by early aeronautical pioneers. This chapter continues with offering a brief introduction to aircraft fundamentals and aircraft flight mechanics, which form the basics of aircraft performance. The chapter also presents the issues involved with units and dimensions in this context.

1.2   Brief Historical Background

Many books cover the broad sweep of aeronautical history, while others discuss specific accomplishments and famous people's achievements in aeronautics. References [1] to [4] are good places to start your exploration. Innumerable web sites on historical topics and technological achievements exist; simply enter keywords such as Airbus, Boeing, or anything that piques your curiosity, and you will find a wealth of information.

1.2.1   Flight in Mythology

People's desire to fly is ancient – every civilization has their early imaginations embedded in mythologies. In human efforts there are the well-known examples such as Daedalus/Icarus, vimanas (aircraft), flying carpets, flying chariots, and so on. In creatures, there are the bird-men (Garuda), flying horses (Pegasus/Sleipnir), flying dragons – our imagination of flight is universal.

IIistory is unfortunately more "down to earth" than mythology, with stories about early pioneers who leapt from towers and cliffs, only to leave the Earth in a different but predictable manner because they did not respect natural laws. Our dreams and imagination became reality only a little over a century ago on 17 December 1903, when the Wright brothers succeeded with the first powered heavier-than-air flight. It only took 65 years from that date to land a man on the Moon.

1.2.2   Fifteenth to Nineteenth Centuries

Tethered kites are recorded to have flown in China as long ago as 600 BC. However, the first scientific attempts to design a mechanism for aerial navigation are credited to Leonardo da Vinci (1452–1519). He was the true "grandfather" of modern aviation, even if none of his

Theory and Practice of Aircraft Performance, First Edition. Ajoy Kumar Kundu, Mark A. Price and David Riordan.
© 2016 John Wiley & Sons, Ltd. Published 2016 by John Wiley & Sons, Ltd.

machines ever defied gravity (Figure 1.1), because he sketched many contraptions in his attempt to make a mechanical bird. Birds possess such refined design features that the initial human path into the skies could not take that route, but today's micro-air devices are increasingly exploring natural designs. After da Vinci, there was an apparent lull for more than a century until Sir Isaac Newton (1642–1727), who computed the power required to make sustained flight. Perhaps we lack the documentary evidence, but we are convinced that the human fascination with and endeavour for flight did not abate. Flight is essentially a practical matter, so real progress paralleled other industrial developments (e.g. isolating gas required for buoyancy).

While it appears that Bartolomeu de Gusmao may have demonstrated balloon flight in 1709 [4] in Portugal, information on this event is still lean. So we credit Jean-François Pilâtre de Rozier and François Laurent d'Arlandes as the first people to effectively defy gravity, using a Montgolfier balloon (Figure 1.1) in France in 1783. For the first time, it was possible to sustain and somewhat control flight above the ground at will. However, these balloon pioneers were subject to the prevailing winds and were thus limited in their navigational options. To become airborne was an important landmark in human history. The Montgolfier brothers (Joseph and Etienne) should be considered among the "fathers" of aviation. In 1784, Jean-Pierre Blanchard (France) with Dr John Jeffries (USA) added a hand-powered propeller to a balloon and made the first aerial crossing of the English Channel on 7 January 1785. (Jules Verne's fictional balloon trip around the world in 80 days became a reality when the late Steve Fossett circumnavigated the globe in fewer than 15 days in 2002.) In 1855, Joseph Pline was the first to use the word *aeroplane* in a paper he wrote proposing a gas-filled dirigible glider with a propeller.

It was not until 1804 that the first recorded controllable heavier-than-air machine to stay freely airborne was recorded when Englishman Sir George Cayley constructed and flew a kite-like glider (Figure 1.2) with movable control surfaces. In 1842, the English engineer Samuel Henson secured a patent on an aircraft design that was driven by a steam engine.

With his brother Gustav, Otto Lilienthal was successfully flying gliders (Figure 1.2) in Berlin more than a decade (1890) before the Wright Brothers' first experiments. His flights

■ **FIGURE 1.1**
Early concepts and reality of flying: Leonardo da Vinci's flying machine, and the Montgolfier Balloon (reproduced with permission of NASA)

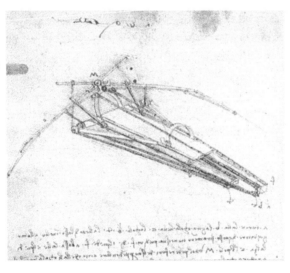 

were controlled but not sustained. The early flight machine designs were hampered by an overestimation of the power requirement needed for sustained flight. This mistake (based in part on Newton's, among others, calculations) may have discouraged attempts of the best German engine-makers of the time to build aircraft engines because they would have been too heavy. Sadly, Lilienthal's aerial developments ended abruptly and his experience was lost when he died in a crash in 1896.

1.2.3   From 1900 to World War I (1914)

The question of who was first in flight is an important event to remember. The Wright Brothers (United States) are recognized as the first to achieve sustained, controlled flight in a heavier-than-air manned flying machine (Wright Flyer, Figure 1.3). Before discussing their achievement, some "also-rans" deserve mention. John Stringfellow accomplished the first powered flight of an unmanned heavier-than-air machine in 1848 in England. In France, Clement Ader also made a successful flight in his "Eole". Gustav Weisskopf (Whitehead), a Bavarian who migrated to the US, claimed to have made a sustained, powered flight [3] on 14 August 1901, in Bridgeport, Connecticut. Karl Jatho of Germany made a 200-foot hop (longer than the Wright Brothers first flight) powered (10-HP Buchet engine) flight on 18 August 1903. At what distance a "hop" becomes a "flight" could be debated. Perhaps most significant are the efforts of

■ **FIGURE 1.2**
Early heavier-than-air unpowered aircraft: Cayley's kite plane, and one of Lilienthal's gliders (reproduced with permission of NASA)

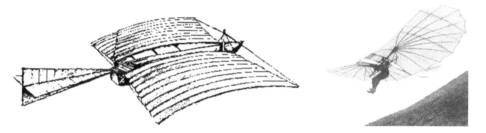

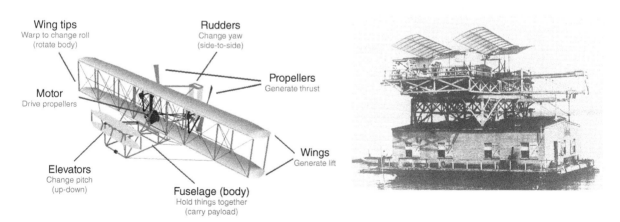

■ **FIGURE 1.3**   **Early heavier-than-air powered aircraft: the Wright Flyer, and Langley's Aerodrome (reproduced with permission of NASA)**

Samuel P. Langley, who made three attempts to get his designs (Aerodrome) airborne with a pilot at the controls (Figure 1.3). His designs were aerodynamically superior to the Wright flyer, but the strategy to ensure pilot safety resulted in structural failure while catapulting from a ramp toward water. His model aircraft were flying successfully in 1902. (To prove the capability, subsequently in 1914 Curtiss made a short flight with a modified Aerodrome.) The failure of his aircraft also broke Professor Langley; a short time afterwards, he died of a heart attack. Professor Langley, a highly qualified scientist, had substantial government funding, whereas the Wright brothers were mere bicycle mechanics without any external funding.

The Wright Brothers' aircraft was inherently unstable, but good bicycle mechanics that they were, they understood that stability could be sacrificed if sufficient control authority was maintained. They employed a foreplane (canard) for pitch control, which also served as a stall-prevention device. Modern designs have reprised this solution as seen in the Burt Rutan-designed aircraft. Exactly a century later, a flying replica model of the Wright Flyer failed to lift off on its first flight. A full-scale non-flying replica of the Wright Flyer is on display at the Smithsonian Museum in Washington, DC. This exhibit and other similar museums are well worth a trip. Strangely, the Wright Brothers did not exploit their invention; however, having been shown that sustained and controlled flight was possible, a new generation of aerial entrepreneurs quickly arose. Newer inventions followed in rapid succession, from pioneers such as Alberto Santos Dumas, Louis Bleriot, and Glenn Curtiss to name but a few. The list grew rapidly. Each inventor presented a new contraption, some of which demonstrated genuine design improvements. Fame, adventure, and *"Gefühl"* (feelings) were the drivers, since the early years saw little financial gain from selling "joy rides" and air shows – spectacles never seen before then and still appealing to the public today.

It did not take long to demonstrate the advantages of aircraft for mail delivery and military applications. At approximately 100 miles per hour (mph), on average, aircraft were travelling three times faster than any surface vehicle – and in straight lines. Mail was delivered in less than half the time. The potential for military applications was dramatic and well demonstrated during World War I. About a decade after the first flight in 1903, aircraft manufacturing had become a lucrative business. The Short Brothers and Harland (now part of the Bombardier Aerospace group) was a company that started aircraft manufacturing by contracting to fabricate the Wright designs. The company is now the oldest surviving aircraft manufacturer still in operation. In 2008, it celebrated its centenary, the first aircraft company to do so.

1.2.4   World War I (1914–1918)

Balloons were the earliest (second half of nineteenth century) airborne military vehicle, but controlled aircraft replaced their role as soon their effectiveness were demonstrated just before World War I. Their initial role was as an observation platform, and soon their military offensive capabilities (bombing, dogfights, etc.) were established. Their combat effectiveness became a decisive factor for military strategy. This rapidly attracted entrepreneurs in both private and public sectors. On both sides of the Atlantic the number of aircraft and engine designs and manufacturing establishments exceeded more than 100 organizations. With the growing recognition of the potential of military aircraft applications, the actual demand was in Europe. Serious military aircraft design activities began after war broke out. German aeronautical science and technologies made rapid advances.

This section shows how quickly the aircraft industry grew within a decade of the first flight, initially driven by military application. This is the period that lay the foundations of what was to come subsequently. The section is kept brief by giving only a few aircraft examples here.

In the US: In 1908, the US Army accepted tender for military aircraft, and after extensive tests the Signal Corps accepted Wright Model A, powered by a 35 HP engine (Figure 1.4) in 1909. In 1912 the Wright Model B was used for the first time to demonstrate the firing of a machine gun from a airplane. Soon after, Glenn Curtiss became the dominant US aircraft designer. Curtiss aircraft introduced naval carrier-based flying during 1910–11. The company became early pioneers of producing military flying boats: planes that could take off and land in water. One of the earlier designs was the Curtiss F4 (Figure 1.4). The Boeing Company was started around this time. Among the famous names of early aviation are Martin, Packard,

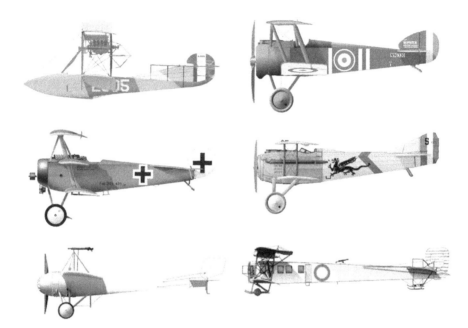

■ **FIGURE 1.4** Very early powered aircraft (World War I). Left panel: top, Curtiss F4 (US) (reproduced with permission of www.wp.scn.ru/); middle, Fokker Dr1 (Germany) (reproduced with permission of www.fokkerdr1.com/); bottom, Caproni Ca.20 (Italy) (reproduced with permission of www.airlinepicture.blogspot.com). Right panel: top, Sopwith Camel (UK) (reproduced with permission of www.worldac.de/); middle, SPAD S VII (France) (reproduced with permission of www.greatwarflyingmuseum.com); bottom, Sikorsky Ilya Muromets Bomber (reproduced with permission of www.aviastar.org/air/russia). See Table 1.1 for their performance summary

■ **TABLE 1.1**
Performance summary of the aircraft in Figure 1.4

	Curtiss F4 flying boat	Sopwith Camel	Fokker Dr1	SPAD S VII	Caprioni Ca.20	Ilya Mouromets
Engine, HP	2×275	130	110	150	110	4×148
Wing area, ft^2	1216	231	201	192	144	1350
MTOM, lb	10,650	1455	1292	1632	≈1290	12,000
Max. speed, knots	85	115	185	119	100	110

Vaught, and so on; possibly in excess of two dozen aircraft and engine design and manufacturing companies emerged in the US during this period. Despite this, America introduced arguably superior European-designed military aircraft into their armed forces.

In the UK: Upon the recommendation of the British Defence Ministry in 1911, the Royal Flying Corps (RFC) was formed in 1912. In 1918 it merged with the Royal Naval Air Service to form the Royal Air Force (RAF). The Royal Aircraft Factory B.E.2 was a single-engined two-seat biplane, in service with the RFC in 1912. They were used as fighters, interceptors, light bombers, trainers and reconnaissance aircraft. A more successful design with better capabilities was the single-seat Sopwith Pup. It entered service in the autumn of 1916. The Avro 504 (100–130 HP) and Sopwith Camel (1913, 110 HP – Figure 1.4) are some of the well-known aircraft of the time. Some of the other famous UK aircraft of the time bore the names of Armstrong-Whitworth, A.V. Roe, Blackburn, Bristol, Boulton/Paul, De Havilland, Fairey, Handley Page, Short Brothers, Supermarine, Vickers, and Westland.

In Germany: *Die Fliegertruppen des Deutschen Kaiserreiches* (the Flier Troops of the German Kaiser Empire) of the Imperial German Army Air Service was formed in 1910, and changed its name to the Luftstretkräfte in 1916 (this became the Luftwaffe in the mid-1930s). Advances made by German aeronautical science and technologies produced many types of relatively high performance aircraft at the time. These saw action during World War I. The triplane Fokker Dr1 (Figure 1.4) was perhaps the most famous fighter of the period. The triplane was flown by the famous "Red Baron", Rittmeister Manfred Freiherr von Richthofen, the top-scoring ace of World War 1 with 80 confirmed kills. Another successful German military airplane, the Albatross III, served on the Western Front until the end of 1917. The Junkers D.I was the first ever cantilever monoplane design to enter production. It utilized corrugated metal wings and front fuselage, with a fabric covering being used only on the rear fuselage. The Friedrichshafen FF.33 was one of the earliest German single-engine amphibious reconnaissance biplanes (1914). Some of the other famous German aircraft of the time bore the names of A.E.G., Aviatik, D.F.W., Fokker, Gotha (Gothaer Waggonfabrik), Halberstadt, Hannoversche, Junkers, Kondor, Roland, L.V.G., and Zeppelin.

In France: The French Air Force (*Armée de l'Air*, ALA) is the air force of the French Armed Forces. It was formed in 1909 as the *Service Aéronautique*, as part of the French Army, and was made an independent military branch in 1933. The first Bleriot XIs entered military service in France in 1910. Other famous French military aircraft are Nieuport 10 (1914, 80 HP) and their subsequent designs. The SPAD S VII (Figure 1.4) was a successful French fighter aircraft of World War I used by many countries. The Caudron G.4 series was the first French-built twin-engine bomber biplane platform introduced in the early years of World War I. Some of the other famous French aircraft of the time bore the names of Hanriot, Maurice Farman, Moraine-Saulnier, and Salmson. Many countries, such as the UK, the US, Italy and Russia, bought French military aircraft for their Air Force.

Other European Countries: Aircraft design and manufacturing activities in other European countries, such as Italy, Russia, the Scandinavian countries, Spain and Portugal, were also vigorously pursued. Only Italian and Russian designs are briefly given below.

Italy could claim to be amongst the earliest to experiment with military aviation. As early as 1884, before powered heavier-than-air vehicles, the *Regio Esercito* (Italian Royal Army) operated balloons as observation platforms. During the early World War I period, Caproni developed a series of successful heavy bombers. The Caproni Ca.20 (1914) was one of the first real fighter planes (Figure 1.4). It is a monoplane that integrated a movable, forward-firing drum-fed Lewis machine gun two feet above the pilot's head, firing over the propeller arc. Some of the other famous Italian aircraft of the time bore the names of Società Italiana Aviazione and Ansaldo.

The Russian Empire under the Czar had the Imperial Russian Air Force possibly before 1910. Russian aeronautical sciences had advanced research of the time through famous names like Tsiolkovsky and Zhukovsky. The history of military aircraft in Imperial Russia is closely associated with the name of Igor Sikorsky. He emigrated to the US in 1919; aircraft bearing his name are still produced. In 1913–14 Sikorsky built the first four-engine biplane, the Russky Vityaz. His famous bomber aircraft, the Ilya Muromets, is shown in Figure 1.4. Other famous aircraft of Russian origin of the time had the names Anade, Antara, Anadwa, and Grigorvich.

1.2.5   The Inter-War Period: the Golden Age (1918–1939)

The urgent necessity for military activities during World War I advanced aeronautical science and technology to the point where it presented an attractive proposition for business growth. The aeronautical activities in the peace period were deployed to increase industrial and national growth. The enhanced understanding of aerodynamics, aircraft control laws, thermodynamics, metallurgy, structural and system analyses ensured that aircraft and engine size and performance grew in rapid strides. A wide variety of innovative new designs emerged to cover wide applications in both military and civil operations. Records for speed, altitude and payload capabilities were updated at frequent intervals. This period is seen as the Golden Age of aeronautics.

With enhanced aeronautical knowledge to increase aircraft capabilities, availability of experienced pilots and public awareness offered the ideal environment to make commercial aviation a reality. Surplus post-war experienced pilots were available who could easily adapt to newer designs. They kept them engaged with performing air-shows and offering joy rides. In this period, aircraft industries geared up in defence applications and in civil aviation, with financial gain as the clear driver. The free market economy of the West contributed much to aviation progress; its downside, possibly reflecting greed, was under-regulation. The proliferation showed signs of compromise with safety issues, and national regulatory agencies quickly stepped in, legislating for mandatory compliance with airworthiness requirements (US, 1926). Today, every nation has its own regulatory agency.

One of the earliest applications of commercial operation with passenger flying was done on the modified Sikorsky Ilya Muromets (Figure 1.4). It had an insulated cabin with heating and lighting, comfortable seats, lounge and toilet. Fokker was a Dutch aircraft manufacturer named after its founder, Anthony Fokker. The company operated under several different names, starting out in 1912 in Schwerin, Germany, moving to the Netherlands in 1919. In the 1920s, Fokker entered its glory years, becoming the world's largest aircraft manufacturer. Its greatest success was the F.VIIa/3 m trimotor passenger aircraft, which was used by 54 airline companies worldwide. It shared the European market with the Junkers all-metal aircraft, but dominated the American market until the arrival of the Ford Trimotor, which copied the aerodynamic features of the Fokker F.VII, and Junkers structural concepts. In May 1927, Charles Lindberg won the Ortega Prize for the first individual non-stop transatlantic flight.

Early aircraft design was centred on available engines, and the size of the aircraft depended on the use of multiple engines. The combination of engines, materials, and aerodynamic technology enabled aircraft speeds of approximately 200 mph; altitude was limited by human physiology. In the 1930s, Durener Metallwerke of Germany introduced *duralumin*, with higher strength-to-weight ratios of isotropic material properties, and dramatic increases in speed and altitude resulted.

1.2.6   World War II (1939–1945)

The introduction of duralumin brought a new dimension to manufacturing technology. Structure, aerodynamics, and engine development paved the way for substantial gains in speed, altitude, and manoeuvring capabilities. These improvements were seen predominantly in World War II designs such as the Supermarine Spitfire, the North American P-51, the Focke-Wolfe 190, and the Mitsubishi Jeero-Sen. Multi-engine aircraft also grew to sizes never before seen.

The invention of the jet engine (independently by Whittle in the UK and von Ohain in Germany) realised the potential for unheard-of leaps in speed and altitude, resulting in parallel improvements in aerodynamics, materials, structures, and systems engineering. Heinkel He 178 was the first jet-powered aircraft (27 August 1939), followed by the Gloster E.28 on 15 May 1941.

1.2.7   Post World War II

A better understanding of supersonic flow and a suitable rocket engine made it possible for Chuck Yeager to break the sound barrier in a Bell X1 in 1949 (the aircraft is on show at the Smithsonian Air and Space Museum in Washington, DC). Tens of thousands of the Douglas C-47 Dakota and Boeing B17 Flying Fortress were produced. Post-war, the De Havilland Comet was the first commercial jet aircraft in service; however, plagued by several tragic crashes, it failed to become the financial success it promised.

The 1960s and 1970s saw rapid progress, with many new commercial and military aircraft designs boasting ever-increasing speed, altitude and payload capabilities. Scientists made considerable gains in understanding the relevant branches of science: in aerodynamics [4], concerning high lift and transonic drag; in materials and metallurgy, improving the structural integrity; and in solid-state physics. Some of the outstanding designs of those decades emerged from the Lockheed Company, including the F104 Starfighter, the U2 high-altitude reconnaissance aircraft, and the SR71 Blackbird. These three aircraft, each holding a world record of some type, were designed in Lockheed's Skunk Works, under the supervision of Clarence (Kelly) Johnson. I recommend that readers study the design of the nearly half-century-old SR71, which still holds the speed-altitude record for aircraft powered by air-breathing engines.

During the late 1960s, the modular approach to gas-turbine technology gave aircraft designers the opportunity to match aircraft requirements (i.e. mission specifications and economic considerations) with "rubberized" engines (see Section 7.2). This was an important departure from the 1920s and 1930s, when aircraft sizing was based around multiples of fixed-size engines. Chapter 7 describes the benefits of modular engine design. This advancement resulted in the development of families of aircraft design. Plugging the fuselage and, if necessary, allowing wing growth accessed a wider market area at a lower development cost because considerable component commonality could be retained in a family: a significant cost-reduction design strategy. Capitalistic objectives render designers quite conservative, forcing them to devote considerably more time to analysis. Military designs emerge from more extensive analysis – for example, the strange-looking Lockheed F117 is configured using stealth features to minimize radar signatures. Now, more mature stealth designs look conventional (e.g. the Lockheed F22).

1.3   Current Aircraft Design Status

A major concern that emerged in the commercial aircraft industry from the market trend and forecast analysis of the early 1990s was the effect of inflation on aircraft manufacturing costs. Since then, all major manufacturers and the subcontracting industries have implemented cost-cutting measures. It became clear that a customer-driven design strategy is the best approach for survival in a fiercely competitive marketplace. The paradigm of "better, farther, and cheaper to market" replaced, in a way, the old mantra of "higher, faster, and farther" [5]. Manufacturing considerations came to the forefront of design, and new methodologies were developed, such as DFM/A and Six Sigma.

With rising airfares, air travellers have become cost-sensitive. In commercial aircraft operations, the direct operating cost (DOC) depends more on the acquisition cost (i.e. unit price) than on the fuel cost (year 2000 prices) consumed for the mission profile. Today, for the majority of mission profiles, fuel consumption constitutes between 15% and 30% of the DOC, whereas the aircraft unit price contributes between three and four times as much, depending on the payload range [6]. For this reason, manufacturing considerations that can lower the cost of aircraft production should receive as much attention as the aerodynamic saving of drag counts. The situation would change if the cost of fuel exceeds the current airfare sustainability limit (see Chapter 17), when drag-reduction efforts regain ground.

The conceptual phase of aircraft design is now conducted using a multidisciplinary approach (i.e. concurrent engineering), which must include manufacturing engineering and an appreciation for the cost implications of early decisions; the "buzzword" is *integrated product and process development* (*IPPD*). Section 1.8 briefly describes typical project phases as they are practised currently. Margins of error have shrunk to the so-called zero tolerance so that tasks are done correctly the first time; the Six Sigma approach is one management tool used to achieve this end. The importance of environmental issues has emerged, forcing regulatory authorities to impose limits on noise and engine emission levels. Recent terrorist activities are forcing the industry and operators to consider preventative design features.

1.3.1   Current Civil Aircraft Trends

Current commercial transport aircraft in the 100 to 300 passenger classes all have a single slender fuselage, backward-swept low-mounted wings, two under-slung wing-mounted engines, and a conventional *empennage* (i.e. a horizontal and a vertical tail); this conservative approach is revealed in the similarity of configuration. The similarity in larger aircraft is the two additional engines; there have been three-engine designs, but only on a few aircraft, because the configuration was rendered redundant by variant engine sizes that cover the in-between sizes and extended twin operations (ETOPS). The largest commercial jet transport aircraft, the Airbus 380 (Figure 1.5), made its first flight on 27 April 2005, and is currently in service. The Boeing 787 Dreamliner (Figure 1.5) is the replacement for its successful Boeing 767 and 777 series, aiming at competitive economic performance.

The last three decades witnessed a 5–6% average annual growth in air travel, exceeding 2×10^9 revenue passenger miles (RPM) per year. Publications by the International Civil Aviation Organization (ICAO), the National Business Aviation Association (NBAA), and other journals provide overviews of civil aviation economics and management. The potential market for commercial aircraft sales is of the order of billions of dollars per year. However, the demand for air travel is cyclical and – given that it takes about four years from the introduction of a new aircraft design to market – operators must be cautious in their approach to new acquisitions. They do not

■ **FIGURE 1.5   Current wide-body large commercial transport aircraft: the Airbus 380 (reproduced with permission of Airbus), and the Boeing 787 Dreamliner (reproduced with permission of Boeing)**

want new aircraft to join their fleet during a downturn in the air-travel market. Needless to say, market analysis is important in planning new purchases.

Deregulation of airfares has made airlines compete more fiercely in their quest for survival. The growth of budget airlines compared with the decline of established airlines is another challenge for operators. Boeing introduced its 737 twin-jet aircraft (derived from the three-engine B727, the bestseller at the time), and after nearly four decades of production to this day, has become the bestseller in the history of the commercial aircraft market. Of course, in that time, considerable technological advancements have been incorporated, improving the B737's economic performance by about 50%.

The gas-turbine turboprop offers better fuel economy than current turbofan engines. However, because of propeller limitations, the turboprop-powered aircraft's cruising speed is limited to about two-thirds of the high-speed subsonic turbofan-powered aircraft. For lower operational ranges (e.g., less than 1000 nautical miles (nm), the difference in sortie time would be of the order of less than half an hour, yet there is a saving in fuel costs of approximately 20%. If a long-range time delay can be tolerated (e.g. for cargo or military heavy-lift logistics), then large turboprop aircraft operating over longer ranges become meaningful. Advances in propeller technology are pushing turboprop powered aircraft cruising speeds close to the turbofan-powered aircraft high subsonic cruise speeds.

1.3.2   Current Military Aircraft Trends

Military aircraft designs have the national interest as a priority over commercial considerations. While commercial aircraft can earn self-sustaining revenue, military operations depend totally on taxpayers' money with no cash flow back, other than export sales that carry the risk of disclosure of tactical advantages. The cost frame of a new design has risen sufficiently to strain the economy of single nations. Not surprisingly, the number of new designs has drastically reduced, and military designs are moving towards multinational collaborations among allied nations, where the retention of confidentiality in defence matters is possible.

There are differences between civil and military design requirements (see Section 1.14.1). However, there are some similarities in their design processes up to the point when a new breakthrough is introduced – one thinks instinctively of how the jet engine changed in design in the 1940s. Consider the F117 Nighthawk (Figure 1.6); the incorporation of stealth technology appeared to be an aerodynamicist's nightmare, but it now conforms to something familiar in the shape of the F35 Lightning II (its prototype X35 is shown in Figure 1.6). We must not forget that military roles are more than just combat; they extend to transportation and surveillance (reconnaissance, intelligence gathering and electronic warfare). F35, Eurofighter, Rafale, Gripen, and Sukhoi 30 are the current frontline fighter aircraft. In strategic bombin, the B52 served for four decades and is to continue for another two decades – some design! The latest B2 bomber (Figure 1.6) looks like an advanced flying wing without the vertical tail.

Combat roles are classified as interdiction, air superiority, air defence and, when missions overlap, multi-role (see Section 10.4 for details). Action in hostile environments calls for special attention to: design for survivability; systems integration for target acquisition and weapons management; and design considerations for reliable navigation and communication. All told, it is a complex system, mostly operated by a single pilot – an inhuman task if the workload was not relieved by microprocessor-based decision-making. Fighter pilots are a special breed of aircraft operators with the best emotional and physical conditioning to cope with the stresses involved. Aircraft designers have a deep obligation to ensure combat pilot survivability. Unmanned aerial vehicle (UAV) technology is in the offing – the Middle-Eastern conflicts saw successful use of the Global Hawk for surveillance. Of late, UAVs are used as a weapon delivery system.

■ **FIGURE 1.6** **Current combat aircraft. Top left: F117 Nighthawk (reproduced with permission of the US Airforce/ Sgt Aaron Allmon; top right: X-35 (F35 experimental) (reproduced with permission of the US Airforce/Dana Russo); bottom, B2 Bomber (reproduced with permission of the US Airforce/Sgt Jeremy Wilson)**

1.4   Future Trends

It is clear that in the near future, vehicle capabilities will be pushed to the extent permitted by economic and defence factors and infrastructure requirements (e.g. navigation, ground handling, support, etc.). It is no exception from past trends that speed, altitude and payload will be expanded in both civilian and military capabilities. Coverage of the aircraft design process in the next few decades is given in [7]. In technology, smart materials (e.g. adaptive structure) will gain ground, microprocessor-based systems will advance to reduce weight and improve functionality, and manufacturing methodology will become digital. However, unless the price of fuel increases beyond affordability, investment in aerodynamic improvement will be next in priority.

1.4.1   Trends in Civil Aircraft

Any extension of payload capability will remain subsonic for the foreseeable future, and will lie in the wake of gains made by higher-speed operational success. High-capacity operations will remain around the size of the Airbus 380. Some well-studied futuristic designs (Figure 1.7) have the possibility of further size increases. A blended-wing body (BWB) can use the benefits of the wing-root thickness being sufficiently large to permit merging (Figure 1.7, top) with the fuselage, thereby benefiting from the fuselage's contribution to lift and additional cabin volume. Another alternative would be that of the joined-wing concept (Figure 1.7, bottom). Studies of twin-fuselage, large transport aircraft also indicate potential. A joined fuselage (Figure 1.7) is also a well studied concept.

The speed–altitude extension will progress initially through supersonic transports (SSTs) and then hypersonic transport (HST) vehicles. SST technology is well proven by three decades of the Anglo-French-designed Concorde, which operated above Mach 2 at altitudes of 50,000 feet carrying 128 passengers.

The next-generation SST will have about the same speed-altitude capability (possibly less in speed capability, around Mach 1.8), but the size will vary from as few as ten business passengers to approximately 300 passengers (Figure 1.7) to cover at least transatlantic and transcontinental operations. Transcontinental operations would demand sonic-shock strength reduction through aerodynamic gains rather than speed reduction; anything less than Mach 1.6 has less to offer in terms of time savings. The real challenge would be to have HST (Figure 1.7) operating at approximately Mach 6 that would require operational altitudes above 100,000 ft. Speeds above Mach 6 offer diminishing returns in time saved because the longest distance necessary is only about 12,000 nm (i.e. about 3 hours of flight time). Military applications for HST vehicles are likely to precede civilian applications, and small-scale HSTs have been flown recently.

The concept of rocket propulsion in modern application came from Von Braun's V2 rocket, an idea taken from Tippu's success in using rockets against the British-led Indian army at the Battle of Srirangapatna in 1792 [8]. The experience of Tippu Sultan's rockets led the British to develop missiles at the Royal Laboratory of Woolwich Arsenal, under the supervision of Sir William Congreve, in the late eighteenth century. A new type of speed-altitude capability will come from suborbital space flight (tourism) using rocket powered aircraft, as demonstrated by Rutan's Space Ship Two that hitchhikes with the White Knight to altitude (Figure 1.8), from where it makes the ascent. Interest in this aircraft has continued to grow; the prize of $10 million offered could be compared with that of a transatlantic prize followed by commercial success.

■ **FIGURE 1.7**
Current combat aircraft. Top left: blended wing aircraft; top right: joined twin fuselage; bottom left, supersonic transport aircraft; bottom right, hypersonic aircraft (all photos reproduced with permission of NASA)

■ **FIGURE 1.8**
White Knight carrying Space Ship Two

Both operators and manufacturers will be alarmed if the price of fuel continues to rise to a point where the air-transportation business finds it difficult to sustain operations. The industry would demand that power plants use alternative fuels such as biofuel, liquid hydrogen (LOH), and possibly nuclear power for large transport aircraft covering long ranges. Aircraft fuelled by LOH have been used in experimental flying for some time, and fossil fuel mixed with biofuel is currently being flight-tested.

A new type of vehicle known as a ground-effect vehicle is a strong candidate for carrying a large payload (e.g. can be bigger than Airbus 380 aircraft) and flying close to the surface, almost exclusively over water. (A ground-effect vehicle is not really new: the Russians built a similar vehicle called the *Ekranoplan*, but it did not appear in the free-market economy.)

Smaller Bizjets and regional jets will morph, and unfamiliar shapes may appear on the horizon, but small aircraft in personal ownership used for utility and pleasure flying are likely to revolutionize the concept of flying through their popularity, similar to how the automobile sector grew. The revolution will occur in short-field capabilities, with vertical takeoffs, and safety issues in both design and operation. Smaller aircraft used for business purposes will see more private ownership to stay independent of the more cumbersome airline operations. There is a good potential for airparks to grow. Various "roadable" aircraft (flying cars) have been designed. The major changes would be in system architecture through miniaturization, automation, and safety issues for all types of aircraft.

1.4.2 Trends in Military Aircraft

Progress in military aircraft would defy all imaginations. Size and shape would be as small as insects (micro-aircraft – dragonfly drones) for surveillance, to larger than any existing kind [9]. Vehicles as small as 15 cm and 1 kg mass have been successfully built for operation. Prototypes much smaller have been successfully flown.

As system-processing power grows, the capability to make weapon delivery decisions advances to an accuracy that could eliminate an onboard human interface, and thereby at one stroke the question of pilot survivability is taken out of the design process, which in turn permits the aircraft to operate at higher load, improving combat capability. Reliance on in-built intelligence would certainly make more remotely piloted vehicles (RPVs) come into in operation. Other terminologies are *unmanned*, *unoccupied*, and *pilotless*. However, unmanned aerial vehicle (UAV) is the prevalent terminology. Nations who can afford it have already entered the race to develop UAVs. Figure 1.9 shows an operational UAV, Ikhana, used for imaging. Futuristic concepts are the Boeing X45A and US Navy X47B (Northrop), as JUCAS (Joint Unmanned Combat Air System).

Once again it is the electronics that would play the main role, although aerodynamic challenges on stealth, manoeuvre and improved capability/efficiency would be as important as structural/material considerations. Engine development would also be a parallel development with all of these discoveries/inventions.

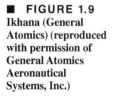

■ **FIGURE 1.9**
Ikhana (General Atomics) (reproduced with permission of General Atomics Aeronautical Systems, Inc.)

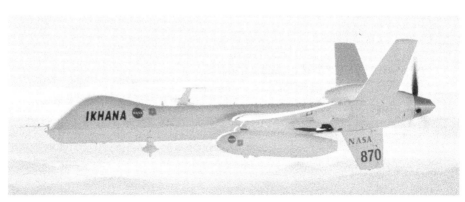

1.4.3   Forces and Drivers

This section discusses the current status of forces and drivers that control design activities. The current aircraft design strategy is linked to industrial growth, which in turn depends on national infrastructure, governmental policies, workforce capabilities, and natural resources; these are generally related to global economic-political circumstances. More than any other industry, the aerospace sector is linked to global trends. A survey of any newspaper provides examples of how civil aviation is affected by recession, fuel price increases, spread of infectious diseases and international terrorism. In addition to its importance for national security, the military aircraft sector is a key element in several of the world's largest economies. Indeed, aerospace activities must consider the national infrastructure as an entire system. A skilled labour force is an insufficient condition for success if there is no harmonization of activity with national policies; the elements of the system must progress in tandem. Because large companies affect regional health, they must share socioeconomic responsibility for the region in which they are located.

The current status stems from the 1980s when returns on investment in classical aeronautical technologies such as aerodynamics, propulsion, and structures began to diminish. Around this time, however, advances in microprocessors enabled the miniaturization of control systems and the development of microprocessor-based automatic controls, which gave additional weight-saving benefits. Dramatic but less ostensible changes in aircraft management began to be embedded in design. At the same time, global political issues raised new concerns as economic inflation drove man-hour rates to a point at which cost-cutting measures became paramount. In the last three decades of the twentieth century, man-hour rates in the West rose four to six times (depending on the country), resulting in aircraft price hikes (typically by about six times for the Boeing 737 – of course, accompanied by improvements in design and operational capabilities.) Lack of economic viability resulted in the collapse or merger/takeover of many well-known aircraft manufacturers. The number of aircraft companies in Europe and North America shrank by nearly three-quarters; currently, only two aircraft companies (Boeing and Airbus) in the West are producing large commercial-transport aircraft. Bombardier Aerospace and Embraer of Brazil have recently entered the large-aircraft market, joining the Russians, the Chinese and the Japanese. Over time, aircraft operating cost terminologies have evolved, and currently, the following standardized definitions are used in this book:

IOC (indirect operating cost):	Comprises costs not directly involved with the sortie (trip).
COC (cash operating cost):	Comprises the trip (sortie) cost elements.
FOC (fixed operating cost):	Comprises cost elements even when not flying but related to trip cost.
DOC (direct operating cost):	= COC + FOC.
TOC (total operating cost):	= IOC + DOC.

1.5   Airworthiness Requirements

From the days of barnstorming and stunt flying in the 1910s, it became obvious that commercial interest had the potential to short-circuit safety considerations. Government agencies quickly stepped in to safeguard people's security and safety without deliberately harming commercial interest. Western countries developed and published thorough systematic rules – these are in the public domain (see relevant websites). In civil applications, they are the Federal Aviation

Regulations or FAR and Certification Standards published by the European Aviation Safety Agency, EASA (formerly Joint Aviation Requirements (JARs) defined by the Joint Aviation Authorities, JAA); both are quite close. The author prefers to work with the established FAR at this point. In military applications, the standards are Milspecs (US) and Defense Standard 970 (earlier AvP 970 – UK); they do differ in places.

The US Government have 50 titles of Code of Federal Regulations (CFRs) published in the Federal Register, covering wide areas subject to federal regulations. The Federal Aviation Regulations (or FARs), are rules prescribed by the Federal Aviation Administration (FAA) governing all aviation activities in the US under title 14 of the CFRs, which covers wide varieties of aircraft-related activities in many parts, of which this book deals mainly with Parts 23, 25, 33 and 35. However, another set of regulations in Title 48 of CFRs is the Federal Acquisitions Regulations, and this has led to confusion with the use of the acronym "FAR". Therefore, the FAA began to refer to aerospace-specific regulations by the term "14 CFR part XX" instead of FAR. There is a growing tendency in the industry to adapt to using 14 CFR part XX. However, to retain the use of FAR meaning Federal Aviation Regulation is still acceptable, and in this book the authors continue with the use of the older practice of the term FAR.

Safety standards were developed through multilateral discussions between manufacturers, operators and government agencies, which continue even today. These minimum standards come as regulations and are mandatory. The regulatory aspects have two kinds of standards, as follows.

- **Airworthiness Standards**: These concern aircraft design by the manufacturers complying with regulatory requirements to ensure design integrity for the limiting performance. These are outlined in FAR 25/JAR 25 in extensive detail in a formal manner, and are revised when required. After substantiating the requirements through extensive testing, an Aircraft Flight Manual (AFM) is issued by the manufacturers for each type of aircraft designed.
- **Operating Standards**: These concern the technical operating rules to be adhered to by the operators, are outlined in FAR 121/JAR-OPS-1 in extensive detail in a formal manner, and are revised when required. The aircraft operational capabilities are substantiated by the manufacturer through extensive flight tests and are certified by the government certification agencies (e.g. FAA/JAA). The contents of the AFM are recast in a Flight Crew Operating Manual (FCOM) that outlines the aircraft limitations and procedures, along with the full envelope of aircraft performance data. Today, with the integration of computers in aircraft operation, it is possible to monitor aircraft performance (APM) for optimum operations. Today, the operational aspects require full understanding of operating microprocessor-based aircraft design.

In civil aviation, every country requires safety standards to integrate with their national infrastructure and climatic conditions for aircraft operation, as well as to relate to their indigenous aircraft designs. Therefore, each country started with their own design and operations regulations. As aircraft started to cross international borders, the standards for foreign-designed aircraft had to be re-examined and possibly re-certified to allow safe operation within their country. To harmonize the diverse nature of the various demands, the International Civil Aviation Organization (ICAO) was formed in 1948, to recommend the international minimum recommended standards. It has now become legal for international practice. However, within each country their own operational regulations might still apply; while countries in North America and some European countries adopt FAR 121, some other European countries follow JAR-OPS-1.

Aircraft operation is prone to litigation, as mishaps do occur. To avoid ambiguity as well to ensure clarity to design, FAA documentations are written in a very elaborate and articulated manner, demanding in-depth study in order to understand and apply them. It is for this reason this book does not exactly copy the FAR lines, but instead quotes the relevant Part number, outlining the requirements with explanations and supported by worked examples. The authors recommend that readers access the latest FAA publication; their web site at http://www.faa.gov/regulations_policies/faa_regulations/should prove useful. Most academic/aeronautical institute libraries necessarily keep FAR documents. For those in industry, these documents will be available there. Aeronautical engineering does not progress without these documents to guarantee a minimum safety in design and operation.

The FAR (14CFT) Part 25 has the most stringent airworthy compliance requirements. The FAR 23 (general aviation aircraft) and the FAR Part 103 (ultra-light aircraft) have considerably lower levels of requirements and use the same performance equations for analyses. This book deals only with the FAR (14CFR) Part 25.

1.6 Current Aircraft Performance Analyses Levels

Aircraft performance analysis is needed at the very early stages of the conceptual design phase and continues in every phase of the programme, updating capabilities as more accurate data are available until it is substantiated through flight tests. At the conceptual stage the performance prediction has to be sufficiently accurate to obtain management "go-ahead" for a programme that bears promise of eventual success. In the next phase the performance figures are fine-tuned to give a guarantee to potential operators. Industry must be able to perform aircraft performance analysis to a high degree of accuracy.

The analyses of aircraft performance cascade down from the preliminary study to final refinement by design engineers, followed by flight test substantiation, and eventually the engineers preparing the aircraft flight manuals (AFM) and the flight crew operating manual (FCOM) for operational usage. The various levels where aircraft performances are evaluated are briefly given below.

By the designers

(i) *At research level (feasibility study)*: In this stage, engineers examine new technologies and their capabilities to advance new aircraft designs, and examine possible modifications to improve existing designs. At this level, researchers explore newer aircraft performance capabilities, and optimize operational procedures using close-form equations that yield quick results for comparison and selection.

(ii) *Conceptual design level (Phase I of a project)*: This is the outcome of the feasibility study showing potential to progress the design towards market launch. In this phase, the study needs to be done in a specific manner to fix configurations in a family framework by sizing the aircraft with matched engines. In this phase, full aircraft analysis is not required. It only covers what is required to speak with potential customers with promising performance specifications sufficient to make comparisons with the competition. If successful, go-ahead for the programme is given at the end of the study.

(iii) *Detailed design level (Phases II and III)*: This is the post go-ahead phase analysis to give guarantees to potential customers. By now, more aerodynamic information is available through wind tunnel testing and CFD analyses. More detailed and accurate aircraft performance estimations are now possible.

(iv) *Final level (Phase IV)*: This is the final design phase, and aircraft performance tests are carried out to obtain certification of airworthiness. All technical/engineering and ground/flight-tested substantiation data are then passed to a dedicated group to prepare the aircraft flight manual (AFM) and the flight crew operating manual (FCOM)

(FCOM) for operational usage. The format of presenting aircraft performance in the AFM and the FCOM is different from the format of aircraft performance documents used by the designers; the former are derived from the latter. The performance documents prepared during Phases I, II and III, and used by engineers, contain predictive data that are substantiated through ground/flight tests. The full set of engineering data is given to experienced performance engineers at the dedicated customer support group who prepare the AFM and FCOM manuals for the airline operators. Typically, the design office uses the prevailing terminologies, but the AFM and the FCOM must incorporate standard formal terminologies specified by the airworthiness agencies to avoid any ambiguity. While preparing the operational manuals does not involve extensive computation, it requires articulated presentation since errors or lack of clarity are not acceptable. This book follows the typical terminologies used by engineers, along with introducing the synonymous formal terminologies to keep the readers informed.

By the operators

Using the manufacturer-supplied AFMs and FCOMs, the operators have their own performance engineers conduct analyses; the extent depends on the operators' strategic plans. Typically, these analyses are not as extensive as what the designers do, as it is not required. They cover: (i) market comparison to make selection of aircraft type; (ii) city-pair route planning (new or old routes); (iii) performance revisions on account of repairs; (iv) design modifications to improvement performance; and (v) accident/incidence analyses. Operating cost analyses form one of the core aspects of the study.

1.7   Market Survey

In a free market economy, industry cannot survive unless it grows; governmental subsistence can only be seen as a temporary relief. The starting point to initiate a new aircraft design project is to establish the key drivers – the requirements and objectives based on market, technical, certification and organizational requirements. These drivers are systematically analysed and then documented by the aircraft manufacturers (Table 1.2). Documents, in several volumes, describing details of the next layer of design specifications (requirements), are issued to those organizations involved with the project. Market surveys determine customer requirements, and user feedback guides the product. In parallel, the manufacturers incorporate the latest, but proven, technologies to improve design and stay ahead of the competition, always constrained

■ **TABLE 1.2**
The drivers leading to the final design

Market drivers from operators	Regulatory drivers from government	Technology drivers from industry
Payload-range, speed	Airworthiness regulations	Aerodynamics
Field performance	Policies (e.g. fare deregulation)	Propulsion
Comfort level	Route permission	Structures
Functionality	Airport fees	Materials
Maintenance	Interest rates	Avionics/electrical
Support	Environmental issues	Systems, fly-by-wire
Aircraft family	Safety issues	Manufacturing philosophy

by the financial viability of what the market can afford. Dialogue between manufacturers and operators continues all the time to bring out the best in the design.

Military product development has a similar approach, but would require some modifications to Table 1.2. Here, government is both the single customer and the regulatory body. Therefore, competition is only between the bidding manufacturers. The market is replaced by the operational requirements arising from perceived threats from potential adversaries. Column 1 of Table 1.2 then becomes "Operational Drivers", which includes weapons management and counter intelligence. Hence, this section on market survey is divided into civil and military customers as shown in Table 1.3. "Customer" is a broad-based term and is defined in this book in the manner given in Table 1.3.

In the UK military, the Ministry of Defence (MoD), as the single customer, searches for a product and floats a Request for Proposal (RFP) to the national infrastructure, where most manufacturers are run privately. It is nearly the same in the US under different terminologies. Product search is a complex process – the MoD must know the potential adversary's existing and future capabilities, and administrate national RD&D infrastructures to be ready with discoveries/innovations to supersede the adversary's capabilities. The Air Staff Target (AST) is an elaborate aircraft specification as customer requirement. A military project is of national interest and in today's practice the capable companies are invited first to produce a 'Technology Demonstrator' as proof of concept. The loser in the competition gets paid by the government for the demonstrator and learns a lot about very advanced technology for the next RFP or civilian design (so that, in a true sense, there is no loser) and the nation hones its technical manpower.

Although used, the authors do not think that RFP is an appropriate terminology in civil applications – here, who is making the request? It is important for the aircraft manufacturers to know the requirements of many operators and supply a product that meets the market's demands in performance, cost and time frames. Airline, cargo and private operators are the direct customers of the aircraft manufacturers, who do not have direct contact with the next level of customers (i.e. passengers and cargo handlers). Airlines do their market surveys of passenger and freight requirements and pass the information to the manufacturer. These are often established by extensive studies of target city pairs, current market coverage and growth trends, and passenger input. Their feedback comes with diverse requirements that need to be coalesced to a marketable product. A large order from a single operator could start a project, but manufacturers must cater for many operators to enlarge and stabilize their market share. The civil market is searched through a multitude of queries to various operators (airlines), nationally and internationally. In civil aviation, the development of national infrastructure must be run in coordination with the aircraft manufacturers and operators to ensure national growth. Airlines generate revenue by carrying passengers and freight; these provide the cash flow that supports the maintenance and development of the civil aviation infrastructure. Cargo generates important revenues for airlines and airports, and its market should not be underestimated, even if it means modifying older airplanes. Manufacturers and operators remain constantly in touch with each other in order to develop product lines with new and/or modified aircraft. The aircraft manufacturers need to

■ **TABLE 1.3**
Customers of aircraft manufacturers

	Civil customers	Military customers
Top level	Airline/cargo/private operators	MoD (single)
Next level	Passengers and cargo handlers	Foreign MoD
Revenue	Cash flows back through revenue earned	No operational revenue, possible export revenue

harmonize diversity in requirements in order to arrive at a point where the management decides to undertake a conceptual study to obtain "go ahead". This is nothing close to the route taken by the MoD to initiate a RFP with a single customer demand.

The private or executive aircraft market is driven by operators who are closely connected to business interests and cover a wide spectrum of types, varying from four passengers to specially modified mid-sized jets.

Military aircraft utilization in peacetime is approximately 7500 hours, about a tenth that of commercial transport aircraft ($\approx$75,000 hours), in its life-span. Peacetime military aircraft yearly utilization is very low (around 600 hours) compared with civil aircraft yearly utilization, which can exceed 3000 hours.

1.8 Typical Design Process

The typical aircraft design process follows the classical pattern of a systems approach. The official definition of a system adopted by the International Council of Systems Engineering (INCOSE – [10]) is that "a 'system' is an interacting combination of elements, viewed in relation to function". The design "system" has an input (a specification/requirement), which undergoes a process (phases of design) to obtain an output (certified design through substantiated aircraft performance), as shown in Figure 1.10.

1.8.1 Four Phases of Aircraft Design

Aircraft manufacturing organizations conduct round-the-year exploratory work on research, design and technology development, as well as market analysis to search for a product, and when it is found, the project gets formally initiated in the four phases as shown in Figure 1.11, which is valid for both civil and military projects.

From organization to organization, the terminologies of the phases vary. The difference between the terminologies is trivial, as the task breakdown covered in various phases is about the same. For example, some may see Market Study and Specification Requirements as Phase 1, making Conceptual Study as Phase 2; some may define the Project Definition Phase (Phase 2) and Detailed Design Phase (Phase 3) as the Preliminary Design Phase and Full Scale Development Phase, respectively. Some would prefer to invest early on risk analysis in Phase 1, but it could be done at Phase 2 when the design is better defined, saving Phase 1 budgetary provision in case the project fails to get a "go-ahead". Military programmes may require early risk analysis as they would be incorporating technologies yet to be proven in operation. Some may see disposal of aircraft at the end as part of the design phase of a project. Figure 1.11 offers a typical/generic pattern prevailing in industry.

Aircraft performance analyses are carried out in all four phases at various levels, as given below. The formulation of physics is the same for all; the difference lies in the extent of coverage and rigour for the performance evaluation required.

■ **FIGURE 1.10**
The aircraft design process

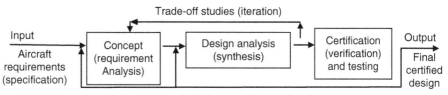

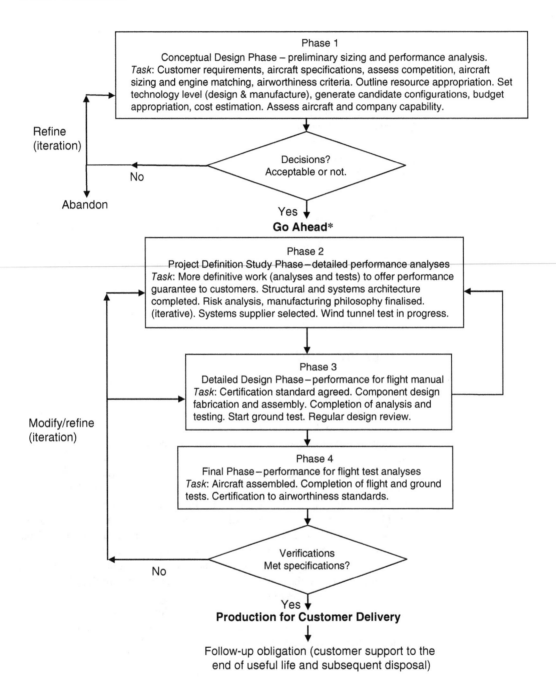

The four phases of aircraft design and development are shown in the figure, including:

Phase 1 — Conceptual Design Phase – preliminary sizing and performance analysis. Task: Customer requirements, aircraft specifications, assess competition, aircraft sizing and engine matching, airworthiness criteria. Outline resource appropriation. Set technology level (design & manufacture), generate candidate configurations, budget appropriation, cost estimation. Assess aircraft and company capability.

Refine (iteration) — No — Abandon

Decisions? Acceptable or not. — Yes — Go Ahead*

Phase 2 — Project Definition Study Phase – detailed performance analyses. Task: More definitive work (analyses and tests) to offer performance guarantee to customers. Structural and systems architecture completed. Risk analysis, manufacturing philosophy finalised. (iterative). Systems supplier selected. Wind tunnel test in progress.

Phase 3 — Detailed Design Phase – performance for flight manual. Task: Certification standard agreed. Component design fabrication and assembly. Completion of analysis and testing. Start ground test. Regular design review.

Modify/refine (iteration)

Phase 4 — Final Phase – performance for flight test analyses. Task: Aircraft assembled. Completion of flight and ground tests. Certification to airworthiness standards.

Verifications Met specifications? — No — Yes — Production for Customer Delivery

Follow-up obligation (customer support to the end of useful life and subsequent disposal)

* Some companies may delay "go ahead" until more information is available. Some Phase 2 tasks (e.g. risk analysis) may be carried out as a Phase 1 task to obtain "go ahead".

■ **FIGURE 1.11**   **Introduction to the four phases of aircraft design and development**

Phase 1 *Conceptual Design Phase*: In this phase, preliminary performance studies are conducted to size aircraft for a family of variants and find matched engines to meet market specifications: takeoff and landing, speeds at initial climb and cruise, and meeting the payload-range. These evaluations are primarily used by management to arrive at the "go-ahead" decision and also by potential customers. Expected accuracy of the results should be within less than ±5%.

Phase 2 *Project Definition Study Phase*: After "go-ahead" is obtained, more definitive work (analyses and tests) are carried out in this phase to offer aircraft performance guarantees to customers. This also offers an opportunity to refine sizing and engine matching before metal cutting starts. Expected accuracy should be within less than ±3%.

Phase 3 *Detailed Design Phase*: This is the time when accurate aircraft performances for the flight manual are carried out to some agreed certification standards. The equations of performance analyses are the same, but evaluated in detail for the full flight envelope for the allowable climatic conditions. Expected accuracy should be within less than ±2%.

Phase 4 *Final Phase*: Aircraft performance for flight test analyses to calibrate with estimation. This is to ensure that aircraft performance does not fall short of the guaranteed values.

The methodologies presented in this book should cater for aircraft performance analyses for all the four stages. Details of activities of the various phases are described in the next sections.

Table 1.4 suggests a generalized functional envelope of aircraft design architecture, which is in line with the index given for commercial transport aircraft by the Aircraft Transport Association (ATA) [11], which recently changed name to A4A, Airlines for America. Further breakdowns of subsystems are given in the respective chapters.

The components of the aircraft as subsystems exist interdependently in a multidisciplinary environment, even if they have the ability to function on their own. For example, wing flap deployment on the ground is inert, while in flight it affects the vehicle motion. Individual components, such as the wing, nacelle, undercarriage, fuel system, air-conditioning, and so on, can also be seen as subsystems. Components are supplied for structural and system testing in conformance with airworthiness requirements in practice. Close contact is maintained with

■ **TABLE 1.4**
Aircraft systems

Design	Operation
Aerodynamics	Training
Structure	Product support
Power Plant	Facilities
Electrical/avionics	Ground/office
Hydraulic/pneumatic	Flight operations
Environmental control	
Cockpit/interior design	
Auxiliary systems	
Production engineering feedback	
Testing and certification	

planning engineers to ensure that production costs are minimized and to ensure that build tolerances are consistent with design requirements.

Extensive wind tunnel, structure and systems testing will be required early in the design cycle to ensure safe flight tests, leading to airworthiness certification approval. The multidisciplinary "systems" approach to aircraft design is carried out within the context of IPPD. The generic methodology has four phases (see next section) to get a new aircraft conceived, designed, built and certified. Civil projects usually proceed to pre-production build aircraft, which will be flight-tested and sold subsequently.

Military projects proceed with technology demonstrators as prototypes before "go-ahead" is given. Military technology demonstrators are normally scaled-down aircraft meant to substantiate untried cutting-edge technology. These are not sold for operational usages.

Company management sets up a "design built team" (DBT) to meet at regular intervals to conduct design reviews and make decisions on the best compromises through multidisciplinary analysis (MDA) and multidisciplinary optimization (MDO) as shown in Figure 1.12 – this is what is meant by IPPD (concurrent engineering) environment.

Specialist areas may optimize their design goals, but in the IPPD environment, compromise has to be sought. *Optimization of individual goals through separate design considerations may prove counterproductive and usually prevent the overall (global) optimization of ownership cost.* MDO offers good potential, but to obtain global optimization is not easy; it is still evolving. In a way, a global MDO, involving large numbers of variables, is still an academic pursuit. Industries are in a position to use sophisticated algorithms in some proven areas. One such situation is to reduce manufacturing cost by reshaping component geometry as a compromise to lower cost (i.e. to minimize complex component curvature). To offer a family of

■ **FIGURE 1.12**
Multidisciplinary analysis (MDA) and optimization (MDO) flowchart (IPPD)

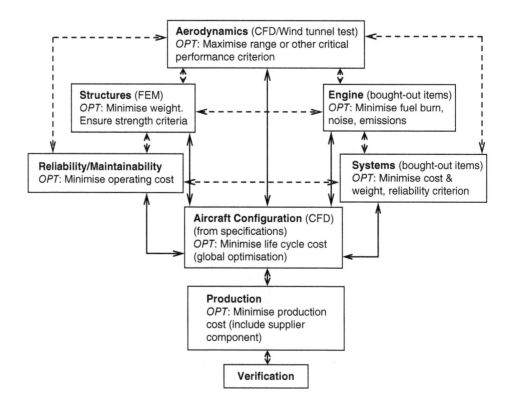

variant aircraft the compromises are evident, as none of the individuals in the family are optimized, but together they offer the best value for money.

Once the aircraft has been delivered to the operators (customers), a manufacturer is not free from obligations. Manufacturers continue with support work on maintenance, design improvements and attention to operational queries, right up to the end of aircraft life. Modern designs are expected to achieve three to four decades of operation. Manufacturers may even face litigation if customers find cause to sue. Compensation payments have crippled some famous general aviation names. Fortunately, the 1990s saw a relaxation of the litigation laws in general aviation – after a certain period of time when a design is established, the manufacturer's liabilities are reduced – which resulted in a revitalization of the general aviation market. Military programmes involve support from "cradle to grave", that is, from delivery of the new aircraft to end of service life.

It is important to emphasize that the product must be "right-first-time". Mid-course changes add needless cost that could hurt the project – a big change may not even prove sustainable. Procedural methodologies such as the Six-Sigma approach have been devised to make sure changes are minimized.

1.9   Classroom Learning Process

To meet our objectives to offer close-to-industrial practice in this book, it would be appropriate here to harmonize some of the recognized gaps between academia and industry as discussed in [12] to [19]. Before we embark on dealing with the aircraft performance analyses, we will lay out our intended classroom learning process, as previously tested in industry and in academia.

It is clear that unless the engineer has sufficient analytical ability, it will be impossible for him/her to convert creative ideas into profitable product. Today's innovators who have no analytical and practical skills must depend on engineers to accomplish routine tasks under professional investigation, and to make necessary decisions to develop an idea into a marketable product.

Traditionally, universities develop analytical abilities by offering the fundamentals of engineering science. Courses are structured with all the material available in textbooks or notes; problem assignments are straightforward with unique answers. This may be termed a "closed-form" education. Closed-form problems are easy to grade and a teacher's knowledge is not challenged (relatively). Conversely, industry requires the tackling of "open-form" problems for which there is no single answer. The best solution is the result of interdisciplinary interaction of concurrent engineering within design built teams (DBTs), in which total quality management (TQM) is needed to introduce "customer-driven" products at the best value. Offering open-ended courses in design education that cover industrial requirements is more difficult and will challenge a teacher, especially when industrial experience is lacking. The associative features of closed- and open-form education are shown in Figure 1.13 [19].

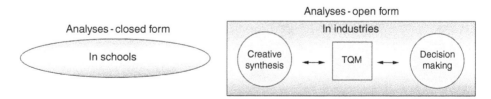

■ FIGURE 1.13   Associative features of "closed" and "open" form education (adapted from [19], American Institute of Aeronautics and Astronautics, Inc.)

To meet industry's needs, newly graduated engineers need a brief transition time before they can become productive, in line with the specialized tasks assigned to them. They must have a good grasp of the mathematics and engineering sciences necessary for analysis and sufficient experience for decision-making. They must be capable of working under minimal supervision, with the creative synthesis that comes from experience that academia cannot offer. The industrial environment will require new recruits to work in a team, with an appreciation of time, cost, and quality under TQM, which is quite different from classroom experience. Today's conceptual aircraft designers must master many trades and specialize in at least one, not ignoring the state-of-the-art "rules of thumb" gained from past experiences; there is no substitute. They need to be good "number-crunchers" with excellent analytical ability. They also need assistance from an equally good support team to encompass wider areas. This is the purpose of the coursework in this book: to provide close-to-industry standard computations and engineering approaches necessary for analysis, and enough experience to work in a team.

For this reason, the authors emphasize that introductory class-work projects should be familiar to students so that they can relate to the examples and subsequently substantiate their work with an existing type. Working on an unfamiliar or non-existent design does not enhance the learning process at the introductory level. In industry, aircraft performance analyses are fully computerized for every phase of project work. However, in the classroom it is recommended to perform manual computation using spreadsheets.

Today, use of computer-aided design (CAD) is an integral part of engineering analyses. As an example, Figure 1.14 gives a 3D CAD drawing of an F16 fighter aircraft. 3D modelling provides fuller, more accurate shapes that are easy to modify, and facilitates maintenance of sequential configuration evolution. Accurate geometric details from CAD can be easily retrieved for drag estimation by the indispensable manual method.

There are considerably more benefits from CAD (3D) solid modelling; it can be uploaded directly into computational fluid dynamics (CFD) analysis to continue with aerodynamic estimations, as one of the first tasks is to estimate loading for structural analysis using the finite element method (FEM). The solid model offers accurate surface constraints for generating internal structural parts. CAD drawings can be uploaded directly to computer-aided manufacture (CAM) operations, ultimately leading to paperless design and manufacture offices. Vastly increased computer power has reached the desktop with parallel processing. Computer-aided engineering (CAE: e.g. CAD, CAM, CFD, FEM and systems analyses) is the accepted practice in the industry. Those who can afford supercomputers will

■ **FIGURE 1.14**
Typical 3D CAD drawing of F16 (reproduced with permission of Pablo Quispe Avellaneda, Naval Engineer, Peru)

have the capability to conduct research in areas hitherto unexplored or facing limitations (e.g. high-end CFD, FEM and multidisciplinary optimization (MDO)).

Finally, the authors recommend that performance engineers have some flying experience, which is most helpful in understanding the flying qualities of aircraft they are trying to analyze. Obtaining a licence requires effort and financial resources, but a few hours of planned flight experience would be instructive. One may discuss with the flight instructor what needs to be demonstrated, for example, aircraft characteristics in response to control input, stalling, "*g*" force in steep manoeuvres, stick forces, and so forth. Some universities offer a few hours of flight-tests as an integral part of aeronautical engineering courses; however, the authors suggest even more: hands-on experience under the supervision of a flight instructor. A driver with a good knowledge of the design features has more appreciation for the automobile.

1.10   Cost Implications

The authors emphasize here that there is a significant difference between civil and military programmes in predicting costs related to aircraft unit-price costing. The civil aircraft design has an international market with cash flowing back from revenues earned from fare-paying customers (i.e. passengers and freight) - a regenerative process that returns funds for growth and sustainability to enhance the national economy. Conversely, military aircraft design originates from a single customer demand for national defence and cannot depend on export potential – it does not have cash flowing back, and it strains the national economy out of necessity. Civil aircraft designs share common support equipment and facilities, which appear as indirect operational costs (IOCs) and do not significantly load aircraft pricing. The driving cost parameter for civilian aircraft design is the DOC, omitting the IOC component. Therefore, using a generic term of life cycle cost (LCC) = (DOC + IOC) in civil applications may be appropriate in context, but would prove to be off track for aircraft design engineers. Military design and operations incorporating discreet advances in technology necessarily have exclusive special support systems, equipment and facilities. The vehicles must be maintained for operation-readiness around the clock. Part of the supply costs and support costs for aircraft maintenance must be borne by manufacturers that know best and are in a position to maintain confidentiality on the high-tech defence equipment. The role of a manufacturer is defined in the contractual agreement to support its product for the entire life cycle of the aircraft "from cradle to grave" Here, LCC is meaningful for aircraft designers in minimizing costs for the support system integral to the specific aircraft design. Commercial transports would have nearly five times more operating hours than military vehicles in peacetime. Military aircraft have relatively high operating costs even when they sit idle on the ground. Academic literature has not been able to address clearly the LCC issues in order to arrive at an applicable standardized costing methodology. Aircraft design strategy is constantly changing. Initially driven by the classical subjects of aerodynamics, structures, and propulsion, the industry is now customer-driven and the design strategies consider the problems for manufacture/assembly that lead the way in reducing manufacturing costs. In summary, an aircraft engineer must be cost-conscious now, and even more so in future projects. Reference [20] addresses cost considerations in detail.

Aircraft design and manufacture are not driven by cost estimators and accountants; they are still driven by engineers. Unlike classical engineering sciences, costing is not based on natural laws; it is derived to some extent from manmade policies, which are rather volatile, being influenced by both national and international origins. The sooner the engineers include costing as an integral part of design, the better will be the competitive edge.

1.11   Units and Dimensions

The postwar dominance of British and American aeronautics has kept the use of the foot-pound-second system (FPS, also known as the Imperial System) current, despite the use of non-decimal fractions and the ambiguity of the word *pound* in referring to both mass and weight. The benefits of the Système International (SI) are undeniable: a decimal system and a distinction between mass and weight. However, there being "nowt so queer as folk," I am presented with an interesting situation in which both FPS and SI systems are used. Operational users prefer FPS (i.e. altitudes are measured in feet); however, scientists and engineers find SI more convenient. This is not a problem if one can become accustomed to the conversion factors. Appendix A provides an exhaustive conversion table that adequately covers the information in this book. However, readers will be relieved to know that in most cases, the text follows current international standards in notation units. Aircraft performance is conducted at the International Standard Atmosphere (ISA) (see Section 2.2). References are given when design considerations must cater to performance degradation in a non-standard day.

1.12   Use of Semi-empirical Relations and Graphs

DATCOM (US) and RAE DATA sheets (UK, recently replaced by ESDU) served many generations of engineers for more than a half century and are still in use. Over time, as technology advanced, new tools using computer-aided engineering (CAE) have somewhat replaced earlier methods. Inclusion of many of DATCOM/ESDU semi-empirical relations and graphs proves meaningless unless their use is shown in worked examples. It is important for instructors to compile as many test data as possible in their resources.

Semi-empirical relations and graphs cannot guarantee exact results; at best it is coincidental if they prove to be error-free. A user of semi-empirical relations and graphs must be aware of the extent of error that can incur. Even when providers of semi-empirical relations and graphs give the extent of the error range, it is difficult to substantiate any errors in a particular application.

If test results are available, they should be used in conjunction with the semi-empirical relations and graphs. Tests (e.g. aerodynamics, structures and systems) are expensive to conduct but they are indispensable to the process. Certifying agencies impose mandatory requirements on manufacturers to substantiate their designs with test results. These test results are archived as a databank to ensure that in-house semi-empirical relations are kept "commercial in confidence" as proprietary information. CFD and FEM are next in priority. The consistency of CFD in predicting drag has to be proven conclusively. At this stage, semi-empirical relations and graphs are used extensively in drag prediction as well as weight prediction.

Data reading from graphs is normal engineering practice, since graphs are readily available and data can be quickly obtained. But not all data are computerized (a good example is the general use of drag polar), and accuracy in reading data from graphs depends on their resolution. The graphs given in this book are small and do not have adequate resolution, therefore any readings are unlikely to be accurate. A good example will be shown in Section 9.9.1. It is recommended that the readers plot graphs with consistent accurate data in high resolution using high-end graph-plotting software.

1.13   How Do Aircraft Fly?

The mechanics of flight stems from the interaction between wind (air) and wing. A special property of air (gas) is the ability to generate lift through *wing–wind* interaction. Nature is conservative. Mass, momentum and energy in airflow is conserved unless there is an external intervention. Static pressure of the system is a form of energy (potential), and velocity is its kinetic energy. Together, the total energy is conserved – if one is changed, then the other is affected. For example, if velocity is increased, then its static pressure drops, and vice versa. This phenomenon is expressed as Bernoulli's Theorem.

A typical bread-slice-like wing section is known as an aerofoil, and its upper surface is more curved than the lower surface. Therefore, airspeed over the wing is faster (reducing its static pressure) than across the lower surface, resulting in a pressure difference directed upward. A sized wing area at a particular speed needs to generate requisite force (lift) to keep an aircraft in sustained flight. There is a minimum aircraft speed (stall speed) below which the wing will stall and will not develop sufficient lift. More details are given in Chapter 2. The entire subject matter comprises flight mechanics and its associated aerodynamic theories.

1.13.1   Classification of Flight Mechanics

The subject of flight mechanics may be divided into four subtopics:

1. **Performance:** The study of how an aircraft performs in terms of its kinematics, which is dependent on aerodynamic characteristics such as lift, drag and moments, and engine characteristics. It involves estimation of the extent of aircraft capability, which can be divided broadly as follows:
   a. Point performance: the aircraft capability at an instant, involving rate capabilities (e.g. speed, climb rate, descent rate, turn rate, etc.).
   b. Integrated performance: the aircraft capability integrated over a time period (e.g. takeoff and landing field length, climb to height, descent, mission range, etc.).
2. **Static stability:** The study of the tendency of the aircraft to remain in steady level flight when slightly perturbed. This leads to the prediction of the control movements and control forces required to change airspeed or load factor. This in turn leads to the idea of *handling qualities*. Longitudinal static stability is introduced in Chapter 6, as this affects aircraft performance.
3. **Dynamic stability:** The study of the motion of the aircraft after it has been disturbed in some fashion. The motions are classically divided into *modes of motion*, and the characteristics of the modes of motion are used to predict the flying qualities of the aircraft. Some modes are more important than others. The five classical modes are described in Chapter 6, but not treated in this book.
4. **Control:** The study of the effect of controls on the flying characteristics of the aircraft. Control is not dealt with in this course. It is treated as a separate subject.

The topic of this book is the first: the performance of aircraft. Aircraft design characteristics influence performance, and so their study is an integral part of understanding performance.

1.14   Anatomy of Aircraft

The study of aeronautics requires familiarity with aircraft configuration and the relevance of its components. The conventional aircraft configuration can be decomposed into the following eight sections.

1. **Fuselage:** This might crudely be regarded as the part of the aircraft that performs the function for which the aircraft was designed – carrying passengers, freight, electronic communications or munitions.

2. **The main wing:** The wing generates most of the lift required for flight. *Dihedral angle* or *sweep* on the wings provides *lateral (roll) stability*. *Flaps* on the trailing edge operate in sympathy and are used to enhance the lifting ability of the wing at low airspeeds. *Leading edge slats* and *wing spoilers* might be found on more complex wings.

3. **Ailerons:** These outboard flaps operate differentially and are used to control the roll rate of the aircraft. (See below for a little more detail.)

4. **Empennage:** The empennage comprises the horizontal (*tailplane*) and vertical (*fin*) surfaces at the rear of the aircraft.

5. **Tailplane:** The tail plane provides *longitudinal (pitch) stability*. The *elevator* is the flap on the tail plane and is used to control the aircraft in pitch. The *trim tab* is a small flap on the trailing edge of the elevator; it is used to balance the aerodynamic loads on the elevator in order to reduce the effort required of the pilot to maintain airspeed.

6. **Fin:** The fin provides *directional (yaw) stability*. The flap on the fin is called the *rudder* and is used to control the sideslip angle of the aircraft.

7. **Powerplant:** The power plant provides the thrust that balances the drag of the aircraft. Without the power plant, the aircraft is a glider. Power plants may be piston-props, turboprops or turbofans (see Chapter 4).

8. **Undercarriage:** The undercarriage or landing gear allow for safe operations on the ground (taxiing, takeoff, landing).

Both civil and military aircraft have different categories of design to cater for specific mission roles, and therefore aircraft performance capabilities will show wide variation. Section 10.4 describes in detail the various types of mission profiles for both civil and military aircraft.

The typical aircraft components of large aircraft are shown in Figure 1.15. The obvious components are generic (e.g. wing, fuselage, nacelle, empennage, control surfaces, etc.) for all types. Less obvious ones are typically winglets and strakes, but they play vital roles – otherwise they would not be there. There are many options. For convenience, components are associated in groups as described below. Not shown in the figure are the trimming surfaces used to reduce control forces experienced by the pilot.

Fuselage group: This starts with the nose cone, and then the constant mid-section fuselage, followed by the tapered aft fuselage, and at the end is the tail cone. The fuselage belly fairing (shown in Figure 1.15 as several sub-assembly components below the fuselage) may be used to house equipment at the wing–fuselage junction, such as the undercarriage wheels.

Wing group: This comprises the main wing, high lift devices, spoilers, control surfaces, tip devices and the structural wing box that passes through the fuselage. High lift devices include leading edge slats or trailing edge flaps; in Figure 1.15, the leading edge slats are shown attached with the main wing but the trailing edge flaps and spoilers are shown detached from the port wing. Spoilers are used to decelerate aircraft on descent and as the name suggests they spoil lift over the wing and are useful as "lift dumpers" on touchdown; thereby the undercarriage more rapidly absorbs the aircraft's weight, allowing more effective application of brakes. In some aircraft, small differential deflections of spoilers with or without the use of ailerons are used to stabilize aircraft rolling tendencies in disturbances. The wing is shown with winglets at the tip: winglets are one of a set of tip treatments that can reduce the induced drag of the aircraft.

Empennage group: The empennage is the set of stability and control surfaces at the back of the aircraft. In Figure 1.15 it is shown as a vertical tail split into the fin in front and the rudder

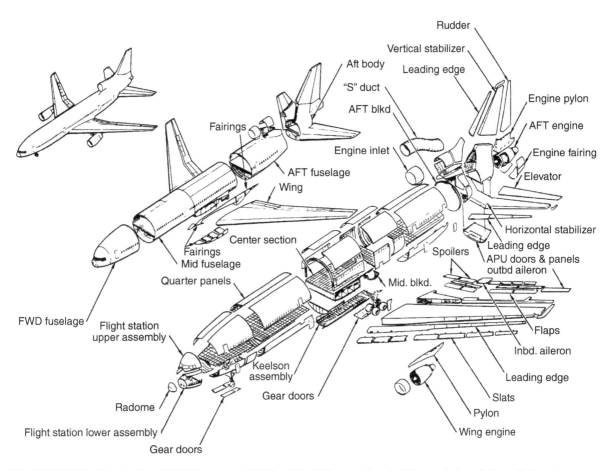

at the back, and an end cap on the top; the horizontal tail, shown as a T-tail set at the top of the vertical tail, comprises the stabilizer and the elevator.

Nacelle group: The podded nacelles are shown mounted on either side of the aft fuselage; pylons effect the attachment.

Undercarriage group: Undercarriage, or landing gear, usually comprises a nose wheel assembly and two sets of main wheels, forming a tricycle configuration. Tail dragging, bicycle and even quad configurations are possible, depending on the application of the aircraft. Wheels are usually retracted in flight, and the retraction mechanism and stowage bay all form part of the undercarriage group.

Military aircraft statistics and geometric details need to be looked at differently on account of the very different mission role. Combat aircraft do not have passengers and the payloads have wide variation in armament type to carry internally and/or externally. Military configurations are more diverse than civil designs. Figure 1.16 shows a blowout diagram for the General Dynamics (now Boeing) F16. The component groups are similar to what is described above, except modern fighters do not have nacelles as the engine is housed inside the fuselage.

■ **FIGURE 1.16**
Military aircraft configuration (courtesy of Michael Niu [20], reproduced with permission of Commlit Press)

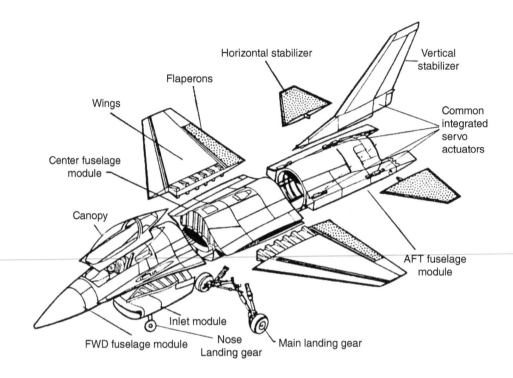

However, in this book a military trainer of the class of RAF Hawk is dealt with. An example of an advanced jet trainer (AJT) with close air support is worked out as a military trainer aircraft performance, greatly simplifying the objective on military aircraft design.

Military aircraft carry mostly externally mounted combat equipment and weaponry, which would affect aircraft performance.

1.14.1 Comparison between Civil and Military Design Requirements

This section compares the two classes of aircraft design: civil and military (Table 1.5). It can be seen how different military aircraft design is compared with civil design.

Readers should consider what might be the emerging design trends within each class of aircraft. In general, new commercial aircraft designs are extensions of existing designs incorporating proven newer technologies (some are fallouts from declassified military applications) in a very conservative manner. Currently, dominant aerodynamic design trends show diminishing returns on investment. Structures technologies are seeking the introduction of suitable new materials (composites, metal alloys, smart adaptive materials) if these can reduce cost and/or weight (or aerodynamic gains). Engine design is still showing aerodynamic improvements to save fuel burn and weight.

1.15 Aircraft Motion and Forces

An aircraft is a vehicle in motion; in fact, it must maintain a minimum speed above the stall speed. The resultant pressure field around the aircraft body (wetted surface) is conveniently decomposed into a usable form for the designers and analysts. The pressure field alters with changes in speed, altitude and orientation (attitude). This book deals primarily with a steady

■ **TABLE 1.5**
Comparison between civil and military design requirements

Issue	Civilian aircraft	Military aircraft
Certification standards	Civil (FAR – US)	Military (Milspecs – US)
Operational environment	Friendly	Hostile
Safety issues	Uncompromised No ejection	Survivability requires ejection
Mission profile	Routine and monitored by ATC	As situation demands and could be unmonitored
Flight performance	Near steady-state operation and scheduled Gentle manoeuvres	Large variation in speed and altitudes – pilot is free to change briefing schedule Extreme manoeuvres
Flight speed	Subsonic and scheduled	Have supersonic segments; in combat unscheduled
Engine performance	Set throttle dependency No afterburner (subsonic)	Varied throttle usage With afterburner
Field performance	Mostly metalled runways generous in length with ATC support	Could have different surfaces with restricted lengths Marginal traffic control
Systems architecture	Moderately complex High redundancies No threat analysis	Very complex Lower redundancies Threat acquisition
Environmental issues	Strictly regulated – legal minimum standards	Relaxed – peacetime operation in restricted zones
Maintainability	High reliability with low maintenance cost in mind	Also with high reliability but at a considerable higher cost
Ground handling	Extensive ground-handling support with standard equipment	Specialized ground support equipment and complex
Economics	Minimize DOC Cash flow-back through revenue earned	Minimize LCC No cash flow-back
Training	Routine	Specialized and more complex

level flight pressure field; the unsteady situation is taken as transient in manoeuvres. Section 2.25 deals with certain unsteady cases (gusty) and reference will be made when occasion demands some design consideration. This section is primarily meant for information on some of the parameters concerning motion and force used in this book.

1.15.1   Motion – Kinematics

Unlike an automobile, which is constrained by the surface of the road, an aircraft is the least restricted vehicle, having all the six degrees of freedom (Figure 1.17) – three linear and three rotational motions along and about the three axes of any coordinate system. In this book, these are represented in the right-handed Cartesian coordinate system. Controlling motion in six degrees of freedom is a complex matter. Very careful aerodynamic shaping of all components of aircraft is of paramount importance, but the wing takes the top priority. Aircraft attitude is measured using Eulerian angles, ψ, θ and ϕ, which are treated in Chapter 4.

In classical flight mechanics, there are many kinds of Cartesian coordinate systems in use. Aircraft have a symmetrical shape, where the left side (port side) is a mirror image of the right side (starboard side). The X and Z axes are in the plane of symmetry (see Figure 1.17).

■ **FIGURE 1.17**
The six degrees of freedom
of aircraft motion

■ **TABLE 1.6**
**Summary of the components of the six degrees of freedom in the
body-fixed coordinate system, F_B**

Linear velocities:	u along X-axis (+ve forward)
	v along Y-axis (+ve right)
	w along Z-axis (+ve down)
Angular velocities:	p about X-axis, known as roll (+ve)
	q about Y-axis, known as pitching (+ve nose up)
	r about Z-axis, known as yaw (+ve)
Angular acceleration:	$\dot{p}$ about X-axis, known as roll (+ve)
	$\dot{q}$ about Y-axis, known as pitching (+ve nose up)
	$\dot{r}$ along Z-axis, known as yaw (+ve)

There are a few asymmetrical aircraft not treated here – they do not present any difficulty to analyse once the axes system is established. The three important kinds of axes systems are as follows.

1. **Body-fixed axes, F_B,** is a system with its origin at the aircraft centre of gravity (CG) with the X-axis pointing forward, the Y-axis going over the right wing and the Z-axis pointing downwards. It is nailed to the aircraft, and normally the X-axis aligns relative to the aircraft zero lift line; for aircraft with a constant section fuselage it is convenient to keep the X-axis running parallel to the constant section. The body axes system F_B is nailed to the aircraft at its CG and is fixed (see Table 1.6).

2. **Wind axes system, F_W,** also has its origin at the CG with the X-axis aligned with the relative direction of airflow to the aircraft and pointing forwards; the Y and Z axes follow the right-handed system. The wing axes F_W is gimballed to the aircraft at its CG and can rotate about it. Wind axes vary corresponding to the airflow velocity vector in relation to the aircraft. Aircraft motion in the vertical plane or in the horizontal plane has the Z-axis in the plane of symmetry. If aircraft have both angle of attack and yaw angle then the Z-axis is not in the plane of symmetry. From the difference between F_B and F_W, the angles of attack (α), yaw (β), and roll (φ) can be established (see Figure 4.2).

3. **Inertial axes, F_I,** are fixed on the ground. For speed–altitudes below Mach 3 and 80,000 ft, the Earth can be considered as flat and not rotating with little error, so the origin of the inertial axes is pegged on the ground, with the X-axis pointing eastwards, which makes the Z-axis point vertically downward in a right-handed system.

If the parameters of one coordinate system are known, then parameters in other coordinate systems can be found out through transformation relationships.

1.15.2   Forces – Kinetics

In a steady-state straight level flight, an aircraft is in equilibrium under the applied forces (lift, weight, thrust and drag) as shown in Figure 1.18. This is the final outcome of the pressure field around the aircraft. Lift is measured perpendicular to aircraft velocity (free stream flow) and drag is opposite to the direction of aircraft velocity. In a steady level flight, lift and weight are opposite to each other. Opposite forces may not be collinear.

Forces and moments are associated with any body moving through fluid. In steady level flight in equilibrium, Σ **Force = 0**; that is, in the vertical direction, lift = weight, and in the horizontal direction, thrust = drag. In a steady-state straight level flight there is no side force.

The aircraft weight is exactly balanced by the lift produced by the wing (the fuselage and other bodies could share a small part of lift – to be discussed later). Thrust provided by the engine is required to overcome the drag arising from viscous, pressure and other forces.

Moments arising from various aircraft components are summed to zero to keep the flight level and straight; that is, Σ **Moment = 0**.

When not in equilibrium, the accelerating forces are taken into account at the instant of computation to find its resultant net force affecting the aircraft flight condition. If there were any force/moment imbalance, it would show up in the aircraft flight profile. That is how aircraft is manoeuvred – through force and/or moment imbalance – even for simple actions of climb and descent. The different axes systems are defined in Section 3.2. A summary of forces and moments in body axes F_B is given in Table 1.7.

■ **FIGURE 1.18**
Equilibrium flight

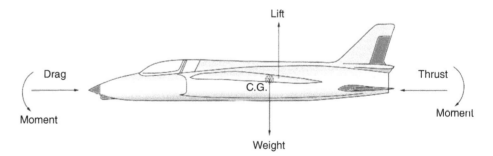

■ **TABLE 1.7**
Summary of the forces and moments in the body-fixed coordinate system, F_B

Axis	Force	Moment	Velocity component	Angle	Angular velocity component
x	X	L	U	φ	p
y	Y	M	V	θ	q
z	Z	N	W	ψ	r

In wind axes F_w the forces and moments are transformed to different magnitudes and directions where the force components X, Y and Z are resolved to lift (L), drag (D) and side force (C).

1.15.3   Aerodynamic Parameters – Lift, Drag and Pitching Moment

This section gives other useful non-dimensional coefficients and derived parameters frequently used in this book. The most common nomenclature, without any conflicts between both sides of the Atlantic, are listed here; these are internationally understood. (See Section 2.6 for the definition of aerofoil chord and Section 2.16 for the definitions of wing area S_w.)

$$q = \tfrac{1}{2}\rho V_\infty^2 = \text{dynamic head}$$

The subscript ∞ represents free stream conditions and is sometimes dropped.

q is a parameter extensively used to non-dimensionalize lumped parameters.

The coefficients of 2D aerofoils and 3D wings differ as shown below (note that for the subscripts, lowercase letters represent 2D aerofoil cases and the capital letters are for 3D wings).

2D aerofoil section:

C_l = Sectional aerofoil lift coefficient = *Section Lift/qc*
C_d = Sectional aerofoil drag coefficient = *Section Drag/qc*
C_m = Aerofoil pitching moment coefficient = *Section Pitching Moment/qc²* (+ nose up)

For wing (3D), replace chord c with wing area S_w:

C_L = Lift coefficient = *Lift/qS$_w$*
C_D = Drag coefficient = *Drag/qS$_w$*
C_M = Pitching moment coefficient = *Lift/qS$_w$²* (+ nose up)

Figure 3.12 gives the pressure distribution at any point over the surface in terms of pressure coefficient, C_p, which is defined as:

$$C_p = \left(p_{local} - p_\infty \right) / q$$

1.15.4   Basic Controls – Sign Convention

Conventional aircraft have four basic controls: the elevators, ailerons, rudder and throttle. The elevator and the throttle are longitudinal controls, in that they affect the three longitudinal degrees of freedom: changes of speed along the Ox axis; heave in the Oz direction; and pitch about the Oy axis. Likewise, the ailerons and the rudder are called lateral controls because they affect the three lateral degrees of freedom: sideslip along the Oy axis; roll about the Ox axis; and yaw about the Oz axis.

1. **Throttle:** The throttle is used to control the thrust of the engines of the aircraft. Its principal purpose is to control the rate of climb or descent of the aircraft.
2. **Elevator:** The elevator is used to control the angle of attack of the aircraft, and therefore its airspeed. Note that positive elevator angle, η, generates negative pitching moment, M, and that this is achieved by pushing the control column forward (Figure 1.19).

■ **FIGURE** 1.19
**Control deflection
and sign convention**

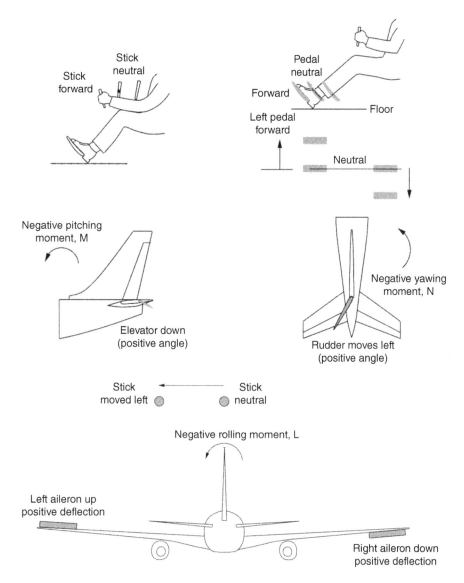

3. **Aileron:** The ailerons are used to control the aircraft in roll. More specifically, the ailerons apply a rolling moment to the aircraft and so are used to demand *roll rate*. Roll rate is used to achieve bank angle, and bank angle is used to initiate a turn. Note that positive (right-hand side down) aileron angle, ξ, generates negative rolling moment, L (Figure 1.19).

4. **Rudder:** The rudder is used to control the side-slip angle of the aircraft. This might be of particular importance when landing in a crosswind. Rudder may also be used to *balance* a banked turn. Note that positive rudder angle, ζ, generates negative yawing moment, N.

References

1. Anderson, J.D., *The Airplane: A History of Its Technology.* AIAA, Reston, VA, 2002.

2. *Jane's Fighting Aircraft of World War I.* Random House Group, Coulsdon, Surrey, 2001.

3. Taylor, M.J.H., *Milestones of Flight.* Chancellor Press, Jane's Information Group, 1983.

4. Anderson, J.D., *History of Aerodynamics.* Cambridge University Press, Cambridge, 1998.

5. Murman, M., Walton, M., and Rebentisch, E., Challenges in the better, faster, cheaper era of aeronautical design, engineering and manufacturing. *The Aeronautical Journal,* 2000, pp. 481–488.

6. Kundu, A.K., Watterson, J., *et al.,* Parametric optimization of manufacturing tolerances at the aircraft surface. *Journal of Aircraft,* 39 (2), 2002, pp. 271–279.

7. McMasters, J., and Cummungs, R., *Rethinking the airplane design process: An early 21st-century perspective.* 42nd Aerospace Science Meeting, 8th January.

8. Von Braun, B., and Ordway, F.I., *History of Rocketry and Space Travel.* Crowell, New York, 1967.

9. McMichael, J.M. and Francis, M.S., *Micro Air Vehicles – Toward a New Dimension in Flight.* Defense Advanced Research Projects Agency, 1997.

10. Jackson, S., *Systems Engineering for Commercial Aircraft.* Ashgate, Aldershot, 1997.

11. Aircraft Transport Association of America (ATA), *Specifications for Manufacturers' Technical Data.* Specification 100, 1989.

12. Kundu, A.K., *et al.,* A proposition in design education with a potential in commercial venture in small aircraft manufacture. *Journal of Aircraft Design* 3, 2000, pp. 261–273.

13. Yechout, T.R., Degrees of expertise: A survey of aerospace engineering programs. *Aerospace America,* April, 1992.

14. Walker, B.K., *et al., Educating Aerospace Engineers for the Twenty-First Century: Results of a Survey.* ASEE Conference Proceedings, 1995.

15. Williams, J.C. and Young, R.L., Making the grade with ABET. *Aerospace America,* April, 1992.

16. McMasters, J.H., *Paradigms Lost, Paradigm Regained: Paradigm Shift in Engineering Education.* SAE Technical Paper Series No. 911179, 1991.

17. Moulton, A.E., *et al., Engineering Design Education.* Design Council UK, April 1976.

18. Engineering Design: *The Fielding Report.* Council of Scientific and Industrial Research. HMSO, London, 1963.

19. Nicolai, L.M., Designing a better engineer. *Aerospace America,* April, 1992.

20. Niu, M., *Airframe Structural Design.* Commlit Press Ltd., Hong Kong, 1999.

CHAPTER 2

Aerodynamic and Aircraft Design Considerations

2.1   Overview

Aircraft configuration as shaped through design considerations is mission-specific. To estimate aircraft performance, a good understanding of aircraft configuration is essential in estimating aircraft drag. Experience engineers can guess, to an extent, the aircraft capability by merely observing the configuration. Only broad descriptive discussions are given here on aerodynamics and aircraft design aspects, to appreciate how these affect aircraft performance. For detailed work, readers may refer to related textbooks. There are many excellent textbooks on aircraft design [1–19] and aerodynamics [20–31] available in the public domain. Because the subject matter is mature, some older introductory books still serve the purpose of this course; however, more recent books relate better to current examples. The design book by the author [15] complements this book, with aircraft performance information for worked examples. The next chapter deals with systems design considerations that affect aircraft performance.

In this book, a preliminary aircraft configured at the conceptual stage is given to performance engineers for sizing and engine matching to finalize the external geometry. It is then followed by complete aircraft performance estimation for the production version.

The information in this chapter is essential for aircraft performance engineers. A set of problems related to this chapter is given in Appendix E.

2.2   Introduction

Any object moving through air interacts with the medium at each point of the wetted (i.e. exposed) surface, creating a pressure field around the aircraft body. An important part of aircraft design is to exploit this pressure field by shaping its geometry to arrive at the desired performance of the vehicle, including shaping of the wing to generate lifting surfaces, shaping of the fuselage to accommodate payload, shaping of the nacelle to house a suitable engine, and tailored control surfaces. This chapter provides an overview and definitions of various parameters

Theory and Practice of Aircraft Performance, First Edition. Ajoy Kumar Kundu, Mark A. Price and David Riordan.
© 2016 John Wiley & Sons, Ltd. Published 2016 by John Wiley & Sons, Ltd.

integral to configuring aircraft mould-lines and shaping of those components normally kept retracted within the streamlined mould-lines.

Aerodynamicists deal with fluid mechanics (the interaction of air with bodies) and performance engineers deal with equations of motion of vehicles in atmosphere; these disciplines are interdependent on each other. Both aerodynamicists and performance engineers work together in the same department and often share work obligations. Therefore knowledge of aerodynamic characteristics of bodies by performance engineers is essential.

This chapter starts with the atmosphere (Section 2.3), the interacting medium that makes flight possible. Understanding of the atmosphere as the environment in which aircraft perform is essential to aerospace engineers. Section 2.4 describes the properties of air and airflow behaviour, explaining the role of viscosity. How the shaping of aircraft components affects their aerodynamics, which in turn contributes to aircraft performance, is discussed in groups, mainly wing (Section 2.11), empennage (Section 2.18), fuselage (Section 2.19) and nacelle (Section 2.20). The last section describes the special purpose devices that contribute to aircraft performance, such as speed brakes.

Aircraft conceptual design starts with shaping an aircraft, finalizing geometric details through aerodynamic considerations in a multidisciplinary manner to arrive at the technology level to be adopted. These decisions affect aircraft performance. To continue with sustained flight, an aircraft requires a lifting surface in the form of a plane – hence, *aeroplane* (the term *aircraft* is used synonymously in this book). Resultant forces and moments are obtained from the pressure field around the wetted surface of an aircraft. Aerodynamic forces of lift and drag (see Sections 1.15.2 and 2.6) are the resultant force components of the pressure field around an aircraft. The aircraft components exhibit inherent moments, the extent depending on the shape of components.

The available engine power propels the aircraft. These forces, along with the engine-generated thrust force, develop moments about the aircraft centre of gravity (CG). Aircraft designers seek to obtain the maximum possible lift-to-drag ratio (i.e., a measure of minimum fuel burn) for an efficient design (this simple statement is complex enough to configure). Aircraft stability and control are the result of harnessing these aerodynamic forces and moments. Control of aircraft is applied through the use of aerodynamic forces modulated by the control surfaces (e.g., elevator, rudder and aileron). In fact, the sizing of all aerodynamic surfaces should lead to meeting the requirements for the full flight envelope without sacrificing safety.

The secret of lift generation is in the sectional characteristics (i.e., *aerofoil*) of the lifting surfaces that serve as wings, similar to birds. This chapter explains how the differential pressure between the upper and lower surfaces of the wing is the lift that sustains the aircraft weight. Details of aerofoil characteristics and the role of the empennage that comprises the lifting surfaces are also explained. Empennages are like small wings that contribute to aircraft stability and control of an aircraft. Section 2.8 summarizes the desirable characteristics in selecting aerofoil. The same aerodynamic logic prevails for other bodies, such as the fuselage and nacelle, except that they serve different roles for the design.

Minimizing aircraft drag is one of the main obligations of aerodynamicists. Viscosity contributes to approximately two-thirds of the total subsonic aircraft drag. The effect of viscosity is apparent in the wake of an aircraft as disturbed air flow behind the body. Its thickness and intensity are indications of the extent of drag and can be measured.

One way to reduce aircraft drag is to shape the body such that it will result in a thinner wake. The general approach is to make the body in a teardrop shape with the aft end closing gradually, as compared to the blunter front-end shape for subsonic flow. (Behaviour in a supersonic flow is different, but it is still preferable for the aft end to close gradually.) The smooth contouring of teardrop shaping is called *streamlining*, which follows the natural airflow lines around the body – it is for this reason that aircraft have attractive smooth contour lines. Streamlining is synonymous with speed, and its aerodynamic influence in shaping is revealed in any object in a relative moving airflow (e.g. boats and automobiles).

Performance engineers must have a full understanding of engine performances, which are given in Chapter 8. Once aircraft drag and engine performance is known, full aircraft performance evaluation for the full flight envelope is possible. The formal aircraft sizing and engine matching exercise is carried out to freeze configuration and precedes the detailed aircraft performance estimation. In this book, aircraft sizing is dealt with in Chapter 16, after studying the aircraft performance in chapter 10 to 15.

2.3   Atmosphere

Since interaction with air is essential for flight, aircraft engineers must have a full understanding of the environment (the atmosphere) and the properties of air. Knowledge of the atmosphere is an integral part of aerospace engineering. The science of atmosphere is quite complex; however the aircraft performance engineers only use standardized tables. This section gives the background of important fundamentals, perhaps a little more than what the book uses.

The atmosphere around the Earth is never uniform nor in a steady state – nature exhibits variance in properties at all times in all places. The scientific community needed some standardization to compare aircraft performance. After substantial data generation, an international consensus was reached to obtain the International Standard Atmosphere (ISA), which follows hydrostatic relations at static, dust-free and zero humidity conditions. ISA data (Appendix B) for up to 32 miles altitude is nearly identical to US Standard Atmosphere Data.

The atmosphere, in the classical definition up to 40 kilometres (km) altitude, is dense (continuum): its homogeneous constituent gases are nitrogen (78%), oxygen (21%), and others (1%). At sea-level (with subscript 0), the ISA condition gives the following properties:

Pressure, p_0 $= 101,325 \text{ N/m}^2 \left(14.7 \text{lb} / \text{in}^2\right)$

Density, ρ_0 $= 1.225 \text{ kg/m}^3 \left(0.02378 \text{ slugs/ft}^3\right)$

Temperature, T_0 $= 288.16 \text{ K} \left(518.69°\text{R}\right)$

Viscosity, μ_0 $= 1.789 \times 10^{-5} \text{ N-s/m}^2 \left(3.62 \times 10^{-7} \text{ lb-sec /ft}^2\right)$

Acceleration due to gravity, $g_0 = 9.807 \text{ m/s}^2 \left(32.2 \text{ ft/s}^2\right)$

2.3.1   Hydrostatic Equations and Standard Atmosphere

The hydrostatic equation for a vertical column of air of density ρ in the elemental height dh standing on one unit area is derived as follows (Figure 2.1a). At any cross-section of the column, the force is the pressure.

Force on height h_1 (lower cross-section) = weight of cube + force on height h_2

$$p_2 + \rho g dh = p_1$$

i.e.

$$p_2 - p_1 = -\rho g dh \text{ (pressure decreases with altitude)}$$

■ **FIGURE 2.1**
**Hydrostatic equation
and altitude effect.
(a) Force diagram.
(b) Aircraft distance
from the centre of
the Earth**

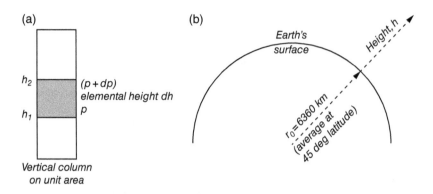

or

$$dp = -\rho g \, dh \tag{2.1}$$

In Equation 2.1, ρ and g vary with altitude, h. To simplify integration using perfect gas laws, the variable ρ is replaced by $p/(RT)$ and g is held constant at the sea-level value of g_0. The justification for holding g constant is given below. Then Equation 2.1 can be written as follows:

$$\frac{dp}{p} = -g_0 \frac{dh}{RT} \tag{2.2}$$

On integrating:

$$\int_0^h \frac{dp}{p} = -g_0 \int_0^h \frac{dh}{RT} \tag{2.3}$$

Relations in Equation 2.1 show dependency on acceleration due gravity, g, which is inversely proportional to the square of the radius (average radius $r_0 = 6360$ km) from the centre of the Earth. If h is the altitude of an aircraft from the Earth's surface, then it is at a distance $(r_0 + h)$ from the centre of the Earth, as shown schematically in Figure 2.1b.

Acceleration due to gravity, g, at height h is expressed as:

$$g = g_0 \left(\frac{r_0}{r_0 + h} \right)^2 \tag{2.4}$$

where $g_0 = 9.807\,\text{m/s}^2$ is the acceleration due to gravity at sea-level. Substituting the values gives:

$$g = 9.807 \times \left(\frac{6360 \times 10^3}{6360 \times 10^3 + h} \right)^2$$

For terrestrial flights, h is much less than r_0; there is less than a 1% change in g up to 30 km. To simplify integration of Equation 2.2, the acceleration due to gravity is held constant at the sea-level value up to 30 km altitude. For keeping g_0 constant, the potential energy at a slightly

EXAMPLE 2.1

Find g at 30 km.

SOLUTION

$$g = 9.807 \times \left(\frac{6360 \times 10^3}{6390 \times 10^3} \right)^2 = 9.807 \times 0.9906 \text{ (less than 1\% change in } g)$$

lower altitude equals the potential energy of pressure altitude h with actual varying g. The small error arising from keeping g_0 constant results in geopotential altitude, h_p, which is slightly lower than the pressure (geometric) altitude, h. This book uses the ISA table that gives geopotential pressure altitude, h_p. The relation between h and h_p can be derived as follows.

For the same potential energy, PE, the following relationship can be set up:

$$PE = m \int_0^h g \, \mathrm{d}h = mg_0 h_p \quad \text{or} \quad \int_0^h g \, \mathrm{d}h = g_0 h_p$$

Substituting the value of g from Equation 2.3, the above equation can be written as:

$$g_0 \int_0^h \left(\frac{r_0}{r_0 + h} \right)^2 \mathrm{d}h = g_0 \left(\frac{r_0 h}{r_0 + h} \right) = g_0 h_p \tag{2.5}$$

This shows $h > h_p$, that is, a geopotential altitude that is slightly lower than the pressure (geometric) altitude. At 25 km altitude, the difference between them is less than half a percent. This book uses the geopotential altitude from the ISA table.

The centripetal force due to Earth's rotation is collinear with gravity vector $\boldsymbol{g}$. The centripetal force is very small compared to gravitational force and is ignored.

Since it is impossible to measure tapeline geometric altitude, pressure is used to compute altitude. In an ideal situation on an ISA day, properties of geometric altitudes in which gravity varies are identical with the hydrostatic equation. The hydrostatic Equation 2.3 shows that with altitude gain, the pressure decreases. The temperature profile with altitude is known from observation and is used to obtain other properties using the hydrostatic equation. In reality, an ISA day is difficult to find; nevertheless, it is used to standardize aircraft performance to a reference condition for assessment and comparison.

Atmospheric temperature variation behaves strangely with altitude – it is influenced by natural phenomena. It decreases linearly up to 11 km at a lapse rate (λ) of –6.5 K/km, reaching 216.66 K (in FPS, $\lambda = 0.00357$ R/ft up to 36,089 ft). Temperature gradient breaks at 11 km altitude, known as the *tropopause*; below the tropopause is the *troposphere* and above it is the *stratosphere*, extending up to 54 km. The initial temperature profile in the stratosphere up to 20 km remains constant at 216.66 K. From 20 km upwards the temperature increases linearly at a rate of 1 K/km up to 32 km. Figure 2.2 shows the typical variation of atmospheric properties

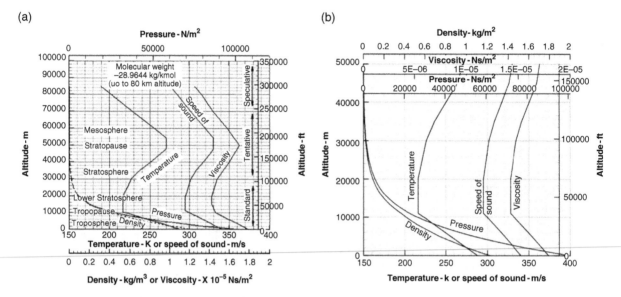

■ FIGURE 2.2 Standard atmosphere (use Appendix B for accurate values). (a) Up to 100,000 m. (b) Magnified up to 50,000 m

with altitude up to 100 km (Figure 2.2b is enlarged by reducing the y-axis scale). From 32 km to 76 km, the atmospheric data are currently considered tentative. Above 76 km, variations in atmospheric data are considered speculative. It is found that up to 90 km the mean molecular mass is nearly homogeneous and termed as *homosphere*, while above it is termed the *heterosphere*. Nearly half the air mass is within 18,000 ft altitude.

Currently, no atmospheric-dependent aircraft fly above 32 km, and therefore the atmospheric properties above this altitude are beyond the scope of this book. Typical atmospheric stratification (based primarily on temperature variation) is as follows (the applications in this book do not exceed 65,000 ft ($\approx$20 km).

Troposphere	up to 11 km (36,089 ft)
Tropopause	11 km (36,089 ft)
Lower stratosphere (isothermal)	from 11 to 20 km (36,089 ft to 65,600 ft)
Upper stratosphere	from 20 to 47 km (65,600 ft to 154,300 ft) – break at 32 km
Stratopause	47 km (154,300 ft)
Upper stratosphere (isothermal)	from 47 km (154,300 ft) to 54 km
Mesosphere	from 54 to 90 km
Mesopause	90 km
Thermosphere	from 100 to 550 km (can extend and overlap with ionosphere)
Ionosphere	from 550 to 10,000 km
Exosphere	above 10,000 km

Enough data are collected up to the stratopause to have an ISA day. Above that altitude, data are being collected using sounding rockets and are treated as tentative.

Clearly, based on the temperature profile for the atmosphere up to 32 km, there are three distinct zones as follows:

Zone 1 (from sea-level to 11 km altitude):	**Troposphere** in negative temperature gradient with lapse rate of –0.0065°K/m, reaching 216.66 K at tropopause.
Zone 2 (from 11 to 20 km altitude):	**Stratosphere** (**lower**) at constant isothermal temperature of 216.66 K.
Zone 3 (from 20 to 32 km altitude):	**Stratosphere** (**upper**) in positive temperature gradient with lapse rate of 0.001°K/m.

From the temperature distribution at distinct altitude zones as shown in Figure 2.2, the following hydrostatic equations are obtained which give the related properties for the given altitude, h, in metres. Computation of each zone uses the properties at the base of the zone and need to be computed out at the beginning. The ISA table is obtained from these equations, which can be used if ISA table is not available.

Using hydrostatic equations up to 32 km, the properties of air in SI units are derived below (for simplicity, in these equations the symbol for geopotential altitude have h instead of h_p, i.e. the subscript p is dropped). The following equations will derive the pressure, p, temperature, T, density, ρ, and viscosity, μ.

The perfect gas law is:

$$\rho = \frac{p}{(T \times R)} \tag{2.6a}$$

and the Sutherland's equation for viscosity is given as:

$$\mu = \mu_0 \frac{(T_0 + C)}{(T + C)} \left(\frac{T}{T_0}\right)^{1.5} = \frac{\lambda T^{1.5}}{(T + C)} = \frac{1.458 \times T^{1.5}}{(T + 110.4)} \,\text{N-s}/\text{m}^2 \tag{2.6b}$$

In SI units, the constants $\lambda = 1.458 \times 10^{-6} (\text{kg}/(\text{msK}^{0.5}))$ and $C = 114.4$.

In FPS:

$$\mu = \mu_0 \frac{(T_0 + C)}{(T + C)} \left(\frac{T}{T_0}\right)^{1.5} \,\text{lb-s}/\text{ft}^2 \tag{2.6c}$$

The following three non-dimensional parameters, δ, σ and θ, are useful in aircraft performance analyses. All atmospheric properties in the three zones can be expressed in terms of these three ratios.

Relative pressure	$\delta = p/p_0$
Relative density	$\sigma = \rho/\rho_0$
Relative temperature	$\theta = T/T_0$

Troposphere (from sea-level to 11 km altitude)   Nature exhibits a temperature drop with altitude gain, and the rate of decay is known as the lapse rate, λ. An international standard lapse rate in SI is taken as $\lambda = -0.0065$°K/m (in FPS, $\lambda = 0.00357$ R/ft) reaching 216.66 K at 11 km. In SI, the constants $g_0 = 9.807$ m/s^2 and R $= 287.053$ J/kgK. At sea-level standard day, $p_0 = 101{,}325$ N/m^2 and $T_0 = 288.16$ K.

Temperature

Temperature T, in $K = T_0 - (0.0065 \times h)$ up to 11,000 m in the tropopause.

$$h = \frac{1}{0.0065}(T_0 - T) = \frac{T_0}{0.0065}\left(1 - \frac{T}{T_0}\right), \text{ where } T_0 = 288.16 \text{ K} \qquad \textbf{(2.7a)}$$

That gives:

$$dh = \frac{T}{-0.0065} \qquad \textbf{(2.7b)}$$

Pressure

Substituting Equation 2.7b in Equation 2.3, and integrating with the limits of sea-level ($p_0 = 101,325 \text{ N/m}^2$ and $T_0 = 288.16 \text{K}$) and h (p and T), the following is obtained.

$$\int_0^h \frac{dp}{p} = -g_0 \int_0^h \frac{dh}{RT} = -g_0 \int_0^h \frac{dT}{0.0065RT} \qquad \textbf{(2.8a)}$$

or

$$\frac{p}{p_0} = \left(\frac{T}{T_0}\right)^{\frac{g_0}{0.0065R}}$$

or

$$\frac{T}{T_0} = \left(\frac{p}{p_0}\right)^{\frac{0.0065R}{g_0}} \qquad \textbf{(2.8b)}$$

In SI, pressure p in $\text{N/m}^2 = 101325 \times \left(\frac{T}{288.16}\right)^{\frac{g_0}{0.0065R}} = 101325 \times \left(\frac{T}{288.16}\right)^{5.2562}$ **(2.8c)**

And Equation 2.7a gives:

$$h = \frac{T_0}{0.0065}\left[1 - \left(\frac{p}{p_0}\right)^{\frac{0.0065R}{g_0}}\right] = \frac{T_0}{0.0065}\left[1 - \left(\frac{p}{p_0}\right)^{0.1906}\right] \qquad \textbf{(2.8d)}$$

Density

Using gas laws (Equation 2.5), Equation 2.8b becomes:

$$\frac{\rho}{\rho_0} = \frac{pT_0}{p_0T} = \left(\frac{T}{T_0}\right)^{\frac{g_0}{0.0065R}}\left(\frac{T_0}{T}\right) = \left(\frac{T}{T_0}\right)^{\frac{g_0}{0.0065R}-1} = \left(\frac{T}{T_0}\right)^{\frac{g_0-0.0065R}{0.0065R}} \qquad \textbf{(2.8e)}$$

or

$$\left(\frac{T}{T_0}\right) = \left(\frac{\rho}{\rho_0}\right)^{\frac{0.0065R}{g_0-0.0065R}}$$

In SI units:

$$\rho = 1.225 \times \left(\frac{T}{288.16}\right)^{\frac{9.807-0.0065 \times 287.053}{0.0065 \times 287.053}} = 1.225 \times \left(\frac{T}{288.16}\right)^{4.256}$$

And from Equation 2.7a:

$$h = \frac{T_0}{0.0065}\left[1-\left(\frac{\rho}{\rho_0}\right)^{\frac{0.0065R}{g_0-0.0065R}}\right] = \frac{T_0}{0.0065}\left[1-\left(\frac{\rho}{\rho_0}\right)^{0.235}\right] \qquad \textbf{(2.8f)}$$

Kinematic viscosity ν in $m^2/s = \mu/\rho$ in m^2/s.

EXAMPLE 2.2

Find the atmospheric properties at a geopotential height of 8 km on an ISA day.

SOLUTION

Temperature at 8 km is $T = 288.16 - (0.0065 \times 8000)$.
$$= 288.16 - 52 = 236.16\,K.$$

Pressure, $p = 101325 \times \left(\frac{236.16}{288.16}\right)^{\frac{9.807}{0.0065 \times 287.053}}$

$$= 101325 \times (0.8195)^{5.2562} = 35598.98\,N/m^2$$

From Equation 2.8c, *Density,* $\rho = 1.225 \times \left(\frac{236.16}{288.16}\right)^{4.256}$

$$= 1.225 \times 0.4286 = 0.525\,kg/m^3$$

Sutherland's Equation 2.6b for viscosity in SI units gives:

$$\mu = \frac{1.458 \times 10^{-6} \times 236.16^{1.5}}{236.16 + 110.4}$$
$$= 1.458 \times 1.0471 \times 10^{-5}\,N\text{-}s/m^2$$
$$= 0.000015267\,N\text{-}s/m^2$$

Stratosphere (lower isothermal) – from 11 km to 20 km
Temperature, $T = 216.66K$ K stays constant up to 20 km.

Pressure – lower stratosphere
Equation 2.3 gives:

$$\int_{h_b}^{h} \frac{dp}{p} = -g_0 \int_{h_b}^{h} \frac{dh}{RT} = -\frac{g_0}{RT} \int_{h_b}^{h} dh$$

Integrating within the limits of the base pressure at $h_p = 11$km and altitude h:

$$p_b = 22{,}633 N/m^2 \text{and } T_b = 216.66K \text{ to altitude } h$$

$$Pressure,\ p = 22633 \times e^{\frac{-g_0}{216.66R}(h-11000)}N/m^2$$

Density – lower stratosphere

$$Density,\ \rho\left(kg/m^3\right) = \rho_1 e^{-\left[\frac{g_0}{RT}\right](h-h_1)}, \text{or use gas law } \rho = \frac{p}{\left(216.66 \times R\right)} \qquad \textbf{(2.8g)}$$

where p is obtained from the above equation.

EXAMPLE 2.3

Find the atmospheric properties at a geopotential height of 15 km.

SOLUTION

Temperature at 15 km is constant at $T = 216.65$ K.

$Pressure, p$ in N/m^2 $= 22633 \times e^{\frac{-9.807}{216.66 \times 287.053}(15000-11000)}$
$= 22633 \times e^{-0.6307} = 22633 \times 0.5322$
$= 12045.7\,N/m^2$

or $\rho = 0.36392 \times e^{-\left[\frac{9.807}{287.053 \times 216.66}\right](15000-11000)}$
$= 0.36392 \times e^{-0.6307}$
$= 0.36392 \times 0.5322$
$= 0.1937\,kg/m^3$

$Density, \rho$ rin $kg/m^3 = \dfrac{p}{(216.66 \times R)} = \dfrac{12045.7}{(216.66 \times 287.053)}$
$= \dfrac{12045.7}{62192.9} = 0.1937\,kg/m^3$

Stratosphere (upper) – from 20 km to 32 km
Temperature
There is a positive temperature gradient in the upper stratosphere from 20 km to 32 km.
Here the lapse rate in SI, $\lambda = 0.001 K/m$ (in FPS, $\lambda = 0.000549$ R/ft).
Temperature, $T = 216.65 + (0.001 \times h)$ up to 32,000 m altitude.
That gives $dh = T/(0.001)$, and substituting in Equation 2.3, the following is obtained.

Pressure
Equation 2.8a gives:

$$\int_0^h \frac{dp}{p} = -g_0 \int_0^h \frac{dh}{RT} = -g_0 \int_0^h \frac{dT}{0.001RT}$$

or
$$\frac{p}{p_0} = \left(\frac{T}{T_0}\right)^{\frac{-g_0}{0.001R}}$$

or
$$\frac{T}{T_0} = \left(\frac{p}{p_0}\right)^{\frac{-0.001R}{g_0}}$$

Integrating within the limits of the base pressure at $h_p = 20$ km and altitude h:

$$p_b = 5475.21\,N/m^2 \text{ and } T_b = 216.66K \text{ to altitude } h$$

$$\text{Pressure, } p = 5475.21 \times \left(\frac{T}{216.66}\right)^{\frac{-g_0}{0.001R}} N/m^2 \tag{2.8h}$$

Density

$$\text{Density, } \rho = 0.08803 \times \left(\frac{T}{216.66} \right)^{\frac{-g_0}{0.001R}-1} \text{kg/m}^3 \qquad \textbf{(2.8i)}$$

Viscosity

$$\mu = \frac{1.458 \times 10^{-6} T^{1.5}}{(T+110.4)} \text{kg/ms} \ \left(\text{from Equation 2.6} \right)$$

Kinematic viscosity $\nu = \mu/\rho$ in m²/s.

EXAMPLE 2.4

Find the atmospheric properties at a geopotential height of 25 km.

SOLUTION

Temperature, T $= 216.66 + \left[0.0047 \times (h - 20,000) \right]$
$\qquad = 216.66 + (0.001 \times 5000) = 221.66 \text{ K}.$

Pressure, p in N/m² $= 5475.21 \times \left(\dfrac{221.66}{216.66} \right)^{\frac{-9.807}{0.001 \times 287.053}}$

$\qquad = 5475.21 \times (1.0231)^{-34.164}$

$\qquad = 5475.21 / 2.1819$

$\qquad = 2510.34 \text{ N/m}^2$

Density, ρ in kg/m³ $= \dfrac{p}{(T \times R)} = \dfrac{2510.34}{(221.66 \times 287.053)}$

$\qquad = \dfrac{2510.34}{63628.17} = 0.0395 \text{ kg/m}^3$

or

$$\rho = 0.08803 \times \left(\frac{221.66}{216.66} \right)^{\frac{-9.807}{0.001 \times 287.053}-1}$$

$\qquad = .08803 \times (1.0231)^{-34.1644-1}$

$\qquad = 0.08803 \times 0.448 = 0.0394 \text{ kg/m}^3$

Viscosity, μ (kg/ms) $= \dfrac{1.458 \times 10^{-6} T^{1.5}}{(T+110.4)}$ (from Equation 2.6b)

Appendix B gives the ISA table up to 25 km altitude, which is sufficient for this book because all aircraft described (except rocket-powered special-purpose aircraft) would be flying below that height. Linear interpolation of properties may be carried out between altitudes.

2.3.2 Non-standard/Off-standard Atmosphere

While there is an internationally accepted standard atmosphere (ISA), there are no accepted standards for atmospheric conditions different from an ISA day, which is the typical situation. Typically, the hydrostatic equations are used to obtain the off-standard day atmospheric properties. The difference between non-standard atmosphere and off-standard atmosphere is that the former has no standards, while the latter takes only the incremental temperature ΔT effects on the ISA atmosphere. This book is concerned with the off-standard atmosphere expressed as $(T_{ISA} \pm \Delta T)$.

As mentioned previously, an aircraft rarely encounters the ISA day. Wind circulation over the globe is always occurring. Surface wind currents such as *doldrums* (slow winds in equatorial regions), *trade winds* (predictable wind currents blowing from subtropical to tropical zones), *westerlies* (winds blowing in the temperate zone), and *polar easterlies* (year-round cold winds blowing in the polar regions) are well known. In addition, there are characteristic winds in typical zones – for example, *monsoon* storms; wind-tunnelling effects of strong winds blowing in valleys and ravines in mountainous regions; steady up and down drafts at hill slopes; and daily coastal breezes. At higher altitudes, these winds have an effect. Storms, twisters, and cyclones are hazardous winds that must be avoided. There are more complex wind phenomena such as wind-shear, high-altitude jet streams, and vertical gusts. Some of the disturbances are not easily detectable, such as clear air turbulence (CAT). Humidity in the atmosphere is also a factor to be considered. The air-route safety standards have been improved systematically through round-the-clock surveillance and reporting. In addition, modern aircraft are fitted with weather radars to avoid disturbed areas. Flight has never been safer apart from manmade hazards. This book deals primarily with the ISA day, with the exception of *gust load*, which is addressed in Chapter 7, for structural integrity affecting aircraft weights. For aircraft performance the critical issues related to non-standard atmosphere is discussed with worked examples. Aircraft performance engineers must ensure aircraft capabilities on hot or cold day (off-standard day) operations; a hot day can be critical for aircraft performance, especially at takeoff. More details of flying in adverse environments is given in Section 18.6.

Unlike ISA, in off-standard atmosphere, ambient pressure at the geopotential altitude is not the same as pressure altitude. Appendix B gives an off-standard chart in which only the change in temperature is given as input. The results from the hydrostatic equations for the geopotential altitude and its corresponding pressure show the changes in density, speed of sound, viscosity and coefficient of friction as a consequence of a change in ambient temperature.

On a hot day, the density of air decreases, resulting in degradation of aircraft and engine performances. Certification authorities (FAA and CAA) require that aircraft demonstrate the ability to perform as predicted in hot and cold weather and in gusty wind. The certification process also includes checks on the ability of the environmental control system (ECS, e.g. anti-icing/de-icing, and air-conditioning) to cope with extreme temperatures. In this book, performance degradation on an off-standard day is addressed to an extent. The procedure to address off-standard atmospheres is identical to the computation using ISA conditions, except that the atmospheric data are different. The atmosphere is not fully dry; there is some humidity always present, affecting ISA conditions. This book does not fully deal with humidity effects, but see Section 2.3.4.

2.3.3   Altitude Definitions – Density Altitude (Off-standard)

It is not possible to measure geometric (tapeline) altitude of aircraft. Aerospace applications have to depend on the hydrostatic relationship between altitude and pressure (Equation 2.2). *Density altitude* is formally defined as "geopotential pressure altitude corrected for off-standard temperature variations". Only temperature variation is considered for computing density altitude.

The ISA day is hard to find, and therefore off-standard day atmospheric behaviour needs also to be looked into from the point of view of aircraft performance, which depends primarily on air density. Aircraft lift and engine performance depends on ambient air density. For example, on a hot day the sea-level air density will be thinner and have lower than standard day pressure. The flight deck altimeter (see Section 3.3.1) will then indicate a higher altitude corresponding to lower pressure. At the same time the engine will produce less thrust and the aircraft generates less lift at the same true air speed (see Section 3.2). Therefore, density altitude is what is used for off-standard atmosphere. Density altitude is the pressure altitude corrected for off-standard conditions. For example, if the top level of a hilly area is 3000 ft and if the altimeter indicates

4000 ft in a hot off-standard day when actually he is below 3000 ft, then a pilot may suddenly realise s/he is flying at the hill and may hit it for not having enough reaction time to pull up.

With an off-standard day the atmospheric properties are denoted by the subscript "off". For off-standard atmospheric conditions let $T_{off} = T_{ISA} \pm \Delta T$. Gas laws give:

$$\rho_{off} = \frac{p_{off}}{RT_{off}},$$

noting $p = p_{off}$. The off-standard density can be obtained as:

$$\rho_{off} = \frac{\rho_0 p T_0}{p_0 \left(T \pm \Delta T\right)}$$

Noting that density altitude is only the function of density, the density altitude h (geopotential) can be computed using Equation 2.8b:

$$h = \frac{T_0}{0.0065}\left[1 - \left(\frac{\rho_{off}}{\rho_0}\right)^{\frac{0.0065R}{g_0 - 0.0065R}}\right]$$

$$= \frac{T_0}{0.0065}\left[1 - \left(\frac{p T_0}{p_0 \left(T \pm \Delta T\right)}\right)^{0.235}\right] \tag{2.9a}$$

A density altitude chart can be generated using Equation 2.9a and is given in Appendix B.

However, if only off-standard temperature at an altitude is known, then the density ρ_{off} can be computed without the ISA table to get the pressure. However, its application in practice is low as it is not in demand.

$$p = p_0\left(\frac{T}{T_0}\right)^{\frac{g_0}{0.0065R}} = p_0\left(\frac{T_0 - 0.0065h}{T_0}\right)^{5.2562} = p_0\left(1 - \frac{0.0065h}{288.16}\right)^{5.2562} \tag{2.9b}$$

Substituting T_{off}, it becomes:

$$\rho_{off} = \frac{101325\left(1 - 0.0000256h\right)^{5.2562}}{287.053\left(T_{ISA} \pm \Delta T\right)} = \frac{353.3\left(1 - 0.0000256h\right)^{5.2562}}{\left(T_{ISA} \pm \Delta T\right)} \tag{2.9c}$$

EXAMPLE 2.5

In Example 2.2 the atmospheric properties at 8 km in STD were computed. In this problem, compute the geopotential density altitude for Example 2.2 for a $T_{off} = ISA + 10°C$ and a $T_{off} = ISA + 40°C$ day.

$T_{off} = ISA + 10°C$
Example 2.2 gives $p = 35598.98 \text{N/m}^2$, $\rho = 0.525 \text{kg/m}^3$, and $T = 236.16$ K. $T_{off} = ISA + 10°C = 246.16$ K. Using Equation 2.9a, the following is obtained:

$$h = \frac{T_0}{0.0065}\left[1 - \left(\frac{p T_0}{p_0 \left(T \pm \Delta T\right)}\right)^{0.235}\right]$$

$$= \frac{288.16}{0.0065}\left[1 - \left(\frac{35598.98 \times 288.16}{101325 \times 246.16}\right)^{0.235}\right]$$

or $h = 44332.3 \times [1 - 0.4112^{0.235}] = 8335 \text{m}$, indicating 335 m higher altitude.

$T_{off} = ISA + 40°C$
$T_{off} = ISA + 40°C = 276.16K$. Using Equation 2.9a, the following is obtained:

$$h = \frac{T_0}{0.0065}\left[1-\left(\frac{pT_0}{p_0\left(T\pm\Delta T\right)}\right)^{0.235}\right]$$

$$= \frac{288.16}{0.0065}\left[1-\left(\frac{35598.98\times288.16}{101325\times276.16}\right)^{0.235}\right]$$

or $h = 44332.3\times\left[1-0.3666^{0.235}\right] = 44332.3\times\left[1-0.79\right]$
$=9310m$, indicating 1310m higher altitude.

The same procedure is adopted for zones 2 and 3 in the stratosphere using relevant equations. In the stratosphere from 11 km to 20 km:

$$\text{Density altitude,}\, h_\rho = 11000 - \frac{(RT_s)}{g_0}\ln\frac{\rho}{\rho_s} \tag{2.9d}$$

On a hot day a pilot could hit the ground while the altimeter is showing altitude, and vice versa for cold days. Serious errors can occur if the altimeter is not adjusted to the conditions of the territory. Density altitude is more used by pilots than engineers. Pilots communicate with the ground station to adjust the altimeter accordingly.

2.3.4   Humidity Effects

The ISA is modelled for dry atmosphere, but there is always some humidity present; in a monsoon climate, it can be very high humidity. The molecular mass ratio of water vapour to dry air is 0.62:1. Humidity affects the density of air. Presence of water vapour decreases the air density to a small extent, so an increase of the relative humidity of the air decreases the air density. Therefore it is considered in computing density altitude. The relative humidity also affects aircraft lift and engine performance. Readers are referred to humidity charts to obtain air density to compute altitude. The humidity effect is not used in this book.

2.3.5   Greenhouse Gases Effect

Greenhouse gas effects concern aerospace engineers. Atmospheric gas compositions greatly influence climatic temperature, affecting the environment. About 99% of the atmosphere is composed of diatomic gases, such as N_2 and O_2, which are transparent to radiation. But the very small percentages of triatomic and higher gases, such as carbon dioxide (CO_2), methane (CH_4), ozone (O_3) and chlorofluorocarbons (CFCs – industrial gases like aerosols and refrigerants) are responsible for atmospheric warming, and are known as greenhouse gases. These are seen as pollutants and exist mostly within 32 km of the Earth. CO_2, CH_4 and CFCs produce a blanket effect, which increases ambient temperature. On the other hand, depletion of O_3 thins out the shield for solar ultraviolet radiation, harming the population. The silent group of aerospace engineers is no less concerned than vociferous environmentalists, and act to minimize pollution, working directly to resolve the industrial challenges. Currently, the aerospace industry produces less than 2% of total manmade pollution. Airworthiness authorities have imposed strict limits to contain pollution from gas emissions. Aeronautical infrastructure deserves credit for it.

2.4   Airflow Behaviour: Laminar and Turbulent

This section deals with air (gas) flow. Air is compressible and its effect is realized when it is flowing. Aircraft design requires an understanding of incompressible, compressible, viscous and inviscid airflow. Characteristics of matter are given in Figure 2.3.

Understanding the role of the viscosity of air is important to aircraft engineers. The simplification of considering air as inviscid may simplify mathematics, but it does not represent the reality of design. Inviscid fluid does not exist, but is a close approximation to air. It provides much useful information rather quickly. Subsequently, the inviscid results are improvised. To incorporate the real effects of viscosity, designs must be tested to substantiate theoretical results.

The fact that airflow can offer resistance due to viscosity has been understood for a long time. Navier in France, and Stokes in England, independently arrived at the same mathematical formulation; their equation for momentum conservation embedding the viscous effect is known as the Navier-Stokes equation. It is a non-linear partial differential equation still unsolved analytically except for some simple body shapes. In 1904, Ludwig Prandtl presented a flow model that made the solution of viscous flow problems easier [23]. He demonstrated by experiment that the viscous effect of flow is realized only within a thin layer over the contact surface boundary; the rest of the flow remains unaffected. This layer is called the boundary layer. Figure 2.4 shows a boundary layer of airflow over a flat surface aligned with the flow direction (i.e. X-axis). Today, numerical methods (i.e. CFD) can address viscosity problems to a great extent.

The best way to model a continuum (i.e. densely packed) airflow is to consider the medium to be composed of very fine spheres of molecular scale (i.e. diameter 3×10^{-8} cm and intermolecular space 3×10^{-6} cm). Like sand, these spheres flow one over another, offering friction in between while colliding with one another. Air flowing over a rigid surface (i.e. acting as a flow boundary) will adhere to it, losing velocity; that is, there is a depletion of kinetic energy of the air molecules as they are trapped on the surface, regardless of how polished it may be. On a molecular scale, the surface looks like the crevices shown in Figure 2.4, with air molecules trapped within to stagnation. The contact air layer with the surface adheres and it is known as the "no-slip" condition. The next layer above the stagnated no-slip layer slips over it – and, of course, as it moves away from the surface, it will gradually reach the airflow velocity. The pattern within the boundary-layer flow depends on how fast it is flowing.

■ **FIGURE 2.3**
Characteristics of matter

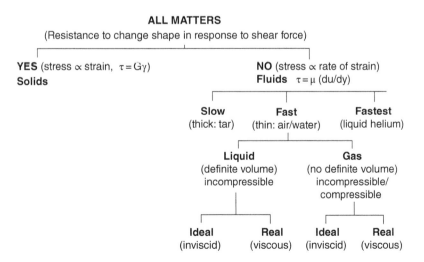

ALL MATTERS
(Resistance to change shape in response to shear force)

YES (stress $\propto$ strain, $\tau = G\gamma$)
Solids

NO (stress $\propto$ rate of strain)
Fluids $\tau = \mu \, (du/dy)$

Slow
(thick: tar)

Fast
(thin: air/water)

Fastest
(liquid helium)

Liquid
(definite volume)
incompressible

Gas
(no definite volume)
incompressible/
compressible

Ideal
(inviscid)

Real
(viscous)

Ideal
(inviscid)

Real
(viscous)

■ **FIGURE 2.4**
Boundary layer effect. Top: boundary layer over a flat plate. Bottom: magnified view of airflow over a rigid surface (boundary)

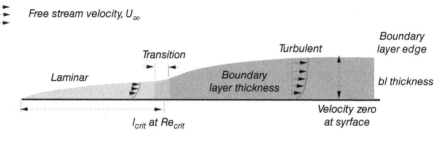

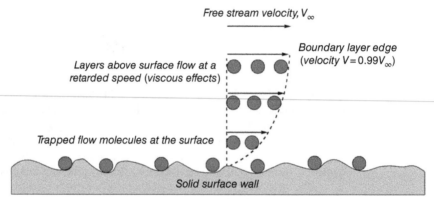

Here is a good place to define the parameter called the *Reynolds Number* (Re). Re is a useful and powerful parameter – it provides information on the flow status with the interacting body involved. Reynolds number increases with length.

$$\text{Re} = \frac{\rho_\infty U_\infty l}{\mu_\infty} \qquad (2.10)$$

= (density × velocity × length)/coefficient of viscosity = (inertia force)/(viscous force), where μ_∞ = coefficient of viscosity.

It represents the degree of skin friction depending on the property of the fluid. The subscript infinity, ∞, indicates the condition (i.e. undisturbed infinite distance ahead of the object). Re is a grouped parameter, which reflects the effect of each constituent variable, whether they vary alone or together. Therefore, for a given flow, characteristic length, l, is the only variable in Re. Re increases along the length. In an ideal flow (inviscid approximation), Re becomes infinity – not much information is conveyed beyond that. However, in real flow with viscosity, it provides vital information: for example, on the nature of flow (turbulent or laminar), on separation, and on many other characteristics.

Initially, when the flow encounters the flat plate at the leading edge (LE), it develops a boundary layer that keeps growing thicker until it arrives at a critical length, when flow characteristics then make a transition and the profile thickness suddenly increases. The friction effect starts at the LE and flows downstream in an orderly manner, maintaining the velocity increments of each layer as it moves away from the surface, much like a sliding deck of cards (in laminae). This type of flow is called *laminar flow*. Surface skin-friction depletes the flow energy transmitted through the layers until at a certain distance (i.e., critical point) from the LE, flow can no longer hold an orderly pattern in laminae, breaking down and creating turbulence.

■ **FIGURE 2.5**
Viscous effect of air on a flat plate (C_f is defined in the text)

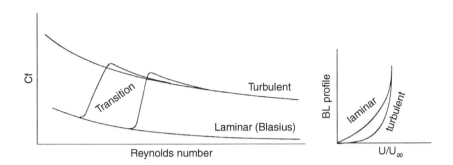

The boundary layer thickness is shown as δ at a height where 99% of the free-stream flow velocity is attained. The region where the transition occurs is called the *critical point*. It occurs at a predictable distance l_{crit} from the LE, having a critical Re of Re_{crit} at that point. At this distance along the plate, the nature of the flow makes the transition from laminar to *turbulent flow*, when eddies of the fluid mass randomly cross the layers. Through mixing between the layers, the higher energy of the upper layers energizes the lower layers. The physics of turbulence that can be exploited to improve performance (e.g. dents on a golf ball force a laminar flow to a turbulent flow) is explained later.

With turbulent mixing, the boundary-layer profile changes to a steeper velocity gradient and there is a sudden increase in thickness, as shown in Figure 2.5 (use Figure 9.20 for C_f values). For each kind of flow situation, there is a Re_{crit}. As it progresses downstream of l_{crit} the turbulent flow in the boundary layer is steadily losing its kinetic energy to overcome resistance offered by the surface. If the plate is long enough, then a point may be reached where further loss of flow energy would fail to negotiate the surface constraint and would leave the surface as a *separated flow* (see Section 2.4.1). Separation can also occur early in the laminar flow.

The extent of the velocity gradient, du/dy, at the boundary surface indicates the tangential nature of the frictional force; hence, it is a shear force. At the surface where u = 0, du/dy is the velocity gradient of the flow at that point. If F is the shear force on the surface area, A, due to friction in the fluid, then shear stress is expressed as follows:

$$\frac{F}{A} = \tau = \mu\frac{du}{dy} \tag{2.11}$$

where μ is the coefficient of viscosity. $\mu = 1.789 \times 10^{-5}\,kg/ms$ or Ns/m^2 (1g/ms = 1 poise) for air at sea-level ISA.

Kinematic viscosity, $v = \mu/\rho\,m^2/s$ ($1m^2/s = 10^4$ stokes), where ρ is the density of the fluid.

The measure of the frictional shear stress is expressed as a *coefficient of friction*, C_f, at the point:

$$C_f = \frac{\tau}{q_\infty} = \left(\text{shear stress/dynamic head}\right) \tag{2.12a}$$

Friction drag for the length l, $D_f = \int_0^l \tau dx$ (τ varies along the length).

Average coefficient of friction, C_F, is computed for a flat plate of unit length for the length l, that is, over the area ($1 \times l$), and is expressed as:

$$C_F = \frac{1}{l}\int_0^l \frac{\tau dx}{0.5\rho U_\infty^2} = \frac{D_f}{0.5\rho U_\infty^2 A} \tag{2.12b}$$

■ **TABLE 2.1**
Boundary layer thickness, δ, and local skin friction C_f

	Laminar	Turbulent
Boundary layer thickness, δ	$\dfrac{5.2x}{\sqrt{Re_x}}$	$\dfrac{0.37x}{Re^{0.2}}$
Local skin friction C_f	$\dfrac{1.328}{\sqrt{Re_l}}$	$\dfrac{0.074}{Re_l^{0.2}}$

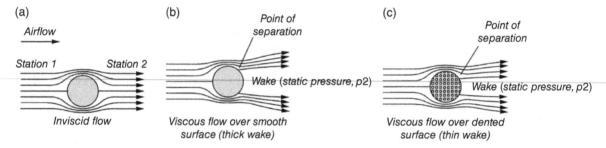

■ **FIGURE 2.6** Airflow past sphere. (a) Inviscid flow ($p_1 = p_2$). (b) Viscous flow ($p_1 > p_2$). (c) Dented surface ($p_1 > p_2$)

The difference of du/dy between laminar and turbulent flow is shown in Figure 2.5; the latter has a steeper gradient and hence a higher C_f. The up arrow indicates an increase and vice versa; for incompressible flow, $temp \uparrow$-$\mu \downarrow$, which reads as viscosity decreases with a rise in temperature, and for compressible flow, $temp \uparrow$-$\mu \uparrow$.

The pressure gradient along a flat plate gives dp / dx = 0. Airflow over curved surfaces (i.e., 3D surface) accelerates or decelerates depending on which side of the curve the flow is negotiating (dp / dx $\uparrow$ 0). Extensive experimental investigations on the local skin-friction coefficient, C_f, on a 2D flat plate are available for a wide range of Re values. The overall coefficient of skin friction over a 3D surface is expressed as C_F and is higher than the 2D flat plate. C_f increases from laminar to turbulent flow, as can be seen from the increased boundary layer thickness. In general, C_F is computed semi-empirically from the flat plate C_f (see Chapter 9).

For flat plate with zero pressure gradient the boundary layer thickness, δ, and local skin friction C_f can be expressed as given in Table 2.1 [22].

To explain the physics of drag, the classical example of flow past a sphere is shown in Figure 2.6. A sphere in inviscid flow will have no drag because it has no skin friction and there is no pressure difference between the front and aft ends – there is nothing to prevent the flow from negotiating the surface curvature. Diametrically opposite to the front stagnation point is a rear stagnation point, equating forces on the opposite sides. This ideal situation does not exist in nature but can provide important information.

In the case of a real fluid with viscosity, the physics changes by the introduction of drag as a combination of skin friction and the pressure difference between fore and aft of the sphere. At low Re, the low-energy laminar flow near the surface of the smooth sphere separates early, creating a large wake in which the static pressure p_2, at the aft, cannot recover to its initial value p_1 at the front of the sphere. The pressure at the front is now higher at the stagnation area, resulting in a pressure difference that appears as pressure drag. It would be beneficial if the flow was made turbulent by denting the sphere surface. In this case, high-energy flow from the upper layers mixes randomly with flow near the surface, reenergizing it. This enables the flow to overcome the sphere's curvature and adhere to a greater extent, thereby reducing the wake. Therefore, a reduction of pressure drag compensates for the increase in skin-friction drag (i.e., C_f increases from laminar to turbulent flow). This concept is applied to golf-ball design (i.e. low Re, low velocity, and small physical dimension). The dented golf ball would go farther than an equivalent smooth golf ball due to reduced drag.

Therefore:

$$\text{drag} = \text{skin friction drag} + \text{pressure drag}$$

The situation changes drastically for a body at high Re (i.e., high velocity and/or large physical dimension, for example, an aircraft wing (or even a golf ball hit at a very high speed that would require more than any human effort), when flow is turbulent almost from the LE. A streamlined aerofoil shape does not have the highly steep surface curvature over a large area like that of a sphere; therefore, separation occurs very late, resulting in a thin wake. Therefore, pressure drag is low. The dominant contribution to drag comes from skin friction, which can be reduced if the flow retains laminarization over more of the surface area (although it is not applicable to a golf ball in low Re). Laminar aerofoils have been developed to retain laminar-flow characteristics over a relatively large part of the aerofoil. These aerofoils are more suitable for low-speed operation (but Re higher than the golf-ball application) such as gliders, and have the added benefit of a very smooth surface made of composite materials.

Clearly, the drag of a body depends on its profile, that is, how much wake it creates. The blunter the body, the greater the wake size will be; it is for this reason that aircraft components are streamlined. This type of drag is purely viscosity-dependent and is termed *profile drag*. In general, in aircraft applications it is also called *parasite drag*, as explained in Chapter 7.

Scientists have been able to model the random pattern of turbulent flow using statistical methods. However, at the edges of the boundary layer, the physics is unpredictable. This makes accurate statistical modelling difficult, with eddy patterns at the edge extremely unsteady and the flow pattern varying significantly. It is clear why the subject needs extensive treatment [23].

2.4.1   Flow Past an Aerofoil

A typical airflow past an aerofoil is shown in Figure 2.7; it is an extension of the diagram of flow over a flat plate (see Figure 2.5).

In Figure 2.7, the front curvature of the aerofoil causes the flow to accelerate, with the associated drop in pressure, until it reaches the point of inflection on the upper surface of the aerofoil. This is known as a region of favourable pressure gradient because the lower pressure downstream favours airflow. Past the inflection point, airflow starts to decelerate, recovering the pressure (i.e., flow in an adverse pressure gradient) that was lost while accelerating. For inviscid flow, it would reach the trailing edge, regaining the original free-stream flow velocity and pressure condition. In reality, the viscous effect depletes flow energy, preventing it from regaining the original level of pressure. Along the aerofoil surface, airflow is depleting its energy due to friction of the aerofoil surface. The viscous effect appears as a wake behind the body.

The result of a loss of energy while flowing past the aerofoil surface is apparent in an adverse pressure gradient; it is like climbing uphill. A point may be reached where there is not enough flow energy left to encounter the adverse nature of the downstream pressure rise; the flow then breaks away and leaves the surface to adjust to what nature allows. Where the flow leaves the surface is called the *point of separation*, and it is critical information for aircraft design. When separation happens over a large part of the aerofoil, it is said that the aerofoil has *stalled* because it has lost the intended pressure field. Generally it happens on the upper

■ **FIGURE 2.7**
Airflow past aerofoil

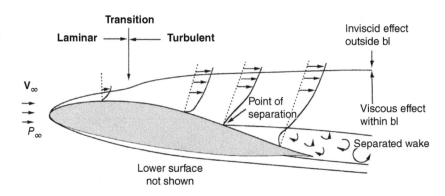

surface; in a stalled condition, there is a loss of low-pressure distribution and, therefore, a loss of lift, and its consequence is described in Section 2.6. This is an undesirable situation for an aircraft in flight. There is a minimum speed below which stalling will occur in every winged aircraft. The speed at which an aircraft stalls is known as the *stalling speed*, V_{stall}. At stall, an aircraft cannot maintain altitude and can even become dangerous to fly; obviously, stalling should be avoided.

For a typical surface finish, the magnitude of skin-friction drag depends on the nature of the airflow. Below Re_{crit}, laminar flow has a lower skin-friction coefficient, C_f, and, therefore lower drag. The aerofoil LE starts with a low Re and rapidly reaches Re_{crit} to become turbulent. Aerofoil designers must shape the aerofoil LE to maintain laminar flow as much as possible. Aircraft surface contamination is an inescapable operational problem that degrades surface smoothness, making it more difficult to maintain laminar flow. As a result, Re_{crit} advances closer to the LE. For high subsonic flight speed (high Re), the laminar flow region is so small that flow is considered fully turbulent.

In summary, as Re increases with length, l, from the LE, the critical point is reached when flow changes from laminar to turbulent. This is the l_{crit} giving the Re_{crit} for transition. For aerofoil application, $Re_{crit} \approx 500,000$.

This section points out that designers should maintain laminar flow as much as possible over the wetted surface, especially at the wing LE. As mentioned previously, gliders that operate at a lower Re offer a better possibility to deploy an aerofoil with laminar-flow characteristics. The low annual utilization in private usage favours the use of a composite material, which provides the finest surface finish. However, although the commercial-transport wing may show the promise of partial laminar flow at the LE, the reality of an operational environment at high utilization does not guarantee adherence to the laminar flow. For safety reasons, it would be appropriate for governmental certifying agencies to examine conservatively the benefits of partial laminar flow. This book considers the fully turbulent flow to start from the LE of any surface of a high-subsonic aircraft.

2.5 Aerofoil

The wing is a 3D surface (i.e. span, chord and thickness). An aerofoil represents 2D geometry (i.e., chord and thickness). The cross-sectional shape of a wing (i.e., the bread-slice-like sections of a wing comprising the aerofoil) is the crux of aerodynamic considerations. Aerofoil characteristics are evaluated at mid-wing to eliminate effects of the finite 3D wing tip. Typically, its characteristics are expressed over the unit chord to allow scaling to fit any suitable size.

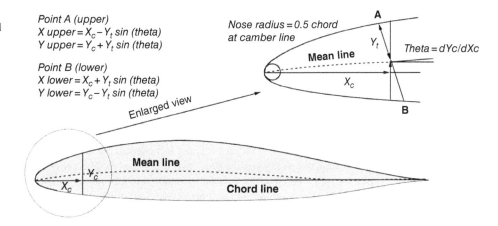

Aircraft performance depends on the type of aerofoil incorporated in its wing design. Aerofoil design is a specialized topic not dealt in this book. There are many well-proven existing aerofoils catering for a wide range of performance demands. NACA/NASA has systematically developed aerofoils in a family of designs and they are widely used by many aircraft. The majority of general aviation aircraft have a NACA aerofoil. However, to stay ahead of the competition, all major aircraft manufacturers develop their own aerofoil kept "commercial in confidence" as proprietary information. These are not available in the public domain. Moreover, unless these aerofoils can be compared, it is not recommended to use them during undergraduate studies. NACA series aerofoils will prove sufficient to make good aircraft and this book depends only on the NACA family of aerofoils.

Nowadays there are specialized application softwares to design aerofoil sections, and can give credible results. The only way to apply them is to verify the analytical results by comparing wind-tunnel test results. Certification of any new aerofoil is an additional task.

This section introduces various types of aerofoil and their geometric characteristics. The next chapter deals with aerofoil characteristics and how to make a selection. The 3D effects of a wing are discussed in Section 2.11.

2.5.1   Subsonic Aerofoil

To standardize aerofoil geometry, Figure 2.8 provides universally accepted definitions that should be well understood [24].

Chord length is the maximum straight-line distance from the leading edge (LE) to the trailing edge (TE). The *mean line* represents the mid-locus between the upper and lower surfaces. The *camber* represents the aerofoil expressed as the percentage deviation of the mean line from the chord line. The mean line is also known as the *camber line*. Coordinates of the upper and lower surfaces are denoted by y_U and x_L for the distance

The chord line is kept aligned to the *x*-axis of the Cartesian coordinate system and measured from the LE. The *thickness (t)* of an aerofoil is the distance between the upper and the lower contour lines at the distance along the chord, measured perpendicular to the mean line and expressed in percentage of the full chord length. Conventionally, it is expressed as the *thickness to chord (t/c)* ratio in percentage. A small radius at the LE is necessary to smooth out the aerofoil contour. It is convenient to present aerofoil data with the chord length non-dimensionalized to unity so that the data can be applied to any aerofoil size.

Aerofoil pressure distribution is measured in a wind tunnel (also in CFD) to establish its characteristics, as shown in [24]. Wind-tunnel tests are conducted at the mid-span of the wing model so that results are as close as possible to 2D characteristics. These tests are conducted at several Re. Higher Re indicates higher velocity; that is, it has more kinetic energy to overcome the skin friction on the surface, thereby increasing the pressure difference between the upper and lower surfaces and, hence, increasing lift.

2.5.1.1   Groupings of Subsonic Aerofoil – NACA/NASA:

From the early days, European countries and the US undertook intensive research to generate better aerofoils to advance aircraft performance. By the 1920s, a wide variety of aerofoils appeared and consolidation was needed. Since the 1930s, NACA-generated families of aerofoils benefited from what was available in the market and beyond. It presented the aerofoil geometries and test results in a systematic manner, grouping them into family series. The generic pattern of the NACA aerofoil family is listed in [24], with well-calibrated wind-tunnel results. The book was published in 1949 and has served aircraft designers (civil and military) for more than half a century, and is still useful. NACA subsequently reorganized to NASA and continued with aerofoil development, concentrating on having better laminar flow characteristics over the aerofoil. Also industry undertook research and development to generate better aerofoils for specific purposes, but they are kept "commercial in confidence".

NACA's earliest attempt (in the 1930s) to make a systemic generic type was the NACA 1-Series (or 16 series) [24]. This new approach to airfoil design had its shape mathematically derived from the desired lift characteristics. Prior to this, airfoil shapes were first created and then had their characteristics measured in a wind tunnel. The 1-series airfoils are described by five digits. Since this type is no longer used, it is not discussed here.

The four-digit, five-digit, and six-digit series of NACA aerofoils are the ones of interest to this book. Many fine aircraft have used the NACA aerofoil. The NACA four- and five-digit aerofoils were created by superimposing a simple camber-line shape with a thickness distribution that was obtained by fitting the following polynomial [24]:

$$y = \pm \frac{t}{0.2} \times \left(0.2969 \times x^{0.5} - 0.126 \times x - 0.3537 \times x^2 + 0.2843 \times x^3 - 0.1015 \times x^4 \right)$$

2.5.1.2   NACA Four-Digit Aerofoil:

The camber line of a four-digit aerofoil section is defined by a parabola from the leading edge to the position of maximum camber, followed by another parabola to the trailing edge (Figure 2.9). This constraint did not allow the aerofoil design to be adaptive. For example, it prevented generations of aerofoil with more curvature towards the leading edge in order to provide better pressure distribution.

Each of the four digits of the nomenclature represents a geometric property, as follows:

First digit: maximum camber, y_c, in % chord
Second digit: location of y_{c-max} in tenths (1/10) of chord from leading edge
Last two digits: maximum thickness to chord.

Explanation for NACA 4415
Here we use the example of the NACA 4415 aerofoil (Figure 2.10).

First digit, **4**: It has 0.02c maximum amount of camber with design lift coefficient = 4 / 10 = 0.4.

Second digit, **4**: Position of maximum camber ratio in percent of chord 4 / 100 = 4% chord length from LE.

Last two digits, **15**: It has maximum thickness to chord ratio = 0.15, i.e. 15% of the chord length.

FIGURE 2.9
Camber line
distributions

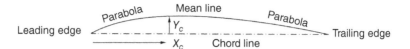

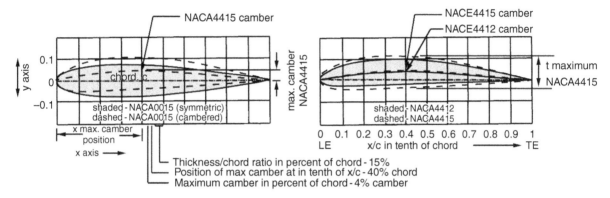

■ **FIGURE 2.10**   Comparing NACA 0015, NACA 4412 and NACA 4415 geometries

2.5.1.3   NACA Five-Digit Aerofoil: After the four-digit sections came the five-digit sections. The first two and the last two digits represent the same definitions as the four-digit NACA aerofoil (Figure 2.9 is still applicable except the camber shape differs as defined by the middle digit). The middle digit stands for the aft position of the mean line, bringing the change in defining camber line curvature. The middle digit has only two options: 0 for straight (i.e. standard) and 1 for inverted cube. These are explained below.

First digit: maximum camber, Y_c, in % chord. The design lift coefficient is 3/2 of it, in tenths
Second digit: maximum thickness of maximum camber in 1/20 of chord X_c
Middle digit: aft position of the mean line: 0 – straight/standard; 1 – inverted cube
Last two digits: maximum thickness to chord, in % of chord.

Explanation for NACA 23012
This example aerofoil is illustrated in Figure 2.11.

First digit, **2**: It has 0.02c maximum amount of camber with design lift coefficient $= 2 \times (3 / 20) = 0.3$.

Second digit, **3**: Position of maximum camber at $3 \times 2 / 200 = 15\%$ chord length from LE.

Third (middle) digit **0**: aft camber shape is straight (standard).

Last two digits, **12**: it has maximum thickness to chord ratio $= 0.12$, i.e. 12% of the chord length.

NACA 23112 is the same as above except that its aft camber shape has an inverted cube shape. Figure 2.11 compares NACA 23012 with NACA 23112 aerofoils.

2.5.1.4   NACA Six-Digit Aerofoil: The five-digit family was an improvement over the four-digit NACA series aerofoil; however, researchers subsequently found better geometric definitions to represent a new family of a six-digit aerofoil. The state of the art for a good aerofoil often follows reverse engineering – that is, it attempts to fit a cross-sectional shape to a

■ **FIGURE 2.11**
Comparison of NACA 23012 aerofoil with NACA 23112 reflex aerofoils

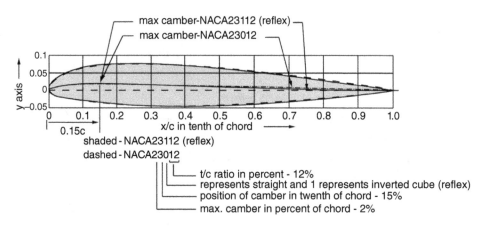

shaded - NACA23112 (reflex)
dashed - NACA23012

- t/c ratio in percent - 12%
- represents straight and 1 represents inverted cube (reflex)
- position of camber in twenth of chord - 15%
- max. camber in percent of chord - 2%

■ **FIGURE 2.12**
Comparison of NACA65₂-415, NACA64₂-415 and NACA63₂-415 aerofoils

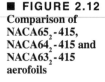

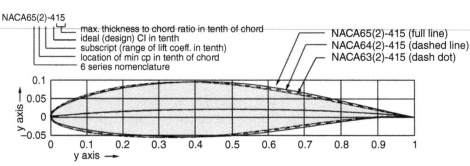

given pressure distribution. The NACA six-digit series aerofoil came much later (it was first used for the P51 Mustang design in the late 1930s) from the need to generate a desired pressure distribution instead of being restricted to what the relatively simplistic four- and five-digit series could offer. The six-digit series aerofoils were generated from a more or less prescribed pressure distribution and were designed to achieve some laminar flow. This was achieved by placing the maximum thickness far back from the LE. Their low-speed characteristics behave like the four- and five-digit series but show much better high-speed characteristics. However, the drag bucket seen in wind-tunnel test results may not show up in actual flight. Some of the six-digit aerofoils are more tolerant to production variation compared with typical five-digit aerofoils.

NACA six-digit aerofoils are possibly the most used ones, widely used in various classes of aircraft. Their success was followed by increased efforts to develop aerofoils with laminar flow characteristics over a wide speed regime.

The definition for the NACA six-digit aerofoil example $63_2 - 212$ is as follows:

First digit: The number "6", indicating the series.
Second digit: location of the minimum c_p (pressure) area in tenths of chord.
Subscript: the range of lift coefficient in tenths above and below the design lift coefficient in which favorable pressure gradients exist on both surfaces (half width of low drag bucket in tenths of C_l).
Fourth digit: ideal aerofoil design lift coefficient C_l in tenths
Last two digits: maximum thickness as % of chord.

Figure 2.12 is an example of the NACA65₃-415 six-series airfoil for which the minimum pressure's position is at the 50% chord location indicated by the second digit. The subscript 3 indicates

FIGURE 2.13
The Seven Series
NACA 747A315
aerofoil

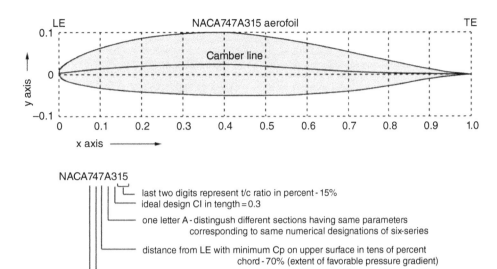

that the minimum drag coefficient (drag bucket) is near its minimum value over a range of C_L of 0.3 above and below the design-lift coefficient. The next digit indicates the design-lift coefficient of 0.4, and the last two digits indicate the maximum thickness as 15% of the chord.

2.5.1.5 *NACA Seven-Digit Aerofoil*:
A seven-digit family of aerofoils have followed, to further maximize laminar flow achieved by separately identifying the low pressure zones on upper and lower surfaces of the airfoil. The airfoil is described by seven digits in the following sequence:

First digit:	The number "7", indicating the series.
Second digit:	location of the minimum c_p (pressure) area on the upper surface in tenths of chord.
Third digit:	location of the minimum c_p (pressure) area on the lower surface in tenths of chord.
Fourth digit:	one letter referring to a standard profile from the earlier NACA series.
Fifth digit:	single digit describing the lift coefficient C_l in tenths.
Last two digits:	maximum thickness as % of chord.

Figure 2.13 is an example of the NACA 747A315, of the seven-digit aerofoil series. The first digit 7 indicates the series. The second digit 4 signifies that it has a favourable pressure gradient on the upper surface to the extent of 40% of chord peaking to a minimum c_p, and thereafter starts the adverse pressure gradient. The third digit 7 says the same for the lower surface, in this case up to 70%. The last three digits are same nomenclature as the six-digit aerofoils, indicating that it has a design lift coefficient of $C_l = 0.3$, and has a maximum thickness of 15% of the chord. The letter A shows that it has a class of different sections.

2.5.1.6 *NACA Eight-Digit Aerofoil*:
The NACA eight-digit aerofoil series are the *supercritical airfoils* designed to independently maximize airflow above and below the wing. It is variation of six-series and seven-series and has the same numbering system as the

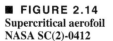

■ **FIGURE 2.14**
Supercritical aerofoil
NASA SC(2)-0412

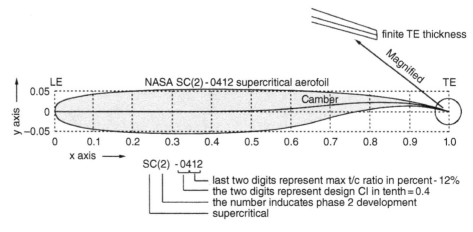

seven-series airfoils, except that it has the first digit as an "8" to identify the series. Application of an NACA eight-digit aerofoil to aircraft is yet to be proven and not discussed here.

2.5.1.7 NASA Supercritical Aerofoil:

In an effort to develop aerofoils to operate at a higher subsonic speed, yet retaining good low-speed characteristics, Richard T. Whitcomb of the Langley Research Center developed supercritical aerofoils during the early 1960s. The goal was to increase the drag-divergence Mach number, thereby reducing drag and allowing for more efficient flight in the transonic regime. It has the characteristic shape of a flat top following a large leading-edge radius and curved tail with thickness.

This distinctive airfoil shape helps the local supersonic flow with isentropic recompression, on account of the reduced curvature over the middle region of the upper surface, and substantial aft camber.

The National Aeronautics and Space Administration (NASA) made systematic studies in three phases during the 1970s to develop a family of aerofoils with thicknesses from 2–18% and design lift coefficients from 0 to 1.0. These were termed *supercritical aerofoil*. The three phases of development are as follows:

Phase 1: Early supercritical airfoils that increased the drag divergence Mach number beyond the six-series NACA family.

Phase 2: The extension of Phase 1 supercritical airfoils with "target pressure distributions".

Phase 3: Airfoils developed for studies to reduce Phase 2 leading-edge radii. The study was eventually abandoned.

The supercritical airfoil number designation is in the form of the example of SC(2)-0412, as shown in Figure 2.14. The aerofoil designation is broken down into two segments – the first segment of 3 characters starts with SC, indicating supercritical airfoil, and then a bracketed number showing the development phase – in this case Phase 2 of three phases of development. The last segment starts with its first two digits as the design lift coefficient in tenths, and the last two are the thickness in percent chord.

2.5.1.8 NASA Low-Speed (LS) Aerofoils – GA(W):

The NASA General Aviation Wing (GAW) series evolved later for low-speed applications and for use by general aviation. In this new research by NASA, low-speed airfoil series for use on light aeroplanes were also developed (e.g. LS(1)-0417 [GA(W)-1], LS(1)-0013 and LS(1)-0413MOD). Although the series

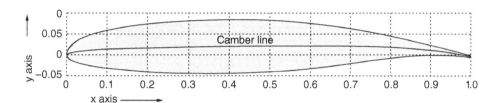

■ **FIGURE** 2.15
NASA/Langley/
Whitcomb
LS(2)-0413
(GA(W)-2) general
aviation airfoil

showed better lift-to-drag characteristics, their performance with flaps deployment, tolerance to production variation, and other issues are still in question. As a result, the GAW aerofoil has yet to compete with some of the older NACA aerofoil designs. However, a modified GAW aerofoil, the Whitcomb LS(2)-0413 (GA(W)-2), has appeared with improved characteristics (Figure 2.15). The numbering system is similar to supercritical aerofoils.

Although these aerofoils showed better lift and drag characteristics than the older NACA series for small aircraft, they did not replace the use of NACA four-, five- and six-digit series for small private aircraft. NACA aerofoils are still popular.

2.5.1.9   *Other Types of Subsonic Aerofoil*:

Subsequently, after the six-series sections, aerofoil design became more specialized, with aerofoils designed for their particular application. In the mid-1960s, Whitcomb's "supercritical" aerofoil allowed flight with high critical Mach numbers (operating with compressibility effects, producing in-wave drag) in the transonic region. In addition to the NACA aerofoil series, there are many other types of aerofoil in use.

To remain competitive, the major aircraft manufacturing companies generate their own aerofoils. One example is the peaky-section aerofoil that were popular during the 1960s and 1970s for the high-subsonic flight regime. Aerofoil designers generate their own purpose-built aerofoil with good transonic performance, good maximum lift capability, thick sections, low drag, and so on – some are in the public domain, but most are held commercial in confidence for strategic reasons. Subsequently, more transonic supercritical aerofoils have been developed, by both research organizations and academic institutions. One such baseline design in the UK is the RAE 2822 aerofoil section, whereas the CAST 7 evolved in Germany.

There are many other types of aerofoil, such as Eppler, Liebeck (used in gliders), and many older types (e.g. Wortmann, Gottingen, Clark Y, RAF aerofoils, etc.) not discussed here. The following websites may provide useful information on various types of aerofoil:

http://aerospace.illinois.edu/m-selig/ads/coord_database.html#N
http://airfoiltools.com/airfoil/details?airfoil=n64212mb-il

2.5.1.10   *Discussion*:

In earlier days, drawing full-scale aerofoils of a large wing, and great effort was required to maintain accuracy to an acceptable level; their manufacture was not easy. Today, CAD/CAM and microprocessor-based numerically controlled lofters have made things simple and very accurate. In December 1996, NASA published a report outlining the theory behind the airfoil sections and computer programs used to generate NACA aerofoil.

Aerofoil characteristics are sensitive to geometry and require hard tooling with tight manufacturing tolerances in order to adhere closely to the profile. Often, a wing design has several airfoil sections varying along the wing span. Appendix F provides six types of aerofoil for use in this book. Readers should note that the 2D aerofoil wind-tunnel test is conducted in restricted conditions and will need corrections for use with real aircraft (see Section 2.12).

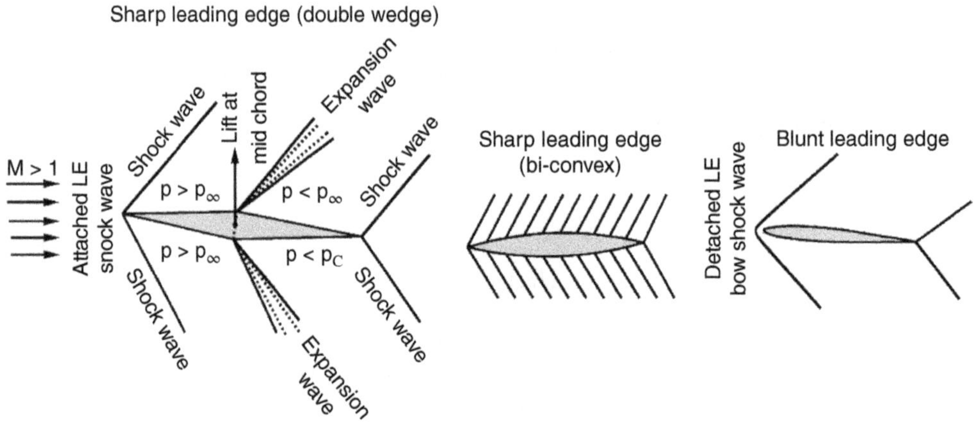

■ **FIGURE 2.16**   Supersonic aerofoil

2.5.2   Supersonic Aerofoil

A supersonic airfoil is a cross-section geometry designed to generate lift efficiently at supersonic speeds. The need for such a design arises when an aircraft is required to operate consistently in the supersonic flight regime. Supersonic aerofoil are necessarily thin, in the range $0.4 \lesssim (t/c) \lesssim 0.7$.

Figure 2.16 shows three main types of supersonic aerofoil. Supersonic airfoils generally have a thin section formed of either angled planes or opposed arcs (called "double wedge airfoils" and "biconvex airfoils" respectively), with very sharp leading and trailing edges. The sharp edges prevent the formation of a detached bow shock in front of the airfoil as it moves through the air. This shape is in contrast to subsonic airfoils, which often have rounded leading edges to reduce flow separation over a wide range of angle of attack. A rounded edge would behave as a blunt body in supersonic flight and thus would form a bow shock, which greatly increases wave drag. The airfoils' thickness, camber, and angle of attack are varied to achieve a design that will cause a slight deviation in the direction of the surrounding airflow.

However, since a round leading edge decreases an airfoil's susceptibility to flow separation, a sharp leading edge implies that the airfoil will be more sensitive to changes in angle of attack. Therefore, to increase lift at lower speeds, aircraft that employ supersonic airfoils also use high-lift devices such as leading edge and trailing edge flaps. The thin six-series and seven-series aerofoils have been used in combat aircraft design.

2.6   Generation of Lift

Figure 2.17a is a qualitative description of the flow field and its resultant forces on the aerofoil. The result of skin friction is the drag force, shown in Figure 2.17b. The lift is normal to the flow. Section 2.4.1 explains that a typical aerofoil has an upper surface more curved than the lower surface, which is represented by the camber of the aerofoil. Even for a symmetrical aerofoil, the increase in the angle of attack increases the velocity at the upper surface and the aerofoil approaches stall, a phenomenon described in Section 2.10. Angle of attack, or incidence, α, is defined as the angle between the freestream direction and the chord line of the aerofoil.

Figure 2.18a shows the pressure field around the aerofoil. The pressure at every point is given by the pressure coefficient distribution, as shown in Figure 2.18b. The upper surface has

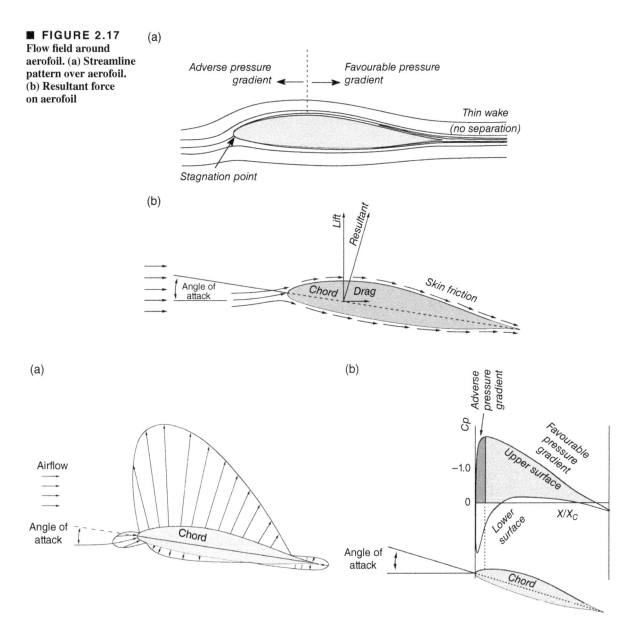

■ **FIGURE 2.17**
Flow field around
aerofoil. (a) Streamline
pattern over aerofoil.
(b) Resultant force
on aerofoil

(a)

Adverse pressure
gradient ◄—————► Favourable pressure
gradient

Thin wake
(no separation)

Stagnation point

(b)

Lift
Resultant

Angle of
attack

Chord | Drag

Skin friction

(a)

Airflow

Angle of
attack

Chord

(b)

Cp | Adverse pressure gradient

Favourable pressure gradient

Upper surface

−1.0

0

X/X_C

Lower surface

Angle of
attack

Chord

■ **FIGURE 2.18**   **Pressure field representations around aerofoil. (a) Pressure field distribution. (b) C$_p$ distribution over aerofoil**

lower pressure, which can be seen as a negative distribution. In addition, cambered aerofoils have moments that are not shown in the figure.

At the LE, the streamlines move apart: one side negotiates the higher camber of the upper surface and the other side negotiates the lower surface. The higher curvature at the upper surface generates a faster flow than the lower surface. They have different velocities when they meet at the trailing edge, creating a vortex sheet along the span. The phenomenon can be decomposed into a set of straight streamlines representing the free-stream flow condition and a set of circulatory streamlines of a strength that matches the flow around the aerofoil. The circulatory flow is

■ **FIGURE 2.19**
Lift generation on aerofoil. Top: Inviscid flow representation of flow around aerofoil. Bottom: Mathematical representation by superimposing free vortex flow over parallel flow

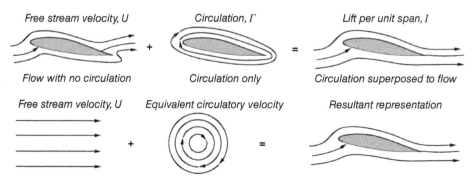

known as the *circulation* of the aerofoil. The concept of *circulation* provides a useful mathematical formulation to represent lift. Circular flow is generated by the effect of the airfoil camber, which gives higher velocity over the upper wing surface. The directions of the circles show the increase in velocity at the top and the decrease at the bottom, simulating velocity distribution over the aerofoil. Either the deflection of the control surface or a change in the angle of attack will alter the pressure distribution. The positive Y-direction has negative pressure on the upper surface. The area between the graphs of the upper and lower surface C_p distribution is the lift generated for the unit span of this aerofoil.

Figure 2.19 shows flow physics around the aerofoil. The flow over an aerofoil develops a lift per unit span of $l = \rho U$ (see textbooks for the derivation). Computation of circulation is not easy. This book uses accurate experimental results to obtain the lift.

Figure 2.20 shows the typical test results of an aerofoil as plotted against a variation of the angle of attack, α. Initially, the variation is linear; it starts to deviate and reaches maximum C_l (C_{lmax} at α_{max}).

Past α_{max}, the C_l drops rapidly – if not drastically – when stall is reached. Stalling starts upon reaching α_{max}. These graphs show aerofoil characteristics. Figure 2.20 depicts the distribution of the pressure coefficient C_p at an angle of attack of 15° on aerofoil with high-lift devices extended.

2.6.1 Centre of Pressure and Aerodynamic Centre

The pressure coefficient at a point, $C_p = (p_{local} - p_\infty)/q$. Then the sectional lift coefficient C_l of an aerofoil can be expressed as:

$$C_l = \frac{1}{c}\int_0^c \left(C_{pl} - C_{pu}\right)dx = \int_0^1 \left(C_{pl} - C_{pu}\right)d\left(x/c\right) \tag{2.13}$$

As stated earlier, an aerofoil is a 2D bread-slice-like wing section with zero thickness and is weightless. The pressure field around an aerofoil develops forces and moment. Aerofoil test results, as shown in Figure 3.20, have viscous consideration with the pressure field that results in resultant force and moment but without any weight.

The *centre of pressure*, *cp*, is defined as a point where moment vanishes and the resultant force of the pressure field acts. In other words, cp is the point where the resultant force represents the equivalent of what the pressure field generates on the aerofoil; the resultant force vector acting at the centre of pressure is the value of the integrated pressure field where there is

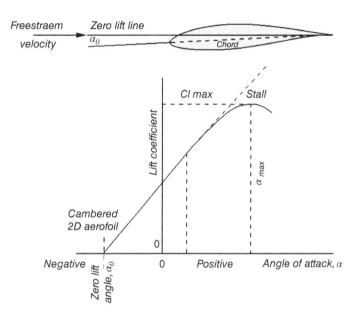

■ **FIGURE 2.20**
Aerofoil
characteristics [22].
C_L **versus alpha.**
Reproduced with
permission from
John Wiley

no moment. The resultant force is resolved into lift and drag. Since the pressure field changes with angle of attack, the position of the cp changes accordingly. This makes the cp a difficult parameter to deal with.

As the angle of attack, α, is increased, the aerofoil gets front-loaded and the cp moves forward. With decrease of α, the cp moves back. When α is reduced to a low value, for a cambered aerofoil the inherent nose-down moment may move the cp rearward outside of the aerofoil. Low resultant force at a low angle of attack requires a large moment arm to balance, pushing the cp further aft. At zero lift, the cp approaches infinity. For a typical aerofoil, the cp stays within the aerofoil for the operating range. Figure 2.21 depicts the movement of the centre of pressure, cp, with change in lift. The figure shows that past α_{max}, the C_l drops rapidly – if not drastically – when stall is reached. Stalling starts upon reaching α_{max}. It will be shown in Section 3.10.1 that cp is always aft of the quarter chord.

The aerodynamic centre, *ac*, is concerned with moments about a point, typically on the chord line (Figure 2.22). It does not deal with forces by themselves. However, at around the quarter-chord point, it is noticed that the moment is invariant to the angle of attack until stall occurs. This point is invariant to α change and is known as the ac, which is a natural reference point through which all forces and moments are defined to act. There could be minor variations of the position of the ac around the quarter-chord among aerofoils. To standardize, the fixed point at the quarter-chord from the leading edge is also measured. In term of coefficients, these are C_{m_ac} and $C_{m1/4}$, which are measured in wind tunnel tests. A symmetric aerofoil has $C_{m\_ac} = 0$, but there is small variation in $C_{m1/4}$ with α variation. The relation between C_{m_ac} and $C_{m1/4}$ can be expressed mathematically as dealt with in Section 3.10.1.

The ac offers much useful information, as discussed later. The ac is a useful parameter in stability analyses when aircraft CG has to be taken into account. The higher the positive camber, the more lift is generated for a given angle of attack; however, this leads to a greater nose-down moment. To counter this nose-down moment, conventional aircraft have a horizontal tail with the negative camber supported by an elevator. For tailless aircraft (e.g. delta-wing designs in which the horizontal tail merges with wing), the trailing edge is given a negative camber as a "reflex". This balancing is known as *trimming* and it is associated with the type of drag known as *trim drag*. Aerofoil selection is then a compromise between having good lift characteristics and a low moment.

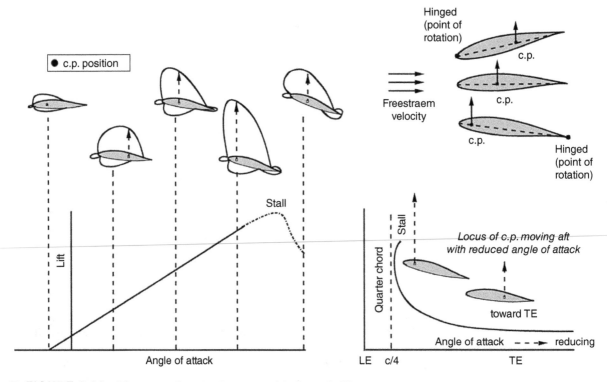

■ **FIGURE 2.21** Movement of centre of pressure with change in lift

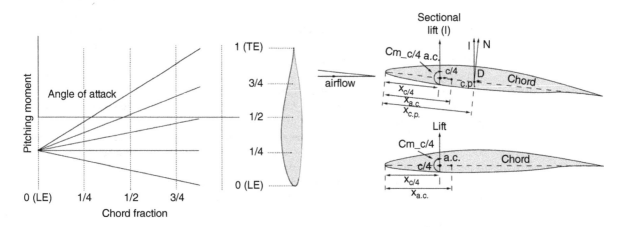

■ **FIGURE 2.22** Aerodynamic centre invariant near quarter-chord (fractional chord position about which moment is taken)

2.6.2 Relation between Centre of Pressure and Aerodynamic Centre

Typically, the ac is within 22% to 27% of the aerofoil chord from the LE. Since the point varies from aerofoil to aerofoil, for standardization the test results are given about a fixed point of quarter-chord, $c/4$ from the LE [24]. From the test results of aerofoil moment around quarter-chord, the aerodynamic centre can be accurately determined (given in the graph in Appendix F).

Figure 2.22 shows a typical aerofoil (NACA six-series) with quarter-chord $c/4$ and the aerodynamic centre ac shown at a distance $x_{c/4}$ and x_{ac} from the LE, respectively. At $x_{c/4}$, the sectional lift, L, and moment $M_{c/4}$ are shown in the diagram (drag is small and its moment contribution is negligibly small).

To estimate the position of the aerodynamic centre, ac, one should proceed as follows. As mentioned above, the quarter chord $x_{c/4}$ is a fixed position and test results of $C_{m\_c/4}$ are available [24]. Also C_{m_ac} is invariant with respect to the angle of attack, α (i.e. C_l). Taking moment about the ac, the moment contribution by drag is small as the moment arm is negligible. The aerofoil is considered weightless and the definitions of the aerodynamic coefficients are given in Equation 1.2.

$$\sum M = 0$$

i.e.

$$M_{ac} = l \times \left(x_{ac} - x_{c/4} \right) + M_{c/4}$$

In coefficient form,

$$C_{m\_ac} = \left(C_l / c \right) \times \left(x_{ac} - x_{c/4} \right) + C_{m\_c/4} \tag{2.14}$$

Differentiating with respect to the angle of attack α (note that C_{m_ac} is constant), it becomes:

$$0 = \left(\frac{dC_l}{d\alpha} \right)\left(\frac{x_{ac}}{c} - \frac{x_{c/4}}{c} \right) + \left(\frac{dC_{m\_c/4}}{d\alpha} \right)$$

Transposing,

$$\left(\frac{x_{ac}}{c} - \frac{x_{c/4}}{c} \right) = -\left(\frac{dC_{m\_c/4}}{d\alpha} \right) \Big/ \left(\frac{dC_l}{d\alpha} \right) = -\frac{m_0}{a_0}$$

where m_0 is the slope of the $C_{M\_c/4}$ curve and a_0 is the lift-curve slope. m_0 and a_0 can be evaluated from the test results (Appendix F), or:

$$\frac{x_{ac}}{c} = \frac{x_{c/4}}{c} - \frac{m_0}{a_0} = 0.25 - \frac{m_0}{a_0} \quad \left(\text{in terms of percentage of chord}\right) \tag{2.15a}$$

Substituting in Equation 3.44:

$$C_{m\_c/4} = C_{m\_ac} + \left[\left(\frac{dC_{m\_c/4}}{d\alpha} \right) \Big/ \left(\frac{dC_l}{d\alpha} \right) \right] C_l = C_{m\_ac} - \left(\frac{m_0}{a_0} \right) C_l \tag{2.15b}$$

Estimating the position of the centre of pressure, cp, is lot more difficult as it moves with the operating range from mid-chord, $c/2$, to ac at aircraft stall. Let the aerofoil section lift l act at the aerofoil cp. Taking moment about quarter-chord, $c/4$:

$$\sum M = 0$$

i.e.

$$0 = l \times (x_{cp} - x_{c/4}) + M_{c/4}$$

In coefficient form:

$$C_l \times \left(x_{cp} - x_{c/4} \right) + C_{m\_c/4} = 0$$

Substituting the value of $C_{m\_c/4}$ from Equation 3.46 into the above equation:

$$C_l \times \left(\frac{x_{cp}}{c} - \frac{x_{c/4}}{c} \right) + C_{m\_ac} - \left(\frac{m_0}{a_0} \right) C_l = 0$$

or

$$\left(\frac{x_{cp}}{c} - \frac{x_{c/4}}{c} \right) = -\frac{C_{m\_ac}}{C_l} + \left(\frac{m_0}{a_0} \right)$$ (2.16a)

C_{m_ac} is negative for conventional aerofoil, which means $x_{cp}/c > x_{c/4}$, always.

Replacing $m_0/a_0 = (x_{a.c.}/c - x_{c/4}/c)$ from Equation 3.46 in the above equation, it becomes:

$$\left(\frac{x_{cp}}{c} - \frac{x_{c/4}}{c} \right) = -\frac{C_{m\_ac}}{C_l} + \left(\frac{x_{ac}}{c} - \frac{x_{c/4}}{c} \right)$$

or

$$\frac{x_{cp}}{c} = \frac{x_{ac}}{c} - \frac{C_{m\_ac}}{C_l}$$ (2.16b)

EXAMPLE 2.6

Find the aerodynamic centre, ac, and centre of pressure, cp, at 6° angle of incidence, α, for the NACA 65-410 aerofoil from the test results given in Appendix F.

SOLUTION

From the test result graph, NACA 65-410 gives:

$dC_l/d\alpha = a_0 = (1.05 - 0.25)/8 = 0.1/\deg$

And $\quad m_0 = (-0.0755 - (-0.062)/[8 - (-4)]$

$\quad\quad\quad = -0.0135/12 = -0.00125/\deg$

Equation 2.15 gives:

$$\frac{x_{ac}}{c} = \frac{x_{c/4}}{c} - \frac{m_0}{a_0} = 0.25 - (-0.00125/0.1)$$

$$\approx 0.2625 \text{ of the chord from the LE.}$$

(The ideal C_l for NACA 65-410 is 0.3.)

To compute the centre of pressure (x_{cp}/c), take $x_{ac}/c = 0.2625$ from above. Then at 6° angle, $C_l = 0.9$ and $C_{m\_c/4} = -0.075$ from the graph in Appendix F. At angle of zero lift:

$$C_{m\_ac} = C_{m\_c/4} + C_l \left(\frac{x_{ac}}{c} - \frac{x_{c/4}}{c} \right)$$

$$= -0.075 - 0.9(0.012)$$

$$= -0.0846$$

Equation 2.16b gives:

$$\frac{x_{cp}}{c} = \frac{x_{ac}}{c} - \frac{C_{m\_ac}}{C_l}$$

$$= 0.2625 + (0.0846/0.9)$$

$$= 0.3565$$

Centre of pressure and aerodynamic centre can be analytically determined if the aerofoil properties can be analytically expressed. Within the operating range, the aerofoil lift characteristics can be expressed as a straight line, and the moment characteristic is not exactly a straight line but can be tolerated as such; however, the drag characteristics may not be amenable to fit as an equation of parabola, resulting in inaccuracies. To obtain industry standard results, it is recommended that test data are used [24].

It is stressed that readers should understand the role of ac and cp. Although they are not much used in aircraft design and often ignored in textbooks, they gives a good insight into how an aerofoil behaves.

2.7   Types of Stall

Aircraft stall will occur below a prescribed speed, and it is an undesirable situation. Stall speeds and stalling characteristics much depend on the airfoil characteristics. Aircraft performance engineers must know the stall speed limits at each configuration and its behaviour.

It is essential that designers understand stalling characteristics. Figure 2.23 shows the general types of stall that can occur. There are two types of stall patterns: (1) gradual stall and (2) abrupt stall, as described below. Aircraft stall is affected by wing stall, which depends on aerofoil characteristics. Section 2.16 addresses wing stall.

1. Gradual stall: This is a desirable pattern and occurs when separation is initiated at the trailing edge of the aerofoil; the remainder maintains the pressure differential. As the separation moves slowly toward the LE, the aircraft approaches stall gradually, giving the pilot enough time to take corrective action. The forgiving and gentle nature of this stall is ideal for an *ab initio* trainee pilot. The type of aerofoil that experiences this type of stall has a generously rounded LE, providing smooth flow negotiation but not necessarily other desirable performance characteristics.

2. Abrupt stall: This type of stall starts with separation at the LE, initially as a small bubble. Then the bubble either progresses downstream or bursts quickly and catastrophically (i.e. abruptly). Aerofoils with a sharper LE, such as those found on higher-performance aircraft, tend to exhibit this type of behaviour.

2.7.1   Buffet

Buffeting is an incipient stage of pre-stall phenomena. If the angle of attack of an aerofoil (i.e. aircraft wing) is gradually increased, then before stall sets in, there is fluctuation of separation point on the upper surface of the wing in an unstable manner, still generating sufficient lift to keep the aircraft in flight. These fluctuations can be felt as aircraft judders and the control stick

■ **FIGURE 2.23**
Stall pattern

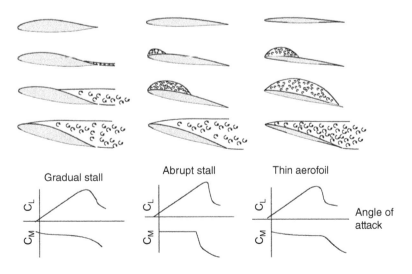

shakes, giving feedback to the pilot. For aerofoils with gentle stall characteristics, buffeting gives sufficient time for the pilot to take corrective measures. This forgiving aerofoil character suits the design for *ab initio* trainer aircraft, although its lift and drag characteristics may be modest. Some of the advanced high-performance aerofoils show abrupt stall where the buffet margin is small, suitable for experienced pilots to fly.

2.8   Comparison of Three NACA Aerofoils

Finally, this section briefly summarizes the important characteristics to be met to select a desirable aerofoil.

The NACA 4412, NACA 23015 and NACA 64_2-415 are three commonly used aerofoils – there are many different types of aircraft that use one of these aerofoils. Figure 2.24 shows their characteristics for comparison purposes [24]. The NACA 4412 is the oldest and, for its time, was the favourite. The NACA 23015 has sharp stalling characteristics; however, it can give a higher sectional lift, C_l, and lower sectional moment, C_m, than others. Drag-wise, the NACA 64_2-415 has a bucket to give the lowest sectional drag. Of these three examples, the NACA 64_2-415 is the best for gentle stall characteristics and low sectional drag, offsetting the small amount of trim drag due to the relatively higher moment coefficient. Designers choose from a wide variety of aerofoils, or generate one suitable for their purposes. Designers look for the following qualities in the characteristics of a 2D aerofoil:

1. The lift should be as high as possible; this is assessed by the C_{Lmax} of the test results.
2. The stalling characteristics should be gradual; the aerofoil should be able to maintain some lift past C_{Lmax}. Stall characteristics need to be assessed for the application. For example, for *ab initio* training, it is better to have aircraft with forgiving, gentle stalling characteristics. For aircraft that will be flown by experienced pilots, designers could compromise with stalling characteristics to achieve better performance.
3. There should be a rapid rise in lift; that is, a better lift-curve slope given by $dC_L/d\alpha$.
4. There should be low drag using a drag bucket, retaining flow laminarization as much as possible at the design C_L (i.e. angle of incidence).
5. C_m characteristics give nose-down moments for a positively cambered aerofoil. It is preferable to have low C_m to minimize trim drag.

An aerofoil designer must produce a suitable aerofoil that encompasses the best of all five qualities – a difficult compromise to make. Flaps are also an integral part of the design.

■ **FIGURE 2.24**
Comparison of three NACA aerofoils (reproduced with permission of NACA)

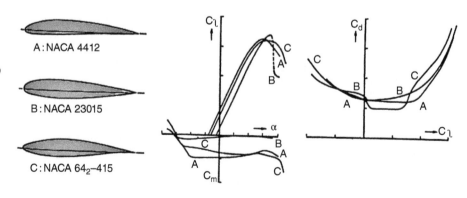

Flap deflection effectively increases the aerofoil camber to generate more lift. Therefore, a designer must also examine all five qualities at all possible flap and slat deflections. From this brief discussion, it is apparent that aerofoil design itself is state of the art and is therefore not addressed in this book. However, experimental data on suitable aerofoils are provided in Appendix F.

2.9   High-Lift Devices

High-lift devices are small aerofoil-like elements that are fitted at the trailing edge of the wing as a *flap*, and/or at the LE as a *slat*. In typical cruise conditions, the flaps and slats are retracted within the contour of the aerofoil. Flaps and slats can be used independently or in combination. At low speed, they are deflected about a hinge line, rendering the aerofoil more curved as if it had more camber. A typical flow field around flaps and slats is shown in Figure 2.25. The entrainment effect through the gap between the wing and the flap allows the flow to remain attached in order to provide the best possible lift.

Higher-performance high-lift devices are complex in construction, and therefore heavier and more expensive. Selection of the type is based on cost versus performance tradeoff studies – in practice, past experience is helpful in making selections.

Considerable lift enhancement can be obtained by incorporating high-lift devices at the expense of additional drag and weight. Figure 2.26 lists the typical 2D values of C_{Lmax} for NACA class aerofoils. These values are representative of other types of NACA aerofoils and may be used if actual data are not available. In 3D application on the wing, C_{Lmax} varies with aspect ratio,

■ **FIGURE 2.25**
Flap and slat flow field

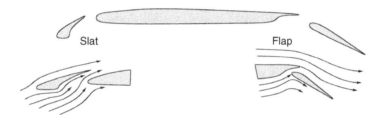

■ **FIGURE 2.26**
High-lift devices and typical values of 2D C_{Lmax}

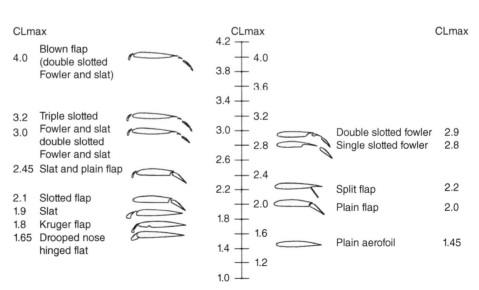

■ **TABLE 2.2**
Correction factors for 3D application of high-lift devices

C_{Lmax} at zero sweep	C_{Lmax} at 20° sweep	C_{Lmax} at 25° sweep	C_{Lmax} at 30° sweep
2.0	0.88	0.85	0.78
2.5	0.88	0.84	0.77
3.0	0.88	0.83	0.76
3.5	0.88	0.82	0.75

■ **FIGURE 2.27**
Transonic aerofoil –
supercritical
Whitcomb aerofoil
(typical flat upper
surface with aft
camberfor rear
loading)

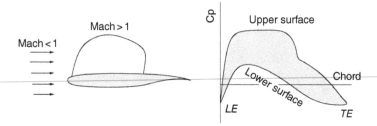

span, sweep, taper, and the high-lift device chord lengths, gap, shape, and so on. It is recommended that experimental values for the particular application be used. If data are not available, then Figure 2.26 values may be used for 2D values, with a correction applied.

Use the following approximate correction factors in Table 2.2 on the 2D values on the zero degree sweep to 20°, 25° and 30° sweep. The factors will vary on application on the 3D wing, depending on the geometries. Intermediate values may be linearly interpolated.

2.10   Transonic Effects – Area Rule

At high subsonic speeds, the local velocity along a curved surface (e.g. on an aerofoil surface) can exceed the speed of sound, whereas flow over the rest of the surface remains subsonic. In this case, the aerofoil is said to be in *transonic flow*. At higher angles of attack, transonic effects can appear at lower flight speeds. Aerofoil thickness distribution along the chord length is the parameter that affects the induction of transonic flow. Transonic characteristics exhibit an increase in wave drag (i.e. the compressibility effect). These effects are undesirable but unavoidable; however, aircraft designers keep the transonic effect to a minimum. Special attention is necessary in generating the aerofoil-section design, which shows a flatter upper surface. Figure 2.27 depicts a typical transonic aerofoil (i.e. the Whitcomb section) and its characteristics.

The Whitcomb section advanced the flight speed by minimizing wave drag (i.e. the critical Mach-number effects); therefore, it is called the *supercritical aerofoil section*. The geometric characteristics are a rounded LE, followed by a flat upper surface and rear-loading with camber; the lower surface at the trailing edge shows the cusp. All modern high-subsonic aircraft have the supercritical aerofoil section characteristics. Manufacturers develop their own section or use any data available to them.

For an aircraft configuration, it has been shown that the cross-sectional area distribution along the body axis affects the wave drag associated with transonic flow. The bulk of this area distribution along the aircraft axis comes from the fuselage and the wing. The best cross-sectional area distribution that minimizes wave drag is a cigar-like smooth distribution

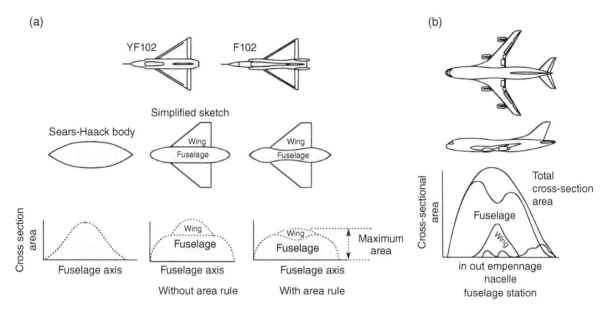

■ **FIGURE 2.28**   Area rule. (a) Transonic area rule. (b) Boeing transonic aircraft

(i.e. uniform contour curvature) known as the *Sears-Haack ideal body* (Figure 2.28a). The fuselage shape approximates it; however, when the wing is attached, there is a sudden jump in volume distribution (Figure 2.28a). In the late 1950s, Whitcomb demonstrated through experiments that waisting of the fuselage in a "coke-bottle" shape could accommodate wing volume. This type of procedure for wing-body shaping is known as the *area rule*. A smoother distribution of the cross-sectional area reduces wave drag (details in [20]). In the late 1960s to early 1970s, Boeing attempted to design a transonic commercial transport aircraft using the area rule, operating above Mach 0.9 (Figure 2.28b). However, it did not get the "go-ahead", possibly because the timing did not favour the market.

2.10.1   Compressibility Correction

Compressibility effects require the application of correction factors (Prandtle-Glauert Rule); high subsonic and supersonic corrections differ slightly. The following are some of the corrections of interest.

Parameter	High subsonic	Supersonic	
2D lift-curve slope, a_0	$a_{0\_comp} = \dfrac{a_{0\_incomp}}{\sqrt{\left(1 - M_\infty^2\right)}}$	$= \dfrac{a_{0\_incomp}}{\sqrt{\left(M_\infty^2 - 1\right)}}$	**(2.17a)**
Centre of pressure, C_p	$C_{p\_comp} = \dfrac{C_{p\_incomp}}{\sqrt{\left(1 - M_\infty^2\right)}}$	$= \dfrac{C_{p\_incomp}}{\sqrt{\left(M_\infty^2 - 1\right)}}$	**(2.17b)**
Lift coefficient, C_L	$C_{L\_comp} = \dfrac{C_{L\_incomp}}{\sqrt{\left(1 - M_\infty^2\right)}}$		**(2.17c)**

■ **FIGURE 2.29**
**Effect of Mach
number on C_{Lmax}**

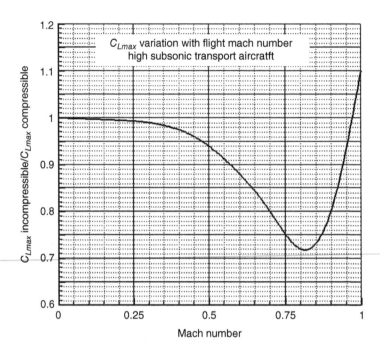

For a low aspect ratio wing, further correction is required to obtain the lift-curve slope [31].

Another effect of compressibility on account of speed gain is the change in C_{Lmax}, as shown in Figure 2.29. In subsonic flow, the ratio of $C_{Lmax\_compressible}/C_{Lmax\_incompressible}$ comes down to around 0.7.

2.11   Wing Aerodynamics

Aircraft performance engineers have to deal with aircraft wing shapes and reference areas. An aircraft's wing is the lifting surface with the chosen aerofoil section, which can vary spanwise. The lift generated by the wing sustains the weight of the aircraft to make flight possible. Proper wing planform shape and size are crucial to improving aircraft efficiency and performance; however, aerofoil parameters are often compromised with the cost involved.

A 3D-finite wing produces vortex flow as a result of tip effects, as shown in Figure 2.30 and explained in Figure 2.31. The high pressure from the lower surface rolls up at the free end of the finite wing, creating the tip vortex.

The direction of vortex flow is such that it generates downwash, which is distributed spanwise at varying strengths. A reaction force of this downwash is the lift generated by the wing. Energy consumed by the downwash appears as lift-dependent induced drag, *Di*, and its minimization is a goal of aircraft designers.

The physics explained thus far is represented in geometric definitions, as shown in Figure 2.32. An elliptical wing planform (e.g. the Spitfire fighter of World War II) creates a uniform span-wise downwash at its lowest magnitude and leads to minimum induced drag. Figure 2.32 shows that the downwash effect of a 3D wing deflects free-stream flow, V_∞, by an angle, ε, to V_{local}. It can be interpreted as if the section of 3D wing behaves as a 2D infinite wing with:

$$\textit{Effective angle of incidence } \alpha_{\textit{eff}} = (\alpha - \varepsilon) \qquad \textbf{(2.18)}$$

where α is the angle of attack at the aerofoil section, by V_∞.

■ **FIGURE 2.30**
Wing tip vortex

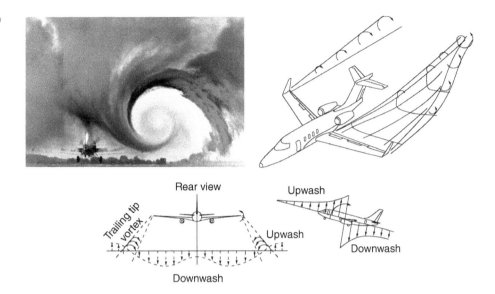

Trailing tip vortex

Rear view

Upwash

Downwash

Upwash

Downwash

■ **FIGURE 2.31**
**Effect of pressure,
flow pattern and
downwash of a finite
3D wing**

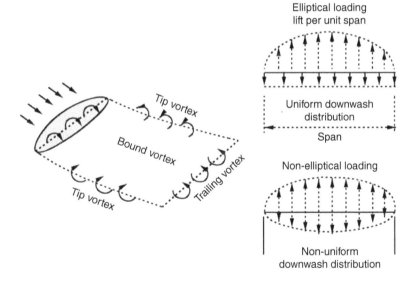

Tip vortex

Bound vortex

Tip vortex

Trailing vortex

Elliptical loading
lift per unit span

Uniform downwash
distribution

Span

Non-elliptical loading

Non-uniform
downwash distribution

■ **FIGURE 2.32**
**Downwash angle and
its distribution on
elliptical wing
planform**

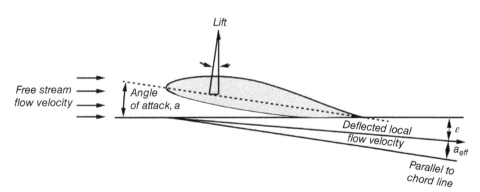

Lift

Free stream
flow velocity

Angle
of attack, a

Deflected local
flow velocity

ε

a_{eff}

Parallel to
chord line

Aerodynamics textbooks may be consulted to derive theoretically the downwash angle:

$$\varepsilon = \frac{C_L}{\pi \text{AR}} \left(\text{in radians}\right) = 57.3 \frac{C_L}{\pi \text{AR}} \left(\text{in degrees}\right) \tag{2.19a}$$

For a nonelliptical wing planform, the downwash will be higher and a semi-empirical correction factor, e, called *Oswald's efficiency factor* (always less than or equal to 1) is applied, as follows:
Average downwash angle,

$$\varepsilon = \frac{C_L}{e\pi \text{AR}} \left(\text{in radians}\right) = 57.3 \frac{C_L}{e\pi \text{AR}} \left(\text{in degrees}\right) \tag{2.19b}$$

The extent of downwash is lift-dependent; that is, it increases with an increase in C_L.

Strictly speaking, Oswald's efficiency factor, e, varies with wing incidence; however, the values used are considered an average of those found in the cruise segment, and remain constant. In that case, for a particular aircraft design, the average downwash angle, ε, is treated as a constant taken at the mid-cruise condition. Advanced wings of commercial transport aircraft can be designed in such a way that at the design point, $e \approx 1.0$.

Equations 2.18 and 2.19 show that downwash decreases with increase in the aspect ratio (AR). When the aspect ratio reaches infinity, there is no downwash and the wing becomes a 2D-infinite wing (i.e. no tip effects), and its sectional characteristics are represented by aerofoil characteristics. The downwash angle, ε, is small – in general less than 5° for aircraft with a small aspect ratio. The aerofoil section of the 3D wing apparently would produce less lift than the equivalent 2D aerofoil. Therefore, 2D-aerofoil test results would require correction for a 3D-wing application, as explained in Section 2.12.

Local lift, L_{local}, produced by a 3D wing, is resolved into components perpendicular and parallel to free streamflow, V_∞. In coefficient form, the integral of these forces over the span gives the following:

$$C_L = \frac{L\cos\varepsilon}{qS_W} \quad \text{and}$$

$$C_{Di} = \frac{L\sin\varepsilon}{qS_W} \quad \left(\text{the induced-drag coefficient}\right)$$

For small angles, ε, it reduces to:

$$C_L = \frac{L}{qS_W} \quad \text{and}$$

$$C_{Di} = \frac{L\varepsilon}{qS_W} = C_L\varepsilon \tag{2.20}$$

C_{Di} is the drag generated from the downwash angle, ε, and is lift-dependent (i.e. induced); hence, it is called the *induced-drag coefficient*. For a wing planform, Equations 2.19 and 2.20 combine to give:

$$C_{Di} = C_L\varepsilon = C_L \times \frac{C_L}{e\pi \text{AR}} = \frac{C_L^{\ 2}}{e\pi \text{AR}} \tag{2.21}$$

Induced drag is lowest for an elliptical-wing planform, when $e = 1$; however, it is costly to manufacture. In general, the industry uses a trapezoidal planform with a taper ratio, $\lambda \approx 0.4$ to 0.5, resulting in an e value ranging from 0.85 to 0.98. A rectangular wing has a ratio of $\lambda = 1.0$ and a delta wing has a ratio of $\lambda = 0$, which results in an average e below 0.8. A rectangular wing with its constant chord is the least expensive planform to manufacture, having the same-sized ribs along the span.

2.11.1   Induced Drag and Total Aircraft Drag

The basic definition of drag is viscosity-dependent. The previous section showed that the tip effects of a 3D wing generate additional drag for an aircraft that appears as induced drag, Di. Therefore, the total aircraft drag in incompressible flow would be as follows:

$$\text{aircraft drag} = \text{skin} - \text{friction drag} + \text{pressure drag} + \text{induced drag}$$
$$= \text{parasite drag} + \text{induced drag}$$

Most of the first two terms does not contribute to the lift and are considered parasitic in nature; hence, they are jointly called the *parasite drag*. In coefficient form, it is referred to as C_{DP}. It changes slightly with lift and therefore has a minimum value. In coefficient form, it is called the *minimum parasite drag coefficient*, C_{DPmin}, or C_{D0}. The induced drag is associated with the generation of lift and must be tolerated. Incorporating this new definition, the above equation can be written in coefficient form as follows:

$$C_D = C_{DP} + C_{Di} \tag{2.22}$$

Chapter 7 addresses aircraft drag in more detail, and the contribution to drag due to the compressibility effect is also presented.

2.12   Aspect Ratio Correction of 2D-Aerofoil Characteristics for 3D-Finite Wing

To incorporate the tip effects of a 3D wing, 2D test data need to be corrected for Re and span. This section describes an example of the methodology. Equation 2.18 indicates that a 3D wing will produce α_{eff} at an attitude when the aerofoil is at the angle of attack, α. Because α_{eff} is always less than α, the wing produces less C_L corresponding to aerofoil C_l, as shown in Figure 2.33. This section describes how to correct the 2D aerofoil data to obtain the 3D wing-lift coefficient, C_L, versus the angle of attack, α, relationship. Within the linear variation, $dC_L/d\alpha$ needs to be evaluated at low angles (e.g. from $-2°$ to $8°$). 2D test data offer the advantage of being able to represent any 3D wing when corrected for its aspect ratio. There exist effects of wing sweep and aspect ratio on $dC_L/d\alpha$.

Let $a_0 = $ 2D aerofoil lift-curve slope. The 2D aerofoil will generate the same lift at a lower α of α_{eff} than the wing will generate at $\alpha(\alpha_{3D} > \alpha_{2D})$. Therefore, using the 2D aerofoil data, the wing-lift coefficient C_L can be worked at the angle of attack, α, as shown here (all angles are in degrees). The wing lift at an angle of attack, α, is as follows:

$$C_L = a_0 \times \alpha_{eff} + \text{constant} = a_0 \times (\alpha - \varepsilon) + \text{constant} \tag{2.23}$$

■ **FIGURE 2.33**
Lift-curve slope
correction for
aspect ratio

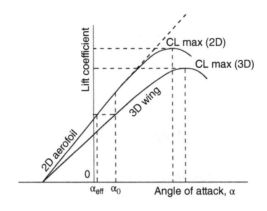

or

$$C_L = a_0 \times \left(\alpha - 57.3 \frac{C_L}{e\pi \text{AR}} \right) + \text{constant}$$

or

$$C_L + \left(57.3\ C_L \frac{a_0}{e\pi \text{AR}} \right) = a_0 \times \alpha + \text{constant}$$

or

$$C_L = \frac{a_0 \times \alpha}{1 + \left(57.3 \dfrac{a_0}{e\pi \text{AR}} \right)} + \frac{\text{constant}}{1 + \left(57.3 \dfrac{a_0}{e\pi \text{AR}} \right)}$$

Differentiating with respect to α, it becomes:

$$\frac{dC_L}{d\alpha} = \frac{a_0}{1 + \left(57.3 \dfrac{a_0}{e\pi \text{AR}} \right)} = a \qquad (2.24)$$

The wing-tip effect delays the stall by a few degrees because the outer-wing flow distortion reduces the local angle of attack; it is shown as α_{max}. Note that α_{max} is the shift of C_{Lmax}; this value α_{max} is determined experimentally. In this book, the empirical relationship of $\alpha_{max} = 2°$, for AR > 5 to 12, $\alpha_{max} = 1°$, for AR > 12 to 20, and $\alpha_{max} = 0°$ for AR > 20.

Evidently, the wing lift-curve slope, $dC_L / d\alpha = a$, is less than the 2D aerofoil lift-curve slope, a_0. Figure 2.33 shows the degradation of the wing lift-curve slope, $dC_L/d\alpha$, from its 2D aerofoil value. The 2D test data offer the advantage of representing any 3D wing when corrected for its aspect ratio. The wing sweep and aspect ratio affects $dC_L/d\alpha$.

If the flight Re is different from the experimental Re, then the correction for C_{Lmax} must be made using linear interpolation. In general, experimental data provide C_{Lmax} for several Re values, to facilitate interpolation and extrapolation.

EXAMPLE 2.7

Given the NACA 2412 aerofoil data (see test data in Appendix F), construct the wing C_L versus α graph for a rectangular wing planform of aspect ratio 7 having an Oswald's efficiency factor $e = 0.75$, at a flight Re $= 1.5 \times 10^6$. From the 2D aerofoil test data at Re $= 6 \times 10^6$, find $dC_l/d\alpha = a_0 = 0.095$ per degree (evaluate within the linear range: -2 to $8°$). C_{lmax} is at $\alpha = 16°$.

Use Equation 2.24 to obtain the 3D wing lift-curve slope:

$$\frac{dC_L}{d\alpha} = a = \frac{a_0}{1 + \left(57.3 \dfrac{a_0}{e\pi \text{AR}}\right)}$$

$$= \frac{0.095}{1 + \left(57.3 \dfrac{0.095}{0.75 \times 3.14 \times 7}\right)}$$

$$= 0.095 / 1.348$$

$$= 0.067$$

From the 2D test data, C_{lmax} for three Res for smooth aerofoils and one for a rough surface, interpolate results for a wing $C_{lmax} = 1.25$ at flight $Re = 1.5 \times 10^6$. Finally, for AR $= 7$, the α_{max} increment is $1°$, which means that the wing is stalling at $(16 + 1) = 17°$.

The wing has lost some lift-curve slope (i.e. less lift for the same angle of attack) and stalls at a slightly higher angle of attack compared with the 2D test data. Draw a vertical line from the 2D stall $\alpha_{max} + 1°$ (the point where the wing maximum lift is reached). Then, draw a horizontal line with $C_{Lmax} = 1.25$.

Finally, translate the 2D stalling characteristic of α to the 3D wing lift-curve slope joining the portion to the C_{Lmax} point following the test data pattern. This demonstrates that the wing C_L versus the angle of attack, α, can be constructed (see Figure 2.27).

2.13   Wing Definitions

This section defines the parameters used in wing design and explains their role. The parameters are the wing planform area (also known as the wing reference area, S_w); wing-sweep angle, Λ; and wing-taper ratio, λ (dihedral and twist angles are given after the reference area is established). Also, the reference area generally does not include any extension area at the leading and trailing edges. Reference areas are concerned with the projected rectangular/ trapezoidal area of the wing.

2.13.1   Planform Area, S_W

The wing planform area acts as a reference area for computational purposes. The wing planform reference area is the projected area, including the area buried in the fuselage shown as a dashed line in Figure 2.34. However, the definition of the wing planform area differs among manufacturers.

In commercial transport aircraft design, there are primarily two types of definitions practised (in general) on either side of the Atlantic. The difference in planform-area definition is irrelevant as long as the type is known and adhered to. This book uses the first type (Figure 2.34a), whichis prevalent in the US and has straight edges extending to the fuselage centreline. Some European definitions show the part buried inside the fuselage as a rectangle (Figure 2.34b); that is, the edges are not straight up to the centreline unless it is a rectangular wing normal to the centreline.

A typical subsonic commercial transport wing is shown in Figure 2.35. An extension at the LE of the wing root is called a *glove*, and an extension at the trailing edge is called a *yehudi* (this is Boeing terminology). The yehudi's low-sweep trailing edge offers better flap characteristics. These extensions can originate in the baseline design or on the existing platform to accommodate a larger wing area. Glove and/or yehudi can be added later as modifications; however, this is not easy because the aerofoil geometry would be affected.

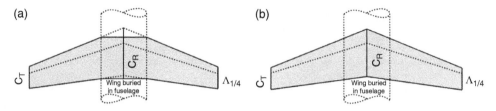

■ **FIGURE 2.34**   **Wing planform definition (half wing shown). (a) Fully tapered (US). (b) Tapered up to fuselage side (Europe)**

■ **FIGURE 2.35**
Wing sweep and twist for a Boeing 737 half wing

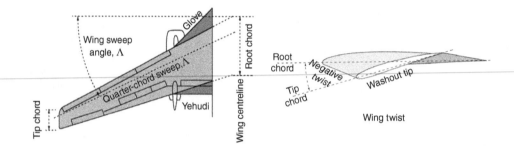

2.13.2   Wing Aspect Ratio

In the simplest rectangular wing-planform area, the aspect ratio is defined as AR = (span, b) / (chord, c). For a generalized trapezoidal wing-planform area:

$$AR = (b \times b) / (b \times c) = (b^2) / (S_W) \qquad (2.25)$$

2.13.3   Wing-Sweep Angle

The wing quarter-chord line is the locus of one quarter of the chord of the reference wing-planform area measured from the LE, as shown in Figure 2.35. Wing sweep angle, Λ, is the angle of the quarter chord measured from the line normal to the wing centreline (X-axis).

2.13.4   Wing Root (c_{root}) and Tip (c_{tip}) Chords

These are the aerofoil chords parallel to the aircraft at the centreline and the tip, respectively, of the trapezoidal reference area.

2.13.5   Wing-Taper Ratio, λ

This is defined as the ratio of the wing-tip chord to the wing-root chord (c_{tip}/c_{root}). The best taper ratio is in the range from 0.3 to 0.6. The taper ratio can affect the wing efficiency by giving a higher Oswald's efficiency factor (see Section 2.10).

2.13.6   Wing Twist

The wing can be twisted (see Figure 2.35) by making the wing-tip nose down (i.e. *washout*) relative to the wing root, which causes the wing root to stall earlier (i.e. retain aileron effectiveness). Typically, 1–2° washout twist is sufficient. Twisting the wing tip upward is known as *wash-in*.

(a)   (b)   (c)

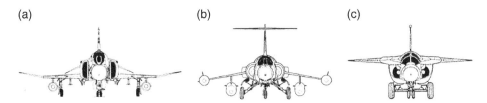

■ **FIGURE 2.36**   Positioning of wing with respect to fuselage. (a) Low wing (F4). (b) Mid wing (T-tail – F104). (c) High wing (MIG31) (reproduced with permission of Richard Ferrier)

2.13.7   High/Low Wing

The position of the wing with respect to the fuselage has a strong influence on aircraft drag generation; the mid-wing position offers nearly twice as much interference drag compared to high or low winged configurations (see Section 9.8.2). Aircraft performance engineers have to make drag estimations and it is important for them to study various configurations. Depending on the design drivers, an aircraft configuration can place the wing anywhere from the top (i.e. high wing) to the bottom of the fuselage (i.e. low wing) or in between (i.e. mid wing), as shown in Figure 2.36 for military aircraft.

Structural considerations of the wing attachment to the fuselage comprise a strong design driver, although in the civilian aircraft market, the choice could be dictated by user requirements and/or preference. The dominant configuration for civilian transport aircraft has been a low wing, which provides a wider main undercarriage wheel track, allowing better ground manoeuvring. A low wing also offers a better crashworthy safety feature in the extremely rare emergency situation of a belly landing.

Design trends shows that military transport aircraft have predominantly high wings with large rear-mounted cargo doors. Aircraft with a high wing have better ground clearance and the fuselage can be closer to the ground, which makes cargo-loading easier, especially with a rear-fuselage cargo door. The main undercarriage is mounted on the fuselage sides, with the bulbous fairing causing some additional drag. Turboprops favour a high-wing configuration to allow sufficient ground clearance for the propeller.

The authors believe that more high-winged commercial transport aircraft/Bizjets can gain a competitive edge as these offer better aerodynamics (e.g. the BAe RJ100 and Dornier 328 are successful high-wing designs).

The wing centre section should not interfere with the cabin passage-height clearance – especially critical for smaller aircraft. A fairing is shown for low-wing aircraft (Cessna Citation) or high-wing aircraft (Dornier 328), where the wing passes under or over the fuselage, respectively. Both cases have a generous fairing that conceals the fuselage mould-line kink (a drag-reduction measure), which would otherwise be visible. Mid-wing (or near-midwing) designs are more appropriate to larger aircraft with a passenger cabin floor-board high enough to allow the wing box to be positioned underneath it.

2.13.8   Dihedral/Anhedral Angles

Aircraft in a yaw/roll motion have a cross-flow over the wing affecting the aircraft roll stability. The *dihedral angle* (i.e. the wing-tip chord raised above the wing-root chord) assists roll stability. A typical dihedral angle is between 2–3° and rarely exceeds 5°. Figure 2.37a shows that the dihedral angle with a low-wing configuration also permits more ground clearance for the wing tip. The opposite of a dihedral angle is an anhedral angle, which lowers the wing tip with respect to the wing root and is typically associated with high-wing aircraft (Figure 2.37b). The dihedral or anhedral angle also can be applied to the horizontal tail.

■ **FIGURE 2.37**
Wing dihedral and
anhedral angles. (a)
Dihedral (mid
wing–low tail).
(b) Anhedral
(high wing–T-tail)

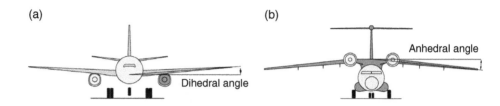

(a)

(b)

Anhedral angle

Dihedral angle

2.14   Mean Aerodynamic Chord

Various wing-reference geometries and parameters are used in aerodynamic computations. A most important parameter is the mean aerodynamic chord (MAC), which is the chord-weighted average chord length of the wing, defined as follows:

$$\text{MAC} = \frac{2}{S_W} \int_0^{b/2} c^2 dy \qquad\qquad (2.26\text{a})$$

Position of MAC from the root chord,

$$y_{\text{MAC}} = \frac{2}{S_W} \int_0^{b/2} c(y)\ y dy \qquad\qquad (2.26\text{b})$$

where c is the local wing chord as a function of y and S_W is the wing reference area.

For the trapezoidal wing reference area, S_W (with sweep), see Figure 2.38 (note $C_R = C_{root}$ and $C_T = C_{tip}$). For the full wing when span $b = 2B$:

$$\text{Wing area, } S_W = \tfrac{1}{2}\left(C_R + C_T\right) \times b$$

From the proportional properties of a triangle, from Figure 2.38 at any distance y from C_R the chord length c can be written as: $c = C_R - 2(C_R - C_T)y/b$.

or

$$c^2 = \left[C_R - 2\left(C_R - C_T\right)y/b\right]^2$$

On substituting in the integral, Equation 2.26 becomes

$$\text{MAC} = \frac{2}{S_W} \int_0^{b/2} c^2 dy = \frac{2}{S_W} \int_0^{b/2} \left[C_R - 2\left(C_R - C_T\right)\left(y/b\right)\right]^2 dy$$

$$= \frac{2}{S_W}\left[\left(yC_R^{\,2}\right)_0^{b/2} - \left\{\frac{2y^2 C_R}{b}\left(C_R - C_T\right)\right\}_0^{b/2} - \left\{4\left(C_R - C_T\right)^2 \frac{y^3}{3b^2}\right\}_0^{b/2}\right]$$

$$= \frac{2}{S_W}\left[\frac{b}{2}C_R^{\,2} - \frac{bC_R}{2}\left(C_R - C_T\right) - \frac{b}{6}\left(C_R - C_T\right)^2\right]$$

■ FIGURE 2.38
Trapezoidal wing planform – mean aerodynamic chord (MAC). In the diagram, $AC' = C_{root} + C_{tip} = B'D$

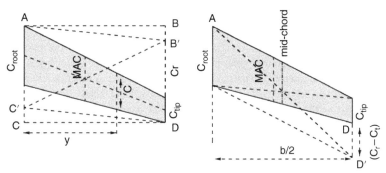

On substituting for $S_W = \frac{1}{2}\left(C_R + C_T\right) \times b$:

$$\text{MAC} = \left[\frac{2}{\left(C_R + C_T\right)}\right]\left[C_R^{\,2} - C_R^{\,2} + C_R C_T + \frac{C_R}{3} + \frac{C_T^{\,2}}{3} - \frac{2C_R C_T}{3}\right]$$

$$= \left[\frac{2}{\left(C_R + C_T\right)}\right]\left[\frac{C_R^{\,2}}{3} + \frac{C_T^{\,2}}{3} - \frac{C_R C_T}{3}\right]$$

$$= \frac{2}{3}\left[\frac{\left(C_R + C_T\right)^2}{\left(C_R + C_T\right)} - \frac{C_R C_T}{\left(C_R + C_T\right)}\right]$$

For a linearly tapered (trapezoidal) wing, this integral is equal to:

$$\text{MAC} = \frac{2}{3}\left[C_R + C_T - \frac{C_R C_T}{\left(C_R + C_T\right)}\right]$$

$$= \frac{2C_R}{3}\left[\frac{1 + \lambda + \lambda^2}{1 + \lambda}\right] \tag{2.27a}$$

From Equation 2.26b:

$$y_{\text{MAC}} = \frac{b}{6}\left(\frac{C_R + 2C_T}{C_R + C_T}\right) = \frac{b}{6}\left(\frac{1 + 2\lambda}{1 + \lambda}\right) \tag{2.27b}$$

where $\lambda = C_T / C_R$

For wings with chord extensions, the MAC may be computed by evaluating the MAC of each linearly tapered portion, then taking an average, weighted by the area of each portion. In many cases, however, the MAC of the reference trapezoidal wing is used. The MAC is often used in the non-dimensionalization of pitching moments, and is also used to compute the reference length for calculation of the Reynolds number as part of the wing drag estimation. The MAC is preferred for computation instead of the simpler mean geometric chord for the aerodynamic quantities whose values are weighted more strongly by local chord than is reflected by their contribution to the area.

2.15   Compressibility Effect: Wing Sweep

Section 2.10 explains the transonic effect resulting from the thickness distribution along an aircraft body. On the wing, the same phenomenon can occur, most importantly along the wing chord, but altered due to the 3D wing-tip influence. A local shock interacting with the boundary layer can trigger early separation, resulting in unsteady vibration and in extreme cases even causing the wing to stall. A typical consequence is a rapid drag increase due to the compressibility effect resulting from the transonic-flow regime. Military aircraft in hard manoeuvre can enter into such an undesirable situation even at a lower speed. As much as possible, designers try to avoid, delay, or minimize the onset of flow separation over the wing due to local shocks.

One way to delay the M_{crit} is to sweep the wing either backward (Figure 2.39a) or forward, which thins the aerofoil t/c ratio and delays the sudden drag rise (Figure 2.39b). The swept-back wing is by far the most prevalent because of structural considerations. Wing slide (i.e. in which the chord length remains the same) is different from wing sweep, in which the chord length is longer by the secant of the sweep angle. Shown here is the relationship between the sweep angle and wing geometries. The chord length of a swept wing increases, resulting in a decrease in the t/c ratio:

$$\text{chord}_{swept} = \left(\text{chord}_{unswept}\right)/\cos\Lambda \tag{2.28}$$

This results in:

$$\left(\text{thickness}/\text{chord}_{swept}\right) < \left(\text{thickness}/\text{chord}_{unswept}\right)$$

This directly benefits the drag divergence Mach number dividing by the cosine of sweep angle, $\Lambda_{\frac{1}{4}}$:

$$\text{Mach}_{div\_swept} = \text{Mach}_{div\_unswept}/\cos\Lambda_{\frac{1}{4}} \tag{2.29}$$

The sweep will also degrade C_{Lmax} by the cosine of sweep angle, $\Lambda_{\frac{1}{4}}$:

$$C_{Lmax\_swept} = C_{Lmax\_unswept} \times \cos\Lambda_{\frac{1}{4}} \tag{2.30}$$

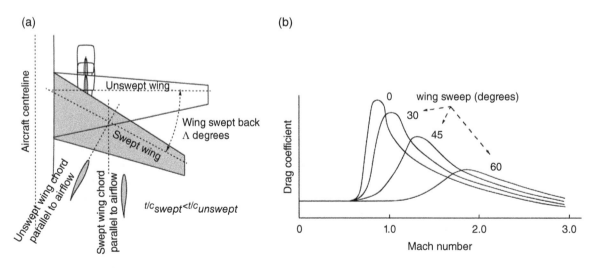

■ **FIGURE 2.39**   Sweeping of wing. (a) Backward sweep definition. (b) Drag comparison

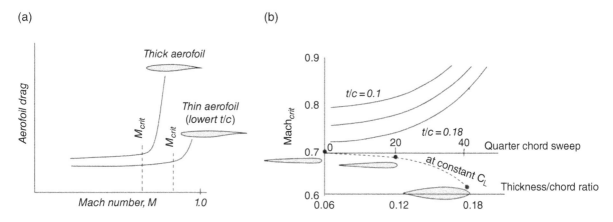

(a)

(b)

■ **FIGURE 2.40**   **Compressibility effects. (a) Drag divergence Mach. (b) Effects of sweep and *t/c***

Drag divergence is a sudden increase in drag. A 20-count drag rise (CD = 0.002) at the Mach number is known as the *drag divergence Mach* (M_{DD}), shown in Figure 2.40a. The critical Mach (M_{crit}) is the onset of the transonic-flow field and is lower than the M_{DD}. Some texts use M_{crit} with a 20-count drag increase. Structural engineers prefer aerofoil sections to be as thick as possible, which favours structural integrity at lower weights and allows the storage of more fuel onboard. However, aerodynamicists prefer the aerofoil to be as thin as possible to minimize the transonic-flow regime in order to keep the wave drag rise lower. To standardize drag-rise characteristics, the flow behaviour is considered to be nearly incompressible up to M_{crit} and can tolerate up to M_{DD}, allowing a 20-count drag increase $\left(\Delta C_D = 0.002\right)$. Qualitative characteristics between the wing sweep and the *t/c* ratio variation are shown in Figure 2.40b. If the trailing edge can remain unswept, then flap effectiveness is less degraded due to a quarter-chord sweep.

Designers require this information for the aerofoil selection. The choice decides the extent of wing sweep required to lower the *t/c* ratio to achieve the desired result (i.e. to minimize the compressible drag increase for the cruise Mach number), while also satisfying the structural requirements.

2.16   Wing-Stall Pattern and Wing Twist

The lower the speed at landing, the safer is the aircraft in case of any inadvertent mishap. An aircraft landing occurs near the wing-stall condition when the aileron effectiveness should be retained to avoid a wing tip hitting the ground. In other words, when approaching the stall condition, its gradual development should start from the wing root, which allows the aileron at the wing tip to retain its ability to maintain level flight. Figure 2.41 shows typical wing-stall propagation patterns on various types of wing planforms.

Because a swept-back wing tends to stall at the tip first, twisting of the wing-tip nose downward (i.e. washout) is necessary to force the root section to stall first, thereby retaining roll control during the landing. A good way to ensure the delay of the wing-tip stall is to twist the wing about the Y-axis so that the tip LE is lower than the wing-root LE (see Figure 2.35). Typical twist-angle values are 1–2° and rarely exceed 3°.

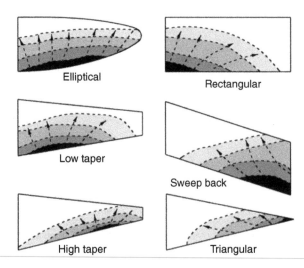

■ **FIGURE 2.41**
Wing stall patterns:
stall progressing
from trailing edge as
angle of attack is
increased

Elliptical

Rectangular

Low taper

Sweep back

High taper

Triangular

2.17   Influence of Wing Area and Span on Aerodynamics

For a given wing loading (i.e. the wing area and maximum takeoff mass (MTOM) invariant), aerodynamicists prefer a large wingspan to improve the aspect ratio in order to reduce induced drag at the cost of a large wing-root bending moment. Structural engineers prefer to see a lower span resulting in a lower aspect ratio.

2.17.1   The Square-Cube Law

Increasing the wing span (i.e. linear dimension, b, maintaining geometric similarity) would increase its volume faster than the increase in surface area, although not at the same rate as for a cube. Volume increase is associated with weight increase, which in turn would require stiffening of the structure, thereby further increasing the weight in a cyclical manner. If the fuselage is considered, then it would be even worse with the additional weight. This is known as the *square-cube law* in aircraft design terminology. This logic was presented half a century ago by those who could not envisage very large aircraft. It gives:

Weight, $W \propto \text{span}^3$
Wing planform area, $S_w \propto \text{span}^2$
Then wing loading, $W/S_w \propto \text{span}$

This indicates that because of excessive weight growth, there should be a size limit beyond which aircraft design may not be feasible.

Yet aircraft keep growing – the size of the Airbus A380 would have been inconceivable to earlier designers. In fact, a bigger aircraft provides better structural efficiency, as shown in Figure 4.6, for operating empty weight fraction (OEWF) reduction with maximum takeoff weight (MTOW) gain. Researchers have found that advancing technology with newer materials (with considerably better strength-to-weight ratios), weight reduction by the miniaturization of systems, better high-lift devices to accommodate higher wing loadings, better fuel economy, and so forth, has defied the square-cube law. Strictly speaking, there is no apparent limit for further growth (up to a point) using the current technology.

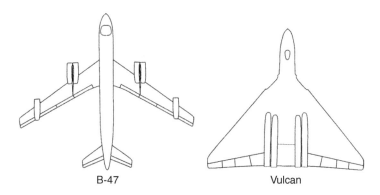

B-47   Vulcan

■ **TABLE 2.3**
Comparison of figures between B47 and Vulcan

	B47	Vulcan
Gross wing area (sq. ft)	1430	3446
Total wetted area (sq. ft)	11300	9500
Span (ft)	116	99
Max. wing loading (W/S_w)	140	43.5
Max. span loading (W/b)	1750	1520
Aspect ratio	9.43	2.84
CD_0	0.0198	0.0069
$L/D_{max}/C_{Lopt}$	17.25/0.682	17.0/0.235
C_{Lmax} at maximum cruise	0.48	0.167

The authors believe that the square-cube law needs better analysis to define it as a law. Currently, it indicates a trend and is more applicable to weight growth with an increase in aspect ratio. What happens if the aspect ratio does not change? The following section provides an excellent example of how a low aspect ratio can compete with a high aspect ratio design.

2.17.2 Aircraft Wetted Area (A_w) versus Wing Planform Area (S_w)

The previous section raised an interesting point on aircraft size, especially related to wing geometry. This section discusses another consideration on how the aircraft wing planform area and the entire aircraft wetted-surface areas can be related. Again, the wing planform area, S_w, serves as the reference area and does not account for other wing parameters (e.g. dihedral and twist). Aerodynamicists desire a high wing aspect ratio, which may make the wing heavy, while structural engineers wish to reduce the wing weight. The conflicting interests between aerodynamicists and structural/stress engineers on the wing aspect ratio presents a challenge for aircraft designers engaged in conceptual design studies (this is an example of the need for concurrent engineering). Both seek to give the aircraft the highest possible *lift-to-drag* (L/D) ratio as a measure of efficient design.

Torenbeek [4] made a fine comparison (Figure 2.42) to reveal the relationship between the aircraft wetted area, A_w, and the wing planform area, S_w (see Table 2.3). Later, Roskam [5] presented his findings to reinforce Torenbeek's point. This is an informative parameter to show how close the configuration is to wing-body configuration (Figure 2.42).

■ **FIGURE 2.43**
Additional vortex
lift. (a) Additional
vortex lift (half
wing shown).
(b) Additional lift
by strake

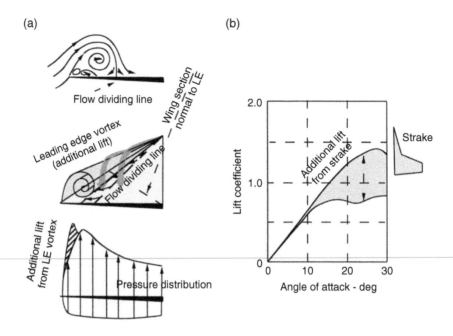

Torenbeek compared an all-wing aircraft (Avro Vulcan bomber) to a conventional design (Boeing B47B bomber) with a similar weight of approximately 90,000 kg and a similar wing span of about 35 m. It was shown that these designs can have a similar L/D ratio despite the fact that the all-wing design has an aspect ratio less than a third of the conventional design. This was possible because the all-wing aircraft precludes the need for a separate fuselage, which adds extra surface area, thereby generating more skin-friction drag. Lowering the skin-friction drag by having a reduced wetted area of the all-wing aircraft compensates for the increase in induced drag for having the lower aspect ratio.

All-wing aircraft provide the potential to counterbalance the low aspect ratio by having a lower wetted area. Again, the concept of blended-wing body (BWB) gains credence. The BWB design for larger aircraft has proven merits over conventional designs, but awaits technological and market readiness. Interesting deductions are made in the following sections.

2.17.3   Additional Wing Surface Vortex Lift – Strake/Canard

Stalling of conventional wings, such as those configured for high-subsonic civilian aircraft, occurs around the angle of attack, α, anywhere from 14–18°. Difficult manoeuvring demanded by military aircraft requires a much higher stall angle (i.e. 30–40°). This can be achieved by having a carefully placed additional low aspect-ratio lifting surface: for example, having a LE strake (e.g. F16 and F18) or a canard (e.g. Eurofighter and Su37). At high angles of attack, the LE of these surfaces produces a strong vortex tube, as shown in Figure 2.43, which influences the flow phenomenon over the main wing. The vortex flow has low pressure at its core, where the velocity is high.

The vortex flow sweeping past the main wing reenergizes the streamlines, delaying flow separation at a higher angle of attack. At airshows during the early 1990s, MIG-29 s demonstrated flying at very high angles of attack (i.e. above 60°); their transient "cobra" movement had never been seen before by the public.

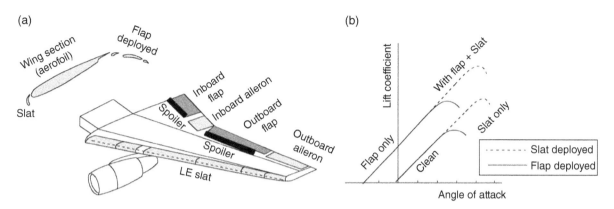

■ **FIGURE 2.44**   High-lift devices. (a) High-lift device layout. (b) Lift characteristics of high-lift devices

2.17.4   Additional Surfaces on Wing – Flaps/Slats and High-Lift Devices

Flaps and slats on a 2D aerofoil were described in Section 2.9. This section describes their installation on a 3D wing (Figure 2.44a). Flaps comprise about two-thirds of an inboard wing at the trailing edge and are hinged on the rear spar (positioned at 60–66%; the remaining third by the aileron) of the wing chord, which acts as a support. Slats run nearly the full length of the LE. The deployment mechanism of these high-lift devices can be quite complex. The associated lift-characteristic variation with incidence is shown in Figure 2.44b. Slat deployment extends the wing maximum lift, whereas flap deployment offers incremental lift increase at the same incidence.

The aileron acts as the roll-control device and is installed at the extremity of the wing for about a third of the span at the trailing edge, extending beyond the flap. The aileron can be deflected on both sides of the wing to initiate roll on the desired side. In addition, ailerons can have trim surfaces to alleviate pilot loads. A variety of other devices are associated with the wing (e.g., spoiler, vortex generator, and wing fence).

Spoilers (Figure 2.44a) are flat plates that can be deployed nearly perpendicular to the airflow over the wing. They are positioned close to the CG (i.e., X-axis) at the MAC to minimize the pitching moment, and they also act as air brakes to decrease the aircraft speed. They can be deployed after touchdown at landing, when they would "spoil" the flow on the upper wing surface, which destroys the lift generated (the US terminology is *lift dumper*). This increases the ground reaction for more effective use of wheel brakes.

2.17.5   Other Additional Surfaces on Wing

Many types of wing-tip device reduce induced drag by reducing the intensity of the wing-tip vortex. Figure 2.45 shows the prevalent types of *winglets*, which modify the tip vortex to reduce induced drag. At low speed, the extent of drag reduction is minimal and many aircraft do not have a winglet. At higher speeds, it is now recognized that there is some drag reduction, no matter how small, and it has started appearing in many aircraft – even as a styling trademark on some. The blended winglet and Whitcomb types are is seen in high subsonic aircraft. The Hoerner-tip and sharp-raked winglet are used in lower-speed aircraft. The Whitcomb-type wing tip and its variants without the lower extension are popular with high-subsonic turbofan aircraft.

■ **FIGURE 2.45**
Types of winglets

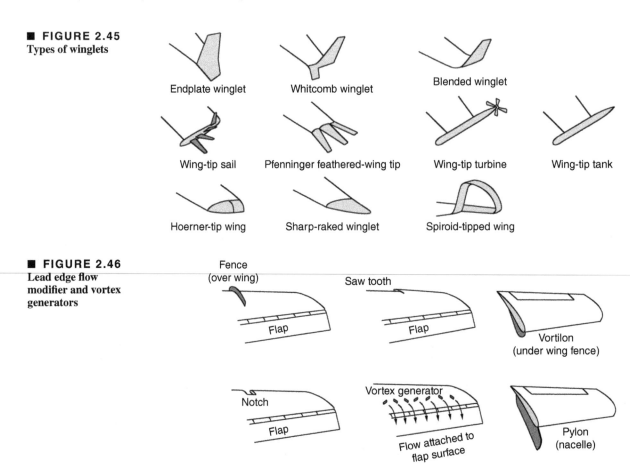

Endplate winglet

Whitcomb winglet

Blended winglet

Wing-tip sail

Pfenninger feathered-wing tip

Wing-tip turbine

Wing-tip tank

Hoerner-tip wing

Sharp-raked winglet

Spiroid-tipped wing

■ **FIGURE 2.46**
Lead edge flow modifier and vortex generators

Fence
(over wing)

Saw tooth

Flap

Flap

Vortilon
(under wing fence)

Notch

Flap

Vortex generator

Flow attached to
flap surface

Pylon
(nacelle)

Extensive analyses and tests indicate that approximately 1% of induced-drag reduction may be possible with a carefully designed winglet. Until the 1970s and 1980s, the winglet was not prominent in aircraft.

Wing flow modifier devices (Figure 2.46) are intended to improve the flow quality over the wing. In the figure, a *fence* is positioned at about half the distance of the wingspan. The devices are carefully aligned to prevent airflow that tends to move spanwise (i.e. outward) on swept wings. Figure 2.46 also shows examples of *vortex generators*, which are stub wings carefully placed in a row to generate vortex tubes that energize flow at the aft wing. This enables the flow to remain attached; however, additional drag increase due to vortex generators must be tolerated to gain this benefit. Vortex generators and/or a fence also can be installed on a nacelle to prevent separation.

2.18 Empennage

Typically, the empennage consists of horizontal and vertical tails for aircraft stability and control. The vertical tail (*V-tail*; UK terms are *fin* and *rudder*) is in the plane of symmetry with a symmetrical aerofoil. A horizontal tail (*H-tail*; UK terms are *stabilizer* and *elevator*) is like a small wing at the tail (i.e. the aft end of the fuselage). The last two decades have seen the return of aerodynamic surfaces placed in front of the wing (Figure 2.53); these are called *canards* and

(a)   (b)   (c)

■ **FIGURE 2.47**   **The dominant options for the wing and nacelle positions. (a) Low tail, under-wing pods. (b) High tee-tail, fuselage-mounted pods. (c) Mid-tail, over-wing pods**

■ **FIGURE 2.48**
Types of empennage configuration

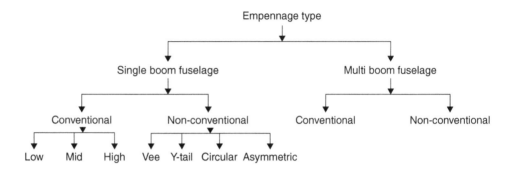

are discussed in subsequent chapters. The canard surface can share some lift (civil aircraft designs) with the wing. This section addresses the definitions associated with the empennage and canard as well as the tail-volume coefficients. The V-tail of a single-engine, propeller-driven aircraft may have an offset of 1–2° to counter the effects of the rotating propeller slipstream.

By far the majority of civil aircraft designs have two surfaces almost orthogonal to each other as V-tail and H-tail. The V-tail is kept symmetric to the aircraft centreline and can be swept (there are exceptions). The H-tail can be positioned low at the fuselage (Figure 2.47a has a dihedral arrangement) or at the top as a tee-tail (Figure 2.47b has an anhedral arrangement) or anywhere in between (Figure 2.47c) as a mid-tail configuration. Any combination of the schemes is feasible but it is decided by consideration of the various aerodynamic, stability and control factors.

Figure 2.48 shows the variety of empennage configuration options available in a systematic way, including some of the unconventional types (as shown in Figure 2.49). The Beechcraft Bonanza 35 has a Vee-shaped tail and in some designs it can be an inverted Vee-tail. One of the early Lear designs had a Y-shaped empennage (Figure 2.49) – a Vee tail along with vertical fin extending below the fuselage. In the past, a circular duct empennage has appeared (Figure 2.49). The unconventional empennages have their merits, but by far the majority of designs have horizontal and vertical surfaces. Vee-tail/Y-tail/circular tail designs are more complex but follow the same routine as for the conventional ones, that is, resolving the forces on the surfaces into vertical and horizontal directions. This book deals with conventional designs. If the V-tail size is large on account of a short tail arm, then the area could to be split up into two or three V-tails (Figure 2.49) from structural and aerodynamic considerations.

The H-tail position relative to V-tail is a serious consideration. Figure 2.50 shows the options available. It can be anywhere from the lowest position going through the fuselage to the other extreme on the top as a tee-tail. Anywhere in between is seen as mid-tail position.

A designer must ensure that the H-tail does not shield the V-tail. The wake (dashed lines) from H-tail should not cover more than 50% of the V-tail surface and also keep more than 50%

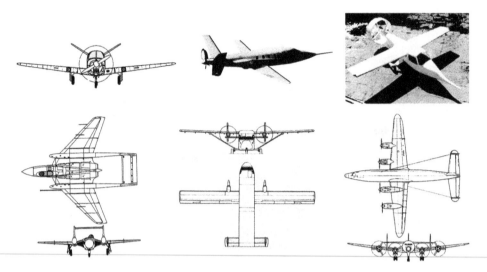

■ **FIGURE 2.49** **Other types of civil aircraft empennage designs. Top left: Vee-tail Beech Bonanza; top centre: Y-tail Lear Fan (reproduced with permission of www.aviastar.org/air/usa/learavia_learfan.php); top right: circular tail (reproduced with permission of Mississippi State University); bottom left: twin V-tail on twin boom, De Havilland Sea Vixen; bottom centre: twin V-tail on fuselage, Short Sky van; bottom right: three V-tail, Lockheed Constellation**

■ **FIGURE 2.50**
Positioning of the
horizontal tail

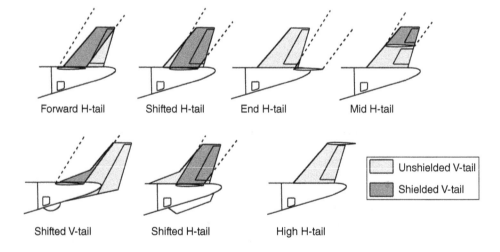

of the rudder area free from its wake to maintain its control effect, especially during spin/stall recovery. Shifting the V-tail aft with the rudder extending below the fuselage will bring the fin and rudder adequately outside the wake. A dorsal and ventral fin can bring more fin surface outside the wake, but the rudder has to be larger to retain some effect. Lowering the H-tail would move the wake aft, but if it is too low then it may hit the ground at rotation, especially if suffering a sudden bank due to gust. At a high angle of attack the wing wake should avoid the H-tail at near-stall conditions so that pitch control remains adequate.

Care must be taken that the H-tail is not within the entrainment effects of the jet exhaust situated at the aft end, typically for aft mounted or within-fuselage jet engines. Military aircraft having an engine inside the fuselage may require a pen-nib type extension to shield jet efflux effects on the H-tail. In that case, the H-tail is moved up to mid level or as a tee-tail.

■ FIGURE 2.51
Geometric
parameters for the
tail volume
coefficients

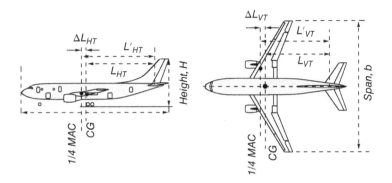

2.18.1   Tail-arm

The *tail-arms* are important parameters to size the empennage for aircraft stability and control [24]. For conventional two-surface aircraft configurations, the tail-arm is defined as the length between the quarter-point of wing $(MAC)_{\frac{1}{4}\_wing}$ and quarter point of tail $(MAC)_{\frac{1}{4}\_H\_tail}$, denoted as L'_{HT}, and the $(MAC)_{\frac{1}{4}\_V\_tail}$ denoted as L'_{VT} (see Figure 2.51). At the conceptual design stage, the wing position with respect to fuselage requires an iterative process (wing chasing) to find the best location.

Wing moments are balanced by the empennage moments about the aircraft CG; the respective tail-arms are denoted as L_{HT} and L_{VT}. The aircraft CG position changes. Therefore, the reference position of CG between wing and empennage proves convenient. However, at the conceptual design stage when the aircraft wing position relative to the empennage is yet to be settled, it is more appropriate to estimate the CG position to compute the tail-arms, thereby avoiding determining the CG position every time a new wing position is tested. It is for this reason that in this book the tail-arms are measured from the aircraft CG as derived in the quoted reference.

Typically, CG variation in civil aircraft is less than 10%, and for military aircraft around 5% of MAC, keeping its quarter-chord point within the margin of movement. Therefore aircraft CG position and $(MAC)_{\frac{1}{4}\_wing}$ are very close (Figure 2.51), hence the discrepancy can be neglected (i.e. $\Delta L_{HT} = L_{HT} - L'_{HT} \approx 0$, and $\Delta L_{VT} = L_{VT} - L'_{VT} \approx 0$) and available statistics can be used. Even for control configured aircraft, the tail-arms can be adjusted to the changes. Stability analysis in Chapter 12 is based on L_{HT} and L_{VT}, and tail-sizing is based on L'_{HT} and L'_{VT}. In this book, dealing with conventional aircraft, we consider that $L_{HT} = L'_{HT}$ and $L_{VT} = L'_{VT}$, removing ambiguity as they are the same. This is a convenient way to start conceptual aircraft configuration, increasing in refinement as more information is available. Strictly speaking, statistics of tail volume coefficients both from CG and $MAC_{\frac{1}{4}}$ should be considered separately, remembering that these do not give exact values.

However, for aircraft configuration with canard (noting the sign convention as negative), the statistical values are to be evaluated appropriately. Whichever method is used, the final result will be the same so long as appropriate interpretation is followed in the computational process.

2.18.2   Horizontal Tail (H-Tail)

The H-tail consists of the stabilizer (fixed or moving) and the elevator (moving) for handling the pitch degree of freedom (Figure 2.52a). The H-tail can be positioned low through the fuselage, in the middle cutting through the V-tail, or at the top of the V-tail to form a T-tail.

■ **FIGURE 2.52**
(a) Horizontal tail
and (b) vertical tail

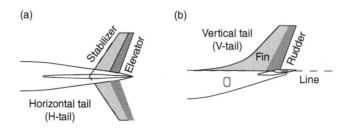

■ **FIGURE 2.52**
(a) Horizontal tail
and (b) vertical tail

Military aircraft can have all moving H-tails with emergency splitting in case there is failure, and there are several choices for positioning it. Figure 2.52a shows the geometrical definition of conventional-type H-tail surfaces.

Like the wing-planform definition, the H-tail reference area, S_H, is the planform area including the portion buried inside the fuselage or V-tail for a low- or mid-tail location, respectively. The T-tail position at the top has a fully exposed planform.

2.18.3   Vertical Tail (V-Tail)

The V-tail (note that the Vee-tail has a different configuration) consists of a fin (fixed) and a rudder (moving) to control the roll and yaw degrees of freedom (Figure 2.52b).

The figure shows the geometric definition of a conventional V-tail surface reference area, S_V. The projected trapezoidal/rectangular area of the V-tail up to this line is considered the reference area, S_V. Depending on the closure angle of the aft fuselage, the root end of the V-tail is fixed arbitrarily through a line drawn parallel to the fuselage centre-line, passing through the point where the mid-chord of the V-tail intersects the line.

2.18.4   Tail-Volume Coefficients

Tail-volume coefficients are used to determine the empennage reference areas, S_H and S_V. The definition of the tail-volume coefficients is derived from the aircraft stability equations provided herein. The CG position is shown in Figure 2.51. The distances from the CG to the aerodynamic centre at the MAC of the V-tail and H-tail (i.e., MAC_{VT} and MAC_{HT}) are designated L_{HT} and L_{VT}, respectively, as shown in Figure 2.51. The ac is taken at the quarter-chord of the MAC.

H-Tail Volume Coefficient, C_{HT}
From the pitching-moment equation for steady-state (i.e. equilibrium) level flight, the H-tail volume coefficient is given as the H-tail plane reference area:

$$S_{HT} = (C_{HT})(S_W \times MAC)/L_{HT}$$ (2.31)

where C_{HT} is the H-tail volume coefficient, $0.5 < C_{HT} < 1.2$; a good value is 0.8. L_{HT} is the H-tail-arm = distance between the aircraft CG to the ac of MAC_{HT}. In general, the area ratio $S_{HT} / S_W \approx 0.25$ to 0.35.

V-Tail Volume Coefficient, C_{VT}
From the yawing-moment equation for steady-state level flight, the V-tail volume coefficient is given as the V-tail plane reference area:

$$S_{VT} = (C_{VT})(S_W \times wing\ span)/L_{VT}$$ (2.32)

where C_{VT} is the V-tail volume coefficient, $0.05 < C_{VT} < 0.1$; a good value is 0.07. L_{VT} is the H-tail-arm = distance between the aircraft CG to the ac of MAC_{VT}. In general, the area ratio $S_{VT} / S_W \approx 0.15$ to 0.25.

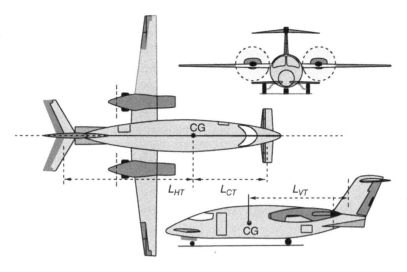

Canard Configuration

Canard is French for "duck", a bird which, in flight, stretches out its long neck with its bulbous head in front. When a horizontal surface is placed in front of the aircraft, it looks like a duck or goose in flight; hence, this surface is sometimes called a canard (Figure 2.53, note the difference in tail-arm definitions).

The Wright brothers' Flyer had a control surface at the front (with a destabilizing effect), which resulted in a sensitive control surface. Military aircraft use a canard to enhance pitch control. However, the use of a canard in civil aircraft serves a different purpose.

In general, the inherent nose-down moment (unless a reflex trailing edge is employed) of a wing requires a downward force by the H-tail to maintain level flight. This is known as the trimming force, which contributes to trim drag. For an extreme CG shift (which can happen as fuel is consumed), high trim drag can exist in a large portion of the cruise sector. The incorporation of a canard surface can reduce trim drag as well as the H-tail area, S_H. However, it adds to the manufacturing cost and, until recently, the benefit from the canard application in large transport aircraft has not been marketable.

Many small civilian aircraft have a canard design (e.g. Rutan designs). A successful Bizjet design is the Piaggio P180 Avanti shown in Figure 2.53. It has achieved a very high speed for its class of aircraft through careful design considerations, embracing not only superior aerodynamics but also the use of composite materials to reduce weight.

Canard-Volume Coefficient, C_{CT}

This is also derived from the pitching-moment equation for steady-state level flight. The canard reference area, S_{CT}, has the same logic for its definition as that of the H-tail. Its tail-arm is L_{CT}. The canard reference area is given as:

$$S_{CT} = (C_{CT})(S_W \times MAC)/L_{CT}, \text{from CG}$$

and

$$S_{CT} = (C'_{CT})(S_W \times MAC)/L'_{CT}, \text{from MAC}_{1/4} \qquad (2.33)$$

where C_{CT} and C'_{CT} are the Canard volume coefficients. In general, $S_{CT}/S_W \approx 0.2$ to 0.3.

2.19   Fuselage

A civil aircraft fuselage is designed to carry revenue-generating payloads, primarily passengers, and the cargo version can also carry containers or suitably packaged cargo. It is symmetrical to a vertical plane and maintains a constant cross-section, with front and aft end closures in a streamlined shape. The aft fuselage is subjected to adverse pressure gradients and therefore is prone to separation. This requires a shallow closure of the aft end so that the low-energy boundary layer adheres to the fuselage, minimizing pressure drag (see Section 2.3). The fuselage also can produce a small amount of lift, but this is typically neglected in the conceptual stages of a configuration study. The following definitions are associated with the fuselage geometry (Figure 2.54) and are used in drag estimation.

2.19.1   Fuselage Axis/Zero-Reference Plane

Fuselage axis is a line parallel to the centreline of the constant cross-section part of the fuselage barrel. It typically passes through the farthest point of the nose cone, facilitating the start of reference planes normal to it. The fuselage axis may or may not be parallel to the ground. The principal inertia axis of the aircraft can be close to the fuselage axis. In general, the zero-reference plane is at the nose cone, but designers can choose any station for their convenience, within or outside of the fuselage. This book considers the fuselage zero-reference plane to be at the nose cone, as shown in Figure 2.54.

2.19.2   Fuselage Length, L_{fus}

This is along the fuselage axis, measuring the length of the fuselage from the tip of the nose cone to the tip of the tail cone (which is unlikely to be on the axis). This is not the same as the aircraft length, L, shown in Figure 2.54a.

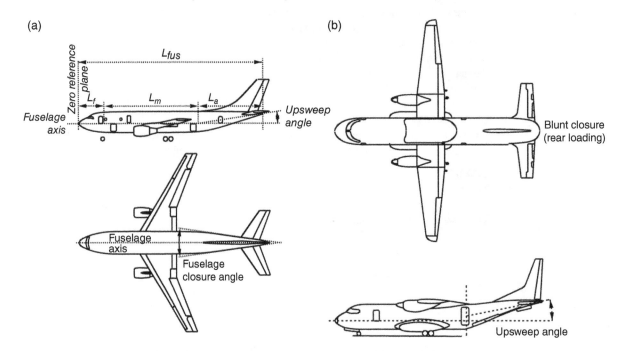

■ **FIGURE 2.54**   Fuselage geometric parameters – lengths associated with the fuselage. **(a)** Conventional aft end.
(b) Rear loading aft end – blunt closure

2.19.3   Fineness Ratio, FR

This is the ratio of fuselage length to average diameter, $FR = L/D_{ave}$. A good value for commercial transport aircraft design is from 8 to 10. It can be as low as 7 and as high as 15.

2.19.4   Fuselage Upsweep Angle

In general, the fuselage aft end incorporates an upsweep (Figure 2.54b) for ground clearance at rotation on takeoff. The upsweep angle is measured from the fuselage axis to the mean line of aft fuselage height. It may not be a straight line if the upsweep is curved like a banana; in that case, it is segmented to smaller straight lines. The rotation-clearance angle is kept to 12–16°; however, the slope of the bottom mould line depends on the undercarriage position and height. Rear-loading aircraft have a high wing with the undercarriage located close to the fuselage belly. Therefore, the upsweep angle for this type of design is high. The upsweep angle can be seen in the elevation plane of a three-view drawing. There is significant variation in the upsweep angle among designs. A higher upsweep angle leads to more separation and, hence, more drag.

2.19.5   Fuselage Closure Angle

The closure angle is the aft fuselage closure seen in a plan view of the three-view drawing, and it varies among designs. The higher the closure angle, the greater the pressure-drag component offered by the fuselage. In rear-loading aircraft the fuselage closes at a blunt angle; combined with a large upsweep, this leads to a high degree of separation and, hence, increased pressure drag.

2.19.6   Front Fuselage Closure Length, L_f

This is the length of the front fuselage from the tip of the nose cone to the start of the constant cross-section barrel of the mid-fuselage (Figure 2.54a). It encloses the pilot cockpit/flight deck and the windscreen – most of which is associated with a kink in the mould lines to allow for a better vision polar and to enable the use of flat windscreens to reduce cost. In general, it includes the front door and passenger amenities, and may have a row or two of passenger seating.

2.19.7   Aft Fuselage Closure Length, L_a

This starts from the end of the constant cross-section barrel of the mid-fuselage up to the tip of the tail cone (Figure 2.54a). It encloses the last few rows of passenger seating, rear exit door, toilet, and – for a pressurized cabin – the aft pressure bulkhead, which is an important component from a structural design perspective ($L_a > L_f$).

2.19.8   Mid-Fuselage Constant Cross-Section Length, L_m

This is the constant cross-section mid-barrel of the fuselage, where passenger seating and other facilities are accommodated (including windows and emergency exit doors, if required).

2.19.9   Fuselage Height, H

This is the maximum distance of the fuselage from its underside (not from the ground) to the top in the vertical plane (see Figure 2.55).

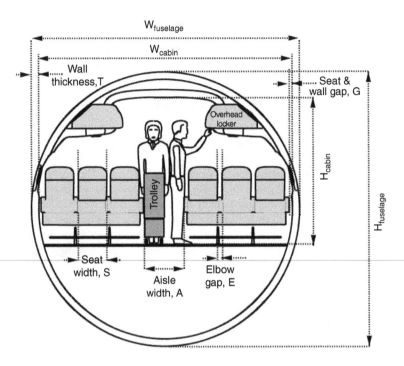

2.19.10   Fuselage Width, *W*

This is the widest part of the fuselage in the horizontal plane. For a circular cross-section, it is the diameter shown in Figure 2.55.

2.19.11   Average Diameter, D_{ave}

For a non-circular cross-section, this is the average of the fuselage height and width at the constant cross-section barrel part ($D_{ave} = (H + W)/2$). Sometimes this is defined as $D_{effective} = (H * W)$; another suitable definition is $D_{equivalent} = \text{perimeter}/2\pi$. This book uses the first definition. Figure 2.55 also shows cabin height, H_{cab} and cabin width, W_{cab}. These dimensions are not in demand by the performance estimators.

2.20   Nacelle and Intake

Air-breathing engines for conventional aircraft must have an intake housed in the aircraft. A *nacelle* is the structural housing for an aircraft engine. In addition to housing the engine, the main purpose of the nacelle is to facilitate the internal air flow reaching the engine face (or the fan of gas turbines) with minimum distortion over a wide range of aircraft speeds and attitudes. The intake acts as a diffuser with an acoustic lining to abate noise generation. The inhaled air-mass flow demanded by an engine varies considerably: at idle, just enough is required to sustain combustion, whereas at maximum thrust, the demand is many times higher. A rigid intake must be sized such that during critical operations (i.e. takeoff, climb and cruise), the engine does not suffer and generates adequate thrust. Supersonic intakes are even more complex and are designed to minimize loss resulting from shock waves. The installation effects of engines to aircraft contribute a sizeable percentage of aircraft drag, affecting aircraft performance.

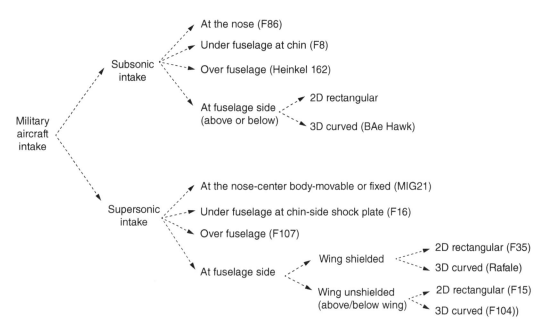

■ **FIGURE 2.56** Types of intake configuration

The purpose of this section is to make engineers aware of the different intake/nacelle design features to compute intake/nacelle drag. Chapters 7 and 8 discuss intake/nacelle and engine performance in detail.

In civilian aircraft, nacelles are invariably externally pod-mounted, either slung under or mounted over the wing or attached to the fuselage (see Figure 2.57). The front part of the nacelle is the intake and the aft end is the nozzle. Pod mounts not only offer lower intake drag as the internal airflow duct length is smaller, but also relieve the fuselage volume for payload accommodation. The military fighter aircraft engines are invariably buried in the fuselage; the single engine installation has no other option, while two engines are also housed close to aircraft centreline, but may bulge out considerably, or even may be just outside the fuselage. The internal intake duct length along the fuselage is relatively long, contributing to higher intake drag. However, bombers and military logistics aircraft are multi-engine and are generally wing-mounted. Note that a supersonic side intake has a side plate acting both as boundary layer bleeder and to adjust oblique shock. Broader classifications of military fighter aircraft engine intake configuration are given in Figure 2.56 (examples given in the bracket).

Over the period of design evolution, there have been many different kinds of intake/nacelle configuration; some are no longer used in current designs. This section only deals with the current types, with passing mention of older, unconventional and futuristic intake/nacelle configurations.

2.20.1 Large Commercial/Military Logistic and Old Bombers Nacelle Group

Large aircraft are heavy, requiring more than one engine (turboprop or turbofan) which will invariably be pod-mounted on the wing or aft fuselage. The dominant options in civil aircraft design are shown in Figure 2.57.

■ **FIGURE 2.57** Options for large aircraft nacelle positions (reproduced with permission of Richard Ferrier). **Top left: four engines on high wing (BAE 146). Top centre: four engines on low wing (Boeing 747). Top right: eight engines on high wing (Boeing B52). Middle left: centre S-duct three-engine (Boeing 727). Middle centre: centre straight-duct (Lockheed 1011). Middle right: over-wing engine (Boeing YC12). Bottom left: buried engine (Comet). Bottom centre: two engines over fuselage (Beriev 200)**

Larger aircraft have pods mounted under the wing (Figure 2.57). The Boeing B52 bomber has eight engines in four pods slung under the wing. Over-wing fuselage-mounted nacelle pods (Berive 200) and over-wing mounted nacelle for small low-wing aircraft (Honda Jet) are gaining ground.

For an odd number of engines, the odd one is placed in the centre line (Douglas DC10) and if buried in fuselage then its intake may require an S-duct type intake (Boeing 727). Over-wing slipper-nacelle design has been flown both by Boeing (Figure 2.57) and Douglas for short takeoff and landing (STOL) performance. Single-engine aircraft will have the engine at the centreline (except special-purpose aircraft), mostly buried in the fuselage. If propeller driven, then it can either be a tractor (majority of designs) or a pusher propeller mounted at the rear.

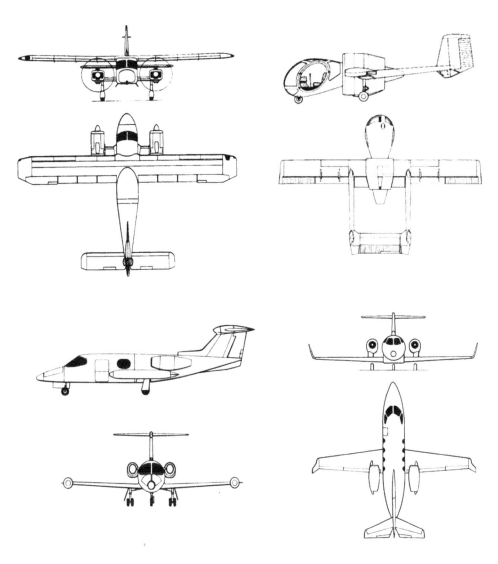

■ **FIGURE 2.58**
**Small aircraft nacelle
positions. From left
to right: engines on
twin mounts
(DO-28); shrouded
propeller (Edgley
Optica); fuselage
mounted (Lear 35);
over-wing
(Honda jet)**

The first commercial jet transport aircraft, the de Havilland Comet, had engines buried in the wing root (Figure 2.57). Such engines generate higher intake/interference drag and have not been pursued in the newer civil aircraft designs, but military aircraft designs may have to override this.

2.20.2   Small Civil Aircraft Nacelle Position

Figure 2.58 shows some of the small multi-engine aircraft with unconventional single and twin engine positions. The first two are propeller-driven aircraft. The Dornier DO-28 has nacelles on the undercarriage housing structure. The Edgley Optica has a unique shrouded propeller as pusher, which offers a helicopter-like all-round view for the occupants. Jet-powered small aircraft with low wings have fuselage-mounted (Lear 35) or over-wing mounted (Honda Jet) nacelle pods for ground clearance reasons.

■ **FIGURE 2.59** Fighter aircraft intake positions (reproduced with permission of Richard Ferrier). Top left: centre-body pitot intake (MIG 21). Top right: slant side-body intake (Dassualt Rafale). Bottom left: chin intake (Boeing F16). Bottom right: fuselage side intake (SAAB Grippen)

2.20.3   Intake/Nacelle Group (Military Aircraft)

Engine nacelle options are shown in Figure 2.59. Combat aircraft do not have pod-mounted nacelles. Single or multi-engines (so far mostly two) are kept side-by-side buried in the fuselage. The fighter aircraft intake path is longer and curvier. Older designs had a forward intake at the nose (MIG 21). A centre body is required for aircraft with a speed capability above Mach 1.8, otherwise it has pitot intake. These have the longest intake duct with low curvature to pass under the pilot, and incur high losses, and hence are no longer pursued. Instead chin (Boeing F16) or side (Rafale and SAAB Grippen) mounted intakes have shorter and carefully designed ducts with comparable curvature, hence with lower losses. A plate is kept above the fuselage boundary layer on which the intake is placed. The B2 has over-wing/fuselage intakes (Figure 1.6) more suited to a BWB type of configuration.

A central forward intake (invariably circular (MIG 21) – can be near-circular (F100)) is an older design, no longer in practice. Considerable duct loss is associated with its long length from nose to tail and the curve to pass underneath the pilot. For supersonic operation, a centre body arrangement produces diffusion through a series of oblique shocks and a normal shock. A centre intake does not have fuselage shielding at yaw and pitch attitudes.

Side intakes (semi-circular, rectangular, e.g. F18 and F35) cut down the internal duct length by nearly half, but are associated with bends (reduced for two side-by-side engines). For supersonic operation, there is a splitter plate which bleeds the fuselage boundary layer to keep it outside the intakes. It needs to be carefully sized for flying in a yawed attitude.

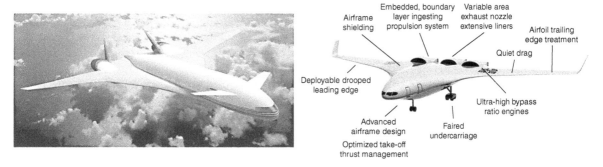

■ FIGURE 2.60   Futuristic designs. Left: futuristic rear engine mount from Boeing (reproduced with permission of Boeing Corp.). Right: BWB rear engine mount (reproduced with permission of Massachusetts Institute of Technology)

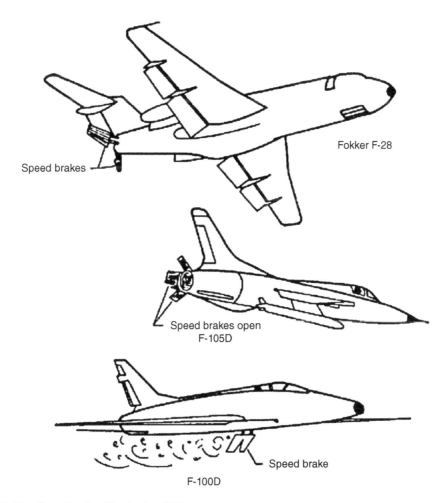

■ FIGURE 2.61   Speed brakes/dive brakes [20]

A chin-mounted intake (near-elliptical/kidney-shaped like Falcon F16, near-rectangular like Eurofighter) can be found in later designed aircraft. At yaw, a chin intake does not have fuselage shielding as for side intakes, but being close to the ground there could be ground ingestion problems, especially during combat operations. This concept has proven very successful and can handle high incidence flying.

A central over-fuselage mounted intake (opposed to chin-mounted intake) configuration is not prevalent – high incidence flying may create serious flow distortion affecting engine performance. But this type of configuration is gaining ground on account on stealth considerations.

Bomber aircraft need to reserve fuselage bay space for bombs and keep engines outside the fuselage. The B52, with eight pod-mounted engines, is shown in Figure 2.57.

2.20.4   Futuristic Aircraft Nacelle Positions

Some futuristic nacelle positions are shown in Figure 2.60. These are proposed designs yet to be built. The figure shows the Boeing Sonic Cruiser, and the silent aircraft BWB proposal by MIT. Future designs will have stealth features, and F22/B2/JUCAS type nacelles will become common shapes of intake design.

Academic establishments such as Virginia Tech Aerospace and Ocean Engineering, Massachusetts Institute of Technology, Michigan University and others have studied some interesting futuristic aircraft configurations.

2.21   Speed Brakes and Dive Brakes

Speed brakes and *dive brakes* have the same definition. They are mounted specifically on the fuselage for military aircraft, and as spoilers on the wings for civilian aircraft (Figure 2.61). However, there are civilian aircraft that use this type of device mounted on the fuselage.

Speed brakes are specifically designed to reduce speed rapidly, typically on approach and in military combat manoeuvres. Speed and dive brakes are primarily drag-producing devices positioned in those areas that will create the smallest change in moments (i.e. kept symmetrical to the aircraft axis with the least moment arm from the CG). Figure 2.61 shows fuselage-mounted devices. The Boeing F22 does not have a separate dive brake. It uses the two rudders of the canted V-tail deflected in opposite directions, along with spoilers and flaps deflected upward and downward, respectively.

References

1. Wood, K.D., *Aircraft Design*. Johnson Publishing Co., Boulder, CO, 1966.
2. Corning, G., *Supersonic and Subsonic Airplane Design*. Published by the author, 1960.
3. Nicolai, L.M., *Fundamentals of Aircraft Design*. METS, Inc., San Jose, CA, 1975.
4. Torenbeek, E., *Synthesis of Subsonic Airplane Design*. Delft University Press, Amsterdam, 1982.
5. Roskam, J., *Aircraft Design*. Published by the author as an 8-volume set, 1985–1990.
6. Stinton, D., *The Design of the Airplane*. Van Nostrand Reinhold, New York, 1983.
7. Raymer, D., *Aircraft Design – A Conceptual Approach*. AIAA, Reston, VA, 1992.
8. Thurston, D., *Design for Flying*, 2nd edn. Tab Books, New York, 1995.
9. Fielding, J., *Introduction to Aircraft Design*. Cambridge University Press, Cambridge, 1999.
10. Jenkinson, L.R., Simpson, P. and Rhodes, D., *Civil Jet Aircraft Design*. Arnold, London, 1999.
11. Jenkinson, L.R. and Marchman, J.F., *Aircraft Design Project Studies*. Butterworth, Oxford, 2003.
12. Howe, D., *Aircraft Conceptual Design Synthesis*. Professional Engineering Ltd, London, 2000.
13. Whitford, R., *Fundamentals of Fighter Design*. Airlife, UK, 2000.
14. Schaufele, R.D., *The Elements of Aircraft Preliminary Design*. ARIES Publications, Santa Ana, CA, 2000.
15. Kundu, A.K., *Aircraft Design*. Cambridge University Press, Cambridge, 2010.
16. Gudmundsson, S., *General Aviation Aircraft Design*. Butterworth-Heinemann, Oxford, 2014.

17. Huenecke, K., *Modern Combat Aircraft Design*. Airlife, UK, 1987.

18. Schmitt, D., *Flugzeugentwerf Umdruck zur Vorlesung [Aircraft Design Handout for Reading]*. Technische Universitat Munchen, 2000.

19. McMasters, J.H., Boeing Commercial Aircraft General Engineering Division, classroom notes, Summer Intern Training Program, 1994.

20. Talay, T.A., *Introduction to the Aerodynamics of Flight*. NASA SP-367, 1975.

21. Holt, A., *Engineering Analyses of Flight Vehicles*. Addison-Wesley, Menlo Park, 1974.

22. Kuethe, A.M. and Chow, C.-Y., *Foundations of Aerodynamics*. John Wiley and Sons Ltd., Chichester, 1986.

23. Schlichting, H., and Gersten, H., *Boundary Layer Theory*, 8th edn. McGraw-Hill, New York, 2003.

24. Abbott, I.H. and von Doenhoff, A.E., *Theory of Wing Sections, Including a Summary of Airfoil Data*. McGraw Hill, New York, 1949.

25. Perkins, C.D., and Hage, R.E., *Aircraft Performance Stability and Control*. John Wiley and Sons Ltd., Chichester, 1949.

26. Liepmann, H.W. and Rosko, A., *Elements of Gas Dynamics*. John Wiley and Sons Ltd., Chichester, 1957.

27. Henne, P.A., *Applied Computational Aerodynamics*. AIAA, Reston, VA, 1990.

28. Bertin, J.J. and Cummings, R.M., *Aerodynamics for Engineers*, 5th edn. Pearson, Upper Saddle River, NJ, 2009.

29. Dommasch, D.O., *et al.*, *Airplane Aerodynamics*, 2nd edn. Pitman, New York, 1958.

30. Anderson, J.D., *Introduction to Flight*, 8th edn. McGraw-Hill, New York, 2015.

31. Anderson, J.D., *Aircraft Performance and Design*. McGraw-Hill, New York, 1999.

CHAPTER 3

Air Data Measuring Instruments, Systems and Parameters

3.1 Overview

Aircraft systems and parameters associated with and affecting aircraft performance are covered in this chapter. First, the fundamentals of flight deck air data instruments, for example, the airspeed indicator (ASI), Mach meter, altimeter and turn/slide-slip indicator, are described. Next, the aircraft weight and centre-of-gravity issues are covered. Performance compromises because of environmental and stealth considerations are covered. Aircraft manoeuvres are carried out by controls, and these are explained.

3.2 Introduction

This chapter covers a wide range of topics, for example, aircraft velocities, air data measuring instruments, aircraft systems, and parameters associated with aircraft performance. It benefits aircraft performance engineers to be conversant with these topics.

Aircraft performance engineers deal with those parameters that comprise air data, displayed on the flight deck for the pilots to fly the aircraft safely and as required. Aircraft speed and altitude are the most important parameters that engineers and pilots have to deal with, along with many other parameters such as aircraft attitude, turn rates, vertical speeds, and so on, and the propulsion (engine) data. An understanding of how aircraft velocity and altitude are obtained is gained in the chapter. It explains the fundamentals of the ASI, Mach meter, altimeter and turn/sideslip indicators.

Flight deck pilot display instruments have changed from a multitude of older analogue dial gauges to an integrated electronic flight information system (EFIS) with multifunctional displays (MFDs), all in one flat LCD panel.

Theory and Practice of Aircraft Performance, First Edition. Ajoy Kumar Kundu, Mark A. Price and David Riordan.
© 2016 John Wiley & Sons, Ltd. Published 2016 by John Wiley & Sons, Ltd.

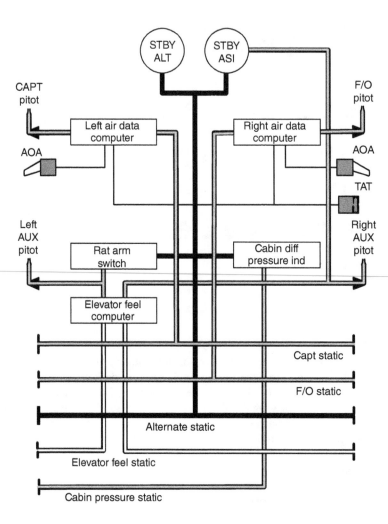

Transducers are used to convert air pressure from probes to digital or analogue electrical signals. Digital data are processed by a centralized air data computer (ADC) system, mostly backed up by independent systems for reliability. Of late, distributed systems are replacing the centralized system (Figure 3.1).

The chapter continues with other topics on aircraft systems, for example engine off-takes for the environmental control system (ECS), as these affect aircraft performance. Finally, stealth and other environmental considerations affecting aircraft performance are also briefly explained.

3.3   Aircraft Speed

One of the most important parameters in computing aircraft performance is its speed or velocity, both terms used interchangeably. Unlike land-based vehicle speed measured directly, aircraft speed measurement is an involved affair, as can be seen in the following sections.

3.3.1   Definitions Related to Aircraft Velocity

Aircraft speed is a vital parameter in computing performance. It is measured using the difference between the total pressure, p_t, and the static pressure, p, expressed as $(p_t - p)$. Static pressure p is the ambient pressure in which an aircraft is flying.

$$\text{Dynamic head} = q = \rho V^2 / 2$$

or

$$V = \sqrt{(2q / \rho)} \tag{3.1a}$$

where V is the true airspeed (short form of V_{TAS}) relative to airflow and is valid for any altitude and at any speed (incompressible and compressible). In still air (zero wind), V_{TAS} is the aircraft ground speed.

Air density ρ changes with altitude – it decreases because atmospheric pressure decreases with a gain in altitude. Therefore, flying at constant dynamic head, aircraft true airspeed has to be increased at altitude. Equivalent airspeed, V_{EAS}, is the aircraft speed maintaining a constant q of the true airspeed V at sea-level on an International Standard Atmosphere (ISA) day. This gives:

$$\text{Equivalent airspeed, } V_{EAS} = V \sqrt{\sigma} = 2q / \rho_0 \tag{3.1b}$$

where σ is the density ratio (ρ/ρ_0) in terms of sea-level ISA day.

The aircraft flies $\sqrt{\sigma}$ times faster than the V at sea-level, which gives the ground speed in zero wind – important information to compute the distance covered.

EXAMPLE 3.1

An aircraft climbs from sea-level to 35,000 ft at constant $V_{EAS} = 280$ kt. Find the V_{TAS}, the Mach number and the dynamic head q at sea-level, 10,000, 20,000, 30,000 and 35,000 ft.

SOLUTION

Altitude (ft)	V_{EAS}(kt)	ρ (slug/ft³)	$\sqrt{\sigma}$	V_{TAS}(kt)	V_{TAS}(ft)	q	a (ft/s)	Mach no.
Sea-level	280	0.00238	1.0	280	472.6	265.5	1116.5	0.423
10,000	280	0.00175	0.858	326	550.3	265.5	1077.38	0.511
20,000	280	0.00126	0.7276	385	650.0	266.0	1036.95	0.627
30,000	280	0.00088	0.608	461	778.2	266.0	994.66	0.782
35,000	280	0.00073	0.554	505	852.4	265.5	977.28	0.872

There are negligible rounding discrepancies.

It can be seen that at constant V_{EAS} climb schedule the dynamic head q remains constant, but V_{TAS}, and hence the Mach number, increases and can become supersonic. It is for this reason there is a cross-over altitude ($\approx$30,000 ft) when the climb schedule changes from constant V_{EAS} to a constant Mach number ($\approx$0.78 Mach) to stay within the design limitations.

3.3.2 Theory Related to Computing Aircraft Velocity

In low-speed flight below Mach 0.3, the aircraft speed can be computed with the simplified assumption that air is incompressible. At higher speeds, compressibility effects have to be considered. Two cases are derived below.

Incompressible Flow. The steady-state momentum conservation equation gives: $dp = -\rho V dV$. On integrating it gives Bernoulli's equation for incompressible isentropic flow:

$$(p_t - p) = \rho V^2 / 2$$

or

$$p + \rho V^2 / 2 = \text{constant} = p_t \tag{3.2a}$$

$$p_t - p = \Delta p = \rho V^2 / 2 = q$$

$$V = \sqrt{(2\Delta p / \rho)}$$

In incompressible flight speed, the value of $(p_t - p)$ gives the dynamic head, which depends on the ambient air density, ρ. The true aircraft speed V_{TAS} must be computed from $(p_t - p)$.

Compressible Flow (Subsonic and Supersonic). Recall that the Mach number $= V / a$, where V is the true airspeed, and $a =$ speed of sound $= \sqrt{(\gamma RT)} = \sqrt{(\gamma p/\rho)}$.

For compressible isentropic flow, Euler's equation (seen as the compressible Bernoulli's equation) is as follows (where $V = V_{TAS}$):

$$\left[\frac{(p_t - p)}{p} + 1\right]^{\frac{\gamma-1}{\gamma}} - 1 = \frac{(\gamma-1)V^2}{2a^2}$$

Rearranging the above equation gives:

$$\frac{p_t}{p} = \left(1 + \frac{\gamma-1}{2}M^2\right)^{\frac{\gamma}{\gamma-1}}$$

By substituting the value of $\gamma = 1.4$ for air, it becomes:

$$p_t - p = \Delta p = p\left[\left(1 + 0.2M^2\right)^{3.5} - 1\right] \tag{3.2b}$$

where $(1 + 0.2M^2)^{3.5}$ is the isentropic pressure rise due to compressibility.
Note that, by definition, the dynamic head in compressible flow is given by:

$$q = \rho V^2 / 2 = \rho a^2 M^2 / 2 = \left(\frac{\gamma p}{2}M^2\right)$$

It is not the same as in the case of incompressible flow expressed in Equation 3.2a.

EXAMPLE 3.2

Compare the dynamic head q and Δp in incompressible and compressible flow at Mach 0.3 (323.2 ft/s) and Mach 0.75 (808 ft/s) at 10,000 ft $(p=1455.33 \text{lb}/\text{ft}^2,\ \rho=0.00175 \text{slug}/\text{ft}^3)$ on an ISA day.

SOLUTION

	$q=\rho V^2/2$	$q=\gamma pM^2/2$	Δp_{incomp} (lb/ft²)	Δp_{comp} (lb/ft²)
At Mach 0.3	91.4	91.7	91.4	93.8
At Mach 0.75	571.3	533.5	571.3	658.2

The example demonstrates that at higher speeds (say, above Mach 0.3), compressibility effects have to be considered.

Exploring this further, the following can be derived:

$$\left[\frac{(p_t-p)}{p}+1\right]^{\frac{\gamma-1}{\gamma}}-1=\frac{(\gamma-1)V^2}{2a^2}=\frac{(\gamma-1)\rho V^2}{2\gamma p} \tag{3.3a}$$

or

$$V^2=\left(\frac{2a^2}{(\gamma-1)}\right)\left[\left(\frac{\Delta p}{p}+1\right)^{\frac{(\gamma-1)}{\gamma}}-1\right] \tag{3.3b}$$

To obtain the expression for $V_{EAS}=V\sqrt{\sigma}$, multiply Equation 3.3a by ρ_0/ρ_0, and then substituting $\sigma=\rho/\rho_0$ the following is obtained.

$$\left[\frac{(p_t-p)}{p}+1\right]^{\frac{\gamma-1}{\gamma}}=\frac{(\gamma-1)\rho\rho_0 V^2}{2\gamma p\rho_0}+1=1+\frac{(\gamma-1)\sigma\rho_0 V^2}{2\gamma p}$$

or

$$\left[\frac{(p_t-p)}{p}\right]+1=\left(1+\frac{(\gamma-1)}{2\gamma}\left(\frac{\rho_0}{p}\right)(V^2\sigma)\right)^{\frac{\gamma}{\gamma-1}}$$

or

$$\left[\frac{(p_t-p)}{p}\right]=\left(1+\frac{(\gamma-1)}{2\gamma}\left(\frac{\rho_0}{p}\right)(V^2\sigma)\right)^{\frac{\gamma}{\gamma-1}}-1 \tag{3.4}$$

or

$$\left(\frac{(\gamma-1)}{2\gamma}\left(\frac{\rho_0}{p}\right)(V^2\sigma)\right)=\left(\frac{(p_t-p)}{p}+1\right)^{\frac{\gamma-1}{\gamma}}-1$$

or

$$\left(V^2\sigma\right)=V_E^{\ 2}=\left[\frac{2\gamma p}{(\gamma-1)\rho_0}\right]\left[\left(\frac{p_t-p}{p}+1\right)^{\frac{\gamma-1}{\gamma}}-1\right]$$ (3.5a)

$$=\left(5\gamma p\,/\,\rho_0\right)\left\{\left[\left(p_t-p\right)/\,p+1\right]^{0.286}-1\right\}$$

or

$$V\sqrt{\sigma}=\sqrt{\frac{2\gamma p}{(\gamma-1)\rho_0}\left[\left(\frac{p_t-p}{p}+1\right)^{0.286}-1\right]}$$ (3.5b)

$$=V_{EAS}\left(\text{defined in Equation 3.1b}\right)$$

This is the mathematical relationship between the V_{TAS} and the V_{EAS}, incorporating density changes with altitude.

Supersonic Velocity – Mach-Meter.   Aircraft velocity at supersonic speed is indicated by the Mach-meter, which does not give aircraft velocity directly. The advantage is that the Mach-meter directly reads the aircraft flight Mach number, and compressibility effects are taken into account. From the aircraft flight Mach number, V_{TAS} can be computed.

For supersonic flow, the Saint-Venant formula gives (see Equation 3.3b):

$$\frac{V^2}{a^2}=M^2=\left(\frac{2}{(\gamma-1)}\right)\left[\left(\frac{\Delta p}{p}+1\right)^{\frac{(\gamma-1)}{\gamma}}-1\right]$$ (3.6)

To compute $\Delta p/p$, the following is presented. Subscript "1" indicates properties of air ahead of the shock wake, that is, the free-stream condition when the subscript can be omitted, and subscript "2" indicates properties behind the shock.

Total pressure behind the normal shock is p_{t2}. Static pressure p is sensed separately. Sensing of static pressure just behind shock is affected, therefore it is measured separately a distance away from the shock wave. $\Delta p=(p_{t2}-p_2)$ is sensed using a pitot-static probe.

From Equation C10 in Appendix C,

$$\frac{p_{t2}}{p_2}=\left(1+\frac{\gamma-1}{2}M_2^2\right)^{\frac{\gamma}{\gamma-1}}\quad\left(\text{behind the shock}\right)$$

Equation C15 gives,

$$M_2^2=\frac{M_1^2+\dfrac{2}{(\gamma-1)}}{\dfrac{2\gamma}{(\gamma-1)}M_1^2-1}$$

and Equation C17 gives

$$\frac{p_2}{p_1}=\frac{2\gamma}{(\gamma+1)}M_1^2-\frac{(\gamma-1)}{(\gamma+1)}$$

Combining the above three equations, the following is obtained:

$$\frac{p_{t2}}{p_1} = \frac{p_{t2}p_2}{p_2 p_1} = \left(1 + \frac{\gamma-1}{2}M_2^2\right)^{\frac{\gamma}{\gamma-1}}\left[\frac{2\gamma}{(\gamma+1)}M_1^2 - \frac{(\gamma-1)}{(\gamma+1)}\right]$$

$$= \left[1 + \frac{(\gamma-1)}{2}\left\{\frac{2+(\gamma-1)M_1^2}{2\gamma M_1^2 - (\gamma-1)}\right\}\right]^{\frac{\gamma}{(\gamma-1)}}\left[\frac{2\gamma}{(\gamma+1)}M_1^2 - \frac{(\gamma-1)}{(\gamma+1)}\right]$$

or

$$\frac{p_{t2}}{p_1} = \left[\frac{(\gamma+1)^2 M_1^2}{4\gamma M_1^2 - 2(\gamma-1)}\right]^{\frac{\gamma}{(\gamma-1)}}\left[\frac{(1-\gamma)+2\gamma M_1^2}{(\gamma+1)}\right] \tag{3.7a}$$

which is known as the Rayleigh Pitot-tube formula.

Substituting the value of $\gamma = 1.4$, Equation 3.7a can be written as

$$\frac{p_{t2}}{p_1} = \left[\frac{2.4^2 M_1^2}{2(2.8M_1^2 - 0.4)}\right]^{3.5}\left[\frac{2.8M_1^2 - 0.4}{2.4}\right]$$

$$= \left[\frac{458.6471 M_1^7}{2^{3.5}(2.8M_1^2 - 0.4)^{2.5}}\right]\left[\frac{1}{2.4}\right]$$

$$= \left[\frac{191.103 M_1^7}{2^{3.5} \times 0.4^{2.5}(7M_1^2 - 1)^{2.5}}\right] = \left[\frac{191.103 M_1^7}{1.1449 \times (7M_1^2 - 1)^{2.5}}\right]$$

$$= \left[\frac{166.9168 M_1^7}{(7M_1^2 - 1)^{2.5}}\right]$$

The above equation is written as

$$\frac{p_{t2}}{p_1} - 1 = \frac{\Delta p}{p_1} = \left[\frac{166.9168 M_1^7}{(7M_1^2 - 1)^{2.5}}\right] - 1 \tag{3.7b}$$

(At Mach 1.0, $\Delta p / p_1 = 166.9168 / 88.18 - 1 = 0.893$. For subsonic speeds, $\Delta p / p_1 < 0.893$.)

The equation is set up for normal shock, that is, at a lowest speed of Mach 1.0 when $\Delta p / p_1 = 0.893$. Therefore for $\Delta p / p_1 < 0.893$, the subsonic Equation 3.1b is used. Being non-dimensional, both Equations 3.1b and 3.2b are valid for both SI (Système International) and FPS (foot-pound-second) systems.

Calibrated at sea-level, Equation 3.7b in terms of V_c becomes:

$$\frac{\Delta p}{p_0} = \left[\frac{166.9168\left(\frac{V_c}{a_0}\right)^7}{\left\{7\left(\frac{V_c}{a_0}\right)^2 - 1\right\}^{2.5}}\right] - 1 \tag{3.7c}$$

3.3.3 Aircraft Speed in Flight Deck Instruments

A pilot reads the airspeed indicator (ASI) installed in the flight deck, which converts the sensed parameter $(p_t - p)$ from the externally mounted pitot probes (see Section 3.4.2). Airspeed indicators have inherent errors on account of their design and manufacture limitations. The readings taken by the external probes are influenced by where they are mounted on the external surface of aircraft. These require corrections to the instrument readings to get the accurate speed.

Given below are the various forms of aircraft airspeed or velocity (both terms used interchangeably) used by engineers and pilots.

V_{TAS}: *True Airspeed (or V)* This is the actual airspeed with respect to the ambient air. In still air (zero wind), V_{TAS} is the ground distance covered in a unit time, that is, the ground speed. This is not the case when the atmosphere has some wind speed (WS), as shown subsequently.

V_i: *Instrument Indicated Airspeed* This is the pneumatic airspeed the gauge will indicate when manufactured, which will have inherent errors because of manufacturing limitations.

V_I: *Indicated Airspeed (or V_{IAS})* The built-in manufacturing error, ΔV_i, for each instrument is calibrated by test. Typically, ΔV_i is small, but an important consideration when an aircraft is close to stall speed. Manufacturer-supplied calibrated charts show ΔV_i for each instrument which is then applied to the instrument readings. This is the bare instrument yet to be installed on an aircraft and calibrated to give the correct ground speed at the sea-level standard day. When corrected, the instrument reads:

$$V_{IAS} = V_i + \Delta V_i \qquad \textbf{(3.8a)}$$

V_C: *Calibrated Airspeed (or V_{CAS})* Once the corrected bare instrument is installed on an aircraft, and depending upon where it is installed, the local flow field around the aircraft distorts the instrument readings, which need to be corrected. The error arising from aircraft flow field distortion is known as the *position error*, ΔV_p. It is not an instrument error, but comes from design consideration depending on the location of its placement on the aircraft. Given below is the position-error correction:

$$V_C = V_I + \Delta V_p = V_i + \Delta V_i \pm \Delta V_p \qquad \textbf{(3.8b)}$$

This is what the pilot uses after making all the corrections.

It is calibrated to give $V = V_C = V_{EAS}$ at sea-level ISA day. At altitude, because of density change, it gives $V_C = V_{EAS}$. In compressible flow, further correction ΔVc has to be applied on account of compressibility effects. Note that position error is also a function of aircraft attitude, that is, its angle of attack (AOA), which is dependent on aircraft weight for the same speed. The aircraft weight slightly affects the position error. Also, flap and undercarriage deployments have their influence on position error correction. Ground proximity also affects position error. Aircraft manufacturers determine ΔV_p for the full flight envelope from flight test for each design, and then supply the correction chart/table for the flight crew to get the gauge reading right. Nowadays these corrections are computerized and shown in the flight deck; a pilot need not make corrections from charts, as done in earlier days.

ΔV_c: *Compressibility Correction* At high subsonic speeds the total pressure sensed by the pitot tube magnifies the reading on account of the compressibility of air. The reading needs to be adjusted by incorporating an adiabatic compressibility correction, ΔV_c. The instrument reads an increment of ΔV_c which has to be subtracted to obtain V_{EAS}, as shown below.

$$V_{EAS} = V_C - \Delta V_c = V_i + \Delta V_i + \Delta V_p - \Delta V_c \qquad \textbf{(3.8c)}$$

At Mach number speeds below 0.3, ΔV_c is small (less than 2%) and may be ignored, thus simplifying to $V_{EAS} \approx V_C$. But at high subsonic speeds, ΔV_c is high (typically less than 7%) and it has to be applied to obtain V_{EAS}. Nowadays onboard microprocessors integrate all the corrections and display the correct speed on the flight-deck instruments for pilot usage.

Engineers prefer to work with V_{EAS} but also have to deal with all kinds of velocities in the process of design. Subsequently, at the later design phases the errors are incorporated and pilots' flight manuals are prepared. Errors are calibrated by ground and flight tests.

Pilots have no option but to observe the ASI for their operation. A pneumatic ASI shows V_I. In earlier days, all the errors had to be added to manually to obtain V_C. Nowadays, the multifunctional EFIS displays the V_{EAS}, having all errors incorporated into it, as well as the V_{TAS} (i.e. V).

3.3.4   Atmosphere with Wind Speed (Non-zero Wind)

The angle between aircraft true airspeed (TAS) and the relative WS is φ, as shown in Figure 3.2. The tail wind has $\varphi = 0$ and head wind has $\varphi = 180°$. The drift angle θ is the angle between the aircraft heading of TAS and the deviated path over the ground as the ground speed (GS), on account of the presence of wind.

$$GS = \sqrt{\left(TAS + WS \times (\cos\varphi)\right)^2 + \left(WS \times (\sin\varphi)\right)^2} \qquad \textbf{(3.9a)}$$

Then the drift angle (DF), θ, is as follows:

$$\theta = \tan^{-1}\left[\frac{WS \times (\sin\varphi)}{TAS + WS \times (\cos\varphi)}\right] \qquad \textbf{(3.9b)}$$

■ **FIGURE 3.2**
Aircraft speed in atmospheric wind

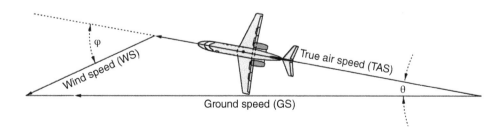

EXAMPLE 3.3

(a) A small aircraft at TAS of 100 kt is flying in atmospheric WS = 10 kt with relative angle $\varphi = 30°$. Find the GS.

(b) If a small aircraft flies with wind speed, WS = 10 kt, TAS of 100 kt and relative angle $\varphi = 25°$, what will be the drift angle, θ? Compare this with the drift angle of part (a) of this example.

SOLUTION

$$GS = \sqrt{\left(100 + 10 \times (\cos 30)\right)^2 + \left(10 \times (\sin 30)\right)^2}$$

$$= \sqrt{(108.66)^2 + 5^2} = \sqrt{11832} = 108.8 \text{ kt}$$

And drift angle,

$$\theta = \tan^{-1}\left[\frac{10 \times (\sin 25)}{100 + 10 \times (\cos 25)}\right]$$

$$= \tan^{-1}\left[\frac{10 \times 0.422}{100 + 9.06}\right] = \tan^{-1}(0.0387) = 2.22°$$

Part (a) gives:

$$\theta = \tan^{-1}\left[\frac{10 \times (\sin 30)}{100 + 10 \times (\cos 30)}\right]$$

$$= \tan^{-1}\left[\frac{10 \times 0.5}{100 + 8.66}\right] = \tan^{-1}(0.046) = 2.63°$$

3.3.5   Calibrated Airspeed

The V_C is the airspeed that the pilot reads in the flight deck. It is the instrument supplied by manufacturers, calibrated at sea-level conditions, which gives V_{TAS} (or V) $= V_C = V_E$. (On installation to the aircraft the position error is added and ΔV_c incorporated to improve accuracy.) Since Equation 3.2b is valid for any altitude, it can be written terms of V_c as follows.

$$p_t - p = \Delta p = p\left[\left(1 + 0.2M^2\right)^{3.5} - 1\right] = p_0\left[\left\{1 + 0.2\left(V_c / a_0\right)^2\right\}^{3.5} - 1\right] \quad \textbf{(3.10)}$$

where subscript "0" indicates sea-level conditions. The compressibility effect is included in the term $(1 + 0.2M^2)$. The above equation can be rearranged to express Mach number M at any altitude in terms of sea-level conditions and what the flight-deck instrument reads.

The instrument only measures $(p_t - p)$. The above equation gives:

$$\left(1 + 0.2M^2\right)^{3.5} = \left(p_0 / p\right)\left[\left\{1 + 0.2\left(V_c / a_0\right)^2\right\}^{3.5} - 1\right] + 1$$

or

$$M^2 = 5\left\{\left[\left(p_0 / p\right)\left[\left\{1 + 0.2\left(V_c / a_0\right)^2\right\}^{3.5} - 1\right] + 1\right]^{0.286} - 1\right\} \quad \textbf{(3.11a)}$$

This gives the Mach number at any altitude in terms of ISA day sea-level conditions.

At sea-level the speed of sound is $a_0 = 661.5$ kt. Substituting $(p/p_0) = \delta$, the above equation can be written as.

$$M^2 = 5[1 / \delta[\left\{1 + 0.2\left(V_C / 661.5\right)^2\right\}^{3.5} - 1] + 1]^{0.286} - 1] \quad \textbf{(3.11b)}$$

■ **TABLE 3.1**
True Mach number versus calibrated airspeed, V_C (kt)

V_C(kt)	V_C/a	$(V_C/a)^2$	Terms of Equation 3.11b				M^2	M
100	0.1512	0.0229	1.0046	0.0542	1.0542	1.0152	0.076	0.276
200	0.3023	0.0914	1.0183	0.22	1.221	1.05863	0.293	0.541
300	0.4535	0.2057	1.041	0.5103	1.51	1.1252	0.62574	0.791
400	0.6047	0.3657	1.073	0.944	1.944	1.2093	1.047	1.023

■ **FIGURE 3.3**
True Mach number
versus calibrated
airspeed

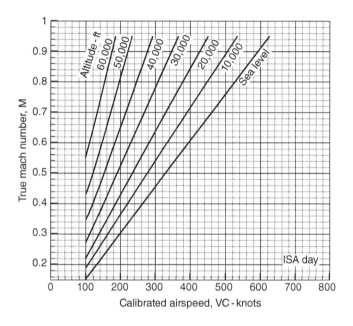

An example of Equation 3.11b at 30,000 ft altitude is tabulated in Table 3.1 and plotted in Figure 3.3.

Equation 3.3b is a function of ambient pressure p varying with altitude. The equations can be standardized to sea-level pressure p_0 in a scale that can be used to define calibrated airspeed V_C.

$$V_C^2 = \left(\frac{2a_0^2}{(\gamma-1)}\right)\left[\left(\frac{\Delta p}{p_0}+1\right)^{\frac{(\gamma-1)}{\gamma}}-1\right] \tag{3.12}$$

Substituting $(p_t - p)$ of Equation 3.10 in Equation 3.5a:

$$V^2\sigma = (5\gamma p/\rho_0)\left\{[(p_0[\{1+0.2(V_C/661.5)^2\}^{3.5}-1])/p+1]^{0.286}-1\right\} \text{ where } V_C \text{ is in knots}$$

$$= (5\gamma p_0 p/\rho_0 p_0)\left\{\left[\left[\{1+0.2(V_C/661.5)^2\}^{3.5}-1\right]\Big/(p/p_0)+1\right]^{0.286}-1\right\} \tag{3.13}$$

$$= (5a_0^2 p/p_0)\left\{\left[\left[\{1+0.2(V_C/661.5)^2\}^{3.5}-1\right]\Big/(p/p_0)+1\right]^{0.286}-1\right\}$$

3.3.6 Compressibility Correction (ΔV_c)

To obtain V_{EAS}, at high subsonic flight speed, the compressibility correction (ΔV_c) is applied. Recalling in Section 3.3 that high subsonic speeds require adiabatic compressibility corrections (ΔV_c) which change with altitude, the following is derived. Taking the square root of Equation 3.13, note V and V_C are in knots and by definition of equivalent airspeed $V_E = V\sqrt{\sigma} = V_C - \Delta V_C$. Substituting $a_0 = 661.5\,kt$ in Equation 3.5b, it becomes:

$$V\sqrt{\sigma} = V_E = V_C - \Delta V_C = 1479 \sqrt{(p/p_0)\left\{\left(\left[\frac{\left\{1+0.2(V_C/661.5)^2\right\}^{3.5}-1}{(p/p_0)}+1\right]^{0.286}-1\right\}} \qquad (3.14)$$

An example of Equation 3.14 at 30,000 ft altitude is tabulated in Table 3.2 and plotted in Figure 3.4.

Since this book deals mainly with subsonic aircraft performance, Figure 3.4 covers speeds up to Mach 0.95. The reader may extend the graph for supersonic application. In any case, from the Mach-meter readings the V_{EAS} can be easily computed. The advantage is that the Mach-meter directly reads aircraft flight Mach number, and compressibility effects are taken into account. From the aircraft flight Mach number, V_{TAS} can be computed.

■ **TABLE 3.2**
True compressibility correction ΔV_C of calibrated airspeed, V_C (kt)

V_C(kt)	$(V_C/a)^2$	Terms of Equation 3.14			V_E	ΔV_C
100	0.0228	0.27572	1.0046	1.0542	0.00452	99.38	0.6206
200	0.0914	0.54144	1.0183	1.22	0.01741	195.15	4.846
300	0.2057	0.791	1.041	1.51	0.0372	285.12	14.88
400	0.366	1.023	1.073	1.9436	0.0622	368.74	31.26

■ **FIGURE 3.4**
Compressibility correction

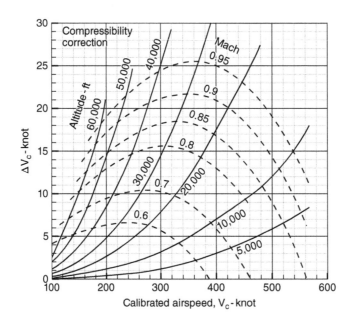

EXAMPLE 3.4

Find the V_{TAS} values and Mach numbers for an aircraft flying at $V_C = 200$ and 300 kt at $40,000$ ft ($p = 391.68$ lb/ft^2, $\sqrt{\rho} = 0.497$, $a = 968.08$ ft/s). Also find V_C when the aircraft becomes just sonic. At sea-level, $a_0 = 1116.5$ ft/s (340 m/s) and $p_0 = 2116.22$ lb/ft^2 ($101,325$ N/m^2).

SOLUTION

Since the flight is subsonic, we must apply the compressibility correction ΔV_C (Figure 3.4), to obtain V_{EAS}.

Use Figure 3.4 and Equations 3.1b and 3.8c.

V_C(kt)	ΔV_C(kt)	V_{EAS}(kt)	$\sqrt{\rho}$	V_{TAS}(kt)	Mach no.
200	8.5	191.5	0.497	385.3	0.672
300	25	275	0.497	553.3	0.965
312[a]	26	286	0.497	575.4	1.003 ($\approx$sonic)

[a] Evidently, sonic speed will be close to $V_C = 300$ kt. Try 312 kt.

Check the values of $\Delta p/p_1$ (use Equation 3.12).

V_C(kt)	(V_C/a_0)	$\dfrac{(\gamma-1)V_C^2}{2a_0^2}$	$\left(\left(\Delta p/p_0\right)+1\right)^{0.2857}$	$\left(\Delta p/p_0\right)+1$	$\Delta p/p_0$	Δp (N/m^2)	$\Delta p/p_1$
200	0.3026	0.0183	1.0183	1.0655	0.0655	6636	0.3538
300	0.4539	0.0412	1.0412	1.1518	0.1518	15,381.135	0.8202

Evidently, it has $\Delta p/p_1 < 0.893$, as expected for subsonic flight.

EXAMPLE 3.5

Find the Mach number of an aircraft flying at $V_C = 500$ kt at $40,000$ ft.

SOLUTION

Here is the computation from Equation 3.12.

V_C(kt)	(V_C/a_0)	$\dfrac{(\gamma-1)V_C^2}{2a_0^2}$	$\left(\left(\Delta p/p_0\right)+1\right)^{0.2857}$	$\left(\Delta p/p_0\right)+1$	$\Delta p/p_0$	Δp (N/m^2)	$\Delta p/p_1$
500	0.7565	0.1145	1.1145	1.4614	0.4614	46 751.355	2.4929

Clearly, $\Delta p/p_1 = 2.4929$, which indicates that $V_C = 500$kt is supersonic, and therefore we use Equation 3.7b to obtain flight M_1. Sonic speed is at $V_C = 312$kt as shown below. Check both the equations to show that it is at $M = 1.0$.

V_C(kt)	$\Delta p/p$	Equation no.	Equation	M_1
312	0.893	3.2b	$\left(1+0.2M^2\right)^{3.5}-1$	1.00 (sonic)
312	0.893	3.7b	$\left[\dfrac{166.9168M_1^7}{\left(7M_1^2-1\right)^{2.5}}\right]-1$	1.00 (sonic)
500	2.4929	3.7b	$\left[\dfrac{166.9168M_1^7}{\left(7M_1^2-1\right)^{2.5}}\right]-1$	1.52 (supersonic)

At Mach 1.52, it gives $V_{TAS} = 871$kt.

Note that on a hot day, say at ISA +10°C, the speed of sound a_0 changes but pressure remains unaffected. Equation 3.5b and other affected equations change, and the Mach number also changes. Working on $V_C = 300$kt for the example of ISA +10°C, $a_0 = 346.14$m/s.

V_C(kt)	(V_C/a_0)	$\dfrac{(\gamma-1)V_C^2}{2a_0^2}$	$\left(\left(\Delta p/p_0\right)+1\right)^{0.2857}$	$\left(\Delta p/p_0\right)+1$	$\Delta p/p_0$	Δp (N/m²)	$\Delta p/p$
300	0.4458	0.0398	1.0398	1.1464	0.1464	14,830.82	0.7908

Using Equation 3.2b, this gives $M_1 = 0.9067$, reduced from 0.966 on an ISA day.

If both ambient pressure and temperature change from ISA conditions, then a_0 has to be worked out using aero-thermodynamic relations.

3.3.7 Other Position Error Corrections

It was mentioned earlier that position error is a function of many factors at low speed (ground effects, flap and undercarriage deployment), at high speed (compressibility effects), and at all speeds (attitude changes). These are established through flight testing and theory. Aircraft manufacturers supply position errors for the full flight envelope. Normally, at any stage the incremental change in position errors arising from these effects are small. These are not given in this book – only the following are mentioned to maintain readers' awareness.

At low speeds during takeoff and landing, ground effects modify the flow field which affects position error. Flap and undercarriage deployment will affect position error. At high speed, the aircraft is in clean configuration, but compressibility effects influence position error. Aircraft attitude can change at any speed. One example is that aircraft weight changes will alter C_L, hence attitude changes in turn affect the position error. Change of speed also changes aircraft attitude.

3.4 Air Data Instruments

This section describes the following analogue air data instruments (Figure 3.5): (a) airspeed indicator (ASI), (b) altimeter, (c) attitude indicator, (d) turn-slip indicator, (e) vertical speed indicator (VSI), (f) Mach number indicator, (g) heading indicator, (h) horizontal

■ **FIGURE 3.5**
Analogue air data
indicating instruments.
Top row: airspeed
indicator; altimeter;
attitude indicator;
turn-slip indicator;
vertical speed indicator.
Bottom row: Mach
meter; heading
indicator; H-S indicator;
angle of attack; g-meter

■ **FIGURE 3.6**
Altimeter

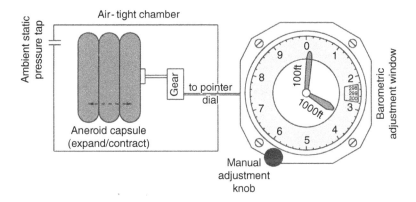

situation indicator (H-S indicator), (i) AOA indicator and (j) g-meter. These are the older types, each displaying a single parameter; as a result there is a clutter of instruments (see Section 3.5) in the flight deck.

All of them use some form of tapping from outside the aircraft. Pitot and static pressure probes and vanes measure most of these parameters. Navigation and communication instruments are beyond the scope of this book.

3.4.1 Altitude Measurement – Altimeter

The altimeter makes use of ambient static pressure taken from a separate probe and not from the pitot-static tube. It records height above the ground corresponding to the static pressure. The altimeter consists of an air-tight chamber encasing a differential pressure-sensitive hollow vacuumed diaphragm (see Figure 3.6). It is a corrugated bronze aneroid capsule. The chamber is connected to the static pressure probe on the aircraft's exterior which exerts differential pressure on the hollow diaphragm, which then expands or contracts as altitude increases or decreases, respectively. The differential pressure is calibrated to ISA conditions, but needs manual adjustment to conditions as explained below. A knob is provided for the flight crew to make adjustments using the Kollsman window shown in the figure. The radio altimeter (rad-alt) and GPS are used in conjunction to read altitude; these are not discussed here.

Section 2.3 gave the mathematical relations to derive the ISA table by holding acceleration due to gravity, g, constant. The pressure of the ISA table is based on the geopotential pressure altitude, h_p. It is an extreme rarity to have a proper ISA day; pilots invariably fly in atmosphere

■ **TABLE 3.3**
Explanation of ICAO standard altitude codes

ICAO altitude code	Description
QFE height	FE stands for field elevation. The QFE pressure level gives height H_{EI} above the surface. It is used for takeoff and landing at the same airfield. Aeronautical maps give QFE heights above AGL (average ground level). QFE heights are important for landing at different airfields, say in cross-country flying.
QNH height	Aeronautical charts give elevation of the terrain above mean sea-level, and QNH heights give the possibility of determining the actual height above the terrain.
QNE height	QNE height is given in terms of flight levels (FLs) in units of 100 ft. It does not depend on the atmospheric height of the takeoff airfield. The altitude is calculated from the ISA sea-level pressure datum of 101,325 Pa (flight level zero, FL0, altimeter at 1013 mb). ATC-controlled traffic uses QNE height so that all aircraft can fly at the designated FL with altimeters showing the FL. There is a transition altitude above which QNH height is changed over to QNE height. The transition level at climb is lower than the transition level at descent, giving a transition layer as shown in Figure 3.7.

■ **FIGURE 3.7**
Typical example altitude schedules. Left-hand panel: altitude schedules at FL300 on an ISA day. Right-hand panel: Altitude schedules on a non-ISA day

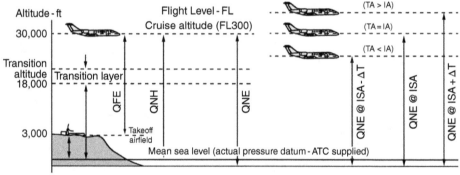

outside ISA conditions. Therefore, h_p, as read in the altimeter gauge, gives an approximate geometric height which may not prove safe as errors can occur, especially when sea-level pressure is lower than ISA conditions. In low visibility, aircraft can hit the ground even when the altimeter shows sufficient altitude. Over elevated terrain it is dangerous, and accidents have happened. To make corrections, all altimeters have an adjustable knob to set zero altitude and ground level pressure at the airfield at the time of takeoff. At landing, pilots should ask for the airfield altitude and ground level pressure at that time to set the altimeter. At ISA sea-level the altimeter is calibrated to show 1013 mb (representing 101,325 Pa as the pressure datum). There are International Civil Aviation Organization (ICAO) codes to follow as given in Table 3.3 below (Figure 3.7).

To maintain accuracy, at lower levels the QNH height is used to make sure that QFE is cleared. At cruise, QNE height is used to maintain the prescribed FL for which an altimeter setting is obtained from air traffic control or other issuing stations. There is a transition altitude above which QNH height changes to QNE height. In North America it is 18,000 ft and in Europe it varies. The example of FL300 is shown in Figure 3.7.

On a non-standard day the indicated altitude (IA) will be different from true altitude (TA) as shown in Figure 3.7. On a hot day, the true altitude is higher than the indicated altitude and vice versa. The relationship between true altitude and indicated altitude is as follows.

$$\text{True altitude}\left(\text{TA}\right) = \text{indicated altitude}\left(\text{IA}\right) \times \frac{T}{T_{ISA}} \qquad \textbf{(3.15)}$$

3.4.2 Airspeed Measuring Instrument – Pitot-Static Tube

A pitot tube is a hollow concentric tube (Figure 3.8), sealed at the front end and aligned to airflow, in which a stream tube is captured in the inner tube to stagnation, converting the kinetic energy of the flow to potential energy of increased total pressure, p_t, as shown in Equations C9 and C10 in Appendix C.

A pitot-static tube also reads the static pressure, p, of the flow by tapping a hole perpendicular to airflow on the hollow tube. The difference in pressure, $\Delta p = (p_t - p)$, is the velocity head, which is calibrated for the indicating gauges. The following subsections are devoted to the theory of calibrating the pitot-static tube reading. Figure 3.8 shows the basic concept of a pitot-static tube to measure airspeed. It consists of two separate concentric hollow tubes sealed from each other. The inner tube is open and aligned to oncoming airflow, which stagnates inside. The outside tube has holes transverse to airflow to measure static pressure, as it does not read the normal component of dynamic head of airflow.

The positions and types of pitot-static probe installation on aircraft are shown in Figure 3.9. Probes should be near to the nose of the aeroplane where the airflow is relatively clean and the boundary layer is thin, minimizing the required probe height. Civil aircraft typically have small probes, with several installed to give redundancy, while older military installations (Figure 3.9 shows F104 installation) have a large boom at the nose to suit supersonic flight.

The positions of static holes have to be carefully chosen on the fuselage sides where the static pressure matches the ambient pressure, to give the minimum pressure error. There are

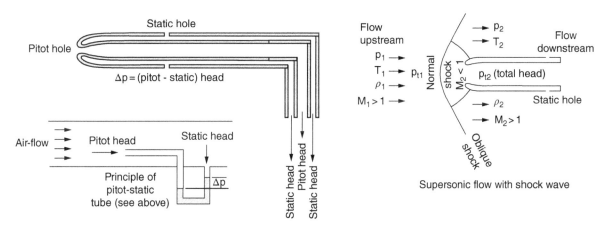

■ **FIGURE 3.8** Pitot-static tube

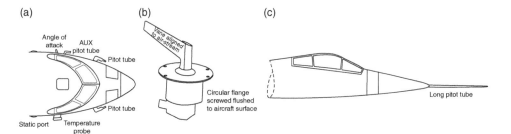

■ **FIGURE 3.9** Pitot-static probe installations on aircraft. (a) Typical positions for civil aircraft. (b) Pitot-static probe with integrated angle-of-attack probe. (c) F104 pitot-static probe installation

■ **FIGURE 3.10**
(a) Angle-of-attack
probe at the centre in
between the pitot-
static probes
and pilot-static
probe. (b) Probe
locations on aircraft
fuselage

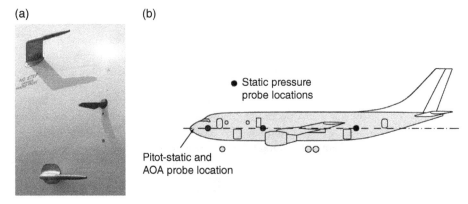

(a)   (b)

Static pressure
probe locations

Pitot-static and
AOA probe location

several holes along the side (Figure 3.10) to give a mean value, as pressure distribution over the fuselage changes with attitude changes.

3.4.3 Angle-of-Attack Probe

The angle of attack (AOA) is one of the important parameters related to aeroplane performance and handling. Takeoff and landing take place close to aircraft stall speed, with an AOA close to wing stall. Information on AOA also helps the pilot to cruise at close-to-maximum lift to drag ratio, for best fuel economy. For the same change in ΔAOA, change in ΔV is higher at high speed than at close to stall. (Stall warning devices are discussed below.)

Lift increases with an increase in AOA at a constant speed. Lift is also weight-dependent. Flaps and/or slats increase lift, while spoilers/speed brakes spoil lift; all affect aircraft stall characteristics. Combat manoeuvres can take place at exceptionally high AOA, for which special design features, for example the strake/canard configuration, may be incorporated. But when the AOA exceeds a certain limit then the aircraft stalls, an undesirable situation that can endanger safety. Figure 3.10 shows an AOA probe, which is a vane that aligns with flow velocity. The instrument is calibrated to show the angle at the flight deck.

Lack of a stall warning device has caused many accidents. Stalls/spins are most likely to occur in manoeuvring flights, when the warning could help the pilot to take preventative steps. At low altitudes, especially on approach to landing, it is critical to know what AOA the aircraft is maintaining, and its indication allows the pilot to control the aircraft safely.

Airworthiness requirements mandate some form of stall warning device to provide safety in operation. There are many forms of stall warning, for example, audio and/or voice warnings when approaching stall speed, a stick shaker or pusher to improve stall avoidance, and so on. Fly-by-wire (FBW) ensures that aircraft have sufficient speed margins, controlled by built-in microprocessors, to meet aircraft control laws. Typically, a pilot is trained to recognize pre-stall buffet characteristics, along with the type of stall warning device installed. As speed reduces and approaches stall speed, the control stick becomes sloppier.

3.4.4 Vertical Speed Indicator

The vertical speed indicator (VSI) gives the rate of climb or descent by the aircraft during flight. Change of altitude records rate of change of ambient (static) pressure, which is sensed by a diaphragm inside the instrument, and the data are calibrated and indicated. Basically, it records differential static pressure between the instantaneous pressure as it changes and the existing trapped pressure in the instrument. It records best when flight is in the vertical plane. It should be maintained to show zero when on ground.

■ **FIGURE 3.11**
Total and static air
temperature probe

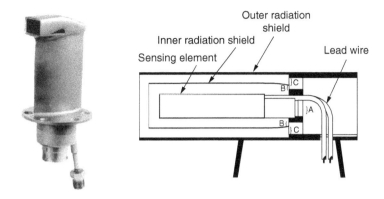

3.4.5 Temperature Measurement

Total air temperature (TAT) and static air temperature (SAT) are part of the air data used to compute aircraft performance. SAT is the atmospheric ambient temperature. TAT (in Equation 3.16 as T_t) is measured by stagnating air velocity which has ambient temperature SAT (in Equation 3.16 as T). Stagnation is through adiabatic compression, converting the kinetic energy of airflow to increase temperature, which makes TAT > SAT. The isentropic relationship is as follows:

$$\left(\frac{T_t}{T}\right)_{isen} = \left(1 + \frac{\gamma - 1}{2}M^2\right) \tag{3.16a}$$

In reality the energy conversion through compression is not perfect and reaches a lower value of $T_r < T_t$. An empirical recovery factor, r, is used to modify the above equation to a realistic value.

$$\frac{T_t}{T} = \left(1 + \frac{\gamma - 1}{2}rM^2\right) \tag{3.16b}$$

where $r = \dfrac{(T_r - T)}{(T_t - T)}$, typically with a value between 0.75 and 1.0, depending on the design.

A TAT probe performs accurate temperature measurement under all flight conditions, including aircraft ground operations and atmospheric icing conditions (see Figure 3.11). The TAT measurement is used to determine static temperature, true airspeed computation, and so on. The sensor consists of two sensing elements that are insulated and hermetically sealed within the sensor housing. A compensating heater used to de-ice the sensor is integrated into the housing. There are two types of TAT: (i) aspirated and (ii) unaspirated. Aspirated TAT design includes the addition of an air-to-air ejector, which induces airflow past the sensing element. The increased airflow helps to eliminate temperature soaking inaccuracies caused by bright sunshine or hot ramp heat radiation.

3.4.6 Turn-Slip Indicator

The turn-slip indicator is a gyroscope-driven instrument. The dial shows a black ball sealed in a curved, liquid-filled glass tube with two wires at its centre, and a needle or aircraft symbol seen from the aft end, as shown in Figure 3.12. The ball measures gravitational and inertial force

■ **FIGURE 3.12**
Turn-slip indicator
showing a right turn

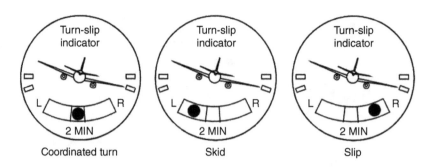

| Coordinated turn | Skid | Slip |

caused by the turn. A right turn is shown in Figure 3.12. Aircraft manoeuvres give readings as follows:

Straight level flying: Aircraft wings level and ball in the centre.

Coordinated turn: In this case when the bank angle is correct then the gravity and inertial forces are equal and the ball stays in the centre but the right wing drops to show the turn. This is executed by coordinated application of right rudder and right aileron.

Skid turn: When the bank is less, then inertial force is greater than gravitational force, and the aircraft skids outward. The ball moves left but a right bank angle is shown by the aircraft symbol.

Slip turn: When the bank is more, then inertial force is less than gravitational force, and the aircraft slips inward. The ball moves right and a right bank angle is shown by the aircraft symbol.

3.5   Aircraft Flight-Deck (Cockpit) Layout

The aircraft flight deck is a better term than the older usage of the word *cockpit*. The flight deck serves as a human–machine interface by providing (i) an outside reference of topography through the cabin windows, (ii) onboard instruments to measure flight parameters, (iii) control facilities to operate an aircraft safely for the mission role, and (iv) management of aircraft systems (e.g. the internal environment). Future designs could result in a visually closed flight deck (i.e. a screen replacing the windows).

Both civil and military pilots have common functions as listed below.

1. mission management (planning, checks, takeoff, climb, cruise, descent and landing);
2. flight path control;
3. systems management;
4. communication;[1]
5. navigation;[1]
6. routine post-flight debriefing;
7. emergency action when required (some drills differ between civil and military).[1]

---

[1] Civil aircraft are assisted by ground control.

In addition, military pilots have an intense workload as listed below:

1. mission planning and management;
2. target acquisition;
3. weapons management/delivery;
4. defensive measures/manoeuvres;
5. counter threats – use of tactics;
6. manage situation when hit;
7. in-flight re-fuelling where applicable; and
8. detailed post-flight debriefing in special situations.

The military aircraft flight deck has more stringent design requirements. The civil aircraft flight deck design is in the wake of military standards, and the provision of space is not as constrained. Figure 3.13 shows that aircraft flight deck design has changed dramatically from the early analogue dial displays (four engine gauges fill the front panel) to modern microprocessor-based data management in an integrated all-glass multifunctional display (MFD). The MFD is also known as the electronic flight information system.

Figure 3.13a shows the old-type panel for two pilots, with duplicated analogue dial gauges for redundancy. The latest Airbus 380 flight-deck panel replaces myriad gauges by EFISs, which are MFD units. The minimum generic layout of a modern flight-deck panel is shown in Figure 3.13b. Numerous redundancies are built into the display with independent circuits. An *EFIS* consists of primary flight displays (PFDs) and navigational displays (NDs) having several pages that display significant data (see next section).

3.5.1 Multifunctional Displays and Electronic Flight Information Systems

Section 3.4 dealt with the older type analogue air data instruments (Figure 3.5) where each one displays specific information, requiring a large number of separate instruments in total. The clutter of instruments in the flight deck is shown in Figure 3.13a. With the advent of microprocessor-based data processing, all information can be integrated and displayed on a liquid crystal display (LCD) monitor with options of flicking through pages for the specific group of data. These are MFDs incorporated into an EFIS – in short known as a "*glass cockpit*".

To reduce clutter in displays for larger aircraft usage, the data are divided into primary and secondary displays. For convenience, displays are usually given two EFIS monitors. The two are the primary air-data system display (SD), and the navigational display (ND); each type of system has several pages and each display screen can be changed for specific information. They include pages for the engine, cruise, flight control, fuel, electrical systems, avionics, oxygen, air-bleed, air-conditioning, cabin pressurization, hydraulics, undercarriage, doors and the auxiliary power unit (APU). Military aircraft also have weapons-management pages. Figure 3.14 shows the two EFIS displays: the primary system display (SD) and the navigational display (ND).

The primary flight display (PFD) consists of air-data systems (SD), including aircraft speed, altitude, attitude, aircraft reference and ambient conditions. The secondary system consists of the ND, which provides directional bearings (i.e. GPS and inertial system), flight plan, route information, weather information, airport information and so forth. Each type has some duplication. In a separate panel, the SD shows the engine data and all other system data, including those required for the environmental control system (ECS). Forward-looking weather

■ **FIGURE 3.13**
**Civil aircraft flight
deck. (a) Old-type
(BAE146) flight deck
with analogue
head-down display.
(b) Outline of a
generic modern
flight deck panel
with head-up display
(HUD)**

(a)

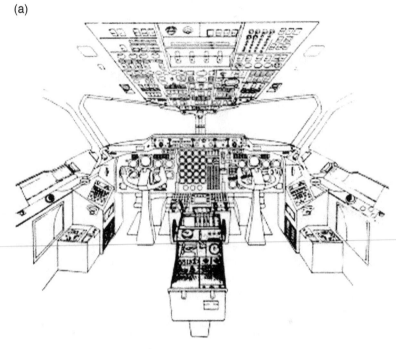

(b)

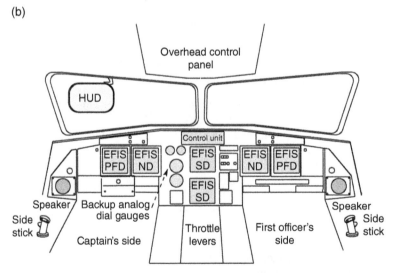

radar can have the ND or a separate display unit. Initially, such flight decks also had basic
analogue gauges showing air-data as redundancies in case the EFIS failed. Currently, with
vastly improved reliability in the EFIS, older analogue gauges are gradually being removed,
except for important ones like the altimeter and ASI.

In some designs, the engine display (ED) is shown separately. Figure 3.15 shows an EFIS
for a General Aviation piston engine aircraft.

■ **FIGURE 3.14**
Multifunctional
display (EFIS).
(a) Primary system
display (SD).
(b) Navigational
display (ND)
(Courtesy of Dynon
Avionics)

(a)

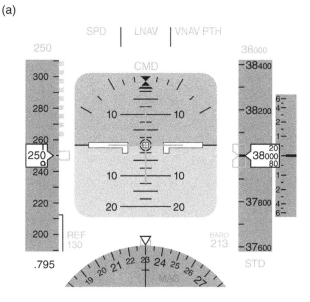

(b)

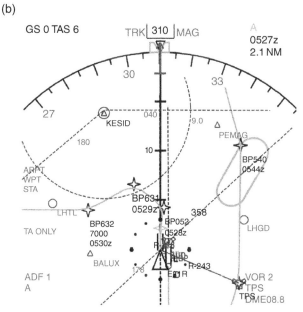

3.5.2   Combat Aircraft Flight Deck

Figure 3.16 shows a typical modern flight deck for military aircraft. Backup analogue gauges are provided as well as the MFD-type EFIS. The left-hand side is the throttle and the right-hand side is the side-stick controller known as the hands-on throttle and stick (HOTAS). The figure indicates which type of data and control a pilot requires. A single pilot's workload is exceptionally high, so computer assistance is required.

■ **FIGURE 3.15**
Integrated display on
an EFIS screen
(General Aviation
piston engine
aircraft)

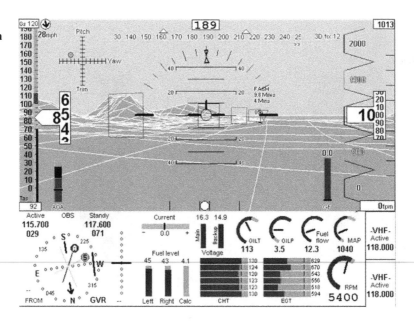

■ **FIGURE 3.16**
Schematic fighter
aircraft flight deck
(flight data EFIS
to the left and
navigational data
EFIS to the right)

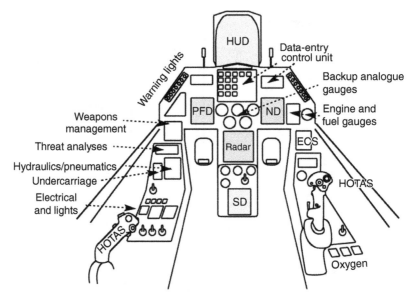

The head-up display (HUD) (see next section) was first seen installed in combat aircraft, but the technology has recently trickled into civil aviation as well. HUDs are being installed on most new medium and large-sized commercial-transport aircraft, if requested by operators.

3.5.3   Head-Up Display (HUD)

The flight-deck display as shown in Figure 3.13 is with the instrument panel in front of the pilot, who must look down for flight information – more frequently in critical situations. When flying close to the ground or chasing a target, however, pilots should keep their head up, looking for external references. A pilot will have to move their head position up and down to read the instruments – this inflicts strain. Engineers have solved the problem to a great extent by projecting the most important flight information (both primary and navigational data) in bright green light on transparent glass mounted in front of the windscreen. With a HUD, pilots can see all necessary information without moving their head, and at the same time, they can see through the HUD for external references. A commercial flight deck with a HUD is shown in Figure 3.13b.

3.6   Aircraft Mass (Weights) and Centre of Gravity

Performance engineers must know various categories of weight groups to decide on the aircraft loading of fuel and payload to meet mission payload-range. The CG position of the aircraft is important to ensure safe operation from engine start to stop. Proper distribution of mass (weight) over the aircraft geometry is the key to establishing the aircraft CG. It is of importance to locate wing, undercarriage, engine and empennage in relation to each other for aircraft stability and control. The convenient way is to estimate each component weight separately and then position them to satisfy the CG location with respect to the overall geometry. This book assumes that aircraft component masses and CG location are supplied. Weight estimation is shown in [1–7]. The examples given here are taken from [1]. Given below is the breakdown of weights of interest to the performance engineer. In this book weight and mass are used interchangeably in pounds for the FPS system. In the SI system, mass is in kilograms and weight is in newtons (mass multiplied by acceleration due to gravity, 9.81 m/s^2).

3.6.1   Aircraft Mass (Weights) Breakdown

Definitions of various groups of aircraft masses (weights) are given below (Figure 3.17).

MEM (manufacturer's empty mass):

.This is the mass that rolls out of the factory before being taken to the flight hangar to make the first flight.

OEM (operating empty mass) = MEM + crew + consumables:

The aircraft is now ready for operation (residual fuel from the previous flight remains).

[Some operators prefer to separate crew and cabin equipment load from consumables by introducing the definition of *dry operational mass* (**DOM**) which replaces OEM as defined above. Since the book has no use for this distinction, we use the standard definition of OEM.]

MTOM (maximum takeoff mass) = OEM + payload + fuel:

MTOM is the reference mass loaded to the rated maximum. This is also known as the brake release mass (**BRM**).

Aircraft are allowed to carry a measured amount of additional fuel for taxiing to reach the end of the runway ready for take-off at BRM (maximum takeoff weight (MTOW)). This additional fuel mass would make the aircraft exceed MTOW to maximum ramp weight (**MRW**).

MRM (maximum ramp mass) $\approx 1.0005 \times$ MTOM (very large aircraft) to $1.001 \times$ MTOM = MTOW + fuel to taxi to end of runway for takeoff

This is also known as the maximum taxi mass (**MTM**) and is heavier than MTOM.

ZFM (zero fuel mass): MTOM minus all fuel (non-usable residual fuel remains).

MLM (maximum landing mass): 0.95 MTOM to 0.98 MTOM, depending on the design.

In case of engine failure at takeoff, the aircraft must be able to return to land immediately, almost fully loaded. Fuel burn to return is taken as 2–5% of MTOM. This is to ensure the structural integrity as designed.

■ **FIGURE 3.17**
**Breakdown of
aircraft weights
(percentages vary
from design to
design)**

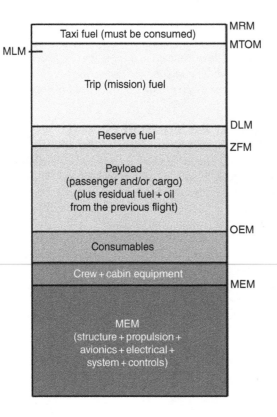

DLM (design landing mass):

This is the aircraft mass when all the trip (mission) fuel is consumed (only reserve fuel left) at touchdown.

Taxiing fuel for mid-sized aircraft could be around 100 kg and it must be consumed before takeoff roll is initiated – the extra fuel for taxiing is not available for range calculation. On busy runways, the waiting period in queue for takeoff could extend to more than an hour in extreme situations.

Military aircraft mass breakdown uses similar nomenclature; the difference is in the type of payload they carry, mostly armament-related loads.

3.6.2 Desirable CG Position

The aircraft aerodynamic centre moves backwards on the ground on account of the flow field being affected by ground constraints. There is also movement of CG location depending on loading (fuel and/or passengers). Typical CG range between forward and aft CG is less than 25% of the mean aerodynamic chord (MAC). Make sure that the pre-flight aftmost CG location is still forward of the in-flight aerodynamic centre by a convenient margin. This should be as low as possible to minimize trim. Also note where the main wheel contact point (and strut line) is aft of the aftmost CG, the subtending angle, β, should be more than the fuselage rotation angle, γ. The main wheel is positioned to ease rotation as well to assist good ground handling. Typical aircraft CG margins affecting aircraft operation are shown in Figure 3.18.

■ **FIGURE 3.18**
Aircraft CG position showing stability margin (axes system meant for CG estimation)

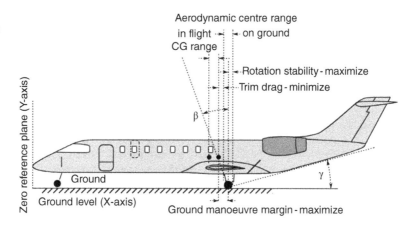

■ **FIGURE 3.19**
Aircraft CG limits of travel – potato curve (shaded area).
(a) Range of CG variation – horizontal limits. (b) Range of CG variation – vertical limits

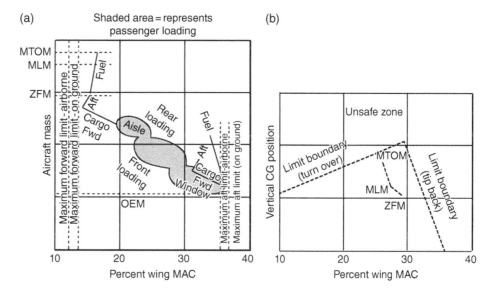

Advanced combat aircraft could have relaxed static stability to give quicker responses. In other words, the margin between the aftmost CG and the in-flight aerodynamic centre is reduced (may even go slightly negative), but the rest of the design considerations with respect to undercarriage position are still the same.

There is some flexibility available to fine-tune the CG position by moving payload positions. It is desired that payload be positioned around the CG so that its variation will have the least effect on CG movement. Fuel storage should be distributed to ensure the least CG movement – if this is not possible, then in-flight fuel transfer is necessary to shift weight to maintain the desired CG position (this was done in supersonic Concorde).

Note that fuel load and payload combinations are variable, and hence the CG position would vary. Figure 3.19a shows the variation of CG position for the full range of fuel and payload combinations. It resembles the shape of a potato, and some loosely call this a "potato" curve for CG variation for all loading conditions. Designers must ensure that at no time during loading up to MTOM does the CG position cross the loading limits, creating the danger that the aircraft might tip over in any direction. Loading is done under supervision. While passengers have a free choice of seating order, cargo and fuel loading is done in prescribed sequences (with options).

For static stability reasons, make sure that the aircraft has a static margin at all loading conditions. With the maximum number of passengers, the CG is not necessarily at the aftmost position. Typically for civil aircraft, the CG should be at around 18% of MAC when fully loaded and at around 22% when empty. The CG is always forward of the neutral point (aircraft aerodynamic centre – established through CFD/wind tunnel test). It is observed that passengers choose to take the window seats first. Figure 3.19a shows the window seating first and the aisle seating last. Note the boundaries of front and aft limits in Figure 3.19b. The cargo and fuel loading is done in an order such that the CG locus of travel gives a line as shown. In the diagram the CG of the operating empty mass (OEM) is at the rear, indicating that the aircraft has aft-mounted engines. For wing-mounted engines the CG at OEM moves forward making, the potato curve more upright.

Military aircraft CGs obey similar logic, but need not consider passengers. Combat aircraft have a smaller stability margin to achieve manoeuvrability.

3.6.3 Weights Summary – Civil Aircraft

Table 3.4 below gives the weights summary obtained from reference [2] for the classroom example in this book.

Clearly within the aircraft family there will be variants that need to be considered. For simplification, linear variation is considered – readers should work them out formally.

Longer Variant: Increase ΔPayload $= 400$ kg, ΔFuselage $= 300$ kg, ΔFurnish $= 200$ kg, ΔFuel $= 300$ kg and ΔOthers $= 200$ kg. Total increase $= +1400$ kg. $\text{MTOM}_{\text{Long}} = 10,900$ kg (no structural changes).

■ **TABLE 3.4**
Bizjet mass (weight) summary

Components	Semi-empirical mass (kg)	Mass fraction (%)
Fuselage group	930	10
Wing group	864	9.2
H-tail group	124	1.32
V-tail group	63	0.67
Undercarriage group	380	4
Nacelle + pylon group	230	2.245
Structures group total	*2591*	*27.56*
Power plant group	1060	11.28
Systems group	1045	11.12
Furnishing group	618	6.57
Contingencies	143	0.7
Manufacturer's empty mass, MEM	*5457*	*58.05*
Crew	180	1.92
Consumables	$\approx$163	1.73
Operational empty mass, OEM	*5800*	*61.7*
Payload (as positioned)	1100	11.7
Fuel (as positioned)	2500	26.6
Maximum take-off mass, MTOM	*9400*	*100*
Maximum ramp mass, MRM	*9450*	*100.53*

Smaller Variant: Decrease ΔPayload = 500 kg, ΔFuselage = 350 kg, ΔFurnish = 250 kg, ΔFuel = 350 kg and ΔOthers = 250 kg (lightening of structures). Total decrease = -1700 kg. $\text{MTOM}_{\text{Small}} = 7800$ kg.

3.6.4   CG Determination – Civil Aircraft

Aircraft CG can be located after obtaining component masses (weights). A reference coordinate system is essential to locate the CG position with respect to the aircraft. A suggested coordinate system is to take the X-axis along the ground level (or at a suitable level) and the Y-axis passing through the furthest point of the nose cone (tip), as shown in Figure 3.18. Typically the fuselage axis is parallel or nearly parallel to the X-axis. In the example it is taken parallel. x is measured from the nose of the aircraft and then converted to MAC_w. Table 3.5 may be used to determine the CG location.

The first task is to estimate the CG position for each component group from statistical data. Figure 3.20 gives generic information to locate component CG positions. Subsequently, when more details of components emerge, then the component CG positions are fine-tuned and aircraft CG estimation is iterated. Typical ranges of the CG position with respect to the component itself are given in Table 3.5. CG height from the ground is represented by z coordinates. Note that the CG should lie in the plane of symmetry (although there are unsymmetrical aircraft).

It is to be emphasized that at the conceptual phase, the aircraft design relies on the experience of the designer. At such an early stage it is not possible to obtain accurate component weights, and their CG location is yet to evolve.

Expression for x, y and z coordinates are given below:

$$\bar{x} = \sum_i^n \frac{\text{component weight} \times \text{distance from nose reference point}}{\text{aircraft weight}} \tag{3.17a}$$

■ **TABLE 3.5**
Typical values of component CG locations – civil aircraft

Component	CG and typical % of component characteristic length
Fuselage group	45%
Wing group	No slat – 30% of MAC
	With slat – 25% of MAC
H-tail and V-tail group	30% (each separately)
Undercarriage group	At wheel centre (nose and main wheels taken separately)
Nacelle + pylon group	35%
Miscellaneous	As positioned – use similarity
Power plant group	50%
Systems group	As positioned – use similarity (typically 40% of fuselage)
Furnishing group	As positioned – use similarity
Contingencies	As positioned – use similarity
MEM	*(Need not compute for CG at this moment)*
Crew	As positioned – use similarity
OEM	*Compute*
Payload	As positioned (distribute around CG)
Fuel	As positioned (distribute around CG)
MTOM and MRM	*To compute*

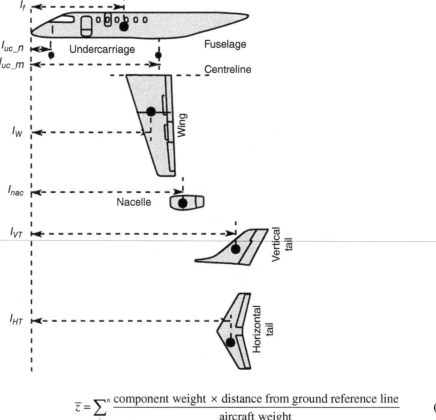

$$\bar{z} = \sum_{i}^{n} \frac{\text{component weight} \times \text{distance from ground reference line}}{\text{aircraft weight}} \tag{3.17b}$$

$$\bar{y} = 0 \quad \left(\text{CG is at the plane of symmetry}\right) \tag{3.17c}$$

3.6.5 Bizjet Aircraft CG Location – Classroom Example

Use Table 3.6 and Equations 3.18a and b to calculate the CG location – SI units are used.

3.6.6 Weights Summary – Military Aircraft

Table 3.7 gives the weights summary of the advanced jet trainer (AJT) classroom examples.

3.6.7 *CG* Determination – Military Aircraft

Table 3.8 gives the CG positions of aircraft components of the classroom examples: x and z are measured from the nose of the aircraft and then convert to MAC_w. Coordinate origin $x = 0$ is at the nose tip, and $z = 0$ is at ground level which is kept horizontal.

The CG should be at around 18% of MAC when fully loaded and at around 22% when empty, and in between when loaded only with pilots. CG is always forward of the neutral point for static stability reasons.

3.6.8 Classroom Worked Example – Military *AJT CG* Location

Table 3.9 gives the CG locations of aircraft components of the advanced jet trainer (AJT), in SI units (kg and m).

■ TABLE 3.6
Determination of Bizjet CG location

Item group	Mass (kg)	X (m)	Moment	Z (m)	Moment
Fuselage	930	6.8	4284	1.6	1488
Wing	864	7.8	6739.2	1	864
H-tail	124	14	1736	8	992
V-tail	63	15	945	3	189
Undercarriage (nose)	110	1.2	132	0.4	44
Undercarriage (main)	270	8.4	2268	0.5	135
Nacelle/pylon	230	10.2	2346	2	460
Power plant	1060	11	11660	1.9	2014
Systems	1045	6.5	6792.5	1	1045
Furnishing	618	6	3708	2	1236
Contingencies	146	3	438	1.2	175.2
MEM	*5460*				
Crew	180	3	540	1.4	252
Consumables	160	5	800	1.5	240
OEM	*5800*				
Payload	1100	6	6600	1.1	1210
Fuel	2500	8.5	21,250	1	2500
Total moment at MTOM of 9400 (20,680 lb)			$\approx$70,239		$\approx$12,779
		$\bar{x}=7.44$m		$\bar{z}=1.357$m	
MRM	*9450 (20,790 lb)*				

■ TABLE 3.7
Advanced jet trainer components and weights summary

Component	% fuselage	Military AJT aircraft kg (lb)
Fuselage group	45%	849
Wing group	30% wing MAC	480
H-tail group	30% H-tail MAC	112
V-tail group	30% V-tail MAC	43
Undercarriage group	$\approx$ wheel centre	300
Nacelle + pylon group	75%	Not applicable
Structures group total		*1784*
Power plant group	Engine centre	750
Systems group	As positioned	780
Furnishing group	As positioned	65
Contingencies	As positioned	98
Manufacturer's empty mass, MEM		*3477*
Crew		180
Operational empty mass, OEM		*3657*
Fuel	As positioned	1100
Normal training configuration (NTC) mass		*4757*
Payload	As positioned	1800
Maximum take-off mass, MTOM		*6557*
Maximum ramp mass, MRM		*6600*

■ **TABLE 3.8**
Typical values of component CG locations – military AJT aircraft

Component	Typical % of component characteristic length
Fuselage group	45–50%
Wing group	No slat – 30% of MAC
	With slat – 25% of MAC
H-tail and V-tail group	30% (each separately)
Undercarriage group	At wheel centre
Nacelle + pylon group	Generally not applicable
Power plant group	70–80%
Systems group	As position – use similarity (typically 35% of fuselage)
Furnishing group	As position – use similarity
Contingencies	As position – use similarity
MEM	*Compute*
Crew	As position – use similarity
OEM	*Compute*
Payload	As position – use similarity
Fuel	As position – use similarity
MTOM and MRM	*Compute*

■ **TABLE 3.9**
Typical values of component CG locations for the AJT

Item group	Mass (kg)	x (m)	Moment	z (m)	Moment
Fuselage	849	5.2	4414.8	1.5	1273.5
Wing	480	6.8	3264	1.9	912
H-tail	112	11.31	1266.72	1.95	218.4
V-tail	43	10.66	458.38	2.6	111.8
Undercarriage (nose)	80	2.08	166	0.33	26
Undercarriage (main)	220	7.4	1628	0.5	110
Nacelle+pylon	None	–	–	–	–
Power plant	754	9.1	6861.4	1.6	1206.4
Systems	780	5	3900	1.3	1014
Furnishing	65	4	260	1.2	78
Contingencies	97	4	388	1.2	116
MEM	*3480*				
Crew	180	4	720	1.7	306
OEM	*3660*				
Fuel	1100	7.5	8250	1.4	1540
Total NTC mass	*4760*		*≈ 30,667.5*		*≈ 6912*
CG at NTC mass		$\bar{x}=6.64$m		$\bar{z}=1.45$m	
This gives the CG angle, $\beta=\tan^{-1}(7.4-6.64)/1.45=\tan^{-1}0.524=27.65$					
Armament payload	1800	6.8	12,240	1.4	2520
Total MTOM	*6560*		*42,915.5*		*9432*
CG at *MTOM*		$\bar{x}=6.55$m		$\bar{z}=1.44$m	

3.7   Noise Emissions

Noise is perceived as environmental pollution [8–9]. In the 1960s, litigation for damage from aircraft noise caused the government regulatory agencies to reduce noise and impose limits for various aircraft classes. Noise reduction affects aircraft performance. Through research and engineering, significant noise reduction has been achieved despite the increase in engine size that produces several times more thrust.

The US airworthiness requirements on noise are governed by Federal Aviation Regulation (FAR) Part 36 (14CFR36). An aircraft MTOM of more than 12,500 lb must comply with FAR Part 36 (14CFR36). The procedure was immediately followed by the international agency governed by ICAO. The differences between the two standards are minor, and there has been an attempt to combine the two into one uniform standard.

Noise is produced by pressure pulses in air generated from any vibrating source. The pulsating energy is transmitted through the air and is heard within the audible frequency range (i.e. 20–20,000 Hz). The intensity and frequency of pulsation determine the physical limits of human tolerance. In certain conditions, acoustic (i.e. noise) vibrations can affect an aircraft structure. The intensity of sound energy can be measured by the sound pressure level (SPL); the threshold of hearing value is 20 µPa. The response of human hearing can be approximated by a logarithmic scale. The advantage of using a logarithmic scale for noise measurement is to compress the SPL range extending to well over a million times. The unit of noise measurement is a decibel, abbreviated to dB, and is based on a logarithmic scale. One "Bel" is a 10-fold increase in the SPL; that is, $1\,\text{Bel} = \log_{10}10$, $2\,\text{Bel} = \log_{10}100$, and so on. A reading of 0.1 Bel is a dB, which is antilog100.1 = 1.258 times the increase in the SPL (i.e. intensity). A two-fold increase in the SPL is $\log_{10}2 = 0.301$ Bel or 3.01 dB.

Technology required a meaningful scale suitable to human hearing. The units of noise continued to progress in line with technological demands. First was the "A-weighted" scale, expressed in dB(A), that could be read directly from calibrated instruments (i.e. sound meters). Noise is more a matter of human reaction to hearing than just a mechanical measurement of a physical property. Therefore, it was believed that human annoyance is a better measure than mere loudness. This resulted in the "perceived-noise" scale expressed in perceived noise decibel (PNdB), which was labelled as the associated "perceived noise level" (PNL). Aircraft in motion present a special situation, with the duration of noise emanating from an approaching aircraft passing overhead and continuing to radiate rearward after passing. Therefore, for aircraft applications, it was necessary to introduce a time-averaged noise – that is, the effective perceived noise level (EPNL), expressed in EPNdB.

Because existing larger aircraft caused the noise problem, the FAA introduced regulations for its abatement in stages; older aircraft required modifications within a specified period to remain in operation. In 1977, the FAA introduced noise-level standards in three tiers, as follows:

Stage I: Intended for older aircraft already flying and soon to be phased out (e.g. the B707 and DC8). These are the noisiest aircraft but least penalized because they are soon to be grounded.

Stage II: Intended for recently manufactured aircraft that have a longer lifespan (e.g. the B737 and DC9). These aircraft are noisy but must be modified to a quieter standard than Stage I. If they are to continue operating, then further modifications are necessary to bring the noise level to the Stage III standard.

Stage III: Intended for new designs with the quietest standards. ICAO standards are in Annexure 16, Volume I, in Chapters 2–10, with each chapter addressing different aircraft classes. This book is concerned with Chapters 3 and 10 of Annexure 16, which are basically intended for new aircraft (i.e. first flight of a jet aircraft after 6 October 1977, and a propeller-driven aircraft after 17 November 1988). This book is concerned with Chapters 4 as the current rule for jet aeroplanes and large propeller aeroplanes, since it has applied since 2006. The cumulative noise level allowed in Chapter 4 is 10 EPNdB less than for Chapter 3. Also, more stringent rules come into effect in 2017 for MTOW>55 tonnes and 2020 for MTOW<55 tonnes.

Stage IV: Further increased stringency was applied for new aircraft certification during 2006.

Readers may refer to FAR Part 36 (14CFR36) and ICAO Annexure 16 for further details.

3.7.1 Airworthiness Requirements

To certify an aircraft's airworthiness, there are three measuring points in the airport's vicinity to ensure that the neighbourhood is within the specified noise limits. Figure 3.21 shows the distances involved in locating the measuring points, which are as follows:

1. Takeoff reference point: 6500 m (3.5 nm) from the brake release (i.e. starting) point and at an altitude given in Table 3.10.
2. Approach reference point: 2000 m (1.08 nm) before the touchdown point, which should be within 300 m of the runway threshold line and maintain at least a 3° glide slope with an aircraft at least at 120 m altitude.
3. Sideline reference point: 450 m (0.25 nm) from the runway centreline. At the sideline, several measuring points are located along the runway. It is measured on both sides of the runway.

■ **FIGURE 3.21**
Noise measurement points at takeoff and landing

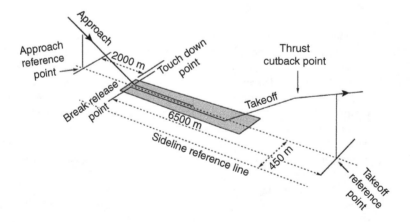

■ **TABLE 3.10**
EPNdB takeoff limits

No. of engines	Two engines	Three engines	Four engines
MTOM (kg)	$\leq 48,100 - 385,000$	$\leq 28,600 - 385,000$	$\leq 20,200 - 385,000$
EPNdB limit	89–101	89–104	89–106
Cutback altitude[a] (m)	300	260	210

[a] Notes: In certain airports, engine throttle cutback (i.e. a lower power setting) is required to reduce the noise level after reaching the altitude shown. At cutback, an aircraft should maintain at least 4% climb gradient. In the event of an engine failure, it should be able to maintain altitude. Make linear interpolations for intermediate aircraft masses.

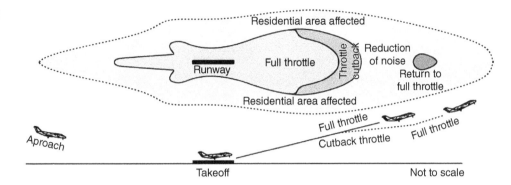

■ **FIGURE 3.22**
Typical noise footprints (≈10km long) of aircraft showing the engine cutback profile

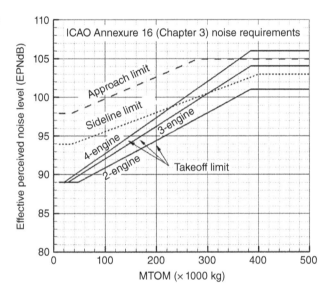

■ **FIGURE 3.23**
ICAO Annexure 16

A typical footprint of the noise profile around a runway is shown in Figure 3.22. The engine cutback area is shown with the reestablished rated thrust for an en-route climb. Residential developments should avoid the noise-footprint areas.

The maximum noise requirements at takeoff in EPNdB, from ICAO, Annexure 16, Volume I, Chapter 3, are listed in Table 3.10 and plotted in Figure 3.23. For the approach, the EPNdB limits do not vary by number of engines, but do vary by aircraft MTOM, ranging from 98 EPNdB for MTOM ≤ 35,000 kg, to 105 EPNdB for MTOM ≥ 280,000 kg. Similarly, for the sidelines the limits are 94 EPNdB for MTOM ≤ 35,000 kg, to 103 EPNdB for MTOM ≥ 400,000 kg. In both cases, linear interpolations should be used to find the limits for intermediate sizes.

It is not only the engines that are responsible for the noise. The airframe also produces a significant amount of noise, especially when an aircraft is in a "dirty" configuration. The noise is aggravated when the undercarriage, flaps and slats are deployed, creating a considerable vortex flow and unsteady aerodynamics; the fluctuation frequencies appear as noise.

Typical noise levels from various sources are shown in Figure 3.24 at both takeoff and landing. Aircraft engines contribute the most noise, which is reduced at landing when the engine power is set low and the jet efflux noise is reduced substantially. There is more noise emanating from the airframe at landing due to higher flap and slat settings, and the aircraft altitude is lower

■ **FIGURE 3.24**
Relative noise
distributions

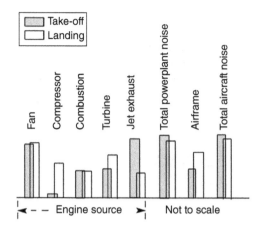

■ **FIGURE 3.25**
In-flight turbofan
noise radiation
profile

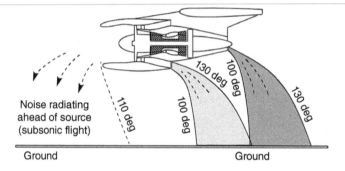

at the measuring point than at the takeoff measuring point. Because the addition of noise level is in a logarithmic scale, the total noise contribution during takeoff and landing is almost at the same level.

The engine and the nacelle are the main sources of noise at takeoff when an aircraft is running at maximum power. All of the gas-turbine components generate noise: fan blades, compressor blades, combustion chamber walls and turbine blades. With an increase in the bypass ratio (BPR), the noise level decreases because a low exhaust velocity reduces the shearing action with ambient air. The difference in noise between an afterburner (AB) turbojet and a high-BPR turbofan can be as high as 30–40 EPNdB. In subsonic flight, noise radiation moves ahead of an aircraft.

To reduce noise levels, engine and nacelle designers must address the sources of noise (Figure 3.25). The goal is to minimize radiated and vibration noise. Candidate areas in engine design are: the fan, compressor and turbine-blade; gaps in rotating components; and, to an extent, the combustion chamber. Power plants are bought-out items for aircraft manufacturers, who must make compromises between engine cost and engine performance in selecting what is available on the market.

Propeller-driven aircraft need to consider noise emanating (radiation and reflection) from propellers. Here the noise reflection pattern depends on the direction of propeller rotation, as can be seen in Figure 3.26. The spread of reflected noise also depends on propeller position with respect to the wing and fuselage.

Inside the aircraft cabin, noise comes from the ECS and must be maintained at the minimum level. These problems are addressed by specialists.

■ **FIGURE 3.26**
Noise considerations
for propeller-driven
aircraft

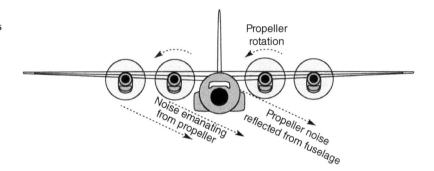

3.7.2   Summary

Aircraft noise reduction is achieved through carefully shaping aircraft and designing engines, but this will affect aircraft performance. Penalties (drag increase and/or loss of thrust) associated with noise reduction should be minimized.

In the near future (i.e. gradually evolving in about two decades), remarkable improvements in noise abatement will be achieved using a multidisciplinary design approach, taking the benefits from various engineering considerations leading to a blended-wing body (BWB) shape. The University of Cambridge and the Massachusetts Institute of Technology have undertaken feasibility studies that show a concept configuration in Figure 2.60 for an Airbus 320 class of subsonic jet commercial-transport aircraft. The engineers predict that the aircraft will be 25 dB quieter than current designs – so quiet as to name it a "silent aircraft". The shaping of the aircraft is not based solely on noise reduction; it is also driven by general aerodynamic considerations (e.g. drag reduction and handling qualities). Noise reduction results from the aircraft body shielding the intake noise, minimizing two-body junctions by blending the wing and the fuselage, and eliminating the empennage. Of course, reduction in engine noise is a significant part of the exercise. However, to bring the research to a marketable product will take time, but the authors believe it will come in many sizes; heavy-lift cargo aircraft are good candidates.

3.8   Engine-Exhaust Emissions

Currently, the civil aviation sector burns about 12% of the fossil fuels consumed by the worldwide transportation industry. It is responsible for an approximate 2–3% annual addition to greenhouse gases and pollutant oxide gases. The environmental debate has become intense on issues such as climate change and depletion of the ozone layer, leading to the debate on long-term effects of global pollution. Smog consists of nitrogen oxide, which affects the pulmonary and respiratory health of humans. The success of the automobile industry in controlling engine emissions is evident by dramatic improvements achieved in many cities. The sustainability of air travel and growth of the industry depends on how technology keeps up with the requirements of human health.

Combustion of air (i.e. oxygen plus nitrogen) and fuel (i.e. hydrocarbon plus a small amount of sulfur) ideally produces carbon dioxide (CO_2), water (H_2O), residual oxygen (O_2) and traces of sulfur particles. In practice, the combustion product consists of all of these plus an undesirable amount of pollutants, such as carbon monoxide (CO, which is toxic), unburned hydrocarbons (UHCs), carbon soot (i.e. smoke, which affects visibility), oxides of sulfur and various oxides of nitrogen (NOX, which affect ozone concentration). The regulations aim to reduce the level of undesirable pollutants by improving combustion technology.

The US Environmental Protection Agency (EPA) recognized these problems decades ago. In the 1980s, the need for government agencies to tackle the engine-emissions issue was emphasized. The early 1990s brought a formal declaration (the Kyoto agreement) to limit pollution (specifically around airports). Currently, there are no regulations for an aircraft's cruise segment. In the US, FAR Part 35, and internationally the ICAO (Annexure 16, Volume II), outline the emissions requirements. The EPA has worked closely with both the FAA and the ICAO to standardize the requirements. Although military aircraft emissions standards are exempt, they are increasingly being scrutinized for MILSPECS standards. Emissions are measured by an emission index (EI).

Lower and slower flying reduces the EI; however, this conflicts with the market demand for flying higher and faster. Designers must make compromises. Reduction of the EI is the obligation of the engine manufacturer; therefore, details of the airworthiness EI requirements are not provided herein (refer to the respective FAA and ICAO publications). Aircraft performance engineers must depend on engine designers to supply airworthiness-certified engines.

3.9   Aircraft Systems

The aircraft control system, engine bleed system and power off-take system all affect aircraft performance, and are introduced in this section. Thrust loss associated with engine bleed and power off-take systems are quantified in Chapter 6.

3.9.1   Aircraft Control System

Aircraft performance is control-dependent. Figure 1.17 is a Cartesian representation of the six degree of freedom, consisting of three linear and three rotational motions. This section describes the associated control hardware and design considerations.

The three axes (i.e. pitch, yaw and roll) of aircraft control have evolved considerably. The use of trim tabs and aerodynamic and mass balances alleviates hinge moments of the deflecting control surfaces, which reduces a pilot's workload. Some operational types are as follows:

1. *Wire-pulley type*: This is the basic type. Two wires per axis act as tension cables, moving over low-friction pulleys to pull the control surfaces in each direction. Although there are many well-designed aircraft using this type of mechanism, it requires frequent maintenance to check the tension level and the possible fraying of wire strands. If the pulley has improper tension, the wires can jump out, making the system inoperable. Other associated problems include dirt in the mechanism, the rare occasion of jamming, and the elastic deformation of support structures leading to a loss of tension. Figure 3.27 shows the wire-pulley (i.e. rudder and aileron) and push-pull rod (i.e. H-tail) types of control linkages.

2. *Push-pull rod type*: The problems of the wire-pulley type are largely overcome by the use of push-pull rods to move the control surfaces. Designers must ensure that the rods do not buckle under a compressive load. In general, this mechanism is slightly heavier and somewhat more expensive, but it is worth installing for the ease of maintenance. Many aircraft use a combination of push-pull rod and wire-pulley arrangements (see Figure 3.27).

3. *Mechanical control linkage boosted by a power control unit (PCU)*: As an aircraft size increases, the forces required to move the control surfaces increase to a point where a pilot's workload exceeds the specified limit. Power assistance by a PCU resolves this

■ **FIGURE 3.27**
Wire-pulley and
push-pull rod control
systems. Reproduced
with permission of
the Europa Aircraft
Company

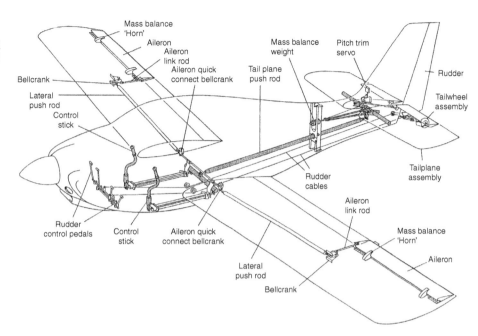

problem. However, one problem of using a PCU is that the natural feedback "feel" of control forces is obscured. Therefore, an artificial feel is incorporated for finer adjustment, leading to smoother flights. PCUs are either hydraulic or electric motors driven by linear or rotary actuators (there are several types).

4. *Electromechanical control system*: In larger aircraft, considerable weight can be saved by replacing mechanical linkages with electrical signals to drive the actuators. Aircraft with FBW use this type of control system. Currently, many aircraft routinely use secondary controls (e.g. high-lift devices, spoilers and trim tabs) driven by electrically signalled actuators.

5. *Optically signalled control system*: This latest innovation uses an optically signalled actuator. Advanced aircraft already have fibre-optic lines to communicate with the control system.

Modern aircraft systems, especially the combat-aircraft control system, have become very sophisticated. A FBW architecture is essential to these complex systems so that an aircraft can fly under relaxed stability margins. Enhanced performance requirements and safety issues have increased the design complexities by incorporating various types of additional control surfaces. Figure 3.28 shows the typical subsonic aircraft control surfaces.

Figure 3.29 shows the various control surfaces and areas as well as the system retractions required for a three-surface configuration. As shown in the figure, there is more control than that required by most modern civil aircraft. Military aircraft control requirements are at a higher level due to the demand for difficult manoeuvres and a possible negative stability margin. The F117 is incapable of flying without FBW. Additional controls are the canard, intake-scheduling and thrust-vectoring devices.

Fighter aircraft may use stabilators (e.g. the F15) in which the elevators can move differentially to improve roll capability. Stabilators are used collectively for pitch control and differentially for roll control. Also, the aileron and rudder can be interconnecting. There can also be automatic control that parallels the basic system.

■ **FIGURE 3.28**
Civil aircraft control surfaces

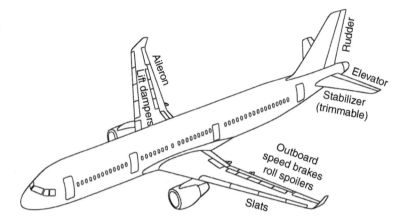

■ **FIGURE 3.29**
Military aircraft control surfaces

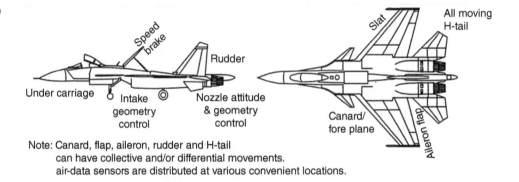

Note: Canard, flap, aileron, rudder and H-tail
      can have collective and/or differential movements.
      air-data sensors are distributed at various convenient locations.

3.9.2   ECS: Cabin Pressurization and Air-Conditioning

At cruise altitude, the atmospheric temperature drops to −50°C and below, and the pressure and density reduce to less than 20% and 25% of sea-level values, respectively. Above 14,000 ft, the aircraft interior environment must be controlled for crew and passenger comfort as well as equipment protection. The cabin ECS consists of cabin pressurization and air-conditioning. Smaller, unpressurized aircraft flying below 14,000 ft suffice with air-conditioning only; the simplest form uses engine heat mixing with ambient cold air supplied under controlled conditions. Hot air from the turbofan compressor bleed controls the cabin environment (Figure 3.30), for example, cabin pressure and air-conditioning. Reduction of air mass from engine aspiration reduces the net thrust and degrades specific fuel consumption (sfc).

Maintaining the cabin interior pressure at sea-level conditions is ideal but expensive. Cabin pressurization is like inflating a balloon; the fuselage skin bulges. The major differential between the outside and the inside pressure requires structural reinforcement, which makes an aircraft heavier and more expensive. For this reason, the aircraft cabin pressure is maintained no higher than 8000 ft, and a maximum differential pressure is maintained at 8.9 psi. During ascent, the cabin is pressurized gradually; during descent, cabin depressurization is also gradual in a prescribed schedule acceptable to the average passenger. Passengers feel it in their ears as they adjust to the change in pressure.

Cabin air-conditioning is an integral part of the ECS, along with cabin pressurization. Supplying a large passenger load at a uniform pressure and temperature is a specialized

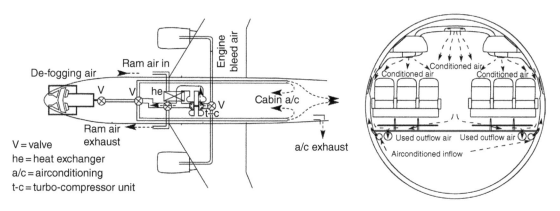

■ **FIGURE 3.30** Engine bleed – environmental control system

■ **FIGURE 3.31**
Military aircraft ECS

design obligation. The engine compressor, which is bled at an intermediate stage with sufficient pressure and temperature, becomes contaminated and must be cleaned, with moisture removed to an acceptable level. Maintaining a proper humidity level is also part of the ECS. The bled-air is then mixed with cool ambient air. In addition, there is a facility for refrigeration.

The avionics black boxes heat up and must be maintained at a level that keeps equipment functioning. The equipment bay is below the floor-boards. Typically, a separate cooling system is employed to keep the equipment cool. Ram-air cooling is a convenient and less expensive way to achieve the cooling. Scooping ram air increases the aircraft drag. The cargo compartment also requires some heating.

An advanced military-aircraft ECS differs significantly (Figure 3.31), using a boot-strap refrigeration system, which has recently also been deployed in civil aircraft applications.

3.9.3 Oxygen Supply

If there is a drop in cabin pressure while an aircraft is still at altitude, the oxygen supply for breathing becomes a critical issue. The aircraft system supplies oxygen to each passenger by dropping masks from the overhead panel. Military aircraft have fewer crew members, and the oxygen supply is directly integrated into a pilot's mask, as shown in Figure 3.31.

3.9.4 Anti-icing, De-icing, Defogging and Rain Removal System

Icing is a natural phenomenon that occurs anywhere in the world depending on weather conditions, operating altitude and atmospheric humidity. Ice accumulation on the wing, empennage and/or engine intake can have disastrous consequences. Icing increases the drag and weight, decreases

■ **FIGURE 3.32**
Anti-icing envelopes

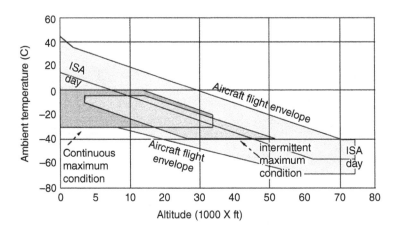

the lift and thrust, and even degrades control effectiveness. On the wing and empennage, icing alters the profile geometry, leading to loss of lift. Ice accumulation at the intake degrades engine performance and can damage the engine if large chunks are ingested. It is a mandatory requirement to keep an aircraft free from icing degradation. This can be achieved by either anti-icing, which never allows ice to form on critical areas, or by de-icing, which allows ice buildup to a point and then sheds it before it becomes harmful. De-icing results in blowing away chunks of ice, which could hit or be ingested into an engine.

Figure 3.32 shows the typical anti-icing envelopes. Section 18.6.2.1 describes several methods for anti-icing and de-icing.

3.10 Low Observable (LO) Aircraft Configuration

A fighter aircraft with low observable (LO) configuration characteristics for stealth design will have to make compromises to performance (aerodynamic and manoeuvre) affecting weapon-carrying capability, thereby limiting combat effectiveness. Aircraft performance engineers will have to make tradeoffs to maximize combat capability. A weapon load of conventional design will not offer stealth. Stealth parameters to be considered include heat and noise emissions, a radar signature and simple visibility.

3.10.1 Heat Signature

An infra-red seeking homing device is a potent method as long as there is a single, clear, identifiable target emanating sufficient signal, such as from the engine exhaust. The drawback is that a missile can be easily fooled by spraying out many heat flares. Missiles aiming downward at targets or facing the sun could lock on to a stationary target elsewhere within its capture angle.

Aircraft designers should aim to reduce the aircraft heat signature below 350°C by mixing engine exhaust with cool atmospheric air through entrainment at the exhaust. Shielding of engine exhaust by its wing is an effective method against ground-launched heat-seeking missiles.

3.10.2 Radar Signature

The radar system works by capturing the reflected returns of its transmitting radio waves from an object. If the object is a moving aircraft, the radar technology adjusts with the Doppler shift phenomenon to give an accurate position of the aircraft; its speed, altitude and range are given in real time, with an alarm and/or homing signal. The fundamental objective of radar stealth is to reduce the radar cross-section (RCS) area; that is, to reduce the echo strength so that it is noticed much later, thereby reducing the reaction time of the adversary.

RCS area is defined as the projected area of an equivalent perfect reflector with uniform properties in all directions, such as from a sphere, and which returns the same amount of power per unit solid angle in steradians as the object under consideration.

The intensity of the reflected radar beam (echo) depends on the surface on which it is reflecting. The parameters influencing reflection are area, orientation, and the nature of the surface. The maximum is when the surface is normal to it and has a large areal extent. Figure 3.33 compares echoing from a sphere to a pointed sharp corner of inclined surfaces.

Even a small sphere would offer a larger normal (and near-normal) surface compared with a point such as the tip of a nose cone. The inclined surfaces would deflect the reflected beam away. In addition, if the surface is coated with radiation-absorbing paints, then the echo strength would further reduce. Radar-absorbing coatings are heavy, difficult to maintain, and increase costs.

Earlier designs, for example the F117 Nighthawk, had inclined flat plate-like surfaces with sharp edges which succeeded in radar signature reduction, but evidently was not sufficient as an F117 was shot down by a missile in the Kosovo conflict. The stealth configuration of the time was aerodynamically inefficient – nicknamed an "aerodynamicists' nightmare". The B2 bomber showed improvement with a more streamlined shape, with the engine intake and exhaust over the wing shielding the hot zones against heat signature. The latest F22 Raptor is a fine example of improvement in shaping, incorporating better streamlining, cutting down on drag. Figure 3.34 compares configurations, and Table 3.11 compares a few combat aircraft RCS values.

■ **FIGURE 3.33**
Typical comparisons of radar signatures (sphere versus stealth aircraft)

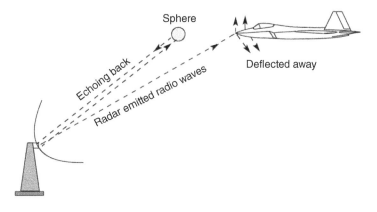

■ **FIGURE 3.34**
Three stealth aircraft configurations

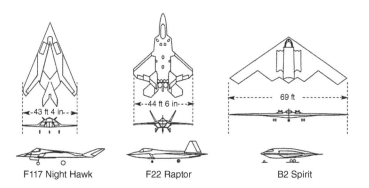

■ **TABLE 3.11**
RCS values of combat aircraft

Aircraft type (older designs)	RCS area (m²)	Aircraft type (newer designs)	RCS area (m²)
F15 Eagle	40.5	F117 Nighthawk (1970s)	0.003
B1 Bomber	10	B2 Spirit (1980s)	0.0014
SR-71	0.01	F22 Raptor (1990s)	0.0065

All modern combat aircraft designs will have a specification for maximum RCS area. Therefore, design of modern combat aircraft without assessing the RCS area requirement is meaningless. Computing RCS area is not difficult but time-consuming, making it unsuitable for undergraduate classroom work. There are software which can measure RCS area and inter-actively tailor aerodynamic surfaces with minimum compromise. Currently, there is no such software available in the public domain.

The downing of the F117 in Kosovo opened an argument on the extent of stealth effectiveness. In addition, stealth features degrade aircraft performance and handling. Advances in missile technology would make stealth technology less effective, but currently stealth is still a desirable feature that designers must exploit. The F117 is an older design, and the stealth technology is also advancing. Aircraft designers have a difficult task to compromise between stealth, performance and cost in a changing environment with newer technologies. The results from future combats will resolve some of the controversies.

References

1. Kundu, A.K., *Aircraft Design*. Cambridge University Press, New York, 2010.
2. Society of Allied Weight Engineers, *Introduction to Aircraft Weight Engineering*. Los Angeles, CA, 1996.
3. Society of Allied Weights Engineers, *Weight Engineer's Handbook*. Los Angeles, CA, 2002.
4. Niu, M., *Airframe Structural Design*. Commlit Press Ltd., Hong Kong, 1999.
5. Torenbeek, E., *Synthesis of Subsonic Airplane Design*. Delft University Press, Amsterdam, 1982.
6. Roskam, J., *Aircraft Design*, 1985–1990. University of Kansas, Lawrence, Kansas, 1990.
7. Jenkinson, L.R., Simpson, P., and Rhodes, D., *Civil Jet Aircraft Design*. Arnold Publishers, London, 1999.
8. Smith, M.J.T., *Aircraft Noise*. Cambridge University Press, Cambridge, 1989.
9. Ruijgrok, G.J.J., *Elements of Aviation Acoustics*. Eburon, Delft, 2003.

CHAPTER 4

Equations of Motion for a Flat Stationary Earth

4.1   Overview

A conventional fixed wing aircraft must have some motion, a minimum velocity to sustain flight (see Section 2.6). Aircraft motion has six degrees of freedom (see Section 1.15), and high-performance combat aircraft motion can be very complex. This chapter is devoted to deriving pertinent equations of motion for use in aircraft performance. Aircraft with a large aspect ratio show elastic deformation and are designed in such a way that deformation at normal cruise conditions does not penalize the aircraft performance to significant extent. Deformation affects structural and control issues.

The mathematical relationships are established using vector/matrix algebra (the basics are reviewed in Appendix C). This chapter adheres closely to the well-established textbook *Dynamics of Atmospheric Flight* by Bernard Etkin [1], meant for aircraft stability and control analysis. Since this book is meant for aircraft performance, some changes are incorporated to suit the context.

Terrestrial flights take place over the round and rotating Earth. Introducing equations of motion for a round and rotating earth during the first course of aircraft performance to under-graduates may prove to be an arduous task; these equations include terms for Coriolis and centripetal forces resulting from the Earth's shape and motion [1]. The following reasoning justifies that aircraft rotational and curvature effects can be approximated to a flat stationary Earth for conventional aircraft without incurring appreciable error.

The Earth's angular velocity is about 7×10^{-10} rad/s, which is very small compared with aircraft rotation capability. Because the Earth's rotation is relatively slow, it yields a low level of Coriolis force. The Coriolis force at aircraft speeds of about 2600 ft/s (this is above Mach 3.0 and at altitude above 80,000 ft) is of the order of 1% of the aircraft's weight. The Earth's curvature is also relatively small compared with aircraft rotation capability about its own centre of gravity (CG). The other contribution is the centripetal force from aircraft angular velocity arising from negotiating Earth's curvature (given aircraft positional latitude and longitude). The centripetal force at aircraft speed of at about 2600 ft/s is also of the order of 1% of the aircraft's weight. These contribute to small errors for speeds less than Mach 2.5 and

Theory and Practice of Aircraft Performance, First Edition. Ajoy Kumar Kundu, Mark A. Price and David Riordan.
© 2016 John Wiley & Sons, Ltd. Published 2016 by John Wiley & Sons, Ltd.

altitude less than 80,000 ft, which are hence ignored. Currently, all aircraft powered by air-breathing engines fly at speeds below these limits, except for the Lockheed Blackbird (YF12) and MIG 31.

In this chapter the simple equations of motion for a flat and stationary Earth are derived.

4.2   Introduction

The motion of an object has a time-varying position, measured as a vector quantity from a point of reference – normally the origin of a reference coordinate frame. The motion of planets, satellites, spacecraft and aircraft require some knowledge of reference frames with respect to which their position vector is measured.

To keep derivation manageable, the following assumptions/considerations are necessary to formulate the equations. The error involved through these assumptions is low, sufficient for industry-standard practice and applicable to high-performance aircraft. This book uses a right-handed Cartesian coordinate system. Vector quantities are shown in bold italics, for example, the unit vectors are i, j and k along X, Y and Z axes, respectively. The six degrees of motion consist of three linear and three angular motions along the three axes of the Cartesian coordinate system. The assumptions are:

1. All manmade vehicles have a plane of symmetry, in which the CG lies. The position vector joins the CG and the origin of the reference frames.
2. Aircraft are reduced to a point mass at its CG that can rotate and translate. The aircraft shape is retained to indicate attitude. (For point mass, $I_{xx} = I_{yy} = I_{zz} = 0$.)
3. All aircraft are treated as rigid bodies – aeroelastic deformations are not considered at this stage. Also not considered are the moving components, such as the propeller. These are primarily concerned with aircraft control and therefore not dealt with in this context.
4. Zero wind velocity, that is, the aircraft moves in stationary air.
5. Section 2.2 showed that acceleration due to gravity g changes less than 1% within 30 km altitude. Therefore, the value of g is kept constant at 9.81 m/s² (32.2 ft/s²) and pressure is maintained up to 30 km ($\approx$100,000 ft) altitude. The Earth is considered spherical.

4.3   Definitions of Frames of Reference (Flat Stationary Earth) and Nomenclature Used

Given below are the pertinent definitions and nomenclature for the frames of reference. The first subscripts denote the frame and can have more than one letter in subscripts. The last subscripts denote the axis. Figure 4.1 shows the frames of reference.

1. *Inertial frame, F_I:* The root frame of any system is an inertial frame, which is stationary in deep space. Every system must have an inertial frame (it is sometimes not shown). Newton's Laws are evaluated in the inertial frame. For terrestrial flight, the origin of the inertial frame is taken at the centre of the Earth with the Z-axis passing through the North Pole. It does not rotate with the Earth; it is stationary in space. For a flat and stationary Earth, the inertial frame is on the surface with the Z-axis vertically upward.

2. *Earth fixed frame, F_E and F_{EC} (Figure 4.1a):* For a round, rotating Earth, the Earth fixed reference frame F_{EC} has its origin at the centre of the Earth with the Z-axis pointing towards the North Pole and the XY plane in the equatorial plane. F_E is fixed on the surface of the Earth with the Z-axis passing downwards collinear to the *g* vector; the X-axis points east and the Y-axis points north. Both F_{EC} and F_E have the angular velocity of the Earth, Ω in F_I. For a flat and stationary Earth, the Earth fixed reference frame, F_E, is the same as the inertial frame, F_I, but the Z-axis passes downward along the *g* vector (F_{EC} is not applicable in this case). Typically, the origin of F_E is placed at the airport from which the aircraft takes off; the X-axis points North and the Y-axis points east. Then the axes of F_E are parallel to the inertial frame, F_I, and have the Z-axis in the opposite direction. If required it can conveniently point along a suitable direction, such as a runway or a flight direction.

3. *Vehicle carried frame, F_V (seen as vertical frame):* Fixed (gimballed) to a vehicle CG with the Z-axis pointing to the centre of the Earth, that is, along the *g* vector, the X-axis pointing north and the Y-axis pointing east, irrespective of aircraft attitude. The X-Y plane is the local horizontal plane and moves with the aircraft. Therefore it is also seen as the vertical frame. Note that when the vehicle is directly above F_E, the frames become parallel to each other. An observer in F_V is external to the aircraft; together with F_E they assist aircraft navigation and guidance. To note that for a flat stationary Earth, F_E and F_V are parallel to each other, the latter moving with the aircraft while the former is fixed to the Earth.

4. *Body fixed frame, F_B:* The origin is fixed (Figure 4.1b) to the vehicle CG with the X-axis pointing forward. A suitable choice for the X-axis is to align with the fuselage axis. Vehicles with a plane of symmetry will have the Z-axis pointing downwards of the vehicle (it turns upward when the vehicle is flying inverted), still in the plane of symmetry. Being fixed (nailed) to the aircraft, it has a fixed relationship with the wing mean aerodynamic chord (MAC). Typically, the angle between the MAC and the X-axis in F_B is small and about the same as the wing incidence angle with the aircraft fuselage. Movement of flap, elevator, and so on, can be measured in F_B. Aircraft mechanical and geometric properties and attitudes are evaluated in this frame. These are useful to

■ **FIGURE 4.1**
Frames of reference.
(a) Earth frames, F_{EC}
and F_E. (b) Body fixed
frame, F_B

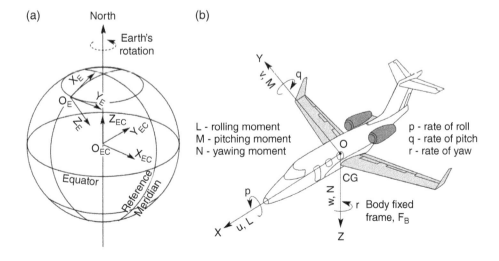

(a)

North

Earth's rotation

Z_{EC}

O_E

O_{EC}

X_{EC}

Y_{EC}

X_E

Y_E

Equator

Reference Meridian

(b)

Y
v, M
q

L - rolling moment
M - pitching moment
N - yawing moment

p - rate of roll
q - rate of pitch
r - rate of yaw

O

CG

p

w, N

r Body fixed
frame, F_B

X u, L

Z

compute vehicle kinetics. The difference between F_B and F_V are the Eulerian angles (Figure 4.2, see next section) of azimuth (ψ), elevation (θ) and bank (ϕ), to be executed in the prescribed order. An observer in the F_B frame is inside the aircraft like the pilot, who can see the horizon (Eulerian angles) changes with aircraft manoeuvre.

5. *Principle frame, F_P:* This is a special set of body axes when the body fixed frame, F_B, coincides with the principle axes (moment of inertia) of the aircraft. Aircraft manufacturers supply the principle axes for a few known weights. There is a fixed relationship between F_B and F_P

6. *Stability frame, F_S:* This is a special set of body axes used to analyse small perturbations from equilibrium flight. At symmetrical (steady level) flight, F_S and F_W are the same. In this book, aircraft F_S and F_P are not used.

7. *Wind frame, F_W:* In this frame, the origin is at the CG, the X-axis is along the vehicle velocity vector, the Z-axis is pointing down, and the Y-axis is in the right-handed system (over the right wing). This is also carried with the aircraft, but can swivel around the CG and therefore should not be seen as a type of body axis. Aircraft motion in the vertical plane or in the horizontal plane has the Z-axis in the plane of symmetry. If the aircraft has both angle of attack and yaw angle, then the Z-axis is not in the plane of symmetry. From the difference between F_B and F_W, the angles of attack (α), yaw (β) and roll (ϕ) can be found out (see Figure 4.2). The origin of F_W traces the aircraft trajectory in the Earth fixed frame. The X-axis is always tangential to the aircraft trajectory. The aircraft aerodynamic characteristics, for example lift, drag, velocities, and so on, are assessed in F_W. The velocity of the aircraft is measured with respect to some stationary frame (e.g. F_I, F_{EC} or F_E). Figure 4.2 gives the angles associated with F_W and F_B.

8. *Atmosphere fixed frame, F_A:* This concerns the case where there is wind velocity – it is not used in this book.

Aircraft with a plane of symmetry have the advantage of keeping the equations of motion uncoupled into longitudinal and lateral motions. Forces (lift, drag and thrust: $\boldsymbol{F} = m\boldsymbol{a}$ in translational motion) are convenient to be dealt in F_W. Moment equations ($\boldsymbol{M} = I\dot{\omega}$ in rotational motion) are more conveniently treated in F_B, in which moments of inertia remain constant. A longitudinal equation of motion takes the same form in both F_W and F_B; therefore F_W is used, while lateral equations of motion are expressed in both F_W and F_B.

In summary: F_E is fixed to the ground. F_V, F_W and F_B have their origins at the aircraft CG and move with it. F_V is the vertical frame and gives the local horizon, that is, its Z-axis is always

■ **FIGURE 4.2**
Angles associated with wind and body axes

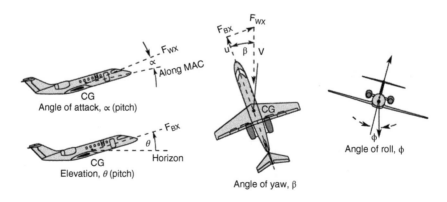

collinear with the g vector; the X-axis of F_B is fixed to the aircraft giving aircraft mechanical characteristics; and for F_W it is aligned with aircraft velocity vector to give aircraft aerodynamic characteristics. These are the four frames of interest in this book – together they assist in estimating aircraft performance.

4.3.1 Notation and Symbols Used in this Chapter

Kinematics and kinetic relationships are derived in the inertial frame. For extended analyses these relationships need to be expressed in different frames of reference. The words "observed" and "expressed" mean the same as equations expressed in the referred frame, and "transformed" means translated from one frame to another. The notations and symbols adhere closely to the well-established textbook *Dynamics of Atmospheric Flight* [1].

One of the difficulties in studying the equations of motion in various frames of reference and their transformation from one frame system to another is the use of a complex notation system. It uses subscripts and superscripts in series and needs to be used with care. For clarity it is outlined below. Vectors are represented by bold letters, matrices in brackets [] and determinants by | |. Given below is the nomenclature used in this chapter. For a flat stationary Earth, the angular velocity relation between F_E and F_V is zero.

Velocities and accelerations (the magnitudes of the vector quantities are given in the scalar form)

V = aircraft inertial velocity along the X-axis of F_W.

u = aircraft inertial velocity in F_B along the X-axis.

v = aircraft inertial velocity in F_B along the Y-axis.

w = aircraft inertial velocity in F_B along the Z-axis.

a = aircraft inertial acceleration.

p = X-component of inertial angular velocity of moving frames F_W and F_B.

q = Y-component of inertial angular velocity of moving frames F_W and F_B.

r = Z-component of inertial angular velocity of moving frames F_W and F_B.

ω = angular velocity of aircraft, with subscripts indicating the reference frame.

Forces and moments

L = aircraft lift along the negative direction of the Z-axis in F_W.

D = aircraft drag vector along the negative direction of the X-axis in F_W.

T = installed engine thrust, in general, approximated along the X-axis in F_B.

W = aircraft weight along the Z-axis in F_V, that is, along the g vector.

L = aircraft rolling moment in F_B along the X-axis (this is not the lift in F_W).

M = aircraft pitching moment in F_B along the Y-axis.

N = aircraft yawing moment in F_B along the Z-axis.

In frames of references

X, Y and Z are reserved to represent axes. The first subscript is in which frame it is expressed. E, V, W and B represent the four frames of reference of interest. The components in the named axis come as the last subscript. For example:

F_{BX} = X-axis of body fixed frame, F_B.

In the inertial frame F_I, the subscript refers to the point of interest, as follows (subscript I may be dropped).

V_C = inertial velocity of point C in the inertial frame.
V_{CW} = inertial velocity of point C expressed in F_W (three components along the axes of F_W).

Invariant and independent vectors such as acceleration due to gravity, $\boldsymbol{g}$, are expressed with respect to a frame of reference as follows:

$\boldsymbol{g}_V$ = acceleration due to gravity in frame F_V, always pointing down along its Z-axis.

The magnitude $\boldsymbol{g}$ is the same in all the frames of reference but the magnitude of the components may differ.
Superscripts are also applied as notation. Their usage is shown below.

ω^B = angular velocity of body frame F_B in inertial space.
V^B = aircraft velocity of body frame F_B in inertial space.

Aircraft velocity in F_V will be different in F_B. The combination of subscripts and superscripts offers four possibilities as shown below. Each has three components with respective subscripts.

$V^V{}_V$ = aircraft velocity of F_V as observed (transformed) in vehicle frame, F_V.
$V^V{}_B$ = aircraft velocity of F_V as observed (transformed) in body frame, F_B.
$V^B{}_V$ = aircraft velocity of F_B as observed (transformed) in vehicle frame, F_V.
$V^B{}_B$ = aircraft velocity of F_B as observed (transformed) in body frame, F_B.

Eulerian angles play an important role in equations of motion. The basic concept of Eulerian angles is given in Section 4.4.

4.4   Eulerian Angles

The Eulerian angles describe aircraft attitudes. Note that the angular displacements are not vector quantities – the order of rotation is important. Angles cannot rotate arbitrarily as they do not obey the commutative law. Angular velocities are angular displacement rates (in the limiting sense) and can be represented as vector quantities. It was mentioned earlier that F_B gives the vehicle attitude with respect to the local horizon of F_V. To coincide with F_B, three distinct rotations of F_V are given in the prescribed manner known as Eulerian angles (Figure 4.3). The prescribed rotations are described below. In Figure 4.3, the dots denote the axes of rotation, with the number in brackets showing the order of rotation.

First rotation about Z-axis (Figure 4.3b): ψ in the horizontal plane ($-\pi \leq \psi < \pi$) gives the azimuth angle (+ right). The rate of change of ψ is the yaw rate r, or $r = \dot{\psi}$.
Second rotation about Y-axis (Figure 4.3c): θ in the vertical plane ($-\pi/2 \leq \theta \leq \pi/2$) gives the elevation of the pitch angle (+ up). The rate of change of θ is the pitch rate q, or $q = \dot{\theta}$.
Third rotation about X-axis (Figure 4.3d): ϕ in the residual plane ($-\pi \leq \phi < \pi$) gives the bank angle (+ right). The rate of change of ϕ is roll rate p or $p = \dot{\phi}$.

Angular velocities will be the same as observed in all non-rotating reference frames. It was mentioned earlier that for the component of any vector along the axis, the first subscript denotes the frame and the last subscript denotes the axes, and can have more than one letter. In general, the Eulerian angles are between any two frames, and not necessarily meant only for attitude indication.

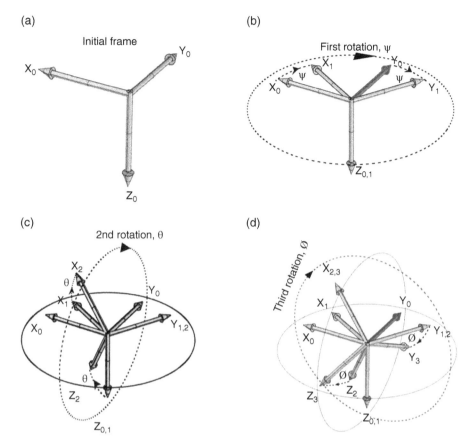

■ FIGURE 4.3   Eulerian angles. (a) Initial frame 0 (subscript, 0). (b) First rotation ψ about axis Z_0 kV to reach frame F_1 with subscript 1. (c) Second rotation θ about axis $Y_{1,2}$ reach frame F_2 with subscript 2. (d) Third and final rotation φ about axis $X_{2,3}$ to reach final frame F_3 with subscript 3

4.4.1   Transformation of Eulerian Angles

Frequently aircraft physical properties (e.g. velocities, forces, and so on,) need to be expressed in different frames of references. Given below is the transformation matrix $[L_{RV}]$ to transform properties from the vehicle frame F_V to the body frame F_B (see textbooks for derivation).

If the initial frame is F_V and the final frame is F_B, then, say, velocity in F_V can be expressed in F_B as

$$[V_B] = [L_{BV}][V_V]$$

where $[L_{BV}]$ is the transformation matrix as follows:

$$[L_{BV}] = \begin{pmatrix} \cos\theta\cos\psi & \cos\theta\sin\psi & -\sin\theta \\ \sin\phi\sin\theta\cos\psi - \sin\psi\cos\phi & \cos\psi\cos\phi + \sin\theta\sin\psi\sin\phi & \cos\theta\sin\phi \\ \sin\psi\sin\phi + \sin\theta\cos\psi\cos\phi & \sin\theta\sin\psi\cos\phi - \sin\phi\cos\psi & \cos\theta\cos\phi \end{pmatrix} \quad (4.1)$$

To transform coordinates the other way, the inverse transformation matrix is used as $[V_V] = [L_{VB}]^{-1}[V_B]$. This comes from the properties of the matrix:

$$[L_{BV}][L_{VB}]^{-1} = [I] \tag{4.2}$$

The benefit of matrix notation and its algebraic manipulations is evident as large expressions can be written compactly. It is the same procedure to transform from wind axes, in the F_W frame.

EXAMPLE 4.1

Take the acceleration vector due to gravity, g, which is positive along the Z axis in the F_V frame. How is it seen in an aircraft (Eulerian angles) at an attitude seen in body axes, F_B?

SOLUTION

In the F_V frame the g vector can be written as: $[g_V] = \begin{pmatrix} 0 \\ 0 \\ g \end{pmatrix}$. This is what an external observer sees. Therefore $[g_B] = [L_{BV}][g_V]$ where

$$[g_B] = \begin{pmatrix} \cos\theta\cos\psi & \cos\theta\sin\psi & -\sin\theta \\ \sin\phi\sin\theta\cos\psi - \sin\psi\cos\phi & \cos\psi\cos\phi + \sin\theta\sin\psi\sin\phi & \cos\theta\sin\phi \\ \sin\psi\sin\phi + \sin\theta\cos\psi\cos\phi & \sin\theta\sin\psi\cos\phi - \sin\phi\cos\psi & \cos\theta\cos\phi \end{pmatrix} \begin{pmatrix} 0 \\ 0 \\ g \end{pmatrix}$$

Therefore g, in body axes, becomes

$$[g_B] = g \begin{pmatrix} -\sin\theta \\ \cos\theta\sin\phi \\ \cos\theta\cos\phi \end{pmatrix} \quad \text{(this is what the pilot feels)} \tag{4.3}$$

EXAMPLE 4.2

This is an example of transformation between wind axes and body axes. An aircraft is in steady level flight bearing due north ($\phi_w = \psi_w = \gamma_w = 0$) at a speed of 100 m/s. The angle of attack $\alpha = \theta_B = 5°$. Calculate u_B, v_B and w_B.

SOLUTION

In wind axes F_w, the velocity vector is given by (100, 0, 0). The body axis system is F_B in which the velocity vector is (u_B, v_B, w_B). The angles are between any two frames and therefore F_B is displaced relative to F_w by rotations $(\phi, \theta, \psi) = (0, 5, 0)$. Hence, using Equation 4.1:

$$\begin{pmatrix} u_B \\ v_B \\ w_B \end{pmatrix} = \begin{pmatrix} \cos5 & 0 & -\sin5 \\ 0 & 1 & 0 \\ \sin5 & 0 & \cos5 \end{pmatrix} \begin{pmatrix} 100 \\ 0 \\ 0 \end{pmatrix} = \begin{pmatrix} 99.6 \\ 0 \\ 8.7 \end{pmatrix} \text{ m/s}$$

EXAMPLE 4.3

A navigational aid shows an aircraft in the vehicle frame F_V is at a steady flight speed of 100 m/s moving north and climbing straight with wings level at a vertical speed of 5 m/s. The pilot is flying the aircraft with Euler angles ($\phi = 6°$, $\theta = 4°$, $\psi = 5°$). Calculate the components of the aircraft velocity along each of the body axes directions.

SOLUTION

Let F_V be the velocity in vehicle axes given by (100, 0, −5). The body axis system F_B has velocity vector (u_B, v_B, w_B). F_V displaced relative to F_B by rotations (ϕ, θ, ψ) = (6, 4, 5).

$$\begin{pmatrix} u_B \\ v_B \\ w_B \end{pmatrix} = \begin{pmatrix} \cos4\cos5 & \cos4\sin5 & \sin4 \\ \sin6\sin4\cos5 - \cos6\sin5 & \sin6\sin4\sin5 + \cos6\cos5 & \sin6\cos4 \\ \cos6\sin4\cos5 + \sin6\sin5 & \cos6\sin4\sin5 - \sin6\cos5 & \cos6\cos4 \end{pmatrix} \begin{pmatrix} 100 \\ 0 \\ 5 \end{pmatrix}$$

$$\begin{pmatrix} u_B \\ v_B \\ w_B \end{pmatrix} = \begin{pmatrix} 0.9938 & 0.0869 & -0.0698 \\ -0.0794 & 0.0014 & 0.1043 \\ 0.0782 & -0.0981 & 0.9921 \end{pmatrix} \begin{pmatrix} 100 \\ 0 \\ -5 \end{pmatrix} \quad \text{or} \quad \begin{pmatrix} u_B \\ v_B \\ w_B \end{pmatrix} = \begin{pmatrix} 99.73 \\ -8.46 \\ 2.86 \end{pmatrix}$$

In fact, vectors can be transformed from any frame to any frame, even transitioning through one or more intermediate frames. In missile applications when a missile is fired from an aircraft at an odd attitude towards a flying target based on data acquisition from another airborne platform (AWACS), the engineer uses various frames of reference for a beyond visual range (BVR) combat engagement. These are more pertinent to control and guidance studies. Vehicle navigation (especially in space application) also uses transformation matrices.

4.5 Simplified Equations of Motion for a Flat Stationary Earth

In this section, the simplified equations of motion are derived for a flat stationary earth. First, the equations of motion in the pitch plane are formulated, followed by derivation in the horizontal plane and finally for 3D motions involving both the planes. The basic equilibrium equations are given in Section 1.15.2. Here is it formally derived in a classical manner. It is suggested readers may go through this section thoroughly as these equations are the basis of performance equations used in this book.

4.5.1 Important Aerodynamic Angles

This section lists the aerodynamic angles associated with aircraft performance (Figure 4.4).

α, angle of attack. This is the angle between the X-axes of F_W and F_B plus the wing-body angle.

γ, pitch (climb) angle. This is the angle between the X-axis of F_W and the horizon, that is, $\theta_W = \gamma$.

θ, elevation angle, an Eulerian angle. This is the angle between the X-axis of F_B and the horizon. When F_B is aligned to the velocity vector, then $\theta = \gamma =$ climb angle.

β, yaw angle. This is the angle between the Y-axes of F_W and F_B.

ψ, azimuth angle of the Eulerian angle. This is the rotation of the X-axis of F_B in horizontal plane.

ϕ, bank angle (Eulerian) with respect to the horizontal plane (vehicle frame F_V).

φ, roll angle, an angle rolling about the X-axis of F_B.

4.5.2 In Pitch Plane (Vertical XZ Plane)

Consider aircraft flight in the plane of symmetry – the generalized attitude is shown in Figure 4.4, with the nomenclature used as given below. For simplification, the flight path is considered along the aircraft axis, that is, $\theta = \gamma$. This is not always the case. The parameters are:

α, angle of attack (wing or body).

θ, flight path angle with respect to the horizon.

λ, thrust line angle with the horizon.

V, vehicle velocity tangent to flight path.

L, D, T, W, are the lift, drag, thrust and weight of the vehicle.

r, instantaneous radius of flight path.

The vehicle acceleration is denoted as $a = dV/dt = \dot{V}$ (it is tangential to a generalized curved trajectory with radius r).

$V = r(d\theta/dt)$, i.e. $rd\theta$ is the distance travelled in time dt, or:

$$V / r = \left(d\theta / dt \right) \tag{4.4}$$

Equating the force equilibrium gives:

$$\sum \text{forces}_{\text{flight path}} = 0 = m\dot{V} - T\cos(\lambda - \theta) + D + mg\sin\theta \quad \left(\text{along flight path}\right) \tag{4.5a}$$

$$\sum \text{forces}_{\text{perpendicular}} = 0 = mV^2 / r - T\sin(\lambda - \theta) - L + mg\cos\theta \quad \left(\text{normal to flight path}\right) \tag{4.5b}$$

Rearranging the equation in terms of coefficients (e.g. $C_L = 2L / \rho V^2 A$, where A is a typical area such as a wing reference area of a conventional aircraft or cross-sectional reference area of a rocket). Combining Equations 4.4 and 4.5a, the following can be derived.

■ **FIGURE 4.4**
Forces on an aircraft in the vertical plane

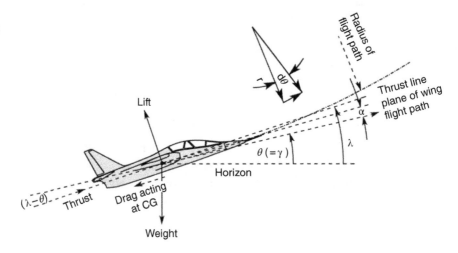

Along the flight path $\quad \dot{V} = (T/m)\cos(\lambda - \theta) - (C_D/2m)\rho V^2 A - g\sin\theta$

$$\text{(4.6)}$$

or

$$m\dot{V} = T\cos(\lambda - \theta) - (C_D/2)\rho V^2 A - mg\sin\theta$$

and perpendicular to the flight path can be obtained from Equations 4.4 and 4.5b (θ is in radian)

$$V(d\theta/dt) = (T/m)\sin(\lambda - \theta) + (C_L/2m)\rho V^2 A - g\cos\theta \qquad \text{(4.7)}$$

Integration of Equations 4.6 and 4.7 is not easy when all aerodynamic variables are to be included. One way to tackle this is to integrate numerically in small steps in which the aerodynamic parameters can be treated as constant to get the trajectory analysis. Typically, thrust line and fuselage axis (X-axis) are close to each other, making $(\lambda - \theta) \approx 0$, when the equations can be simplified to:

$$m\dot{V} = (T) - (C_D/2)\rho V^2 A - mg\sin\theta \qquad \text{(4.8)}$$

$$V(d\theta/dt) = (C_L/2m)\rho V^2 A - g\cos\theta$$

For shallow climb (when $\sin\theta \approx 0$ and $\cos\theta \approx 1$) the equations can be further simplified:

$$dV/dt = (T/m) - \left(\frac{1}{2}\rho V^2 A C_D\right)/m \qquad \text{(4.9a)}$$

$$V(d\theta/dt) = (C_L/2m)\rho V^2 A - g = (C_L/m)qA - g \qquad \text{(4.9b)}$$

For the case of a simplified level trajectory, evidently, in straight and level flight $\theta = 0$ and $d\theta/dt = 0$, which makes the above two equations as follows:

$$dV/dt = (T/m) - \left(\frac{1}{2}\rho V^2 A C_D\right)/m \qquad \text{(4.10)}$$

or

$$T - D = m\dot{V} \;(\text{i.e. thrust - drag} = \text{aircraft acceleration})$$

Also $dV/dt = 0$, when Equation 4.10 gives $T = D$ (i.e. thrust = drag) and Equation 4.9b gives:

$$0 = (C_L/2m)\rho V^2 A - g \qquad \text{(4.11)}$$

or

$$mg = \left(\frac{1}{2}\rho V^2 A C_L\right)$$

or

$$W = L$$

Figure 4.5a shows a yawed aircraft in a linear flight path. Here the yaw angle is β as the side-slip angle. The track angle is τ to the north. Therefore, the heading angle is $\tau = (\psi + \beta)$, not shown in the figure. Equations 4.10 and 4.11 are still valid; note that the aircraft drag in the yawed position is higher than that for aircraft flying without yaw. A crude first-cut estimation may be taken as $C_{D\_yaw} = C_D/\sqrt{\cos\beta}$. All forces have to be resolved along the direction of aircraft velocity and normal to it. Turn equations are derived in Section *4.5.3*.

■ **FIGURE 4.5**
Forces on an aircraft
in the horizontal
plane. (a) Yaw but
linear trajectory. (b)
Pure turn, showing
the bank angle. (c)
Pure turn – no yaw

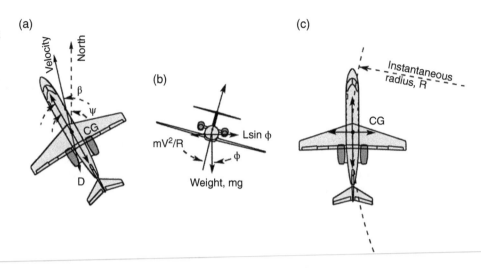

In a trajectory of vertical climb, when $\theta = 90°$, $d\theta/dt = 0$, $\sin\theta = 1$ and $\cos\theta = 0$, then Equation 4.9b vanishes and Equation 4.8 reduces to:

$$dV/dt = (T/m) - \left(\frac{1}{2}\rho V^2 AC_D\right)/m - g \tag{4.12}$$

For vertical acceleration:

$$ma = T - W.$$

or

$$a = T/m - g = (Tg)/w - g \quad \text{i.e.} \quad a/g = T/w - 1$$

which means if thrust is twice the aircraft weight, then body weight also doubles.

4.5.3   In Yaw Plane (Horizontal Plane) – Coordinated Turn

Turning in the horizontal plane, that is, yawing, makes the outer wing move faster than the inner wing (Figure 4.5c). This will cause the outer wing to generate more lift and rise in relation to the inner wing, so the aircraft also rolls; it has coupled motion of yaw and roll. There is a tendency for aircraft to side-slip unless the aircraft is appropriately banked to make a coordinated turn; that is, yaw with the right amount of roll prevents side-slip. Actually, an aircraft may have to keep its nose up to stay level.

The next section tackles the case with nose up (Eulerian angle θ) for aircraft having a helical path either as a climbing turn or a descending turn. The bank angle is the Eulerian angle ϕ, as shown in Figure 4.5b.

The equation of motion is for an instantaneous turn with radius R, for instance. The aircraft velocity, V, is tangential to the flight path. Like in the previous case, the thrust line is taken along the flight path. In a coordinated turn the Eulerian angles ϕ and θ are the same in F_W and F_B. Forces in the wind axes F_W are as follows. (Equation 5.3 gives $L' = L + \Delta L = W + a \times W/g = W(1 + a/g)$.)

Along a steady flight path, thrust=drag. Perpendicular to the flight path, the centrifugal force=centripetal force, i.e. $mV^2/R = L'\sin\phi$, and weight $mg = L'\cos\phi$. From above:

$$\text{Load factor, } n = L'/mg = 1/\cos\phi = \sec\phi \tag{4.13}$$

$$\text{Angular velocity of the turn, } \omega_t = V / R$$

$$\text{Instantaneous radius of turn, } R = V^2 / \left(g \tan \phi \right)$$

or

$$\tan \phi = \omega_t V / g \qquad \textbf{(4.14)}$$

4.5.4 In Pitch-Yaw Plane – Coordinated Climb-Turn (Helical Trajectory)

A more generalized motion of climbing turn (Figure 4.6) follows a helical path in the pitch and yaw plane, with the aircraft turning in climb or descent. In this situation (body axes), unlike in the vertical or horizontal plane, the aircraft motion is in pitch and yaw plane with elevation attitude pitch angle, θ, and bank angle, ϕ. Since it is a coordinated turn, there is no yaw and weight, *mg*, is pointing vertically down.

The climbing-turn formulation of the equations of motion is more complex. To keep the equations of motion simple, the following assumptions are made (refer to the force diagram in Figure 4.6).

1. Elevation, θ, is small, that is, $\sin\theta \approx 0$ and $\cos\theta \approx 1$. The other Eulerian angle is the bank angle, ϕ.
2. The Z-axis of the body axes (F_B) and the wind axes (F_W) are collinear. This is equivalent to having small θ.
3. Components of velocities, V (in F_W) $= u$ (in F_B) and $v = w = 0$ (in F_B).
4. The aircraft is in a coordinated turn, hence there is no side-slip or skidding; that is, $\beta = 0$.
5. Turn angular velocity ω_t is about the vertical axis and steady. Angular velocity $\omega_t = V / R$.
6. All aerodynamic forces (lift, drag and thrust) are in the aircraft plane of symmetry.

Forces involved in a helical turn are as follows (in body axes F_B). To develop the angle in a helical turn, first the elevation angle θ is given when the wings stay level, and then the bank angle ϕ is given. The gravitational force components on account of elevation, θ only (no turning yet), are:

$$\text{In the vertical plane} \quad mg = L' \cos\theta$$

$$\text{In the horizontal plane} \quad L' \sin\theta \approx 0 \quad \left(\text{no turning yet}\right)$$

■ **FIGURE 4.6**
Angles of flight plane

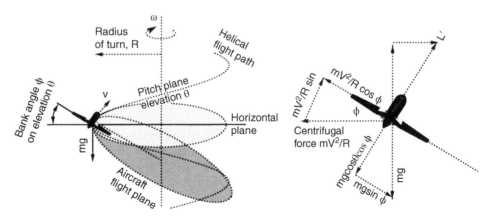

Next is the bank angle ϕ, given after having the elevation, θ, to accomplish the coordinated climb turn in helical trajectory. Force components in body axes F_B on account of coordinated helical climb turn are as follows:

Along X-axis:

$$T - D - mg\sin\theta = m\dot{V} \tag{4.15}$$

Along Y-axis:

Component of weight along Y-axis = component of centrifugal force along Y-axis

That is:

$$mg\sin\phi = m\left(V^2 / R\right)\cos\phi = mV\omega_t \cos\phi$$

or

$$\tan\phi = V\omega_t / g \tag{4.16}$$

Along Z-axis:

Lift, L' = component of weight along Z-axis − component of centrifugal force along Z-axis

or

$$L' = mg\cos\theta\cos\phi + m\left(V^2 / R\right)\sin\phi = mg\cos\phi + mV\omega_t \sin\phi \quad \left(\text{since } \cos\theta \approx 1\right)$$

Note that Equation 4.15 gives $V\omega_t = g\tan\phi$, and substituting in the above equation

$$L' = mg\cos\phi + mg\tan\phi \sin\phi = mg\left(\cos\phi + \sin^2\phi / \cos\phi\right)$$

This gives load factor

$$n = L' / mg = 1 / \cos\phi = \sec\phi \tag{4.17}$$

Instantaneous radius of turn R can be obtained from Equation 4.15:

$$mg\tan\phi = m\left(V^2 / R\right)$$

or

$$R = V^2 / g\tan\phi \tag{4.18}$$

4.5.5 Discussion on Turn

Turn performance analyses can be complex. The above derivations are simplified, so only applicable in a restricted manner to aircraft operations. The climbing turn equations could be applied to commercial transport in en-route climb when the elevation angle θ is less than 10°. These relaxed simplifications are not valid for military aircraft applications in hard manoeuvres when θ is not small and its sine and cosine components are to appear in the equations, making them more complex.

Aircraft turning performance is control-dependent. Turning need not be only in a circular path. It can tighten inside or loosen outside the circular/helical trajectory. The computation for a turn is a point performance evaluated at an instant. Integrated turn performance is carried out in small time steps in which the variables are treated as constants. If the bank angle Φ is less than what is required for a coordinated turn, then the aircraft will side-slip (skid) outwards, and if it is higher than it will side-slip inward of the intended trajectory.

EXAMPLE 4.4

Consider an aircraft performing a coordinated banked turn at 250 m/s and 60 degrees of bank. Calculate the turn rate $\dot{\psi}$, the yaw rate r and the pitch rate q. (This is a good example to understand where to use degrees or radians for the angle.)

SOLUTION

We know from the balance of forces on the aircraft that the radius of turn is:

$$R = \frac{V^2}{g\tan\phi} = \frac{250^2}{9.81\times\tan 60°} = 3678 \text{ m}$$

Therefore the time required for a complete turn will be:

$$t = \frac{2\pi R}{V} = \frac{2\pi\times 3678}{250} = 92.45 \text{ s}$$

Hence the turn rate is

$$\dot{\psi} = \frac{2\pi}{t} = \frac{V}{R} = 0.0680 \text{ rad/s} = 3.9 \text{ deg/s}$$

The only non-zero Euler angle involved in the turning manoeuvre is the bank angle $\Phi = 60°$. Hence we can write

$$\begin{pmatrix} p \\ q \\ r \end{pmatrix} = \begin{pmatrix} 1 & 0 & -\sin 0 \\ 0 & \cos 60 & \sin 60\cos 0 \\ 0 & -\sin 60 & \cos 60\cos 0 \end{pmatrix}\begin{pmatrix} \Phi \\ \dot{\theta} \\ \dot{\psi} \end{pmatrix} \text{ or } \begin{pmatrix} p \\ q \\ r \end{pmatrix} = \begin{pmatrix} 1 & 0 & 0 \\ 0 & 0.5 & 0.866 \\ 0 & -0.866 & 0.5 \end{pmatrix}\begin{pmatrix} 0 \\ 0 \\ 0.068 \end{pmatrix} = \begin{pmatrix} 0 \\ 0.059 \\ 0.034 \end{pmatrix} \text{ rad/s}$$

Pitch rate $q = 0.059$ rad/s and yaw rate $r = 0.034$ rad/s. These would be measured by rate gyros fixed in the aircraft, with their sensitive axes aligned with the Oy and Oz axes of the aircraft, respectively.

Reference

1. Etkin, B., *Dynamics of Atmospheric Flight*. Dover Publications Inc., New York, 2005.

CHAPTER 5

Aircraft Load

5.1   Overview

An aircraft's performance depends on its structural integrity to withstand the design load level. Aircraft structures must withstand the imposed load during operations; the extent depends on what is expected from the intended mission role. The higher the load, the heavier is the structure for its integrity; hence the maximum takeoff weight (MTOW) affecting aircraft performance. Aircraft designers must comply with the mandatory certification regulations to meet the minimum safety standards.

The information provided herein is essential for understanding performance considerations that affect aircraft mass (i.e. weight), and hence its performance. The *V-n* diagram depicts the loading conditions and the limits of the aircraft flight envelope. Only the loads and associated *V-n* diagram in symmetrical flight are discussed herein. Estimation of load is a specialized subject covered in focused courses and textbooks [1]. However, this chapter does outline the key elements of aircraft loads.

5.2   Introduction

Loads are the external forces applied to an aircraft in its static or dynamic state of existence, in flight or on the ground. In-flight loads are due to symmetrical flight, unsymmetrical flight, or atmospheric gusts from any direction. On-ground loads result from ground handling and field performance (e.g. in static, takeoff and landing). Aircraft designers must be aware of the applied aircraft loads, ensuring that configurations are capable of withstanding them. The subject matter concerns the interaction between aerodynamics and structural dynamics (i.e. deformation occurring under load), a subject that is classified as *aeroelasticity*. Even the simplified assumption of an aircraft as an elastic body requires study beyond the scope of this book. This book addresses rigid aircraft, as explained in Section 4.1.

User specifications define the manoeuvre types and speeds that influence aircraft weight, which then dictates aircraft design, affecting aircraft performance. In addition, enough margin must be allocated to cover inadvertent excessive load encountered through pilot-induced manoeuvres, or sudden severe atmospheric disturbances (i.e. external input), or a combination of the two scenarios. The limits of these inadvertent situations are derived from historical statistical data, and pilots must avoid exceeding the margins. To ensure safety, governmental regulatory agencies have mandatory requirements for structural integrity specified with

Theory and Practice of Aircraft Performance, First Edition. Ajoy Kumar Kundu, Mark A. Price and David Riordan.
© 2016 John Wiley & Sons, Ltd. Published 2016 by John Wiley & Sons, Ltd.

minimum margins. *Load factor* (not to be confused with the passenger load factor) is a term that expresses structural-strength requirements. The structural regulatory requirements are associated with *V-n* diagrams, which are explained in Section 5.7. In fact, they also require mandatory strength tests to determine ultimate loads, which must be completed before the first flight, with the exceptions of homebuilt and experimental categories of aircraft. In the following sections, buffet and flutter are introduced.

5.2.1 Buffet

Section 2.7.1 described the buffet phenomenon at the incipient phase of stall at low-speed 1-*g* level flight. Actually buffet can occur at any speed, whenever flow instability over the wing takes place, for example during extreme manoeuvres, at higher *g* loads ($n > 1$) when the angle of attack increases to high values. Buffet can also occur at high speed (called Mach buffet) when transonic flow over the wing (or any other lifting surface). When transonic flow occurs over the wing (or any other lifting surface), it develops local shocks that interact with the boundary layer to cause unstable separation, resulting in unsteady airflow. The separation line over the wing keeps fluctuating forwards and backwards. This causes the aircraft to shudder, and it is a warning to the pilot. The aircraft structure is not necessarily at its maximum loading.

It is clear that every class of aircraft has two stall speeds and hence two buffet speeds: one at low speed and the other at high speed, the magnitude depending on its weight, altitude and ambient conditions. At higher altitudes, because of lower air density, the low-speed buffet goes up while the Mach buffet speed goes down. At a ceiling these two buffet speeds merge, loosely called "coffin corner", as it is very difficult to fly at that point (see Section 12.7.1). The buffet boundary for each class of aircraft has to be established for the full flight envelope. At the design stage, simulation analyses can give a preliminary indication of the buffet boundary, followed by wind-tunnel tests and finetuned through flight tests for operational usage. Typically, combat aircraft are allowed to tolerate higher levels of buffet compared with transport aircraft, for operational reasons. The aircraft design must be free from Mach buffet within the operational envelope.

In the Federal Aviation Administration (FAA) regulations, buffeting is a flight-test task. The pass/fail assessment is made by the approved test pilot and is solely the result of the pilot's experience and judgement. These data are proprietary information kept commercial-in-confidence. It is difficult to find examples to give a worked example. The readers may contact aircraft manufacturers to get a proper buffet boundary with explanations.

5.2.2 Flutter

This is the vibration of the structure that may cause deformation: primarily the wing, but also any other component depending on its stiffness. At transonic speeds, the load on the aircraft is high while the shock–boundary layer interaction could result in an unsteady flow, causing vibration of the wing, for example. The interaction between aerodynamic forces and structural stiffness is the source of flutter. A weak structure enters into flutter; in fact, if it is too weak, flutter could happen at any speed because the deformation would initiate the unsteady flow. If it is in resonance, then it could be catastrophic – such failures have occurred. Flutter is an aeroelastic phenomenon, while buffet is not. Aircraft flutter is dangerous and must be avoided. At the design stage, simulation analyses can give preliminary indications of flutter speed, followed by wind-tunnel tests to improve accuracy, followed by flight test substantiation. Being dangerous, flight tests are carried out with the utmost caution.

5.3   Flight Manoeuvres

Although throttle-dependent linear acceleration would generate flight load in the direction of the flight path, pilot-induced control manoeuvres could generate extreme flight loads that may be aggravated by inadvertent atmospheric conditions. Aircraft weight is primarily determined by the air load generated by manoeuvres in the pitch plane. The associated *V-n* diagram described in Section 5.7 is useful information for aircraft performance analyses. Section 1.15 describes the six degrees of freedom for aircraft motions – three linear and three angular. Given herein are the three Cartesian coordinate planes of interest.

5.3.1   Pitch Plane (X-Z) Manoeuvre

The pitch plane is the symmetrical vertical plane (i.e. X-Z plane) in which elevator/canard-induced motion occurs with angular velocity q about the Y-axis, in addition to the linear velocities in the X-Z plane. Changes in the pitch angle due to angular velocity q results in changes in C_L. The most severe aerodynamic loading occurs in this plane.

5.3.2   Roll Plane (Y-Z) Manoeuvre

Aileron-induced motion generates the roll manoeuvre with angular velocity p about the X-axis, in addition to velocities in the Y-Z plane. Roll-plane loading is not discussed herein.

5.3.3   Yaw Plane (Y-X) Manoeuvre

Rudder-induced motion generates the yaw (coupled with the roll) manoeuvre with angular velocity r about the Z-axis, in addition to the linear velocities in the Y-X plane. Aerodynamic loading of an aircraft due to yaw is also necessary for structural design. It is not discussed herein.

5.4   Aircraft Loads

An aircraft is subject to load at any time. The simplest case is an aircraft stationary on the ground experiencing its own weight. Under heavy landing, an aircraft can experience severe loading, and there have been cases of structural collapse. Most of these accidents showed failure of the undercarriage, but breaking of the fuselage has also occurred.

In flight, aircraft loading varies with manoeuvres and/or when gusts are encountered. Early designs resulted in many structural failures in operation. In-flight loading in the pitch plane is the main issue considered in this chapter. The aircraft structure must be strong enough at every point to withstand the pressure field around the aircraft, along with the inertial loads generated by flight manoeuvres. The *V-n* diagram is the standard way to represent the most severe flight loads that occur in the pitch plane (i.e. X-Z plane).

Civil aircraft designs have conservative limits; there are special considerations for the aerobatic category aircraft. Military aircraft have higher limits for hard manoeuvres, and there is no guarantee that, under threat, a pilot would be able to adhere to the regulations. Survivability requires widening the design limits and strict maintenance routines to ensure structural integrity. Typical human limits are currently taken at $9\,g$ in sustained manoeuvres and can reach $12\,g$ for instantaneous loading. Continuous monitoring of the statistical database retrieved from aircraft-mounted "black boxes" provides feedback to the next generation of aircraft design or for midlife modifications. A g-meter in the flight deck records the g-force and a second needle remains at the maximum g reached in the sortie. If the prescribed limit is exceeded, then the aircraft must be grounded for a major inspection. An important aspect of design is to know what could happen at the extreme points of the flight envelope (i.e. the V-n diagram).

5.5   Theory and Definitions

In steady level flight, an aircraft is in equilibrium; that is, the lift, L, equals the aircraft weight, W and the thrust, T, equals drag, D. It is understood that the wing produces all the lift with a spanwise distribution (see Section 2.11). In equation form, for steady level flight:

$$L = W \quad \text{and} \quad T = D \tag{5.1}$$

5.5.1   Load Factor, n

Newton's law states that change from an equilibrium state requires an additional applied force, which is associated with some form of acceleration, a. When applied in the pitch plane to increase the angle of incidence, α, initiated by rotation of the aircraft, the additional force appears as an increment in lift, ΔL, resulting in gain in height (Figure 5.1).

$$\Delta L = \text{acceleration} \times \text{mass} = a \times W / g \tag{5.2}$$

The resultant force equilibrium gives:

$$L' = L + \Delta L = W + a \times W / g = W\left(1 + a / g\right) \tag{5.3}$$

Load factor, n, is defined as:

$$n = \left(1 + a / g\right) = L / W + \Delta L / W = 1 + \Delta L / W \tag{5.4}$$

The load factor, n, indicates the increase in force contributed by the centrifugal acceleration, a. For example, the load factor $n = 2$ indicates a twofold increase in weight; that is, a 90-kg person would experience a 180-kg weight. The load factor, n, is loosely termed as the *g-load*; in this example, it is the $2\,g$ load.

A high g-load damages the human body, with the human limits of the instantaneous g-load higher than for continuous g-loads. For a fighter pilot, the limit (i.e. continuous) is taken as $9\,g$; for the civil aerobatic category, it is $6\,g$. Negative g-loads are taken as half of the positive g-loads. Fighter pilots use pressure suits to control blood flow (i.e. delay blood starvation) to the brain to prevent blackouts. A more inclined pilot seating position reduces the height of the carotid arteries to the brain, providing an additional margin on the g-load that causes a blackout.

■ **FIGURE 5.1**
Equilibrium flight

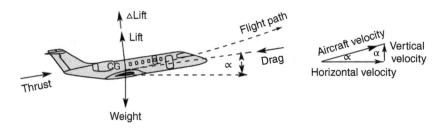

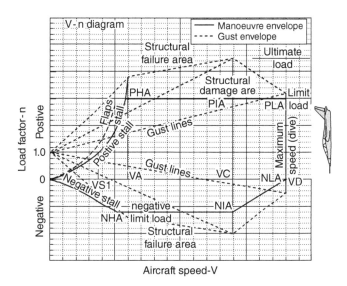

■ **FIGURE 5.2**
**Typical *V-n*
diagram showing
load and speed
limits (pitch plane)**

Because they are associated with pitch-plane manoeuvres, pitch changes are related to changes in the angle of attack, α and the velocity, V. Hence, there is variation in C_L, up to its limit of C_{Lmax}, in both the positive and negative sides of the wing incidence to airflow. The relationship is represented in a *V-n* diagram (Figure 5.2). Atmospheric disturbances are natural causes that appear as a gust load from any direction. Aircraft must be designed to withstand this unavoidable situation up to a statistically determined point that would encompass flights in almost all weathers except extremely stormy conditions. Based on the sudden excess in loading that can occur, margins are built in, as explained in the next section.

5.6 Limits – Loads and Speeds

The *limit load* is defined as the maximum load that an aircraft can be subjected to in its lifecycle. Under the limit load, any deformation recovers to its original shape and would not affect structural integrity. Structural performance is defined in terms of stiffness and strength. Stiffness is related to flexibility and deformation and has implications for aeroelasticity and flutter. Strength concerns the loads that an aircraft structure is capable of carrying, and is addressed within the context of the *V-n* diagram.

To ensure safety, a margin (factor) of 50% increase is enforced through regulations as a *factor of safety* to extend the limit load to the *ultimate load*. A flight load exceeding the limit load but within the ultimate load should not cause structural failure but could affect integrity with permanent deformation. Aircraft are equipped with *g*-meters to monitor the load factor – the *n* for each sortie – and, if exceeded, the airframe must be inspected at prescribed areas and maintained by prescribed schedules that may require replacement of structural components. For example, an aerobatic aircraft with a 6 *g* limit load will have an ultimate load of 9 *g*. If an in-flight load exceeds 6 *g* (but is below 9 *g*), the aircraft may experience permanent deformation but should not experience structural failure. Above 9 *g*, the aircraft would most likely experience structural failure.

The factor of safety also covers inconsistencies in material properties and manufacturing deviations. However, aerodynamicists and stress engineers should calculate for load and

■ **TABLE 5.1**
Typical permissible *g*-load for civil aircraft

Type	Maximum positive *n*	Maximum negative *n*
FAR25 (14CFR25)		
Transport aircraft below 50,000 lb	3.75	−1 to −2
Transport aircraft above 50,000 lb	2.1 + 24000/(*W* + 10000) [should not exceed 3.8]	−1 to −2
FAR23 (14CFR23)		
Aerobatic category (FAR23 only)	6	−3
Military aircraft (military specification MIL-A-8860, MIL-A-8861 and MIL-A-8870)		
In general, factor of safety = 1.5 but can be modified through negotiation		

■ **TABLE 5.2**
Typical *g*-loads for different classes of aircraft (instantaneous limits based on typical human capability)

Club flying	Sports aerobatic	Transport	Fighter	Bomber
+4 to −2	+6 to −3	3.8 to −2	+9 to −4.5	+3 to −1.5

component dimensions such that their errors do not erode the factor of safety. Geometric margins, for example, should be defined such that they add positively to the factor of safety:

$$\text{ultimate load} = \text{factor of safety} \times \text{limit load}$$

For civil aircraft applications, the factor of safety equals 1.5 (FAR 23 (14CAR23) and FAR 25 (14CAR 25), Vol. 3).

5.6.1 Maximum Limit of Load Factor

This is the required manoeuvre load factor at all speeds up to V_C. (The next section defines speed limits.) Maximum elevator deflection at V_A and pitch rates from V_A to V_D also must be considered. Table 5.1 gives the *g*-limit of various aircraft classes as limited by regulation. Table 5.2 shows the typical *g*-levels for various types of aircraft.

5.7   *V-n* Diagram

To introduce the *V-n* diagram, the relationship between load factor, *n*, and lift coefficient, C_L, must be understood. The fundamental flight operational domain is termed the "*manoeuvre*" envelope. The *V-n* envelope also includes the domain termed the "*gust envelope*", covering the atmospheric disturbances from statistical weather data which is seen as all-weather conditions except for some extreme situations which must be avoided. An aircraft must be able to perform safely within the boundaries of speed and load it encounters.

Pitch-plane manoeuvres result in the full spectrum of angles of attack at all speeds within the prescribed boundaries of limit loads. Since C_L varies with changes in attitude and/or aircraft weight, the *V-n* diagram for each altitude and/or weight will be different. Typically, specification is given at the maximum takeoff mass (MTOM) and sea-level at ISA conditions. The *V-n* diagram is established at the specified weight and altitude for detailed analyses. The *V-n* diagram is constructed for the critical altitude, typically 20,000 ft at its maximum weight at that altitude. In principle, it may be necessary to construct several *V-n* diagrams representing different altitudes.

Depending on the direction of pitch-control input, at any given aircraft speed, positive or negative angles of attack may result. The control input would reach either the $C_{L\mathrm{max}}$ or the maximum load factor *n,* whichever is the lower of the two. The higher the speed, the greater is the load factor. Compressibility has an effect on the *V-n* diagram. This chapter explains the role of the *V-n* diagram to understand aircraft manoeuvre performance only in the pitch plane.

Figure 5.2 represents a typical *V-n* diagram showing varying speeds within the specified structural load limits. The figure illustrates the variation in load factor with airspeed for manoeuvres. Some points in a *V-n* diagram are of minor interest to configuration studies – for example, at the point $V = 0$ and $n = 0$ (e.g. at the top of the vertical ascent just before the tail slide can occur). Inadvertent situations may take aircraft from within the limit-load boundaries to conditions of ultimate-load boundaries. The points of interest are explained in the remainder of this section.

5.7.1   Speed Limits

The *V-n* diagram (Figure 5.2) described in the next section uses various speed limits as defined below. More details of the speed definitions are given in Section 5.8.

V_S: Stalling speed during normal level flight.
V_A: Design manoeuvring speed. FAR (14CFR) specifies that $V_A = V_S \sqrt{n}$, for the stalling speed at limit load. In pitch manoeuvre, the aircraft stalls at higher speeds than V_S. In accelerated manoeuvre of pitching up, the angle of attack, α, reduces and hence stalls occur at higher speeds. The tighter is the pitch manoeuvre, the higher the stalling speed until it reaches V_A.
V_B: Stalling speed at maximum gust velocity. It is the design speed for maximum gust intensity. The speed should not be less than $V_B = V_S \sqrt{(n_{gust} \mathrm{at} V_C)}$.
V_C: Maximum design cruising (level) speed. Typically, $V_C \geq V_B + 43$ knots. This serves as a safety margin above V_B. V_C is the specified operating condition. See Table 5.3 for the relationship between V_C and the positive/negative vertical gust velocity at altitudes.
V_D: maximum permissible speed (occurs in dive – sometimes known as the placard speed). This reflects the structural limitations.
V_F: Design flap speeds (separate *V-n* diagram.)

An aircraft can fly below the stall speed if it is in a manoeuvre that compensates for loss of lift or the aircraft attitude is below the maximum angle of attack, α_{max}, for stalling.

Bizjet Example (Figure 5.3)

$$V_S = 82.6 \mathrm{K_{EAS}} \quad V_C = 328.6 \mathrm{K_{EAS}} \quad V_A = 180 \mathrm{K_{EAS}} \ \left(\text{see below}\right)$$

$$V_{MO} = 330 \mathrm{K_{EAS}} < 24,000 \,\mathrm{ft\ altitude,\ } 0.8 \,\mathrm{Mach} \geq 23,500 \,\mathrm{ft.}$$

$$V_D = 380 \mathrm{K_{EAS}} < 21,000 \,\mathrm{ft\ altitude,\ } 0.86 \,\mathrm{Mach} \geq 20,500 \,\mathrm{ft.}$$

5.7.2   Extreme Points of the *V-n* Diagram

Following corner points of the flight envelope is of interest for stress and performance engineers. Beefing up of structures would establish aircraft weight that must be predicted at the conceptual design phase. It also indicates the limitation points of the flight envelope in pitch plane.

■ **FIGURE 5.3**
Bizjet speed limits

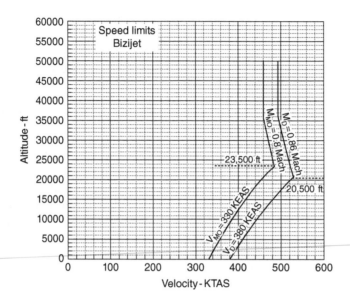

■ **FIGURE 5.4**
Aircraft angles of attack in pitch-plane manoeuvres (see text for explanation)

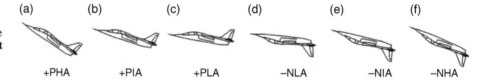

Figure 5.4 shows various attitudes in pitch-plane manoeuvres associated with the *V-n* diagram, each of which are explained below. Note that manoeuvre is a transient situation. The various positions of Figure 5.4 from (a) to (f) (as listed below) can occur under more than one situation. Only the attitudes associated with the predominant cases in pitch plane are outlined below. Negative *g* is when the manoeuvre force is directed in the opposite direction towards the pilot's head, irrespective of his/her orientation with respect to the Earth.

5.7.2.1   *Positive Loads*   This is when the aircraft (and occupants) experiences more force than its normal weight. The aircraft stalls during the manoeuvre reaching α_{max}. The following are the various possibilities. In level flight at 1-*g*, the aircraft angle of attack, α, increases with reduction of speed and reaching its maximum value, α_{max}, the aircraft would stall at a speed V_S.

(a) Positive high angle of attack (+PHA): This occurs during a pull-up manoeuvre raising the aircraft nose with a high pulling *g*-force, reaching the limit. The aircraft could stall if pulled harder. At the limit load of *n* the aircraft reaches +PHA at aircraft speeds of V_A.

(b) Positive intermediate angle of attack (+PIA): This occurs during high-speed level flight when control is actuated to set wing incidence at an angle of attack. Note that the aircraft has a maximum operating speed limit of V_C, when +PIA reaches the maximum limit load of *n* in manoeuvre – it is in transition or held at an intermediate level of manoeuvre.

(c) Positive low angle of attack (+PLA): This occurs when the aircraft gains the maximum allowable speed, sometimes in a shallow dive (dive speed, V_D) and then at very small elevator pull (low angle of attack) would hit the maximum limit load of n. Some high-powered military aircraft can reach V_D at level flight. The higher the speed, the lower would be the angle of attack, α, needed to reach limit load – at highest speed, it would have $+\text{PLA}$.

5.7.2.2   Negative Loads   Negative loads occur when the aircraft (and occupants) experiences force less than its weight. In an extreme manoeuvre in bunt (developing negative g in a nose-down curved trajectory), the centrifugal force pointing away from the centre of the earth can cancel weight such that the pilot feels weightless during the manoeuvre. The corner points follow the same logic of the positive load description, except that the limit load of n is in the negative side, which is lower for not being in the normal flight regimes. It can occur during aerobatic flight, in combat, or in an inadvertent situation through atmospheric gusts.

(d) Negative high angle of attack (−NHA): It is the inverted situation of case (a) explained above. With $-g$ it has to be in a manoeuvre.

(e) Negative intermediate angle of attack (−NIA): It is the inverted situation of case (b). The possibility of negative α was mentioned when the elevator is pushed down, and is called a "bunting" manoeuvre. Negative α can classically occurs at inverted flight at the highest design speed, V_C (coincided with PIA). When it reaches the maximum negative limit load of n, the aircraft takes NIA.

(f) Negative low angle of attack (−NLA): It is the inverted situation of case (c). At V_D, the aircraft should not exceed zero g.

5.7.3   Low Speed Limit

At low speeds, the maximum load factor is constrained by the aircraft C_{Lmax}. The low-speed limit in a *V-n* diagram is established at the velocity at which the aircraft stalls in an acceleration flight load of n until it reaches the load factor limit. At higher speeds, the manoeuvre-load factor may be restricted to the load factor limit, as specified by the regulatory agencies. A *V-n* diagram is constructed with V_{EAS} representing the altitude effects.

Let V_{S1} be the stalling speed at $1\,g$, then $V_{S1}^{2} = \left(\dfrac{1}{0.5\rho C_{Lmax}}\right)\left(\dfrac{W}{S_W}\right)$ or $L = W = (0.5\rho V_{S1}^{2} S_W) C_{Lmax}$

or

$$W = \frac{\rho C_{Lmax} S_W V_{S1}^{2}}{2} \tag{5.5}$$

This gives

$$V_{S1} = \sqrt{\frac{2(W/S_W)}{\rho C_{Lmax}}} \tag{5.6}$$

In manoeuvre, let V_{Sn} be the stalling speed at load factor n,

then

$$nW = 0.5\rho V_{Sn}^{2} S C_{Lmax} = \frac{\rho C_{Lmax} S_W V_{Sn}^{2}}{2} \tag{5.7}$$

or

$$n = \frac{\rho C_{Lmax} S_W V_{Sn}^{2}}{2W} = \frac{\rho C_{Lmax} V_{Sn}^{2}}{2\left(W/S_W\right)} \tag{5.8}$$

Using Figure 5.5 to make a Mach number correction on C_{Lmax}, this gives

$$V_{Sn} = \sqrt{\frac{2n(W/S_W)}{\rho C_{Lmax}}} \qquad (5.9)$$

Equating for W in Equations 5.6 and 5.7

$$\frac{\rho C_{Lmax} S_W V_{S1}^2}{2} = \frac{\rho C_{Lmax} S_W V_{Sn}^2}{2n}$$

or $\qquad\qquad V_{S1} = \dfrac{V_{Sn}^2}{2n} \quad$ i.e., $\quad V_{Sn} = V_{S1}\sqrt{n} \qquad (5.10)$

In terms of Imperial units, Equation 5.8 becomes

$$n = \frac{1.688^2 \times \rho_0 C_{Lmax} V_{sn\_KEAS}^2}{2\left(\dfrac{W}{S_W}\right)} = \frac{C_{Lmax} V_{sn\_KEAS}^2}{296 \times \left(\dfrac{W}{S_W}\right)} \qquad (5.11)$$

V_A is the speed at which the positive stall and maximum load factor limits are simultaneously satisfied, that is $V_A = V_{S1}\sqrt{n_{limit}}$.

The negative side of the boundary can be estimated in a similar fashion. For a cambered aerofoil, C_{Lmax} at a negative angle is less than at a positive angle.

V_C is the design cruise speed. For transport aircraft, V_C must not be less than $V_B + 43$ knots. The Joint Airworthiness Regulations (JAR) contain more precise definitions, plus definitions of several other speeds. In civil aviation, for aircraft weighing less than 50,000 lb, the maximum manoeuvre load factor is usually +2.5. The appropriate expression to calculate the load factor is given by (FAR(14CAR) 25.337b):

$$n = 2.1 + 24000/(W + 10000), \text{up to a maximum of } 3.8 \qquad (5.12)$$

This is the required manoeuvre load factor at all speeds up to V_C, unless the maximum achievable load factor is limited by stall.

Within the limit load, the negative value of n is −1.0 at speeds up to V_C, decreasing linearly to 0 at V_D (Figure 5.2). Maximum elevator deflection at V_A and pitch rates from V_A to V_D must also be considered.

5.7.4   Manoeuvre Envelope Construction

The computational procedure is as follows:

1. Assume V_{KEAS} and determine the flight Mach number.
2. Apply the Mach number correction (experimental data, Figure 5.5 – refer to Section 2.10.1) to revise the value of C_{Lmax}.
3. Compute $q = 0.5\rho V_{Sn}^2$.
4. Estimate n using Equation 5.11 (Imperial units).

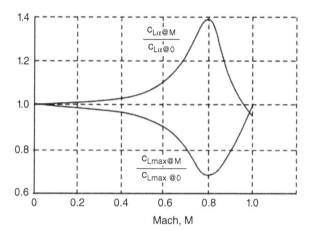

■ **FIGURE 5.5**
Effect of Mach
number on
C_{Lmax} **– experimental**
data. Reproduced
with permission of
CommLit Press

5.7.5   High Speed Limit

V_D is the maximum design speed. It is limited by the maximum dynamic pressure that the airframe can withstand. At high altitude, V_D may be limited by the onset of high-speed flutter.

5.8   Gust Envelope

Encountering unpredictable atmospheric disturbance is unavoidable. Weather warnings help, but full avoidance is not possible. Gusts can hit aircraft from any angle and the gust envelope can be shown in a separate set of diagrams. The most serious type is the vertical gust (Figure 5.2) affecting load factor n. Vertical gust increases the angle of attack, α, developing ΔL. Regulatory bodies have specified vertical gust rates which must be superimposed on V-n diagrams to describe limits of operation. It is usual practice to combine manoeuvre and gust envelopes in one diagram as shown in Figures 5.2 and 5.6.

The FAR (14CFR) gives a detailed description to establish required gust loads. To keep within the maximum load, the limits of vertical gust speeds are reduced with an increase in aircraft speed. Pilots should fly at a lower speed if high turbulence is encountered. Examining Equation 5.6 it can be seen that aircraft with low wing loading (W/S_w), flying at high speed are more affected by gust load. From statistical observations, the regulatory bodies (FAR (14CFR) Part 25.341a) have established maximum gust load at level flight as follows.

V_{MO}/M_{MO} (maximum operating limit speed - airspeed or Mach Number), whichever is critical at a particular altitude, may not be deliberately exceeded in any regime of flight (climb, cruise, or descent), unless a higher speed is authorized for flight test or pilot training operations.

Linear interpolation is used to get appropriate velocities between 20,000 and 50,000 ft. Construction of V-n diagrams is relatively easy from the aircraft specifications. The corner points of V-n diagrams are specified. Computation to superpose gust lines is more complex, for which FAR has given semi-empirical relations.

Flight speed, V_B, is determined by the gust loads (see Table 5.3). V_B is the design speed for maximum gust intensity. This definition assumes that the aircraft is in steady level flight at speed V_B when it enters an idealised upward gust of air, which instantaneously increases the aircraft angle of attack and, hence, load factor. The increase in angle of attack must not stall the aircraft, that is, take it beyond the positive and negative stall boundaries.

■ **FIGURE 5.6**
Composite *V-n*
diagram, typical of
Boeing 707 type
(clean configuration
at cruise speed and
altitude).
Reproduced with
permission of Boeing
Corp.

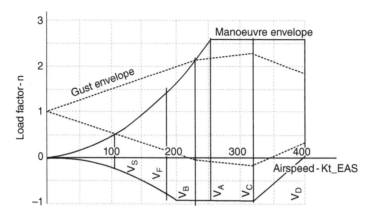

■ **TABLE 5.3**
FAR (14CFR) Part 25.341a specified gust velocity (in EAS)

Altitude	Altitudes 20,000 ft and below (ft/s)	Altitudes 20,000 ft and above
V_B (rough air gust – maximum)	66	Linear reduction to 38 ft/s at 50,000 ft
V_C (gust at max. design speed)	50	Linear reduction to 25 ft/s at 50,000 ft
V_D (gust at max. dive speed)	25	Linear reduction to 12.5 ft/s at 50,000 ft

Except for extreme weather conditions, this gust limit amounts to almost all-weather flying. In gusts, the aircraft load may cross the limit load but must not exceed ultimate load, as shown in Figure 5.2. If the aircraft crossed the limit load, then appropriate action through inspection is carried out.

Given below is a brief outline for constructing a *V-n* diagram superposed with gust load. The starting point for developing the gust envelope is when an aircraft flies a straight and level flight at $n=1$. A factor is used to derive the gust load relationship [1]. A typical *V-n* diagram, with the gust speeds intersecting the lines, is illustrated in Figures 5.2 and 5.6. The *V-n* diagram is plotted with the X-axis in V_{EAS}. Separate *V-n* diagrams are required for changes in aircraft weight and/or operating altitude.

5.8.1   Gust Load Equations

The simplest way to consider gust load is to assume that a gust hits sharply with a vertical velocity U_z (for extensive treatment, the readers may refer to [1]). The increment of the angle of attack due to this vertical velocity is $\Delta\alpha$:

$$\Delta\alpha = U_z / V \tag{5.13}$$

If V is in knots, then $\Delta\alpha = U_z / (1.688 \times V)$, U_z remains in ft/s, and U_{ze} is the equivalent airspeed (EAS) in ft/s. Note that U_z and V are in the same units, typically kept in EAS.

Considerable simplification is made here to obtain satisfactory estimation. The simplification is made in establishing the value of $C_{L\alpha\_aircraft}$ (in brief, $C_{L\alpha\_ac}$). Typically, take $C_{L\alpha\_ac}$ as 10% higher than $C_{L\alpha\_wing}$.

Incremental lift, ΔL, caused by the gust, is then:

$$\Delta L = \Delta nW = 0.5 \times \rho \times V^2 \times \Delta C_L S_W = \frac{\left(\Delta \alpha C_{La\_ac} \rho V^2 S_W\right)}{2} \tag{5.14a}$$

Noting that $V_{EAS} = \sqrt{\sigma} V_{TAS}$, Equation 5.13 becomes:

$$\Delta L = \Delta nW = \frac{\left(\Delta \alpha C_{La\_ac} \rho_0 V_{EAS}^2 S_W\right)}{2} = \frac{\left(\Delta \alpha C_{La\_ac} V_{EAS}^2 S_W\right)}{2 \times 421.9} \tag{5.14b}$$

When V_{EAS} is in knots, then the above equation reduces to

$$\Delta L = \Delta nW = \frac{\left(\Delta \alpha C_{La\_ac} \times 1.688^2 \times V_{KEAS}^2 S_W\right)}{2 \times 421.9} = \frac{\left(\Delta \alpha C_{La\_ac} \times V_{KEAS}^2 S_W\right)}{296} \tag{5.14c}$$

Modification to these equations are required as vertical gusts may not hit suddenly and sharply, but develop gradually. In addition to the gust gradient, other factors (Kussner-Wagner effect) affecting the incremental load Δn are the aircraft response characteristics (mass and geometry dependent) and a lag in lift-increment due to the gradual increment of angle of attack $\Delta \alpha$ (aeroelastic effect). Incremental load Δn becomes less intense compared with if it encountered a sudden and sharp gust. An empirical factor K_g, known as the *gust alleviation factor*, is applied and the equations become as follows.

Using Equation 5.13 in Equation 5.14a for $\Delta \alpha$ (velocities in ft/s TAS, i.e. U_Z in ft/s), it becomes

$$\Delta L = \Delta nW = \frac{\left(\Delta \alpha C_{La\_ac} \rho V^2 S_W\right)}{2} = \frac{\left(U_Z C_{La\_ac} \rho V S_W\right)}{2} \tag{5.15a}$$

Equation 5.14c, in K_{EAS} becomes,

$$\begin{aligned}
\Delta L = \Delta nW &= \frac{\left(U_{Ze} C_{La\_ac} \times \rho_0 \times 1.688 \times V_{KEAS} S_W\right)}{2 \times \sqrt{\sigma}} \\
&= \frac{\left(U_{Ze} C_{La\_ac} \times V_{KEAS} S_W\right)}{421.94 \times 1.185 \times \sqrt{\sigma}} = \frac{\left(U_{Ze} C_{La\_ac} \times V_{KEAS} S_W\right)}{498 \times \sqrt{\sigma}}
\end{aligned} \tag{5.15b}$$

Using Equation 5.15a,

$$\Delta n = \frac{\left(K_g U_Z C_{La\_ac} \rho V\right)}{2\left(W / S_W\right)} \tag{5.16a}$$

where $K_g = (0.88\mu) / (5.3 + \mu)$ and where $\mu = 2W / (g \times \text{MAC} \times \rho \times C_{La\_ac} \times S_W)$.

When V_{EAS} is in knots (V_{KEAS}) then Equation 5.15b reduces to

$$\Delta n = \frac{\left(K_g U_{Ze} C_{La\_ac} \times V_{KEAS}\right)}{498\left(W / S_W\right)} \tag{5.16b}$$

The higher the wing aspect ratio, the higher is $C_{L\hat{a}\_ac}$, and the higher the wing sweep, the lower is C_{La_ac}. This means that a straight wing aircraft with high wing aspect ratio will be more sensitive to loading than a swept wing with low aspect ratio. Military aircraft are more tolerant to gust loads compared with a turboprop drive civil transport aircraft.

5.8.2   Gust Envelope Construction

The computational procedure is as follows:

1. Assume V_{KEAS} and determine the flight Mach number.
2. Apply Mach number correction (Figure 5.5) to revise the value of C_{Lmax} and $C_{L\hat{a}}$
3. Compute $q = 0.5\rho V_{Sn}^2$.
4. Estimate n using Equation 5.11 (Imperial units).

Taking as a brief example the Bizjet specification at sea-level and 20,680 lb, the manoeuvre load limit = 3.0 to −1 and gust load limit = 4.5 to −1.5. The n range will increase at lighter weights and higher altitudes. The maximum design cruising (level) speed V_C at 20,000 ft altitude is Mach 0.8 (TAS = 767.3 ft/s, 328.6 K_{EAS}). The maximum speed at dive is Mach 0.86 (380 K_{EAS}).

EXAMPLE 5.1

Find the manoeuvre load n and the gust load, n_{gust}, with vertical velocity, $U_{ZE} = 66$ ft/s on an aircraft of 20,000 lb flying at 200 K_{EAS} (337.6 ft/s) at 20,000 ft altitude ($\rho = 0.00126$ slug, $\sigma = 0.312E$ and $\sqrt{\sigma} = 0.729$), on an ISA day. Maximum operating speed is Mach 0.80 ($V_{MO} = 380 K_{EAS}$), $W = 19,900$ lb, $S_W = 323$ ft² ($W/S_W = 61.92$ lb/ft²), MAC = 7 ft, $C_{Lmax} = 1.59$. The aerofoil used is NACA 65-410.

SOLUTION

At sea-level on an ISA day, $V_{EAS} = V_{TAS}$.

$$V_S = 82.6 \ K_{EAS}$$
$$V_C = 328.6 \ K_{EAS}$$
$$V_A = 180 \ K_{EAS} \left(\text{see below}\right)$$
$$V_D = 380 \ K_{EAS}$$

To estimate manoeuvre load, n, FAR (14CFR) 25.337b gives:

$$n = 2.1 + 24000/(W+10000) = 2.1 + 0.83 = 2.93$$

Take $n = 3.0$ for the additional safety. A negative manoeuvre load can be estimated in a similar manner.

Use Equation 5.11 in the Imperial system: $n = \dfrac{C_{Lmax}V_{Sn\_KEAS}^2}{296 \times \left(W/S_W\right)}$

Use Figure 5.5 to correct C_{Lmax} (at low speeds, a Mach number correction is not applied). To keep things simple, the effect of relatively small amount of wing sweep on C_{Lmax} is ignored. This gives:

$$n = \frac{C_{Lmax}V_{Sn\_KEAS}^2}{296 \times \left(W/S_W\right)} = \frac{1.59 \times 200^2}{296 \times 61.92} = 63600/18328.32 = 3.47$$

Therefore, V_A is at 180 K_{EAS} to the limit of $n = 3.0$ (this is determined by plotting the locus of stall airspeed versus n).

To estimate gust load, n_{gust}, use Equations 5.15 and 5.16.

$$\text{Gust load } n_{gust} = \left(1 + \Delta n\right)$$

For NACA 65-410, given in Appendix F, take the lines for $Re = 9 \times 10^6$, being close to the flight Reynolds number. It gives a 2D lift-curve slope, $dC_l/d\alpha = a_0 = (1.05 - 0.25)/8 = 0.11/$deg.

For a 3D wing,

$$\left(dC_L / d\alpha\right)_w = a = a_0 / \left[1 + \left(57.3 a_0 / e\pi AR\right)\right]$$

$$= 0.11 / \left[1 + \left(57.3 \times 0.11\right) / 0.96 \times 3.14 \times 7.5\right]$$

$$= 0.11 / \left[1 + / \left(7.3 / 22.6\right)\right] = 0.11 / 1.32$$

$$= 0.0833$$

$$\left(57.296 \times 0.833 = 4.77 \text{ is for the wing.}\right)$$

For aircraft increase by 10%, that is $\left(dC_L / d\alpha\right)_a = 1.1 \times 4.77 = 5.25 / \text{rad}$.

Mach number correction (Figure 5.5) at $200\,\text{K}_{EAS}$:

$$C_{L\alpha\_ac} = 1.04 \times 5.25 = 5.46 / \text{rad}$$

From Equation 5.16,

$$\mu = 2W / \left(g \times \text{MAC} \times \rho \times C_{L\alpha\_ac} \times S_W\right)$$

or

$$\mu = \left(2 \times 20\ 000\right) / \left(32.2 \times 7 \times 0.00126 \times 5.46 \times 323\right)$$

$$= 40000 / 500.9 = 79.86$$

and

$$K_g = \left(0.88\mu\right) / \left(5.3 + \mu\right) = \left(0.88 \times 79.86\right) / \left(5.3 + 79.86\right)$$

$$= 70.28 / 85.16 = 0.825$$

Using Equation 5.16b

$$\Delta n = \frac{\left(K_g U_{Ze} C_{L\alpha\_ac} \times V_{KEAS}\right)}{498 \left(W / S_W\right)} = \frac{\left(0.825 \times 66 \times 5.46 \times 200\right)}{498 \times 61.92} = 1.93$$

Gust load $n_{gust} = \left(1 + \Delta n\right) = 1 + 1.93 = 2.93$

Readers may construct the full flight envelope (see problem assignments in Appendix E).

Reference

1. Niu, M., *Airframe Structural Design*. Commlit Press Ltd., Hong Kong, 1999.

CHAPTER 6

Stability Considerations Affecting Aircraft Performance

6.1 Overview

Aircraft performance depends on aircraft stability characteristics. Aircraft stability analysis forms a subject of its own (more details on the subject are given in [1–10]), but it is necessary to explain some of the pertinent points that affect aircraft performance. Substantiation of aircraft performance by itself would not prove sufficient if the aircraft stability characteristics do not offer satisfactory handling qualities and safety. These characteristics concern aircraft *flying qualities* (NASA has codified these flying qualities). Many good designs had to go through considerable modifications to substantiate aircraft performance and handling criteria. The position of the aircraft centre of gravity (CG) plays an important role in inherent aircraft motion characteristics, which can be seen in how an aircraft performs. This chapter gives a qualitative understanding of the geometric arrangement of aircraft components that affect aircraft performance associated with stability. Only the equations governing static stability are given to explain some of the important effects on aircraft performance.

6.2 Introduction

Inherent aircraft performance and stability are a result of CG location, wing and empennage sizing/shaping, fuselage and nacelle sizing/shaping, and their relative locations. This chapter highlights some of the lessons learnt on how aircraft components affect aircraft performance. Designers have to ensure that there is adequate trim authority at any condition.

The pitching motion of an aircraft is in the plane of aircraft symmetry (about the Y-axis – elevator actuated) and is uncoupled from any other type of motion. Directional (about Z-axis – rudder actuated) and lateral (about X-axis – aileron actuated) motions are not in the plane of symmetry. Activating any of the controls, for example rudder or ailerons, causes coupled aircraft motion in both the directional and lateral planes.

Theory and Practice of Aircraft Performance, First Edition. Ajoy Kumar Kundu, Mark A. Price and David Riordan.
© 2016 John Wiley & Sons, Ltd. Published 2016 by John Wiley & Sons, Ltd.

Flight tests would reveal whether the aircraft satisfies the flying quality and safety considerations. Practically all projects require some sort of minor tailoring and/or rigging of control surfaces to improve the flying qualities as a consequence of flight tests.

6.3   Static and Dynamic Stability

It is pertinent here to review briefly the terms *static* and *dynamic stability*. Stability analyses examine what happens to an aircraft when it is subjected to forces and moments applied by a pilot and/or induced by external atmospheric disturbances. There are two types of stability, as follows:

1. *Static stability*: This refers to the instantaneous tendency of an aircraft when disturbed during equilibrium flight. The aircraft is statically stable if it has restoring moments when disturbed; that is, it shows a tendency to return to the original equilibrium state. However, this does not cover what happens in the due course of time. The recovery motion can overshoot into oscillation, which may not return to the original equilibrium flight.

2. *Dynamic stability*: This is the time history of an aircraft response after it has been disturbed, which is a more complete picture of aircraft behaviour. A statically stable aircraft may not be dynamically stable, as explained in subsequent discussions. However, it is clear that a statically unstable aircraft also is dynamically unstable. Establishing static stability before dynamic stability is for procedural convenience.

Pitch-plane stability may be compared to a spring–mass system, as shown in Figure 6.1a. The oscillating characteristics are represented by the spring-mass system, with the resistance to the rate of oscillation as the damping force (i.e. proportional to pitch rate, $\dot{q}$) and the spring compression proportional to pitch angle θ. Figure 6.1b shows the various possibilities of the vibration modes. Stiffness is represented by the stability margin, which is the distance between the CG and the neutral point (NP). The higher the force required for deformation, the greater is the stiffness. Damping results from the rate of change and is a measure of resistance (i.e. how fast the oscillation fades out); the higher the H-tail area, the greater is the damping effect. An aircraft only requires adequate stability; making it more stable than required poses other difficulties in the overall design.

Aircraft motion in 3D space is represented in the three planes of the Cartesian coordinate system (Section 2.4). Aircraft have six degrees of freedom of motion in 3D space. They are decomposed into the three planes; each exhibits its own stability characteristics, as listed herein.

■ **FIGURE 6.1**
Aircraft stability compared with spring–mass system. (a) Aircraft as spring–mass system. (b) Typical response characteristics

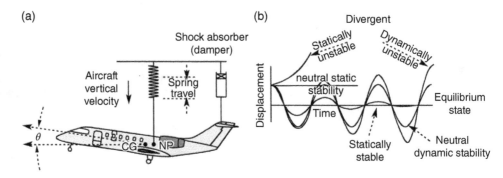

The sign conventions associated with the pitch, yaw and roll stabilities need to be learned (they follow the right-handed rule). The brief discussion of the topic herein is only for what is necessary in this chapter. The early stages of stability analyses are confined to small perturbations – that is, small changes in all flight parameters.

1. *Longitudinal stability in the pitch plane*: The pitch plane is the XZ plane of aircraft symmetry. The linear velocities are u along the X-axis and w along the Z-axis. Angular velocity about the Y-axis is $\dot{q}$, known as *pitching* (+ve nose up). Pilot-induced activation of the elevator changes the aircraft pitch. In the plane of symmetry, the aircraft motion is uncoupled; that is, motion is limited only to the pitch plane.

2. *Directional stability in the yaw plane*: The yaw plane is the XY plane and is not in the aircraft plane of symmetry. Directional stability is also known as *weathercock stability* because of the parallel to a weathercock. The linear velocities are u along the X-axis and v along the Y-axis. Angular velocity about the Z-axis is $\dot{r}$, known as *yawing* (+ve nose to the left). Yaw can be initiated by the rudder; however, pure yaw by the rudder alone is not possible because yaw is not in the plane of symmetry. Aircraft motion is coupled with motion in the YZ plane.

3. *Lateral stability in the roll plane*: The roll plane is the YZ plane and also is not in the aircraft plane of symmetry. The linear velocities are v along the Y-axis and w along the Z-axis. Angular velocity about the X-axis is $\dot{p}$, known as *rolling* (+ve when right wing drops). Rolling can be initiated by the aileron, but a pure roll by the aileron alone is not possible because roll is associated with yaw. To have a pure rolling motion in the plane, the pilot must activate both the yaw and roll controls.

It is convenient now to explain static and dynamic stability in the pitch plane using diagrams. The pitching motion of an aircraft is in the plane of symmetry and is uncoupled; that is, motion is limited only to the pitch plane. The static and dynamic behaviour in the other two planes has similar characteristics, but it is difficult to depict the coupled motion of yaw and roll. These are discussed separately in Sections 6.3.2 and 6.3.4.

Figure 6.2 depicts the stability characteristics of an aircraft in the pitch plane, which provide the time history of aircraft motion after it is disturbed from an initial equilibrium. It shows that aircraft motion is in an equilibrium level flight – here, motion is invariant with time. Readers may examine what occurs when forces and moments are applied.

A statically and dynamically stable aircraft tends to return to its original state even when it oscillates about the original state. An aircraft becomes statically and dynamically unstable if the pitching motion diverges outward – it neither oscillates nor returns to the original state. The third diagram of Figure 6.2 provides an example of neutral static stability – in this case, the aircraft does not have a restoring moment. It remains where it was after the disturbance and requires an applied pilot effort to force it to return to the original state. The tendency of an aircraft to return to the original state is a good indication of what could happen in time: Static stability makes it possible but does not guarantee that an aircraft will return to the equilibrium state.

As an example of dynamic stability, Figure 6.2 also shows the time history for when an aircraft returns to its original state after a few oscillations. The time taken to return to its original state is a measure of the aircraft's damping characteristics – the higher the damping, the faster the oscillations fade out. A statically stable aircraft showing a tendency to return to its original state can be dynamically unstable if the oscillation amplitude continues to increase, as shown in the last diagram of Figure 6.2. When the oscillations remain invariant in time, the aircraft is statically stable but dynamically neutral – it requires an application of force to return to the original state.

■ **FIGURE 6.2**
Stability in the pitch plane

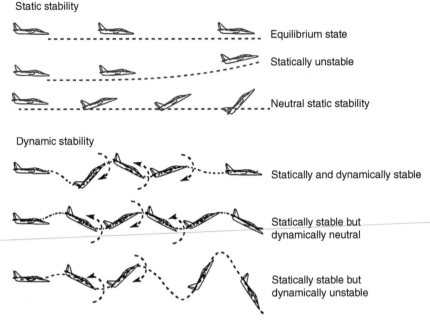

6.3.1   Longitudinal Stability – Pitch Plane (Pitch Moment, M)

Figure 6.3 depicts the conditions for aircraft longitudinal static stability. In the pitch plane, by definition, the angle of attack (AOA), α, is positive when an aircraft nose is above the direction of free-stream velocity.

A nose-up pitching moment is considered positive. Static-stability criteria require that the pitching-moment curve exhibits a negative slope, so that an increase in the AOA, α, causes a restoring negative (i.e. nose-down) pitching moment. At equilibrium, the pitching moment is equal to zero ($C_m=0$) when it is in trimmed condition (α_{trim}). The higher the static margin (see Figure 6.3), the greater is the slope of the curve (i.e. the greater is the restoring moment). Using the spring analogy, the stiffness is higher for the response.

The other requirement for static stability is that at zero AOA, there should be a positive nose-up moment, providing an opportunity for equilibrium at a positive AOA ($+\alpha_{trim}$), typical in any normal flight segment.

6.3.2   Directional Stability – Yaw Plane (Yaw Moment, N)

Directional stability can be compared to longitudinal stability, but it occurs in the yaw plane (i.e. the XY plane about the Z-axis) as shown in Figure 6.4. By definition, the angle of sideslip, β, is positive when the free-stream velocity vector, V, relative to aircraft is from the right (i.e. the aircraft nose is to the left of the velocity). V has component aircraft velocities u along the X-axis and v along the Y-axis, subtending the side-slip angle $\beta=\tan^{-1}(v/u)$.

The V-tail is subjected to an angle of incidence ($\beta+\sigma$), where σ is the sidewash angle generated by the wing vortices (like the downwash angle in longitudinal stability). Static stability criteria require that an increase in the side-slip angle, β, should generate a restoring moment, C_n, that is positive when turning the nose to the right. The moment curve slope of C_n is positive for stability. At zero β, there is no yawing moment (i.e. $C_n=0$).

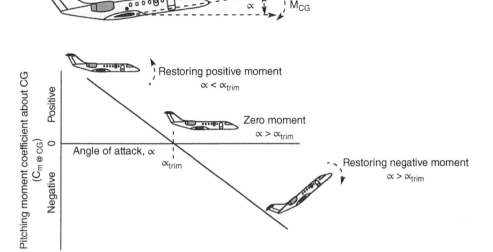

■ **FIGURE 6.3**
Longitudinal static stability – nose-up gives +α and C_m.

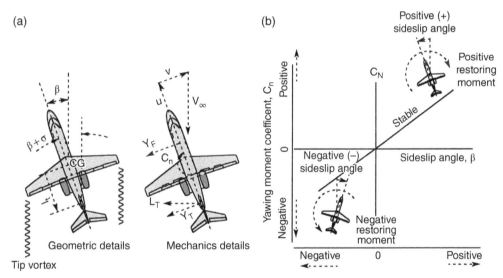

■ **FIGURE 6.4**
Direction stability.
(a) Side-slip angle β.
(b) Directional stability criteria

Yaw motion invariably couples with roll motion because it is not in the plane of symmetry. In yaw, the windward wing works more to create a lift increase while the lift decreases on the other wing, thereby generating a rolling moment. Therefore, a pure yaw motion is achieved by the use of compensating, opposite ailerons. The use of an aileron is discussed in the next section.

6.3.3   Lateral Stability – Roll Plane (Roll Moment, *L*)

Roll stability is more difficult to analyse compared with longitudinal and lateral stabilities. A banked aircraft attitude through pure roll keeps the aircraft motion in the plane of symmetry and does not provide any restoring moment. However, roll is always coupled with a yawing motion, as explained previously. As a roll is initiated, the side-slip velocity, *v*, is triggered by the weight component towards the downwing side, as shown in Figure 6.5. Then (see the previous

■ **FIGURE 6.5**
Lateral stability – rear view, right wing drop

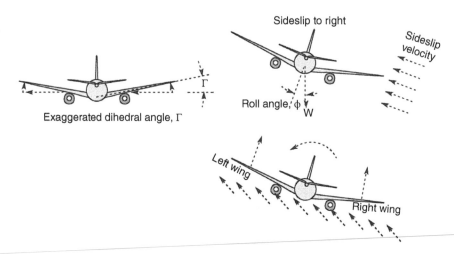

■ **FIGURE 6.6**
Lateral stability – fuselage contributions

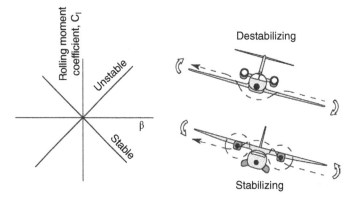

section) the side-slip angle is $\beta = \tan^{-1}(v/u)$. The positive angle of roll, φ, is when the right wing drops as shown in the figure (the aircraft is seen from the rear showing the V-tail but no windscreen). A positive roll angle generates a positive side-slip angle, β.

Recovery from a roll is possible as a result of the accompanying yaw (i.e. coupled motion), with the restoring moment contributing by increasing the lift acting on the wing that has dropped. Roll static-stability criteria require that an increase in the roll angle, φ, creates a restoring moment coefficient, C_l (not to be confused with the sectional aerofoil-lift coefficient). The restoring moment has a negative sign.

Having a coupled motion with the side-slip, Figure 6.6 shows that C_l is plotted against the side-slip angle β, not against the roll angle φ, because it is β that generates the roll stability. The sign convention for restoring the rolling moment with respect to β must have $C_{l\beta}$ negative; that is, with an increase of roll angle φ, the sideslip angle β increases to provide the restoring moment. An increase in β generates a restoring roll moment due to the dihedral. At zero φ, there is no β, and hence zero rolling moment ($C_l = 0$).

The wing dihedral angle, Γ, is one way to increase roll stability, as shown in Figure 6.5. The dropped wing has an airflow component from below the wing generating lift, while at the other side, the airflow component reduces the AOA (i.e. the lift reduction creates a restoring moment).

■ **FIGURE 6.7**
Effect of wing sweep on roll stability

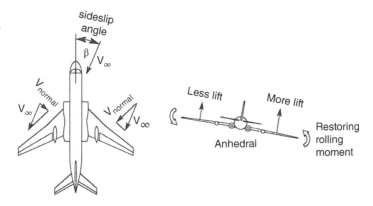

■ **FIGURE 6.8**
Vertical tail contribution to roll, viewed from the rear (modified and reproduced from [7])

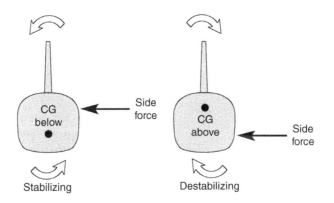

The position of the wing relative to the aircraft fuselage has a role in lateral stability, as shown in Figure 6.6. At yaw, the relative airflow about the low wing has a component that reduces the AOA; that is, the reduction of lift and the other side act in opposite ways: a destabilizing effect that must be compensated for by the dihedral shape, as explained previously. Conversely, a high-wing aircraft has an inherent roll stability that acts opposite to a low-wing design. If it has too much stability, then an anhedral angle (−ve dihedral) is required to compensate for it. Many high-wing aircraft have an anhedral angle (e.g. the Harrier and the BAe RJ series).

An interesting situation occurs with a wing sweepback on a high-speed aircraft, as explained in Figure 6.7. At side-slip, the windward wing has an effectively reduced sweep; that is, the normal component of air velocity increases, creating a lift increment, whereas the leeward wing has an effectively increased sweep with a slower normal velocity component, thereby losing lift. This effect generates a rolling moment, which can be quite powerful for high-swept wings; even for low-wing aircraft, it may require some anhedral angle to reduce the excessive roll stability (i.e. stiffness), especially for military aircraft, which require a quick response in a roll. The Tu-104 in Figure 6.7 is a good example of a low-wing military aircraft with a high sweep coupled with an anhedral shape.

The side force by the fuselage and V-tail contributes to the rolling moment, as shown in Figure 6.8. If the V-tail area is large and the fuselage has a relatively smaller side projection, then the aircraft CG is likely to be below the resultant side force, thus increasing the stability. Conversely, if the CG is above the side force, then there is a destabilizing effect.

Table 6.1 shows a summary of the forces, moments and sign conventions for stability analyses.

TABLE 6.1

Forces, moments and sign conventions

	Longitudinal static stability	Directional stability	Lateral stability
Angles	Pitch angle α	Side-slip angle, β	Roll angle, φ
Positive angle	When nose up	Nose to left	Right wing down
Moment coefficient	C_m	C_n	C_l
Positive moment	Nose up	Nose left	Right wing down

6.4   Theory

Forces and moments affect aircraft motion. In a steady level flight (in equilibrium), the summation of all forces is zero; the same applies to the summation of moments. When not in equilibrium, the resultant forces and moments cause the aircraft to manoeuvre. The following sections provide the related equations for each of the three aircraft planes. A sense of these equations helps with an understanding of aircraft performance.

6.4.1   Pitch Plane

In equilibrium, Σforce $=0$, when drag $=$ thrust and lift $=$ weight. An imbalance in drag and thrust changes the aircraft speed until equilibrium is reached. Drag and thrust act nearly collinearly; if they did not, pitch trim would be required to balance out the small amount of pitching moment it can develop. The same is true with the wing lift and weight, which are rarely collinear and generate a pitching moment. This scenario also must be trimmed, with the resultant lift and weight acting collinearly.

Together in equilibrium, Σmoment $=0$. Any imbalance results in an aircraft rotating about the Y-axis. Figure 6.9 shows the generalized forces and moments (including the canard) that act in the pitch plane. The forces are shown normal and parallel to the aircraft reference lines (i.e. body axes). Lift and drag are obtained by resolving the forces of Figure 6.9 into perpendicular and parallel directions to the free-stream velocity vector (i.e. aircraft velocity). The forces can be expressed as lift and drag coefficients, dividing by qS_w, where q is the dynamic head and S_w is the wing reference area. Subscripts identify the contribution made by the respective components. The arrowhead directions of component moments are arbitrary – they must be assessed properly for the components. With its analysis, Figure 6.9 gives a good idea of where to place aircraft components relative to the aircraft CG and NP. The static margin is the distance between the NP and the CG.

The generalized expression for the moment equation can be written as given in Equation 6.1. The equation sums up all the moments about the aircraft CG. In the trimmed condition the aircraft moment about the aircraft CG must be zero ($M_{ac\_cg} = 0$).

$$M_{ac\_cg} = \left(N_c \times l_c + C_c \times z_c + M_c\right)_{canard} + \left(N_w \times l_w + C_w \times z_w + M_w\right)_{wing} + \left(N_t \times l_t + C_t \times z_t + M_t\right)_{tail}$$
$$+ M_{fus} + M_{nac} + \left(thrust \times z_{th} + Nac\_Drag \times z_{th}\right) + \text{any other item} \quad \textbf{(6.1)}$$

The forces and the moments of each of the components are estimated first. When assembled together as aircraft, each component would be influenced by the flow fields of the others (e.g. flow over the horizontal tail is affected by the wing flow). Therefore a correction factor, η, is applied. This is shown in Equation 6.2 written in coefficient form, dividing by $qS_w c$, where q

■ **FIGURE 6.9**
Generalized force
and moment in pitch
plane

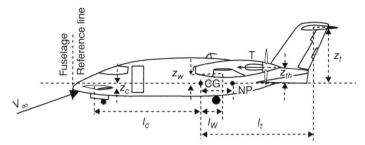

Geometry, velocity and force details

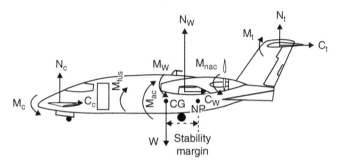

Force and moment details

is the dynamic head, c is the wing mean aerodynamic chord (MAC) and S_w is the wing reference area. Subscripts identify the contribution made by the respective components. The component moment coefficients are initially computed as isolated bodies and then converted to the reference wing area.

$$C_{mcg} = \left[C_{Nc}\left(S_c / S_w\right)\left(l_c / c\right) + C_{Cc}\left(S_c / S_w\right)\left(z_c / c\right) + C_{mc}\left(S_c / S_w\right)\right]_{\text{for canard}}$$
$$+ \left[C_{Nw}\left(l_a / c\right)\eta_w + C_{Cw}\left(z_a / c\right)\eta_w + C_{mw}\eta_w\right]_{\text{for wing}}$$
$$+ \left[C_{Nt}\left(S_t / S_w\right)\left(l_t / c\right)\eta_t + C_{Ct}\left(S_t / S_w\right)\left(z_t / c\right)\eta_t + C_{mt}\left(S_t / S_w\right)\eta_t\right]_{\text{for tail}} \quad \textbf{(6.2)}$$
$$+ C_{mfus} + C_{nac} + \left(\text{Thrust} \times z_{th} + Nac\_Drag \times z_{th}\right) / qS_w c$$

where $\eta \; (= q_i/q_\infty)$ represents the wake effect of lifting surfaces behind another lifting surface producing downwash, q_i is the incident dynamic head, and q_∞ is the free stream dynamic head. The vertical distances (z) of each component could be above or below the CG, depending on configuration as described below:

1. For fuselage-mounted engines, z_{th} is likely to be above the aircraft CG and its thrust would generate nose-down moment. Under-wing slung engines have z_{th} below the CG, generating nose-up moment. For most military aircraft, the thrust line is very close to the CG and hence for a preliminary analysis the z_{th} term can be ignored, that is, no moment is generated with thrust (unless vectored).

2. Drag of low wing below CG (z_a) will have nose-down moment, and vice versa for high wing. Mid-wing positions have to be noted in terms of which side of CG it lies, and z_a could be small enough to be ignored for a preliminary analysis.

3. Position of the H-tail shows the same effect as for the wing, but would invariably be above the CG. For a low H-tail, z_t can be ignored. In general, the drag generated by the H-tail is small and can be ignored. It is also true for a T-tail design.

4. It is for the same reason that the contribution by the canard vertical distance, z_c, can be ignored.

In summary, Equation 6.2 can be further simplified by comparing the order of magnitude of the contributions of the various terms. At first, the following simplifications are suggested.

1. The vertical distance of the canard and the wing from the CG is small. Therefore the terms with z_c and z_a can be dropped.

2. The canard and H-tail reference areas are much smaller that the wing reference area, and their C ($\approx$ drag) component force is less than a tenth of their lifting forces. Therefore the terms with C_{Cc} (S_c/S_w) and C_{Ct} (S_t/S_w) can be dropped (and even for a T-tail, but it is better to check its overall contribution).

3. High or low wing configurations will have z_a with opposite signs. For mid-wing, z_a may be small enough to be ignored.

Then Equation 6.2 can be simplified as given below:

$$C_{mcg} = \left[C_{Nc}\left(S_c/S_w \right)\left(l_c/c \right) + C_{mc}\left(S_c/S_w \right) \right] + \left[C_{Nw}\left(l_a/c \right)\eta_w + C_{mw}\eta_w \right]$$
$$+ \left[C_{Nt}\left(S_t/S_w \right)\left(l_t/c \right)\eta_t + C_{mt}\left(S_t/S_w \right)\eta_t \right]$$
$$+ C_{mfus} + C_{nac} + \left(\text{thrust} \times z_{th} + Nac\_Drag \times z_{th} \right)/qS_wc \qquad (6.3)$$

Conventional aircraft do not have a canard, so Equation 6.3 can be further simplified to Equation 6.4. The conventional aircraft CG is possibly ahead of the wing MAC. In this case, the H-tail needs to have negative lift to trim the moment generated by the wing and body.

$$C_{mcg} = \left[C_{Nw}\left(l_a/c \right) + C_{mw} \right] + \left[C_{Nt}\left(S_t/S_w \right)\left(l_t/c \right)\eta_t + C_{mt}\left(S_t/S_w \right)\eta_t \right]$$
$$+ C_{mfus} + C_{nac} + \left(\text{thrust} \times z_{th} + Nac\_Drag \times z_{th} \right)/qS_wc \qquad (6.4)$$

Normal forces can now be resolved in terms of lift and drag, and for small angles of α, the cosine of the angle is taken as 1. The drag components of all the C_N terms are very small and can be neglected.

Then the first term,

$$C_{Nw}\left(l_a/c \right) + C_{mw} \approx C_{Lw}\left(l_a/c \right) + C_{mw}$$

and the second term,

$$C_{Nt}\left(S_t/S_w \right)\left(l_t/c \right)\eta_t + C_{mt}\left(S_t/S_w \right)\eta_t \approx C_{Lt}\left(S_t/S_w \right)\left(l_t/c \right)\eta_t + C_{mt}\left(S_t/S_w \right)\eta_t$$

where C_{mt} (S_t/S_w) $\eta_t \ll C_{Lt}$ $(S/S_w)(l_t/c)$ η_t, hence the moment contribution by the horizontal tail is represented as $C_{m\_HT} = C_{Lt}$ $(S_H/S_w)(l/c)$ η_{HT}. Then Equation 6.4 is rewritten as (note the sign):

$$C_{mcg} = \left[C_{Lw}\left(l_a/c \right) + C_{mw} \right] + C_{m\_HT} + C_{mfus} + C_{nac} + \left(\text{thrust} \times z_{th} + Nac\_Drag \times z_{th} \right)/qS_wc \qquad (6.5)$$

$$C_{m\_HT} = -C_{LHT} \times \left[\left(l_t/S_{HT} \right)/\left(S_wc \right) \right]\eta_{HT} = -V_H\eta_{HT}C_{LHT} \qquad (6.6)$$

■ **FIGURE 6.10**
Pitch stability.

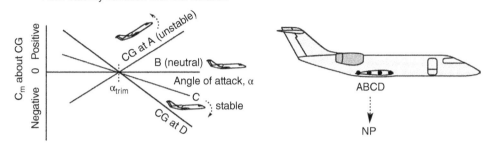

For conventional aircraft, C_{LHT} has a downwards direction to keep the nose up, and hence carries a negative sign.

Here, $$V_H = \text{horizontal tail volume coefficient} = (l_t / S_{HT}) / (S_w c) \tag{6.7}$$

Then Equation 6.5 without engines becomes as follows:

$$C_{mcg} = \left[C_{Lw} \left(l_a / c \right) + C_{mw} \right] - C_{m\_HT} + C_{mfus} \tag{6.8}$$

The right-hand panel in Figure 6.10 shows the stability effects of different CG positions on a conventional aircraft. The stability margin is the distance between the aircraft CG and the NP (i.e. a point through which the resultant force of the aircraft passes). When the CG is forward of the NP, then the static margin has a positive sign and the aircraft is statically stable. The stability increases as the CG moves further ahead of the NP. There is a convenient range from the CG margin in which the aircraft design exhibits the most favourable situation. In Figure 6.10, position B is where the CG coincides with the NP and shows neutral stability (i.e. at a zero stability margin) – the aircraft can still be flown with the pilot's efforts controlling the aircraft attitude. In fact, an aircraft with relaxed stability, outside the usual limits, can have a small negative margin that requires little force to make rapid manoeuvres – these aircraft invariably have a FBW control architecture in which the aircraft is continuously controlled by a computer.

Engine thrust has a powerful effect on stability. If it is placed above and behind the CG such as in an aft-fuselage-mounted nacelle, it causes an aircraft to have a nose-down pitching moment with thrust application. For an underslung wing nacelle ahead of the CG, the pitching moment is with the aircraft nose up. It is advisable for the thrust line to be as close as possible to the aircraft CG (i.e. a small z_e to keep the moment small). High-lift devices also affect aircraft pitching moments, and it is better that these devices be close to the CG.

In summary, designers must carefully consider where to place components to minimize the pitching-moment contribution, which must be balanced by the tail at the expense of some drag affecting aircraft performance – this is unavoidable but can be minimized.

6.4.2   Yaw Plane

The equation of motion in the yaw plane can be set up in a similar manner as in the pitch plane. The weathercock stability of the vertical tail contributes to the restoring moment.

Figure 6.4 depicts moments in the yaw plane. In the diagram the aircraft is yawing to the left with positive yaw angle β. This would generate a destabilizing moment by the fuselage with the moment $(N_F = Y_F \times l_f)$, where Y_F is the resultant side force by the fuselage and l_f is the distance of L_f from the CG. Contributions from wing, H-tail and nacelle are small (small projected areas and/or shielded by fuselage projected area). The restoring moment is positive when it tends to turn the nose to the right to realign with the airflow. The weathercock stability of the vertical tail causes the restoring moment ($N_{VT} = Y_T \times l_t$), where Y_T is the resultant side force on the V-tail (for small angles of ($\beta + \sigma$), it can be approximated as the lift generated by the V-tail, L_{VT}) and l_t is the distance of L_T from the CG. Therefore, the total aircraft yaw moment, N (conventional aircraft), is the summation of N_F and N_{VT} as given in Equation 6.9.

$$N_{ac\_cg} = N_F + N_{VT} \tag{6.9}$$

At equilibrium flight,

$$N_{ac\_cg} = 0 \quad \text{i.e.,} \quad N_{VT} = -N_F \tag{6.10}$$

In coefficient form, the fuselage contribution can be written as:

$$C_{nf} = -k_n k_{Rl} N_F \left[\left(S_f l_f \right) / \left(S_w b \right) \right] \beta \tag{6.11}$$

where k_n is an empirical wing–body interference factor, k_{Rl} is an empirical correction factor, S_f is the projected side area of the fuselage, l_f is the fuselage length, and b is the wing span.

In coefficient form, the V-tail contribution can be written as (L_{VT} in coefficient form is C_{LVT}):

$$C_{nVT} = \left[\left(l_t / S_{VT} \right) / \left(S_w c \right) \right] \eta_{VT} C_{LVT} = L_{VT} V_V \eta_{VT} C_{LVT} \tag{6.12}$$

where

$$V_V = \text{vertical tail volume coefficient} = \left(l_t / S_{VT} \right) / \left(S_w c \right) \tag{6.13}$$

Then Equation 6.13 in coefficient form becomes:

$$C_{n\_cg} = -k_n k_{Rl} N_F \left[\left(S_f l_f \right) / \left(S_w b \right) \right] \beta + L_{VT} V_V \eta_{VT} C_{LVT} \tag{6.14}$$

6.4.3   Roll Plane

As explained earlier, roll stability derives mainly from the following three aircraft features:

1. *Wing dihedral* Γ: Side-slip angle β will increase the AOA, α, on the windward wing:

$$\Delta\alpha = \left(V sin\Gamma \right) / u, \text{generating } \Delta Lift$$

For small dihedral and side-slip angle perturbations, $\beta = v/u$, which would approximate to $\Delta\alpha = \beta\Gamma$. Restoring moment is the result of $\Delta Lift$ generated by $\Delta\alpha$. It is quite powerful – normally, for low wing Γ it is between $1°$ and $3°$, depending on the wing sweep. For straight wing aircraft, the maximum dihedral angle rarely exceeds $5°$.

2. *Wing position relative to the fuselage* (Figure 6.10): Section 6.4.1 explains the contribution to rolling moment caused by different wing positions relative to the fuselage. Semi-empirical methods are used to determine the extent of the rolling moment contribution.

3. *Wing sweep at quarter chord,* $\Lambda_{¼}$: The lift produced by a swept wing is a function of the component of velocity V_n, normal to the $c_¼$ line; for steady rectilinear flight,

$V_n = V \cos \Lambda_{\frac{1}{4}}$. When the aircraft side-slips with angle β, the component of velocity normal to the $c_{\frac{1}{4}}$ line becomes (small β):

$$V'_n = V \cos \left(\Lambda_{\frac{1}{4}} - \beta \right) = V \left(\cos \Lambda_{\frac{1}{4}} + \beta \sin \Lambda_{\frac{1}{4}} \beta \right)$$

For the leeward wing,

$$V'_{n\_lw} = V \cos \left(\Lambda_{\frac{1}{4}} + \beta \right) = V \left(\cos \Lambda_{\frac{1}{4}} - \beta \sin \Lambda_{\frac{1}{4}} \beta \right)$$

Evidently, the windward wing has $V'_n > V_n$ and vice versa, hence will offer $\Delta Lift$ as the restoring moment in conjunction with a lift decrease on the leeward wing. As $\Lambda_{\frac{1}{4}}$ increases, the restoring moment becomes powerful enough so that it must be compensated for by the use of a wing anhedral configuration.

6.5 Current Statistical Trends for Horizontal and Vertical Tail Coefficients

Figure 6.11 provides additional statistics for current aircraft (21 civil and 9 military aircraft types), plotted separately for the H-tail and the V-tail. It is advised that readers create separate plots to generate their own aircraft statistics for the particular aircraft class in which they are interested, to obtain an appropriate average value. For civil aircraft designs, the typical H-tail area is about a quarter of the wing reference area. The V-tail area varies from 12% of the wing reference area, S_w, for large, long aircraft, to 20% for smaller, short aircraft.

Military aircraft require more control authority for greater manoeuvrability, and they have shorter tail-arms that require larger tail areas. The H-tail area is typically about 30–40% of the wing reference area. The V-tail area varies from 20% to 25% of the wing reference area. Supersonic aircraft have a movable tail for control. If a V-tail is too large, then it is divided into two halves.

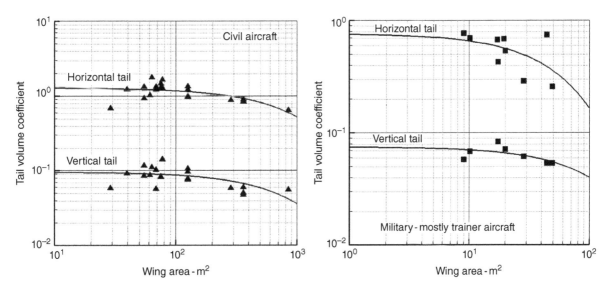

■ **FIGURE 6.11** Statistics of aircraft tail volume coefficients [11, 12]

Modern aircraft with FBW technology can operate with more relaxed stability margins, especially for military aircraft designs; therefore, they require smaller empennage areas compared with older conventional designs (see Section 6.8.3).

6.6   Inherent Aircraft Motions as Characteristics of Design

Once an aircraft is built, its flying qualities (how the aircraft performs) are the result of the effects of its mass (i.e. inertia), CG location, static margin, wing geometry, empennage areas and control areas. Flying qualities are based on a pilot's assessment of how an aircraft behaves under applied forces and moments. The level of ease or difficulty in controlling an aircraft is a subjective assessment by a pilot. In a marginal situation, recorded test data may satisfy airworthiness regulations yet may not prove satisfactory to the pilot. Typically, several pilots evaluate aircraft flying qualities to resolve any debatable points.

Practically all modern aircraft incorporate active control technology (FBW) to improve flying qualities. This is a routine design exercise and provides considerable advantage in overcoming any undesirable behaviour, which is automatically and continuously corrected. Described herein are six important flight dynamics of particular design interest. They are based on fixed responses associated with small disturbances, making the rigid-body aircraft motion linearized. Military aircraft have additional considerations as a result of nonlinear, hard manoeuvres. The six flight dynamics (seen as aircraft performance) are as follows:

- Short-period oscillation
- Phugoid motion (long-period oscillation)
- Dutch roll
- Slow spiral
- Roll subsidence
- Spin

6.6.1   Short-Period Oscillation and Phugoid Motion

The diagrams in Figure 6.12 show an exaggerated aircraft flight path (i.e. altitude changes in the pitch plane). In the pitch plane, there are two different types of aircraft dynamics that result from the damping experienced when an aircraft has a small perturbation. The two longitudinal modes of motion are as follows:

1. *Short-period oscillation* (SPO) is associated with pitch change (α change) in which the H-tail acts as a powerful damper (see Figure 6.1). If a disturbance (e.g. a sharp flick of the elevator and return) causes the aircraft to enter this mode, then recovery is also quick for a stable aircraft. The H-tail acts like an aerodynamic spring that naturally returns to equilibrium. The restoring moment comes from the force imbalance generated by the AOA, α, created by the disturbance. Damping (i.e. resistance to change) comes as a force generated by the tail plane, and the stiffness (i.e. force required) comes from the stability margin. The heavy damping of the H-tail resists changes to make a quick recovery.

    The bottom diagram of a short period in Figure 6.12 plots the variation of the AOA, α, with time. All aircraft have a short-period mode and it is not problematic for pilots. In a well-designed aircraft, oscillatory motion is almost unnoticeable because it damps out in about one cycle. Although aircraft velocity is only slightly affected, the AOA, α and the vertical height are related. Minimum α occurs at maximum vertical

■ **FIGURE 6.12**
**Short-period
oscillations and
phugoid motion
(modified from [7],
reproduced with
permission from
NASA)**

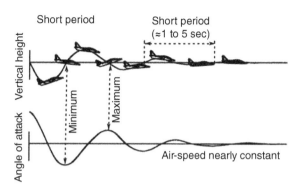

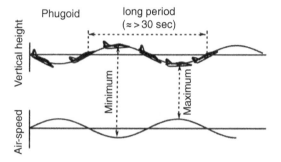

displacement, and maximum α occurs at about the original equilibrium height. The damping action offered by the H-tail quickly smoothes out the oscillation; that is, one oscillation takes a few seconds (typically, 1–5 seconds). The exact magnitude of the period depends on the size of the aircraft and its static margin. If the H-tail plane area is small, then damping is minimal and the aircraft requires more oscillations to recover.

2. *Phugoid motion* is the slow, oscillatory aircraft motion in the pitch plane, as shown in the bottom diagram in Figure 6.12. It is known as the *long-period oscillation* (LPO) – the period can last from 30 seconds to more than 1 minute. Typically, a pilot causes the LPO by a slow up-and-down movement of the elevator. In this case, the AOA, α, remains almost unchanged while in the oscillatory motion. The aircraft exchanges altitude gain (i.e. increases in potential energy, PE) for decreases in velocity (i.e. decreases in kinetic energy, KE). The phugoid motion has a long period, during which time the KE and PE exchange. Because there is practically no change in the AOA, the H-tail is insignificant in the spring–mass system. Here, another set of springs and masses is activated but is not shown in schematic form (it results from the aircraft configuration and inertia distribution – typically, it has low damping characteristics). These oscillations can continue for a considerable time and fade out comparatively slowly.

The frequency of a phugoid oscillation is inversely proportional to an aircraft's speed. Its damping also is inversely proportional to the aircraft lift-to-drag (L/D) ratio. A high L/D ratio is a measure of aircraft performance efficiency. Reducing the L/D ratio to increase damping is not preferred; modern designs with a high L/D ratio incorporate automatic active control (e.g. FBW) dampers to minimize a pilot's workload. Conventional designs may have a dedicated automatic damper at a low cost. Automatic active control dampers are essential if the phugoid motion has undamped characteristics.

All aircraft have an inherent phugoid motion. In general, the slow motion does not bother a pilot – it is easily controlled by attending to it early. The initial onset, because it is in slow motion, sometimes can escape a pilot's attention (particularly when instrument-flying), which requires corrective action and contributes to pilot fatigue.

6.6.2   Directional/Lateral Modes of Motion

Aircraft motion in the directional (i.e. yaw) and the lateral (i.e. roll) planes is coupled with side-slip and roll; therefore, it is convenient to address the lateral and directional stability together. These modes of motion are relatively complex in nature. FAR 23 (14CFR23), Sections 23.143 to 23.181, address airworthiness aspects of these modes of motion. Spinning as a post-stall phenomenon is discussed separately in Section 6.7.

The four typical modes of motion are (i) directional divergence, (ii) spiral, (iii) Dutch roll and (iv) roll subsidence. The limiting situation of directional and lateral stability produces two types of motion. When yaw stability is less than roll stability, the aircraft can enter directional divergence. When roll stability is less than yaw stability, the aircraft can enter spiral divergence. Figure 6.13 shows the two extremes of directional and spiral divergence. The Dutch roll occurs along the straight initial path, as shown in Figure 6.14. The wing acts as a strong damper to the roll motion; its extent depends on the wing aspect ratio. A large V-tail is a strong damper to the yaw motion. It is important to understand the role of damping in stability. When configuring an aircraft, designers need to optimize the relationship between the wing and V-tail geometries. The four modes of motion are as follows:

1. *Directional divergence*: This results from directional (i.e. yaw) instability. The fuselage is a destabilizing body, and if an aircraft does not have a sufficiently large V-tail to provide stability, then side-slip increases accompanied by some roll, with the extent depending on the roll stability. The condition can continue until the aircraft is broadside to the relative wind, as shown in Figure 6.13.

2. *Spiral*: If the aircraft has a large V-tail with a high degree of directional (i.e. yaw) stability but is not very stable laterally (i.e. roll), for example, a low-wing aircraft with no dihedral or sweep, then the aircraft banks as a result of rolling while side-slipping. This is a non-oscillatory motion with characteristics that are determined by the balance of directional and lateral stability. In this case, when an aircraft is in a bank

■ **FIGURE 6.13**
Spiral mode of motion showing divergence (from [7], reproduced with permission from NASA)

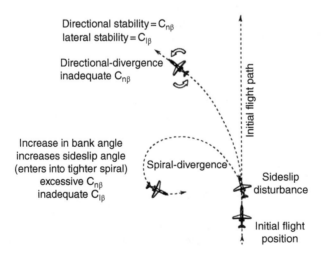

Directional stability = $C_{n\beta}$
lateral stability = $C_{l\beta}$

Directional-divergence
inadequate $C_{n\beta}$

Initial flight path

Increase in bank angle increases sideslip angle (enters into tighter spiral) excessive $C_{n\beta}$ inadequate $C_{l\beta}$

Spiral-divergence

Sideslip disturbance

Initial flight position

and side-slipping, the side force tends to turn the plane into the relative wind. However, the outer wing is travelling faster, generating more lift, and the aircraft rolls to a still higher bank angle. If poor lateral stability is available to negate the roll, the bank angle increases. The aircraft continues to turn into the sideslip in an ever-increasing (i.e. tighter) steeper spiral, which is *spiral divergence* (Figure 6.13). Spiral divergence is strongly affected by C_{lr}.

The initiation of a spiral is typically very slow and is known as a *slow spiral*. The time taken to double the amplitude from the initial state is long – 20 seconds or more. The slow build-up of a spiral motion can cause high bank angles before a pilot notices an increase in the *g*-force. If a pilot does not notice the change in horizon, this motion may become dangerous. Night-flying without proper experience in instrument-flying has cost many lives due to spiral divergence. Trained pilots should not experience the spiral mode as dangerous – they would have adequate time to initiate recovery actions. A Boeing 747 has a non-oscillatory spiral mode that damps to half amplitude in 95 seconds under typical conditions; many other aircraft have unstable spiral modes that require occasional pilot input to maintain a proper heading.

3. *Dutch roll*: A Dutch roll is a combination of yawing and rolling motions, as shown in Figure 6.14. It can happen at any speed, developing from the use of the stick (i.e. aileron) and rudder, which generate a rolling action when in yaw. If a sideslip disturbance occurs, the aircraft yaws in one direction and, with strong roll stability, then rolls away in a countermotion. The aircraft "wags its tail" from side to side, so to speak. The term *Dutch roll* derives from the rhythmic motion of Dutch ice skaters swinging their arms and bodies from side to side as they skate over wide frozen areas.

When an aircraft is disturbed in yaw, the V-tail performs a role analogous to the H-tail in SPO; that is, it generates both a restoring moment proportional to the yaw angle, and a resisting, damping moment proportional to the rate of yaw. Thus, one component of the Dutch roll is a damped oscillation in yaw. However, lateral stability responds to the yaw angle and the yaw rate by rolling the wings of the aircraft. Hence, the second component of a Dutch roll is an oscillation in a roll. The Dutch-roll period is short – of the order of a few seconds. In other words, the main contributors to the

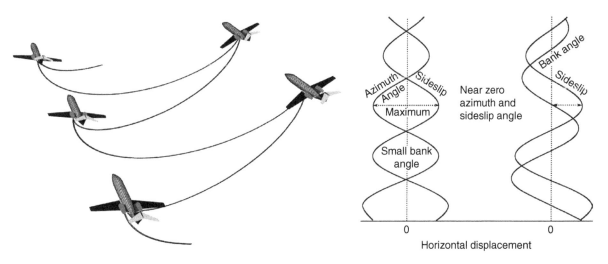

■ **FIGURE 6.14**   **Dutch roll motion**

Dutch roll are two forms of static stability: the directional stability provided by the V-tail and the lateral stability provided by the effective dihedral and sweep of the wings – both forms offer damping. In response to an initial disturbance in a roll or yaw, the motion consists of a combined lateral-directional oscillation. The rolling and yawing frequencies are equal but slightly out of phase, with the roll motion leading the yawing motion.

Snaking is a pilot term for a Dutch roll, used particularly at approach and landing when a pilot has difficulty aligning with the runway using the rudder and ailerons. Automatic control using yaw dampers is useful in avoiding the snaking/Dutch roll. Today, all modern transport aircraft have some form of yaw damper. The FBW control architecture serves the purpose well.

All aircraft experience the Dutch-roll mode when the ratio of static directional stability and dihedral effect (i.e. roll stability) lies between the limiting conditions for spiral and directional divergences. A Dutch roll is acceptable as long as the damping is high; otherwise, it becomes undesirable. The characteristics of a Dutch roll and the slow spiral are both determined by the effects of directional and lateral stability; a compromise is usually required. Because the slow-spiral mode can be controlled relatively easily, slow-spiral stability is typically sacrificed to obtain satisfactory Dutch-roll characteristics. High directional stability ($C_{n\beta}$) tends to stabilize the Dutch-roll mode but reduces the stability of the slow-spiral mode. Conversely, a large, effective dihedral (rolling moment due to sideslip, $C_{l\beta}$) stabilizes the spiral mode but destabilizes the Dutch-roll motion. Because sweep produces an effective dihedral and because low-wing aircraft often have excessive dihedral to improve ground clearance, Dutch-roll motions often are poorly damped on swept-wing aircraft.

4. *Roll subsidence*: The fourth lateral mode is also non-oscillatory. A pilot commands the roll rate by application of the aileron. Deflection of the ailerons generates a rolling moment, but the aircraft has a roll inertia and the roll rate builds up. Very quickly, a steady roll rate is achieved when the rolling moment generated by the ailerons is balanced by an equal and opposite moment proportional to the roll rate. When a pilot has achieved the desired bank angle, the ailerons are neutralized and the resisting rolling moment very rapidly damps out the roll rate. The damping effect of the wings is called roll subsidence.

6.7   Spinning

Spinning of an aircraft is a post-stall phenomenon. An aircraft stall occurs in the longitudinal plane; its nose drops as a result of loss of lift at stall. Unavoidable manufacturing asymmetry in geometry and/or asymmetric load application makes one wing stall before the other. This creates a rolling moment and causes an aircraft to spin around the vertical axis, following a helical trajectory while losing height – even though the elevator has maintained in an up position. The vertical velocity is relatively high (i.e. descent speed on the order of 30–60 m/s), which maintains adequate rudder authority, whereas the wings have stalled, losing aileron authority. Therefore, recovery from a spin is by the use of the rudder, provided it is not shielded by the H-tail (Section 2.18). Spinning is different from spiralling; it occurs in a helical path and not in a spiral. In a spiral motion, there is a large bank angle; in spinning, there is only a small bank angle. In a spiral, the aircraft velocity is sufficiently high and recovery is primarily achieved by using opposite ailerons. Spin recovery is achieved using the rudder and then the elevator. After straightening the aircraft with the rudder, the elevator authority is required to bring the aircraft nose down in order to gain speed and exit the stall.

There are two types of spin: a steep and a flat-pitch attitude of an aircraft. The type of spin depends on the aircraft inertia distribution. Most general aviation aircraft have a steep spin with the aircraft nose pointing down at a higher speed, making recovery easy – in fact, the best aircraft recover on their own when the controls are released (i.e. hands off). Conversely, the rudder authority in a flat spin may be low. A military aircraft with a wider inertia distribution can enter into a flat spin from which recovery is difficult and, in some cases, impossible. A flat spin for transport aircraft is unacceptable. Records show that the loss of aircraft in a flat spin is primarily from not having sufficient empennage authority in the post-stall wake of the wing.

The prediction of aircraft-spinning characteristics is still not accurate. Although theories can establish the governing equations, theoretical calculations are not necessarily reliable because too many variables are involved, and they require accurate values that are not easily obtainable. Spin tunnels are used to predict spin characteristics, but proper modelling on a small scale raises questions about its accuracy. In particular, the initiation of the spin (i.e. the throwing technique of the model into the tunnel) is a questionable art subjected to different techniques. On many occasions, spin-tunnel predictions did not agree with flight tests; there are only a few spin tunnels in the world.

The best method to evaluate aircraft spinning is in the flight test. This is a relatively dangerous task for which adequate safety measures are required. One safe method is to drop a large 'dummy' model from a flying 'mother' aircraft. The model has onboard, real-time instrumentation with remote-control activation. This is an expensive method. Another method is to use a drag chute as a safety measure during the flight test of the piloted aircraft. Spin tests are initiated at a high altitude; if a test pilot finds it difficult to recover, the drag chute is deployed to pull the aircraft out of a spin. The parachute is then jettisoned to resume flying. If a test pilot is under a high g-strain, the drag chute can be deployed by ground command, where the ground crew maintains real-time monitoring of the aircraft during the test. Some types of military aircraft may not recover from a spin once it has been established. If a pilot does not take corrective measures in the incipient stage, then ejection is the routine procedure. FBW technology avoids entering spins because air data systems recognize the incipient stage and automatic recovery measures take place.

6.8   Summary of Design Considerations for Stability

6.8.1   Civil Aircraft

Positioning of the wing relative to the fuselage depends on the mission role, but it is sometimes influenced by a customer's preference [1–3,13]. A high- or low-wing position affects stability in opposite ways (see Figure 6.6). The wing dihedral is established in conjunction with the sweep and position relative to the fuselage. Typically, a high-wing aircraft is anhedral and a low-wing aircraft is dihedral, which also assists in ground clearance of the wing tips. In extreme design situations, a low-wing aircraft can be anhedral (see Figure 6.7) and a high-wing aircraft can be dihedral. There are case-based "gull-wing" designs, which are typically for flying boats. Passenger-carrying aircraft are predominantly low-winged, but there is no reason why they should not have high wings; a few successful designs exist. Wing-mounted, propeller-driven aircraft favour a high wing for ground clearance, but there are low-wing, propeller-driven aircraft with longer undercarriage struts. Military transport aircraft invariably have a high wing to facilitate the rear loading of bulky items.

A conventional aircraft H-tail has a negative camber, the extent depending on the moment produced by an aircraft's tail-less configuration, as described previously. For larger, wing-mounted turbofan aircraft, the best position is a low H-tail mounted on the fuselage, the robust structure of which can accommodate the tail load. A T-tail on a swept V-tail increases the tail-arm but should be avoided unless it is essential, such as when dictated by an aft-fuselage-mounted engine. T-tail drag is destabilizing and requires a larger area if it is in the wing wake at nearly stalled attitudes. The V-tail requires a heavier structure to support the T-tail load. Smaller turbofan aircraft are constrained with aft-fuselage-mounted engines, which force the H-tail to be raised up from the middle to the top of the V-tail. The canard configuration affords more choices for the aircraft CG location. In general, if an aircraft has all three surfaces (i.e. canard, wing and H-tail), then they can provide lift with a positive camber of their sectional characteristics. It is feasible that future civil aircraft designs of all sizes may feature a canard.

Typically, a V-tail has a symmetric aerofoil, but for propeller-driven aeroplanes it may be offset by 1° or 2° to counter the skewed flow around the fuselage (as well as gyroscopic torque).

6.8.2   Military Aircraft – Non-linear Effects

Military aircraft often perform extreme manoeuvres involving large disturbances, and hence require non-linear stability analyses. Take as an example the early 1990s demonstrations by MIGs doing the spectacular "cobra" manoeuvres. Aircraft roll rate could be as high as 200 deg/s. Military aircraft flying qualities are tackled in MIL-STD-1797A, which supersedes MIL-STD-8785. In studying the stability of military aircraft, similar design considerations as for civil aircraft (e.g. small disturbances involving linear treatment) are initially used; but additional features associated with large disturbances and involving non-linear treatment must also be considered as indicated below (although these topics are beyond the scope of this book).

1. Inertial pitch and yaw divergence in roll manoeuvres.
2. Aerodynamic yaw departure at high angles of attack.
3. Wing rock.

There are yet other problems arising from weapons release simultaneously or asymmetrically, causing a sudden CG shift that could severely affect aircraft stability. Provision has to be made for quick recovery by fuel transfer in a short time – this is microprocessor-based management incorporated into FBW technology. Stealth of aircraft gives additional constraints to aircraft configuration. The F117 Nighthawk is a classic example of such considerations – it is an unstable aircraft and cannot fly without FBW.

Two modern combat aircraft configurations are shown in Figure 6.15: a delta wing design with one large V-tail, and a typical swing-wing twin tail configuration.

Supersonic flights would require an all-moving H-tail as shown in the figure. Also, at high AOA it is immersed in its wing trailing vortex system and becomes ineffective. In one

■ **FIGURE 6.15**
Typical modern fighter aircraft configurations

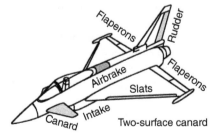
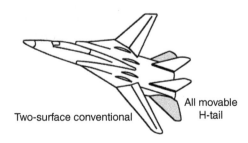

or two rare examples the fins (V-tail) are all moving surfaces. In many designs the all-moving surfaces are split with some elevator and rudder authority primarily serving as redundancy to protect against failures. A single large V-tail is not desirable for high performance combat aircraft. It cannot be canted and does not offer stealth. A tall single fin would also generate higher rolling moments in yaw, and its stability would depend on the *CG* position.

The use of twin canted fins (strictly speaking not a vertical tail) for military aircraft is common nowadays for the following reasons. A twin canted tail is not a Vee tail as there is a separate horizontal tail. (A Vee tail has to combine the work of both pitch and directional control, and the required size becomes very large in order to achieve the required authority.)

1. It is desirable that a vertical tail (fin) should be canted for stealth reasons (deflects radar signals), and hence two such tails are required. A twin fin also halves the size, easing structural considerations with very little weight penalty.
2. When the aircraft is yawed, the upwind canted fin is less effective than the downwind fin, but together they provide the desired authority.
3. Twin-tail aircraft do not need to have a separate speed brake. To achieve the braking action, the two rudders are deflected in opposite directions, and similarly for spoiler and flaps.

At supersonic speeds the aerodynamic centre moves aft, making the aircraft more stable. At low speeds as the aerodynamic centre moves forward, thrust vectoring in the pitch plane ($\pm 20°$) is helpful. Thrust vectoring is mainly used in low-speed extreme manoeuvres. In high Mach or high q (low altitude) conditions, the AOA is low.

Delta or all-wing designs use wing reflex. If the V-tail is eliminated to avoid a radar signature, then splitting of the ailerons for directional control is necessary (as in the B2 bomber). Such features will invariably require FBW designs.

6.8.3 Active Control Technology (ACT) – Fly-by-Wire

It is clear that stability considerations are important in aircraft design configurations. Although the related geometric parameters are from statistical data of past designs and subsequently sized, this chapter provides a rationale for their role in the conceptual design stage. To control inherent aircraft motions, feedback control systems such as a *stability augmentation system* (SAS) (e.g. a yaw damper) and a *control augmentation system* (CAS) have been routinely deployed for some time. The rationale continues with a discussion on how the feedback control system has advanced to the latest technologies, such as FBW and fly-by-light (FBL), known collectively as active control technology (ACT). Today, almost all types of larger aircraft incorporate some form of ACT.

FBW is basically a feedback control system based on the use of digital data. Earlier SASs and CASs had mechanical linkage from the pilot to the controls; FBW does not have the direct linkage. It permits the transmission of several digital signal sources through one communications system, known as *multiplexing*. A microprocessor is in the loop that continuously processes air data (i.e. flight parameters) to keep an aircraft in a preferred motion with or without pilot commands. Aircraft control laws – algorithms relating a pilot's command to the control-surface demand and aircraft motion, height and speed, which involve equations of motion, aircraft coefficients and stability parameters – are embedded in the computer to keep the aircraft within the permissible flight envelope. Under the command of a human pilot, the computer acts as a subservient flier. The computer continuously monitors aircraft behaviour and acts accordingly, ensuring a level of safety that a human pilot cannot match.

The flight-control computer takes the pilot's steering commands, which are compared to the commands necessary for aircraft stability to ensure safety and that control surfaces are activated accordingly. Air data are continually fed to the computers (i.e. speed, altitude and attitude). Built into the computer are an aircraft's limitations, which enable it to calculate the optimum control-surface movements. Steering commands are no longer linked mechanically from the cockpit to the control surface, but rather via electrical wiring. FBW flight-control systems seem to be the ideal technology to ensure safety and reduce a pilot's workload.

FBW reacts considerably faster than a conventional control system and does not encounter fatigue problems. A strong driver for incorporating FBW in military aircraft design is the ability to operate at relaxed stability (even extending to a slightly unstable condition) used for rapid manoeuvre (increased agility) as a result of minimal stiffness in the system. It is difficult for a typical pilot to control an unstable aircraft without assistance; a computer is needed and a regulator supplies the necessary stability. This system does not generate the

■ **TABLE 6.2**

Conventional and CCV comparison

	Conventional	CCV
MTOM (kg)	38,000	38,000
OEM (kg)	27,490	26,764
S_W (m²)	130	130
S_H (m²)	26	15.8
Payload (kg)	5000	5730
CG range (%MAC)	15–35	32–53

■ **FIGURE 6.16**
Comparison between a conventional (upper half) and a CCV design (lower half) [13].

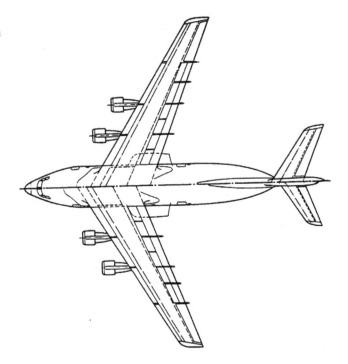

natural stability of a conventional aircraft, but automatically trims the aircraft to the preferred flight conditions. Progress in FBW systems depends to a great extent on the progress of onboard computer power.

An aircraft flying under relaxed stability using FBW does not have the same requirement for geometric features to provide low stiffness and damping. Hence stability and control-surface sizing is different than in a conventional design: they are smaller and lighter with less drag. This is what is meant by a *control configured vehicle* (CCV).

Stable designs already have a down-pitching force because of the position of the NP aft of the CG. Any balancing force must be generated by a larger downward lift of the H-tail. Again, this decreases the maximum possible lift and increases the trim drag. In an unstable layout (e.g. the CG moving aft), the elevator's lift is directed upward to counterbalance the moment. In this way, the aircraft's total lift is increased; the aircraft wing can therefore be designed to be smaller and lighter and still provide the same performance. There is another benefit from the use of an unstable design: in addition to the aircraft's increased agility, there is a reduction in drag and weight.

The difference between a conventional and a CCV design is shown in Table 6.2 and Figure 6.16 [13] for longitudinal stability. The wing area and maximum takeoff mass (MTOM) of both designs were unchanged; the CCV design yielded a smaller H-tail area with a larger CG range. The directional stability exhibits similar gains with a smaller V-tail area, thereby further reducing the operating empty mass (OEM) and permitting a bigger payload. This can be translated to performance gain as the incremental payload can be traded for fuel to increase range.

References

1. Perkins, C.D., and Hege, R.E., *Airplane Performance, Stability and Control*. John Wiley & Sons, Inc., New York, 1988.
2. Etkin, B., *Dynamics of Flight*. John Wiley & Sons, Inc., New York, 1959.
3. Nelson, R.C., *Flight Stability and Automatic Control*. McGraw-Hill, New York, 1989.
4. Phillip, W.F., *Mechanics of Flight*. John Wiley & Sons, Inc., Hoboken, NJ, 2004.
5. Watterson, J., Class Lecture Notes at the Queen's University Belfast, Belfast, UK. 2000.
6. US DATCOM. The USAF Stability and Control Digital DATCOM. 1979.
7. Talay, T.A., *Introduction to the Aerodynamics of Flight*. NASA SP-367. NASA, Langley, 1975.
8. Roskam, J., *Airplane Design*, vol. 1–8. University of Kansas, Lawrence, KS, 1990.
9. McCormick, B.W., *Aerodynamics, Aeronautics and Flight Mechanics*. John Wiley & Sons, Inc., New York, 1995.
10. Russell, J.B., *Performance and Stability of Aircraft*. Butterworth-Heinemann, Oxford, 1996.
11. Munroe, A., Statistics and Trends for Vertical and Horizontal Tails of Commercial Transport and Military Aircraft. Undergraduate project, with author as one of the supervisors.
12. Murphy, S., Statistics and Trends for Vertical and Horizontal Tails of Commercial Transport and Military Aircraft. Undergraduate project, with author as one of the supervisors.
13. Klug, H.G., *Transport Aircraft with Relaxed/Negative Longitudinal Stability – Results of a Design Study*. AGARD CP-157. 1975.

CHAPTER 7

Aircraft Power Plant and Integration

7.1   Overview

The engine may be considered as the heart of any powered aircraft as a system. Because of its importance, this largish topic is divided into two chapters; Chapter 7 gives the fundamentals of the associated theories and installation details when integrated with aircraft. Chapter 8 deals purely with engine performance, without which aircraft performance analyses cannot be progressed; it includes uninstalled and installed thrust/power and fuel flow data for various types of engines.

This chapter starts with a brief introduction to the evolutionary past of engines, followed by classification of the types of engines available and their domain of application, some fundamentals of engine theory, installation details, nacelles and thrust reversers (TRs). Primarily, this chapter deals with gas turbines (both jet and propeller driven) and to a lesser extent piston engines, which are used only in small general aviation aircraft. Therefore, propeller theory performance is also included (the next chapter computes thrust developed by propellers).

7.2   Background

Gliders were flying long before the Wright brothers flew, but they could not install engines even when automobile piston engines were available – they were simply too heavy. Gustav Weisskopf made his own engine. The Wright brothers made their own light gasoline engine with the help of Curtiss. Up till World War II, aircraft were designed around available engines. Aircraft sizing was a problem – the design was based on the number and/or the size of existing engine types that could be installed.

During the late 1930s, Frank Whittle in the UK and Hans von Ohain in Germany were working independently and simultaneously on reaction-type engines using vane/blade type pre-compression before combustion. Their efforts resulted in today's gas turbine engines. The end of World War II saw gas turbine powered jet aircraft in operation. A good introduction to jet engines is given in [1].

Post-war research led to the rapid development of gas turbines to a point where, from a core gas generator module, a family of engines could be produced in a modular concept (Figure 7.1) that allowed engine designers to offer engines as specified by the aircraft

Theory and Practice of Aircraft Performance, First Edition. Ajoy Kumar Kundu, Mark A. Price and David Riordan.
© 2016 John Wiley & Sons, Ltd. Published 2016 by John Wiley & Sons, Ltd.

■ **FIGURE 7.1**
Modular concept of
gas turbine design
(core module is also
known as the gas
generator module)

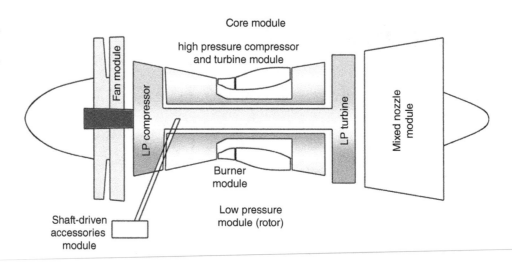

Figure 7.1 labels: Core module; high pressure compressor and turbine module; Fan module; LP compressor; LP turbine; Mixed nozzle module; Burner module; Low pressure module (rotor); Shaft-driven accessories module

designers. Similarity laws in thermodynamic design parameters allow power plants to be scaled (rubberized) to the requisite size around the core gas generator module, to meet the demands of the mission requirements. The size and characteristics of the engine are determined by matching with the aircraft mission. Now both the aircraft and the engine can be sized to the mission role, thereby improving operational economics. Modular engine design also favours less down-time for maintenance.

Potential energy locked in fuel is released through combustion in the form of heat energy. In gas turbine technology, the high energy of the combustion product can be used in two ways: (i) converted into an increase in kinetic energy(KE) of the exhaust to produce the reactionary thrust (turbojet/turbofan); or (ii) further extracted through an additional turbine to drive a propeller (turboprop) to generate thrust

Initially, the reactionary type engines came as simple straight-through airflow turbojets (see Section 7.4.1). Subsequently, turbojet development improved with the addition of a fan (long compressor blades visible from the outside) in front of the compressor – termed a turbofan.

The gas turbine operating environment demands more complex aerodynamic considerations than aircraft. Very high stress levels at considerably elevated temperatures, but the need to make it as light as possible, imposes stringent design considerations. Gas turbine parts manufacture is also a difficult task – very tough material has to be machined into a complex 3D shape to a very tight tolerance level. All these considerations make gas turbine design a very complex technology, possibly second to none. Engine control also involves very complex microprocessor-based management.

Gas turbine engines have a wide range of applications, from land-based large prime movers for power generation and shipping (civil and military) to weight-critical airborne applications. The theory behind all the types has a common base – the hardware design differs depending on the application requirements and technology level adopted. For example, land-based applications are not weight-critical and so have fewer constraints. Surface-based gas turbines have to run economically for days/months generating a very large power output, compared with stand-alone light aircraft engines running for hours with varying power, altitude, g-load and airflow demands. Even the biggest aircraft gas turbine is small compared with surface-based turbines.

Gas turbine design has advanced to incorporate sophisticated microprocessor-based control systems with automation, called FADEC(full authority digital engine control), working in conjunction with FBW(fly-by-wire) control of aircraft. Computer-aided design, computer-aided manufacture, computational fluid dynamics and finite element method (CAD/CAM/CFD/

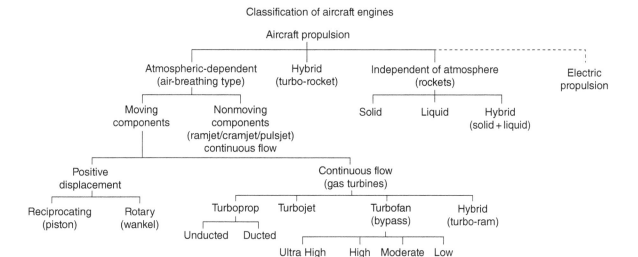

Classification of aircraft engines

■ **FIGURE 7.2**   **Classification of aircraft engines**

FEM) are now the standard tools for engine design. Current developments focus on having laminar flow in the intake duct, noise reduction and emission reduction.

Liquid cooled aircraft piston engines of over 3000 horse power (HP) have been built. Except for a few types, they are no longer in production as they are too heavy for the power generated; in their place, gas turbines have taken over. Gas turbine engines have a better thrust to weight ratio. Two of the successful piston engines were World War II types – the Rolls Royce Merlin and Griffon. They produced 1000–1500 HP and weighed around 1500 lb dry. Another factor is that AVGAS(aviation gasoline – petrol) is considerably more expensive than AVTUR(aviation turbine – kerosene). Today, the biggest piston engine in production is around 5000 HP. Of late, diesel fuel piston engines (less than 2500 HP) have entered the market for general aviation usage. In the home-built market, MOGAS(motor gasoline – petrol) powered engines have appeared, approved by certification authorities. Battery-powered small engines for very light aircraft are gaining ground.

Figure 7.2 classifies all types of aircraft engines in current usage. Electric propulsion is not included in this book, which will only cover the air-breathing types.

Application domains for the types dealt with in this book are shown in Figure 7.3 (excluding electric propulsion). High bypass ratio(BPR) turbofans are meant for high subsonic speeds. At supersonic speeds the bypass ratio comes down to less than 3. Typically, turboprop-powered aircraft reach speeds of around Mach 0.5 or below (higher speeds exceeding Mach 0.7 have been achieved). Piston engine powered aircraft are at the low speed end.

Turbofans (bypass turbojets) start to compete with turboprops for ranges over 1000 nm on account of time saved as a consequence of higher subsonic flight speed. Fuel cost is not the only consideration, depending on the sector of operation. Combat aircraft power plant uses lower bypass turbofans; in earlier days they used straight-through (no bypass) turbojets.

Figure 7.4a gives the thrust to weight ratio of various kinds of engine. Figure 7.4b gives the specific fuel consumption(sfc) at sea-level static take-off thrust (T_{SLS}) rating on an ISA day for various classes of current engines. At cruise, sfc would be higher.

Typical levels of specific thrust ($F / \dot{m}_a$ – lb/lb/s) and sfc (lb/h/lb) of various types of gas turbines are shown Figure 7.5.

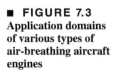

■ **FIGURE 7.3**
**Application domains
of various types of
air-breathing aircraft
engines**

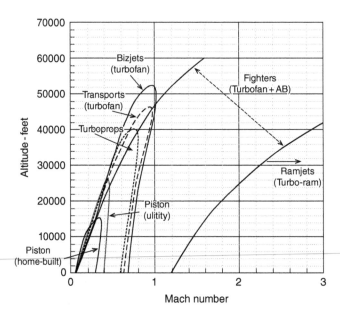

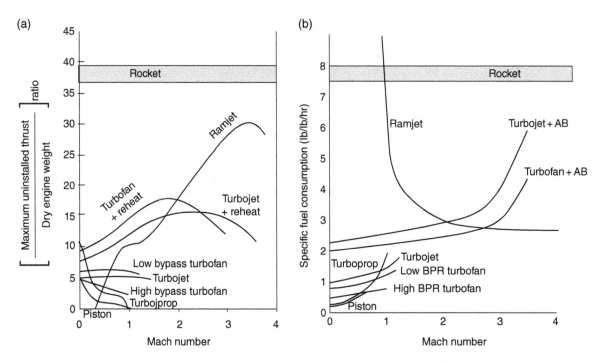

■ **FIGURE 7.4** Engine performance. **(a) Thrust to weight ratio. (b) Specific fuel consumption** (modified from [2])

Table 7.1 gives various kinds of efficiencies of the different classes of aircraft engine. Table 7.2 gives progress made in the last half a century. It indicates considerable advances made in engine weight savings. In the 1970s the engine noise levels came into force as a requirement to comply with certification. Pollution levels of noise and emissions are steadily being lowered.

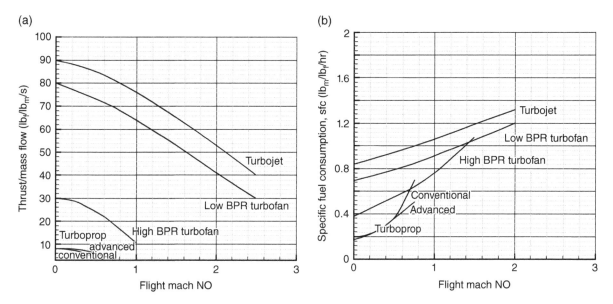

■ **FIGURE 7.5** Typical performance levels of various gas turbine engines. (a) Specific thrust. (b) Specific fuel consumption

■ **TABLE 7.1**
Efficiencies of engine types

	Thermal efficiency	Propulsive efficiency	Overall efficiency
Current types	0.60–0.65	0.75–0.85	0.50–0.55
Propfan	0.52–0.55	0.80–0.85	0.54–0.55
Advanced propfan	0.52–0.55	0.70–0.76	0.46–0.50
High bypass ratio	0.48–0.55	0.62–0.68	0.40–0.42
Low bypass ratio	0.40–0.50	0.55–0.60	0.35–0.38
Turbojet	0.40–0.45	0.45–0.52	0.28–0.32

■ **TABLE 7.2**
Progress in jet engines

	Thrust/weight ratio (lbf/lbm)
1950s (J69 class)	2.8–3.2
1960s (JT8D, JT3D class)	3.2–3.6
1970s (J79 class)	4.5–7.0
1980s (TF34 class)	6.0–6.5
1990s (F100, F404 class)	7.0–8.0
Current	8.0–10.0

7.3 Definitions

Below are the definitions of various terminologies used in jet engine performance analysis.

Specific fuel consumption (sfc) is the fuel flow rate required to produce one unit of thrust or SHP (shaft horse power).

$$\text{sfc} = (\text{fuel flow rate}) / (\text{thrust or power}) \tag{7.1}$$

Units of sfc are lb/h per pound of thrust produced (in SI units, gm/s/N) – the lower the better. To be more precise, the reaction-type engines use "tsfc" and propeller-driven ones use "psfc", where t and p stand for thrust and power, respectively. For turbofan engines (see Section 7.4.2):

$$\text{Bypass ratio, BPR} = \frac{\text{secondary airmass flow over the core engine}}{\text{primary airmass flow through the core (combustion)}} = \dot{m}_s / \dot{m}_p \tag{7.2}$$

Given below are the definitions of various kinds of efficiencies of jet engines. Subscript numbers indicate gas turbine component station numbers as shown in Figure 7.5 (in the figure, subscript 5 represents e).

Thermal efficiency:

$$\eta_t = \frac{\text{mechanical energy produced by the engine}}{\text{heat energy of (air + fuel)}} = \frac{W_E}{Q}$$

$$\text{or} \quad \eta_t = \frac{V_e^2 - V_\infty^2}{2C_p(T_3 - T_2)} = 1 - PR^{\frac{1-\gamma}{\gamma}} \tag{7.3}$$

For a particular aircraft speed, $V_\circ$, the higher the exhaust velocity V_e, the better would be the η_t of the engine.

Heat addition at the combustion chamber, $q_{2-3} = C_p(T_3 - T_1) \approx C_p(T_{t3} - T_{t1})$.

Propulsive efficiency:

$$\eta_p = \frac{\text{useful work done on airplane}}{\text{mechanical energy produced by the engine}}$$

$$\text{or} \quad \eta_p = \frac{W_A}{W_E} = \frac{2V_\infty}{V_e + V_\infty} \tag{7.4}$$

For subsonic aircraft, $V_e \gg V_\circ$. Clearly, for a given engine exhaust velocity, V_e, the higher the aircraft speed, then the better would be the propulsion efficiency, η_p. Jet aircraft flying below Mach 0.5 is not desirable – it is better to use propeller-driven aircraft flying at Mach 0.5 and below.

Overall efficiency:

$$\eta_o = \frac{\text{useful work done on airplane}}{\text{heat energy of (air + fuel)}} = \frac{W_A}{Q}$$

$$\text{or} \quad \eta_o = \frac{W_A}{Q} = \eta_t \times \eta_p = \frac{(V_e - V_\infty)V_\infty}{q_{2-3}} \tag{7.5}$$

It can be shown [3–5] that for non-afterburning engines, ideally the best overall efficiency, η_o, would be when the engine exhaust velocity, V_e, is twice the aircraft velocity, V_∞. Bypassed turbofans offer such an opportunity at high subsonic aircraft speed.

Recovery Factor, RF:

Intake Pressure Recovery Factor, RF, is defined as follows.

RF = (total pressure head at engine fan face)/(free stream air total head) = $p_{t1}/p_{t\infty}$

where subscript '∞' represents free stream condition and subscript '1' represent fan face

Pod mounted nacelles generally have short intake ducts that reduce the level of perturbed free-stream air entering the engine, thereby ensuring minimal loss of flow energy and yielding high RF. The RF during cruise speed is 0.98. Military aircraft with long intake ducts incur higher loss, hence the value of RF is lower, typically of the order of 0.92 to 0.96 at the design Mach number. In supersonic speed, associated shock waves further degrade the RF.

7.4 Air-Breathing Aircraft Engine Types

This section starts with describing various types of gas turbines followed by introducing piston engines. Aircraft propulsion depends on the extent of thrust produced by the engine. Chapter 8 presents the available thrust and power from various types of engine. Some statistics of various kinds of aircraft engine are given at the end of this chapter. Gas turbine sizes are progressing in making both larger and smaller engines than the current sizes, giving an expanded application envelope.

7.4.1 Simple Straight-through Turbojets

The most elementary form of gas turbine is a simple straight-though turbojet as shown schematically in Figure 7.6. In this case, the intake airflow goes straight through the whole length of the engine and comes out at a higher velocity and temperature after going through the processes of compression, combustion and expansion. It burns like a stove under a pressurized environment. The readers may note the *waisting* (narrowing) of the airflow passage as a result of the compressor reducing the volume while the turbine expands. Typically, at long-range cruise(LRC) conditions, the free-stream tube far upstream is narrower in diameter than at the compressor face. As a result, airflow ahead of the intake plane slows down during the pre-compression phase.

Components associated with the thermodynamic processes within the engine are assigned station numbers as shown in Table 7.3.

Overall engine efficiency can be improved if the higher energy of the exhaust gas of a straight-through turbojet is extracted through an additional turbine which can drive a fan in front of the compressor (when it is called turbofan engine) or a propeller (when it is called turboprop engine).

■ **FIGURE 7.6**
Sketch of a simple straight-through turbojet (bare engine does not have intake and exhaust nozzle)

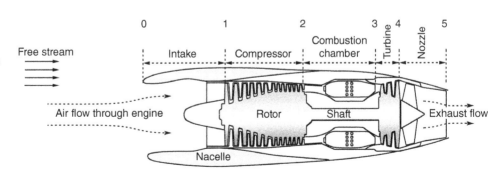

■ **TABLE 7.3**
Gas turbine station numbers

Number	Station	Description
∞	Free stream	Far upstream (if pre-compression is ignored than it is the same as 0)
0–1	Intake	A short divergent duct as a diffuser to compress inhaled air mass
1–2	Compressor	Active compression to increase pressure – temperature rises
2–3	Burner (combustion chamber)	Fuel is burned to release the heat energy
3–4	Turbine	Extracts the power from heat energy to drive the compressor
4–5	Nozzle (exhaust duct)	Generally convergent to increase flow velocity
5	e	Station 5 is also known as e, representing the exit plane

7.4.2 Turbofan – Bypass Engine

Energy extraction through the additional turbine lowers the rejected energy at the exhaust, resulting in lower exhaust velocity, pressure and temperature. The additional turbine drives a fan in front of the compressor. The large amount of air-mass flow through the fan provides thrust. The intake air-mass flow is split into two streams (Figure 7.7b); the core airflow passes through the combustion chamber of the engine as primary (or hot) flow and is made to burn, while the secondary flow through the fan is bypassed (hence the term bypass engine) around the engine and remains as cold flow. Figure 7.7a gives a schematic diagram of a bare PW 2037 turbofan engine. Lower exhaust velocity reduces engine noise. Also the lower exhaust pressure allows the nozzle exit area to be sized to make the exit pressure equal to ambient pressure (perfectly expanded nozzle), unlike simple turbojets which can have higher exit pressure.

Readers may note that the component station numbering system follows the same pattern as for the simple straight-through turbojet. The combustion chamber in the middle maintains the same numbers of 2–3. The only difference is the fan exit has a subscript of f. Intermediate stages of the compressor and turbine are primed ($'$).

Typically, the bypass ratio (BPR – Equation 7.2) for commercial jet aircraft turbofans (high subsonic flight speeds of less than 0.98 Mach) is around 4–6. Of late, turbofans for the new B787 have exceeded a BPR of 8. For military applications (supersonic flight speeds up to Mach 2.5), the BPR is around 1–3. Lower BPR keeps the fan diameter smaller and hence lowers frontal drag. Multi-spool drive shafts offer better efficiency and response characteristics, mostly with two concentric shafts. The shaft driving the low-pressure(LP) section runs inside the hollow shaft high-pressure(HP) section – see Figure 7.7b. Three shaft turbofans have been designed, but most of the current designs are with twin spools. The recent advent of geared turbofans offers better fuel efficiencies.

A lower fan diameter compared with a propeller permits higher rotational speeds and offers scope for having a thinner aerofoil section to extract better aerodynamic benefits. The higher the BPR, the better is the fuel economy. A higher BPR demands a larger fan diameter, and reduction gears may be required to keep the RPM at a desirable level. Ultra-high bypass ratio(UHBPR) turbofans approach the class of ducted fan or ducted propeller or propfan engines. Such engines have been built, but their cost versus performance means they are yet to break into the market arena.

7.4.3 Afterburner Jet Engines

An afterburner (AB) is another form of thrust augmentation exclusively meant for supersonic combat category aircraft (Concorde is the only civil aircraft that has used an afterburner). Figure 7.8 gives a cutaway diagram of a modern AB engine meant for combat aircraft.

(a)

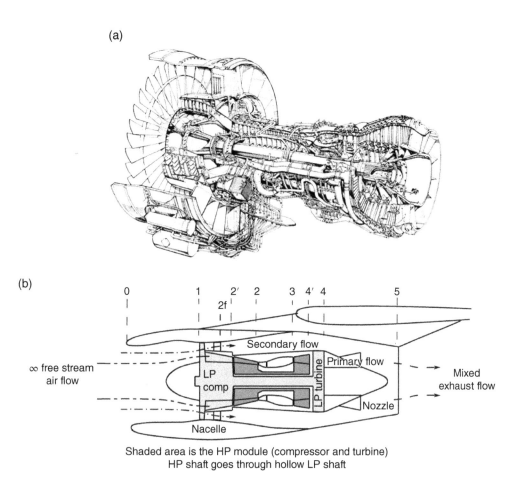

(b)

Secondary flow

∞ free stream
air flow

LP
comp

LP turbine

Primary flow

Mixed
exhaust flow

Nozzle

Nacelle

Shaded area is the HP module (compressor and turbine)
HP shaft goes through hollow LP shaft

■ **FIGURE 7.7**   Turbofan engine. (a) Pratt and Whitney 2037 turbofan. Reproduced with permission from Flightglobal.
(b) Schematic diagram of a pod-mounted long duct two-shaft turbofan engine

■ **FIGURE 7.8**
**Afterburning
engine – Volve-Flygmotor
RM8. Reproduced with
permission from
Flightglobal**

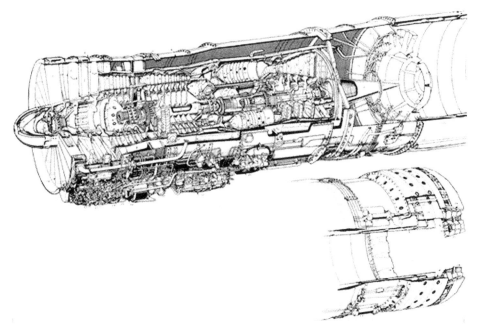

The simple straight-through turbojets or low bypass ratio turbofans have a relatively small frontal area to give low drag, and have excess air in the exhaust flow. If additional fuel can be burnt in the exhaust nozzle beyond the turbine exit plane, then additional thrust can be generated to propel aircraft at a considerably higher speed and acceleration, thereby also improving propulsive efficiency. However, the reason for using an AB arises from the mission demand, for example at takeoff with high payload, or acceleration to engage/disengage during combat/evasion manoeuvres. Mission demand overrides the fact that there is a very high level of energy rejection in the high exhaust velocity. The fuel economy with an AB degrades – it takes 80–120% higher fuel burn to gain 30–50% increase in thrust. Nowadays, most supersonic aircraft engines have some BPR when afterburning can be done within the cooler mixed flow past the turbine section of the primary flow.

7.4.4 Turboprop Engines

Lower speed aircraft can use propellers for thrust generation. Instead of driving a smaller encased fan (turbofan), a large propeller (hence known as turboprop) can be driven by a gas turbine to improve efficiency as the exhaust energy can be further extracted to very low exhaust velocities (nearly zero nozzle thrust). There could be some residual jet thrust left at the nozzle exit plane when it needs to be added to the propeller thrust. The nozzle thrust is converted into HP and together with the SHP(shaft horse power) generated; it becomes the equivalent shaft horse power(ESHP).

However, a large propeller diameter would limit rotational speed on account of both aerodynamic (transonic blade tips) and structural considerations (centrifugal force). Heavy reduction gears are required to bring down the propeller RPM to a desirable level. Propeller efficiency drops when aircraft operate at flight speeds above Mach 0.7. For shorter-range flights, the turboprop's slower speed does not become time-critical to the users, and yet it offers better fuel economy. Military transport aircraft are not as time-critical as commercial transport and may use turboprop aircraft (e.g. A400). Figure 7.9b gives a schematic diagram of a typical turboprop engine. Modern turboprops have up to eight blades (see Section 7.10), allowing reduction of the diameter and so a relatively high RPM. Advanced propeller designs allow aircraft to fly above Mach 0.7.

7.4.5 Piston Engines

Most of the aircraft piston engines are reciprocating types (positive displacement – intermittent combustion); the smaller ones have an air-cooled two-stroke cycle, and the bigger ones (typically over 200 HP) have a liquid-cooled four-stroke cycle. There are a few rotary-type positive displacement engines (Wankel), which in principle are very attractive but have some sealing problems. Costwise, the rotary-type positive displacement engines are not popular yet. Figure 7.10 shows an aircraft piston engine with its installation components.

To improve high-altitude performance (having low air density), supercharging is used. Figure 7.10 shows vane supercharging for pre-compression. Also aviation fuel (AVGAS – petrol) differs slightly from automobile petrol (MOGAS). Of late, some engines in the home-built category are able to use MOGAS. Recently, small diesel engines have been introduced into the market.

Piston engines are the oldest type to power aircraft. Over the life cycle of aircraft, gas turbines prove more cost-effective for engine sizes over 500 HP. Currently, general aviation aircraft and recreational small aircraft are the main users of piston engines.

■ **FIGURE 7.9**
Aircraft turboprop engines. (a) General Electric CT7. Reproduced with permission from Flightglobal. (b) Schematic drawing of a turboprop engine

(a)

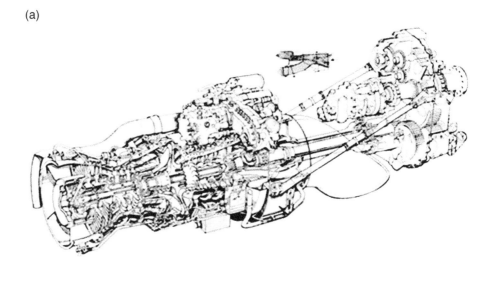

(b)

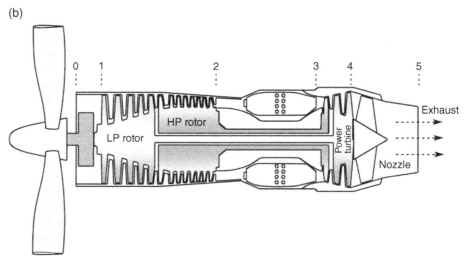

7.5   Simplified Representation of Gas Turbine (Brayton/Joule) Cycle

Figure 7.11a depicts a standard schematic diagram representing a simple straight-through turbojet engine as shown in Figure 7.6, with the appropriate station numbers. The thermodynamic cycle associated with gas turbines is known as the Joule cycle(also known as the Brayton cycle). Figure 7.11b gives the corresponding temperature–entropy diagram for an ideal Joule cycle in which compression and expansion takes place isentropically.

Real engine processes are not isentropic, and there are losses associated with increased entropy. A comparison of real and ideal cycles is shown in Figure 7.12.

■ **FIGURE 7.10**
Schematic of an aircraft piston engine and the supercharged scheme

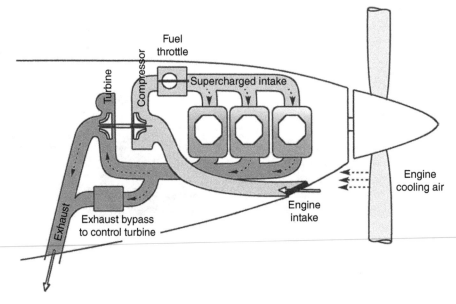

■ **FIGURE 7.11**
Simple straight-through jet.
(a) Generic schematic diagram.
(b) Ideal Joule cycle

(a)

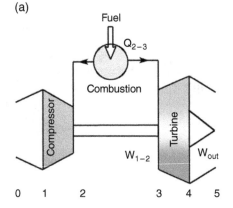

(b)

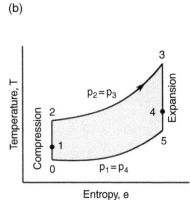

■ **FIGURE 7.12**
Real and ideal Joule (Brayton) cycle comparison of a straight-through jet

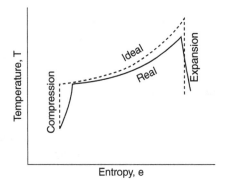

7.6   Formulation/Theory – Isentropic Case

The basic thermodynamic relationships and related gas turbine equations are given in Appendix C. These are valid for all types of processes. For more details, see [4–9].

7.6.1   Simple Straight-through Turbojets

Consider a control volume(CV– dashed line – note the waist-like shape of the simple turbojet) representing a straight-through axi-symmetric turbojet engine as shown in Figure 7.13. The CV and the component station numbers are as per the sketch and convention shown in Figure 7.6. Gas turbine intake starts with subscript 0 or ∞ and ends at nozzle exit plane with subscript 5 or e. Free-stream air mass flow rate(MFR), $\dot{m}_a$, is inhaled into the CV at the front face perpendicular to the flow, fuel MFR $\dot{m}_f$ (taken from an onboard tank) is added at the combustion chamber, and the product flow rate $(\dot{m}_a + \dot{m}_f)$ is exhausted out of the nozzle plane perpendicular to flow. It is assumed that the inlet face static pressure is p_∞, which is fairly accurate. Pre-compression exists, but for ideal consideration it has no loss.

No flow crosses the other two lateral boundaries of the CV as it is aligned with the walls of the engine. Force experienced by this CV is the thrust produced by the engine. Consider cruise conditions with an aircraft velocity of V_∞. At cruise, the demand for air inhalation is considerably lower than at takeoff. Intake area is sized in between the two demands. At cruise the intake stream tube cross-sectional area is smaller than the intake face area – it is closer to that of exit plane area, A_e (gas exiting at very high velocity). Since in ideal conditions there is no pre-compression loss, station 0 may be considered as having free-stream properties with subscript ∞.

From Newton's second law:

Applied force, F = rate of change of momentum + net pressure force

Note that the momentum rate is given by the MFR.

Where inlet momentum rate $= \dot{m}_a V_\infty$ exit momentum rate $= (\dot{m}_a + \dot{m}_f)V_e$

Therefore the rate of change of momentum $= (\dot{m}_a + \dot{m}_f)V_e - \dot{m}_a V_\infty$ **(7.6)**

Net pressure force between the intake and exit planes $=(p_e A_e - p_\infty A_\infty)$. The axi-symmetric side pressure at the CV walls cancels out. Typically, at cruise and sufficiently upstream, $A_e \approx A_\infty$. Therefore:

$$F = (\dot{m}_a + \dot{m}_f)\, V_e - \dot{m}_a V_\infty + A_e (p_e - p_\infty) = \text{net thrust}$$ **(7.7)**

■ **FIGURE 7.13**
Control volume representation of a straight-through turbojet

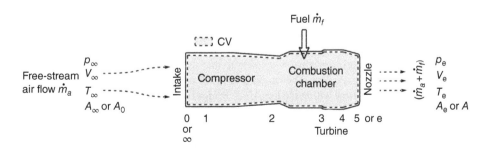

Then

$$\left(\dot{m}_a + \dot{m}_f\right)V_e + A_e\left(p_e - p_\infty\right) = \text{gross thrust}$$

and

$$\dot{m}_a V_\infty = \text{ram drag}\left(\text{with} - \text{ve sign, it has to be drag}\right).$$

It is the loss of energy seen as drag on account of slowing down of the income air as the ram effect. This gives:

$$\text{Net thrust} = \text{gross thrust} - \text{ram drag}$$
$$A_e\left(p_e - p_\infty\right) = \text{pressure thrust}$$

In general, subsonic commercial transport turbofans have a convergent nozzle, with its exit area sized in such a way that in cruise, $p_e \approx p_\infty$ (known as a perfectly expanded nozzle). It is different for military engines, especially with afterburning, when $p_e > p_\infty$ requiring a convergent–divergent nozzle.

For a perfectly expanded nozzle ($p_e - p_\infty$):

$$\text{Net thrust } F = \left(\dot{m}_a + \dot{m}_f\right)V_e - \dot{m}_a V_\infty \tag{7.8}$$

Further simplification is possible by ignoring the effect of fuel flow, $\dot{m}_f$, since $\dot{m}_a \gg \dot{m}_f$. Then the thrust for a perfectly expanded nozzle is:

$$F = \dot{m}_a V_e - \dot{m}_a V_\infty = \dot{m}_a\left(V_e - V_\infty\right) \tag{7.9}$$

At sea-level static take-off thrust (T_{SLS}) ratings $V_\infty = 0$, which makes:

$$F = \dot{m}_a V_e + A_e\left(p_e - p_\infty\right) \tag{7.10}$$

The expression indicates that the thrust increase can be achieved by increasing intake air MFR and/or increasing exit velocity.

Equation 7.4 gave propulsive efficiency: $\eta_p = \dfrac{2V_\infty}{V_e + V_\infty}$.

Clearly, jet propelled aircraft with low flight speeds will have poor propulsive efficiency, η_p. Jet propulsion is favoured for aircraft flight speeds above Mach 0.6.

The next question is, where does the thrust act? Figure 7.14 shows a typical gas turbine where the thrust is acting. It is all over the engine, and the aircraft realises the net thrust transmitted through the engine-mount bolts.

7.6.2   Bypass Turbofan Engines

Typically, in this book a long duct nacelle is preferred to obtain better thrust and fuel economy, offsetting the weight gain compared with short duct nacelles (see Section 7.7.1). Note that the pressure rise across the fan (secondary cold flow) is substantially lower than the pressure rise of the primary airflow. Also note that the secondary airflow does not have heat addition as in the primary flow. The cooler and lower exit pressure of the fan exit, when mixed with the primary hot flow within the long duct, brings the final pressure lower than the critical pressure, favouring a perfectly expanded exit nozzle ($p_e = p_\infty$). Through mixing, there is a reduction in jet velocity, which offers a vital benefit of noise reduction to meet airworthiness requirements. A long duct nacelle exit plane can be sized to expand perfectly.

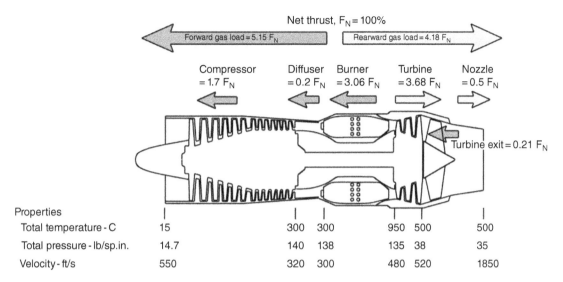

Net thrust, F_N = 100%

Forward gas load = 5.15 F_N

Rearward gas load = 4.18 F_N

Compressor = 1.7 F_N Diffuser = 0.2 F_N Burner = 3.06 F_N Turbine = 3.68 F_N Nozzle = 0.5 F_N

Turbine exit = 0.21 F_N

Properties						
Total temperature - C	15	300	300	950	500	500
Total pressure - lb/sp.in.	14.7	140	138	135	38	35
Velocity - ft/s	550	320	300	480	520	1850

■ **FIGURE 7.14**   **Where does the thrust act? [1]. Reproduced with permission from Rolls Royce**

The primary flow has a subscript designation of p and the secondary flow has a subscript designation of s. Therefore, F_p and V_{ep} stand for primary flow thrust and exit velocity, respectively, and F_s and V_{sp} stands for bypass flow thrust and its exit velocity, respectively. Thrust (F) equations of perfectly expanded turbofans are separately computed for primary and secondary flows, and then added to obtain the net thrust, F, of the engine (perfectly expanded nozzle, i.e. $p_e = p_\infty$).

$$F = F_p + F_s = \left[\left(\dot{m}_p + \dot{m}_f \right) \times V_{ep} - \dot{m}_p \times V_\infty \right] + \left[\dot{m}_s \times \left(V_{es} - V_\infty \right) \right]$$

Specific thrust in terms of primary flow becomes f (= fuel to air ratio) or

$$F / \dot{m}_p = \left[\left(1 + f \right) \times V_{ep} + BPR \times V_{es} - V_\infty \times \left(1 + BPR \right) \right] \tag{7.11}$$

If fuel flow is ignored, then

$$F / \dot{m}_p = \left[V_{ep} - V_\infty \right] + BPR \times \left(V_{es} - V_\infty \right) \tag{7.12}$$

Kinetic energy,

$$KE = \dot{m}_p \left[\frac{1}{2} \left(V_{ep}^2 - V_1^2 \right) \right] + \dot{m}_s \left[\frac{1}{2} \left(V_{sp}^2 - V_\infty^2 \right) \right]$$

or

$$KE / \dot{m}_p = \left[\frac{1}{2} \left(V_{ep}^2 - V_1^2 \right) \right] + BPR \times \left[\frac{1}{2} \left(V_{sp}^2 - V_\infty^2 \right) \right] \tag{7.13}$$

At a given design point, that is, flight speed $V_\circ$, BPR, fuel consumption and $\dot{m}_p$ are held constant. Then the best specific thrust and KE can be found by varying the fan exit velocity for a given V_{ep}. This is when setting the differentiation with respect to V_{es} is equal to zero. (This may be taken as trend analysis for ideal turbofan engines. The real engine analysis is more complex.)

Then by differentiating Equation 7.12,

$$d\left(F/\dot{m}_p\right)/d\left(V_{es}\right)=0=d\left(V_{ep}\right)/d\left(V_{es}\right)+\text{BPR} \tag{7.14}$$

and Equation 7.13 becomes

$$d\left(\text{KE}/\dot{m}_p\right)/d\left(V_{es}\right)=0=V_{ep}d\left(V_{ep}\right)/d\left(V_{es}\right)+\text{BPR}\times V_{sp} \tag{7.15}$$

Combining Equations 7.14 and 7.15, $-\text{BPR}\times V_{ep}+\text{BPR}\times V_{sp}=0$.
Since BPR ↑ 0, then the optimum has to be when

$$V_{ep}=V_{sp} \tag{7.16}$$

so the best specific thrust is when the primary (hot core flow) exit flow velocity equals the secondary (cold fan flow) exit flow velocity.

Equation 7.4 gave turbojet propulsive efficiency, $\eta_p=\dfrac{2V_\infty}{V_e+V_\infty}$, for a simple turbojet engine, but in the case of a turbofan there are two exit plane velocities: one each for the hot core primary flow (V_{ep}) and cold fan secondary flow (V_{es}). Therefore an equivalent mixed turbofan exit velocity (V_{eq}) could substitute for V_e in the above equation. Fuel flow rates are small and can be ignored. The equivalent turbofan exit velocity (V_{eq}) is obtained by equating the total thrust (perfectly expanded nozzle) as if it is a turbojet engine with total mass flow $\dot{m}_p+\dot{m}_s$. Thus

$$\left(\dot{m}_p+\dot{m}_s\right)\times\left(V_{eq}-V_\infty\right)=\dot{m}_p\times\left(V_{ep}-V_\infty\right)+\dot{m}_s\times\left(V_{es}-V_\infty\right)$$

or

$$\left(1+\text{BPR}\right)\times\left(V_{eq}-V_\infty\right)=\left(V_{ep}-V_\infty\right)+\text{BPR}\times\left(V_{es}-V_\infty\right)$$

or

$$\left(1+\text{BPR}\right)\times V_{eq}=\left(V_{ep}-V_\infty\right)+\text{BPR}\times\left(V_{es}-V_\infty\right)+V_\infty\times\left(1+\text{BPR}\right)=V_{ep}+\text{BPR}\times V_{es}$$

or

$$V_{eq}=\left[V_{ep}+\text{BPR}\times V_{es}\right]/\left(1+\text{BPR}\right) \tag{7.17}$$

Then turbofan propulsive efficiency:

$$\eta_{pf}=\frac{2V_\infty}{V_{eq}+V_\infty} \tag{7.18}$$

Large engines could benefit from weight savings by installing short-duct turbofans.

7.6.3   Afterburner Jet Engines

Figure 7.15 gives a schematic diagram with the station numbers for the afterburning jet engine. To keep numbers consistent with turbojet numbering system, there is no difference between stations 4 and 5 representing turbine exit conditions. Station 5 is the start and station 6 is the end of afterburning. Station 7 is the final exit plane. Figure 7.15 also shows the real cycle of afterburning in a T-s diagram.

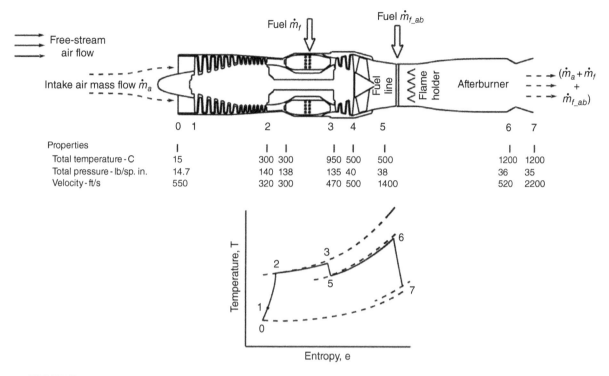

FIGURE 7.15 Afterburning turbojet and its *T-s* diagram (real cycle)

Afterburning is deployed only in military vehicles as a temporary thrust augmentation device, to meet mission demand at takeoff and/or for fast acceleration/manoeuvre in combat. Afterburning is applied at full throttle by flicking a fuel switch. The pilot can feel its deployment with a sudden increase in g levels in the flight direction. A ground observer would notice a sudden increase in noise levels. The afterburning glow is visible at the exit nozzle, and in darkness appears as a spectacular plume with supersonic expansion diamonds. In the absence of any downstream rotating machines, the afterburning temperature limit can be raised to around 2000–2200 K at the expense of a large increase in fuel flow (richer fuel/air ratio than in the core combustion).

The afterburning exit nozzle would invariably run choked and would require a convergent-divergent nozzle to give a supersonic expansion to increase momentum for the thrust augmentation. Typically, to gain a 50% thrust increase, fuel consumption increases by about 80–120%. That is why it is used for short periods, not necessarily in one burst. Interestingly enough, afterburning in bypass engines is an attractive proposition because the afterburning inlet temperature is lower. In fact, all modern combat category engines use low bypass of 1–3.

Losses in the afterburning exit nozzle are high – the flame holders, and so on, act as an obstruction. It is desirable to diffuse the flow speed at the afterburner to around 0.2–0.3 Mach – this results in a small bulge in the jet pipe diameter around that area, and a combat aircraft fuselage has to house this.

7.6.4 Turboprop Engines

Section 7.4.4 described turboprops. They have considerable similarities with turbojets/turbofans except that the high energy of the exhaust jet is utilized to drive a propeller by incorporating additional low pressure turbine stages as shown in Figure 7.9. Thrust developed by the propellers is the propulsive force for the aircraft. There could be a small amount of residual thrust left at the nozzle exit plane that should be added to the propeller thrust. Thrust power (T_p) and gas turbine shaft power (SHP) are related to propeller efficiency, η_{prop} as:

$$T_P = \text{SHP} \times \eta_{prop} + F \times V_\infty \tag{7.19}$$

ESHP is a convenient way to define the combination of shaft and jet power as follows:

$$\text{ESHP} = T_P / \eta_{prop} = \text{SHP} + \left(F \times V_\infty \right) / \eta_{prop} \tag{7.20}$$

Evidently, aircraft at a static condition will have ESHP = SHP as the small thrust at the exit nozzle is not utilized. As speed increases, ESHP > SHP as long as there is some thrust at the nozzle; sfc and specific power are expressed in terms of ESHP.

The formulae in Section 7.6 give the background for the gas turbine domain of application as shown in Figure 7.3. Turboprops would offer best economy for design flight speeds of Mach 0.5 and below, and are well suited for shorter ranges of operation. At higher speeds up to Mach 0.98, turbofans with high BPR offer the better efficiencies (see comments following Equation 7.5). At supersonic speed, the bypass ratio is reduced and in most cases use an afterburner. Small aircraft have piston engines up to " 500 HP, above which size gas turbines can compete.

7.7   Engine Integration to Aircraft – Installation Effects

Engine manufacturers supply bare engines to the aircraft manufacturers who install them on to aircraft. The same type of engine could be used by different aircraft manufacturers, each with their own integration requirements. Engine installation on to aircraft is a specialized technology that aircraft designers must understand, and is carried out by the aircraft manufacturers in consultation with the engine manufacturer.

In civil aircraft, a *nacelle* is the structural pod into which the supplied bare engine is installed. Multi-engine civil aircraft nacelles are invariably externally pod-mounted, either slung under or mounted over the wing or attached to the fuselage supported by pylons. Trijet aircraft configuration has its odd engine at the centreline placed at the aft of the aircraft. The front part of the nacelle is the intake and the aft end is the nozzle for the engine exhaust flow. Thrust reversers are installed mainly at the aft end around the nozzle. A typical nacelle pod with associated station numbers is shown in Figure 7.16.

A bare engine on a ground test bed inhales free-stream air directly and performs the best. In contrast, an engine installed in an aircraft can only inhale the airflow diffused through the nacelle/intake duct, incurring losses in pressure head, expressed as recovery loss as defined below:

Intake pressure recovery factor,

$$\text{RF} = \left(\text{total pressure head at engine fan face} \right) / \left(\text{free-stream air total head} \right)$$
$$= p_{t1} / p_{t\infty}$$

where subscript ∞ represents free-stream conditions and subscript 1 represent the fan face.

■ **FIGURE 7.16**
**Installation effects on
an engine**

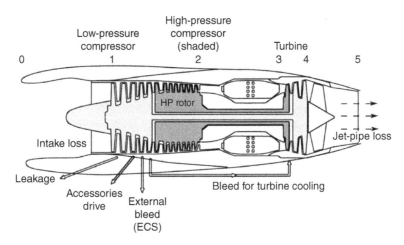

■ **FIGURE 7.16**
**Installation effects on
an engine**

Pod-mounted nacelles stand isolated on the wing or fuselage, and hence have the least perturbed free-stream air to flow through, offering minimal flow energy loss and high range factor (RF). The RF during cruise speed is around 0.98. During takeoff, the RF starts at a low value and gradually builds to high values as the aircraft speed increases to cruise speed.

Aircraft performance engineers use installed engine performance. Installation effects on the engine are those arising from having a nacelle, the losses at intake (RF) and exhaust, plus off-takes of power (to drive motor/generators) and air bleeds (anti-icing, environmental control, etc.). Total loss of thrust at takeoff can be as high as 8–10% of what is generated by a bare engine on the test bed. At cruise, it can be brought down to less than 5%.

Aircraft performance engineers must be given the following data to generate installed engine performance.

1. Intake and jet pipe losses.
2. Compressor air-bleed for the environmental control systems(ECS), for example, cabin air-conditioning and pressurization, de-icing/anti-icing, and any other purpose.
3. Power off-takes from the engine shaft to drive the electrical generator, accessories, and so on.

Propeller-driven engine nacelles also have similar considerations as podded nacelles, modified by the presence of a propeller (Section 7.7.2). Aircraft with one engine have it aligned in the plane of aircraft symmetry (engines with propellers can have a small lateral inclination of a degree or two about the aircraft centreline to counter the slipstream and gyroscopic effects from the rotating propeller). It was mentioned earlier that wing-mounted nacelles are the best to relieve wing-bending under flight load. Military aircraft have engines buried into the fuselage, and so do not have nacelles unless designers choose to have pods (some older designs). Military aircraft designers only consider intake design as treated in Section 7.8.2. The position of the nacelle with respect to the aircraft and their shaping to reduce drag are important considerations (Section 7.8).

7.7.1   Subsonic Civil Aircraft Nacelle and Engine Installation

A nacelle is a multifunctional system comprising: (i) inlet; (ii) exhaust nozzle; (iii) thrust reverser (TR), if required; and (iv) noise suppression system. The design aim of the nacelle is to minimize associated drag and noise, and to provide airflow smoothly to the engine in all flight conditions.

As of today, except for Concorde, all civil aircraft are subsonic with maximum speeds less than Mach 0.98. All subsonic aircraft use some form of pod-mounted nacelles, and in a way have become generic in design. Figure 7.17 shows a turbofan installed in a civil aircraft nacelle pod. Over-wing nacelles like those of the VFW614 is a possibility yet to be explored properly (the Honda jet aircraft has reintroduced this). Under-wing nacelles is the current best practice, but for smaller aircraft the ground clearance issues force nacelles to be fuselage-mounted.

There are two types of podded nacelles. Figure 7.18a figure shows a long-duct nacelle where both the primary and secondary flows mix within the nacelle. The mixing increases the thrust and reduces the noise level compared with the short-duct nacelles, possibly compensating

■ **FIGURE 7.17**
Installed turbofan an aircraft. Reproduced with permission from FlightGlobal

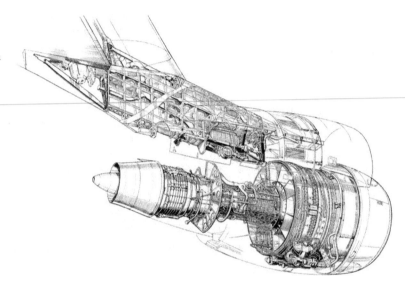

(a)

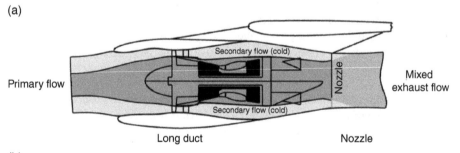

(b)

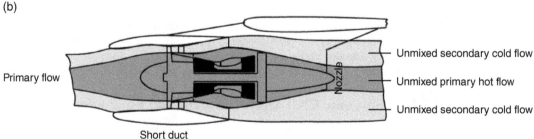

■ **FIGURE 7.18** Podded nacelle types. (a) Long duct. (b) Short duct. Reproduced with permission from Bombardier-Aerospace Shorts

for the weight gain through fuel and cost savings. Figure 7.18b shows a short-duct nacelle in which the bypassed cold flow does not mix with the hot core flow. The advantage is that there is considerable weight saving by cutting down the length of the outside casing of the nacelle by not extending up to the end. Short-duct nacelle length can vary.

The aircraft performance engineers are to substantiate to the certification authorities that the thrust available from the engine after deducting the losses are sufficient to cater for the full flight envelope as specified. It can become critical at aircraft takeoff if (i) the runway is not sufficiently long and/or at high altitude, (ii) if there is an obstruction to clear or (iii) ambient temperature is high. In that case, the aircraft may take off with less weight. Airworthiness requirements require that the aircraft has to maintain a minimum gradient (Chapter 11) at takeoff with one engine inoperative. Customer requirements could demand more than the minimum performance.

7.7.2   Turboprop Integration to Aircraft

The turboprop nacelle is also a multifunctional system comprising: (i) inlet; (ii) exhaust nozzle; and (iii) noise suppression system. Thrust reversal can be achieved by changing the propeller pitch angle sufficiently. There are primarily two types of turboprop nacelles as shown in Figure 7.19. The scoop intake could be above or below (as chin) the propeller spinner. Interestingly, quite a few turboprop nacelles have integrated undercarriage mounts with storage space in the same nacelle housing, as can be seen in Figure 7.19a. The other kind is with an annular intake as shown in Figure 7.19b. Installation losses are of the same order as discussed in the turbofan installation.

Turboprop nacelle position is dictated by the propeller diameter. The key geometric parameters for turboprop installation are shown in Figure 7.20. Figure 7.20a shows a wing-mounted turboprop installation. The overhang should be as much forward as the design can take, like that of the turbofan overhang to reduce interference drag. For high-wing aircraft, the turboprop nacelle is generally under-wing like the Bombardier Q400 aircraft. For low-wing aircraft nacelles are generally over the wing to get propeller ground clearance. Both types can house the undercarriage. The propeller slipstream assists lift and has a strong effect on static stability, while flap deployment aggravates the stability changes. Depending on the extent of wing incidence with respect to the fuselage, there is some angle between the wing chord line and the thrust line – typically from 2° to 5°.

A fuselage-mounted propeller-driven system arrangement is shown in Figure 7.20b. Note the angle between the thrust line and the wing chord line as it is with wing-mounted propeller nacelle. Sometimes the propeller axis is given about a degree down-inclination with respect to the fuselage axis. This assists with longitudinal stability. To counter the propeller slipstream effect, an inclination of a degree or two in the yaw direction can be given. Otherwise, the V-tail can be given such inclination to counter the effect.

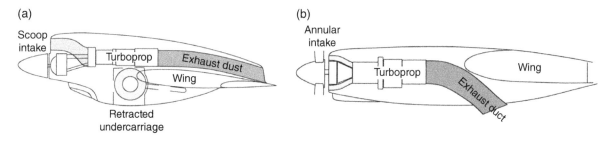

■ **FIGURE 7.19**   **Typical wing-mounted turboprop installation [8]. (a) Scoop intake. (b) Annular intake. Reproduced with permission from Cambridge University Press**

(a)

(b)

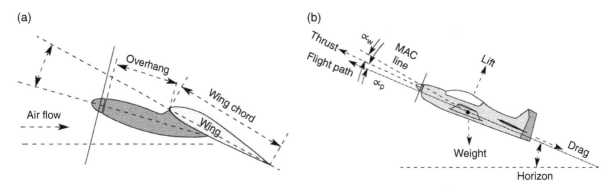

■ **FIGURE 7.20** Typical turboprop installation parameters. (a) Wing-mounted turboprop (note overhang). (b) Fuselage-mounted turboprop

■ **FIGURE 7.21**
Installed engine in a combat aircraft [8]. Reproduced with permission from Cambridge University Press

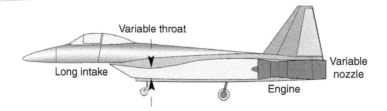

Piston engine nacelles on the wing follow the same logic. The older designs had a more closely-coupled installation.

7.7.3 Combat Aircraft Engine Installation

Combat aircraft have engines integral with the fuselage, mostly buried inside, but in some cases with two engines, they can bulge out to the sides. Therefore pods do not feature, unless it is required to have more than two engines on large aircraft. Figure 7.21 shows a turbofan installed on a supersonic combat aircraft.

The external contour of the engine housing is integral to fuselage mould-lines. Early designs had an intake at the front of the aircraft, pitot-type for subsonic fighters (Sabrejet F86, etc.), and with a movable centre body for supersonic fighters (MIG 21). The long intake duct snaking inside the fuselage below the pilot seat incurs high losses. Side intakes overtook the nose intake designs. A possible choice for side intakes is given in Section 2.20.3– primarily, they are side-mounted or chin-mounted.

Aircraft performance engineers must be given the following data to generate installed engine performance.

1. Intake losses plus losses arising from supersonic shock waves and the duct length.
2. Exit nozzle losses (military aircraft nozzle design is complex).
3. Additional losses at the intake and nozzle on account of suppression of exhaust temperature for stealth.
4. Compressor air-bleed for the ECS, for example, cabin air-conditioning and pressurization, de-icing/anti-icing, and any other purpose.
5. Power off-takes from the engine shaft to drive the electrical generator, accessories, for example pumps, and so on.

Military aircraft have excess thrust (with or without afterburner) to cater for hot and high-altitude conditions and operating from short airfields, and are capable of climbing at a steeper angle than civil aircraft. When the thrust/weight ratio is more than 1, then the aircraft is capable of climbing vertically.

7.8   Intake/Nozzle Design

7.8.1   Civil Aircraft Intake Design

Engine mass flow demand varies a lot, as shown in Figure 7.22. To size the intake area, the reference cross-section of incoming air-mass stream tube is taken at the maximum cruise condition as shown in Figure 7.22a, when it has a cross-sectional area almost equal to that of the highlighted area (i.e. $A_\infty = A_1$). The ratio of MFR in relation to the reference condition (air-mass flow at maximum cruise) is a measure of spread the intake would encounter. At the maximum cruise condition, MFR = 1 as a result of $A_\infty = A_1$. At the normal cruise condition the intake air mass demand is lower (MFR < 1, shown in Figure 7.22b). At the maximum takeoff rating (MFR > 1), intake air-mass flow demand is high and the streamline patterns are as shown in Figure 7.22c. Variation in intake mass flow demand is quite high.

If the takeoff airflow demand is high enough, then a blow-in door can be provided which closes automatically when the demand drops off. Figure 7.22d shows a typical flow pattern at incidence at high demand, when an automatic blow-in door may be necessary. At idle, the engine is kept running with very little thrust generation ($MFR \ll 1$). At inoperative conditions the rotor is kept windmilling to minimize drag. If the rotor is seized due to mechanical failure, then there will be a considerable increase in drag.

The engine (fan) face Mach number should not exceed Mach 0.5 to avoid degradation due to compressibility effects. At fan face Mach number above 0.5, the relative velocity at the fan tip region approaches near sonic speed on account of high rotational speeds.

The purpose of intake is to serve engine airflow demand as smoothly as possible – there should be no flow distortion at the compressor face on account of any separation and/or flow asymmetry. Nacelle intake lip cross-section is designed with similar logic to designing aerofoil leading-edge cross-sections – to avoid flow separation within the flight envelope. More details on subsonic pod-mounted intake design are given in references [7, 8].

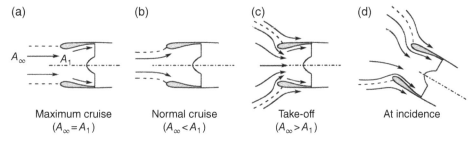

(a) Maximum cruise ($A_\infty = A_1$) (b) Normal cruise ($A_\infty < A_1$) (c) Take-off ($A_\infty > A_1$) (d) At incidence

Dashed line represents captured stream tube
A_∞ = Captured stream tube area far upstream
A_1 = Intake highlight area

■ **FIGURE 7.22**   **Subsonic intake airflow demand at various conditions (valid both for civil and military intakes). (a) MFR ratio = 1. (b) MFR ratio <1. (c) MFR ratio >1. (d) MFR ratio >1 (at climb)**

7.8.2   Military Aircraft Intake Design

Military aircraft have supersonic capabilities [9, 10] and therefore must manage the shock losses associated with the intake. Ideally, at design point (at supersonic cruise) the bow shock wave just attaches to the intake lip, which is sharp compared with a subsonic intake lip. For aircraft operating above Mach 2, a movable centre body keeps the oblique wave to the lip as the shock angle changes with speed change. The simplest centre body is a cone (or half cone for side-mounted nacelles). The cone could be in steps to give multiple oblique shocks at reduced intensity, that is, lowering shock loss. The best design is an Oswatitsch curved contour design, which generates infinite weak shocks with minimum loss.

Figure 7.23 gives four kinds of flow regimes associated with ideal supersonic intake with a fixed centre-body. Four types of operating situation arise as follows.

1. At the design flight speed seen as critical operation when it is desirable that the oblique shock wave just touches the lip, followed by other shocks culminating with a normal shock at the throat, beyond which airflow becomes subsonic (Figure 7.23). The captured free stream tube is shown by the highlighted area. Intake area is sized to inhale the air mass required at critical operation.

2. If the back pressure is lower than the critical operation (on account of throttle action), then more air mass is inhaled and the throat area remains supersonic, pushing back a stronger normal shock. This is known as supercritical operation (Figure 7.23). The captured free stream tube is still the same as the highlighted area and the oblique shock position depends on the aircraft speed.

3. When the back pressure is high, especially when below critical speed but still at supersonic level, the oblique shock is wider and is followed by a normal shock which pops out ahead of the nacelle lip and is known as subcritical operation (Figure 7.23). The captured free stream tube is smaller than the highlighted area. It is not as efficient as in critical operation, but loss is less than supercritical operation.

4. The last situation could happen at a particular combination of aircraft speed (below critical) and air mass inhalation when the normal shock outside keeps oscillating, which starves the engine and makes it run erratically, possibly leading to flameout. This is known "buzz". Aircraft must avoid this situation, and modern designs have FBW/FADEC to keep clear of it.

Modern combat aircraft with advanced missiles have BVR (beyond visual range) capabilities. Rapid manoeuvres at high speed is the combat specification, for which typical speeds are of the order of Mach 1.8 and the requirement for a movable centre body is not stringent. For side intakes, boundary layer bleed plates serve as the centre body to position oblique shock at the lip (critical design point operation).

■ **FIGURE 7.23**
Types of ideal supersonic intake demand conditions [9]. Reproduced with permission from the American Institute of Aeronautics & Astronautics

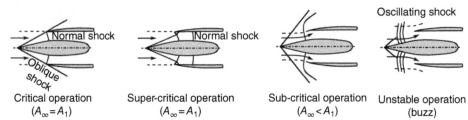

Critical operation
$(A_\infty = A_1)$

Super-critical operation
$(A_\infty = A_1)$

Sub-critical operation
$(A_\infty < A_1)$

Unstable operation
(buzz)

Dashed line represents captured stream tube
A_∞ = captured stream-tube area far upstream
A_1 = Intake highlight area

7.9   Exhaust Nozzle and Thrust Reverser

Thrust reversers (TRs) are not required by the regulatory authorities (FAA/CAA). These are expensive components, heavy and only applied on the ground, yet their impact on aircraft operation is significant on account of having additional safety through better control, reduced time to stop, and so on, especially at aborted takeoffs and other related emergencies. Airlines want to have TRs, even with increased DOC(direct operating cost).

A TR is part of the exhaust nozzle, so they are treated together in this section. While this section offers an empirical sizing method for the nozzle, it will not size or design the TR, which is a state of technology by itself. To tackle exhaust nozzles, it is desirable to know something about TRs beforehand.

The role of the TR is to retard aircraft speed by applying thrust in the forward direction, that is, in a reversed application. The rapid retardation by TR application reduces the required landing field length. In civil applications it is only used on the ground. Because of its severity, in-flight application is not permitted. It reduces the wheel brake load, hence reducing wear and heating-up hazards. The TR is very effective on slippery runways (ice, water, etc.) when braking is less effective. A typical example of the benefits to stopping distance for TR application during landing on icy runway is that it cuts down the required field length to less than half. A mid-sized jet transport aircraft would stop in about 4000 ft with TR, but 12,000 ft without it. Without the TR, the energy depleted to stop the aircraft is absorbed by the wheel brakes and aerodynamic drag. Application of the TR also offers additional intake momentum drag (at full throttle), contributing to energy depletion. TR is useful for manoeuvring aircraft backwards on the ground for parking, and so on, but most aircraft with TR still use a specialized vehicle for ground manoeuvres.

The TR is integrated on the nacelle, and is the obligation of aircraft manufacturers to design. Sometimes the design is subcontracted to specialist organizations devoted to TR design. Basically there are two types of civil aircraft TR design: those operating on both fan and core flow, and those operating on fan flow only. Their choice depends on BPR, nacelle location and customer specification.

7.9.1   Civil Aircraft Exhaust Nozzles

Civil aircraft nozzles are conical, into which the TR is integrated. Small turbofan aircraft may not need a TR, but aircraft of regional jet(RJ) size and above use a TR. Inclusion of a TR may slightly elongate nozzle length – this will be ignored in this book.

In general, the nozzle exit area is sized as a perfectly expanded nozzle ($p_e = p_\infty$) at LRC conditions. At higher engine ratings it has $p_e > p_\infty$. The exit nozzle of a long-duct turbofan does not run choked at cruise ratings. At takeoff ratings, the back pressure is high at lower altitude, and thereby a long-duct turbofan could escape from running choked (low-pressure secondary flow mixes within the exhaust duct). The exhaust nozzle runs with a favourable pressure gradient, and hence in its shaping it is relatively simple to establish geometric dimensions. However, it is not simple engineering at elevated temperatures and with the need to suppress noise levels.

The nozzle exit plane is at the end of the engine. Its length from the turbine exit plane is about 0.8–1.5 fan face diameter. The nozzle exit area diameter may be taken roughly as half to three-quarters of the intake throat diameter.

7.9.2   Military Aircraft *TR* Application and Exhaust Nozzles

The afterburning military engine TR is integral to the nozzle design and is positioned at the end of the fuselage. Afterburning engine nozzles always run choked at maximum cruise rating, and has a variable convergent–divergent (de Laval) nozzle to match with the throttle demands.

■ **FIGURE 7.24**
Supersonic nozzle area adjustment and thrust vectoring [9, 10]. Reproduced with permission from the American Institute of Aeronautics & Astronautics, and Patrick Stephens, Ltd.

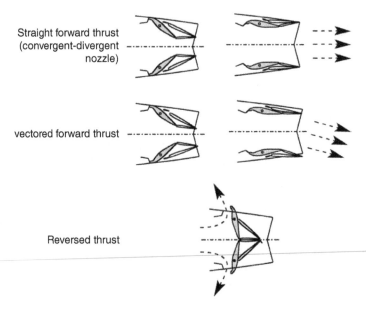

Straight forward thrust (convergent-divergent nozzle)

vectored forward thrust

Reversed thrust

Of late, all combat aircraft have thrust vectoring capabilities by deflecting the exhaust jet to the desired angles, affecting aircraft manoeuvrability capabilities.

The lower diagram of Figure 7.24 shows that the integral mechanism can even provide a mild form of in-flight thrust reversing to spoil thrust (air-braking action) to wash out high speed at the approach to land or in combat to. The full extent of TR deployment is shown in the last figure. This has an integral mechanism capable of adjustment for all demands.

7.10   Propeller

Aircraft flying at speeds less than Mach 0.5 are propeller-driven, with bigger aircraft powered by gas turbines and smaller ones by piston engines. More advanced turboprops have pushed the flight speed to over Mach 0.7 (Airbus A400). This book deals with the conventional types of propellers operating at flight speeds of Mach ≤ 0.7. After a brief introduction to the basics of propeller theory, this section concentrates on the engineering aspects of what is required by the aircraft designers. References [11–17] may be consulted for more details. It is recommended to use certified propellers manufactured by well-known companies.

Propellers are twisted wing-like blades that rotate in a plane normal to the aircraft flight path. Thrust generated by the propeller is the lift component produced by the propeller blades in the flight direction. It acts as a propulsive force and is not meant to lift weight unless the thrust line is vectored. It has aerofoil sections that vary from being thickest at the root and to thinnest at the tip chord (Figure 7.25). In rotation, the tip experiences the highest tangential velocity. The three important angles are blade pitch angle, β, angle subtended by the relative velocity, φ, and angle of attack, $\alpha = (\beta - \varphi)$. Shown in the figure is the effect of coarse and fine blade pitch, β. The definition of propeller pitch p is given in Section 7.10.1.

The propeller types are shown in Figure 7.26. There can be anywhere from two or three blades for small aircraft, to as high as seven/eight blades for bigger aircraft. When the propeller is placed in front of the aircraft, then it is called a tractor(Figure 7.19) and when placed at aft, it is called a pusher. The majority of propellers serve as tractors.

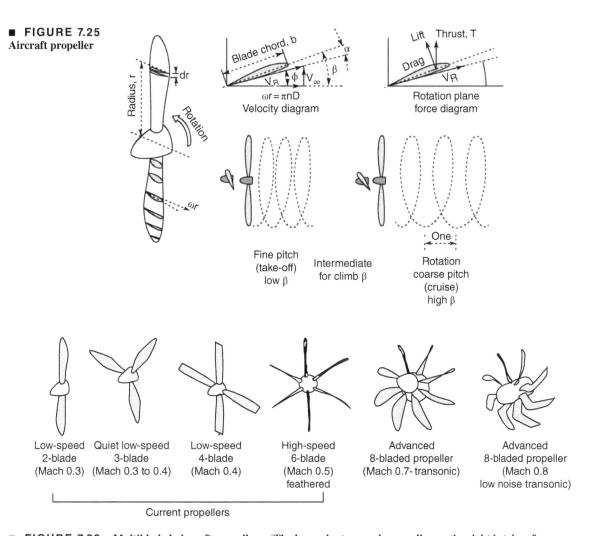

■ **FIGURE 7.25**
Aircraft propeller

■ **FIGURE 7.26**   **Multibladed aircraft propellers. (The low noise transonic propeller on the right is taken from unpublished work by Dr R. Cooper at Queen's University, Belfast)**

Blade pitch should match with aircraft speed, V, to keep the blade angle of attack α to produce best lift. To cope with aircraft speed change, it is best if the blade is rotated (varying the pitch) about its axis through the hub to maintain favourable α at all speeds. It is then called a *variable pitch* propeller. Typically for pitch variation, the propeller is kept at constant RPM with the help of a *governor*, when it is termed a constant speed propeller. Almost all aircraft flying at higher speed will have a constant speed, variable pitch propeller (when done manually it is β-controlled). The smaller low-speed aircraft are with fixed pitch propellers, which would run best at one combination of aircraft speed and propeller RPM. If the fixed pitch is meant for cruise, then at takeoff (low aircraft speed and high propeller revolution) the propeller would be less efficient. Typically, aircraft designers would like to have a fixed pitch propeller matched for the climb, a condition in between cruise and takeoff, to minimize the difference between the two extremes. Obviously, for high-speed performance it should match the high-speed cruise condition. Figure 7.27 shows the benefit of a constant speed, variable pitch propeller over the speed range.

■ **FIGURE 7.27**
Comparison of fixed pitch and constant speed variable pitch propellers (≈200 HP)

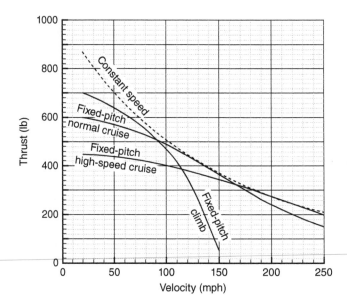

■ **FIGURE 7.27**
Comparison of fixed pitch and constant speed variable pitch propellers (≈200 HP)

The β control can extend to reversing of the propeller pitch. A full reverse thrust acts with all the benefits of a TR described in Section 7.9. The pitch control can be made to "fine-pitch" to produce zero thrust when the aircraft is static. This could assist aircraft to *wash out* speed, especially at approach to land.

When the engine fails, that is, the system senses insufficient power, the pilot or the automatic sensing device elects to *feather* the propeller. Feathering is changing β to 75–85° (maximum coarse) when the propeller slows down to 0 RPM, producing zero net drag/thrust (part of propeller has thrust and the rest drag).

Windmilling of the propeller is when engine has no power and is free to rotate, driven by the relative airspeed to the propeller when the aircraft is in flight. The β angle is in a fine position.

7.10.1 Propeller-Related Definitions

Industry uses propeller charts, which incorporate some special terminologies. Definitions of the necessary terminologies/parameters are given below.

D,	propeller diameter $= 2 \times r$.
n,	revolutions per second (RPS).
ω,	angular velocity.
N,	number of blades.
b,	propeller blade width (varies with radius, r).
P,	propeller power.
C_p,	power coefficient (not to be confused with pressure coefficient) $= P/(\rho n^3 D^5)$.
T,	propeller thrust.
C_{Li},	integrated design lift coefficient (C_{Ld} = sectional lift coefficient).
C_T,	propeller thrust coefficient $= T/(\rho n^2 D^4)$.
β,	blade pitch angle subtended by the blade chord and its rotating plane (also known as geometric pitch).
p,	propeller pitch = no slip distance covered in one rotation $= 2\pi r \tan\beta$ (explained above).

V_R, relative velocity to the blade element $= \sqrt{(V^2 + \omega^2 r^2)}$ (blade Mach No. $= V_R / a$).

φ, angle subtended by the relative velocity $= \tan^{-1}(V/2\pi nr)$ or $\tan\phi = V/\pi nD$. It is the pitch angle of the propeller in flight and is not the same as blade pitch, which is independent of aircraft speed.

α, angle of attack $= (\beta - \varphi)$.

J, advance ratio $= V / (nD) = \pi \tan\varphi$ (a non-dimensional quantity – analogous to α).

AF, activity factor $= (10^5/16) \int_0^{1.0} (r/R)^3 (b/D) d(r/R)$.

TAF, total activity factor $= N \times AF$ (N is number of blades – it gives an indication of power absorbed).

It is found that at around $0.7r$ (tapered propeller) to $0.75r$ (square propeller), the blades give an aerodynamically average value that can be applied uniformly over the entire radius to obtain the propeller performance.

Figure 7.25 shows a two-bladed propeller along with a blade elemental section, dr, at radius r. The propeller has diameter D. If ω is the angular velocity, then the blade element linear velocity at radius r is $\omega r = 2\pi nr = \pi nD$, where n is the number of revolutions per unit time. An aircraft with true airspeed of V and with propeller angular velocity of ω will have a blade element moving in a helical path. At any radius, the relative velocity V_R has angle $\varphi = \tan^{-1}(V/2\pi nr)$. At the tip, $\varphi_{tip} = \tan^{-1}(V/\pi nD)$.

However, irrespective of aircraft speed, the inclination of blade angle from the rotating plane can be seen as the solid-body screw thread inclination and is known as the *pitch angle*, β. The solid-body screw-like linear advancement through one rotation is called the *propeller pitch*, p. The pitch definition has a problem since, unlike mechanical screws, the choice of inclination plane is not standardized. It can be the zero-lift line, which is aerodynamically convenient, or the chord line, which is easy to locate, or the bottom surface – each plane will give a different pitch. All these planes are interrelated by fixed angles. This book takes the chord line as the reference line for the pitch, as shown by blade pitch angle, β, in Figure 7.25. This gives propeller pitch $p = 2\pi r \tan\beta$.

Since the blade linear velocity ωr varies with radius, its pitch angle needs to be varied as well to make the best use of the blade element aerofoil characteristics. When β is varied in such a way that the pitch does not change along the radius, then the blade has *constant pitch*. This means that β decreases with increase in r (variation in β is about $40°$ from root to tip).

$$\text{Blade angle of attack, } \alpha = (\beta - \varphi) = \tan^{-1}(p/2\pi r) - \tan^{-1}(V/2\pi nr) \qquad (7.21)$$

This brings out an analogous non-dimensional parameter:

$$J = \text{advance ratio} = V/(nD) = \pi\tan\varphi.$$

7.10.2   Propeller Theory

The fundamentals of propeller performance start with the idealized consideration of momentum theory. Its practical application in industry is based on subsequent "blade-element" theory. Both are presented in this section, followed by the engineering considerations relevant to the aircraft designers. Industry still uses manufacturer-supplied propellers, with wind-tunnel tested generic charts/tables to evaluate their performance. There are various forms of propeller charts; the two prevailing ones are the NACA (National Advisory Committee for Aeronautics) method and the Hamilton Standard (propeller manufacturer) method. This book uses the Hamilton Standard method as used in industry [13]. For designing advanced propellers/propfans to operate at speeds greater than Mach 0.6, computational fluid dynamics(CFD) plays an important role, substantiated by wind tunnel tests. CFD employs more advanced theories, for example, vortex theory.

7.10.2.1   Momentum Theory – Actuator Disc

The classical incompressible inviscid momentum theory provides the basis of propeller performance [17]. In this theory, the propeller is represented by a thin *actuator disc* of area A, placed normal to a free-stream flow of velocity V_0. This captures a stream tube within a CV having the front surface sufficiently upstream represented by subscript 0, and sufficiently downstream by subscript 3 (Figure 7.28). It is assumed that thrust is uniformly distributed over the disc and the tip effects are ignored. Whether the disc is rotating or not is redundant as flow through it is taken without any rotation. Station numbers just in front and aft of the disc are designated as 1 and 2.

The impulse given by the disc (propeller) increases the velocity from the free-stream value of V_0, smoothly accelerating to V_2 behind the disc and continuing to accelerate to V_3 (station 3) until the static pressure equals the ambient pressure p_0. The pressure and velocity distribution along the stream tube is shown in Figure 7.28. There is a jump of static pressure across the disc (from p_1 to p_2) but there is no jump in velocity change.

Newton's law gives that the rate of change of momentum is the applied force; in this case it is the thrust, T. Consider station 2 of the stream tube immediately behind the disc producing the thrust. It has *MFR*, $\dot{m} = \rho A_{disc} V_2$. The change of velocity is $\Delta V = (V_3 - V_0)$. This is the reactionary thrust experienced at disc through the pressure difference times its area, A.

Thrust produced by the disc:

$$T = \text{pressure across the disc} \times A_{disc} = A_{disc} \times \left(p_2 - p_1\right)$$
$$= \text{rate of change of momentum} \tag{7.22}$$
$$= \dot{m}\Delta V = \rho A_{disc} \times V_2 \times \left(V_3 - V_0\right)$$

Equating, Equation 7.22 can be rewritten as

$$\rho \times \left(V_3 - V_0\right) \times V_2 = \left(p_2 - p_1\right) \tag{7.23}$$

■ **FIGURE 7.28**
Control volume showing the stream tube of the actuator disc

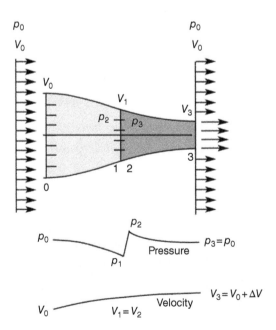

For incompressible flow, Bernoulli's equation cannot be applied through the disc imparting energy. Instead, two equations are set up, one for conditions ahead of the disc and the other aft of it. Ambient pressure p_0 is the same everywhere.

Ahead of the disc:
$$p_0 + \tfrac{1}{2}\rho V_0^{\,2} = p_1 + \tfrac{1}{2}\rho V_1^{\,2} \tag{7.24}$$

Aft of the disc:
$$p_0 + \tfrac{1}{2}\rho V_3^{\,2} = p_2 + \tfrac{1}{2}\rho V_2^{\,2} \tag{7.25}$$

Subtracting the front relation from the aft relation:
$$\tfrac{1}{2}\rho\left(V_3^{\,2} - V_0^{\,2}\right) = \left(p_2 - p_1\right) \times \tfrac{1}{2}\rho\left(V_2^{\,2} - V_1^{\,2}\right) \tag{7.26}$$

Since there is no jump in velocity across the disc, the last term drops out.
Next, substitute the value of $(p_2 - p_1)$ from Equations 7.23 to 7.26:
$$\tfrac{1}{2}\left(V_3^{\,2} - V_0^{\,2}\right) = \left(V_3 - V_0\right) \times V_2 \quad \text{or} \quad \left(V_3 + V_0\right) = 2V_2 \tag{7.27}$$

Note that $(V_3 - V_0) = \Delta V$, which when subtracted from Equation 7.27 gives $2V_3 = 2V_2 + \Delta V \text{ or } V_3 = V_2 + \Delta V / 2$. This implies that:
$$V_1 = V_0 + \Delta V / 2 \tag{7.28}$$

Using conservation of mass, $A_3 V_3 = A V_1$, Equation 7.22 becomes
$$T = \rho A_{disc} V_1 \times \left(V_3 - V_0\right) = A_{disc}\left(p_2 - p_1\right)$$

or
$$\left(p_2 - p_1\right) = \rho V_1 \times \left(V_3 - V_0\right) \tag{7.29}$$

It says that half of the added velocity, $\Delta V/2$, is added ahead of the disc and the rest, $\Delta V/2$, is added aft of the disc.

Using Equations 7.28 and 7.29, the thrust Equation 7.22 can be rewritten as:
$$T = A_{disc}\rho V_1 \times (V_3 - V_0) = A_{disc}\rho(V_0 + \Delta V/2) \times \Delta V \tag{7.30}$$

Applying to aircraft, V_0 may be seen as aircraft velocity, V, by dropping the subscript. Then the useful work rate (power, P) done on the aircraft is:
$$P = TV \tag{7.31}$$

For ideal flow without tip effects, the mechanical work produced in the system is the power, P_{ideal}, generated to drive the propeller, which is force (thrust, T) times velocity, V_1, at the disc. The maximum possible value in an ideal situation is thus given by:
$$P_{ideal} = T\left(V + \Delta V/2\right) \tag{7.32}$$

Therefore, ideal efficiency,
$$\eta_i = P/P_{ideal} = TV / \left[T\left(V + \Delta V / 2\right)\right] = 1 / \left[1 + \left(\Delta V / 2V\right)\right] \tag{7.33}$$

The real effects include viscous effects, propeller tip effects and other installation effects. In other words, to produce the same thrust, the system has to provide more power. In the case of a piston engine it is seen as brake horse power(BHP), and in the case of a turboprop as ESHP, the equivalent shaft horse power that converts the residual thrust at the exhaust nozzle to HP, dividing by an empirical factor 2.7. The propulsive efficiency as given in Equation 7.33 can be written as:

$$\eta_p = TV/[\text{BHP or ESHP}] \tag{7.34}$$

This gives

$$\eta_p / \eta_i = \{TV/[\text{BHP or SHP}]\}/\{1/[1+(\Delta V/2V)]\}$$
$$= \{TV[1+(\Delta V/2V)]/[\text{BHP or SHP}]\} \tag{7.35}$$
$$= 85-86\% \text{ (typically)}$$

7.10.2.2 *Blade Element Theory* Practical application of propellers is obtained through blade element theory, as given below. A propeller blade cross-sectional profile has the same functions as that of a wing aerofoil, that is, to operate at the best *L/D*.

Figure 7.25 shows a blade elemental section, dr, at radius r, is valid for any number of blades at any radius, r. As blades are rotating elements, its properties would vary along the radius.

Figure 7.25 gives the velocity diagram showing an aircraft with a flight speed of V, having its propeller rotating at n revolutions per second, would have its blade element advancing in a helical manner. V_R is the relative velocity to the blade with angle of attack α. Here, β is the propeller pitch angle as defined above. Strictly speaking, each blade rotates in the wake (downwash) of the previous blade, but the current treatment ignores this effect to use propeller charts without appreciable error. Figure 7.25 gives the force diagram at the blade element in terms of lift, L, and drag, D, normal and parallel respectively to V_R. Then the thrust, dT, and force, dF (producing torque), on the blade element can be easily obtained, by decomposing lift and drag in the direction of flight and in the plane of propeller rotation, respectively. Integrating over the entire blade length (non-dimensionalized as r/R – an advantage to be applicable to different sizes) would give the thrust, T, and torque-producing force, F, by the blade. The root of the hub (with spinner or not) does not produce thrust, and integration is normally carried out from 0.2 to the tip, 1.0, in terms of r/R. When multiplied by the number of blades, N, it gives the propeller performance.

Therefore,

$$\text{Propeller thrust, } T = N \times \int_{0.2}^{1.0} dT\, d(r/R) \tag{7.36}$$

and

$$\text{Force that produces torque, } F = N \times \int_{0.2}^{1.0} dF\, d(r/R) \tag{7.37}$$

From the definition of the advance ratio, $J = V/(nD)$.

It can be shown that the thrust to power ratio is best when the blade element works at the highest lift to drag ratio (L/D_{max}). It is clear that a fixed pitch blade works best at a particular aircraft speed for the given power rating (RPM) – typically the climb condition is matched for the compromise. It is for this reason that constant speed, variable pitch propellers have better performance over a wider aircraft speed range. It is convenient to express thrust and torque in non-dimensional form as given below. From dimensional analysis (note that the denominator does not have the ½):

Non-dimensional thrust $\qquad T_C = \text{thrust}/(\rho V^2 D^2)$

Thrust coefficient

$$C_T = T_C \times J^2 = \text{thrust} \times \left[V/(nD) \right]^2 / \left(\rho V^2 D^2 \right)$$
$$= \text{thrust} / \left(\rho n^2 D^4 \right)$$

(7.38)

In FPS system,

$$C_T = 0.1518 \times \left[\frac{(T/1000)}{\sigma \times (N/1000)^2 \times (D/10)^4} \right]$$

(7.39)

where σ is the ambient density ratio for altitude performance.

Non-dimensional force (for torque), $T_F = F/(\rho V^2 D^2)$

Force coefficient

$$C_F = T_F \times J^2 = F \times \left[V/nD \right]^2 / \left(\rho V^2 D^2 \right) = F / \left(\rho n^2 D^4 \right)$$

(7.40)

Torque

$$Q = \text{force} \times \text{distance} = Fr = C_F \times \left(\rho n^2 D^4 \right) \times D/2 = \left(C_F/2 \right) \times \left(\rho n^2 D^5 \right)$$

Torque coefficient

$$C_Q = Q / \left(\rho n^2 D^5 \right) = C_F / 2$$

(7.41)

Power consumed

$$P = 2\pi n \times Q$$

Power coefficient

$$C_P = P / \left(\rho n^3 D^5 \right) = 2\pi n \times Q / \left(\rho n^3 D^5 \right) = 2\pi C_Q = \pi C_F$$

(7.42)

In the FPS system,

$$C_P = 0.5 \times \left[\frac{(\text{BHP}/1000)}{\sigma \times (N/1000)^3 \times (D/10)^5} \right]$$
$$= \left[\frac{(237.8 \times \text{SHP})}{2000 \times (6/100)^3 \rho n^3 D^5} \right] = \left[\frac{(550 \times \text{SHP})}{\rho n^3 D^5} \right]$$

(7.43)

The wider the blade, the higher would be the power absorbed to a point when any further increase would offer diminishing returns in increasing thrust.

A non-dimensional number defined as the activity factor expresses the integrated capacity of the blade element to absorb power. It indicates that increase of blade width outwardly is more effective than in the hub direction.

Activity factor,

$$\text{AF} = \left(10^5/16 \right) \int_0^{1.0} (r/R)^3 (b/D) d(r/R)$$

Total activity factor,

$$\text{TAF} = N \times \left(10^5/16 \right) \int_0^{1.0} (r/R)^3 (b/D) d(r/R)$$

The TAF expresses the integrated capacity of the total number of blades of a propeller to absorb power. (Typical propeller charts give the propeller properties taking into account the number of blades, as given in Figures 7.29–7.31 [13].)

A piston engine or a gas turbine would drive the propeller. Propulsive efficiency η_p can be computed by using Equations 7.34, 7.38 and 7.43.

$$\text{Propulsive efficiency,} \, \eta_p = (TV)/[\text{BHP or ESHP}]$$
$$= \left[C_T \times \left(\rho n^2 D^4 \right) \times V \right] \Big/ \left[C_P \times \left(\rho n^3 D^5 \right) \right] \quad \textbf{(7.44)}$$
$$= \left(C_T / C_P \right) \times \left[V / (nD) \right] = \left(C_T / C_P \right) \times J$$

The theory determines that geometrically similar propellers can be represented in a single non-dimensional chart (propeller graph) by combining the above non-dimensional parameters such as shown in Figures 7.29 and 7.30 for three-bladed propellers and Figures 7.31 and 7.32 for four-bladed propellers. A considerable amount of classroom work can be conducted with these graphs. All these graphs and the procedure to estimate propeller performance are taken from [13] as a courtesy from Hamilton Standard who kindly allowed the use of the graphs in this book. All Hamilton Standard graphs are replotted retaining maximum fidelity. The reference gives a full range of graphs for other types and has charts for propellers with higher AF. The charts use the following relations.

In flight, thrust

$$T = \left(550 \times \text{BHP} \times \eta_p \right) / V, \text{ where } V \text{ is in ft/s}$$
$$= \left(375 \times \text{BHP} \times \eta_p \right) / V, \text{ where } V \text{ is in mph} \quad \textbf{(7.45)}$$

For static performance (takeoff)

$$T_{TO} = \left[\left(C_T / C_P \right) \times \left(550 \times \text{BHP} \right) \right] / nD \quad \textbf{(7.46)}$$

Static computation gives a problem since when V is zero, then $\eta_p = 0$. Different sets of graphs are required to obtain the values of (C_T / C_P) to compute takeoff thrust as given in Figures 7.29

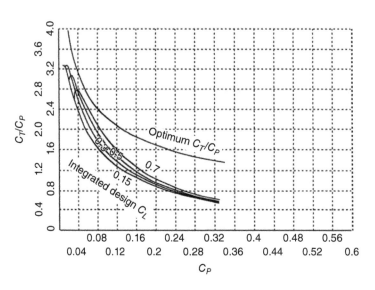

■ **FIGURE 7.29**
Static performance: three-bladed propeller performance chart – AF100 (for piston engines). (Adapted from Hamilton Standard [13] with permission)

■ **FIGURE 7.30**
Three-bladed
propeller perfor-
mance chart – AF100
(for piston
engines). (Adapted
from Hamilton
Standard [13] with
permission)

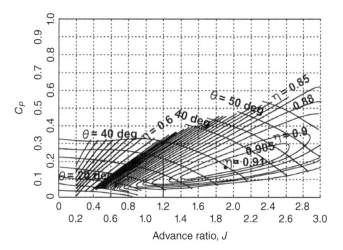

Advance ratio, J

■ **FIGURE 7.31**
Static performance:
four-bladed
propeller perfor-
mance chart – AF180
(for high-performance
turboprop). (Adapted
from Hamilton
Standard [13] with
permission)

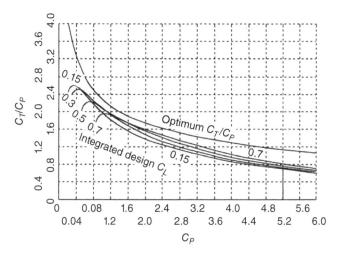

C_P

and 7.31. Finally, Figure 7.31 is meant for selecting the design C_L for the propeller to avoid compressibility loss.

7.10.3 Propeller Performance – Practical Engineering Applications

The book does not deal with propeller design. Aircraft designers select a propeller offered by a manufacturer, mostly off-the-shelf, unless specially designed propellers are required, in consultation with aircraft designers. This section gives the considerations necessary and appropriate for the aircraft designers to select an appropriate propeller to match with the sized engine to produce thrust for the full flight envelope.

Readers may note that the propeller charts for the number of blades only use three variables, C_p, β and η (subscript p is dropped). They do not specify propeller diameter and RPM. Therefore, similar propellers with the same AF and C_{Li} can use the same chart. Aircraft designers will have to choose AF or C_{Li} based on the critical phase of operation. Propeller selection would require some compromises since optimized performance for the full flight envelope is not possible, especially for fixed pitch propellers.

■ **FIGURE 7.32**
Four-bladed
propeller performance
chart – AF180 (for
high-performance
turboprop). (Adapted
from Hamilton
Standard [13] with
permission)

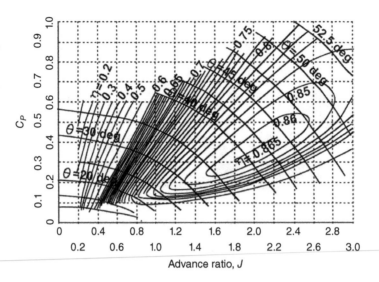

Of late, certification requirements for noise have affected these factors, especially for high-performance propeller design. High tip Mach number is detrimental to noise, and to reduce it, η is compromised by reducing RPM and/or diameter, hence increasing J and/or the number of blades. Increasing the number of blades would increase cost and weight. Propeller curvature suits transonic operation and helps to reduce noise.

Equation 7.21 gave the aerodynamic incidence, that is, the blade angle of attack, $\alpha = (\beta - \varphi)$, where φ is determined from the aircraft speed and propeller RPM, which is a function of $J = V/nD$. It is desirable to keep α constant along the blade radius to get the best C_L, that is, α is maintained at around 6–8°, the values at $0.7r$ or $0.75r$ being used as the reference point (the propeller chart mentions reference radius).

The combination of the designed propeller RPM is matched with its diameter to keep operation parameters below those where compressibility effects would be experienced. A suitable reduction gear ratio brings the engine RPM down to the desired propeller RPM. Figure 7.33 is used to obtain the integrated design C_L for the combination of propeller RPM and diameter. The factor of $ND \times$ (ratio of speed of sound at ISA day sea-level to altitude) would establish the integrated design C_L. A spinner at the propeller root is recommended to reduce loss. Typically, the majority of production propellers have integrated design C_L within the range of 0.35–0.6. Figure 7.33 gives the limits of integrated design C_L, which should not be exceeded.

The following step-by-step observations/information may prove helpful in using the propeller charts given in Figures 7.29–7.32 (Hamilton Standards):

1. In this book, the integrated design C_L is taken as 0.5. Checking of the integrated design C_{Li} using Figure 7.33 is not used.
2. Typical blade activity factor, AF, is of the following order:
   - Low power absorption, two- to three-bladed propellers for home-built flying = $80 < \text{AF} < 90$.
   - Medium power absorption, three- to four-bladed propellers for piston engines (utility) = $100 < \text{AF} < 120$.
   - High power absorption, four-bladed and above propellers for turboprops = $140 < \text{AF} < 200$.
3. Keep the tip Mach number around 0.85 at cruise and make sure that at takeoff RPM it exceeds the value at the second segment climb speed.

■ **FIGURE 7.33**
Limits of the integrated design C_L to avoid compressibility loss. (Adapted from Hamilton Standard [13] with permission)

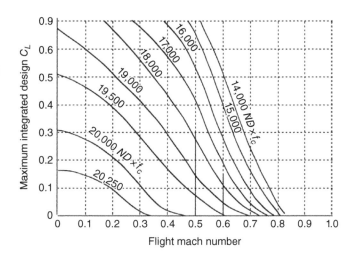

■ **FIGURE 7.34**
Engine power versus propeller diameter [16]. (Reproduced with permission from John Wiley & Sons, Ltd.)

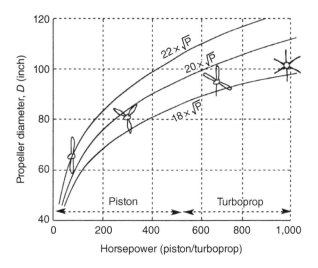

4. Typically, for constant speed, variable pitch propellers, β is kept low for takeoff, gradually increasing at climb speed, reaching an intermediate value at cruise and a high value at the maximum speed.

5. Propeller diameter in inches can be coarsely determined by an empirical relation $D = K(P)^{0.25}$ where $K = 22$ for two blades, 20 for three blades and 18 for four blades. Power P is the installed power, which is less than the bare engine rating supplied by the engine manufacturer. Figure 7.34 gives the typical relationship between engine power and propeller diameter. It is a useful graph to initially size the propeller empirically. If n and J are known beforehand, then propeller diameter can be determined using $D = 1056 V/(NJ)$ in the FPS system.

6. Maintain at least about 0.5 m (1.6 ft) propeller tip clearance from the ground, which in extreme demand can be slightly reduced. This should take care of nose wheel tyre bursts and undercarriage oleo collapse.

7. At maximum takeoff static power, the thrust developed by the propellers is about four times the power.

Continuing (in FPS) with propeller performance for static takeoff and in-flight cruise separately:

Static performance (Figures 7.29 and 7.31):

1. Compute the power coefficient, $C_p = (550 \times \mathrm{SHP})/(\rho n^3 D^5)$, where n is in RPS. Note ρ for f_c.
2. From propeller chart, find C_T/C_P.
3. Compute static thrust, $T_s = (C_T/C_P)(33,000 \times \mathrm{SHP})/nD$, where n is in RPM.

In-flight performance (Figures 7.30 and 7.32):

1. Compute the advance ratio, $J = V/(nD)$.
2. Compute the power coefficient, $C_p = (550 \times \mathrm{SHP})/(\rho n^3 D^5)$, where n is in RPS.
3. From the propeller chart, find efficiency, η_p.
4. Compute thrust, T, from $\eta_p = (TV)/(550 \times \mathrm{SHP})$, where V is in ft/s.

If necessary, off-the-shelf propeller blade tips could be slightly chopped off to meet geometric constraints. Typical penalties are that 1% reduction of diameter would reduce thrust by 0.65% – for small changes, linear interpolation may be used.

7.10.4 Propeller Performance – Three- to Four-Bladed

Using these graphs, linear extrapolations could be made for two- and five-bladed propellers with similar AF. Reference [15] treats the subject in detail, with propeller charts for other AFs.

Installed turboprop thrust is obtained by reducing the thrust by the various loss factors. In addition, fuselage factors such as the blockage factor f_b and excrescence factor f_h, are to be applied, further reducing the effective thrust level.

References

1. Rolls Royce, *The Jet Engine*. John Wiley & Sons, Inc., New York, 2015.
2. Harned, M., *The Ramjet Power Plant*. Aero. Digest, 69 1, 43, 1954.
3. Cohen, H., Rogers, G.F.C., and Saravanamuttoo, H.I.H., *Gas Turbine Theory*, 5th edn. Pearson, Harlow, 2001.
4. Mattingly, J.D., *Elements of Gas Turbine Propulsion*. McGraw-Hill, 1996.
5. Oates, G., *Aerodynamics of Gas Turbines and Rocket Propulsion*, 3rd edn. AIAA, Reston, US, 1997.
6. Swavely, C.E., *Propulsion System Overview*. Pratt and Whitney Short Course, 1985.
7. Farokhi, S., et al., *Propulsion System Integration Course*. University of Kansas, 1991.
8. Kundu, A.K., *Aircraft Design*. Cambridge University Press, Cambridge, 2010.
9. Seddon, J., and Goldsmith, E., *Practical Intake Aerodynamic Design*. AIAA, Reston, US, 1993.
10. Gunston, B., *The Development of Jet and Turbine Aeroengines*, 4th edn. PSL-Haynes Publishing, Cambridge, 2006.
11. McCormick, B.W., *Aerodynamics, Aeronautics and Flight Mechanics*. John Wiley & Sons, Inc., Chichester, 1994.
12. Perkins, C.D., and Hage, R.E., *Aircraft Performance Stability and Control*. John Wiley & Sons, Inc., New Jersey, 1949.
13. *Propeller Performance Manual*. Hamilton Standard.
14. Society of British Aircraft Constructors, *SBAC Standard Method of Performance Estimation*. London, 1930.
15. Lan, C.T.E., and Roskam, J., *Aeroplane Aerodynamics and Performance*. Darcorporation, Lawrence, 2000.
16. Stinton, D., *The Design of the Airplane*. Blackwell Science, Ltd., Oxford, 2001.
17. Glauert, H., *Elements of Aerofoil and Airscrew Theory*. Cambridge University Press, Cambridge, 1947.

CHAPTER 8

Aircraft Power Plant Performance

8.1 Overview

This chapter covers both uninstalled and installed engine performance. The aircraft performance engineers require installed thrust and fuel flow (FF) data to estimate aircraft performance. In most industries, aircraft performance engineers generate the installed engine performance using engine manufacturer-supplied computer programs amenable to input of the various losses arising from taking engine bleed/off-takes for aircraft systems and intake/nozzle losses.

Uninstalled gas-turbine engine performances in non-dimensional form are given in Sections 8.3–8.5. From the non-dimensional engine data, matched engine installed performances are generated. These are validated against industry standard data. Performances are given in non-dimensional form for the following types of engine (except the piston engine performance):

- Turbofan engine of BPR (bypass ratio) ≤ 4 (civil engine of the class $T_{SLS} \approx 3500 \, \text{lb}$).
- Turbofan engine of BPR ≥ 4 (civil engine of the class $T_{SLS} \approx 40,000 \, \text{lb}$).
- Medium-sized turbofan engine of BPR ≤ 1 (military engine of the class $T_{SLS\_dry} \approx 6000 \, \text{lb}$).
- Turboprop engine of the class of 1000 SHP at sea-level static conditions.
- 180 HP Lycoming piston engine.

Industry standard engine performance data for classroom usage are not easy to obtain – these have proprietary value. Readers need to be careful in applying engine data – an error in engine data would degrade/upgrade the aircraft performance and corrupt the design. Verification/substantiation of aircraft design is carried out through flight tests of its performances. It becomes difficult to locate the source of any discrepancy between predicted and tested performance, whether the discrepancy stems from the aircraft or from the engine or from both. Appropriate engine data may be obtained beyond what is given in this book. It was mentioned earlier that the US contribution to aeronautics is indispensable, and their data are generated in the foot-pound-second (FPS) (Imperial) system. Many of the data and worked examples are kept in the FPS system. An extensive list of conversion factors is given in Appendix A.

Theory and Practice of Aircraft Performance, First Edition. Ajoy Kumar Kundu, Mark A. Price and David Riordan.
© 2016 John Wiley & Sons, Ltd. Published 2016 by John Wiley & Sons, Ltd.

8.2   Introduction

Chapter 7 mentioned that post WWII research led to the rapid advancement of gas turbine development to a point where, from a core gas generator module, a family of engines could be produced in a modular fashion (Figure 7.1) that allows engine designers to offer engines as specified by the aircraft designers. Similarity laws in thermodynamic design parameters allow power plants to be scaled (rubberized) to the requisite size around the core gas generator module to meet the demands of the mission requirements.

Gas turbine engines operate at variable power demands; they are rated in a band of power settings, that is in ratings (in this book in terms of engine RPM) maximum at takeoff, and then at the next level at climb, and then at cruise demand, and finally at descent at part-throttle to the idle power level. Below the various power ratings are explained.

8.2.1   Engine Performance Ratings

All power plants have prescribed power ratings as given below. Power settings are decided by the engine RPM and/or by the exhaust pressure ratio (EPR) at the jet pipe temperature (JPT) which should remain below prescribed levels. An engine is identified at its sea-level ISA day at takeoff in static conditions. For turbofans it is abbreviated as T_{SLS}. The following are the typical power settings (ratings) of engines.

Takeoff rating: This is the highest rating, producing the maximum power, and is rated at 100% (at a static run of turbofans, it is the T_{SLS}). At this rating, the engine runs the hottest and therefore is time-limited for longer life and lower maintenance. For civil engines the limit is about 10 minutes, while military engines could extend a little longer. A situation may arise when one engine is inoperative and the operating engine is required to supply more power at the augmented power rating (*APR*). Not all engines have APR, which exceeds 100% power by about 5% for very short periods (say for about 5 minutes).

Flat-rated takeoff rating: For any engine rating running at a constant RPM, variation in ambient temperature would make the engine thrust vary as well. On a cold day, higher air density would mean that the engine inhales more air mass flow and therefore will produce increased thrust, and vice versa. It is most critical ay takeoff rating as the engine is operating at maximum power. To protect the engine from structural failure, the fuel control is set to keep the thrust level at a set value known as the flat-rated thrust, for a limited period of operation (say around 10 minutes) at takeoff. To ensure that 100% thrust is available on a hotter day, the ambient temperature limit for the flat-rated thrust is set to anywhere from ISA + 15°C to ISA + 20°C, depending on the engine design. A typical flat-rated takeoff setting is shown in Figure 8.1. Thrust at ambient temperature above the flat-rated limit will drop but will remain constant at the flat-rated value for ambient temperatures colder than the limiting value. In an emergency, for example, if an engine fails at takeoff phase, some turbofans have the APR capability to extract more power from the running engine, which can prove helpful just to allow the aircraft to continue with go around and land back. Engine manufacturers like to have engines inspected if APR is applied.

Maximum continuous rating: This is the highest engine rating that can operate continuously at around 90–95% of the maximum power. It is more than what is required for climb at a prescribed speed schedule (see Section 12.4.4) for pilot ease and good fuel economy. Operational demand in this rating arises from specific situations, say an engine failure of a multi-engine aircraft, or for a very fast/steep climb to altitude.

■ **FIGURE 8.1**
Flat-rated thrust at takeoff rating

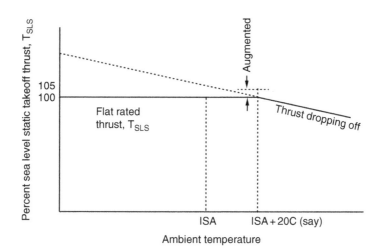

Maximum climb rating: A climb schedule is carried out at around 85–90% of the maximum power. This is to reduce stress on the engine and gain better fuel economy. Typical climb time for civil aircraft is less than 30–40 minutes, but can run continuously. It is for this reason that some engines do not include the maximum continuous rating and merge it with the maximum climb rating.

Maximum cruise rating: This is at around 80–85% of the maximum power matched to the maximum cruise speed capability. Unless there is a need for higher speed, typical cruise is performed at 70–80% power rating, which may be called a *cruise rating*. This gives the best fuel economy for the long range cruise (LRC). At a holding pattern in the vicinity of an airport, engines are run at still lower power, barely to maintain altitude and wait for clearance to proceed. The rating depends on the weight of aircraft; at the end of the cruise (light weight) about 60–70% rating would suffice.

Idle rating: This is at around 40–55% of the maximum power meant for the engine to run without flameout, but produces practically no thrust. This situation arises at descent, at approach or on the ground. During descent, it is found that better economy can be achieved by coming down at part throttle, say around 60% power rating. This gives a shallower glide slope to cover more distance without consuming much fuel.

Representative engine performances of various types at takeoff rating, maximum climb rating and maximum cruise ratings are given in this chapter for an ISA day. Engine manufacturers also supply performance data for non-ISA days; more critical is for hot and high-altitude conditions when engines would produce considerably less power. To protect engines from heat stress, the fuel control system is tuned to cut off power generation to a flat-rated value (ISA day engine rating) for up to 20°C above an ISA day. This book does not deal with non-ISA day performances, but in industry this will need to be accounted for.

8.2.2 Turbofan Engine Parameters

Engine manufacturers supply engine data to aircraft manufacturers, who generate the engine data for in-house use. The format of data transfer from engine to aircraft manufacturer has changed considerably with the advent of powerful desktop computers. In the past, engine manufacturers produced tabulated engine data for the full flight envelope for the aircraft under evaluation. The aircraft designers converted these tabulated data into just a few non-dimensional graphical plots, but still allowed the expression of engine performance characteristics. Such non-dimensional graphs were generated for each type of turbofan for the particular aircraft, incorporating the associated installation losses. Today, instead of supplying data in tabulated format, the engine manufacturers supply a computer deck for aircraft designers to generate installed engine performances in detail for exact off-take demands and installation losses.

The generalized non-dimensional parameters include the fan face diameter, D_{fan} (not the intake diameter), but when expressed for a particular turbofan, then D_{fan} can be omitted, as it is of a known constant value. Also the sea-level ISA atmospheric values are brought in to include atmospheric ratios for total conditions (subscript t) at the fan face. These parameters are also used for gas turbine component performance.

Non-dimensional thrust, $\dfrac{T}{\delta_t}$

Non-dimensional RPM, $\dfrac{N}{\sqrt{\theta_t}}$

Non-dimensional turbofan air mass flow, $\dfrac{W_a\sqrt{\theta_t}}{\delta_t}$

Non-dimensional turbofan fuel mass flow, $\dfrac{W_f\sqrt{\theta_t}}{\delta_t}$

$\dfrac{W_a\sqrt{\theta_t}}{\delta_t}$ and $\dfrac{W_f\sqrt{\theta_t}}{\delta_t}$ are mainly used in gas turbine component performance analyses. Note that T/δ_t is not truly non-dimensional, but when multiplied by p_0/D_{fan} it becomes non-dimensional. It is convenient to use T/δ as the non-dimensional parameter where δ is the ambient pressure ratio p/p_0. Today, the engine manufacturer-supplied computer deck has made non-dimensionally plotted engine performance data almost redundant.

Turbofan characteristics show that net thrust is a function of engine geometry, pressure ratio, inlet and outlet flow conditions. A given engine shows net thrust proportional to engine pressure ratio (EPR, i.e. the ratio of pressure at turbine exit to pressure at compressor face). At a particular ambient condition (temperature and pressure) and aircraft speed, the thrust of a particular turbofan can expressed as a function of EPR. Also engine RPM can also be considered as an index for thrust and can be related to engine ratings (Section 8.3). Although EPR gives more refined thrust values, this book uses RPM as an index for net thrust because it is more convenient.

8.3 Uninstalled Turbofan Engine Performance Data – Civil Aircraft

All thrusts given in this section are uninstalled thrust [1]. There is a loss of power when an engine is installed into an aircraft, as discussed in the last chapter. If exact values are not available, then use 7–10% loss at takeoff rating depending on how the environmental control system (ECS) is managed. At cruise, it would come down to 3–5%. To keep it simple, both military and civil installation losses are kept at similar percentages, although the demands of the off-takes are quite different.

Figures 8.2 and 8.3 give the turbofan powers at the three ratings in a non-dimensional form for civil engines for low and high BPR. Civil turbofan performance is also divided into two

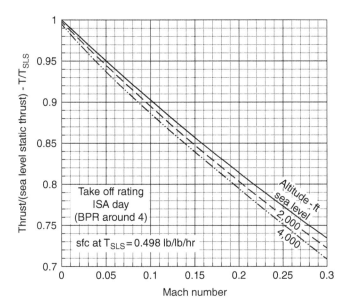

■ **FIGURE 8.2**
Uninstalled takeoff performance (BPR $\lesssim 4$) (fuel flow rate is nearly constant at takeoff rating)

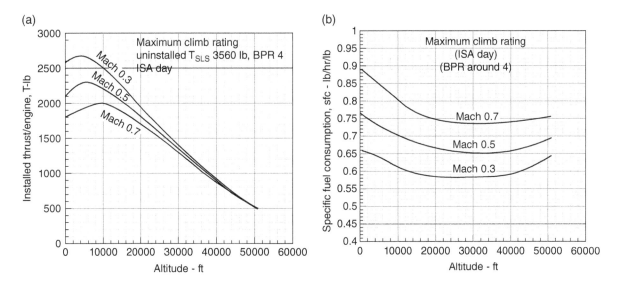

■ **FIGURE 8.3** Uninstalled maximum climb rating (BPR $\lesssim 4$). (a) Non-dimensional thrust. (b) Specific fuel consumption

categories, one for lower BPR of the order of around ≤ 4, and another for BPR ≥ 5. The latest engines (for Boeing 787 and Airbus 350) have achieved BPR above 8, but the author could not obtain realistic data for the latest class of turbofans. The higher the BPR, the lower would be the specific thrust ($T_{SLS}/\dot{m}_a$ – lb/lb/s). A similar trend is found for the specific dry engine weight (T_{SLS}/dry engine weight, non-dimensional). Table 8.1 may be used to compare values for different BPR categories.

■ **TABLE 8.1**
Turbofan parameters, BPR and specific thrust

BPR	$F/\dot{m}_a$, lb/s/lb at T_{SLS}	T_{SLS}/dry engine weight
Around 4	35–40	4–5
Around 5	32–34	5–6
Around 6	30–34	5–7

8.3.1 Turbofans with BPR around 4

8.3.1.1 Turbofan Performance Uninstalled turbofan (BPR around ≤ 4) thrust and fuel flow data in normalized form are given in Figures 8.2–8.4.

8.3.1.2 Takeoff Rating Figure 8.2 gives the uninstalled takeoff thrust in non-dimensional form for the standard day for turbofans with BPR of 4 or less. The fuel flow rate remains nearly invariant for the envelope shown in the graph. The sfc at takeoff rating is taken at the value of the T_{SLS} of 0.498 lb/lb/h per engine.

8.3.1.3 Maximum Climb Rating Figure 8.3 gives the uninstalled maximum climb thrust and fuel flow in non-dimensional form for the standard day up to 50,000 ft altitude for 0.3, 0.5 and 0.7 Mach numbers. Intermediate values may be linearly interpolated. Note that there is a break in thrust generation at about 6000–10,000 ft altitude, depending on Mach number, due to fuel control to keep exhaust gas temperature (EGT) within the specified limit.

Equation 16.14 requires a factor k_{cl} to be applied to T_{SLS} to get the initial climb thrust. At climb, the throttle is cut back to a maximum of about 85% of the takeoff rating. Hence this scaling factor k_2 reduces the thrust level as the throttle is reduced. In the example, the initial climb starts at 800 ft altitude at 250 V_{EAS} (Mach 0.38) which gives the factor $T/T_{SLS} = 0.67 = k_{cl}$, that is, $T_{SLS}/T = 1.5$. At constant equivalent airspeed (EAS), Mach number increases with altitude, and in the example when it reaches 0.7 (depending on the aircraft type) the Mach number is held constant. Fuel flow at climb is obtained from Figure 8.3b. Having varying values with altitude, climb calculations are done in small steps of altitude within which the variation is taken as the mean and kept constant for the step altitude.

8.3.1.4 Maximum Cruise Rating Figure 8.4 gives the uninstalled maximum cruise thrust and fuel flow in non-dimensional form for the standard day from 5000 to 50,000 ft altitude for Mach numbers varying from 0.5 to 0.8, sufficient for this class of engine-aircraft combination. Intermediate values may be linearly interpolated.

The classroom example gives the design initial maximum cruise speed Mach 0.7 at 41,000 ft. At cruise the throttle is further cut back to a maximum of about 75% of the takeoff rating. Hence this scaling factor k_{cr} reduces the thrust level as the throttle is further reduced. From the graph that point reads $T/T_{SLS} = 0.222 = k_{cr}$, which has $T_{SLS}/T = 4.5$. Fuel flow per engine can be computed from Figure 8.4b.

8.3.2 Turbofans with *BPR* around 5–6

8.3.2.1 Turbofan Performance Bigger engines have higher BPR. Currently operational bigger turbofans are at 5–6 BPR, which would have non-dimensional engine performance characteristics slightly different from smaller engines, as can be compared with Figures 8.2–8.4.

8.3.2.2 Takeoff Rating Figure 8.5 gives the uninstalled takeoff thrust in non-dimensional form for the standard day. Typically, at takeoff, the specific fuel consumption, sfc, is nearly invariant; the value is given in the graph.

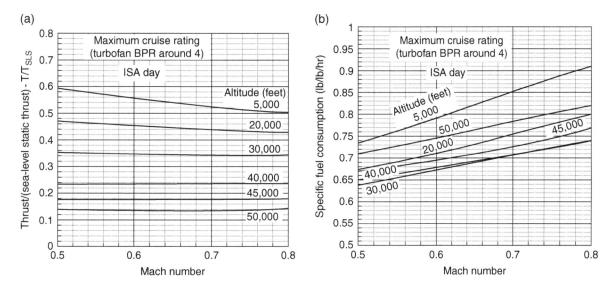

■ **FIGURE 8.4**   Uninstalled maximum cruise rating (BPR $\lesssim$ 4). (a) Non-dimensional thrust. (b) Specific fuel consumption

■ **FIGURE 8.5**
**Uninstalled takeoff
performance
(BPR $\gtrsim$ 5)**

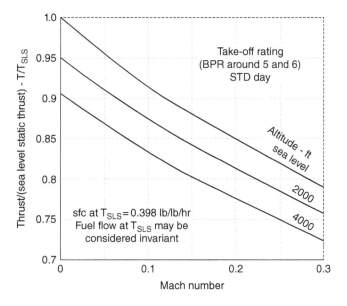

8.3.2.3   *Maximum Climb Rating*   Figure 8.6 gives the uninstalled maximum climb thrust and fuel flow in non-dimensional form for the standard day up to 50,000 ft altitude for three Mach numbers. Intermediate values may be linearly interpolated.

8.3.2.4   *Maximum Cruise Rating*   Figure 8.7 gives the uninstalled maximum cruise thrust and fuel flow in non-dimensional form for the standard day from 5000 to 50,000 ft altitude for Mach numbers varying from 0.5 to 0.8, sufficient for this class of engine-aircraft combination. Intermediate values may be linearly interpolated.

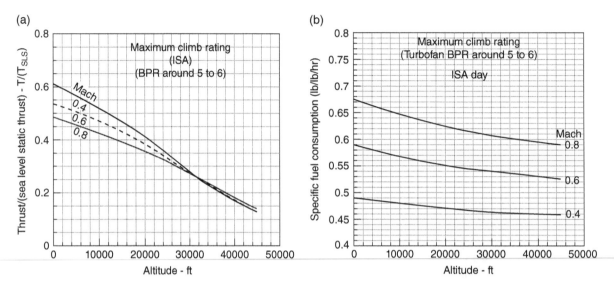

■ **FIGURE 8.6**   Uninstalled maximum climb rating (BPR ≳ 5). (a) Non-dimensional thrust. (b) Specific fuel consumption

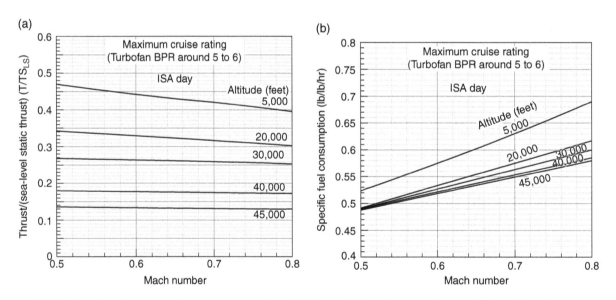

■ **FIGURE 8.7**   Uninstalled maximum cruise rating (BPR ≳ 5). (a) Non-dimensional thrust. (b) Specific fuel consumption

8.4   Uninstalled Turbofan Engine Performance Data – Military Aircraft

Military turbofan ratings are slightly different from civil turbofan ratings [1]. Military engines are allowed to run longer at maximum ratings, not only at takeoff but also for fast acceleration in combat. Of course, these still operate for a limited period of time, for example at takeoff, and the extent depends on the design.

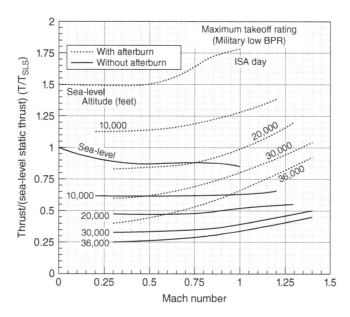

■ FIGURE 8.8
Military turbofan engine with and without reheat (BPR = 0.75). Take 90% of maximum rating as the maximum continuous rating

For continuous operation, the engine RPM is throttled down to a maximum of 90% of T_{SLS} as the maximum continuous rating, mostly applied in a demanding situation during combat. For climb, it is further throttled back to a maximum of 85% of T_{SLS} as the maximum climb rating. Typically, the maximum continuous rating is close to the maximum climb rating. For convenience, some military engine designers have merged the two ratings in one maximum climb rating.

Military engines operate at considerably varying throttle demands. Here, cruise is less meaningful unless it is ferry flight. Flight to operation theatre or return to base is not exactly cruising – but this period can be executed at lower throttle settings below 75% of T_{SLS}, as the mission demands.

Reheat (afterburning) is added at maximum rating when the throttle is set to the fully forward position. Running at a relatively longer duration at this high power (higher combustion temperature) has the price of a shorter engine life span and shorter time between overhaul (TBO). In addition, the hot zone components use more expensive materials to withstand stress at elevated temperatures.

A typical military turbofan engine performance at maximum rating suited to the classroom example of an advanced jet trainer, with a derivative in a close air support (CAS) role, is given in Figure 8.8. The sfc for this engine at T_{SLS} is 0.75 lb/h/lb when operating without reheat, and 1.1 lb/h/lb when under reheat (afterburner lit). Rated (maximum) air mass flow is 95 lb/s.

Currently, not much information can be supplied on military engines. However, for a classroom example this would prove sufficient, as will be shown in subsequent chapters.

8.5   Uninstalled Turboprop Engine Performance Data

Engine performance characteristics vary from type to type [1]. The real engine data are not easy to obtain in the public domain. The graphs in this section are generic in nature, representing typical current turboprop engines (all types up to 100 passenger class). The graphs include the small amount of jet thrust available at a 70% rating and higher. The jet power is converted to SHP and the total equivalent shaft horse power (ESHP) is labelled only "SHP" in the graph.

In industry, engine manufacturers supply performance data incorporating exact installation losses for accurate computation.

The sizing exercise provides the required thrust for the specified aircraft performance requirements. Using the propeller performance given in Section 7.10, the SHP at sea-level static conditions can be worked out. From this information, engine performance at other ratings can be established for the full flight envelope. Takeoff rating maintains constant power but the thrust changes with speed. Figures 8.9 and 8.10 give the typical turboprop thrust and fuel flow (in terms of psfc – turboprop specific fuel consumption) at maximum climb and maximum cruise ratings in a non-dimensional form. Intermediate values may be linearly interpolated from these graphs.

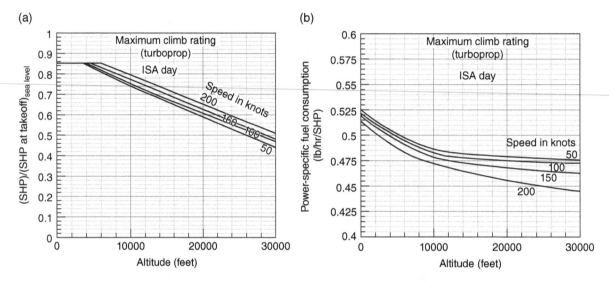

■ **FIGURE 8.9**   Uninstalled maximum climb rating (turboprop). (a) Shaft horse power. (b) Specific fuel consumption

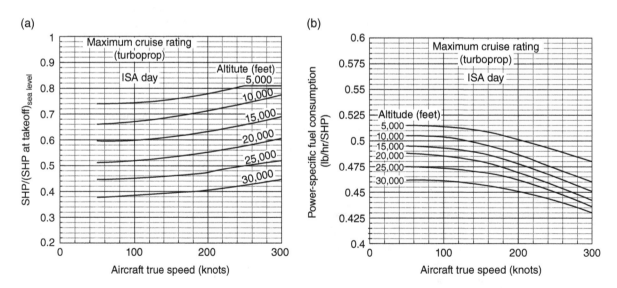

■ **FIGURE 8.10**   Uninstalled maximum cruise rating (turboprop). (a) Shaft horse power. (b) Specific fuel consumption

■ **TABLE 8.2**
Turboprop shaft horse power for sizes (based on uninstalled SHP)

	$SHP/\dot{m}_a$ (SHP/lb/s) at takeoff	SHP_{SLS}/dry engine weight (SHP/lb)
Smaller turboprops, ≤1000 SHP	≈0.012	≈2.2−2.75
Larger turboprops, >1000 SHP	≈0.01	≈2.5−3.0

Typically, the higher the SHP, the lower is the specific SHP ($SHP/\dot{m}_a$ − SHP/ lb/s), and the higher the specific dry-engine weight (SHP_{SLS}/dry engine weight –lb/lb). Table 8.2 may be used for these computations.

8.5.1 Typical Turboprop Performance

8.5.1.1 Takeoff Rating The takeoff rating for turboprops is held constant up to Mach 0.3. In this case, the non-dimensional value is 1.0 and no graph is given. The psfc of the turboprop at takeoff is from 0.475 to 0.6 lb/h/shp. The sizing exercise in Chapter 16 gives the Tucano class military trainer aircraft uninstalled $SHP_{SLS} = 1075$ to get 4000 lb installed thrust.

8.5.1.2 Maximum Climb Rating Figure 8.9 shows the uninstalled maximum climb SHP and fuel flow (sfc) in non-dimensional form for a standard day up to 30,000 ft altitude for four true airspeeds from 50 to 200 kt. The break in SHP generation up to 6000 ft altitude is due to fuel control to keep the EGT low. As in the case of takeoff rating, the SHP remains invariant up to about 5000 ft altitude.

8.5.1.3 Maximum Cruise Rating Figure 8.10 shows the uninstalled maximum cruise SHP and fuel flow in non-dimensional form for a standard day from 5000 to 30,000 ft altitude for true airspeed from 50 to 300 kt.

8.6 Installed Engine Performance Data of Matched Engines to Coursework Aircraft

Section 7.7 describes losses associated with installation effects of engine to aircraft [1]. This section generates the available installed thrust and fuel flow graphs matched for the sized aircraft worked examples: one Bizjet and one advanced jet trainer (AJT). In addition, a 1075 SHP turboprop engine performance is given for the readers to work out aircraft performance. The installed engine data are obtained from the non-dimensional data after applying the installation losses and the sized T_{SLS}.

Since the given sfc graphs are based on uninstalled thrust, the fuel flow rates are computed using uninstalled thrust. Therefore, the installed sfc will be higher than the uninstalled sfc. Installation losses at cruise come down to about half the losses of that at takeoff.

8.6.1 Turbofan Engine (Smaller Engines for Bizjets – *BPR* ≈)

Figures 8.11–8.13 give typical uninstalled turbofan thrust in non-dimensional form in terms of T_{SLS} along with corresponding sfc for the Bizjet class of aircraft. Section 16.5 establishes that it requires an uninstalled matched $T_{SLS\_UNINSTALLED} = 3560$ lb per engine. Table 8.3 summarizes the factors associated with installed thrust and fuel flows for the three engine ratings. The data are sufficient for the example taken in the book. Intermediate values may be linearly interpolated.

■ **FIGURE 8.11**
Installed takeoff performance per engine (BPR ≈ 3 – 4), ISA day

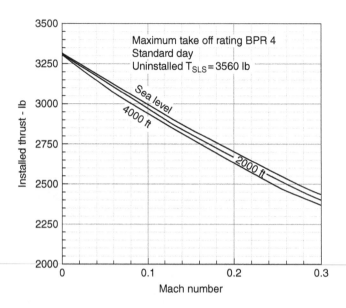

(a)

(b)

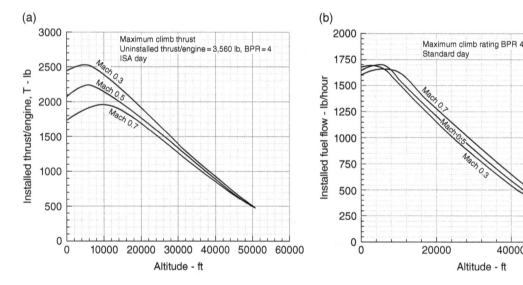

■ **FIGURE 8.12** Installed maximum climb performance per engine (BPR ≲ 4), ISA day. (a) Installed maximum climb thrust per engine. (b) Installed maximum climb fuel flow per engine

8.6.1.1 *Takeoff Rating (Bizjet)*

Depending on how the ECS is managed, typical installation losses vary from 6–8% of the uninstalled sea-level static thrust. If required, the air-conditioning could be shut off for a brief period until the undercarriage is retracted. Taking 7% installation loss at takeoff, the installed $T_{SLS\_INSTALLED} = 0.93 \times 3560 = 3315$ lb per engine for the sized Bizjet. Figure 8.11 gives the installed engine thrust at the takeoff rating.

Fuel flow rate is computed from its sfc of 0.498 lb/h/lb at sea-level static conditions. Using uninstalled $T_{SLS} = 3560$ lb per engine, the fuel flow rate is $3560 \times 0.498 = 1772$ lb/h per engine. Fuel flow is kept nearly constant at takeoff up to the en-route climb segment, when the engine is throttled down to the maximum climb rating computed in the next section.

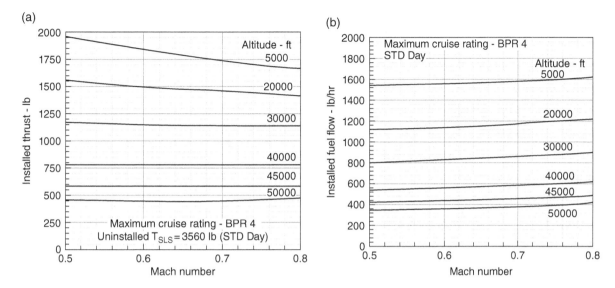

■ **FIGURE 8.13**   Installed maximum cruise performance per engine (BPR $\lesssim$ 4), ISA day. (a) Installed maximum cruise thrust per engine. (b) Installed maximum cruise fuel flow per engine

■ **TABLE 8.3**
Summary example of installed thrust and fuel flow data per engine at the three ratings (all computations are based on $T_{SLS\_UNINSTALLED} = 3560$ lb per engine)

Rating	Altitude (ft)	Mach	Loss, %	Scaling factor, k	sfc (lb/lb/h)	Available thrust (lb) Uninstalled	Available thrust (lb) Installed	Fuel flow (lb/h)
Takeoff	0	0	7	1	0.498	3560	3315	1772
Maximum climb	1000	0.38	6	0.67	0.7	2373	2231	1661
Maximum cruise	41000	0.74	4	0.222	0.73	790	758	578

8.6.1.2   *Maximum Climb Rating (Bizjet)*   Figure 8.3 gives the uninstalled maximum climb thrust in non-dimensional form in terms of T_{SLS} and fuel consumption (sfc) up to 50,000 ft altitude for three Mach numbers on an ISA day. Installation loss during climb is taken as 5% of the uninstalled thrust. Using the uninstalled graphs, the installed thrust and fuel flow rates are plotted in Figure 8.12. Note that this turbofan has a break in fuel flow at around 5000–10,000 ft altitude, depending on the flight Mach number, to keep EGT within the limits, resulting in a corresponding break in thrust generation.

An example to establish thrust and fuel flow at sea-level and Mach 0.3 is given here. The installed climb data can be generated in the same manner using Figure 8.3, which has the applied scaling factor k_{cl}. Figure 8.3 gives $T/T_{SLS} = 0.72 = k_{cl}$. Then the uninstalled initial climb thrust is $3560 \times 0.72 = 2563$ lb per engine. Then the installed thrust becomes $T_{SLS\_INSTALLED} = 0.95 \times 2563 = 2435$ lb/engine. The graph gives uninstalled sfc = 0.68. It gives fuel flow of $0.66 \times 2563 = 1690$ lb/h per engine. The classroom example is checked out in Chapter 12. Estimation of the payload range would require the full aircraft climb performance up to cruise altitude.

8.6.1.3 *Maximum Cruise Rating (Bizjet)*

Figure 8.4 gives the installed maximum cruise thrust in non-dimensional form and fuel consumption (sfc) from 5000 to 50,000 ft altitude for Mach number varying from 0.5 to 0.8 on an ISA day. Figure 8.13 gives the installed engine thrust at maximum cruise rating for the sized Bizjet.

The classroom example specifies an initial maximum cruise speed high-speed cruise (HSC) of Mach 0.74 at 41,000 ft. An example to establish thrust and fuel flow at the initial cruise is given here. The rest of the installed cruise data can be generated in the same manner using Figure 8.4, which has an applied scaling factor k_{cr}. Figure 8.4 gives $T/T_{SLS} = 0.222 = k_{cr}$. Then the uninstalled initial climb thrust is $3560 \times 0.222 = 790$ lb per engine. Taking into account a 4% installation loss at cruise, the installed thrust $T = 0.96 \times 790 = 758$ lb per engine. Chapter 13 checks out whether the thrust is adequate for the aircraft to reach the maximum cruise speed. Fuel flow in Figure 8.13 is $0.73 \times 790 = 577$ lb/h per engine.

8.6.1.4 *Idle Rating (Bizjet)*

Figure 8.14 gives the idle thrust and fuel flow at installed idle engine rating for Mach number varying from 0.3 to 0.6 and from sea-level to 51,000 ft altitude on an ISA day. The idle rating shows patterns of performance variation quite different from climb and cruise ratings. Part-throttle ratings are not given here, but may be interpolated.

8.6.2 Turbofans with BPR around 5–6 (Larger Jets)

Bigger engines have higher BPR. Current large operational turbofans have BPR around 5–6 (new-generation turbofans have achieved BPR greater than 8). These big engines have performance characteristics slightly different from the smaller ones: specifically the maximum climb rating is without break in thrust with altitude gain. Using Figures 8.10–8.13, the installed thrust and fuel flow rates can be calculated in a similar manner as above.

8.6.3 Military Turbofan (Very Low *BPR*)

Military turbofans differ from civil turbofans. Figure 8.15 gives the typical available uninstalled thrust in non-dimensional form for the AJT/CAS class of aircraft. Earlier we stated that only one graph at the maximum ratings would prove sufficient for use in this book. In industry, separate graphs are used for each rating.

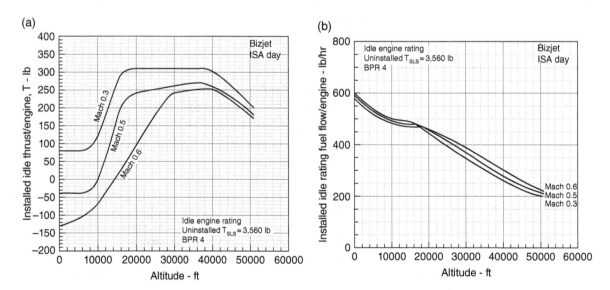

■ **FIGURE 8.14** Installed idle engine rating performance per engine (BPR $\lesssim$ 4), ISA day. (a) Installed idle thrust per engine. (b) Installed idle fuel flow per engine

■ **FIGURE 8.15**
Installed maximum rating, military turbofan (BPR = 0.75)

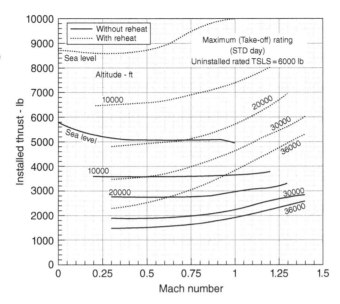

The installation loss of a single-seat military trainer aircraft is low and taken as 3.33% reduction in thrust (combat aircraft can have as high as around 15% loss of thrust). Section 16.7 sizes the AJT that requires installed T_{SLS} of 5300 lb (uninstalled T_{SLS} = 5500 lb) for normal training configuration (NTC). Using Figure 8.8, the installed thrust at the maximum (takeoff) rating for the AJT is shown in Figure 8.15. Maximum continuous rating (primarily for climb) is at 95% of the maximum takeoff rating. High-speed runs are done at 90% of the maximum takeoff rating (combat could demand up to 100% thrust in short bursts). For the CAS role, the thrust values need to be scaled up by 30% with an uprated engine.

The AJT sfc at T_{SLS} is 0.7 lb/h/lb and the fuel flow rate is based on the uninstalled T_{SLS}. Therefore, fuel flow rate = 0.7 × 5500 = 3850 lb/h.

Unlike civil aircraft, the AJT training profile is throttle-dependent. A training profile operates in varying speeds and altitudes. At NTC (meant for airmanship and navigational training with low levels of armament practice) the average throttle setting may be considered as about 75–85% of the maximum rating. At 30,000 ft altitude and Mach 0.7 the average operating installed thrust T = 0.75 × 1980 (Figure 8.8) = 1485 lb (uninstalled thrust of 1538 lb) and the average sfc is given as 0.7 lb/h/lb. It gives the average fuel flow rate = 0.7 × 1536 = 1076 lb/h. On-board fuel carried is 2425 lb.

8.7 Installed Turboprop Performance Data

8.7.1 Typical Turboprop Performance

Given the Tucano class trainer aircraft, the uninstalled SHP_{SLS} = 1075 to get 4000 lb installed thrust [1]. Once the SHP at sea-level static conditions is established from the sizing exercise, the thrust requirement of an installed four-bladed propeller can be computed.

8.7.1.1 Takeoff Rating The psfc of turboprops at takeoff is from 0.475 to 0.6 lb/h/shp. For SHP of less than 1000, use a psfc of 0.6 lb/h/shp; for more than 1000, use a psfc of approximately 0.48 lb/h/shp. Therefore, fuel flow at SHP_{SLS} is 0.5 × 1075 = 538.5 lb/h.

From Table 8.2, the intake air mass flow at SHP_{SLS} is 0.011 × 1075 = 11.83 lb/s. The dry engine weight = 1075/2.75 = 390 lb.

8.7.1.2 ***Maximum Climb Rating*** Figure 8.9 shows the uninstalled maximum climb SHP and fuel flow (sfc) in non-dimensional form for an ISA day up to 30,000 ft altitude for four true airspeeds from 50 to 200 kt. Intermediate values may be linearly interpolated. The break in SHP generation up to 6000 ft altitude is due to fuel control to keep the EGT low.

As before, the turboprop class also requires a factor k_{cl} (varies with speed and altitude) to be applied to the SHP. From Figure 8.9a, a value of 0.85 may be used to obtain the initial climb SHP. In the example, the uninstalled initial climb power is then $0.85 \times 1075 = 914$ SHP. The integrated propeller performance after deducting the installation losses gives the available thrust. Typically, the initial climb starts at a constant EAS of approximately 200 kt, which is approximately Mach 0.3. At a constant EAS climb, Mach number increases with altitude; when it reaches 0.4, it is held constant. Fuel flow at the initial climb (at 800 ft altitude) is obtained from Figure 8.9b. The example gives $0.522 \times 914 = 477$ lb/h. With varying values of altitude, climb calculations are performed in small increments of altitude, within which the variation is taken as the mean and is kept constant for the increment.

8.7.1.3 ***Maximum Cruise Rating*** Figure 8.10 shows the uninstalled maximum cruise SHP and fuel flow in non-dimensional form for an ISA day from 5000 to 30,000 ft altitude for true airspeeds from 50 to 300 kt. Intermediate values may be linearly interpolated. The graph takes into account the factor k_{cr} (varies with speed and altitude).

In the example, the design initial maximum cruise speed is given as 300 kt at 25,000 ft altitude. From Figure 8.10a, the uninstalled power available is $0.525 \times 1075 = 564$ SHP. In Figure 8.10b, the corresponding fuel flow is $0.436 \times 564 = 246$ lb/h. The integrated propeller performance, after deducting the installation losses, gives the available installed propeller performance.

8.7.2 Propeller Performance – Worked Example

In a stepwise manner, thrust from a propeller is calculated as a coursework exercise. The method uses the Hamilton Standard charts intended for constant-speed propellers. These charts can also be used for fixed-pitch propellers when the pitch of the propeller should match the best performance at a specific speed: either cruise speed or climb speed. Two forms are shown: (1) from the given thrust, compute the HP (in turboprop case SHP) required; and (2) from the given HP (or SHP), compute the thrust. The starting point is the C_p. Typically, at sea-level takeoff rating at static conditions, 1 SHP produces about 4 lb on an ISA day. At the first guesstimate, a factor of 4 is used to obtain SHP to compute the C_p. One iteration may prove sufficient to refine the SHP.

8.7.2.1 ***Problem Description*** Consider a single, four-bladed, turboprop military-trainer aircraft of the class RAF Tucano operating with a constant-speed propeller at $n = 2400$ RPM, giving installed $T_{SLS} = 4000$ lb. The specified aircraft speed is 320 mph at 20,000 ft altitude (i.e. Mach 0.421). For the aircraft speed, the blade AF is taken as 180. Establish its rated SHP at sea-level static conditions and thrusts at various speeds and altitudes. All computations are for an ISA day.

8.7.2.2 ***Propeller Diameter*** The first task is to establish the propeller diameter. From Figure 7.34 for a four-bladed propeller, the diameter is taken to be 98 inches. Figure 7.33 establishes the integrated design C_{Li}.

To check, use the empirical relation to determine propeller diameter in Section 7.10. Take $SHP_{SLS} = 4000/4 = 1000$. It gives $D = K(P)^{0.25} = 18 \times (1000)^{0.25} = 101.2$ in, where K for a four-bladed propeller is 18. The empirical relation is close to what is obtained from Figure 7.34. A mean diameter of the two results of 100 inches is accepted.

Aircraft configuration must ensure ground clearance in the event of a collapsed nose-wheel tyre or undercarriage oleo failure. A higher number of blades (i.e. greater solidity) could reduce the diameter, at the expense of higher cost. For this aircraft class, it is best to retain the largest permissible propeller diameter, keeping the number of blades to four or five. To set an example for the ground clearance, a 2 in radius is cut off from the tip. Then the final diameter is 96 in = 8 ft.

8.7.2.3 *Case 1*

The takeoff performance (HP from thrust) is used to compute the SHP at a sea-level takeoff. A rough value of installed SHP = 4000/4 = 1000 SHP. For a clean nacelle for turboprop, an optimistic value of a fuselage loss factor $(f_b \times f_h)$ of 0.98 is taken.

Taking into account 5% installation losses at takeoff, the uninstalled T_{SLS} = 4000/0.95 = 4200 lb. In addition, apply a fuselage loss of 0.98, that is, T_{SLS} = 4200/0.97 = 4300 lb. It may now be summarized that to obtain 4000 lb installed thrust, the uninstalled rated power needs to be 1000/(0.95 × 0.98) ≈ 1075 SHP.

Figure 7.33 gives the factor $ND \times f_c$ = 2400 × 8 × 1.0 = 19,200. It corresponds to the ratio of the speed of sound at ISA sea-level to the altitude, f_c = 1.0, at static conditions (Mach 0).

Figure 7.33 gives the integrated design C_{Li} ≈ 0.5.

Equation 7.43 gives:

$$C_P = (550 \times \text{SHP})/(\rho n^3 D^5)$$

or

$$C_P = (550 \times 1000)/(0.00238 \times 40^3 \times 8^5)$$
$$= 550\ 000/4\ 991\ 222 = 0.11$$

Figure 7.31 (four-blade, AF = 180, C_L = 0.5) gives C_T/C_P = 2.4, corresponding to integrated design C_{Li} = 0.5 and C_P = 0.11. Therefore installed static thrust:

$$T_{SLS} = (C_T/C_P)(33000 \times \text{SHP})/ND$$
$$= (2.4 \times 550 \times 1000)/(40 \times 8)$$
$$= 1320000/320 = 4125\ \text{lb}$$

Therefore the installed SHP is revised to 1000 × (4000/4125) = 970 SHP, giving installed thrust T_{SLS} = 4000 lb. It is close enough to avoid any further iteration. As a safe consideration, it is taken that 4000 lb of thrust is produced by installed 1000 SHP. The corresponding uninstalled SHP is 1075. Readers may carry out the iteration with installed 970 SHP.

8.7.2.4 *Case 2*

Once the installed SHP_{SLS} is known, the thrust for the takeoff rating can be computed. The turboprop fuel control maintains a constant SHP at takeoff rating, keeping it almost invariant. This section computes the thrust available at speeds up to 160 mph to cover lift-off and entering the en-route climb phase. Available thrust is computed at 20, 50, 80, 120 and 160 mph, as shown in Table 8.4.

■ **TABLE 8.4**
Propeller installed thrust results

Velocity, V (mph)	20	50	80	120	160
$J = 0.00463 \times$ mph	0.092	0.23	0.37	0.55	0.74
C_p	0.11	0.11	0.11	0.11	0.11
Installed SHP	1000	1000	1000	1000	1000
From Figure 7.32, η_{prop}	0.19	0.4	0.56	0.7	0.77
Installed thrust, T (lb)	3820	3225	2820	2350	1940

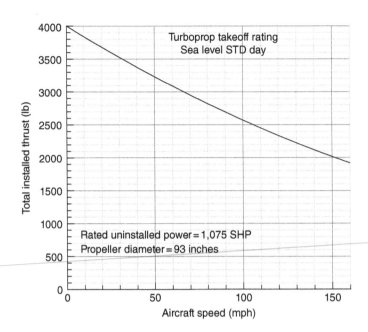

■ **FIGURE 8.16**
Takeoff thrust for the turboprop engine

For integrated design, $C_{Li} = 0.5$ and $C_p = 011$. The propeller η_{prop} corresponding to J and C_p is obtained from Figure 7.32 (four-blade, AF = 180, $C_L = 0.5$), as shown in Table 8.4.

$$J = V/nD = (1.467 \times V)/(40 \times 8) = 0.00458 \times V, \text{ where } V \text{ is in mph}$$

$$\text{Power coefficient}, C_p = (550 \times \text{SHP})/(\rho n^3 D^5)$$

$$= (550 \times 1000)/(0.00238 \times 40^3 \times 8^5) = 0.11$$

For integrated design, $C_{Li} = 0.5$ and $C_p = 011$. The propeller η_{prop} corresponding to J and C_p is obtained from Figure 7.32 (four-blade, AF = 180, $C_L = 0.5$), as shown in Table 8.4.

Thrust, $T = (550\eta_{prop} \times 1000)/V = 550,000 \times \eta_{prop}/V$, where V is in mph (see Table 8.4). Use Equation 7.45, for the FPS system.

Figure 8.16 plots the thrust versus speed at the takeoff rating. In a similar manner, thrust at any speed, altitude and engine rating can be determined from the relevant graphs.

Typically, an engine throttles back from the takeoff rating to the maximum climb rating for an en-route climb. At up to about 4000 ft altitude, it is kept around 85% of the maximum power, and thereafter SHP goes down with altitude.

8.8   Piston Engine

There are several ways to present piston engine performance. Figure 8.17 gives the Lycoming IO-360 series 180 HP rated piston engine performance graph. It may be noted that the power ratings are given by RPM. It consists of two graphs to be used together. The first one is at full throttle and part throttle at ISA sea-level and is known as the *full throttle sea-level* graph, and the second one is at full throttle and part throttle at altitude and is known as the *full throttle altitude* graph. At each altitude there is a unique value of HP for the combination setting of RPM and manifold pressure. The second graph also gives correction factors for a non-standard day.

Piston engine power is dependent on the amount of air mass inhaled, which is indicated by its RPM and manifold pressure, $p_{manifold}$, at a particular ambient condition. Engine RPM is adjusted by throttle control (fuel control). Above 5000 ft altitude, the air/fuel mixture is

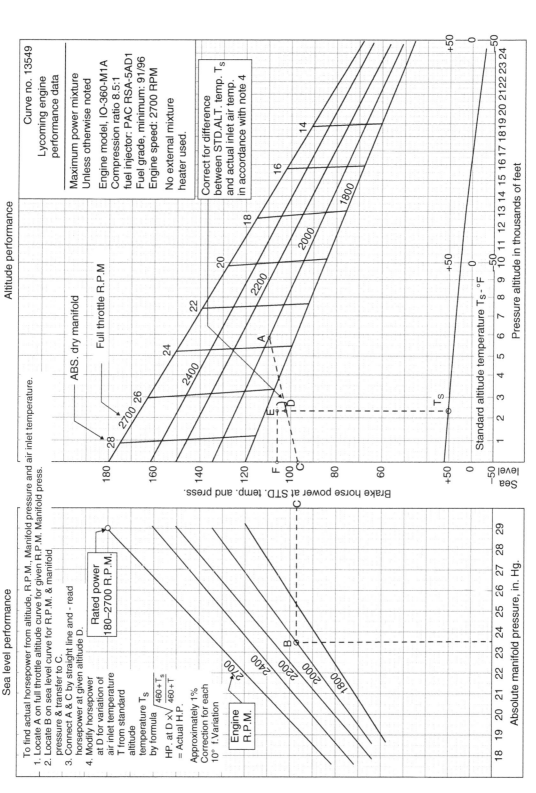

■ **FIGURE 8.17** Textron-Lycoming IO-360 series 180 HP piston engine. (Reproduced with permission of Lycoming Engine)

controlled by a pilot-operated lever to make it leaner. Nowadays, all piston engines are fuel-injected, with the carburettor type disappearing fast.

A valve controls air-mass aspiration, and when closed there is no power, that is, $p_{manifold} = 0$. When the engine is running at full aspiration creating suction, $p_{manifold}$ reads the highest suction values. If there is less propeller load at higher altitudes at the same RPM, then it will generate less power and the valve will be partially closed to inhale less air mass than running at equilibrium. Evidently, at low RPM the aspiration level would be low and there is a limiting $p_{manifold}$ line. Therefore, the variables affecting engine power are RPM, $p_{manifold}$, altitude and atmospheric temperature for non-standard days. If the engine is supercharged, then the graphs will indicate the effect.

The Lycoming IO-360 series power ratings at ISA sea-level are as follows:

Take-off: 180 HP at 2700 RPM consuming 15.6 US gal/h (this is the normal rated power – 100%).
Typical climb: at 2500–2650 RPM (80–95% power).
Performance cruise: 135 HP at 2450 RPM consuming 11.0 US gal/h (75% power).
Economy cruise: 117 HP at 2350 RPM consuming 8.5 US gal/h (60% power).
Low-speed cruise: at 2000–2300 RPM (light weight/holding).
Idle/descent: below 2000 RPM.

Figure 8.17 shows the parameters in graphical form and how to use them. The fuel flow graph is given separately in Figure 8.18. Figure 8.18a is meant for fuel flow. It has two settings, one for *best power* and one for *best economy* – this is adjusted by a mixture ratio lever. It can be seen in Figure 8.18b that for higher power and RPM the mixture setting is at *best power*, and for cruise it is at *best economy*. Evidently, at 2000 RPM, the example is meant for *best economy*.

EXAMPLE 8.1

Find the power and fuel flow at 2000 RPM and at a manifold pressure of 23.75 Hg at 2400 ft altitude on an ISA day. Also find the HP on an ISA–20°F day. Take the inlet manifold temperature as the ambient temperature. (The Textron-Lycoming IO-360 series is a normally aspirated fuel-injected engine).

SOLUTION

On an ISA day:

Step 1: In the first graph locate the point at 2000 RPM and at manifold pressure of 23.75. Draw a horizontal to *HP* axis giving 96 HP. This is at sea level.

Step 2: In the second graph locate the point 2000 RPM and at manifold pressure of 23.75. It gives 110 HP at 6400 ft altitude.

Step 3: At 2400 ft altitude, interpolate by joining the two points by straight line and read 102 HP.

Fuel flow is obtained from Figure 8.18. It gives about 42.5 lb/h = 7.4 US gal/h. The aircraft is at slow speed cruise and possibly at light weight.

At ISA–20°F day

The expression for HP on a non-standard day, $[HP_{act} / HP_{std}] = \sqrt{(T_{std}/T_{act})}$.

Or,

$$HP_{non\_STD} = \sqrt{\{(460 + T_{STD})/(460 + T_{non\_STD})\}} \times HP_{STD}$$

At 2400 ft altitude, $T_{std} = 510°F$, therefore $T_{act} = 490°F$.

$$[HP_{act}/HP_{std}] = \sqrt{(T_{std}/T_{act})} = \sqrt{(510/490)} = 1.02$$

This gives $HP_{act} = 104$ HP.

The expression gives approximately 1% HP variation for a 10° change from standard conditions.

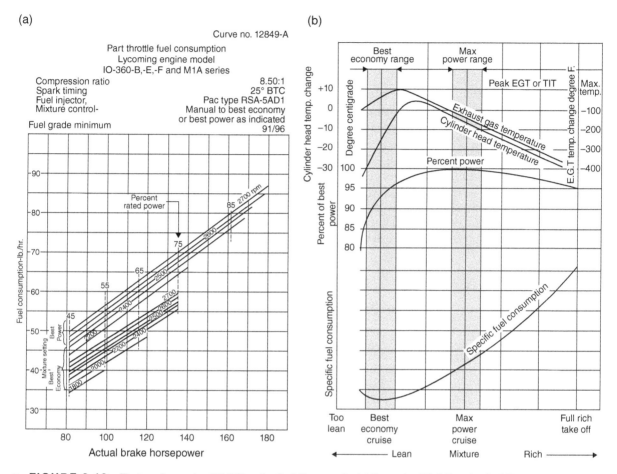

■ **FIGURE 8.18** **Textron-Lycoming IO-360 series fuel flow graph. (a) Lycoming IO-360 series fuel flow graph; and (b) Lycoming piston engine operating range. (Reproduced with permission of Lycoming Engine)**

With old analogue gauges, a pilot would not know at what HP the aircraft is operating. Below 5000 ft and with fixed pitch propeller (typical club/recreational flying), a pilot has only the throttle lever to control power. If he/she wishes to know the power, then he/she will have to record the RPM, manifold pressure, altitude and outside air temperature (OAT), and compute the HP after landing using the method shown. Some modern electronic flight information systems (EFIS) can display the power by processing the engine and air data using an onboard computer.

Readers may obtain appropriate engine charts from the manufacturers for other engines. Failing that, this graph may be scaled for classroom use.

8.9   Engine Performance Grid

Non-dimensional engine performance data as plotted in Figures 8.2–8.10 serve as generalized representations which are converted into actual installed engine performance data as given in Figures 8.11–8.15 for the Bizjet and AJT aircraft. Since engine performance is affected by atmospheric conditions, the thrust changes with altitude as density changes. Therefore, it is convenient to plot engine performance normalized to atmospheric conditions. This is done by using non-dimensional atmospheric property ratios as follows.

Density ratio $=\sigma=$ (density at altitude ρ)/(density at sea-level ρ_0)

Pressure ratio $=\delta=$ (pressure at altitude p)/(pressure at sea-level p_0)

Temperature ratio $=\theta=$ (temperature at altitude T)/(temperature at sea-level T_0)

From thermodynamics, $\delta=\sigma\theta$

Ratio of speed of sound $a/a_0=\sqrt{(T/T_0)}=\sqrt{\theta}$, or $a=a_0\sqrt{\theta}$

Mach number, $M=V/a$ where V is the aircraft velocity

(Note $\rho_0=0.002378\text{slugs/ft}^3$, $p_0=2118.22\text{lb/ft}^2$, $a_0=1118.4\text{ft/s}$ in Imperial system.)

In engineering practice, thrust is plotted as T/δ instead of thrust, T, alone. In the past, engineers used F/δ to represent thrust available and T/δ to represent thrust required by the aircraft; another way is to use subscripts, for example $(T/\delta)_{avail}$ and $(T/\delta)_{reqd}$. Installed $(T/\delta)_{avail}$ can include intake losses for the particular nacelle for the chosen engine, as the intake efficiency is known. Drag of aircraft also depends on atmospheric density as well as aircraft weight. In equilibrium flight, thrust equals drag, giving $(T/\delta)_{reqd}$. This gives $D=T$. Normalized to atmospheric properties, $D/\delta=(T/\delta)_{reqd}$.

Note that $\quad D=0.5\rho V^2 S_W C_D=0.5(\rho_0\delta)(Ma_0)^2 S_W C_D=1481\,C_D M^2 S_W\delta$

or

$$D/\delta=1481\,C_D M^2 S_W \tag{8.1}$$

or

$$(T/\delta)_{reqd}=1481\,C_D M^2 S_W \tag{8.2}$$

Since aircraft drag depends on aircraft weight, $(T/\delta)_{reqd}$ can be tied down to aircraft weight as W/δ, keeping in line with the normalization. From W/δ the lift coefficient C_L can be determined, which in term yields induced drag of the aircraft.

$$L=0.5\rho V^2 S_W C_L=0.5(\rho_0\delta)(Ma_0)^2 S_W C_L=1481\,C_L M^2 S_W\delta$$

or

$$L/\delta=W/\delta=1481\,C_L M^2 S_W \tag{8.3}$$

Finally, by definition, fuel flow is expressed as:

$$FF=\text{sfc}\times T$$

or

$$FF/\delta=\text{sfc}\times T/\delta \tag{8.4}$$

Chapter 13 deals with the specific range, SR, in detail. It is defined as nautical air miles (nam) covered at the expense of a unit of fuel consumption. In still air, nautical air miles (nam) = nautical miles (nm) on the ground travelled by the aircraft. With wind velocity V_w, nam = nm $\pm\ V_w$, where positive for a tail wind and negative for a head wind.

$$\text{Specific range, SR}=\text{nam}/\text{lb}=V/(T\times\text{sfc}) \tag{8.5}$$

It is convenient to plot with Mach number, M, instead of velocity, V:

$$M=V/(a_0\sqrt{\theta})$$

Then

$$SR = nam / lb = M/\left(a_0 \sqrt{\theta}\right)/\left(T \times sfc\right)$$

Equation 8.4 can be rewritten as

$$FF/\left(\delta\sqrt{\theta}\right) = sfc \times T/\left(\delta\sqrt{\theta}\right) \tag{8.6}$$

The two normalized parameters are now $FF/(\delta\sqrt{\theta})$ and $sfc \times T/(\delta\sqrt{\theta})$. Relevant values of δ and θ for different altitudes are given in Table 8.5.

8.9.1 Installed Maximum Climb Rating (TFE 731-20 Class Turbofan)

Tables 8.6 and 8.7 below are data taken from Figures 8.11–8.13 to plot the Bizjet engine performance grid, or simply engine grid. This gives available thrust and fuel flow rate for a TFE731-20 class turbofan engine. In this section, only the maximum cruise rating is plotted.

■ **TABLE 8.5**
Atmospheric properties by altitude

Altitude (ft)	1500	5000	10000	20000	30000	36089[a]	40000	45000	49000
δ	0.971	0.832	0.688	0.49	0.297	0.223	0.185	0.146	0.1201
θ		0.9656	0.931	0.8625	0.7937	0.7518	0.7518	0.7518	0.7518
$\sqrt{\theta}$		0.9826	0.965	0.9287	0.891	0.867	0.867	0.867	0.867

[a] This is the altitude of the tropopause.

■ **TABLE 8.6**
Installed maximum climb thrust: T/engine (lb)

Altitude (ft)	1500	5000	10000	20000	30000	36089[a]	40000	45000
Mach 0.3	2640	2680	2450	1950	1390	1100	910	690
Mach 0.4	2400	2480	2300	1850	1350	1070	880	670
Mach 0.5	2200	2300	2180	1800	1320	1050	860	660
Mach 0.6	1980	2120	2080	1750	1300	1050	860	660
Mach 0.7	1830	1920	2000	1700	1320	1070	870	670

[a] This is the altitude of the tropopause.

■ **TABLE 8.7**
Installed maximum climb thrust: (T/δ)/engine (lb)

Altitude (ft)	1500	5000	10000	20000	30000	36089[a]	40000	45000
Mach 0.4	2719	3221	3561	3980	4680	4933	4919	4726
Mach 0.4	2472	2980	3343	3776	4546	4798	4757	4589
Mach 0.5	2266	2764	3198	3674	4444	4708	4649	4520
Mach 0.6	2039	2548	3023	3572	4377	4708	4648	4520
Mach 0.7	1885	2308	2907	3470	4377	4798	4702	4590

[a] This is the altitude of the tropopause.

8.9.2 Maximum Cruise Rating (TFE731-20 Class Turbofan)

Installed total maximum cruise thrust (T/δ – lb) and fuel flow ($FF/\delta\sqrt{\theta}$ – lb/h) are plotted in Figure 8.19 as the engine grid for the Bizjet application. To obtain constant lines of $FF/(\delta\sqrt{\theta})$, a cross-plot is required using data given in Tables 8.8 to 8.11. These kinds of graphs are useful for evaluating aircraft performance by superimposing on to the aircraft grid as done in Figure 8.19. An example of how to use the graph is shown below.

■ **FIGURE 8.19**
Engine grid for Bizjet (total of two engines) (read fuel flow values at the same point of thrust data)

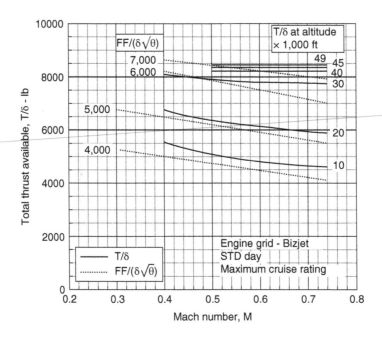

■ **TABLE 8.8**
Installed maximum cruise thrust: T/engine (lb)

Altitude (ft)	5000	10000	20000	30000	36089[a]	40000	45000	49000
Mach 0.5	1950	1740	1550	1170	914	770	615	510
Mach 0.6	1840	1650	1500	1155	910	770	615	510
Mach 0.7	1740	1600	1450	1150	910	770	615	510
Mach 0.74	1700	1580	1440	1150	910	770	615	515

[a] This is the altitude of the tropopause.

■ **TABLE 8.9**
Installed maximum cruise thrust: (T/δ)/engine (lb)

Altitude (ft)	5000	10000	20000	30000	36089[a]	40000	45000	49000
Mach 0.5	2344	2529	3163	3940	4099	4162	4212	4246
Mach 0.6	2212	2398	3061	3889	4081	4162	4212	4246
Mach 0.7	2092	2326	2529	3872	4081	4162	4212	4246
Mach 0.74	2043.3	2297	2939	3872	4081	4162	4212	4288

[a] This is the altitude of the tropopause.

■ **TABLE 8.10**
Installed maximum cruise thrust: corresponding fuel flow *FF*/engine (lb/h)

Altitude (ft)	5000	10000	20000	30000	36089[a]	40000	45000	49000
Mach 0.4	1530	1380	1100	770	620	530	430	370
Mach 0.5	1540	1400	1120	800	640	540	440	380
Mach 0.6	1560	1420	1150	830	660	580	455	390
Mach 0.7	1580	1440	1180	860	690	590	470	400
Mach 0.74	1600	1460	1200	880	710	600	485	410

[a] This is the altitude of the tropopause.

■ **TABLE 8.11**
Installed maximum cruise thrust: corresponding fuel flow $FF/(\delta\sqrt{\theta})$ per engine (lb/h)

Altitude (ft)	5000	10000	20000	30000	36089[a]	40000	45000	49000
Mach 0.4	1871	2079	2417	2910	3206	3304	3397	3552
Mach 0.5	1884	2109	2461	3024	3310	3367	3476	3649
Mach 0.6	1908	2139	2527	3137	3414	3616	3595	3745
Mach 0.7	1932	2169	2593	3250	3568	3678	3714	3842
Mach 0.74	1957	2199	2637	3326	3671	3740	3831	3938

[a] This is the altitude of the tropopause.

EXAMPLE 8.2

Find the total thrust and fuel flow for a Bizjet flying at 45,000 ft ($\delta = 0.146$ and $\sqrt{\theta} = 0.867$) and Mach 0.5 using the engine grid given in Figure 8.19. This is a good example of the comments made in Section 1.12 on using graphs.

SOLUTION

From the graph corresponding to Mach 0.5, T/δ at 45,000 ft = 8400 lb, or T/δ per engine = 4200 lb. This gives total thrust from two engines, $T = 8400 \times 0.146 = 1228.4$ lb, that is, T per engine ≈ 613 lb. Compare this with the tabulated value which gives T/δ per engine = 4212 lb and T per engine = 615 lb.

$FF/(\delta\sqrt{\theta})$ is interpolated between the dashed lines at the same point where T/δ is read. In this case, total fuel flow,

$FF/(\delta\sqrt{\theta}) = 6980$ lb/h, or $FF/(\delta\sqrt{\theta})$ per engine = 3490 lb/h, that is, $FF = 442$ lb/h. Compare with the tabulated value which gives $FF/(\delta\sqrt{\theta})$ per engine = 3476 lb/h and FF per engine = 440 lb/h.

In general, tabulated computerized outputs are better than reading from graphs.

Aircraft performance estimation for a non-ISA day requires the plots to be further normalized for the conditions. Readers may refer to aircraft engine textbooks for this technique. Since the 1990s, almost all aircraft started to use engine manufacturer-supplied engine performance programs (engine deck – computer programs) to obtain engine data at any condition, and the use of generalized engine performance graphs gradually died out, applied only when required, as discussed in Chapter 12.

8.10   Some Turbofan Data

Tables 8.12 to 8.14 show some example data for different models of engine for ISA day conditions.

■ TABLE 8.12
Civil aircraft engine data

Model	T_{SLS} (lb)	Fan diameter (inches)	BPR	OPR	Airflow (lb/s)	Altitude ('000 ft)	Mach	Thrust (lb)	tsfc (lb/lb/h)
		Takeoff				Cruise			
CF6-50-C2	52500	134.1	4.31	30.4	1476	35	0.8	11555	0.63
CF6-80-C2	52500	88.4	4.31	28.4	1450	35	0.8	12000	0.576
GE90-B4	87400	134	8.4	39.3	3037	35	0.8	17500	–
JT8D-15A	15500	49.2	1.04	18.6	327	30	0.8	4920	0.779
JT9D-59A	53000	97	4.9	24.5	1639	35	0.85	11950	0.666
PW2040	41700	84.8	8.0	28.6	1210	35	0.85	6500	0.582
PW4052	52000	97	5.0	28.5	1700	–	–	–	–
PW4084	87500	118.5	8.41	34.4	2550	35	0.83	–	–
CFM56-3	23500	60	5.0	22.6	655	35	0.85	4890	0.667
CFM56-5C	31200	72.3	8.6	31.5	1027	35	0.8	6600	0.545
RB211-524B	50000	85.8	4.5	28.4	1513	35	0.85	11000	0.643
RB211-535E	40100	73.9	4.3	25.8	1151	35	0.8	8495	0.607
RB211-882	84700	8.01	39.0	2640	35	–	0.83	16200	0.557
V2528-D5	28000	63.3	4.7	30.5	825	35	0.8	5773	0.574
ALF502R	6970	41.7	5.7	12.2		35	0.7	2250	0.72
TFE731-20	3500	28.2	3.34	14.4	140	40	0.8	986	0.771
PW300	4750	38.2	4.5	23.0	180	40	0.8	1113	0.675
FJ44	1900	20.9	3.24	12.8	63.3	30	0.7	600	0.75
Olympus593	38000	11.3	410	53	2.0	10030	1.15	–	–

BPR, bypass ratio; OPR; overall pressure ratio.

■ TABLE 8.13
Military engine sea-level static data at takeoff

Model	BPR	Weight (lb)	OPR	Airflow (lb/s)	Without afterburner Thrust (lb)	TSFC (lb/lb/h)	With afterburner Thrust (lb)	tsfc (lb/lb/h)
P&W F119	0.45	3526	35		23600	–	35400	–
P&W F100	0.36	3740	32	254.5	17800	0.74	29090	1.94
GE F110	0.77	3950	30.7	270	17020	–	29000	–
GE F404	0.27	2320	26	146	12000	0.84	17760	1.74
GE F414	0.4	2645	30	170	12600	–	22000	–
Snecma-M88	0.3	1980	24	143	11240	0.78	16900	1.8

BPR, bypass ratio; OPR; overall pressure ratio.

■ **TABLE 8.14**
Turboprop data

Model	SHP$_{SLS}$	Dry weight (lb)	psfc (lb/h/SHP)
RR-250-B17	450	195	–
PT6-A	850	328	–
TPE-331-12	1100	400	0.5
GE-CT7	1940	805	–
AE2100D	4590	1548	–

SHP, shaft horse power.

Reference

1. Kundu, A.K., *Aircraft Design*. Cambridge University Press, Cambridge, 2010.

CHAPTER 9

Aircraft Drag

9.1   Overview

An important task for aircraft performance engineers is to make the best possible estimation of all the different types of drag associated with aircraft aerodynamics. Commercial aircraft design is sensitive to the DOC (direct operating cost), which is drag-dependent. Just one count of drag (i.e. $C_D = 0.0001$) could account for several million US dollars in operating costs over the lifespan of a small fleet of midsized aircraft. This will become increasingly important with increasing fuel costs. Accurate estimation of the different types of drag remains a central theme. (Equally important are other ways to reduce DOC; one of them is reducing manufacturing cost.)

For a century, a massive effort has been made to understand and estimate drag, and the work is still continuing. Possibly some of the best work in the English language on aircraft drag is compiled by NACA/NASA (National Advisory Committee for Aeronautics/National Aeronautics and Space Administration), RAE, AGARD (Advisory Group for Aerospace Research and Department), ESDU, DATCOM, RAeS (Royal Aeronautical Society), AIAA (American Institute for Aeronautics and Astronautics) and others [1–11]. These publications indicate that the drag phenomena are still not fully understood [10], and that the way to estimate aircraft drag is by using semi-empirical relations. CFD (computational fluid dynamics) is gaining ground, but is still some way from supplanting the proven semi-empirical relations. In the case of work on excrescence drag, efforts are lagging.

The 2D surface skin-friction drag, elliptically loaded induced drag and wave drag can be accurately estimated – together, they comprise most of the total aircraft drag. The problem arises when estimating drag generated by the 3D effects of the aircraft body, interference effects and excrescence effects. In general, there is a tendency to underestimate aircraft drag. Accurate assessments of aircraft mass, drag and thrust are crucial in aircraft performance estimation.

9.2   Introduction

The drag of an aircraft depends on its shape and speed, which are design-dependent, as well as on the properties of air, which are nature-dependent. Drag is a complex phenomenon arising from several sources, such as the viscous effects that result in skin friction and pressure differences as well as the induced flow-field of the lifting surfaces and compressibility effects (see Sections 2.10 and 2.15).

Theory and Practice of Aircraft Performance, First Edition. Ajoy Kumar Kundu, Mark A. Price and David Riordan.
© 2016 John Wiley & Sons, Ltd. Published 2016 by John Wiley & Sons, Ltd.

The aircraft drag estimate starts with the isolated aircraft components (e.g. wing and fuselage). Each component of the aircraft generates drag largely dictated by its shape. Total aircraft drag is obtained by summing the drag of all components plus their interference effects when the components are combined. The drag of two isolated bodies increases when they are brought together due to the interference of their flow fields.

The Reynolds number, Re, has a deciding role in determining the associated skin-friction coefficient, C_F, over the affected surface, and the type, extent and steadiness of the boundary layer (which affects parasite drag). Boundary-layer separation increases drag and is undesirable; separation should be minimized.

A major difficulty arises in assessing drag of small items attached to an aircraft surface such as instruments (e.g. pitot and vanes), ducts (e.g. cooling), blisters and necessary gaps to accommodate moving surfaces. In addition, there is the unavoidable discrete surface roughness from mismatches (at assembly joints) and imperfections, perceived as defects, which result from limitations in the manufacturing processes. Together, from both manufacturing and non-manufacturing origins, they are collectively termed *excrescence drag*.

Currently, accurate total aircraft drag estimation by analytical or CFD methods is not possible. Schmidt, in the AGARD AR-256 review publication [8], was categorical about the inability of CFD, analytical methods or even wind-tunnel model testing to estimate drag. CFD is steadily improving and can predict wing-wave drag (C_{Dw}) accurately, but not the total aircraft drag – most of the errors are due to the smaller excrescence effects, interference effects and other parasitic effects. Industrial practices employ semi-empirical relations (with CFD) validated against wind-tunnel and flight tests and generally produce proprietary information. Most of the industrial drag data are not available in the public domain. The methodology given in this chapter is a modified and somewhat simplified version of standard industrial practice [1, 3, 7]. The method is validated by comparing its results with the known drag of existing operational aircraft.

The design criterion for today's commercial high-subsonic jet transport aircraft is that the effects of separation and local shocks are minimized at the LRC (*long-range cruise*) (i.e. compressibility drag almost equal to zero before the onset of wave drag) condition. At HSC (*high-speed cruise*), a 20-count drag increase is allowed, reaching M_{crit}, due to local shocks (i.e. transonic flow) covering small areas of the aircraft. Modern streamlined shapes maintain low separation at M_{crit}; therefore, such effects are small at HSC. The difference in the Mach number at HSC and LRC for subsonic aircraft is small – of the order of Mach 0.05–0.075. Aircraft drag characteristics are plotted as *drag polar* (C_D versus C_L).

Strictly speaking, drag polar at several speeds and altitudes will give better resolution of drag value; however, estimation of the drag coefficient at LRC is sufficient because it has a higher C_f, which gives conservative values at HSC when ΔC_{Dw} is added. The LRC condition is by far the longest segment in the mission profile; the industry standard practice uses the LRC drag polar for all parts of the mission profile (e.g. climb and descent). The Re at LRC provides a conservative estimate of drag for the climb and descent segments. At takeoff and landing, the undercarriage and high-lift device drags must be added.

Supersonic aircraft operate over a wider speed range: the difference between M_{crit} and maximum aircraft speed is of the order of Mach 1.0–1.2. Therefore, estimation of C_{Dpmin} is required at three speeds: (i) at a speed before the onset of wave drag; (ii) at M_{crit}; and (iii) at maximum speed (say, Mach 2.0).

It is difficult for the industry to absorb drag prediction errors of more than 5% (the goal is to ensure errors of less than 3%) for civil aircraft; overestimating is better than underestimating. Practitioners are advised to be generous in allocating drag – it is easy to miss a few of the many sources of drag, as shown in the worked examples in this chapter.

Underestimated drag causes considerable design and management problems; failure to meet customer specifications is expensive for any industry. Subsonic aircraft drag prediction has advanced to the extent that most aeronautical establishments are confident in predicting drag with adequate accuracy. Military aircraft shapes are more complex; therefore it is possible that predictions will be less accurate.

9.3   Parasite Drag Definition

The components of drag due to viscosity do not contribute to lift. For this reason, it is considered "parasitic" in nature. For bookkeeping purposes, parasite drag is usually considered separately from other drag sources. The main components of parasite drag are as follows:

1. Drag due to skin friction.
2. Drag due to the pressure difference between the front and the rear of an object.
3. Drag due to the lift-dependent viscous effect and therefore seen as parasitic (to some extent resulting from the non-elliptical nature of lift distribution over the wing); this is a small but significant percentage of total aircraft drag (at LRC it is about 2–3%).

All of these components vary (to a small extent) with changes in aircraft incidence (i.e. as C_L changes). The minimum parasite drag, C_{Dpmin}, occurs when shock waves and boundary-layer separation are at a minimum, by design, around the LRC condition. Any change from the minimum condition (C_{Dpmin}) is expressed as ΔC_{Dp}. In summary:

$$
\begin{aligned}
\text{Parasite drag } (C_{Dp}) &= \text{drag due to skin friction [viscosity]} + \text{drag due to pressure} \\
&\quad \text{difference [viscosity]} \\
&= \text{minimum parasite drag } (C_{Dpmin}) + \text{incremental parasite drag } (\Delta C_{Dp}) \\
&= C_{Dpmin} + \Delta C_{Dp} \text{ (both terms have friction and pressure drag)} \quad \textbf{(9.1)}
\end{aligned}
$$

Oswald's efficiency factor (see Section 2.11) is accounted for in the lift-dependent parasite drag, ΔC_{Dp}. The nature of ΔC_{Dp} is specific to a particular aircraft. Numerically, it is small and difficult to estimate.

Parasite drag of a body depends on its form (i.e. shape) and is also known as *form drag*. The form drag of a wing profile is known as *profile drag*. These two terms are not used in this book. In the past, parasite drag in the FPS (foot-pound-second) system was sometimes expressed as the drag in pound force (lb_f) at 100 ft/s speed, represented by D100. This practice was useful in its day as a good way to compare drag at a specified speed, but it is not used today.

The current industrial practice using semi-empirical methods to estimate C_{Dpmin} is a time-consuming process. (If computerized, then faster estimation is possible, but the author recommends relying more on the manual method at this stage.) Parasite drag constitutes half to two-thirds of subsonic aircraft drag. Using the standard semi-empirical methods, the parasite-drag units of an aircraft and its components are generally expressed as the drag of the "equivalent flat-plate area" (or "flat-plate drag") placed normal to airflow, as shown in Figure 9.1 (see Equation 9.8). These units are in square feet to correspond with literature in the public domain. This is not the same as air flowing parallel to the flat plate and encountering only the skin friction.

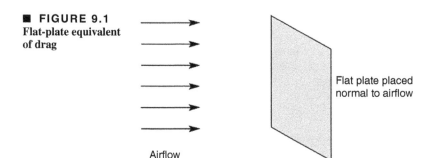

The inviscid idealization of flow is incapable of producing parasite drag because of the lack of skin friction and the presence of full pressure recovery.

9.4 Aircraft Drag Breakdown (Subsonic)

There are many variations and definitions of the bookkeeping methods for components of aircraft drag; this book uses the typical US practice [2, 3]. The standard breakdown of aircraft drag is as follows (see Equation 9.1):

Total aircraft drag = (drag due to skin friction + drag due to pressure difference) +
           drag due to lift generation + drag due to compressibility (wave drag)
       = parasite drag (C_{Dp}) + lift-dependent induced drag (C_{Di}) + wave drag (C_{Dw})
       = minimum parasite drag $[C_{Dpmin}]$ + incremental parasite drag $[\Delta C_{Dp}]$ +
       induced drag (C_{Di}) + wave drag (C_{Dw})

Therefore, the total aircraft drag coefficient is:

$$C_D = C_{Dpmin} + \Delta C_{Dp} + C_L^2/\pi AR + C_{Dw} \tag{9.2}$$

The advantage of keeping pure induced drag separate is obvious because it is dependent only on the lifting-surface aspect ratio (AR) and is easy to compute. The total aircraft drag breakdown is shown in Figure 9.2.

It is apparent that C_D varies with C_L. When the C_D–C_L relationship is shown in graphical form, it is known as a *drag polar*, shown in Figure 9.3 (all components of drag in Equation 9.2 are shown in the figure). The C_D versus the C_L^2 characteristics of Equation 9.2 are rectilinear, except at high and low C_L values (see Section 9.18), because at high C_L (i.e. low speed, high angle of attack), there could be additional drag due to separation; at a very low C_L (i.e. high speed), there could be additional drag due to local shocks. Both effects are nonlinear in nature. Most of the errors in estimating drag result from computing ΔC_{Dp}, three-dimensional effects, interference effects, excrescence effects of the parasite drag, and non-linear range of aircraft drag. Designers should keep $C_{Dw} = 0$ at LRC and aim to minimize to $\Delta C_{Dp} \approx 0$ (perceived as the design point).

An aircraft on a long-range mission typically can have a weight change of more than 25% from the initial to the final cruise condition. As the aircraft becomes lighter, its induced drag decreases. Therefore, it is more economical to cruise at a higher altitude to take advantage of having less drag. In practical terms, this is achieved in the *step-climb* technique, or a gradual climb over the cruise range (Chapter 13).

■ **FIGURE 9.2**
Breakdown of total aircraft drag

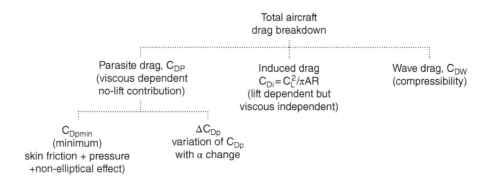

■ **FIGURE 9.3**
Aircraft drag polar (from the classroom Bizjet, Section 9.18)

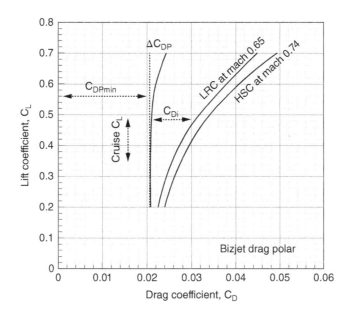

9.5　Aircraft Drag Formulation

A theoretical overview of drag is provided in this section to show that aircraft geometry is not amenable to the equation for an explicit solution. CFD is yet to achieve an acceptable result for the full aircraft.

Recall the expression in Equation 9.2 for the total aircraft drag, C_D, as:

$$C_D = C_{Dparasite} + C_{Di} + C_{Dw} = C_{Dpmin} + \Delta C_{Dp} + C_{Di} + C_{Dw}$$

where

$$C_{Dparasite} = C_{Dfriction} + C_{Dpressure} = C_{Dpmin} + \Delta C_{Dp}$$

At LRC, when $C_{Dw} \approx 0$, the total aircraft drag coefficient is given by:

$$C_D = C_{Dpmin} + \Delta C_{Dp} + C_{Di} \tag{9.3}$$

■ **FIGURE 9.4**
Control volume
approach to
formulate aircraft
drag. (a) 2D body
and wake in a control
volume. (b) 3D
aircraft in a control
volume

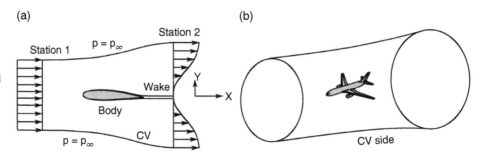

The general theory of drag on a 2D body (Figure 9.4a) provides the closed-form Equation 9.4. A 2D body has infinite span. In the diagram, airflow is in the X direction and wake depth is shown in the Y direction. The wake is formed due to viscous effects immediately behind the body, where integral operation is applied. Wake behind a body is due to the viscous effect in which there is a loss of velocity (i.e. momentum) and pressure (depletion of energy), as shown in the figure. Measurement and computation across the wake are performed close to the body; otherwise, the downstream viscous effect dissipates the wake profile. Consider an arbitrary control volume (CV) large enough in the Y direction where static pressure is equal to free-stream static pressure (i.e. $p = p_\infty$). The subscript ∞ denotes the free-stream condition. Integration over the Y direction on both sides up to the free-stream value gives:

$$D = D_{press} + D_{skin} = \int_{-\infty}^{\infty}(p_\infty - p)dy + \int_{-\infty}^{\infty}\rho u(U_\infty - u)dy$$

$$= \int_{-\infty}^{\infty}\left[(p_\infty - p) + \rho u(U_\infty - u)\right]dy \qquad (9.4)$$

An aircraft is a 3D object (Figure 9.4b) with the additional effect of finite wing-span which will produce induced drag. In that case, the above equation can be written as

$$D = D_{skin} + D_{press} + D_i$$

$$= \int_{-\infty}^{\infty}\int_{-b/2}^{b/2}\left[(p_\infty - p) + \rho u(U_\infty - u)\right]dx\,dy \qquad (9.5)$$

where b is the span of the wing in the X direction (the axis system has changed).

The finite wing effects on the pressure and velocity distributions result in induced drag D_i embedded in the expression on the right-hand side of Equation 9.5. Because the aircraft cruise condition (i.e. LRC) is chosen to operate below M_{crit}, the wave drag, D_w, is absent; otherwise, it must be added to the expression. Therefore, Equation 9.5 can be equated with the aircraft drag expression as given in Equation 9.3. Finally, Equation 9.5 can be expressed in non-dimensional form, by dividing by $\frac{1}{2}\rho_\infty U_\infty^2 S_W$. Therefore

$$C_D = \frac{1}{S_W}\int_{-\infty}^{\infty}\int_{-b/2}^{b/2}\left[-\frac{(p - p_\infty)}{\frac{1}{2}\rho_\infty U_\infty^2} + \frac{2\rho u}{\rho_\infty U_\infty}\left(1 - \frac{u}{U_\infty}\right)\right]dx\,dy$$

$$= \frac{1}{S_W}\int_{-\infty}^{\infty}\int_{-b/2}^{b/2}\left[-C_p + \frac{2\rho u}{\rho_\infty U_\infty}\left(1 - \frac{u}{U_\infty}\right)\right]dx\,dy \qquad (9.6)$$

Unfortunately, the complex 3D geometry of an entire aircraft in Equation 9.6 is not amenable to easy integration. CFD discretizes the flow field into small domains that are numerically integrated, resulting in some errors. Mathematicians have successfully managed the error level with sophisticated algorithms. The proven industrial-standard, semi-empirical methods are currently the prevailing practice, and are backed up by theories and validated by flight tests. CFD assists in the search for improved aerodynamics.

9.6 Aircraft Drag Estimation Methodology

The semi-empirical formulation of aircraft drag estimation used in this book is a credible method based on [1, 3, 7]. It follows the findings from NACA/NASA, RAE and other research establishment documents. This chapter provides an outline of the method used. It is clear from Equation 9.2 that the following four components of aircraft drag are to be estimated:

1. *Minimum parasite drag, C_{Dpmin}* (see Section 9.7)

   Parasite drag is composed of skin friction and pressure differences due to viscous effects that are dependent on the Re. To estimate the minimum parasite drag, C_{Dpmin}, the first task is to establish geometric parameters such as the characteristic lengths and wetted areas and the Re of the discrete aircraft components.

2. *Incremental parasite drag, ΔC_{Dp}* (see Section 9.10)

   C_{Dp} is characteristic of a particular aircraft design and includes the lift-dependent parasite drag variation, 3D effects, interference effects and other spurious effects not easily accounted for. There is no theory to estimate ΔC_{Dp}; it is best obtained from wind-tunnel tests or the ΔC_{Dp} of similarly designed aircraft wings and bodies. CFD results are helpful in generating the C_{Dp} versus C_L variation.

3. *Induced drag, C_{Di}* (see Section 2.11)

   Pure induced drag, C_{Di}, is computed from the expression

   $$C_{Di} = C_L^2 / \pi AR \tag{9.7}$$

4. *Wave drag, C_{Dw}* (see Section 9.11).

   The last component of subsonic aircraft drag is the wave drag, C_{Dw}, which accounts for compressibility effects. It depends on the thickness parameter of the body: for lifting surfaces it is the t/c ratio, and for bodies it is the diameter-to-length ratio. CFD can predict wave drag accurately but must be substantiated using wind-tunnel tests. Transport aircraft are designed so that HSC at M_{crit} (e.g. 320 type ≈ 0.82 Mach) allows a 20 count ($C_{Dw} = 0.002$) drag increase. At LRC, wave-drag formation is kept at zero.

The methodology presented herein considers fully turbulent flow from the LE (leading edge) of all components. Here, no credit is taken for drag reduction due to possible laminar flow over a portion of the body and lifting surface. This is because it may not always be possible to keep the aircraft surfaces clear of contamination that would trigger turbulent flow. The certifying agencies recommend this conservative approach.

9.7 Minimum Parasite Drag Estimation Methodology

The practised method to compute C_{Dpmin} is first to dissect (i.e. isolate) the aircraft into discrete identifiable components, such as the fuselage, wing, V-tail, H-tail, nacelle and other smaller geometries (e.g. winglets and ventral fins). The wetted area and the Re of each component establishes the skin friction associated with each component. The 2D flat-plate basic mean skin-friction coefficient, C_{F_basic}, corresponding to the Re of the component, is determined from the flight Mach number. Sections 2.4.1 and 9.18.2 explain the worked examples carried out in this book for fully turbulent flow.

The C_Fs arising from the 3D effects (e.g. supervelocity) and wrapping effects of the components are added to the basic flat-plate C_{F_basic}. Supervelocity effects result from the 3D nature (i.e. curvature) of aircraft body geometry where, in the critical areas, the local velocity exceeds the free-stream velocity (hence the term *supervelocity*). The axi-symmetric curvature of a body (e.g. fuselage) is perceived as a wrapping effect when the increased adverse pressure gradient increases the drag. The interference in the flow field is caused by the presence of two bodies in proximity (e.g. the fuselage and wing). The flow field of one body interferes with the flow field of the other body, causing more drag. Interference drag must be accounted for when considering the drag of adjacent bodies or components – it must not be duplicated when estimating the drag of the other body.

The design of an aircraft should be streamlined so that there is little separation over the entire body, thereby minimizing parasite drag obtained by taking the total C_F (by adding various C_F to C_{F_basic}). Hereafter, the total C_F will be known as the C_F. Parasite drag is converted to its flat-plate equivalent expressed in "f" square feet. Although it can be easily converted into the SI system, in this book the FPS system is used for comparison with the significant existing data that use the FPS system. The flat-plate equivalent f is defined as:

$$f_{component} = \left(A_w \times C_F\right)_{component} \tag{9.8}$$

where A_w is the wetted area (units of ft^2).

The minimum parasite drag C_{Dpmin} of an aircraft is obtained by totalling the contributing fs of all aircraft components with other sundries. Therefore, the minimum parasite drag of the aircraft is obtained by:

$$C_{Dpmin} = \left(f_{component} + \text{sundries}\right)/S_W = \left(A_w \times C_F\right)_{component}/S_W \tag{9.9}$$

The stepwise approach to computing C_{Dpmin} is described in the following three subsections.

9.7.1 Geometric Parameters, Reynolds Number and Basic C_F Determination

The Re has the deciding role in determining the skin-friction coefficient, C_F, of a component. First, the Re per unit length, speed and altitude are established. Then the characteristic lengths of each component are determined. The characteristic length L_{comp} of each component is as follows (i.e. Re $= (\rho_\infty L_{comp} V_\infty)/\mu_\infty$):

Fuselage: axial length from the tip of the nose cone to the end of the tail cone (L_{fus}).
Wing: the wing MAC (mean aerodynamic chord).
Empennage: the MACs of the V-tail and the H-tail.
Nacelle: axial length from the nacelle-highlight plane to the nozzle exit plane (L_{nac}).

Figure 9.23, in the reference figure section at the end of this chapter, shows the basic 2D flat-plate skin-friction coefficient, C_{F_basic}, of a fully turbulent flow for local and average values. For a partial laminar flow, the C_{F_basic} correction is made using factor f_1, given in Figure 9.24. It has been shown that the compressibility effect increases the boundary layer, thus reducing the local C_F. However, in LRC until the M_{crit} is reached, there is little sensitivity of the C_F change with Mach number variation, and therefore the incompressible C_F line (i.e. the Mach 0 line in Figure 9.23b) is used. At HSC at M_{crit} and above, the appropriate Mach line is used to account for the compressibility effect. The basic C_F changes with changes in the Re, which depends on speed and altitude of the aircraft. Section 9.2 explains that a subsonic aircraft C_{Dpmin} computed at LRC would cater to the full flight envelope, for the purposes of this book.

9.7.2 Computation of Wetted Area

Computation of the wetted area, A_w, of the aircraft component is shown herein. Skin friction is generated on that part of the surface over which air flows, the so-called wetted area. Wetted area computation has to be accurate as parasite drag is directly proportionate to A_w. A 2% error in area estimation will result in about 1% error in overall subsonic aircraft drag.

While 3D CAD (computer-aided design) modelling can give an accurate wetted area, A_w, it may not be available in the early stages of a conceptual design when a 2D CAD or manually drawn three-view diagram on a large sheet (say A0 or A1 size) is generated by draftsmen in a different department, and then given to the aerodynamics group where drag estimation is done. Planimeters are useful in making accurate area measurements. Normally, the three-view 2D drawings have the reference areas of the geometry, which can help in calibrating the planimeter.

9.7.2.1 Lifting Surfaces
These are approximate to the flat surfaces, with the wetted area slightly more than twice the reference area due to some thickness. Care is needed in removing the areas at intersections, such as the wing area buried in the fuselage. A factor k is used to obtain the wetted area of lifting surfaces, as follows:

$$A_w = k \times \left(\text{reference area, } S_W - \text{the area buried in the body} \right)$$

The factor k may be interpolated linearly for other t/c ratios, for example:

$$k = 2.02 \text{ for } t/c = 0.08\%;$$
$$k = 2.04 \text{ for } t/c = 0.12\%;$$
$$k = 2.06 \text{ for } t/c = 0.16\%.$$

9.7.2.2 Fuselage
The fuselage is divided conveniently into sections – typically, for civil transport aircraft, into a constant cross-section mid-fuselage with varying cross-section front and aft fuselage closures. The constant cross-section mid-fuselage barrel has a wetted area of $A_{wfmid} = \text{perimeter} \times \text{length}$. The forward and aft closure cones could be sectioned more finely, treating each thin section as a constant section "slice". A military aircraft is unlikely to have a constant cross-section barrel, and its wetted area must be computed in this way. The wetted areas must be excluded where the wing and empennage join the fuselage or for any other considerations.

9.7.2.3 Nacelle
Only the external surface of the nacelle is considered the wetted area, and it is computed in the same way as the fuselage, taking note of the pylon cut-out area. (Internal drag within the intake duct is accounted for as installation effects in engine performance as a loss of thrust.)

9.7.3 Stepwise Approach to Computing Minimum Parasite Drag

The following seven steps are carried out to estimate the minimum parasite drag, C_{Dpmin}:

Step 1: Distinguish and isolate aircraft components such as the wing, fuselage, nacelle, and so on. Determine the geometric parameters of the aircraft components, such as the characteristic length and wetted areas.

Step 2: Compute the Reynolds number per foot at the LRC condition. Then obtain the component Reynolds number by multiplying its characteristic length.

Step 3: Determine the basic 2D (Figure 9.23b) average skin friction coefficient C_{F_basic}, corresponding to the Reynolds number for each component.

Step 4: Estimate ΔC_F as the increment taking account of 3D effects on each component.

Step 5: Estimate the interference drag of two adjacent components. Avoid duplication.

Step 6: Add results obtained in steps 3–5 to get the minimum parasite drag of the component in terms of flat-plate equivalent area (ft² or m²), that is, $C_F = C_{F\_2D} + \Sigma\Delta C_F$ for the component: $(f)_{comp} = (A_w)_{comp} \times C_F$, where $(C_{Dpmin})_{comp} = (f)_{comp}/S_w$.

Step 7: Add all the component minimum parasite drags. Then add other drags, for example, trim excrescence drag, and so on. Finally add 3% drag due to surface roughness effects. The aircraft minimum parasite drag is expressed in coefficient form as C_{Dpmin}.

The semi-empirical formulation for each component is given in the following subsections.

9.8   Semi-Empirical Relations to Estimate Aircraft Component Parasite Drag

Isolated aircraft components are worked on to estimate component parasite drag. The semi-empirical relations given here embed the necessary corrections required for 3D effects. Associated coefficients and indices are derived from actual flight-test data. (Wind-tunnel tests are conducted at a lower Re and therefore require correction to represent flight-tested results.) The influence of the related drivers is shown as drag increasing by ↑ and drag decreasing by ↓. For example, an increase of the Re reduces the skin-friction coefficient and is shown as Re ($\downarrow$).

9.8.1   Fuselage

The methodology for the fuselage (denoted by the subscript f) is discussed in this section. The Re_f is calculated first using the fuselage length as the characteristic length. The fuselage characteristic length, L_{fus}, is the length from the tip of the nose cone to the end of the tail cone. The wetted area, $A_{wf}(\uparrow)$, and fineness ratio (length/diameter) ($\downarrow$) of the fuselage are computed. Ensure that cutouts at the wing and empennage junctions are subtracted. Obtain the Re_f ($\downarrow$). The corresponding basic C_{Ff} for the fuselage using Figure 9.23b is intended for the flat plate at the flight Mach number.

The semi-empirical formulation is required to correct the 2D skin-friction drag for 3D effects and other influencing parameters, as listed herein. These are incremental values shown by the symbol Δ. There are many incremental effects, and it is easy to miss some of them.

1. 3D effects [1]. 3D effects are on account of surface curvature resulting in changes in local flow speed and associated pressure gradients:

   (a) *Wrapping:*

$$\Delta C_{Ff} = C_{Ff} \times \left[k \times (\text{length / diameter}) \times Re_f^{-0.2} \right] \qquad (9.10)$$

where k is between 0.022 and 0.025 (take the higher value) and Re_f = Reynolds number of fuselage.

■ **FIGURE 9.5**
Canopy types for drag estimation

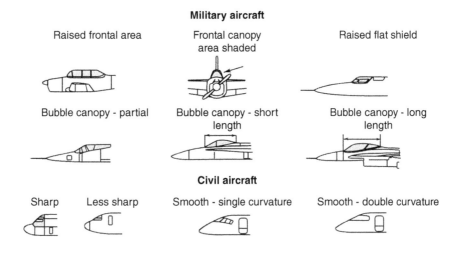

Military aircraft

Raised frontal area   Frontal canopy area shaded   Raised flat shield

Bubble canopy - partial   Bubble canopy - short length   Bubble canopy - long length

Civil aircraft

Sharp   Less sharp   Smooth - single curvature   Smooth - double curvature

(b) *Supervelocity:*

$$\Delta C_{Ff} = C_{Ff} \times \left(\text{diameter / length}\right)^{1.5} \qquad (9.11)$$

(c) *Pressure:*

$$\Delta C_{Ff} = C_{Ff} \times 7 \times \left(\text{diameter / length}\right)^{3}. \qquad (9.12)$$

2. Other effects on the fuselage (increments are given in a percentage of 2D C_{Ff}) are listed herein. The industry has more accurate values of these incremental values of ΔC_{Ff}. Readers in the industry should not use the values given here – they are intended only for coursework using estimates extracted from industrial data.

(a) *Canopy drag:* There are two types of canopy (Figure 9.5), as follows:

(i) *Raised or bubble-type canopy or its variants:* These canopies are mostly associated with military aircraft and smaller aircraft. The canopy drag coefficient $C_{D\pi}$ is based on the frontal cross-sectional area shown in the military-type aircraft in Figure 9.5 (the front view of the raised canopy is shaded). The extent of the raised frontal area contributes to the extent of drag increment and $C_{D\pi}$ accounts for the effects of canopy rise. $C_{D\pi}$ is then converted to $C_{Ffcanopy} = (A_\pi \times C_{D\pi})/A_{wf}$, where A_{wf} is the fuselage wetted area. The dominant types of a raised or bubble-type canopy and their associated $C_{D\pi}$ are summarized in Table 9.1.

(ii) *Windshield-type canopy for larger aircraft*: These canopies are typically associated with payload-carrying commercial aircraft, from a small Bizjet upwards. Flat panes lower the manufacturing cost but result in a kink at the double-curvature nose cone of the fuselage. A curved and smooth transparent windshield avoids the kink that would reduce drag at an additional cost. Smoother types have curved panes with a single or double curvature. Single-curvature panes come in smaller pieces, with a straight side and a curved side. Double-curvature panes are the most expensive, and considerable attention is required during manufacturing to avoid distortion of vision. The values in square feet in Table 9.2 are used to obtain a sharp-edged windshield-type canopy drag.

■ **TABLE 9.1**

Typical $C_{D\pi}$ associated with raised or bubble-type canopies

Canopy type	$C_{D\pi}$
Raised frontal area (older boxy design – has sharp edges)	0.2
Raised flat shield (reduced sharp edges)	0.15
Bubble canopy (partial)	0.12
Bubble canopy (short)	0.08
Bubble canopy (long)	0.06

■ **TABLE 9.2**

Typical $C_{D\pi}$ associated with sharp windshield-type canopies

Aircraft size	Associated $C_{D\pi}$
2 passengers abreast	0.1 ft^2
4 passengers abreast	0.2 ft^2
6 passengers abreast	0.3 ft^2
8 passengers abreast	0.4 ft^2
10 passengers abreast or more	0.5 ft^2

Adjust the values for the following variations:

- Kinked windshield (less sharp) – reduce the value by 10%.
- Smoothed (single curvature) windshield – reduce the value by 20–30%.
- Smoothed (double curvature) windshield – reduce the value by 30–50%.

(b) *Body pressurization:* for fuselage surface waviness, 5–6% (use 5.5%).

(c) *Non-optimum fuselage shape* (interpolate the in-between values).

  (i) *Nose fineness ratio, Fcf* (see Section 2.19.5)

    For Fcf ≤ 1.5 : 8%

    For 1.5 ≤ Fcf ≤ 1.75 : 6%

    For Fcf ≥ 1.75 : 4%

    For military aircraft with high nose fineness: 3%.

  (ii) *Fuselage closure* – above Mach 0.6 (see Section 2.19)

    Less than 10°: 0%

    11–12° closure: 1%

    13–14°: 4%

  (iii) *Upsweep closure* (see Section 2.19.4)

    No upsweep: 0%

    4° of upsweep: 2%

    10° of upsweep: 8%

    15° of upsweep: 15%.

(iv) *Aft-end cross-sectional shape*

Circular: 0%

Shallow keel: 0–1%

Deep keel: 1–2%.

(v) Rear-mounted door (with fuselage upsweep): 5–10%.

(d) Cabin pressurization leakage: 3–5% (if unknown, use higher value).

(e) Excrescence (non-manufacturing types such as windows):

(i) Windows and doors: 2–4% (use higher values for larger aircraft)

(ii) Miscellaneous: 1%.

(f) *Wing-fuselage belly fairing,* if any: 1–5% (use higher value if houses undercarriage).

(g) *Undercarriage fairing* – typically for high-wing aircraft (if any fairing): 2–6% (based on fairing protrusion height from fuselage).

3. The interference drag increment with the wing and empennage is included in the calculation of lifting-surface drag and therefore is not duplicated when computing the fuselage parasite drag. Totalling the C_{Ff} and ΔC_{Ff} from the wetted area A_{wf} of the isolated fuselage, the flat-plate equivalent drag C_{Ff} (see Step 6 in Section 9.7.3), is estimated in square feet.

4. Surface roughness is 2–3%. These effects are from the manufacturing origin, discussed in Section 9.8.4.

Total all the components of parasite drag to obtain C_{Dpmin}, as follows. It should include the excrescence drag increment. Converted into the fuselage contribution to $[C_{Dpmin}]_f$ in terms of aircraft wing area, it becomes:

$$C_{Ff} = 1.03 \times \left(\text{basic } C_{Ff} + \Sigma \Delta C_{Ff} \right) \tag{9.13}$$

$$f_f = \left(C_{Ff} + \Delta C_{Ffwrap} + \Delta C_{Ffsupervel} + \Delta C_{Ffpress} + \Delta C_{Ffother} + \Delta C_{Ffrough} \right) \times A_{wf} \tag{9.14}$$

$$\left[C_{Dpmin} \right]_f = f_f / S_W \tag{9.15}$$

Because surface roughness drag is the same percentage for all components, it is convenient to total them after evaluating all components. In that case, the term $\Delta C_{Ffrough}$ is dropped from Equation 9.14 and it is accounted for as shown in Equation 9.26.

9.8.2 Wing, Empennage, Pylons and Winglets

The wing, empennage, pylon and winglets are treated as lifting surfaces, and we use an identical methodology to estimate their minimum parasite drag. It is similar to the fuselage methodology except that it does not have the wrapping effect. Here, the interference drag with the joining body (e.g. for the wing, it is the fuselage) is taken into account because it is not included in the fuselage *ff*.

The methodology for the wing (denoted by the subscript *w*) is discussed in this section. The Re_w is calculated first using the wing MAC as the characteristic length. Next, the exposed wing area is computed by subtracting the portion buried in the fuselage and then the wetted area,

A_{Ww}, using the k factors for the t/c as in Section 9.7.2. Using the Re_w, the basic C_{Fw_basic} is obtained from the graph in Figure 9.23b for the flight Mach number. The incremental parasite drag formulae are as follows:

1. *3D effects* (taken from [1]):
   (a) *Supervelocity:*

$$\Delta C_{Fw} = C_{Fw} \times K_1 \times \left(\text{aerofoil thickness/chord ratio}\right)_{ave} \qquad (9.16)$$

where K_1 is 1.2–1.5 for a supercritical aerofoil, and 1.6–2 for a conventional aerofoil.
   (b) *Pressure:*

$$\Delta C_{Fw} = C_{Fw} \times 60 \times \left(\text{aerofoil thickness/chord ratio}\right)_{ave}^{4} \times \left(\frac{6}{AR}\right)^{0.125} \qquad (9.17)$$

where aspect ratio $AR \geq 2$ (modified from [1]). The last term of this expression includes the effect of non-elliptical lift distribution.

2. *Interference:*

$$\Delta C_{Fw} = C_B^{2} \times K_2 \times \left\{ \frac{0.75 \times (t/c)_{root}^{3} - 0.0003}{A_w} \right\} \qquad (9.18a)$$

where $K_2 = 0.6$ for high- and low-wing designs and C_B is the root chord at the fuselage intersection. For mid wing, $K_2 = 1.2$. This is valid for t/c ratios up to 0.09. For t/c ratios below 0.07, take the interference drag:

$$\Delta C_{Fw} = 3 - 5\% \text{ of } C_{Fw} \qquad (9.18b)$$

The same relations apply to the V-tail and H-tail.

For pylon interference, add 10–12%. Interference drag is not accounted for in fuselage drag, so it is accounted for in wing drag. (For the pylon, one interference is at the aircraft side and one is with the nacelle.)

3. *Other effects*: Excrescence (non-manufacturing, e.g. control surface gaps, etc.)
   Flap gaps: add 4–5%
   Slat gaps: add 4–5%
   Others: add 4–5%
4. *Surface roughness (to be added later):*

The flat plate equivalent of wing drag contribution is:

$$f_w = \left(C_{Fw} + \Delta C_{Ffw\_supervel} + \Delta C_{Fw\_press} + \Delta C_{Fw\_inter} + \Delta C_{Fw\_other} + \Delta C_{Fw\_rough}\right) \times A_{ww} \qquad (9.19)$$

which can be converted into C_{Dpmin} in terms of aircraft wing area, that is:

$$\left[C_{Dpmin}\right]_w = f_w / S_W \qquad (9.20)$$

(Drop the term ΔC_{Fw_rough} of Equation 9.19 if it is accounted for at the end after computing fs for all components as shown in Equation 9.26.)

The same procedure is used to compute the parasite drag of empennage, pylons, and so on, which are considered as being wing-like lifting surfaces.

$$f_{lifting\_surface} = \left[\left(C_F + \Delta C_{F\_supervel} + \Delta C_{F\_press} + \Delta C_{F\_inter} + \Delta C_{F\_other} + \Delta C_{F\_rough}\right) \times A_w\right]_{lift\,sur}$$

(9.21)

which can be converted into

$$\left[C_{Dpmin}\right]_{lifting\_surface} = f_{lifting\_surface}/S_w$$

(9.22)

As before, it is convenient to total $\Delta C_{Ffrough}$ after evaluating all components. In that case, the term $\Delta C_{Ffrough}$ is dropped from Equation 9.21 and it is accounted for as shown in Equation 9.26.

9.8.3   Nacelle Drag

The nacelle requires different treatment, with the special consideration of throttle-dependent air flowing through as well as over it, like the fuselage. This section provides the definitions and other considerations needed to estimate nacelle parasite drag (see [2, 8, 12, 13]). The nacelle is described in Section 7.7.1.

The throttle-dependent variable of the internal flow passing through the turbofan engine affects the external flow over the nacelle. The dominant changes in the flow field due to throttle dependency are around the nacelle at the lip and aft end. When the flow field around the nacelle is known, the parasite drag estimation method for the nacelle is the same as for the other components, but must also consider the throttle-dependent effects.

Civil aircraft nacelles are typically pod-mounted. In this book, only the long duct is considered. Military aircraft engines are generally buried in the aircraft shell (i.e. fuselage). A podded nacelle may be thought of as a wrapped-around wing in an axisymmetric shape like that of a fuselage. The nacelle section shows aerofoil-like sections in Figure 9.6; the important sources of nacelle drag are listed here (a short-duct nacelle (see Figure 7.18) is similar except for the fan exhaust coming out at high speed over the exposed outer surface of the core nozzle, for which its skin friction must be considered):

Throttle-independent drag (external surface):
- skin friction
- wrapping effects of axisymmetric body
  excrescence effects (includes non-manufacturing types such as cooling ducts).

Throttle-dependent drag:
- inlet drag (front end of the diffuser)
- nacelle base drag (zero for an engine operating at cruise settings and higher)
- boat-tail drag (curvature of the nozzle at the aft end of the nacelle).

Definitions and typical considerations for drag estimation associated with the flow field around an isolated long-duct podded nacelle (approximated to circular cross-section) are shown in Figure 9.6. Although there is internal flow through the nacelle, the external geometry of the nacelle may be treated as a fuselage, except that there is a lip section similar to the LE of an aerofoil. The prevailing engine throttle setting is maintained at a rating for LRC or HSC for the mission profile. The *intake drag* and the *base drag/boat-tail drag* are explained below.

■ **FIGURE 9.6**
Aerodynamic
considerations for an
isolated long-duct
nacelle

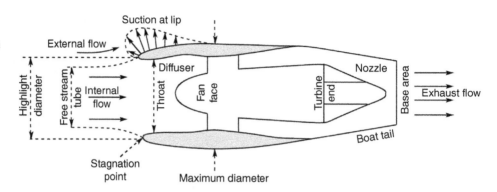

1. *Intake drag*: The intake stream tube at cruise operates in a subcritical condition, which is complex and makes estimation of the intake drag difficult. There is spillage during the subcritical operation due to the stream tube being smaller than the cross-sectional area at the nacelle highlight diameter (see Figure 9.6), where external flow turns around the lip, creating suction (i.e. thrust). This develops pre-compression, ahead of the intake, when the intake velocity is slower compared with the free-stream velocity expressed in the fraction (V_{intake}/V_∞). At $(V_{intake}/V_\infty) < 0.8$. The excess airflow spills over the nacelle lip. The intake lip acts as the LE of a circular aerofoil. The subcritical airflow diffusion ahead of the inlet results in pre-entry drag, termed *additive drag*. The net effect results in *spillage drag,* as described herein. The spillage drag added to the friction drag at the lip results in the *intake drag*, which is a form of parasite drag. (For military aircraft intake drag, see Section 9.16.)

$$\text{spillage drag} = \text{additive drag} + \text{lip suction}\left(\text{thrust sign changes to} - \text{ve}\right)$$

$$\text{intake drag} = \text{spillage drag} + \text{friction drag at the lip}\left(\text{supervelocity effect}\right)$$

   Figure 9.7 shows intake-drag variations with the mass flow rate for both subsonic and supersonic (i.e. sharp LE) intake.

2. *Intake recovery factor*, η_i (refer to Figure 7.6 for the engine station numbers). The energy in freestream air, far ahead of the aircraft, can be expressed in terms of its total pressure $P_{t\infty}$, where the greater is the speed, the higher is the energy entering into the engine intake plane at station 0 without any loss, where $P_{t\infty} = P_{t0}$. At the engine fan face at station 1, there is depletion of flow energy down to a reduced total pressure P_{t1} on account of wall skin friction and/or separation because of aircraft flight attitude, both in subsonic and supersonic flight, where the latter incurs further losses caused by the shock wake. The ratio of P_{t1}/P_{t0} is known as the intake recovery factor, η_i. It is also known as the ram recovery factor, as seen as intake efficiency.

3. *Base drag*: The design criteria for the nozzle-exit area sizing is such that at LRC, the exit-plane static pressure P_e equals the ambient pressure P_∞ (for a perfectly expanded nozzle, $P_e = P_\infty$) to eliminate any base drag. At higher throttle settings, when $P_e > P_\infty$, there is still no base drag. At lower settings, for example idle rating, there is some base drag as a result of the nozzle-exit area being larger than what is required.

■ **FIGURE 9.7**
Throttle-dependent spillage drag

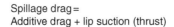

Spillage drag =
Additive drag + lip suction (thrust)

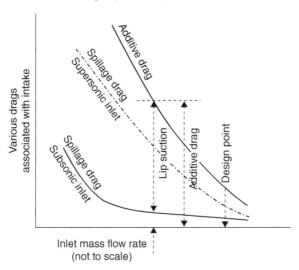

4. *Boat-tail drag*: The long-duct contour for closure (i.e. "boat-tail" shape) at the aft end is shallow enough to avoid separation, especially with the assistance of entrainment effects of the exhaust plume. Hence, the boat-tail drag is kept low. At the idle throttle setting, considerable flow separation can occur and the magnitude of boat-tail drag could be higher, but it is still small compared with the nacelle drag.

For bookkeeping purposes and to avoid conflict with aircraft manufacturers, engine manufacturers generally include internal drag (e.g. ram, diffuser and exhaust-pipe drag) in computing the net thrust of an engine. Therefore, this book only needs to estimate the parasite drag (i.e. external drag) of the nacelle. Intake-duct loss is considered engine-installation losses expressed as intake-recovery factor. Intake- and exhaust-duct losses are approximately 1–3% in engine thrust at LRC (throttle- and altitude-dependent). The net thrust of the turbofan, incorporating installation losses, is computed using the engine manufacturer-supplied program and data. These manufacturers work in close liaison to develop the internal contour of the nacelle and intake. External nacelle contour design and airframe integration remain the responsibility of the aircraft manufacturer.

The long-duct nacelle characteristic length, L_{nac}, is the length measured from the intake highlight plane to the exit-area plane. The wetted area A_{Wn}, Re and basic C_{Fn} are estimated as for other components. The incremental parasite drag formulae for the nacelle are provided herein. The supervelocity effect around the lip section is included in the intake drag estimation; hence, it is not computed separately. Similarly, the pressure effect is included in the base/boat-tail drag estimation. These two items are addressed this way because of the special consideration of throttle dependency. Following is the stepwise approach used to compute the nacelle drag coefficient C_{Dn}.

1. ΔC_{Dn} *effects* (same as fuselage being axisymmetric):

$$\text{Wrapping:} \quad \Delta C_{Fn} = 0.025 \times \left(\text{diameter/length}\right)^{-1} \times \text{Re}^{-0.2} \tag{9.23}$$

■ **TABLE 9.3**
Nacelle interference drag (per nacelle)

Nacelle position	Interference drag
Wing-mounted	
High (long) overhang	0
Medium overhang	4% of C_{Fn}
Low (short) overhang	7% of C_{Fn}
Fuselage-mounted	
Raised	5% of C_{Fn}
Medium	5% of C_{Fn}
Low	5% of C_{Fn}
S-duct	6.5% of C_{Fn}
Straight duct (centre)	5.8% of C_{Fn}

2. *Other incremental effects*

   Drag contributions made by the following effects are given in percentage of C_{Fn}. These are typical of the generic nacelle design.

   (a) Intake drag at LRC – includes supervelocity effects $\approx 40-60\%$ (higher bypass ratio having higher percentage)

   (b) Boat-tail/base drag (throttle dependent) – includes pressure effects $\approx 10-12\%$ (higher value for smaller aircraft)

   (c) Excrescence (non-manufacturing type, e.g. cooling air intakes, etc.) $\approx 20-25\%$ (higher value for smaller aircraft).

3. *Interference drag*

   Podded nacelle in the vicinity of the wing or body will give interference drag. For a wing-mounted nacelle, the higher the overhang forward of the wing, the less would be the interference drag. Typical values of the interference drag by the pylon (each) interacting with the wing or the body are given in Table 9.3.

4. *Surface roughness*: finally add $\approx 3\%$.

A long overhang in front of the wing keeps the nacelle free from interference effects. A short overhang gives the highest interference. However, there is little variation of inference drag of nacelle mounted on different position at the aft fuselage (see Figure 9.8). Much depends on the proximity of other bodies, for example the wing, empennage, and so on. If the nacelle is within one diameter, then interference drag may be increased by another 0.5%. A central engine is close to the fuselage and with a V-tail – they have increased interference.

By adding all the components, the flat plate equivalent of nacelle drag contribution is given by Equation 9.24:

$$f_n = \left(C_{Fn} + \Delta C_{Fn\_wrap} + \Delta C_{Fn\_intake} + \Delta C_{fn\_boat\ tail} + \Delta C_{Fn\_excres} + \Delta C_{Fn\_rough} \right) \times A_{wn} \qquad (9.24)$$

As before, it is convenient to total ΔC_{Ff_rough} after evaluating all components. In that case, the term ΔC_{Ff_rough} is dropped from Equation 9.21 and it is accounted for as shown in Equation 9.26. Converted into the nacelle contribution to C_{Dpmin} in terms of aircraft wing area, it becomes

$$\left[C_{Dpmin} \right]_n = f_n / S_W \qquad (9.25)$$

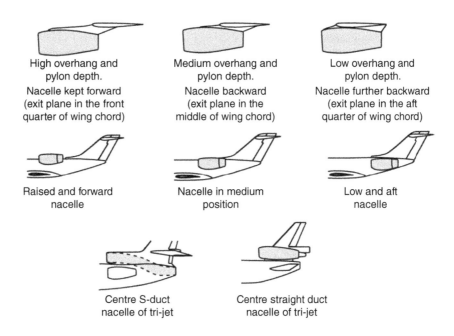

High overhang and
pylon depth.
Nacelle kept forward
(exit plane in the front
quarter of wing chord)

Medium overhang and
pylon depth.
Nacelle backward
(exit plane in the
middle of wing chord)

Low overhang and
pylon depth.
Nacelle further backward
(exit plane in the aft
quarter of wing chord)

Raised and forward
nacelle

Nacelle in medium
position

Low and aft
nacelle

Centre S-duct
nacelle of tri-jet

Centre straight duct
nacelle of tri-jet

In the last three decades, the nacelle drag has been reduced by approximately twice as much as has been achieved with other aircraft components. This demonstrates the complexity of and unknowns associated with the flow field around nacelles. CFD is important in nacelle design and its integration with aircraft. In this book, nacelle geometry is simplified to the axisymmetric shape without loss of methodology.

9.8.4 Excrescence Drag

An aircraft body is not smooth; located all over the body are probes, blisters, bumps, protrusions, surface-protection mats for steps, small ducts (e.g. for cooling) and exhausts (e.g. environmental control and cooling air) – these are unavoidable features. In addition, there are mismatches at subassembly joints – for example steps, gaps and waviness originating during manufacture, and treated as discrete roughness. Pressurization also causes fuselage-skin waviness (i.e. areas ballooning up). In this book, excrescence drag is addressed separately as two types:

1. *Manufacturing origin* [13]. This includes aerodynamic mismatches as discrete roughness resulting from tolerance allocation. Aerodynamicists must specify surface smoothness requirements to minimize excrescence drag resulting from the discrete roughness, within the manufacturing tolerance allocation.
2. *Non-manufacturing origin*. This includes aerials, flap tracks and gaps, cooling ducts and exhausts, bumps, blisters and protrusions.

Excrescence drag due to surface roughness is accounted for by using 2–3% of component parasite drag as roughness drag [1, 7]. As indicated in Equation 9.13, it is factored using 3% after computing all component parasite drag, as follows:

$$f_{comp\ total} = 1.03\left(f_f + f_w + f_n + f_p + f_{other\ comp}\right)$$ **(9.26)**

■ **TABLE 9.4**
Air-conditioning drag

Number of passengers	Drag – f(ft²)	Thrust – f(ft²)	Net drag – f(ft²)
50	0.1	−0.04	0.06
100	0.2	−0.1	0.1
200	0.5	−0.2	0.4
300	0.8	−0.3	0.5
600	1.6	−0.6	1.0

The difficulty in understanding the physics of excrescence drag was summarized by Haines [10] in his review by stating "…one realises that the analysis of some of these early data seems somewhat confused, because three major factors controlling the level of drag were not immediately recognised as being separate effects". These factors are as follows:

1. how skin friction is affected by the position of the boundary layer transition;
2. how surface roughness affects skin friction in a fully turbulent flow;
3. how geometric shape (non-planar) affects skin friction.

Haines' study showed that a small but significant amount of excrescence drag results from manufacturing origin and was difficult to understand.

9.8.5   Miscellaneous Parasite Drags

In addition to excrescence drag, there are other drag increments such as from intake drag of the ECS (environmental control system), which has a fixed value depending on the number of passengers; and aerials and trim drag, which are included to obtain the minimum parasite drag of the aircraft.

9.8.5.1   Air-Conditioning Drag   Air-conditioning air is inhaled from the atmosphere through flush intakes that incur drag. It is mixed with hot air bled from a mid-stage of the engine compressor and then purified. Loss of thrust due to engine bleed is accounted for in the engine thrust computation, but the higher pressure of the expunged cabin air causes a small amount of thrust. Table 9.4 shows the air-conditioning drag based on the number of passengers (interpolation is used for the intermediate sizes).

9.8.5.2   Trim Drag   Due to weight changes during cruise, the centre of gravity could shift, thereby requiring the aircraft to be trimmed in order to relieve the control forces. Change in the trim-surface angle causes a drag increment. The average trim drag during cruise is approximated as shown in Table 9.5, based on the wing reference area (interpolation is used for the intermediate sizes).

9.8.5.3   Aerials   Navigational and communication systems require aerials that extend from an aircraft body, generating parasite drag of the order 0.06–0.1 ft², depending on the size and number of aerials installed. For midsized transport aircraft, 0.075 ft² is typically used. Therefore:

$$f_{aircraft\_parasite} = f_{comp\_total} + f_{aerial} + f_{air\,cond} + f_{trim} \tag{9.27}$$

■ **TABLE 9.5**
Trim drag (approximate)

Wing reference area (ft²)	Trim drag – f(ft²)
200	0.12
500	0.15
1000	0.2
2000	0.3
3000	0.5
4000	0.8

9.9 Notes on Excrescence Drag Resulting from Surface Imperfections

Semi-empirical relations discussed in Sections 9.8.4 and 9.8.5 are sufficient for the purpose. Excrescence drag due to surface imperfections is difficult to estimate; therefore, this section provides background on the nature of the difficulty encountered. Capturing all the excrescence effects over the full aircraft in CFD is yet to be accomplished with guaranteed accuracy.

A major difficulty arises in assessing the drag of small items attached to the aircraft surface, such as instruments (e.g. pitot and vanes), ducts (e.g. cooling) and necessary gaps to accommodate moving surfaces. In addition, there is the unavoidable discrete surface roughness from mismatches and imperfections – *aerodynamic defects* – resulting from limitations in the manufacturing process. Of all of these sources of excrescence drag, of particular interest is that resulting from the discrete roughness, within the manufacturing tolerance allocation, in compliance with the surface-smoothness requirements specified by aerodynamicists to minimize drag.

Mismatches at the assembly joints are seen as discrete roughness (i.e. aerodynamic defects) – for example steps, gaps, fastener flushness and contour deviation – placed normal, parallel or at any angle to the free-stream airflow. In consultation with production engineers, aerodynamicists specify tolerances to minimize the excrescence drag to the order of 1–3% of the C_{Dpmin}.

The "defects" are neither at the maximum limits throughout, nor uniformly distributed. The excrescence dimension is of the order of less than 0.1 in.; for comparison, the physical dimension of a fuselage is nearly 5000–10,000 times larger. It poses a special problem for estimating excrescence drag; that is, capturing the resulting complex problem in the boundary layer downstream of the mismatch.

The methodology involves first computing excrescence drag on a 2D flat surface without any pressure gradient. On a 3D curved surface with a pressure gradient, the excrescence drag is magnified. The location of a joint of a subassembly on the 3D body is important for determining the magnification factor that will be applied to the 2D flat-plate excrescence drag obtained by semi-empirical methods. The body is divided into two zones: Zone 1 (the front side) is in an adverse pressure gradient and Zone 2 is in a favourable pressure gradient [13]. Excrescences in Zone 1 are more critical to magnification than those in Zone 2. At a LRC flight speed (i.e. below M_{crit} for civil aircraft), shocks are local and subassembly joints should not be placed in Zone 1.

Estimation of aircraft drag uses an average skin-friction coefficient C_F (see Figure 9.23b), whereas excrescence drag estimation uses the local skin-friction coefficient C_f (see Figure 9.23a), appropriate to the location of the mismatch. These fundamental differences in drag estimation methods make the estimation of aircraft drag and excrescence drag quite different.

After World War II, efforts continued for the next two decades (especially at the RAE by Gaudet, Winters, Johnson, Pallister, Tillman *et al.*) using wind-tunnel tests to understand and estimate excrescence drag. Their experiments led to semi-empirical methods subsequently compiled by ESDU as the most authoritative source of information on the subject [5]. Aircraft and excrescence drag estimation methods still remain state-of-the-art, and efforts to understand the drag phenomena continue.

Surface imperfections inside the nacelle – that is, at the inlet diffuser surface and at the exhaust nozzle – could affect engine performance as a loss of thrust. Care must be taken so that the defects do not perturb the engine flow field. The internal nacelle drag is accounted for as an engine installation effect.

9.10   Minimum Parasite Drag

The aircraft C_{Dpmin} can now be obtained from $f_{aircraft}$. The minimum parasite drag of the entire aircraft is $C_{Dpmin} = (1/S_w)f_i$, where f_i is the sum of the total fs of the entire aircraft:

$$C_{Dpmin} = f_{aircraft}/S_w \qquad (9.28)$$

9.11 ΔC_{Dp} Estimation

Equation 9.2 shows that ΔC_{Dp} is not easy to estimate. ΔC_{Dp} contains the lift-dependent variation of parasite drags due to changes in pressure distribution with changes in the angle of attack. Although it is a small percentage of the total aircraft drag (it varies from 0 to 10%, depending on the aircraft C_L), it is the most difficult to estimate. There is no proper method available to estimate the ΔC_{Dp} versus C_L relationship; it is design-specific and depends on wing geometry (i.e. planform, sweep, taper ratio, aspect ratio and wing–body incidence) and aerofoil characteristics (i.e. camber and t/c). The values are obtained through wind-tunnel tests and, currently, by CFD.

During cruise, the lift coefficient varies with changes in aircraft weight and/or flight speed. The design-lift coefficient, C_{LD}, is around the mid-cruise weight of the LRC. Let C_{LP} be the lift coefficient when $\Delta C_{Dp} = 0$. The wing should offer C_{LP} at three-quarters of the value of the C_{LD}. This would permit an aircraft to operate at HSC (at M_{crit}; i.e. at the lower C_L) with almost zero ΔC_{Dp}. Figure 9.9a shows a typical ΔC_{Dp} versus C_L variation. This graph can be used for the coursework in Sections 9.18 and 9.19.

For any other type of aircraft, a separate graph must be generated from wind-tunnel tests and/or CFD analysis. The industry has a large databank to generate such graphs during the conceptual-design phase. In general, the semi-empirical method takes a ΔC_{Dp} versus C_L relationship for a tested wing (with sufficiently close geometric similarity) and then corrects it for the differences in wing sweep ($\downarrow$), aspect ratio ($\downarrow$), t/c ratio ($\uparrow$), camber, and any other specific geometric differences.

9.12   Subsonic Wave Drag

Thickness parameters of lifting surfaces (e.g. wing) and bodies (e.g. fuselage) have a strong influence in drag generation. Thickness gives rise to local supervelocities (higher than free-stream velocities) that increase the local Reynolds number (Re), altering the skin friction. If local velocities are sufficiently high, then compressibility effects develop. Semi-empirical formulae to account for the supervelocity effects altering skin friction are given in Section 9.8. This section deals with the compressibility drag, C_{Dw}.

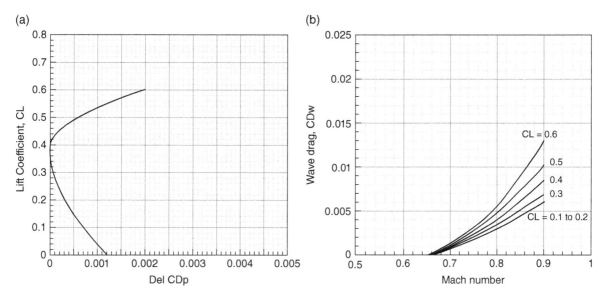

■ FIGURE 9.9   Typical relationships of (a) ΔC_{Dp} and (b) C_{Dw}

Wave drag is caused by compressibility effects of air as an aircraft approaches high subsonic speeds. Local shocks start to appear on a curved surface as aircraft speed increases. This is in the transonic-flow regime, in which a small part of the flow over the body is supersonic while the remainder is subsonic. In some cases, a shock interacting with the boundary layer can cause premature flow separation, thus increasing pressure drag. Initially, it is gradual and then shows a rapid rise as it approaches the speed of sound. The industry practice is to tolerate a 20-count (i.e. $C_D = 0.002$) increase due to compressibility at a speed identified as M_{crit} (Figure 9.9b). At higher speeds, higher wave-drag penalties are incurred.

The extent of compressibility drag rise is primarily dependent on the aerofoil design of the lifting surfaces, and to lesser extent in shaping of the rest of aircraft. In a proper sense, aerofoil design is an iterative process and offers several options to choose from. Once the aircraft configuration is frozen in the conceptual design phase, then the proper C_{Dw} versus Mach graph is obtained and the exact drag rise at LRC can be applied. The difficulty is at the conceptual stage when not much information is available and when approximations are to be made. Aircraft conception design is treated in [7]. In this book aircraft configuration is given and its performance is to be estimated. One of the first tasks is to estimate the aircraft drag from the three-view diagram of the given aircraft. Also supplied are the wing characteristics, including the wave drag characteristics. Initially, C_{Dw} has to be assumed from past data, and as the project progresses with more information, it is fine-tuned.

At the conceptual phase of the project, it may prove convenient to keep compressibility drag $C_{Dw} = 0$ up to LRC. This is permissible, since the drag prediction is constantly updated as more information becomes available. Some industries use this approach as it offers the advantage of expressing drag polar in a *closed-form* equation of the shape of a parabola ($C_D = C_{Dpmin} + kC_L^2$; there is no C_{Dw} term) with little error in the operating range of cruise, climb and descent. The advantage of having a closed-form equation is that it can quickly analyse aircraft characteristics to ascertain the LRC Mach number based on, say, best economical cruise or any other criteria (see chapters 9 and 12). If, however, the proper C_{Dw} versus Mach graph for the aircraft is available (suiting the LRC criteria), then it can be used with the proper C_{Dw} rise when the closed-form equation gets a little more complicated. Otherwise a suitable drag rise to account for the compressibility effect at LRC may be used – say, approximately 5–10 counts. This may fall within the average value but at the conceptual stage it is still a guesstimate. The crux is in the accurate bookkeeping of drag counts for whichever method is used.

A typical wave drag (C_{D_w}) graph is shown in Figure 9.9b, which can be used for course-work (civil aircraft) described in Section 9.19.

9.13   Total Aircraft Drag

Total aircraft drag is the sum of all drags estimated in Sections 9.8–9.12, as follows.

At LRC,
$$C_D = C_{Dpmin} + \Delta C_{Dp} + \frac{C_L^2}{\pi \text{AR}} \qquad (9.29)$$

At HSC (M_{crit}),
$$C_D = C_{Dpmin} + \Delta C_{Dp} + \frac{C_L^2}{\pi \text{AR}} + C_{Dw} \qquad (9.30)$$

At takeoff and landing, additional drag exists, as explained in the next section.

9.14   Low-Speed Aircraft Drag at Takeoff and Landing

For safety in operation and aircraft structural integrity, aircraft speed at takeoff and landing must be kept as low as possible. At ground proximity, a lower speed would provide a longer reaction time for the pilot, easing the task of controlling an aircraft at a precise speed. Keeping an aircraft aloft at low speed is achieved by increasing lift through increasing wing camber and area using high-lift devices such as a flap and/or a slat. Deployment of a flap and slat increases drag; the extent depends on the type and degree of deflection. Of course, in this scenario, the undercarriage remains extended, which also would incur a substantial drag increase. At approach to landing, especially for military aircraft, it may require "washing out" of speed to slow down by using fuselage-mounted speed brakes (in the case of civil aircraft, this is accomplished by wing-mounted spoilers). Extension of all these items is known as a *dirty configuration* of the aircraft, as opposed to a *clean configuration* at cruise. Deployment of these devices is speed-limited in order to maintain structural integrity; that is, a certain speed for each type of device extension should not be exceeded.

After takeoff, typically at a safe altitude of 200 ft, pilots retract the undercarriage, resulting in noticeable acceleration. At about 800 ft altitude with appropriate speed gain, the pilot retracts the high-lift devices. The aircraft is then in the clean configuration, ready for an en-route climb to cruise altitude; therefore, this is sometimes known as the *en-route configuration* or *cruise configuration*.

9.14.1   High-Lift Device Drag

High-lift devices (described in Section 2.9) are typically flaps and slats, which can be deployed independently of each other. Some aircraft have flaps but no slats. Flaps and slats conform to the aerofoil shape in the retracted position. The function of a high-lift device is to increase the aerofoil camber when it is deflected relative to the baseline aerofoil. If it extends beyond the wing LE and trailing edge, then the wing area is increased. A camber increase causes an increase in lift for the same angle of attack at the expense of drag increase. Slats are nearly full span, but flaps can be anywhere from part to full span (i.e. flaperon). Typically, flaps are sized up to about

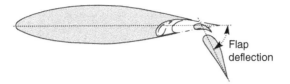

■ **FIGURE 9.10**
Fowler flap, NACA
632-118 aerofoil

Flap
deflection

two-thirds from the wing root. The flap chord to aerofoil chord (cc/c) ratio is of the order of 0.2–0.3. The main contribution to drag from high-lift devices is proportional to their projected area normal to free-stream air. The associated parameters affecting drag contributions are as follows:

- type of flap or slat;
- extent of flap or slat chord to aerofoil chord (typically, flap has 20–30% of wing chord);
- extent of deflection (flap at takeoff is from 7° to 15°; at landing, it is from 25° to 60°);
- gaps between the wing and flap or slat (depends on the construction);
- extent of flap or slat span;
- fuselage width fraction of wing span;
- wing sweep, t/c, twist and AR.

The myriad variables make formulation of semi-empirical relations difficult. References [1, 4, 5] offer different methodologies. It is recommended that practitioners use CFD and test data. Reference [14] gives detailed test results of a double-slotted flap (0.309c) NACA 632-118 aerofoil (Figure 9.10). Both elements of a double-slotted flap move together, and the deflection of the last element is the overall deflection. For wing application, this requires an aspect ratio correction, as described in Section 2.12.

Figure 9.11 is generated from various sources giving averaged typical values of C_L and C_{D_flap} versus flap deflection. It does not represent any particular aerofoil and is intended only for coursework to be familiar with the order of magnitude involved without loss of overall accuracy. The methodology is approximate; practising engineers should use data generated by tests and CFD.

The simple semi-empirical relation for flap drag given in Equation 9.31 is generated from flap-drag data shown in Figure 9.11. The methodology starts by working on a straight wing (Λ_0) with an aspect ratio of 8, flap-span to wing-span ratio (b_f/b) of two-thirds, and a fuselage-width to wing-span ratio of less than a quarter. Total flap drag on a straight wing (Λ_0) is seen as composed of 2D parasite drag of the flap ($C_{Dp\_flap\_2D}$), change in induced drag due to flap deployment (ΔC_{Di_flap}), and interference generated on deflection ($\Delta C_{Dint\,flap}$). Equation 9.32 is intended for a swept wing. The basic expressions are corrected for other geometries, as given in Equations 9.33 and 9.34.

$$\text{Straight wing,}\ C_{D\_flap\_\Lambda 0} = \Delta C_{D\_flap\_2D} + \Delta C_{Di\_flap} + \Delta C_{Dint\_flap} \tag{9.31}$$

$$\text{Swept wing,}\ C_{D\_flap\_\Lambda \frac{1}{4}} = C_{D\_flap\_\Lambda 0} \times \cos \Lambda_{\frac{1}{4}} \tag{9.32}$$

The empirical form of the second term of Equation 9.31 is given by

$$\Delta C_{Di\_flap} = 0.025 \times \left(8/\text{AR}\right)^{0.3} \times \left[\left(2b\right)/\left(3b_f\right)\right]^{0.5} \times \left(\Delta C_L\right)^2 \tag{9.33}$$

where AR is the wing aspect ratio and (b_f/b) is the flap to wing-span ratio.

■ **FIGURE 9.11**
Flap drag

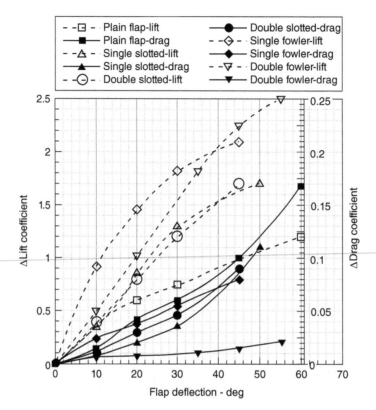

The empirical form of the third term of Equation 9.31 is given by

$$\Delta C_{Dint\_flap} = k \times C_{D\_flap\_2D} \tag{9.34}$$

where k is 0.1 for single-slotted, 0.2 for double-slotted, 0.25–0.3 for single Fowler and 0.3–0.4 for double Fowler flaps (see Figure 2.26). Lower values may be used at lower settings.

Figure 9.11 shows the $C_{D\_flap\_2D}$ for various flap types at various deflection angles, with the corresponding maximum C_L gain given in Figure 2.26. Aircraft fly well below C_{Lmax}, keeping a safe margin. Increase $C_{Di\,flap}$ by 0.002 if the slats are deployed.

9.14.1.1 Worked Example
An aircraft has AR =9.5, quarter-chord sweep $\Lambda_{1/4} = 20°$, $(b_f/b) = 2/3$, and a fuselage to wing-span ratio of less than 1/4. The flap type is a single-slotted Fowler flap and there is a slat. The aircraft has $C_{Dpmin} = 0.019$. Construct its drag polar.

At 20° deflection: This is typical for takeoff with $C_L = 2.2$ (approximate), but could be used at landing.

From Figure 9.11, $\Delta C_{D\_flap\_2D} = 0.045$ and $\Delta C_L = 1.46$.
From Equation 9.33,

$$\Delta C_{Di\_flap} = 0.025 \times (8/9.5)^{0.3} \times \left[(2/3)/(3/2)\right]^{0.5} \times (1.46)^2 = 0.025 \times 1.02 \times 2.13 = 0.054.$$

From Equation 9.34, $\Delta C_{Dint\_flap} = 0.25 \times 0.045 = 0.01125$.

$$C_{D\_flap\_\Lambda 0} = 0.045 + 0.054 + 0.01125 = 0.11.$$

With slat on, $C_{D\_highlift} = 0.112$.

FIGURE 9.12
Typical drag polar with high-lift devices

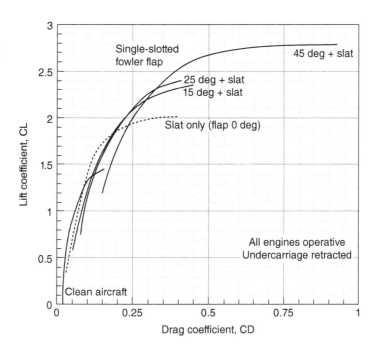

For the aircraft wing, $C_{D\_flap\_\Lambda\frac{1}{4}} = C_{D\_flap\_\Lambda 0} \times \cos \Lambda_0 = 0.112 \times \cos 20 = 0.105$.

Induced drag, $C_{Di} = (C_L^2)/(\pi AR) = (2.2)^2/(3.14 \times 9.5) = 4.48/23.55 = 0.21$.

Total aircraft drag, $C_D = 0.019 + 0.105 + 0.21 = 0.334$.

At 45° deflection: This is typical for landing with $C_L = 2.7$ (approximate).

From Figure 9.11, $\Delta C_{D\_flap\_2D} = 0.08$ and $\Delta C_L = 2.1$.

From Equation 9.33,

$$\Delta C_{Di\_flap} = 0.025 \times (8/9.5)^{0.3} \times [(2/3)/(3/2)]^{0.5} \times (2.1)^2 = 0.025 \times 1.02 \times 4.41 = 0.112.$$

From Equation 9.34, $\Delta C_{Dint\_flap} = 0.3 \times 0.08 = 0.024$.

$$C_{Dp\_flap\_\Lambda 0} = 0.08 + 0.112 + 0.024 = 0.216.$$

With slat on, $C_{Dp\_highlift} = 0.218$.

For the aircraft wing,

$$C_{D\_flap\_\Lambda\frac{1}{4}} = C_{D\_flap\_\Lambda 0} \times \cos \Lambda_0 = 0.218 \times \cos 20 = 0.201 \times 0.94 = 0.205.$$

Induced drag, $C_{Di} = (C_L^2)/(\pi AR) = (2.7)^2/(3.14 \times 9.5) = 9.29/23.55 = 0.31$.

Total aircraft drag, $C_D = 0.019 + 0.205 + 0.31 = 0.534$.

Drag polar with a high-lift device extended is plotted as shown in Figure 9.12 at various deflections. It is cautioned that this graph is intended only for coursework; practising industry-based engineers must use data generated by tests and CFD.

A typical value of C_L/C_D for high-subsonic commercial transport aircraft at takeoff with flaps deployed is of the order of 10–12; at landing, it is reduced to 6–8.

A more convenient method would be as given in Figure 9.13, and this is used for the classroom example (civil aircraft) worked out in Section 9.18.

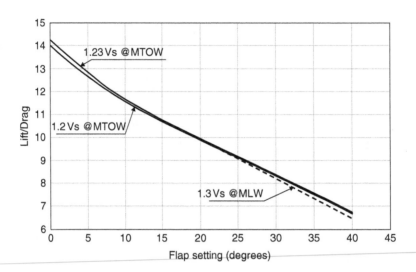

9.14.2 Dive Brakes and Spoilers Drag

To decrease aircraft speed, whether in combat or at landing, flat plates are used which are attached to the fuselage and shaped to its geometric contour when retracted. They could be placed symmetrically on both sides of the wing or on the upper fuselage (i.e. for military aircraft). The flat plates are deployed during subsonic flight. Use $C_{D\pi\_brake} = 1.2\_2.0$ (average 1.6) based on the projected frontal area of the brake to air-stream. The force level encountered is high and controlled by the level of deflection. The best position for the dive brake is where the aircraft moment change is the least (i.e. close to the aircraft CG line).

9.14.3 Undercarriage Drag

Undercarriages, fixed or extended (i.e. retractable type), cause considerable drag on smaller, low-speed aircraft. A fixed undercarriage (not streamlined) can cause up to about a third of aircraft parasite drag. When the undercarriage is covered by a streamlined wheel fairing, the drag level can be halved. It is essential for high-speed aircraft to retract the undercarriage as soon as it is safe to do so (like birds). Below a 200 ft altitude from takeoff and landing, an aircraft undercarriage is kept extended.

The drag of an undercarriage wheel is computed based on its frontal area: A_{π_wheel} (product of wheel diameter and width). For twin side-by-side wheels, the gap between them is ignored and the wheel drag is increased by 50% from a single-wheel drag. For the bogey type, the drag also would increase – it is assumed by 10% for each bogey, gradually decreasing to a total maximum 50% increase for a large bogey. Finally, interference effects (e.g. due to undercarriage doors and tubing) would double the total of wheel drag. The drag of struts is computed separately. The bare single-wheel C_{D_wheel} based on the frontal area is in Table 9.6.

For the smooth side, reduce by half. In terms of an aircraft:

$$C_{Dp\_wheel} = \left(C_{D\pi\_wheel} \times A_{\pi\_wheel} \right) / S_W.$$

A circular strut has nearly twice the amount of drag compared with a streamlined strut in a fixed undercarriage. For example, the drag coefficient of a circular strut based on its cross-sectional area per unit length is $C_{D\pi\_strut} = 1.0$, as it operates at low Re during takeoff and landing. For streamlined struts with fairings, it decreases to 0.5–0.6, depending on the type.

■ **TABLE 9.6**
Bare single-wheel drag with side ridge

Wheel aspect ratio (diameter/width)	3	4	5	6
$C_{D\pi\_wheel}$	0.15	0.25	0.28	0.3

■ **TABLE 9.7**
One-engine inoperative drag

	$\Delta C_{Dengine\ out+rudder}$
Fuselage-mounted engines	0.0035
Wing-mounted twin-engine	0.0045
Wing-mounted four-engine (outboard failure)	0.005

Torenbeek [9] suggests using an empirical formula, if details of undercarriage sizes are not known at an early conceptual design phase. This formula is given in the FPS system as follows:

$$C_{D\_UC} = 0.00403 \times \left(\text{MTOW}^{0.785}\right)/S_W \qquad (9.35)$$

Understandably, it could result in a slightly higher value (see the following example). If aircraft geometry is available, then it is advisable to accept the computed drag values as worked out below.

9.14.3.1 Worked Example We continue with the previous example in Section 9.14.1.1 using the largest in the design (i.e. MTOM (*maximum takeoff mass*) = 24,200 lb and S_W = 323 ft^2) for the undercarriage size. It has a twin-wheel, single-strut length of 2 ft (i.e. diameter of 6 in., $A_{\pi\_strut} \approx 0.2$ ft^2) and a main wheel size of 22 in. diameter and 9.6 in. width (i.e. wheel aspect ratio = 3.33, $A_{\pi\_wheel} \approx 1$ ft^2). From Table 9.6, a typical value of $C_{D\pi\ wheel} = 0.18$, based on the frontal area and increased by 50% for the twin-wheel (i.e. $C_{D0} = 0.27$). Including the nose wheel (although it is smaller and a single wheel, it is better to be liberal in drag estimation), the total frontal area is about 3 ft^2. (A more accurate approach would be to consider nose and main wheels separately.)

$$f_{wheel} = 0.27 \times 3 = 0.81\,\text{ft}^2\,(\text{nose wheel included})$$

$$f_{strut} = 1.0 \times 3 \times 2 \times 0.2 = 1.2\,\text{ft}^2\,(\text{nose wheel included})$$

Total $f_{UC} = 2 \times (0.81 + 1.2) = 4.02$ ft^2 (100% increase due to interference, doors, tubing, etc.), in terms of $C_{Dpmin\_UC} = 4.02/323 = 0.0124$. Checking the empirical relation in Equation 9.35, $C_{D\_UC} = 0.00403 \times (24,200^{0.785})/323 = 0.034$; this may be used if undercarriage geometric details are not available.

9.14.4 One-Engine Inoperative Drag

Mandatory requirements by certifying agencies specify that multi-engine commercial aircraft must be able to climb at a minimum specified gradient with one engine inoperative at "dirty" configuration. This immediately safeguards an aircraft in the rare event of an engine failure. Certifying agencies require backup for mission-critical failures to provide safety regardless of the probability of an event occurring.

Asymmetric drag produced by the loss of an engine would make an aircraft yaw, requiring a rudder to fly straight by compensating for the yawing moment caused by the inoperative engine. Both the failed engine and rudder deflection substantially increase drag, expressed by $C_{Dengine\ out+rudder}$. Typical values for coursework are given in Table 9.7.

9.15   Propeller-Driven Aircraft Drag

Drag estimation of propeller-driven aircraft involves additional considerations. The slip-stream of a tractor propeller blows over the nacelle, which blocks the resisting flow. Also, the faster-flowing slipstream causes a higher level of skin friction over the downstream bodies. This is accounted for as a loss of thrust, thereby keeping the drag polar unchanged. The following two factors arrest the propeller effects with piston engines (see Chapter 8 for calculating propeller thrust):

1. Blockage factor, f_b, for tractor-type propeller: 0.96–0.98 applied to thrust (for the pusher type, there is no blockage; therefore, this factor is not required – that is, $f_b = 1.0$).
2. A factor, f_h, as an additional profile drag of a nacelle: 0.96–0.98 applied to thrust (this is the slipstream effect applicable to both types of propellers).

Turboprop nacelles have a slightly higher value of f_b than piston-engine types because of a more streamlined shape. For the worked example, take $f_b \times f_h = 0.98$ as an optimistic design. Typically, it is about 0.96.

9.16   Military Aircraft Drag

Military aircraft drag estimation requires additional considerations to account for the weapon system, because most are external stores (e.g. missiles, bombs, drop-tanks, and flares and chaff launchers); few are carried inside the aircraft mould lines (e.g. guns, ammunition and bombs inside the fuselage bomb bay, if any). Without external carriages, military aircraft are considered at typical configuration (the pylons are not removed – part of a typical configuration). Internal guns without consumables are considered typical configuration; with armaments, the aircraft is considered to be in a loaded configuration. In addition, most combat aircraft have a supersonic capability, which requires additional supersonic-wave drag.

Rather than drag due to passenger doors and windows as in a civil aircraft, military aircraft have additional excrescence drag (e.g. gun ports, extra blisters, antennae and pylons) that requires a drag increment. To account for these additional excrescences, [3] suggests an increment of the clean flat-plate equivalent drag, f, by 28.4%.

Streamlined external-store drag is shown in Table 9.8, based on the frontal maximum cross-sectional area. Bombs and missiles flush with the aircraft contour line have minor interference drag and may be ignored at this stage. Pylons and bomb racks create interference, and Equation 9.18a is used to estimate interference on both sides (i.e. the aircraft and the store). These values are highly simplified at the expense of unspecified inaccuracy; readers should be aware that these simplified values are not far from reality (see [1, 4, 5] for more details).

■ **TABLE 9.8**
External store drag

External store	$C_{D\pi}$ (based on frontal area)
Drop tanks	0.1–0.2
Bombs (length/diameter <6)	0.1–0.25
Bombs/missiles (length/diameter >6)	0.25–0.35

Military aircraft engines are generally buried into the fuselage and do not have nacelles and associated pylons. Intake represents the air-inhalation duct. Skin-friction drag and other associated 3D effects are integral to fuselage drag, but their intakes must accommodate large variations of intake air-mass flow. Military aircraft intakes operate supersonically; their power plants are low bypass turbofans (i.e. of the order of less than 3.0 – earlier designs did not have any bypass). For speed capabilities higher than Mach 1.9, most intakes and exhaust nozzles have an adjustable mechanism to match the flow demand in order to extract the best results. In general, the adjustment aims to keep the V_{intake}/V_∞ ratio more than 0.8 over operational flight conditions, thereby practically eliminating spillage drag. Supersonic flight is associated with shock-wave drag.

9.17   Supersonic Drag

A well-substantiated reference for industrial use is [3], which was prepared by Lockheed as a NASA contract for the National Information Service, published in 1978. A comprehensive method for estimating supersonic drag that is suitable for classroom work is derived from this exercise. The empirical methodology (called the Delta method) is based on regression analyses of 18 subsonic and supersonic military aircraft (i.e. the T-2B, T37B, KA-3B, A-4F, TA-4F, RA-5C, A-6A, A-7A, F4E, F5A, F8C, F-11F, F100, F101, F104G, F105B, F106A and XB70) and 15 advanced (i.e. supercritical) aerofoils. The empirical approach includes the effects of the following:

- wing geometry (AR, t/c and aerofoil section)
- cross-sectional area distribution
- C_D variation with C_L and Mach number.

The methodology presented herein follows [3], modified to simplify C_{Dp} estimation resulting in minor discrepancies. The method is limited and may not be suitable for analysis of more exotic aircraft configurations. However, this method is a learning tool for understanding the parameters that affect supersonic aircraft drag build-up. Results can be improved when more information is available.

Aircraft with supersonic capabilities require estimation of C_{Dpmin} at three speeds: (i) at a speed before the onset of wave drag; (ii) at M_{crit}; and (iii) at maximum speed. The first two speeds follow the same procedure as for the high-subsonic aircraft discussed in Sections 9.7–9.14. In the subsonic drag estimation method, the viscosity-dependent C_{Dp} varying with C_L is separated from the wave drag, C_{Dw} (i.e. transonic effects), which also varies with C_L but independent of viscosity.

For bookkeeping purposes, in supersonic flight such a division between C_{Dp} and C_{Dw} is not clear with C_L variation. At supersonic speeds, there is little complex transonic flow over the body even when C_L is varied. It is not clear how shock waves affect the induced drag with a change in the angle of attack. For simplicity, however, in the empirical approach presented here, it is assumed that supersonic drag estimation can use the same approach as subsonic drag estimation by keeping C_{Dp} and C_{Dw} separate. The C_{Dp} values for the worked example are listed in Table 9.14. Here, drag due to shock waves is computed at $C_L = 0$, and C_{Dw} is the additional shock-wave drag due to compressibility varying with $C_L > 0$. The total supersonic aircraft drag coefficient can then be expressed as follows:

$$C_D = C_{Dpmin} + C_{Dp} + C^2/\pi AR + \left(C_{DLshock@CL=0}\right) + C_{Dw} \tag{9.36}$$

It is recommended that in current practice, CFD analysis should be used to obtain the variation of C_{Dp} and C_{Dw} with C_L. Reference [3] was published in 1978 using aircraft data before the advent of *CFD*. Readers are referred to [1, 4, 5] for other methods. The industry has advanced methodologies, which are naturally more involved.

The aircraft cross-sectional area distribution should be as smooth as possible, as discussed in Section 2.10 (see Figure 2.28). It may not always be possible to use narrowing of the fuselage with an appropriate distribution of areas.

The stepwise empirical approach to estimating supersonic drag is as follows:

Step 1: Progress in the same manner as for subsonic aircraft to obtain the aircraft component Re for the cruise flight condition and the incompressible $C_{Fcomponent}$.

Step 2: Increase $C_{Fcomponent}$ in Step 1 by 28.4% for the typical military aircraft excrescence drag effect.

Step 3: Compute C_{Dpmin} at the three speeds discussed previously.

Step 4: Compute induced drag using $C_{Di} = C^2/\pi AR$.

Step 5: Obtain C_{Dp} from CFD and tests or from empirical relations.

Step 6: Plot the fuselage cross-sectional area versus the length and obtain the maximum area, S_π and base area, S_b (see Section 9.20).

Step 7: Compute the supersonic wave drag at zero lift for the fuselage and the empennage using graphs; use the parameters obtained in Step 6.

Step 8: Obtain the design C_L and the design Mach number using graphs (see example).

Step 9: Obtain the wave drag, C_{Dw}, for the wing using graphs.

Step 10: Obtain the wing–fuselage interference drag for supersonic flight using graphs.

Step 11: Total all the drags to obtain the total aircraft drag and plot as C_D versus C_L.

The worked example for the North American RA-5C Vigilante aircraft is a worthwhile coursework exercise. Details of the Vigilante aircraft drag are in [3]. The subsonic drag estimation methodology described in this book differs from what is presented in [3], yet is in agreement with it. The supersonic drag estimation follows the methodology described in [3]. A typical combat aircraft of today is not too different from the Vigilante in configuration details and, similar logic can be applied. Exotic shapes (e.g. the F117 Nighthawk) should rely more on information generated from CFD and tests along with the empirical relations. For this reason, the author does not recommend undertaking coursework on exotic aircraft configurations unless the results can be substantiated. Learning with a familiar design that can be substantiated gives confidence to practitioners. Those in the industry are fortunate to have access to more accurate in-house data.

9.18   Coursework Example – Civil Bizjet Aircraft

Figure 9.14a gives a three-view diagram of a Bizjet as the coursework example.

9.18.1   Geometric and Performance Data

The geometric and performance parameters discussed herein were used in previous chapters. Figure 9.14b illustrates the dissected anatomy of the coursework baseline aircraft. Aircraft cruise performance for the basic drag polar is computed as follows:

- Cruise altitude = 40 000 ft.
- LRC Mach = 0.65 (630 ft/s) to 0.7 Mach at high altitudes.

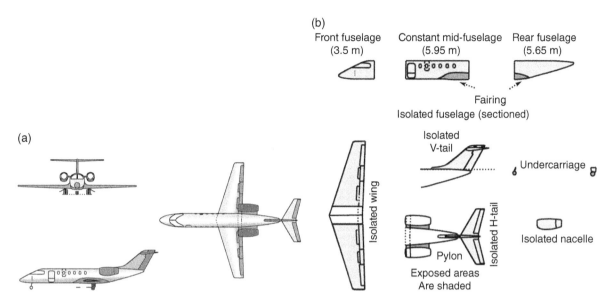

■ **FIGURE 9.14** Bizjet aircraft component diagram. (a) Three-view diagram of the Bizjet. (b) Dissected aircraft components

The Bizjet C_{Dpmin} is evaluated at 0.65 Mach to remain conservative in estimates. The 2D flat-plate C_{F_basic} decreases as Re increases. It is recommended that the readers evaluate C_{Dpmin} at 0.7 Mach to find an insignificant reduction in C_{Dpmin}.

- Ambient pressure $= 391.68 \text{ lb / ft}^2$.
- Ambient temperature $= 390 \text{ K}$.
- Ambient density $= 0.00058 \text{ sl / ft}^3$.
- Ambient viscosity $= 2.96909847 \times 10^{-7} \text{ lbs / ft}^2$.
- Re/ft $= 1.2415272 \times 10^6$ (use incompressible zero Mach line, as explained in Section 9.7.1).
- C_L at LRC (Mach 0.65) $= 0.5$.
- C_L at HSC (Mach 0.7) $= 0.43$.

Fuselage:

- Fuselage length, $L_f = 15.24 \text{ m} (50 \text{ ft})$.
- Average diameter $= 175.5 \text{ cm} (70 \text{ in})$.
- Overall width $= 173 \text{ cm} (68.11 \text{ in})$.
- Overall height (depth) $= 178 \text{ cm} (70 \text{ in})$.
- Average diameter at the constant cross-section barrel, $D_f = 1.75 \text{ m} (5.74 \text{ ft})$.
- $L_f / D_f = 8.71$.
- Fuselage upsweep angle $= 10°$.
- Fuselage closure angle $= 10°$.
- Fineness ratio $= 8.6$.

Wing:

- Aerofoil: NACA 65-410 having 10% thickness to chord ratio for design $C_L = 0.4$.
- Planform reference area, $S_W = 30\,\text{m}^2\,(323\,\text{ft}^2)$.
- Span $= 15\,\text{m}\,(49.2\,\text{ft})$, aspect ratio $= 7.5$.
- Wing MAC $= 2.132\,\text{m}\,(7\,\text{ft})$.
- Dihedral $= 3°$.
- Twist $= 1°$ (washout).
- t/c $= 10\%$.
- Root chord at centreline, $C_R = 2.86\,\text{m}\,(9.38\,\text{ft})$.
- Tip chord, $C_T = 1.143\,\text{m}\,(3.75\,\text{ft})$.
- Quarter chord wing sweep $= 14°$.

V-tail (aerofoil 64-010):

- Planform (reference) area, $S_V = 4.4\,\text{m}^2\,(49.34\,\text{ft}^2)$.
- Height $= 2.13\,\text{m}\,(7\,\text{ft})$, AR $= 2.08$.
- Root chord, $C_R = 2.57\,\text{m}\,(8.43\,\text{ft})$.
- Tip chord, $C_T = 1.54\,\text{m}\,(5.05\,\text{ft})$.
- MAC $= 2.132\,\text{m}\,(7\,\text{ft})$, t/c $= 10\%$.
- Taper ratio, $\lambda = 0.6$, $\Lambda_{1/4} = 40°$.
- Rudder $= 0.75\,\text{m}^2\,(8\,\text{ft}^2)$.

H-tail (Tee tail, aerofoil 64-210, installed with negative camber):

- Planform (reference) area $S_H = 9.063\,\text{m}^2\,(65.3\,\text{ft}^2)$.
- Span $= 5\,\text{m}\,(19.4\,\text{ft})$, AR $= 4.42$.
- Root chord, $C_R = 1.54\,\text{m}\,(5.04\,\text{ft})$.
- Tip chord, $C_T = 0.77\,\text{m}\,(2.52\,\text{ft})$.
- H-tail MAC $= 1.28\,\text{m}\,(4.2\,\text{ft})$, t/c $= 10\%$.
- Taper ratio, $\lambda = 0.5$, $\Lambda_{1/4} = 15°$.
- Dihedral $= 5°$.
- Elevator $= 1.21\,\text{m}^2\,(13\,\text{ft}^2)$.

Nacelle (each, two required):

- Length $= 2.62\,\text{m}\,(8.6\,\text{ft})$.
- Maximum diameter $= 1.074\,\text{m}\,(3.52\,\text{ft})$.
- Nacelle fineness ratio $= 2.62/1.074 = 2.44$.

Bare engine (each, two required):

- Takeoff static thrust at ISA sea-level $= 3800$ lb (17,235 N) per engine with BPR $= 5$.
- Dry weight $= 379$ kg (836 lb).
- Fan diameter $= 0.716\,\text{m}\,(28.2\,\text{in})$.
- Length $= 1.547\,\text{m}\,(60.9\,\text{in})$.

9.18.2 Computation of Wetted Areas, Re and Basic C_F

The aircraft is split into isolated components as shown in Figure 9.14b. The Re, wetted area and basic 2D flat-plate C_{F_basic} of each component are worked out herein. Section 9.18.1 explains that the Bizjet C_{Dpmin} is evaluated at 0.65 Mach to remain conservative in estimates. The summary results from this section are shown in Table 9.9.

Fuselage:

The fuselage is conveniently sectioned into three parts:

1. Front fuselage length, $L_{Ff} = 3.5$ m with a uniformly varying cross-section; wetted area A_{wFf} (no cutout) $= 110$ ft^2.
2. Mid-fuselage length, $L_{Fm} = 5.95$ m with an average constant cross-section diameter $= 1.75$ m; wetted area A_{wFm} (with two sides of wing cutouts) $= 352 - 2 \times 6 = 340$ ft^2.
3. Aft-fuselage length, $L_{Fa} = 5.79$ m, with a uniformly varying cross-section; wetted area A_{wFs} (with empennage cutouts) $= 180 - 10 = 170$ ft^2.

Include an additional wetted area for the wing–body fairing housing the undercarriage ≈ 50 ft^2. Therefore the total wetted area, $A_{wf} = 110 + 340 + 170 + 50 = 670$ ft^2.
Fuselage Re $= 50 \times 1.2415272 \times 10^6 = 9.2 \times 10^7$.
From Figure 9.23b (fully turbulent) at LRC, the incompressible basic $C_{Ff} = \mathbf{0.0022}$.

Wing:

Wing exposed reference area $= 323 - 50$ (area buried in the fuselage) $= 273$ ft^2.
For t / c $= 10\%$ of the wing wetted area, $A_{ww} = 2.024 \times 273 = 552.3$ ft^2 (see Section 9.9.2).
Wing Re $= 7 \times 1.2415272 \times 10^6 = 8.7 \times 10^6$.
From Figure 9.23b at LRC, the incompressible basic $C_{Fw} = \mathbf{0.003}$.

V-tail:

Reference area, $S_V = 4.4$ m^2 (49.34 ft^2).
Exposed reference area $= 49.34 - 9.34$ (area buried in the fuselage) $= 40$ ft^2.
For t / c $= 10\%$, the V-tail wetted area, $A_{w\_VT} = 2.024 \times 40 = 81$ ft^2.
MAC $= 2.132$ m (7 ft).
V-tail Re $= 7 \times 1.2415272 \times 10^6 = 8.7 \times 10^6$.
From Figure 9.23b (fully turbulent) at LRC, the incompressible basic $C_{F\_V\text{-}tail} = \mathbf{0.003}$.

■ **TABLE 9.9**

Summary of Bizjet component Reynolds numbers and 2D basic skin friction C_{Fbasic}

Parameter	Reference area (ft^2)	Wetted area (ft^2)	Characteristic length (ft)	Reynolds number	2D C_{F_basic}
Fuselage	n/a	670	50	9.2×10^7	0.0022
Wing	323	552.3	7 (MAC$_w$)	8.7×10^6	0.003
V-tail	49.34	81	7 (MAC$_{VT}$)	8.7×10^6	0.003
H-tail	65.3	132.2	4.22 (MAC$_{HT}$)	5.24×10^6	0.0032
2×nacelle	n/a	152	8.6	1.07×10^7	0.0029
2×pylon	24	48.6	9.5	9.3×10^6	0.00295

H-tail:

Reference area, $S_H = 9.063$ m^2 (65.3 ft^2). It is a T-tail and it is fully exposed.

For t/c $= 10\%$, the H-tail wetted area, $A_{wHT} = 2.024 \times 65.3 = 132.2$ ft^2.

H-tail Re $= 4.22 \times 1.2415272 \times 10^6 = 5.24 \times 10^6$.

From Figure 9.23b (fully turbulent) at LRC, the incompressible basic $C_{F\_H-tail} = \mathbf{0.003185}$.

Nacelle:

Fineness ratio $= 2.44$.

Nacelle Re $= 8.6 \times 1.2415272 \times 10^6 = 1.07 \times 10^7$.

Two-nacelle wetted area, $A_{wn} = 2 \times 3.14 \times 3.1\,(D_{ave}) \times 8.6 - 2 \times 5$ (two pylon cutouts) $= 158$ ft^2.

From Figure 9.23b (fully turbulent) at LRC, the incompressible basic $C_{F\_nac} = \mathbf{0.0029}$.

Pylon:

Each pylon exposed reference area $= 14$ ft^2.

Length $= 2.28$ m (9.5 ft), t/c $= 10\%$.

Two-pylon wetted area $A_{wp} = 2 \times 2.024 \times 14 = 59.7$ ft^2.

Pylon Re $= 9.5 \times 1.2415272 \times 10^6 = 9.3 \times 10^6$.

From Figure 9.23b (fully turbulent) at LRC, the incompressible basic $C_{Fpylon} = \mathbf{0.00295}$.

9.18.3   Computation of 3D and Other Effects

A component-by-component example of estimating C_{Dpmin} is provided in this section.

Fuselage:

The basic $C_{Ff} = 0.0022$, and likely 3D effects are described earlier with Equations 9.10–9.12.

Wrapping:

$$\Delta C_{Ff} = C_{Ff} \times 0.025 \times (\text{length / diameter}) \times \text{Re}^{-0.2}$$
$$= 0.0022 \times 0.025 \times (8.71) \times (9.2 \times 10^7)^{-0.2}$$
$$= 0.00048 \times 0.0276$$
$$= 0.0000132 \left(0.6\% \text{ of basic } C_{Ff}\right).$$

Supervelocity:

$$\Delta C_{Ff} = C_{Ff} \times (\text{diameter / length})^{1.5}$$
$$= 0.0022 \times (1 / 8.71)^{1.5}$$
$$= 0.0000856 \left(3.9\% \text{ of basic } C_{Ff}\right).$$

Pressure:

$$\Delta C_{Ff} = C_{Ff} \times 7 \times (\text{diameter / length})^3$$
$$= 0.0022 \times 7 \times (0.1148)^3$$
$$= 0.0000233 \left(1.06\% \text{ of basic } C_{Ff}\right).$$

Other effects on the fuselage include (see Section 9.8.1):

- Body pressurization – fuselage surface waviness: 5%
- Non-optimum fuselage shape
    - nose fineness – for $1.5 \leq Fcf \leq 1.75 : 6\%$
    - fuselage closure – above Mach 0.6, less than $10°$: 0
    - upsweep closure – $10°$ upsweep: 8%
    - aft-end cross-sectional shape – circular: 0%.
- Cabin pressurization leakage (if unknown, use higher value): 5%
- Excrescence (non-manufacturing types; e.g. windows)
    - windows and doors (higher values for larger aircraft): 2%
    - miscellaneous: 1%.
- Wing-fuselage-belly fairing, if any (higher value if it houses undercarriage): 5%.

The ECS (see Section 9.8) gives $0.06\,\text{ft}^2$: 3.6%.

Total C_{Ff} increment: 41.16%.

Table 9.10 gives the Bizjet fuselage C_{Ff} components.

Add the canopy drag for two-abreast seating, $f = 0.1\,\text{ft}^2$ (see Section 9.8.1). Therefore, the equivalent flat-plate area, f, becomes $= C_{Ff} \times A_{wF} + $ canopy drag.

$$f_f = 1.416 \times 0.0022 \times 670 + 0.1$$
$$= 2.087 + 0.1 = 2.187\,\text{ft}^2$$

Surface roughness (to be added later): 3%.

Wing:

Basic $C_{FW} = 0.003$. 3D effects are described in Equations 9.16–9.18.

■ **TABLE 9.10**

Bizjet fuselage ΔC_{Ff} correction (3D and other shape effects)

Item	ΔC_{Ff}	% of $C_{Ffbasic}$
Wrapping	0.0000132	0.6
Supervelocity	0.0000856	3.9
Pressure	0.0000233	1.06
Body pressurization	–	5
Fuselage upsweep of $10°$	–	8
Fuselage closure angle of $9°$	0	0
Nose fineness ratio 1.7	–	6
Aft-end cross-section – circular	–	–
Cabin pressurization/leakage	–	5
Excrescence (windows/doors, etc.)	–	3
Belly fairing	–	5
Environmental control system (ECS) exhaust	–	3.6
Total ΔC_{Ff}	0.000906	41.16

Supervelocity:

$$\Delta C_{Fw} = C_{Fw} \times 1.4 \times (\text{aerofoil thickness/chord ratio})$$
$$= 0.003 \times 1.4 \times 0.1$$
$$= 0.00042 \, (14\% \text{ of basic } C_{Fw}).$$

Pressure:

$$\Delta C_{Fw} = C_{Fw} \times 60 \times (\text{aerofoil thickness/chord ratio})^4 \times \left(\frac{6}{\text{AR}}\right)^{0.125}$$
$$= (0.003 \times 60) \times (0.1)^4 \times (6/9.5)^{0.125}$$
$$= 0.0000175 \, (0.58\% \text{ of basic } C_{Fw}).$$

Interference:

$$\Delta C_{Fw} = C_B^2 \times 0.6 \times \left\{ \frac{0.75 \times (t/c)_{root}^3 - 0.0003}{A_w} \right\}$$
$$= 9.38^2 \times 0.6 \times \left[\{0.75 \times 0.1^3 - 0.0003\} / 552.3 \right]$$
$$= 0.000043 \, (1.43\% \text{ of basic } C_{Fw}).$$

Other effects include excrescence (non-manufacturing, e.g. control surface gaps, etc.):

Flap gaps: 5%
Others: 5%

Therefore the total ΔC_{Fw} increment is 26%. In terms of equivalent flat plate area, f, it becomes $= C_{Fw} \times A_{ww}$:

$$f_f = 1.26 \times 0.003 \times 552.3 = 2.09 \, \text{ft}^2$$

Table 9.11 summarizes the Bizjet wing C_{Fw} components.
Surface roughness (to be added later): 3%.

■ TABLE 9.11
Bizjet wing ΔC_{Fw} correction (3D and other shape effects)

Item	ΔC_{FW}	% of $C_{FWfbasic}$
Supervelocity	0.00042	14
Pressure	0.0000175	0.58
Interference (wing–body)	0.000043	1.43
Flaps gap	–	5
Excrescence (others)	–	5
Total ΔC_{Fw}		26

Empennage:
Since the procedure is the same as for the wing, it is not repeated. It may be emphasized here that, in industry, one must compute systematically as shown in the case of the wing.

V-tail: Wetted area, $A_{wVT} = 81$ ft^2, basic $C_{F\_V-tail} = 0.003$.
The T-tail configuration gives interference from the T-tail (add another 1.2%).

$$f_{VT} = 1.262 \times 0.003 \times 81 = 0.307 \text{ ft}^2$$

H-tail: Wetted area, $A_{wHT} = 132.2$ ft^2, basic $C_{F\_H-tail} = 0.0032$.

$$f_{HT} = 1.25 \times 0.0032 \times 132.2 = 0.529 \text{ ft}^2$$

Surface roughness (to be added later): 3%.

Nacelle:
Fineness ratio $= 2.45$, nacelle $Re = 1.07 \times 10^7$.
Wetted area of two nacelles, $A_{wn} = 158$ ft^2, basic $C_{Fnac} = 0.0029$.
3D effects include wrapping:

$$\Delta C_{Fn} = C_{Fn} \times 0.025 \times (\text{length/diameter}) \times Re^{-0.2}$$
$$= 0.025 \times 0.003 \times 2.45 \times (1.07 \times 10^7)^{-0.2}$$
$$= 0.0000072 \, (0.25\% \text{ of basic } C_{Ff}).$$

Other increments are shown in Table 9.12 for one nacelle.
For two nacelles (shown in wetted area), $f_n = 1.8325 \times 0.0029 \times 158 = 0.84$ ft^2.
Surface roughness (to be added later): 3%.

Pylon:
Since the procedure is the same as for the wing, it is not repeated. Interference occurs on both sides of the pylon.
Each pylon exposed reference area $= 14$ ft^2. Length $= 2.28$ m (9.5 ft). Therefore the two-pylon wetted area $A_{wp} = 59.7$ ft^2.
Pylon $Re = 9.5 \times 1.2415272 \times 10^6 = 9.3 \times 10^6$.
Basic $C_{Fpylon} = 0.00295$.
For two pylons (shown in wetted area), $f_{py} = 1.26 \times 0.00295 \times 59.7 = 0.21$ ft^2.
Surface roughness (to be added later): 3%.

■ **TABLE 9.12**
Bizjet nacelle ΔC_{Fn} correction (3D and other shape effects)

Item (one nacelle)	ΔC_{Fn}	% of $C_{Fnbasic}$
Wrapping (3D effect)	0.0000072	0.25
Excrescence (non-manufacture)	–	22
Boat-tail (aft end)	–	11
Base drag (at cruise)	0	0
Intake drag (BPR 4)	–	50
Total ΔC_{Fn}	–	83.25

9.18.4 Summary of Parasite Drag

Table 9.13 gives the aircraft parasite drag build-up summary in a tabular form. The surface roughness effect as a 3% increase (Equation 9.26) in f is added in this table for all surfaces. The parameters are: wing reference area $S_w = 323 \text{ ft}^2$, $C_{Dpmin} = f/S_w$, ISA day, 40,000 ft altitude and Mach 0.65.

9.18.5 ΔC_{Dp} Estimation

Use Figure 9.9 as typical data (requires CFD/testing). Table 9.14 gives a typical estimation.

9.18.6 Induced Drag

The formula used for induced drag is:

$$C_{Di} = C_L^2/(3.14 \times 9.5) = C_L^2/23.55$$

Table 9.15 shows the estimated induced drag for the Bizjet.

9.18.7 Total Aircraft Drag at LRC

Drag polar at LRC can be summed up as shown in Table 9.16. Drag polar at HSC (Mach 0.74) would require adding wave drag from Figure 9.9b.

The drag polar is plotted in Figure 9.3. C_L^2 versus C_D is plotted in Figure 9.15 below. Note the non-linearity at low and high C_L.

■ **TABLE 9.13**
Bizjet parasite drag build-up summary and C_{Dpmin} estimation

	Wetted area, A_w (ft²)	Basic C_F	ΔC_F	Total C_F	f (ft²)	C_{Dpmin}
Fuselage + U/C fairing	670	0.0022	0.000906	0.003106	2.08	0.00644
Canopy	–	–	–	–	0.1	0.00031
Wing	552.3	0.003	0.00078	0.003784	2.09	0.00647
V-tail	81	0.003	0.000786	0.003786	0.302	0.00095
H-tail	132.2	0.0032	0.0008	0.004	0.529	0.00164
2×nacelle	2×79	0.0029	0.002414	0.005314	0.84	0.0026
2×pylon	2×28.35	0.00295	0.000767	0.00372	0.21	0.00065
Rough (3%)	Equation 9.26	–	–	–	0.182	0.00056
Air-conditioning	–	–	–	–	0.1	0.00031
Aerial, lights	–	–	–	–	0.05	0.000155
Trim drag	–	–	–	–	0.13	0.0004
Total					9.61	0.0205

■ **TABLE 9.14**
Bizjet ΔC_{Dp} estimation

C_L	0.1	0.2	0.3	0.4	0.5	0.6	0.7
ΔC_{Dp}	0.0007	0.0003	0.00006	0	0.0006	0.002	0.004

■ **TABLE 9.15**
Bizjet induced drag

C_L	0.2	0.3	0.4	0.5	0.6	0.7	0.8
C_{Di}	0.0017	0.00382	0.0068	0.0106	0.0153	0.0206	0.0272

■ **TABLE 9.16**
Bizjet total aircraft drag coefficient, C_D

C_L	0.2	0.3	0.4	0.5	0.6	0.7
C_{Dpmin}			0.0205 from Table 9.13			
ΔC_{Dp} (Table 9.14)	0.0003	0.00006	0.00004	0.0	0.002	0.003
C_{Di} (Table 9.15)	0.0017	0.00382	0.0068	0.0106	0.0153	0.0206
Total aircraft C_D at LRC	0.0225	0.02438	0.0274	0.0311	0.0378	0.0441
Wave drag, C_{Dw} (Figure 9.9)	0.0014	0.0017	0.002	0.0025	0.0032	0.0045
Total aircraft C_D at HSC	0.024	0.02618	0.0294	0.0336	0.041	0.0486

■ **FIGURE 9.15**
Drag polar of
Figure 9.3, plotting
C_L^2 versus C_D

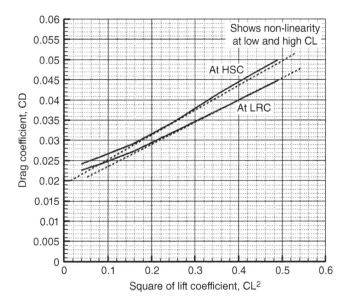

9.19   Classroom Example – Subsonic Military Aircraft (Advanced Jet Trainer)

A classroom example for military aircraft is conducted on the subsonic AJT (advanced jet trainer) type of aircraft, such as the BAe Hawk, using the same method as for civil aircraft drag estimation. To avoid repetition, only the drag polar and other drag details of the AJT is given in Figure 9.16.

Figure 9.17 gives the three-view diagram of the AJT class (e.g. BAe Hawk trainer) and its variant CAS aircraft. Given below is the defence specification for which the design was developed [1]. HSC Mach is 0.8 at 9 km altitude, while LRC Mach is 0.7 at 9 km altitude.

■ **FIGURE 9.16**
AJT drag polar

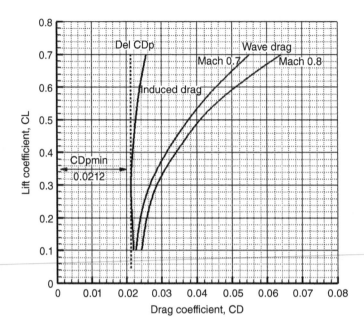

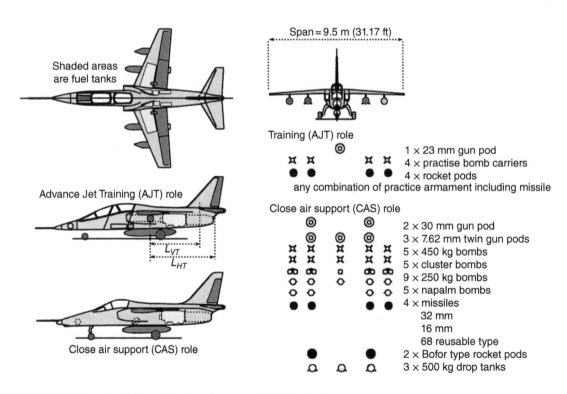

■ **FIGURE 9.17** The AJT and its close air support (CAS) variant

9.19.1   AJT Specifications

The important external dimensions of the baseline *AJT* design are as follows.

Fuselage:

- Length $= 12$ m (39.4 ft).
- Maximum overall width $= 1.8$ m.
- Overall height (depth) $= 4.2$ m.
- Cockpit width $= 0.88$ m.
- Fineness ratio $= 12 / 1.8 = 9.67$.

Wing:

- Planform (reference) area $= 17 \text{m}^2 (183 \text{ ft}^2)$.
- Span $= 9.5$ m (31.17 ft).
- Aspect ratio $= 5.31$.
- Root chord, $C_R = 2.65$ m (8.69 ft).
- Tip chord, $C_T = 0.927$ m (3.04 ft).
- MAC $= 1.928 (9.325$ ft$)$.
- Taper ratio, $\lambda = 0.35$, $\Lambda_{1/4} = 20°$.
- Dihedral $= -2°$ (anhedral – high wing).
- Twist $= 1°$ (washout).
- Flap $= 2.77 \text{m}^2 (29.8 \text{ ft}^2)$.
- Aileron $= 1.06 \text{m}^2 (11.4 \text{ ft}^2)$.

V-tail:

- Planform (reference) area $= 3.83 \text{m}^2 (41.1 \text{ ft}^2)$.
- Span $= 2.135$ m (7 ft).
- Root chord, $C_R = 2.8$ m.
- Tip chord, $C_T = 0.8$ m.
- MAC $= 2.132$ (7 ft).
- Taper ratio, $\lambda = 0.286$, $\Lambda_{1/4} = 35°$.
- Aspect ratio $= 1.2$.
- Tail-arm $= 4$ m (13.1 ft).
- Rudder area $= 0.98 \text{m}^2 (10.5 \text{ ft}^2)$.
- t/c $= 9\%$.

H-tail:

- Planform (reference) area $= 4.78 \text{m}^2 (51.45 \text{ ft}^2)$.
- Span $= 4.2$ m (13.8 ft).
- Root chord, $C_R = 1.652$ m.

- Tip chord, $C_T = 0.6$ m.
- MAC = 2.132 (7 ft).
- Taper ratio, $\lambda = 0.36$, $\Lambda_{\frac{1}{4}} = 25°$.
- Aspect ratio = 3.7.
- Tail-arm = 4 m (13.1 ft).
- Elevator area = 0.956 m^2 (10.3 ft^2).
- t/c = 9%.

Nacelle: None, as the engine is buried into the fuselage.

Engine:

- Take-off static thrust at ISA sea-level = 5390 lb.
- BPR = 0.75.
- Fan diameter = 0.56 m (22 in.).
- Length = 1.956 m (77 in.).
- Maximum depth = 1.04 m (3.4 ft).
- Maximum width = 0.75 m (2.46 ft).

Baseline aircraft mass:

- MTOM = 6500 kg (15,210 lb).
- Normal training configuration mass = 4800 kg (10,800 lb).
- OEM (operating empty mass) = 3700 kg.
- Fuel mass = 1300 kg.

9.19.2 CAS Variant Specifications

The close air support (CAS) role aircraft is the only variant of the AJT aircraft. The configuration of the CAS variant is achieved by splitting the AJT front fuselage, then replacing the tandem seat arrangement with a single seat cockpit. The length can be kept the same, since the nose cone needs to house more powerful acquisition radar. The front-loading of radar and single pilot is placed in such a way that the CG location is kept undisturbed. All other component dimensions except the fuselage length are kept unchanged.

- Wing area = 17 m^2 (183 ft^2).
- Payload = 2500 kg.
- Payload = 2500 kg.
- LRC Mach = 0.7.
- Fuselage length = 12 m (39.4 ft).
- Maximum overall fuselage width = 1.8 m.
- Overall height (depth) = 4.2 m.
- Cockpit width = 0.88 m.
- Fineness ratio = 12/1.8 = 9.67.

■ **TABLE 9.17**
Relevant weights for the AJT and CAS variant

	AJT (kg)	CAS (kg)	Remarks
OEM	3700	3700	
Fuel	1100	1300	Maximum capacity 2390 kg
Clean aircraft	4800 (10,582 lb)	5000 (11,023 lb)	
Wing loading, W/S_W	282 kg/m²(59.8 lb/ft²)	294 kg/m²(60.23 lb/ft²)	
Armament mass	1800	2500	
MTOM, kg (lb)	6600 (14,550 lb)	7500 (16,535 lb)	
Wing loading at MTOM, W/S_W	388.7 kg/m²(79.5 lb/ft²)	441.2 kg/m²(90.36 lb/ft²)	

■ **TABLE 9.18**
AJT total aircraft drag coefficient, C_D

C_L	0.1	0.2	0.3	0.4	0.5	0.6
C_{Dpmin}			0.0212			
ΔC_{Dp}	0.0001	0.0003	0.0006	0.001	0.0014	0.0026
$C_{Di} \cong C_L^2/(3.14 \times 5.3)$	0.0006	0.0024	0.0054	0.00961	0.015	0.0216
Total aircraft C_D at Mach 0.7	0.0219	0.0239	0.0272	0.03181	0.0376	0.0454
Wave drag, C_{Dw}	0.0016	0.0018	0.002	0.00271	0.0033	0.0056
Total aircraft C_D at Mach 0.8	0.0235	0.0257	0.0292	0.03452	0.0409	0.051

9.19.3 Weights

A summary of the relevant weights is given in Table 9.17 below. The CAS mass has been derived by removing one pilot, the instrument ejection seat, and so on (260 kg), and including radar and combat avionics (100 kg). There is an increase of 60 kg in engine mass. Armaments and fuel could be traded for range.

The drag levels of the clean AJT and CAS aircraft may be considered to be about the same. There would be an increase in drag on account of the weapon load. For the CAS aircraft, there is a wide variety, but to give a general idea, the typical drag coefficient increment for armament load is $C_{D\pi} = 0.25$ (includes interference effect) each for five hard points as a weapon carrier. Weapon drag is based on a maximum cross-sectional area (say 0.8 ft²) of the weapons.

Parasite drag increment due to armaments, $\Delta C_{Dpmin} = (5 \times 0.25 \times 0.8)/183 = 0.0055$, where $S_W = 183 \text{ ft}^2$.

9.19.4 AJT Details

The drag polar at Mach 0.7 and at Mach 0.8 are given in Table 9.18 and plotted in Figure 9.16.

9.20 Classroom Example – Turboprop Trainer

This classroom example of a turboprop trainer (TPT) type aircraft, of the class Shorts Tucano, uses the same procedure as for the civil aircraft drag estimation. However, propeller-driven aircraft drag has to account for the propeller slipstream effects on the wetted surface. Propeller wakes have higher velocity than the aircraft free-stream velocity. This gives a different average skin friction coefficient C_F. A V-tail is given a 1° offset to counter the slipstream rotational effect.

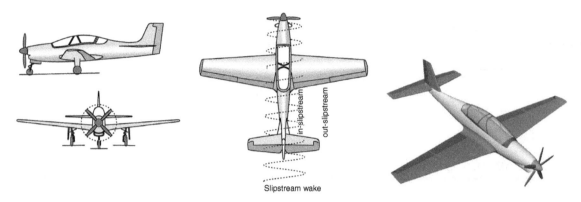

■ **FIGURE 9.18** **Turboprop trainer aircraft (TPT)**

To avoid repetition, the TPT drag evaluation is done using a short-cut method (based on experience). Some of the flat plate equivalent drags *f*s are given, using comparable industrial data. It is suggested that the reader should undergo the full methodology. The results should not be much different from the drag polar given here. The three-view diagram of a TPT is given in Figure 9.18, showing the slipstream effects. The normal training configuration (NTC) is the aircraft without any weapon load, that is, no external stores.

9.20.1 TPT Specification

Given below is the defence specification for which the TPT design was developed [1]. Practice altitude $= 25,000$ ft ($\rho = 0.00106$ slug/ft^3 and $\mu = 0.3216 \times 10^{-6}$ lb/ft^2). $C_{DPmin} = 0.023$. The important TPT details are as follows:

- MTOM $= 6600$ lb (3000 kg).
- NTC-TOM (takeoff mass) $= 5000$ lb (2500 kg).
- OEW $= 4400$ lb (2000 kg).
- Typical mid-training aircraft mass $= 4780$ lb (2400 kg).
- Crew $= 400$ lb (≈ 180 kg).
- Fuel $= 1000$ lb (≈ 450 kg).
- Maximum speed $= 280$ kt.
- Sustained speed $= 240$ kt (405.2 ft/s).

Fuselage:

- Length $= 30$ ft.
- Overall height $= 10$ ft.
- Cockpit width $= 35$ in.
- Fineness ratio $= (30 \times 12) / 35 = 10.3$.

Wing (NACA 63-212):

- Reference area $= 18.5$ m^2 (200 ft^2).
- Span $= 9.5$ m (31.17 ft).

- Aspect ratio $= 9.5$.
- Root chord, $C_R = 2.605\,\text{m}$ (9.94 ft).
- Tip chord, $C_T = 1.04\,\text{m}$ (3.17 ft).
- MAC $= 1.805\,\text{m}$ (5.5 ft).
- Taper ratio, $\lambda = 0.35$, $\Lambda_{\frac14} = 0°$.
- Dihedral $= 5°$ (dihedral – low wing).
- Twist $= 1°$ (washout).
- t/c $= 0.12$.
- Flap $= 2.77\,\text{m}^2$ (29.8 ft^2).
- Aileron $= 1.06\,\text{m}^2$ (11.4 ft^2).

V-tail (NACA 0012):

- Reference area $= 2.04\,\text{m}^2$ (22 ft^2).
- Span $= 1.615\,\text{m}$ (5.3 ft).
- Aspect ratio $= 1.52$.
- Root chord, $C_R = 1.77\,\text{m}$ (5.8 ft).
- Tip chord, $C_T = 0.762\,\text{m}$ (2.5 ft).
- V-tail MAC $= 1.28\,\text{m}$ (4.2 ft).
- Taper ratio, $\lambda = 0$, $\Lambda_{\frac14} = 35°$.
- Rudder area $= 0.743\,\text{m}^2$ (8 ft^2).
- t/c $= 0.12$.

H-tail (NACA 0012):

- Reference area $= 4.1\,\text{m}^2$ (44.2 ft^2).
- Span $= 4.2\,\text{m}$ (13.8 ft).
- Aspect ratio $= 3.5$.
- Root chord, $C_R = 1.28\,\text{m}$ (4.2 ft).
- Tip chord, $C_T = 0.67\,\text{m}$ (2.2 ft).
- H-tail MAC $= 1\,\text{m}$ (3.3 ft).
- Taper ratio, $\lambda = 0$, $\Lambda_{\frac14} = 25°$.
- Elevator area $= 1.12\,\text{m}^2$ (12 ft^2).
- t/c $= 0.12$.

Nacelle: None, as the engine is buried into the fuselage.

9.20.2 TPT Details

ISA day, 25,000 ft altitude. Wing reference area $S_w = 200\,\text{ft}^2$ (18.5 m^2). $C_{Dpmin} = f/S_w$. Cruise $C_L = 0.319$. Outstream velocity $= 405.2$ ft/s ($\approx$ Mach 0.4). Instream velocity $= 569.2$ ft/s (increase 40%).

Outstream Re/ft at 25,000 ft and at 405.2 ft / s $= 1.3355 \times 10^6$.
Instream Re/ft at 25,000 ft and at 569.2 ft / s $= 1.87 \times 10^6$.

Drag estimation follows the Bizjet methodology incorporating slipstream effects. Note that the TPT does not have nacelles. Intake drag is to be estimated accordingly.

9.20.3 Component Parasite Drag Estimation

The shortcut method uses Bizjet ΔC_F percentage increases. No account is taken of partial laminar flow.

Fuselage:
The wetted surface of the fuselage is entirely in slipstream; in short, instream.

- Fuselage instream $Re = 5.6 \times 10^7$.
- Corresponding basic $C_{Ff} = 0.0022$.
- Wetted area $A_{wf} = 280$ ft^2.

Unlike the Bizjet, the TPT does not have (i) fuselage pressurization; (ii) windows and doors; (iii) wing-fuselage-belly fairing; and (iv) upsweep closure reducing ΔC_{Ff} by 20%. Take the 3D effects, for example (i) wrapping, (ii) supervelocity and (iii) pressure together as 5%.
   Other effects on the fuselage are (see Section 9.8.1):

(a) nose fineness – for $1.5 \leq \text{Fcf} \leq 1.75 : 6\%$.
(b) aft-end cross-sectional shape near circular: 0%.
(c) excrescence (non-manufacturing types): 3%.
(d) miscellaneous: 1%.

Total ΔC_{Ff} increment: $\approx 15\%$.
   Fuselage blockage factors of f_b and f_h are included as engine installation losses, reducing thrust by about 3%.
   Add the canopy drag for two-abreast seating $f = 0.1$ ft^2 (see Section 9.8.1). Therefore, the equivalent flat-plate area, f, becomes $C_{Ff} \times A_{wF} +$ canopy drag.

$$f_f = 1.15 \times 0.0022 \times 280 + 0.1 = 0.708 + 0.1 = 0.81 \text{ ft}^2$$

Surface roughness (to be added later): 3%.

Wing:
Of the wing wetted A_{wf} surface, approximately 20% is instream and 80% is outstream.

- Wetted area $A_{ww} = 350$ ft^2.
- Instream $= A_{ww\_in} = 70$ ft^2 and outstream $A_{ww\_out} = 250$ ft^2.
- Instream $Re_{in} = 1.03 \times 10^7$, giving basic $C_{FW\_in} = 0.0029$.
- Outstream $Re_{out} = 9.35 \times 10^6$, giving basic $C_{FW\_out} = 0.003$.

Take the 3D effects, for example (i) interference, (ii) supervelocity and (iii) pressure, together as 16% of the basic C_{Fw}.
   Other effects on the wing include excrescence from flap gaps (5%) and others (5%), giving a total ΔC_{Fw} increment of 26%.

Empennage:
Since this has the same procedure as in the case of the wing, it is not repeated. The same percentage increment as in the case of the wing is used for the classroom exercise. It may be emphasized here that, in industry, one must compute systematically as shown in the case of the wing.

V-tail:
The V-tail is treated as fully immersed in the slipstream (strictly speaking a small portion is outside the slipstream).

- Wetted area, $A_{wVT} = 45$ ft^2.
- Basic $C_{F\_H-tail} = 0.00298$.
- $f_{VT} = 1.26 \times 0.00298 \times 45 = 0.17$ ft^2.

H-tail:
Of the H-tail wetted A_{wf} surface, approximately 45% is instream and 55% is outstream.

- Wetted area $A_{wHT} = 90.6$ ft^2.
- Instream $A_{wHT\_in} = 40.6$ ft^2 and outstream $A_{wHT\_out} = 50$ ft^2.
- Instream $\text{Re}_{in} = 9.1 \times 10^6$, giving basic $C_{FW\_in} = 0.00304$ and $f_{HT\_in} = 1.26 \times 0.00304 \times 40.6 = 0.156$ ft^2.
- Outstream $\text{Re}_{out} = 4.4 \times 10^6$, giving basic $C_{FW\_out} = 0.0035$ and $f_{HT\_out} = 1.26 \times 0.0035 \times 50 = 0.22$ ft^2.

Strakes + dorsal: $f_{strake+dorsal} = 0.05$ ft^2.
Surface roughness (to be added later): 3%.

Engine intake:
There is no nacelle since the engine is integrated with the fuselage. The TPT has one turboprop engine with low intake mass flow. Only spillage drag and friction drag at the lip are to be estimated. Every aircraft requires the generation of a graph of spillage drag versus mass flow for its engine intake, like that given in Figure 9.7. TPT intake is sized for takeoff mass flow that is higher compared with the cruise demand. Hence there will be spillage drag (= additive drag + lip suction). Here f_{intake} is taken as 0.1 ft^2 (from similar industrial data).

Other sources of drag:

- *Exhaust stub*: the turboprop has a relatively large exhaust stub. $f_{stub} - 0.2$ ft^2.
- *Excrescences*: mainly those of non-manufacturing origin. $f_{excr} = 0.108$.
- *Aerial and lights*: $f_{aerial+light} = 0.05$.
- *Trim drag*: $f_{trim} = 0.013$.
- *Miscellaneous*: $f_{misc} = 0.04$ (military aircraft can be dirtier).

Table 9.19 gives the aircraft parasite drag build-up summary in a tabular form. The surface roughness effect as a 3% increase in f is added in this table for all surfaces. Table 9.20 summarizes the components of parasite drag for the TPT, and Table 9.21 gives the total aircraft drag coefficient. The actual drag polar matches closely with the parabolic drag polar expressed as $C_D = 0.0226 + 0.052\,C_L{}^2$. The TPT drag polar is plotted in Figure 9.22.

■ **TABLE 9.19**
Summary of TPT component Reynolds numbers and 2D basic skin friction C_{Fbasic}

Parameter (instream or outstream)	Reference area (ft²)	Wetted area (ft²)	Characteristic length (ft)	Reynolds number	$2DC_{F\_basic}$
Fuselage (in)	n/a	280	30	5.6×10^7	0.0022
Wing (in)	200	70	5.5 (MAC$_w$)^a	1.03×10^7	0.0029
Wing (out)	–	280	5.5 (kept same)^a	9.35×10^6	0.003
V-tail (in)	22	45	4.2 (MAC$_{VT}$)	9.85×10^6	0.00298
H-tail (in)	20	40.6	3.3 (MAC$_{HT}$)^a	9.1×10^6	0.00304
H-tail (out)	24.2	50	3.3 (MAC$_{HT}$)^a	4.4×10^6	0.0035
Strakes (in)	–	–	–	–	–

^{a} Strictly speaking, separate *MAC* should be taken. The error is small.

■ **TABLE 9.20**
TPT parasite drag buildup summary and C_{Dpmin} estimation

	Wetted area, A_w (ft²)	Basic C_F	ΔC_F	Total C_F	f(ft²)	C_{Dpmin}
Fuselage + canopy	280	0.0022	0.00035	0.00266	0.81	0.004
Wing (in)	70	0.0029	0.000754	0.00365	0.256	0.00128
Wing (out)	280	0.003	0.00078	0.00378	1.06	0.0053
V-tail	45	0.00298	0.000775	0.00376	0.17	0.00085
H-tail (in)	40.6	0.00304	0.0008	0.00384	0.156	0.00078
H-tail (out)	50	0.0035	0.00091	0.00441	0.22	0.0011
Strakes + dorsal	–	–	–	–	0.05	0.00025
Engine intake	–	–	–	–	0.1	0.0005
Exhaust stub	–	–	–	–	0.2	0.001
NACA intakes	–	–	–	–	0.8	0.004
Roughness (3%)	–	–	–	–	0.182	0.0006
Excrescences	–	–	–	–	0.106	0.00053
Aerial, lights	–	–	–	–	0.05	0.0004
Trim drag	–	–	–	–	0.13	0.00001
Miscellaneous		–			0.4	0.002
Total					4.5	0.0226

■ **TABLE 9.21**
TPT total aircraft drag coefficient, C_D

C_L	0.1	0.2	0.3	0.4	0.5	0.6
C_{Dpmin}			0.0226			
ΔC_{Dp}	0.0005	0.0002	0.0	0.00005	0.0004	0.001
$C_{Di} = C_L^2/(3.14\times9.5)$	0.0005	0.00196	0.0044	0.00784	0.01225	0.0176
Total aircraft C_D at 0.4 M	0.0236	0.0247	0.027	0.0305	0.03525	0.0412

9.21   Classroom Example – Supersonic Military Aircraft

To show proper supersonic drag estimation method, a three-view diagram of a North American RA-C5 Vigilante aircraft is shown in Figure 9.19, as an example to work out here. Reference [3] gives the Vigilante drag polar for comparison. Subsonic drag estimation of the Vigilante aircraft follows the same procedure as that for the civil aircraft example. Therefore the results of drag at Mach 0.6 (no compressibility) and at Mach 0.9 (at M_{crit}) are worked out in brief and tabulated. The supersonic drag estimation is worked out in detail following the empirical methodology of [3].

9.21.1   Geometric and Performance Data for the Vigilante RA-C5 Aircraft

The following pertinent geometric and performance parameters are taken from Ref. [3]. The RA-C5 has two crew.

- Engine: $2 \times$ turbo-jet GE J-79-8(N), 75.6 kN.
- Length: 22.3 m, height: 5.9 m.
- Wing-span: 19.2 m, wing area: 65.0 m².
- Start mass: 27,300 kg, max. speed Mach 2.0.
- Ceiling: 18,300 m, range: 3700 km.
- Armament: nuclear bombs and missiles (only the clean configuration is evaluated).

Fuselage:

- Fuselage length = 73.25 ft.
- Average diameter at the maximum cross section = 9.785 ft.

■ FIGURE 9.19
North American RA-C5 Vigilante aircraft (very clean wing – no pylon shown)

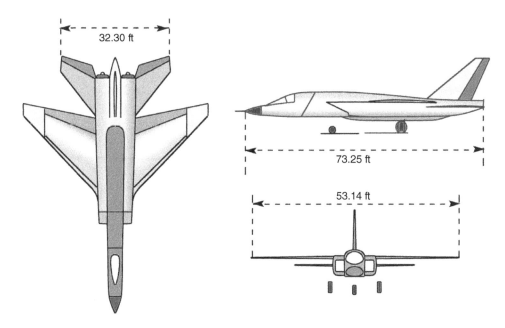

- Fuselage length/diameter $= 9.66$ (fineness ratio).
- Fuselage upsweep angle $= 0°$.
- Fuselage closure angle $\approx 0°$.

Wing:

- Planform reference area, $S_W = 65.03 \, \text{m}^2 \, (700 \, \text{ft}^2)$.
- Span $= 19.2 \, \text{m} \, (53.14 \, \text{ft})$.
- Aspect ratio $= 3.73°$.
- t/c $= 5\%$.
- Taper ratio, $\lambda = 0.19$, camber $= 0$.
- Wing MAC $= 4.63 \, \text{m} \, (15.19 \, \text{ft})$.
- $\Lambda_{\frac{1}{4}} = 39.5°$ and $\Lambda_{LE} = 43°$.
- Root chord at centreline $= 9.65 \, \text{m} \, (20 \, \text{ft})$.
- Tip chord $= 1.05 \, \text{m} \, (3.46 \, \text{ft})$.

Empennage:

- V-tail: $S_V = 4.4 \, \text{m}^2 \, (49.34 \, \text{ft}^2)$, span $= 3.6 \, \text{m} \, (11.82 \, \text{ft})$, MAC $= (8.35 \, \text{ft})$, t/c $= 4\%$.
- H-tail: $S_H = 9.063 \, \text{m}^2 \, (65.3 \, \text{ft}^2)$, span $= 9.85 \, \text{m} \, (32.3 \, \text{ft})$, MAC $= (9.73 \, \text{ft})$, t/c $= 4\%$.

Empennage: There is no nacelle or pylon since the engine is integrated with the fuselage.

Aircraft cruise performance where the basic drag polar has to be computed:
Drag estimated at cruise altitude $= 36,152$ ft and Mach number $= 0.6$ (no compressibility drag). Ambient pressure $= 391.68 \, \text{lb/ft}^2$, Re/ft $= 1.381 \times 10^6$.
Design $C_L = 0.365$, design Mach number $= 0.896$ (M_{crit} is at 0.9), maximum Mach number $= 2.0$.

9.21.2   Computation of Wetted Areas, Re and Basic C_F

The aircraft is first dissected into isolated components to obtain Reynolds numbers, the wetted areas, and the basic 2D flat plate C_F of each component as listed below. Note that there is no correction factor for C_F at Mach 0.6 (no compressibility drag). The C_F compressibility correction factor (computed from Figure 9.25) at Mach 0.9 and Mach 2.0 will be applied later.

Fuselage:

- Fuselage wetted area $= A_{wf} = 1474 \, \text{ft}^2$.
- Fuselage Re $= 69 \times 1.381 \times 10^6 = 9.53 \times 10^7$ (length trimmed to what is pertinent for Re).
- Use Figure 9.23b to obtain basic $C_{Ff} = 0.0021$.

Wing:

- Wing wetted area $= A_{ww} = 1144.08 \, \text{ft}^2$.
- Wing Re $= 15.19 \times 1.381 \times 10^6 = 2.1 \times 10^7$.
- Use Figure 9.23b to obtain basic $C_{Fw} = 0.00257$.

Empennage (same procedure as for the wing):

- V-tail wetted area $= A_{wVT} = 235.33$ ft^2.
- V-tail Re $= 8.35 \times 1.381 \times 10^6 = 1.2 \times 10^7$.
- Use Figure 9.23 to obtain basic $C_{F\_V-tail} = 0.00277$.
- H-tail wetted area $= A_{wHT} = 388.72$ ft^2.
- H-tail Re $= 9.73 \times 1.381 \times 10^6 = 1.344 \times 10^7$.
- Use Figure 9.23 to obtain basic $C_{F\_H-tail} = 0.002705$.

9.21.3 Computation of 3D and Other Effects to Estimate Component C_{Dpmin}

The component examples are given below.

Fuselage:
From above, at Mach 0.6, the basic $C_{Ff} = 0.0021$. 3D effects apply as follows:

Wrapping:

$$\Delta C_{Ff} = C_{Ff} \times 0.025 \times (\text{length/diameter}) \times \text{Re}^{-0.2}$$
$$= 0.025 \times 0.0021 \times (9.66) \times (9.53 \times 10^7)^{-0.2}$$
$$= 0.000507 \times 0.0254 = 0.0000129 \,(0.6\% \text{ of basic } C_{Ff})$$

Supervelocity:

$$\Delta C_{Ff} = C_{Ff} \times (\text{diameter} / \text{length})^{1.5}$$
$$= 0.0021 \times (1 / 9.66)^{1.5}$$
$$= 0.0021 \times 0.033 = 0.0000693 \,(3.3\% \text{ of basic } C_{Ff})$$

Pressure:

$$\Delta C_{Ff} = C_{Ff} \times 7 \times (\text{diameter} / \text{length})^3$$
$$= 0.0021 \times 7 \times (0.1035)^3$$
$$= 0.0000163 \,(0.8\% \text{ of basic } C_{Ff})$$

There are other effects on the fuselage – this time the intake has to be included. It is to be noted here that Ref. [3] suggests applying a factor of 1.284 to include most of the other effects except intake. Therefore, unlike the civil aircraft example, it is simplified to the following only:

Intake (little spillage – rest taken in 3D effects): 2%

The total ΔC_{Ff} increment is given in Table 9.22. In terms of equivalent flat plate area, f, it becomes $= C_{Ff} \times A_{wF}$:

$$f = 1.067 \times 0.0021 \times 1474 = 3.3 \text{ ft}^2$$

■ **TABLE 9.22**
Vigilante fuselage ΔC_{Ff} corrections (3D and other shape effects)

Item	ΔC_{Ff}	% of $C_{Ffbasic}$
Wrapping	0.000015	0.6
Supervelocity	0.0001	3.3
Pressure	0.0000274	0.8
Intake (little spillage)	–	2
Total ΔC_{Ff}	0.00105	9.7

■ **TABLE 9.23**
Vigilante wing ΔC_{Fw} correction (3D and other shape effects)

Item	ΔC_{Fw}	% of $C_{Fwbasic}$
Supervelocity	0.000385	7
Pressure	0.0000136	0.04
Interference (wing-body)	0.0000328	3
Flap/slat gaps	–	2
Total ΔC_{Fw}	–	12.04

Add canopy drag, $C_{D\pi} = 0.08$ (approximated from Table 9.1). Noting $C_{Ffcanopy} = (A_p \times C_{Dp})/A_{wF}$, it can be written in terms of flat plate drag:

$$f_{can} = 0.08 \times 4.5 = 0.36 \text{ ft}^2.$$
$$f_f = 3.3 + 0.36 = 3.66 \text{ ft}^2$$

Wing:
From above, at Mach 0.06, the basic $C_F = 0.00259$. 3d effects are as follows:

Supervelocity:

$$\Delta C_{Fw} = C_{Fw} \times 1.4 \times (\text{aerofoil thickness/chord ratio})$$
$$= 0.00257 \times 1.4 \times 0.05$$
$$= 0.00018 \,(7\% \text{ of basic } C_{Fw})$$

Pressure:

$$\Delta C_{Fw} = C_{Fw} \times 60 \times (\text{aerofoil thickness/chord ratio})^4 \times \left(\frac{6}{\text{AR}}\right)^{0.125}$$
$$= 0.00257 \times 60 \times (0.05)^4 \times (6/3.73)^{0.125}$$
$$= 0.00000102 \,(0.04\% \text{ of basic } C_{Fw})$$

Interference: ΔC_{Fw} for thin high wing, take 3% of C_{Fw}.

Other effects include excrescence drag (non-manufacturing) – for flap/slat gaps, allow 2%. This gives the total ΔC_{Fw} increment $= 12.04\%$ (Table 9.23).

■ **TABLE 9.24**
Vigilante parasite drag summary

Element	Component parasite drag
Fuselage	$3.66\,\text{ft}^2$
Wing	$3.3\,\text{ft}^2$
V-tail	$0.73\,\text{ft}^2$
H-tail	$1.18\,\text{ft}^2$
Total	$8.87\,\text{ft}^2$

Therefore in terms of equivalent flat plate area, f, it becomes $= C_{Fw} \times A_{ww}$:

$$f_w = 1.12 \times 0.00257 \times 1144.08 = 3.3\,\text{ft}^2$$

Empennage:
Since this has the same procedure as in the case of the wing, it is not repeated. The same percentage increment as in the case of the wing is used for the classroom exercise. It may be emphasized here that, in industry, one must compute systematically as shown in the case of the wing.

V-tail: Wetted area, $A_{wVT} = 235.33\,\text{ft}^2$, basic $C_{F\_V-tail} = 0.00277$:

$$f_{VT} = 1.12 \times 0.00277 \times 235.33 = 0.73\ \text{ft}^2$$

H-tail: Wetted area, $A_{wHT} = 388.72\,\text{ft}^2$, basic $C_{F\_H-tail} = 0.002705$:

$$f_{HT} = 1.12 \times 0.002705 \times 388.72 = 1.18\,\text{ft}^2$$

9.21.4 Summary of Parasite Drag

For an ISA day, 40,000 ft altitude and Mach 0.9, wing reference area $S_w = 700\,\text{ft}^2$, $C_{Dpmin} = f/S_w$. Table 9.24 lists the total of component drag elements as calculated above.

As mentioned earlier, [3] offers a correlated factor of 1.284 to include all so-called other effects. Therefore the final flat plate equivalent drag $f_{aircraft} = 1.284 \times 8.87 = 11.39\,\text{ft}^2$. This gives C_{Dpmin} at Mach 0.6 $= 11.39 / 700 = 0.01627$ (Ref. [3] gives 0.1645, close enough).

This is the C_{Dpmin} at flight Mach number before compressibility effects start to show up, seen as C_{Dpmin} for incompressible flow. At higher speeds there is a C_F shift to a lower value. The C_{Dpmin} estimation needs to be repeated with lower C_F at Mach 0.9 and at Mach 2.0. To avoid repetition to account for compressibility, a factor of 0.97 (ratio of values at Mach 0.9 and Mach 0 in Figure 9.23b – a reduction of 3%) is taken at Mach 0.9. A factor of 0.8 (a reduction of 20%) is taken at Mach 2.0, as shown in Table 9.27. For compressible flow, add the wave drag. At supersonic speed, this is contributed by shock waves.

To stay in line with the methodology presented here, the following values of ΔC_{Dp} have been extracted from Ref. [3].

9.21.5 ΔC_{Dp} Estimation

The data for ΔC_{Dp} given in Table 9.25 is extracted from Ref. [3] and is approximate.

■ **TABLE 9.25**
Vigilante ΔC_{Dp} estimation

C_L	0	0.1	0.16	0.2	0.3	0.4	0.5	0.6
ΔC_{Dp}	0.0008	0.00015	0	0.0001	0.0008	0.00195	0.0036	0.006

■ **TABLE 9.26**
Vigilante induced drag

C_L	0.2	0.3	0.4	0.5	0.6	0.7
C_{Di}	0.00342	0.00768	0.0137	0.0214	0.0307	0.0418

9.21.6   Induced Drag

The formula used is $C_{Di} = C_L^2/(3.14 \times 3.73) = C_L^2/11.71$ (see results in Table 9.26).

9.21.7   Supersonic Drag Estimation

Supersonic flight produces a bow shock wave that is a form of compressibility drag, evaluated at zero C_L. Drag increases with change of angle of attack. The difficulty arises in understanding the physics involved with the increase in C_L. Clearly, the increase, though lift dependent, has little to do with viscosity unless shock interacts with the boundary layer to increase pressure drag. Since the very point of design is to avoid such an interaction up to a certain C_L, this book treats the compressibility drag at supersonic speeds to be composed of compressibility drag at zero C_L (that is C_{D_shock}) plus a compressibility drag at higher C_L (that is, ΔC_{Dw}).

To compute compressibility drag at zero C_L, the following empirical procedure is adopted, taken from [3]. The compressibility drag of an object depends on its thickness parameter; in the case of a fuselage it is the fineness ratio, and in the case of a wing it is the thickness to chord ratio. Fuselage (including empennage) and wing compressibility drags are computed separately and then added along with the interference effects. Graphs (Figures 9.24–9.30) are used extensively for the empirical methodology. Compressibility drags for both Mach 0.9 and Mach 2.0 are estimated.

The drag estimation at Mach 0.9 follows the same method as used in earlier examples.

Fuselage compressibility drag (includes empennage contribution) at Mach 2.0:
The thickness parameter is the fuselage fineness ratio.

Step 1: Plot the fuselage cross-section along fuselage length as shown in Figure 9.20, and obtain the maximum cross-section, $S_\pi = 45.25$ ft^2, and the fuselage base $S_b = 12$ ft^2. Find the ratios:

$$(1 + S_b/S_\pi) = 1 + 12/45.25 = 1.27$$
$$S_\pi/S_w = 45.25/700 = 0.065$$

Step 2: Get the fuselage fineness ratio $l/d = 73.3/9.788 = 9.66$ (d is minus intake width). Obtain $(l/d)^2 = (9.66)^2 = 93.3$.

Step 3: Use Figure 9.25 to get $C_{D\pi}(l/d)^2 = 18.25$ at $M_\infty = 2.0$ for $(1 + S_b/S_\pi) = 1.27$. This gives $C_{D\pi} = 18.25/93.3 = 0.1959$. Convert it to fuselage contribution of compressibility drag expressed in terms of wing reference area:

$$C_{Dwf} = C_{D\pi} \times (S_\pi/S_w) = 0.1956 \times 0.065 = 0.01271$$

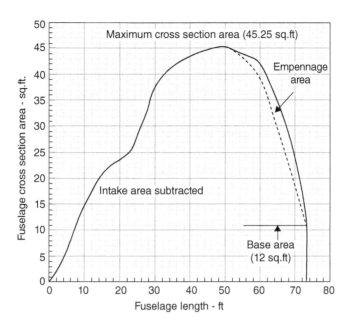

■ FIGURE 9.20
Vigilante RA-C5 fuselage cross-sectional area distributions

Wing compressibility drag at Mach 2.0:

Step 1: Obtain design C_{L_DES} from Figure 9.26 for a supersonic aerofoil for:

$$\text{AR} \times (t/c)^{1/3} = 3.73 \times (0.05)^{1/3} = 1.374$$

This gives $C_{L\_DES} = 0.352$. Test data of C_{L_DES} from [3] gives 0.365, which is close enough. Test data are used.

Step 2: Obtain from Figure 9.27 the two-dimensional design Mach number, $M_{DES\_2D} = 0.784$. Using Figure 9.28 obtain $\Delta M_{AR} = 0.038$ for $1/\text{AR} = 0.268$. Using Figure 9.28, obtain $\Delta M_{DA\frac{1}{4}} = 0.067$ for $\Lambda_{\frac{1}{4}} = 39.5°$.

Step 3: Make correction to get design Mach as:

$$M_{DES} = M_{DES\_2D} + \Delta M_{AR} + \Delta M_{DA\frac{1}{4}}$$

or

$$M_{DES} = 0.784 + 0.038 + 0.067 = 0.889$$

Then

$$\Delta M = M_\infty - M_{DES} = 2.0 - 0.889 = 1.111$$

Step 4: Compute

$$(t/c)^{5/3} \times \left[1 + ((h/c)/10)\right] = (0.05)^{5/3} \times \left[1 + (0/10)\right] = 0.00679$$

Step 5: Compute $\text{AR} \ \tan \Lambda_{LE} = 3.73 \times \tan 43 = 3.73 \times 0.9325 = 3.48$

■ **TABLE 9.27**
Vigilante supersonic drag summary

	C_{Dw} at Mach 2.0
Fuselage/empennage contribution	0.01271
Wing contribution	0.00458
Wing-fuselage interference (supersonic only)	0.00075
Total	0.01804

Step 6: Use Figure 9.29 and the corresponding value of AR $\tan \Lambda_{LE} = 3.48$ in Step 5.

$$\Delta C_{DC\_WING} \Big/ \Big\{ (t/c)^{5/3} \times \big[1 + (h/c)/10 \big] \Big\} = 0.675$$

Compute $\Delta C_{DC\_WING} = 0.675 \times 0.00679 = 0.00458$

Wing-fuselage interference drag:

Finally, interference drag at supersonic speed has to be added to fuselage and wing compressibility drag, using the procedure given below.

Step 1: Compute fuselage diameter at maximum area/wing span = 9.785 / 53.14 = 0.1465.
Step 2: Having taper ratio, $\lambda = 0.19$, compute $(1 - \lambda) \cos \Lambda_{\frac{1}{4}} = (1 - 0.19) \cos 39.5 = 0.643$.
Step 3: Using Figure 9.30 obtain

$$\Delta C_{D\_INT} \times \Big[(1 - \lambda) \cos \Lambda_{\frac{1}{4}} \Big] = 0.00048$$

Compute $\Delta C_{D\_INT} = 0.00048 / 0.643 = 0.00075$.

Summary of compressibility drag of Vigilante at zero lift:

The supersonic drag summary for the Vigilante is given in Table 9.27.

Compressibility drag at Mach 0.6 (3% reduction on account of C_F change) and Mach 0.9 (20% reduction on account of C_F change) has been calculated as in the case of civil aircraft, and is given in Table 9.28 along with drag at Mach 2.0.

9.21.8 Total Aircraft Drag

Total Vigilante drag at the three Mach numbers is tabulated in Table 9.28. Figure 9.21 gives the Vigilante drag polar at the three aircraft speeds. Figures 9.23–9.30 are re-plotted at the end of this chapter from [3].

9.22 Drag Comparison

The following equations give the parabolic drag polar of the aircraft used as classwork examples.

Bizjet (Oswald's efficiency factor, $e = 0.95$, determined in Section 10.6.2):
Bizjet drag may be approximated as parabolic drag polar, as follows.

$$C_D = 0.0205 + 0.0447 C_L^2 \text{ with } S_W = 323 \, \text{ft}^2 \tag{9.37}$$

■ **TABLE 9.28**
Vigilante total aircraft drag coefficient, C_D

C_L	0.0	0.1	0.2	0.3	0.4	0.5	0.6
At Mach 0.6							
C_{Dpmin}				0.01627			
ΔC_{Dp}	0.0008	0.00015	0.0001	0.0008	0.00195	0.0036	0.006
C_{Di}	0	0.000854	0.00342	0.00768	0.0137	0.0214	0.0307
Aircraft C_D	0.01707	0.01727	0.0198	0.02475	0.03192	0.04127	0.0523
At Mach 0.9							
C_{Dpmin}			0.97 × 0.01627 = 0.01582 (Ref. [3] gives 0.01575)				
ΔC_{Dp}	0.0008	0.00015	0.0001	0.0008	0.00195	0.0036	0.006
C_{Di}	0	0.000854	0.00342	0.00768	0.0137	0.0214	0.0307
Wave drag, C_{Dw}[a]	0.0003	0.001	0.002	0.0032	0.0055	0.01	0.02
Aircraft C_D	0.01737	0.01827	0.0218	0.02795	0.03742	0.05127	0.0723
At Mach 2.0							
C_{Dpmin}			0.8 × 0.01627 = 0.013 (Ref. [3] gives 0.01302)				
ΔC_{Dp}	0.0008	0.00015	0.0001	0.0008	0.00195	0.0036	0.006
C_{Di}	0	0.000854	0.00342	0.00768	0.0137	0.0214	0.0307
Shock drag			0.01804 at zero C_L (C_{D_shock})				
Wave drag, ΔC_{Dw}[b]	0.0	0.003	0.011	0.023	0.041	–	–
Aircraft C_D	0.0318	0.035	0.0455	0.0625	0.087	–	–

[a] C_{Dw} versus C_L is to be taken from CFD/test data. Here it is reduced from Ref. [3].
[b] ΔC_{Dw} versus C_L is to be taken from CFD/test data.

■ **FIGURE 9.21**
Vigilante RA-C5 drag polar (not from industry)

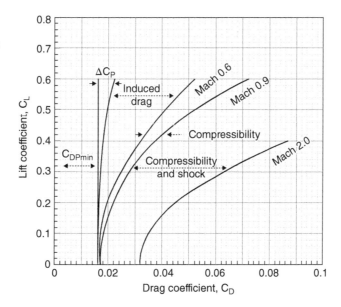

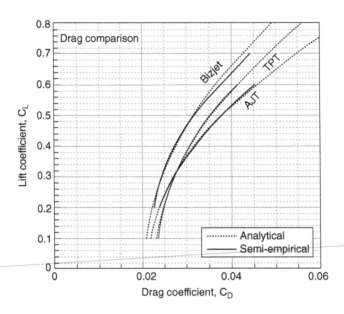

AJT (readers to work out the Oswald efficiency factor, *e*):

AJT drag may be approximated as parabolic drag polar, as follows.

$$C_D = 0.0212 + 0.068C_L^{\,2} \text{ with } S_W = 183\,\text{ft}^2 \tag{9.38}$$

(gives Oswald's efficiency factor, *e* = 0.85).

Turboprop trainer (low speed aircraft – Oswald efficiency factor, *e*, fits closely with actual polar):

Maximum TPT speed is below 0.5 Mach (incompressible flow). Here, parabolic drag is close enough to actual drag polar.

$$C_D = 0.0226 + 0.052C_L^{\,2} \text{ with } S_W = 200\,\text{ft}^2 \tag{9.39}$$

(gives Oswald's efficiency factor, *e* = 0.875).

Figure 9.22 gives a drag comparison between the above three examples: the Bizjet, AJT and TPT.

9.23   Some Concluding Remarks and Reference Figures

Drag estimation is state-of-the-art and covers a very large territory, as can be seen in this chapter. There is a tendency to underestimate drag, primarily on account of missing out some of the multitude of items that need to be considered in the process of estimation. The object of this chapter is to make readers aware of the sources of drag as well as to give a methodology in line with typical industrial practices (without CFD results).

Some of the empirical relations are guesstimates based on industrial data available to the author; these are not in the public domain. The formulation could not possibly cover all aspects of drag estimation methodologies and therefore has to be simplified for classroom usage. For example, the drag calculations for high-lift devices can only give ballpark figures.

■ **FIGURE 9.23**
Flat plate skin-friction coefficients versus Reynolds number (turbulent flow). (a) Local skin friction. (b) Mean skin friction

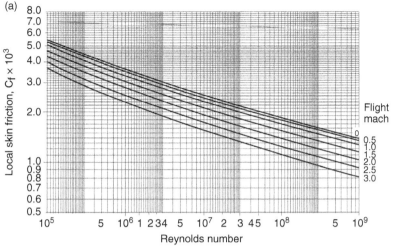

(a)

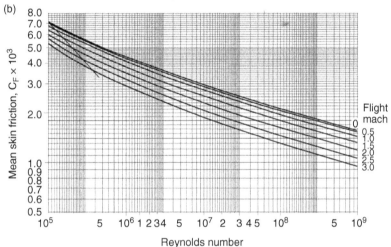

(b)

■ **FIGURE 9.24**
Corrections for laminarization. Reproduced from NASA [3]

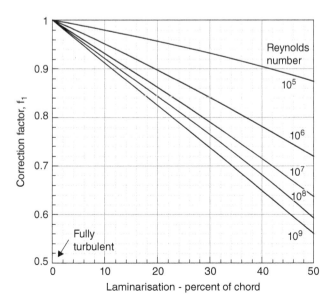

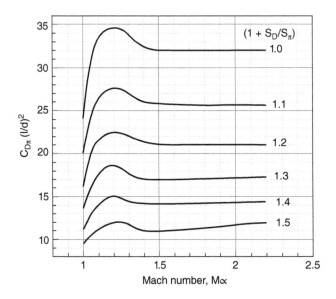

■ FIGURE 9.25
Supersonic fuselage
compressibility drag.
Reproduced from
NASA [3]

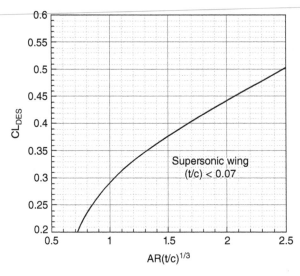

■ FIGURE 9.26
Design lift coefficient.
Reproduced from
NASA [3]

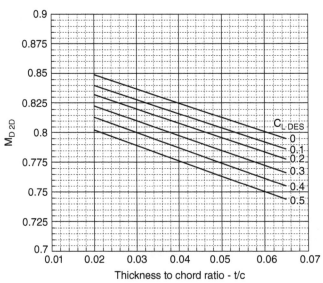

■ FIGURE 9.27
2D drag divergence
Mach number for
supersonic aerofoil.
Reproduced from
NASA [3]

■ **FIGURE 9.28**
Design Mach
number. Reproduced
from NASA [3]

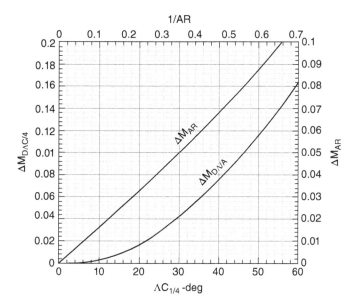

■ **FIGURE 9.29**
Supersonic wing
compressibility drag.
Reproduced from
NASA [3]

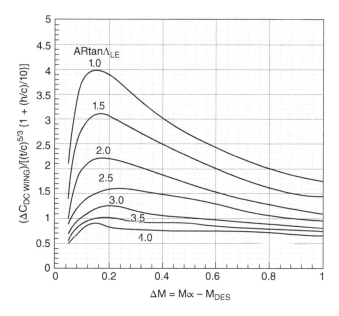

Readers are advised to rely on industrial data or to generate their own databank through CFD/tests. The author would be grateful to receive data and/or substantiated formulation that could improve the accuracy of future editions (with due acknowledgement).

Figures 9.23 to 9.30 are given below to provide reference data to the reader. Most are re-plotted from the NASA report listed as [3].

■ **FIGURE 9.30**
Wing-body zero lift interference drag. Reproduced from NASA [3]

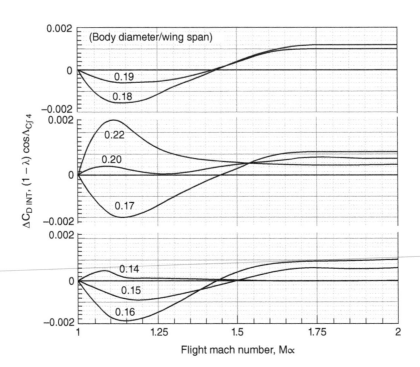

References

1. Hoerner, S.F., *Fluid Dynamic Drag, Practical Information on Aerodynamic Drag and Hydrodynamic Resistance*, Hoerner Fluid Dynamics, Bricktown, NJ, 1965.

2. *Prediction Method for Aircraft Aerodynamic Characteristics*, AGARD LS 67, May 1974.

3. Feagan, R.C., and Morrison, W.D., *Delta Method – An Empirical Drag Build-Up Technique*. NASA Contractor Report 151971. Lockheed California Company, 1978.

4. USAF DATCOM, *The USAF Stability and Control Digital DATCOM*. 1979.

5. ESDU, *An Introduction to Aircraft Excrescence Drag*. ESDU 90029. London, 2012.

6. Roskam, J., *Airplane Design*, vol. 1–8. University of Kansas, Lawrence, KS, 1990.

7. Kundu, A.K., *Aircraft Design*. Cambridge University Press, Cambridge, 2010.

8. Schmidt, W., and Sacher, P., *Technical Status Review on Drag Prediction and Analysis from Computational Fluid Dynamics: State of the Art*. AGARD AR-256, June 1989.

9. Torenbeek, E., *Synthesis of Subsonic Airplane Design*. Delft University Press, Delft, 1982.

10. Haines, A.B. Subsonic aircraft drag: an appreciation of present standards. *Aeronautical Journal* 72 687, 253–266, 1968.

11. Young, A.D., and Paterson, J.H., *Aircraft Excrescence Drag*. AGARD AG264. Cranfield University, Cranfield, 1981.

12. Farokhi, S., *et al. Propulsion System Integration*. Short course. University of Kansas, Lawrence, KS, 1991.

13. Kundu, A.K., Watterson, J., Raghunathan, S., MacFadden, R., Parametric optimisation of manufacturing tolerances at the aircraft surface. *Journal of Aircraft*, 39, 2, 271–279, 2002.

14. Seddon, J., and Goldsmith, E., *Practical Intake Aerodynamic Design*. AIAA, Reston, VA, 1965.

CHAPTER 10

Fundamentals of Mission Profile, Drag Polar and Aeroplane Grid

10.1   Overview

This chapter gives an overall picture of what is required to evaluate aircraft performance. It starts with classifying the types of aircraft and with a brief introduction to progress made in aircraft capabilities. Typical mission profiles for civil and military aircraft are shown. Next, the most important term in aircraft performance, the drag polar is explained in detail. It captures the relationship between aircraft lift and drag, a topic that needs to be well understood. The crux of aircraft performance analyses is to make use of drag polar. Finally, the important aircraft performance-related theories associated with lift and drag are presented; these embrace the scope of the chapters to follow. Good coverage of aircraft performance is given in [1–23].

There is some difference between using close-form parabolic drag polar expressed in equation form and actual drag polar in graphical form obtained through using semi-empirical relations, especially for high subsonic jet aircraft. The difference arises from how the Oswald's efficiency factor "e" is treated (Section 10.9). Easy manipulation of the close-form parabolic drag equation quickly gives aircraft characteristics, and gives valuable information of whether to proceed with more accurate actual drag polar calculations. For low-speed aircraft operating below 0.5 Mach, the parabolic drag polar is, in general, close to actual drag polar on account of low-order non-linear effects.

Sections 10.3.1 and 10.4.1 give the typical mission profiles of commercial transport and military aircraft, respectively. The component mission segments of the profiles are separately analysed in subsequent chapters. Finally, the mission segments are totalled to obtain the overall integrated aircraft performances. There are two types of aircraft performance: (i) point performance (at an instant); and (ii) integrated performance (over a period of time). A performing aircraft could be in steady or unsteady states.

The aircraft flight envelope gives the boundary within which the vehicle should perform. Situations can be experienced inadvertently, for example, deliberate hard manoeuvres such as

Theory and Practice of Aircraft Performance, First Edition. Ajoy Kumar Kundu, Mark A. Price and David Riordan.
© 2016 John Wiley & Sons, Ltd. Published 2016 by John Wiley & Sons, Ltd.

collision avoidance, evasive manoeuvres under adversary threat, encountering gust load, and so on, for which safety margins are provided as shown in the flight envelope. Therefore, flight envelopes of typical civil and military aircraft are also discussed in the context, in addition to what is considered in Chapter 5. To ensure safety, certification authorities stipulate minimum design criteria, but aircraft designers offer more than the minimum requirements as market demands.

There are many types of aircraft in production, serving different types of mission requirement – the civil and military missions differ substantially (Section 1.13.1). It is important to classify the aircraft category to isolate and identify strong trends existing within the class (Section 10.3). Existing patterns of correlation (through regression analysis) within a class of aircraft indicates what to expect from a new design. Nature is the same for all – there are no surprise elements until new research establishes something radical in technology, or designers introduce a new class of aircraft (e.g. Airbus 380). In civil aircraft design, a 10–15% improvement on operating economics from current designs within the class would be considered good. Of course, economic gains are to be supported by gains in safety, environmental considerations, reliability and maintainability, which in turn add to the cost.

A good part of aircraft performance computation uses aeroplane and engine grids (Section 10.10), that is, the thrust available and thrust required for the aircraft performance estimation for the segment of operation. Aircraft drag polar provides the thrust required, and engine performance provides the thrust available.

Readers should study this chapter thoroughly as it serves as the foundation of aircraft performance. Readers should prepare aeroplane and engine grids for the classwork examples. These are required for overall aircraft performance analyses. A problem set for the chapter is given in Appendix E.

10.2   Introduction

There are many types of aircraft in production, serving different sectors of mission requirements – the civil and military missions differ substantially. Readers are advised to examine what could be the emerging design trends within each class of aircraft. In general, new commercial aircraft designs are extensions of the existing designs, incorporating proven newer technologies (some are fallouts from declassified military applications) in a very conservative manner.

Figure 10.1 gives aircraft classification based on its mission role and usage, power plants, configuration type, size, and so on. Here, the first level of classification is based on operational role, that is, civil or military. In the next level, the classification branches into their generic mission roles, which also indicate size. The next level proceeds with classification based on the type of power plant used. The examples used in this book are convenient for a classroom course. It should be noted that there are many to choose from without stretching too far into studying exotic types.

A new design, with commercial considerations, has to be a cautious progression, advancing through the introduction of the latest proven technologies. It is not surprising, therefore, to see a strong statistical correlation with the past. Military designs necessarily have to be bolder, making bigger leaps to stay decisively ahead of potential adversaries, even if it costs a great deal more. Eventually, matured, declassified military technology trickles down to commercial usage; one such example is fly-by-wire (FBW). Fortunately, the aircraft performance equations and methodologies between civil and military designs are the same, irrespective of configuration differences. However, there is some computational difference arising from relatively more

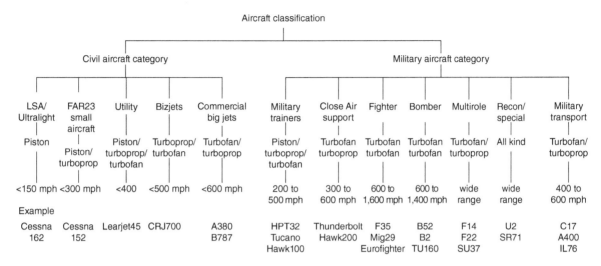

■ FIGURE 10.1   **Aircraft classification**

complex military designs. Extreme aircraft performance capabilities are dictated by the combat aircraft requirements, exceeding the capabilities of aerobatic sports aircraft. Progress in missile technology demands hard aircraft manoeuvres rather than speed-altitude gains. However, with the advent of space-planes and the near future possibility of hypersonic flights, the aircraft performance envelope has stretched again after a pause in speed–altitude gains.

10.2.1   Evolution in Aircraft Performance Capabilities

Figure 10.2 gives the history of growth in speed and altitude capabilities. The impressive growth in a century is astounding – from the first time leaving the Earth's surface in a heavier-than-air vehicle to return from the Moon in less than 66 years!

It is interesting to note that for air-breathing engine-powered aircraft, the speed–altitude record is still held by an over 40 years old design, the SR71 Blackbird, capable of operating at around Mach 3.0 and 100,000 ft altitude. Properties of aluminium alloy allow flight speeds up to Mach 2.5. Above Mach 2.5, a change of material to temperature-tolerant types and/or cooling would be required because stagnation temperature would come close to 600 K, crossing the strength limit of aluminium alloys. Combat aircraft speed–altitude capabilities have remained stagnant since the 1960s. Recently a breakthrough appeared from the success of "Space Ship One", taking aircraft to the fringes of the atmosphere at 100 km altitude. Note that in civil aviation, the supersonic transport (SST) aircraft "Concorde" was designed nearly 40 years ago and has not been exceeded yet. Concorde's speed-altitude capability was Mach 2.2 at around 60,000 ft.

In military scenarios, gone (almost) are the days of dogfighting which demanded high-speed chases to bring an adversary within machine-gun firing range (low projectile speed, low impact energy, and no homing), and if missed the hunter became the hunted. After World War II, say around the late 1960s, air superiority combat required fast acceleration and speed (Lockheed F104 Starfighter) to engage with infra-red homing missiles fired a relatively short distance from the target. As missile capabilities advanced, the current combat aircraft design trend shows a decrease in speed capabilities, reduced to around Mach 1.8. Instead, high turn rates and acceleration, integrated with superior missile capabilities (guided, high-speed and high impact

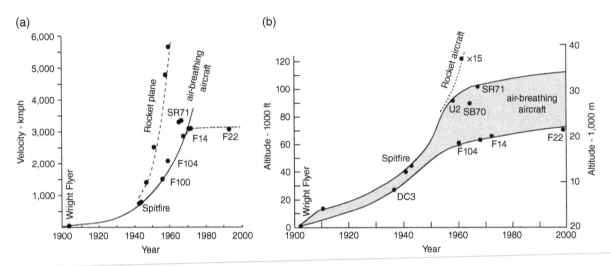

■ **FIGURE 10.2** Aircraft operational envelope. (a) Speed. (b) Altitude

energy, even when detonated in the proximity of the target), are the current trend. Target acquisition beyond visual range (BVR) by the use of advance warning systems from separate platform, and rapid aiming are the combat rules for mission success and survivability. Current military aircraft operate below Mach 2.5. Hypersonic military aircraft are in the offing.

Figure 7.3 indicates the speed–altitude regimes for the type of power plant used. Figure 7.4 gives the thrust to weight ratio and specific fuel consumption (sfc) at sea-level static takeoff thrust (T_{SLS}) rating on an ISA day for various classes of current engines. At cruise, sfc would be higher.

10.2.2   Levels of Aircraft Performance Analyses

There are two main classes of aircraft performance analyses – one done by the designers until it is certified and delivered to operators, and the other by the operators. All use the same fundamental equations but differ in approach as demanded for the purpose; the performance analyses by the operators are less intensive compared with what designers do. Operators typically use the performance manual for the aircraft type prepared by the designers. The format of presenting aircraft performance in the pilot's manual is different from the format of aircraft performance used by the designers. This book deals with what the designers do.

Manufacturers keep a statistical status deck of all aircraft produced; data are extracted from mandatory flight tests before delivery to customers. Companies have their own specification, which is slightly better than what is guaranteed to customers to cover degradation due to ageing during usage. This is the minimum performance a new aircraft must meet. If failed, the aircraft is sent back for rework to meet the minimum performance level.

10.3   Civil Aircraft Mission (Payload–Range)

The real-life experience of past designs has a strong influence on future designs – there is no substitute for it. It is therefore important that past information is properly synthesized through studying statistical trends and examining all aspects of the influencing parameters.

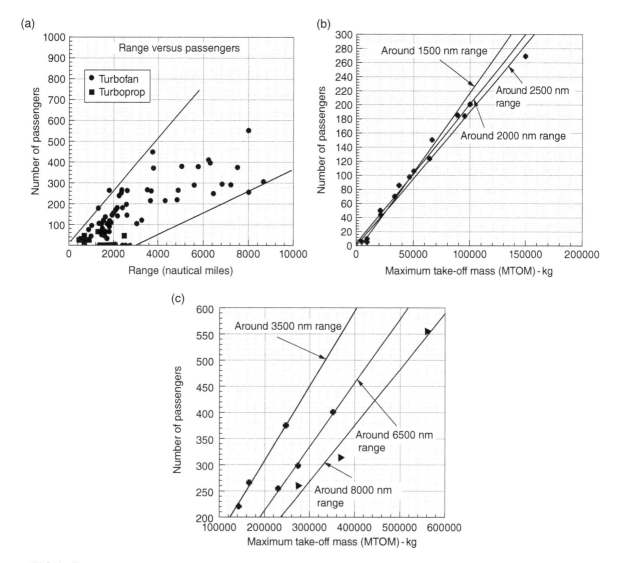

■ **FIGURE 10.3** Passengers versus range and MTOM. (a) Passengers versus range (statistics of 70 aircraft). (b) Passengers and MTOM (lower capacity). (c) Passengers and MTOM (higher capacity)

Payload–range capability combines the two most important parameters in commercial transport aircraft performance. It is the basic requirement for aircraft arising from market studies for a new aircraft design.

Figure 10.3 gives the payload–range capabilities for a number of subsonic transport aircraft (turbofan and turboprop). The figure captures over 80 different kinds of current designs, including: LEAR31A, LEAR45, LEAR60, CESS525A, CESS650, CESS500, CESS550, CESS560, CESS560XL, ERJ135ER, ERJ140, ERJ145ER, CRJ100, CRJ700, ERJ170, DC-9-10, CRJ900, ERJ190, 737-100, 717-200, A318-100, A319-100, A320-100, Tu204, A321-100, 757-200, A310-200, A330-200, L1011, A300-600, A300-100, A320, A340-200, A380, DC-10-10, MD11, B737, B767-200, B777-200, B747-100, Short 330 and 360, ATR 42 and 72, Jet stream 31, SAAB 340A, Dash 7 and 8, Jetstream41, EMB120, EMB120ER, Dornier328-100, Dash8 Q400, and so on.

The trend shows a range increase with payload increase, reflecting the market demand of flying longer distances with higher payloads. Long-range aircraft will have a lower frequency of sorties and need to carry more passengers at one go. Interestingly, there are practically no aircraft carrying a high passenger load for shorter ranges (less than 2000 nm). At the other extreme, the Global Express and Gulfstream V, as high subsonic long-distance executive jets are already in the market (not on the graph), carrying executives/small number of passengers for very long ranges (> 6500 nm) at a considerably higher cost per passenger. Obviously, on account of considerably lower speeds, turboprop-powered aircraft cater for the shorter range market sector – they offer better fuel economy over turbofans. The future may show potential markets in the lean areas. Large countries with substantial populations could bus around passengers using very large aircraft within their borders, such as in China, India, Indonesia, Russia and the US.

Commercial aircraft operation is singularly dependent on revenue earned from fare-paying passengers/cargo. Operating sector passenger load factor (LF) is defined as the ratio of occupied seats to available seats. Typically, for aircraft of medium sizes and above, aircraft operational costs break even at around a third full (varying from airlines to airline – fuel cost at year 2000 level and having regular airfares) – that is a load factor of around 0.33. Of late, the steep rise in fuel costs has raised the *break-even LF* to some higher value. Naturally, the empty seats could be filled up with deregulated cut fares, contributing to the revenue earned. Aircraft direct operating cost (DOC) is dealt with in Chapter 17.

10.3.1 Civil Aircraft Classification and Mission Segments

In general, the civil aircraft category covers five kinds, namely

Small club trainers: up to four occupants including pilot.
Utility aircraft: 4–10 passengers. Other types are for agriculture and ambulance usage.
Business aircraft: around 8–30 passengers. Other types for corporate usage.
Regional transport aircraft: around 30–100 passengers.
Mid-size transport aircraft: around 100–250 passengers. All are narrow-bodied.
Large transport aircraft: around 250–1000 passengers. Majority are wide-bodied, those over 400 passengers often double-deck.

Other than recreational flying, for example in the club/home-build category, all typical Bizjet/transport aircraft have a mission profile as shown in Figure 10.4. The mission fuel consumed, distance covered (range), and time taken can be expressed as follows:

$$
\textit{Mission fuel consumed}\,(block\ fuel) = \sum fuel_{segments}
$$
$$
= \text{taxi}\left(fuel_{taxi}\right) + \text{takeoff}\left(fuel_{TO}\right) + \text{climb fuel}\left(fuel_{climb}\right) + \text{cruise fuel}\left(fuel_{cruise}\right) \quad \textbf{(10.1)}
$$
$$
+ \text{descent fuel}\left(fuel_{descent}\right) + \text{land}\left(fuel_{land}\right) + \text{return taxi}\left(fuel_{return}\right)
$$

$$
\textit{Mission distance covered}\,(range) = \sum distance_{segments}
$$
$$
= \text{climb}\left(distance_{climb}\right) + \text{cruise}\left(distance_{cruise}\right) + \text{descent}\left(distance_{descent}\right) \quad \textbf{(10.2)}
$$

$$
\textit{Mission time taken}\,(block\ time) = \sum time_{segments}
$$
$$
= \text{taxi}\left(time_{taxi}\right) + \text{takeoff}\left(time_{TO}\right) + \text{climb}\left(time_{climb}\right) + \text{cruise}\left(time_{cruise}\right) \quad \textbf{(10.3)}
$$
$$
+ \text{descent}\left(time_{descent}\right) + \text{land}\left(time_{land}\right) + \text{return taxi}\left(time_{return}\right)
$$

■ **FIGURE 10.4**
Typical transport aircraft mission profile

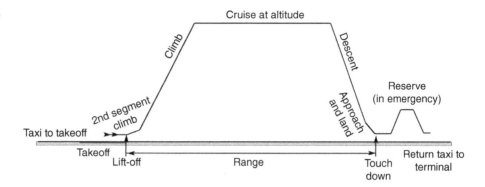

Fuel is consumed during taxiing, takeoff and landing without any range contribution, and this fuel is to be added to the mission fuel, with the total known as *block fuel*. This additional fuel burn and time consumed without contributing to range are taken from the statistics of operation. Minimum reserve fuel quantity depends on the type of mission, mainly on range. It has fuel for holding around the destination airfield plus diversion in case of emergency demands. These are dealt with in Chapter 14. The time taken from the start of the engine at the beginning of the mission to stop of the engine at the end of the mission is known as the *block time*, in which a small part of time is not contributing to a gain in range. Note that there is no credit for distance covered during taxi, takeoff, landing and return taxi. All segments are in a straight line (in a vertical plane), hence the question of *block range* does not arise.

The mission segments as given in Equations 10.1–10.3 can be conveniently grouped as follows:

Field performance (Chapter 11):	(i) Takeoff up to second segment climb
	(ii) Approach and landing
Climb and descent (Chapter 12):	(i) En-route climb to initial cruise altitude
	(ii) Descent to approach from final cruise altitude
Cruise (Chapter 13):	(i) At constant altitude
	(ii) At varying altitude
Overall mission analyses (Chapter 14):	Totals the segment performances.

In addition, manoeuvre and other flight performances are considered in Chapter 15. Aircraft sizing exercises are carried out early during the conceptual stage of the project, but it has to be dealt with at the end in Chapter 16 after learning the equations of performance analyses treated in Chapters 11–15.

10.4   Military Aircraft Mission

Military aircraft operational roles are extremely varied. A preliminary classification of military aircraft is given in Figure 10.1, consisting of fighters, bombers, reconnaissance, transport aircraft, and so on. From time to time, depending on the perceived combat, the mission requirements need specialist roles; nevertheless it has to be borne in mind that there is considerable overlap in the functional capabilities between the roles. The fighter subdivision has many classifications. A multi-role large combat aircraft (F14, 33,000 kg) can be used in air-to-air combat as well as for precision bombing of specific targets, for example, enemy radar stations. On the other hand, an air defence role (F16, 16,500 kg) calls for light, agile aircraft, mostly in defence roles to destroy enemy aircraft. A heavy bomber aircraft like the B52 operates as a strategic bomber with few high *g* manoeuvres. The modern B2 bomber has stealth features to penetrate deep into enemy territory, but not much is known about its all-round capabilities. Given below are the typical terminologies for various kinds of combat aircraft in use. The combat categories use advanced technologies.

Air superiority: its role is to prevent enemy aircraft retaliation over the battlefield in enemy territory, so that ground attack aircraft can carry out their tasks to disable adversaries. The aircraft should be very agile to carry out air-to-air combat in BVR capability. As it has to fly longer distances into enemy territory, and loiter in the vicinity in preparedness, it is a relatively heavier aircraft.

Air defence: its role is to prevent enemy aircraft gaining any superiority of home sky. It has to outmanoeuvre the best adversary. Air defence aircraft are smaller, lighter, very agile and primarily meant for air-to-air combat with BVR capabilities. It requires rapid response.

Ground attack aircraft: caters for the tactical and other specific requirements on the battlefield. It is capable of a close air support (CAS) role (see below).

CAS: Air to ground support (gun/missile/light bombs) on the battlefield. It is a relatively lighter aircraft, highly manoeuvrable, and may be slower than the ground attack type above. Rapid-fire gunships are variants of CAS.

Interdiction: carries heavier bombs (JDAMs), capable of precision bombing on the battlefield. It has deep penetration capability into enemy territory.

Multi-role fighter: heavier aircraft capable of performing a variety of combat roles, for example, air superiority, air defence, ground attack and interdiction.

Advanced tactical support (ATF aircraft): the F22 aircraft clearly illustrates the long-range air superiority mission, which was envisaged for penetrating deep into enemy airspace to destroy enemy air defense aircraft and to disrupt offensive air operations. This is advanced tactics with a multi-role capability – hence this new class.

Strategic bombing: carpet bombing (B52 class).

Air-to-air refuelling: larger tanker aircraft for mid-air refuelling (K135 type).

Maritime patrol: has a special role to cover threats from oceans (anti-submarine role, etc.). In addition they do surveillance and patrolling with long endurance flying.

Reconnaissance: Very high performance aircraft beyond missile range (SR71, U2). Photographs enemy territory.

Airborne early warning (AEW): it is capable of early detection of threats with long-range sensors.

Electronic warfare (EW): capable of electronic countermeasures. RPVs are playing an increasing role in EW.

Military transport aircraft: to serve the logistic requirements, for example, troop movement, equipment ferrying, and so on (C17 type).

Military pilot training: These are specific types of training aircraft normally through two or three types leading up to advanced combat training, ready for operational conversion. Its single-seat variant can serve in a CAS role.

UAVs/RPVs: These are without onboard pilots, and are increasingly appearing in the battlefield in various roles. This type could one day replace advanced manned combat aircraft.

It is unrealistic to assume that any nation will expend vast sums of money to acquire specific weapons systems without seeing how that expenditure will further national interests.

10.4.1   Military Aircraft Performance Segments

Military aircraft missions differ from those of civil aircraft. Military missions depend on the strategy, and need to be looked into from different angles depending on the different mission roles required. Combat aircraft do not have passengers and the payload has wide variation in armament type to be carried internally and/or externally. Military payloads are released, but this is not an option for civil aircraft; in fact commercial transport aircraft do not carry parachutes.

Military aircraft certification standards are also different from those of civil aircraft. The Regulatory Military Specifications are generally available in the public domain. Training mission and combat mission can be planned in a definite pattern as shown in Figures 10.5 and 10.6, respectively.

10.4.1.1   Mission Profile – Military Advanced Jet Trainer   Figure 10.5 gives a typical advanced jet trainer (AJT) training profile to gain airmanship and navigational skills in an advanced aircraft. The training profile segment breakdown is given in Table 10.1.

Block fuel and block time are computed by adding fuel and time consumed in each segment. However, distance covered is meaningless as the aircraft returns to the base. Armament training practice closely follows a combat mission profile.

10.4.1.2   Mission Profile – Combat Aircraft   Combat missions depend on target range and expected adversary defence capability. The study of a combat mission requires complex analysis by a specialist. Defence organizations conduct these studies, which are understandably kept

■ **FIGURE 10.5**
Normal training configuration sortie profile for an AJT (60 minutes)

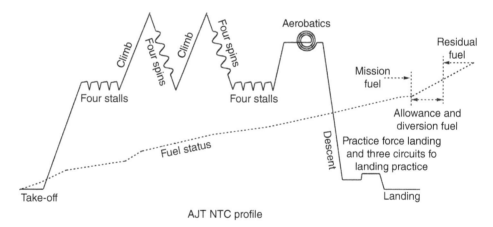

■ **FIGURE 10.6**   **Two typical mission profiles for combat aircraft**

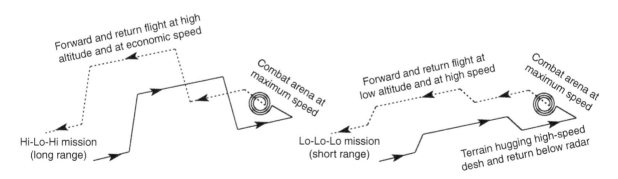

■ **TABLE 10.1**
AJT engine ratings – detailed segments

Segment	Engine rating, % RPM
Taxi and takeoff	60% (idle)
Takeoff and climb to 6 km altitude	TO at 100%, then 95%
Four turns/stalls	1 min at 95% + 3 min at 60%
Climb from 5 to 8 km altitude	95%
Four turn spins	60%
Climb from 5 to 8 km altitude	95%
Four turn spins	60%
Four turns/stalls	1 min at 95% + 3 min at 60%
Climb from 5 to 6 km altitude	95%
Aerobatics practice	95%
Descent and practise force landing	2 min at 95% + 6 min at 60%
Three circuits for landing practice	Average 80%
Approach, land, return taxi	60%
Trainee pilot allowance	95%

■ **TABLE 10.2**
Hi-Lo-Hi mission engine ratings, by segment

Segment	Engine rating, % RPM
Taxi to takeoff	60% (idle)
Takeoff and climb to 12 km altitude	TO at 100%, then 95%
High subsonic cruise to distance to rapid descent	500 nm at 80–85%
Rapid descent	60% (idle)
High-speed dash to combat zone	95% (if required with afterburner)
Combat (extreme manoeuvres)	5 min with afterburner
Rapid climb to 15 km (lighter aircraft)	2 min with afterburner
High-speed return from combat	95%, then reduce to 75% past threat zone
Descent	60% (idle)
Approach and land	60% (idle)

confidential in nature. Game theory, twin dome combat simulations, and so on, are some of the tools for such analysis. Actual combat may yet prove quite different, as not all is known about adversary tactics and capabilities. Detailed study is beyond the scope of this book, and possibly of most academies. However, out of the many types, there are two fundamental typical missions as shown in Figure 10.6.

The Hi-Lo-Hi mission profile segment breakdown is given in Table 10.2. A Lo-Lo-Lo mission profile can be worked out along similar lines. Other types of military aircraft mission profiles can be handled in a similar manner by breaking down the type of segment in use.

10.4.1.3   *Mission Profile – Turboprop Trainer*   Normal TPT training profiles to gain airmanship and navigational skills are identical to Figure 10.5. A training profile segment breakdown is given in Table 10.3. However, there is a difference in their capabilities.

■ **TABLE 10.3**
TPT mission engine ratings, by segment

Segment	Engine rating, % RPM
Taxi and takeoff	60% (idle)
Takeoff and climb to 2000 ft altitude	TO at 100%, then 95%
Four turns/stalls	1 min at 95% + 3 min at 60%
Climb from 2000 to 2500 ft altitude	95%
Four turn spins	60%
Climb from 2000 to 2500 ft altitude	95%
Four turn spins	60%
Four turns/stalls	1 min at 95% + 3 min at 60%
Climb from 2000 to 2500 ft altitude	95%
Aerobatics practice	95%
Descent and practise force landing	2 min at 95% + 6 min at 60%
Three circuits for landing practice	Average 80%
Approach, land, return taxi	60%
Trainee pilot allowance	95%

10.5   Aircraft Flight Envelope

Besides aircraft flying the routine segments of any mission, they have to undergo transient operations. Aircraft transient performance needs to be assessed to make sure that it can execute the operation within the permissible design limits. Such a situation can occur inadvertently, for example encountering a gust, or deliberately, for example, hard manoeuvres such as collision avoidance or evasive manoeuvres under adversary threat. Aircraft have capabilities beyond what mission segments demand. The aircraft flight envelope gives the boundary within which the vehicle should perform. Therefore, flight envelopes of typical civil and military aircraft are outlined below and subsequently dealt with in detail in later chapters.

Figure 10.7a gives a generic depiction of a flight corridor for aircraft in the speed–altitude domain. Each altitude offers high-speed limitation; at low altitude, with higher air density, it is the force limitation arising from an increase in dynamic pressure challenging structural integrity. At higher altitudes the density is lower, allowing higher speed, and this time the temperature rise causes structural degradation.

Next is to examine the flight envelope as described in Chapter 5. Departure from a 1-g flight load can be seen as a manoeuvre. Manufacturers must provide the users with the capability envelope for the aircraft. Aircraft performance engineers estimate the flight envelope represented by the V-n diagram. Figure 10.7b gives the V-n diagram without gust lines. This gives specific aircraft limitations for a particular design within the flight corridor (Figure 10.7a). Here the stall limitation and maximum speed limits for both positive and negative g-loads are shown. The aircraft must fly within the shaded area.

Figure 10.8 gives the speed–altitude capabilities of typical subsonic and supersonic aircraft. Specific excess energy (specific excess power, SEP) is a measure of excess aircraft speed (kinetic energy) that can be converted to potential energy to gain height.

Figure 10.9a compares the speed–altitude capabilities of the three main categories of aircraft. *Agility* means how fast a system can change; in the case of aircraft, manoeuvrability. Turn manoeuvre is more applicable to military aircraft and is part of the MOD specification as early as when a Request for Proposal (RFP) is floated. A typical example of such a capability is given in Figure 10.9b.

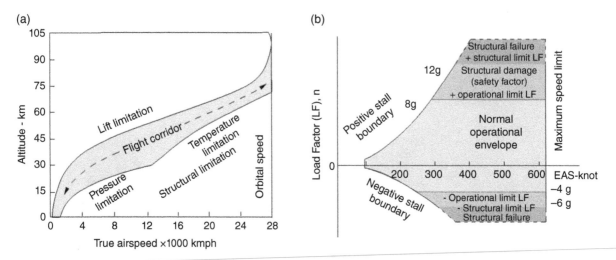

■ **FIGURE 10.7**   Typical flight corridor and envelope. (a) Flight corridor. (b) V–n envelope (typical AJT aircraft)

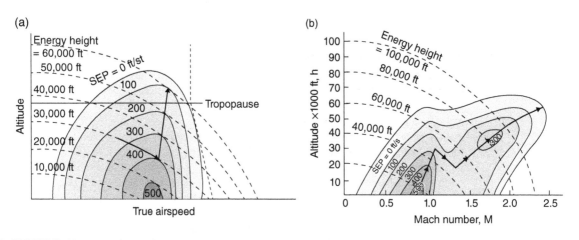

■ **FIGURE 10.8**   (a) Subsonic and (b) supersonic SEP flight envelopes

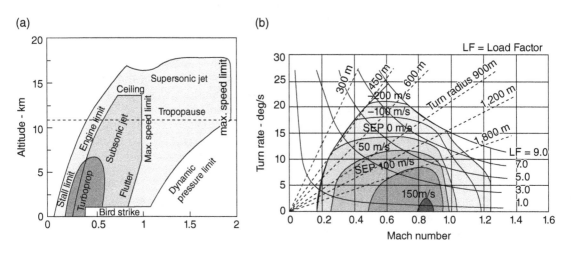

■ **FIGURE 10.9**   (a) Flight envelope and (b) turn capability

10.6   Understanding Drag Polar

Aircraft drag polar is one of the most important pieces of information for evaluating aircraft capabilities. Instead of discussing the properties of drag polar in Chapter 9 dedicated to drag, it is presented here in the context of understanding its characteristics pertaining to aircraft performance. Aircraft drag polar gives the relationship between drag and lift at any instant of flight under consideration. If drag polar could be expressed in a simple analytical form, then it would be easier to obtain close-form solutions rather quickly and avoid the cumbersome computational effort. Engine performance characteristics are not easy to express accurately in analytical form.

10.6.1   Actual Drag Polar

Actual drag polar has the same format as parabolic drag polar, but is not an exact fit to the equation of a parabola, especially at low and high speed ranges. The methodology for estimating actual drag using semi-empirical relations is given in Chapter 9. These are based on experimental data from both wind-tunnel and flight tests. These are the best available drag polar values for industry standard analyses. Drag polar obtained by this methodology is not easily amenable to being represented by *close-form* equations, especially for high subsonic speed aircraft. Generating graphs to fit a parabolic shape may require shaping of graphs with least loss in accuracy. Considerable insight into aircraft performance can be obtained by manipulating parabolic drag polar equations. The industrial methods presented in this book give all the information, but it requires laborious computation effort.

10.6.2   Parabolic Drag Polar

As mentioned earlier, *close-form* solutions from the analytical expressions quickly show important aircraft characteristics and prove useful for trend analyses. Readers may refer to [1–4] to study analytical treatments. At long-range cruise (LRC) when wave drag is simplified to $C_{DW} = 0$ (see Section 9.12), then Equation 9.2 takes the form as follows

$$C_D = C_{Dpmin} + \Delta C_{Dp} + C_L^2 / \pi AR \approx C_{Dpmin} + kC_L^2 \qquad \text{where } k = 1/e\pi AR$$

The difference between the actual aircraft drag polar $\left(C_{Dpmin} + \Delta C_{Dp} + C_L^2 / \pi AR\right)$ and its corresponding parabolic drag polar $\left(C_{Dpmin} + kC_L^2\right)$ is in the approximation of the variable lift-dependent parasite drag ΔC_{Dp} to a constant Oswald's efficiency factor e associated with the induced drag C_{Di}. The variable parameter e is approximated to a constant value, bringing parabolic drag polar close to actual drag polar around the cruise segment, and also giving a good match for the climb and descent segments.

Figure 10.10 shows the variation of Oswald's efficiency factor e with Mach number for a Bizjet aircraft at 18,000 lb weight cruising at 41,000 ft altitude ($C_L \approx 0.512$). It is evaluated using the following equation and plotted in the figure.

$$k = \left(C_D - 0.0205\right) / C_L^2 \qquad \text{where } e = \left(1 / k\pi AR\right)$$

Parabolic drag polar incorporates a value of $k = 0.0447$ ($e \approx 0.95$) representing average values. At the design C_L (typically, at mid-LRC) it approaches to 1. This will cause a discrepancy between parabolic drag polar and actual drag polar values at higher speeds. For close to ground operations, flying at low speeds in a "dirty configuration", extra drag from the high-lift devices will have a deteriorating effect on e. It is for this reason that this book suggests that for high

■ **FIGURE 10.10**
**Oswald's efficiency
factor e variation for
Bizjet at 41,000 ft
altitude (18,000 lb
weight, $C_L \approx 0.512$)**

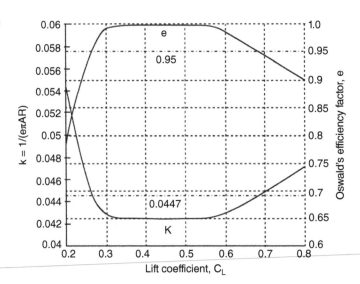

subsonic aircraft the actual drag polar should be used for accurate industry standard results, for example for certification substantiation, preparing the pilot manual, and so on, while parabolic drag polar may be used for exploring and establishing aircraft characteristics.

10.6.3   Comparison between Actual and Parabolic Drag Polar

The parabolic representation of aircraft drag polar offers many advantages. Easy mathematical manipulation gives a quick insight into aircraft characteristics at an early design phase, allowing improvements to be made if required. But a word of caution: performance estimation using approximated parabolic drag polar does not represent the real performance characteristics. The results are not sufficiently accurate to satisfy operational needs, since even 1% discrepancy in drag will show up in operational economics, losing competitive advantage. Industry needs the drag prediction to be as accurate as possible, and therefore results from approximate analytical expressions are not adequate for high subsonic aircraft, especially when a good prediction of drag polar can be obtained through semi-empirical relations substantiated by testing.

This section presents an approximated analytical expression for high subsonic aircraft drag polar. Recall Equation 9.2, which gives the expression for high subsonic aircraft drag as

$$C_D = C_{Dpmin} + \Delta C_{Dp} + C_L^{\,2} / \pi AR + C_{Dw} \qquad (10.4)$$

where $C_{Dpmin} + C_L^{\,2} / \pi AR$ represents a parabola in which C_{Dpmin} is the minimum distance of the drag polar from the Y-axis representing C_L.

Equation 10.4 is not exactly of parabolic shape, depending on the extent contributed by the non-parabolic component of ΔC_{Dp} and the wave drag term C_{Dw}. In an incompressible flow regime, C_{Dw} drops out and ΔC_{Dp} is integrated with the induced drag term with the suitable coefficient $k = 1 / e\pi\,AR$ (where e is the Oswald's efficiency factor) to represent the equation of a parabola as discussed above. A carefully designed aircraft can have $\Delta C_{Dp} \approx 0$ at cruise C_L. The simplification brings Equation 10.4 to a simpler form as in Equation 10.5, making it easier to handle:

$$C_D = C_{Dpmin} + kC_L^{\,2} \qquad (10.5)$$

■ FIGURE 10.11
**Analytical
(parabolic) and
semi-empirical drag
polar comparison**

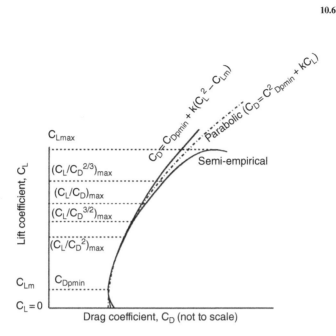

This form of representation can be applied to high-speed subsonic aircraft up to LRC, that is, up to M_{crit}. When the parabolic part of the C_D is plotted against C_L^2, it is a straight line (Figure 10.14b).

Equation 10.5 can be improved by modifying to a more accurate form as shown in Equation 10.6 (plotted in Figure 10.11).

$$C_D = C_{Dpmin} + k\left(C_L - C_{Lm}\right)^2 \tag{10.6}$$

where C_{Dpmin} is at C_{Lm} and not at $C_L = 0$.

In the generalized situation of high-speed subsonic aircraft, C_{Dpmin} is not necessarily at $C_L = 0$. Figure 10.11 typically represents Equations 10.5 and 10.6, with drag polar estimated by the semi-empirical method as in Table 9.16. The various C_L points shown in the graph appear in analytical equations derived in the subsequent sections using the parabolic drag Equation 10.5.

The most generalized form of aircraft drag polar can be expressed in a polynomial expression as given in Equation 10.7:

$$C_D = C_{Dpmin} + k_1 C_L + k_2 C_L^2 + k_3 C_L^3 + k_4 C_L^4 + \ldots \tag{10.7}$$

For high subsonic aircraft, the terms above $k_3 C_L^3$ contribute very little and can be ignored. Even then the polynomial form is not amenable to an easy close-form solution. Supersonic drag expression can be dealt with using a similar rationale.

Figure 10.15 later in this chapter compares actual drag polar with parabolic drag polar for the three classwork examples: the Bizjet, the TPT and the AJT. It may be noted that the three graphs are close enough within the operating range below M_{crit}. Equation 10.5 represents typical high-speed subsonic aircraft operational segments of LRC, en-route climb and descent. The TPT flies in the incompressible flow region, and parabolic drag polar can be used to get good results. At HRC the non-linear effects show up. The Bizjet and AJT operate in high subsonic at M_{crit}, and therefore semi-empirically determined actual drag polar is preferred. This is the standard procedure as practised in industry, which gives credible output as well as offering close-form solutions. In the Bizjet example, Section 10.10 gives more details (see Figure 10.18a,b) on comparing parabolic drag polar with actual drag polar.

10.7   Properties of Parabolic Drag Polar

In equilibrium flight, thrust developed by the engine is adjusted to equate with aircraft drag. Gas turbines produce thrust in two types of engines: turbofans and turboprops. Turbofans produce thrust directly, but turboprop thrust is obtained by extracting turbine power through the propeller, and therefore the related equations require additional manipulation by incorporating propeller characteristics. It is convenient to treat turbofans first, and then the same equations can be used to proceed with turboprop powered aircraft. (Note that piston-engine aircraft can use the turboprop related derivations.)

At any instant of aircraft operation, the relationship between its drag (thrust required) and available engine thrust are of importance. The graphs to show thrust required and thrust available are presented non-dimensionally in a defined manner called an *aircraft and engine grid*, considered in Section 10.10. Aircraft equations of motion have drag and thrust terms, and together they give aircraft performance capabilities.

All the equations derived in this subsection using parabolic drag polar are applicable to both turbofan and turboprop aircraft. In the next subsection, additional relations for turboprop aircraft are separately treated, incorporating the parameters of the installed propeller.

The expression for parabolic drag polar as given in Equation 10.5 is

$$C_D = C_{Dpmin} + C_L{}^2 / e\pi AR = C_{Dpmin} + kC_L{}^2$$

This gives

$$\frac{dC_D}{dC_L} = 2kC_L \quad \text{where } k = 1/e\pi AR \tag{10.8}$$

10.7.1   The Maximum and Minimum Conditions Applicable to Parabolic Drag Polar

Aircraft lift and drag are interrelated, being design dependent. It is important to examine aircraft design characteristics early enough before programme go-ahead is given so that changes may be incorporated to improve design, if required. This subsection derives analytical relationships between aircraft lift and drag, highlighting the main performance characteristics of all types of aircraft. It has 12 main relationships and can get cluttered. The summary at the end of this section will help to take an overall view.

10.7.1.1   Velocity at Minimum Drag, V_{Dmin}   Equation 10.5 can be written as:

$$D = 0.5\rho V^2 S_W C_D = 0.5\rho V^2 S_W C_{DPmin} + 0.5\rho V^2 S_W kC_L{}^2$$
$$= 0.5\rho V^2 S_W C_{DPmin} + kWC_L \tag{10.9}$$

At equilibrium flight, $W = L = 0.5\rho V^2 S_W C_L$ or $C_L = 2\left(\dfrac{W^2}{\rho V^2 S_W}\right)$.

To evaluate the condition for minimum drag, differentiate Equation 10.9 with respect to V and set equal to 0:

$$\frac{dD}{dV} = 0 = \rho V S_W C_{DPmin} + \frac{2d\left(\dfrac{kW^2}{\rho V^2 S_W}\right)}{dV} \tag{10.10}$$

For a particular aircraft, S_W is constant, and when flying at a particular weight (W) and altitude, W and ρ are held constant. Then the above equation reduces to the following form:

$$\rho V S_W C_{DPmin} = \frac{4kW^2}{\rho V^3 S_W} \quad \text{or} \quad V^4 = \frac{4kW^2}{\rho^2 S_W{}^2 C_{DPmin}} \quad \text{or} \quad V^2 = \left(\frac{2W}{\rho S_W}\right)\left(\frac{\sqrt{k}}{\sqrt{C_{DPmin}}}\right)$$

Therefore velocity at minimum drag, $V_{Dmin} = \sqrt{\left(\frac{2W}{\rho S_W}\right)}\left(\frac{k}{C_{DPmin}}\right)^{0.25}$ **(10.11)**

10.7.1.2   *Velocity at Minimum D/V*   Equation 10.9 can be written as

$$D/V = 0.5\rho V S_W C_D = 0.5\rho V S_W C_{DPmin} + k\left(0.5\rho V S_W C_L\right)^2$$

Substituting for $C_L{}^2$, it becomes

$$D/V = 0.5\rho V S_W C_{DPmin} + k\left(\frac{2W^2}{\rho V^3 S_W}\right) \quad\quad\quad \textbf{(10.12)}$$

Differentiating for the minimum condition,

$$\frac{d\left(D/V\right)}{dV} = 0 = 0.5\rho S_W C_{DPmin} - \left(\frac{8kW^2}{\rho S_W V^4}\right)$$

or $0.5\rho S_W C_{DPmin} = \left(\frac{8kW^2}{\rho S_W V^4}\right) \quad \text{or} \quad V^4 = \left(\frac{16kW^2}{\rho^2 S_W{}^2 C_{DPmin}}\right)$

or $V_{@(D/V)min} = \sqrt{\frac{2W}{\rho S_W}}\left(\frac{k}{C_{DPmin}}\right)^{0.25} \quad\quad\quad \textbf{(10.13)}$

Note that this is identical to Equation 10.11: $V_{Dmin} = V_{@(D/V)min} = $ velocity at minimum D/V.

Below in Section 10.7.1.5 it is shown that V_{Dmin} and $V_{@(D/V)min}$ as well as the velocity at maximum L/D are identical. Section 13.7 derives that the best endurance occurs when an aircraft flies at $(L/D)_{max}$, that is, at V_{Dmin}. Velocity at minimum D/V gives the maximum range (see Chapter 12).

10.7.1.3   *Conditions for Maximum Lift to Drag Ratio, $(C_L/C_D)_{max}$*   Differentiating the general expression C_L/C_D with respect to C_L, and setting it equal to zero gives $(C_L/C_D)_{max}$ as follows:

$$d\left(C_L/C_D\right)/dC_L = 0 = \frac{C_D - C_L\left(\dfrac{dC_D}{dC_L}\right)}{C_D{}^2} \quad\quad\quad \textbf{(10.14)}$$

in which drag C_D can never be zero. Therefore

$$C_L\left(\frac{dC_D}{dC_L}\right) = C_D \quad \text{or} \quad \left(\frac{dC_D}{dC_L}\right) = \frac{C_D}{C_L} \quad\quad\quad \textbf{(10.15)}$$

Substituting Equation 10.8 the above equation reduces to

$$\frac{dC_D}{dC_L} = 2kC_L = \frac{C_D}{C_L}\frac{\left(C_{DPmin} + kC_L^2\right)}{C_L}$$

or
$$2kC_L^2 = C_{DPmin} + kC_L^2 \tag{10.16a}$$

or
$$kC_L^2 = C_{DPmin} \tag{10.16b}$$

This gives $C_{Di} = C_{DPmin}$ $\left(\text{since } kC_L^2 = C_{Di}\right)$. Equation 10.16a is the condition for $(C_L/C_D)_{max}$. Substituting in Equation 10.5:

$$C_{D@L/Dmax} = C_{DPmin} + C_L^2 / e\pi AR = C_{DPmin} + C_{DPmin} = 2C_{DPmin} \quad \text{at} \left(C_L / C_D\right)_{max} \tag{10.17a}$$

or

$$C_{L@L/Dmax} = \sqrt{\left(C_{DPmin}/k\right)} \quad \text{where } k = 1/e\pi AR \tag{10.17b}$$

10.7.1.4   Expression for $(C_L/C_D)_{max}$ Using Equations 10.16a and 10.16b, $(C_L/C_D)_{max}$ can be written as

$$\left(\frac{C_L}{C_D}\right)_{max} = \frac{\sqrt{\frac{C_{DPmin}}{k}}}{2C_{DPmin}} = \frac{1}{2}\sqrt{\frac{1}{kC_{DPmin}}} = \frac{1}{2}\sqrt{\frac{e\pi AR}{C_{DPmin}}} = \frac{1}{2kC_L} \tag{10.18}$$

10.7.1.5   Velocity V_{Dmin} at Minimum Drag (Occurs at $(C_L/C_D)_{max}$)   Equation 10.17b gives

$$C_L = \sqrt{\left(C_{DPmin}/k\right)} \quad \text{at} \left(C_L/C_D\right)_{max}$$

Substituting $C_L = \frac{2W}{\rho S_W V^2}$, the above equation can be written as $\frac{2W}{\rho S_W V^2} = \sqrt{\frac{C_{DPmin}}{k}}$

or
$$V_D^2 = \frac{2 \times W \times \sqrt{k}}{\rho \times S_W \times \sqrt{C_{DPmin}}}$$

or
$$V_{D@L/Dmax} = \sqrt{\left(\frac{2W}{\rho S_W}\right)}\left(\frac{k}{C_{DPmin}}\right)^{0.25} = V_{Dmin} \tag{10.19}$$

This is identical to Equation 10.11.

Section 13.7 derives that the best endurance occurs when the aircraft flies at V_{Dmin} giving $(L/D)_{max}$.

Equation 10.16a can be derived in another way using Equation 10.18:

$$\left(\frac{C_L}{C_D}\right)_{max} = \frac{\sqrt{\frac{C_{DPmin}}{k}}}{2C_{DPmin}} = \frac{1}{2}\sqrt{\frac{1}{kC_{DPmin}}} = \frac{1}{2}\sqrt{\frac{e\pi AR}{C_{DPmin}}} = \frac{1}{2kC_L} = \frac{e\pi AR \times \rho V_{Dmin}^2 S_W}{2 \times 2W}$$

$$\text{or } V_{Dmin} = \sqrt{\frac{4W\left(\dfrac{C_L}{C_D}\right)_{max}}{e\pi AR\rho \times S_W}} = \sqrt{\frac{2W}{\rho S_W}}\sqrt{\frac{2\left(\dfrac{C_L}{C_D}\right)_{max}}{e\pi AR}} = \sqrt{\frac{2W}{\rho S_W}}\sqrt{2\frac{1}{2}\sqrt{\frac{e\pi AR}{C_{DPmin}}}}{e\pi AR}}$$

$$= \sqrt{\frac{2\times W}{\rho S_W}}\left(\frac{1}{e\pi AR C_{DPmin}}\right)^{0.25} \tag{10.20}$$

10.7.1.6   *Drag at $(C_L/C_D)_{max}$*   [Note: drag, $D = 0.5\rho V^2 S_W C_D$.]

At $(C_L/C_D)_{max}$, Equation 10.16b gives

$$C_{Di} = C_{DPmin}$$

Since the velocity at minimum drag D_{min} and at $(C_L/C_D)_{max}$ are the same, it can be inferred that D_{min} occurs at $(C_L/C_D)_{max}$ (see Figure 10.14b).

Equation 10.17a gives

$$C_{D@L/Dmax} = C_{DPmin} + C_L^2/e\pi AR = C_{DPmin} + C_{DPmin} = 2C_{DPmin}$$

Multiplying both sides of Equation 10.17a by $(0.5\times\rho\times(V_{Dmin})^2\times S_W)$ gives

$$D_{min} = 2\times 0.5\times\rho\times(V_{Dmin})^2\times S_W\times C_{DPmin}$$

Substituting the expression for V_{Dmin} from Equation 10.11,

$$D_{min} = 2\times 0.5\times\rho\times\sqrt{\frac{2\times W}{\rho S_W}}\left(\frac{k}{C_{DPmin}}\right)^{1/4}\times S_W\times C_{DPmin}$$

$$\text{or } D_{min} = D_{@(L/D)max} = 2\times D_{Pmin} = \rho\times\frac{2W}{\rho S_W}\left(\frac{k}{C_{DPmin}}\right)^{1/2}\times S_W\times C_{DPmin} \tag{10.21a}$$

$$= 2W\left(\frac{C_{DPmin}}{e\pi AR}\right)^{1/2} = 2W\sqrt{kC_{DPmin}}$$

10.7.1.7   *C_L at Maximum Lift/Drag Ratio*

$$C_L \text{ at } V_{Dmin} = C_{L@VDmin} = \frac{2W}{\rho V_{Dmin}^2 S_W} \tag{10.21b}$$

Equation 10.11 gives V_{Dmin}.

10.7.1.8   *Conditions for Maximum $\dfrac{C_L^3}{C_D^2}$*   Conditions for maximum drag $\dfrac{C_L^3}{C_D^2}$ are derived here. Differentiating $\dfrac{C_D^2}{C_L^3}$ with respect to C_L and setting it equal to zero, the velocity at maximum value:

$$\frac{d\left(\dfrac{C_D^2}{C_L^3}\right)}{dC_L} = 0 = \frac{3C_D^2 C_L^2 - 2C_L^3 C_D\left(\dfrac{dC_D}{dC_L}\right)}{C_D^4}$$

or
$$2C_L^{3}C_D\left(\frac{dC_D}{dC_L}\right)=3C_D^{2}C_L^{2}$$

The condition for $\left(\dfrac{C_D^{2}}{C_L^{3}}\right)_{max}$ gives $\left(\dfrac{dC_D}{dC_L}\right)=\dfrac{3C_D}{2C_L}$ **(10.22)**

Using Equation 10.8 for parabolic drag polar and equating to Equation 10.22, the following is obtained:

$$\left(\frac{dC_D}{dC_L}\right)=\frac{d\left(C_{DPmin}+kC_L^{2}\right)}{dC_L}=2kC_L=\frac{3C_D}{2C_L}$$ **(10.23)**

Expanding Equation 10.23,

$$4kC_L^{2}=3\times\left(C_{DPmin}+kC_L^{2}\right)$$

or
$$C_L=\sqrt{(3/k)C_{DPmin}}$$ **(10.24)**

Substituting the condition as obtained in Equation 10.24 into Equation 10.5, the following is obtained:

$$C_D=C_{DPmin}+kC_L^{2}=C_{DPmin}+3C_{DPmin}=4C_{DPmin}$$ **(10.25)**

Using Equations 10.23 and 10.24,

$$\left(\frac{C_L^{3}}{C_D^{2}}\right)_{max}=\frac{\left(3C_{DPmin}/k\right)^{3.5}}{16C_{DPmin}^{2}}=\frac{\left(3C_{DPmin}/k\right)\sqrt{3C_{DPmin}/k}}{16C_{DPmin}^{2}}$$

or
$$\left(\frac{C_L^{3}}{C_D^{2}}\right)_{max}=\left(\frac{3\sqrt{3}}{16k}\right)\sqrt{\frac{1}{kC_{DPmin}}}$$ **(10.26)**

10.7.1.9 *Conditions for Maximum* $\dfrac{C_L}{C_D^{2}}$ Chapter 12 uses $\left(\dfrac{C_L}{C_D^{2}}\right)_{max}$. For convenience

the condition for maximum $\dfrac{C_L}{C_D^{2}}$ is derived here.

Differentiating $\dfrac{C_L}{C_D^{2}}$ with respect to C_L and setting it equal to zero, the following is obtained:

$$\frac{d\left(\frac{C_L}{C_D^{2}}\right)}{dC_L}=\frac{C_D^{2}-2C_LC_D\left(\frac{dC_D}{dC_L}\right)}{C_D^{4}}=0$$

or
$$\frac{dC_D}{dC_L}=\frac{C_D}{2C_L}\text{ as the condition for }\left(\frac{C_L}{C_D^{2}}\right)_{max}$$ **(10.27)**

Substituting Equation 10.27 into the parabolic drag polar of Equation 10.9, the following is obtained

$$\frac{dC_D}{dC_L} = \frac{d\left(C_{DPmin} + kC_L^{2}\right)}{dC_L} = 2kC_L = \frac{C_D}{2C_L} = \frac{C_{DPmin} + kC_L^{2}}{2C_L}$$

or

$$4kC_L^{2} = C_{DPmin} + kC_L^{2}$$

or

$$C_L^{2} = C_{DPmin}/3k \quad \text{or} \quad C_L^{2} = C_{DPmin}/3k \tag{10.28a}$$

or

$$C_L = \sqrt{(C_{DPmin}/3k)} \quad \text{(same as Equation 10.24)} \tag{10.28b}$$

Substituting Equation 10.28a into parabolic drag polar Equation 10.9 to obtain the condition for $\left(\dfrac{C_L}{C_D^{2}}\right)_{max}$:

$$C_D = C_{DPmin} + kC_L^{2} = C_{Dpmin} + C_{DPmin}/3 = (4/3)C_{DPmin} \quad \text{and} \quad kC_L^{2} = C_{DPmin}/3 \tag{10.29}$$

or

$$C_L = \sqrt{(C_{DPmin}/3k)} \quad \text{(same as Equation 10.28b)} \tag{10.30}$$

Therefore using Equations 10.29 and 10.30,

$$\left(\frac{C_L}{C_D^{2}}\right)_{max} = \frac{\sqrt{\left(\dfrac{C_{DPmin}}{3k}\right)}}{\dfrac{16}{9}C_{DPmin}^{2}} \tag{10.31}$$

Note: The same expression of velocity $V = \sqrt{\dfrac{2W}{\rho S_W}} \left(\dfrac{1}{e\pi ARC_{DPmin}}\right)^{0.25}$ can be derived in various ways, stating that it occurs at the minimum drag, D_{min}, that is, at minimum $(D/V)_{min}$ which is at $(L/D)_{max}$. Chapter 12 shows that the maximum range is achieved at this velocity.

10.7.1.10   *Summary of the Sections on Parabolic Drag Polar*   Table 10.4 shows a summary of the equations derived in the parabolic drag polar sections above, which are applicable to both turbofans and turboprops.

10.7.2   Propeller-Driven Aircraft

The equations common to both turbofans and turboprops have been derived in the previous section. As mentioned earlier, turboprop thrust is obtained by extracting turbine power through the propeller, and therefore the related equations require additional manipulation incorporating the propeller characteristics to obtain engine power. This subsection is devoted to deriving the expressions for the power required by the turboprops to produce the required thrust. All derivations in this subsection are also applicable to propeller-driven piston engine aircraft.

To develop thrust for equilibrium flight, $(T_p)_{reqd}$ is the thrust power required from the turboprop engine. It is power $(ESHP)_{reqd}$ and not thrust force. Thrust is obtained by taking into account the propeller efficiency, η_{prop} (propeller characteristic) as given in Equations 7.34 and 7.35.

■ **TABLE 10.4**
Summary of the parabolic drag polar equations

Property	Equation	Equation no.
Velocity at minimum drag, V_{Dmin}	$V_{Dmin} = \sqrt{\left(\dfrac{2W}{\rho S_W}\right)\left(\dfrac{k}{C_{DPmin}}\right)^{0.25}}$	10.11
Velocity at minimum D/V	Same as Equation 10.11	
Conditions for maximum lift to drag ratio $(C_L/C_D)_{max}$	$C_{DPmin} = C_{Di}$	10.16a
	$C_D = 2C_{DPmin}$ at $\left(\dfrac{C_L}{C_D}\right)_{max}$	10.17a
	$C_L = \sqrt{(C_{DPmin}/k)}$	10.17b
Expression for $(C_L/C_D)_{max}$	$\left(\dfrac{C_L}{C_D}\right)_{max} = \dfrac{1}{2kC_L}$	10.18
Velocity V_{Dmin} is at minimum drag (occurs at $(C_L/C_D)_{max}$)	$V_{Dmin} = \sqrt{\dfrac{2 \times W}{\rho S_W}\left(\dfrac{1}{e\pi ARC_{DPmin}}\right)^{0.25}}$ (same as Equation 10.11 but derived differently)	10.20
$D_{(L/D)max}$ at $(C_L/C_D)_{max}$	$D_{min} = 2W\left(\dfrac{C_{DPmin}}{e\pi AR}\right)^{1/2}$	10.21a
C_L at maximum lift/drag ratio $(L/D)_{max}$	C_L at $V_{Dmin} = \dfrac{2W}{\rho V_{Dmin}^{2} S_W}$	10.21b
	$C_L = \sqrt{(3/k)C_{DPmin}}$	10.24
Conditions for maximum $\dfrac{C_L^{3}}{C_D^{2}}$	$\left(\dfrac{C_L^{3}}{C_D^{2}}\right)_{max} = \left(\dfrac{3\sqrt{3}}{16k}\right)\sqrt{\dfrac{1}{kC_{DPmin}}}$	10.26
$C_L^{2} = C_{DPmin}/3k$	or $C_{DPmin} = 3kC_L^{2}$	10.28a
Conditions for maximum $\dfrac{C_L}{C_D^{2}}$	$\left(\dfrac{C_L}{C_D^{2}}\right)_{max} = \dfrac{\sqrt{\dfrac{C_{DPmin}}{3k}}}{\dfrac{16}{9}C_{DPmin}^{2}}$	10.31

Turboprop cruise equations also use the terms $\dfrac{C_L^{3}}{C_D^{2}}$ and $\left(\dfrac{C_L}{C_D^{2}}\right)_{max}$ (Equations 10.26 and 10.27). The subsection has four relationships as derived below, and is summarized at the end of this section.

10.7.2.1   Power Required In equilibrium level flight, power required by propeller driven aircraft is expressed as:

$$\text{Drag}\,(D) \times \text{aircraft velocity}\,(V) = DV = (T_P)_{reqd} \tag{10.32}$$

or

$$(T_P)_{reqd} = DV = \left(\frac{DW}{L}\right)V = W\left(\frac{D}{L}\right)\sqrt{\frac{2W}{(\rho \times S_W \times C_L)}} = W\sqrt{\frac{2W \times C_D^{2}}{(\rho \times S_W \times C_L^{3})}} \tag{10.33}$$

Equation 10.33 is valid for both turboprop and piston engine applications. In imperial units, turboprop is in shaft horse power (SHP) and piston engine is in horse power (HP).

Power generated by turboprop engines to produce the propeller thrust is related by Equation 7.20 incorporating propeller efficiency η_{prop} as

$$(\text{ESHP})_{reqd} = (T_P)_{reqd} / \eta_{prop} \tag{10.34}$$

At steady level flight, $L = W$ (note $k = (1/e\pi\text{AR})$), Equation 10.34 reduces to:

$$
\begin{aligned}
(\text{ESHP})_{reqd} &= DV / \eta_{prop} = \left(0.5 \times \rho \times V^2 \times S_W \times C_D \right) V / \eta_{prop} \\
&= \left[0.5 \times \rho \times V^2 \times S_W \times \left(C_{DPmin} + C_L^2 k \right) \right] V / \eta_{prop} \\
&= 0.5 \times \rho \times V^3 \times S_W \times \left(C_{DPmin} \right) / \eta_{prop} + k \times 0.5 \times \rho \times V^3 \times S_W \times C_L^2 / \eta_{prop}
\end{aligned}
\tag{10.35}
$$

For a particular aircraft, C_{DPmin}, k and η_{prop} are constant, which makes the above equation:

$$(\text{ESHP})_{reqd} = k_1 \times \rho \times V^3 \times S_W \times \left(C_{DPmin} \right) + k_2 \times V^3 \times S_W \times C_L^2 \tag{10.36}$$

where $k_1 = 0.5/\eta_{prop}$ and $k_2 = (k \times 0.5 \times \rho)/\eta_{prop}$.

10.7.2.2 *Minimum Power $(\text{ESHP})_{min\_reqd}$ and Velocity at Minimum Power V_{Pmin}*

At constant altitude, $(\text{ESHP})_{reqd}$ is a function of velocity. Differentiating $(\text{ESHP})_{reqd}$ with respect to V and setting it equal to 0, the velocity for minimum power, V_{Pmin}, can be obtained.

$$\text{d}(\text{ESHP})_{reqd}/\text{d}V = 3k_1 \times \rho \times V^2 \times S_W \times \left(C_{DPmin} \right) - k_2 \times W^2 / \left(\rho \times S_W \times V^2 \right) = 0$$

or

$$3k_1 \times \rho \times V^2 \times S_W \times \left(C_{DPmin} \right) = k_2 \times W^2 / \left(\rho \times S_W \times V^2 \right)$$

or

$$V^4 = k \times W^2 / \left(3\rho^2 \times S_W^2 \times C_{DPmin} \right) \quad \left[\text{note that } \eta_{prop} \text{ cancels out} \right]$$

or

$$V_{Pmin} = \left(\frac{k}{3C_{DPmin}} \right)^{1/4} \sqrt{\frac{2W}{\rho S_W}} = \left(\frac{1}{3e\pi\text{AR}C_{DPmin}} \right)^{1/4} \sqrt{\frac{2W}{\rho S_W}} \tag{10.37}$$

Substituting Equation 10.37 into Equation 10.35, the minimum turboprop power can be found.

$$
\begin{aligned}
(\text{ESHP})_{min\_reqd} &= \left(0.5 \times \rho \times V_{min}^3 \times S_W \times C_D \right) / \eta_{prop} \\
&= \left(0.5 / \eta_{prop} \right) \times \rho \times V_{Pmin}^3 \times S_W \times C_D \\
&= \left(1 / \eta_{prop} \right) \times \rho \times V_{Pmin}^3 \times S_W \times C_{DPmin} \\
&= k_1 \times \rho \times \left(\frac{k_3}{3C_{DPmin}} \right)^{3/4} \left(\frac{2W}{\rho S_W} \right)^{3/2} \times S_W \times C_{DPmin} \\
&= k_1 \times 2W \sqrt{\frac{2W}{\rho S_W}} \times \left(\frac{k_3}{3C_{DPmin}} \right)^{3/4} \times C_{DPmin} \\
&= k_1 \times \left(\frac{1}{3e\pi\text{AR}} \right)^{3/4} 2W \sqrt{\frac{2W}{\rho S_W}} \times \left(C_{DPmin} \right)^{1/4}
\end{aligned}
\tag{10.38}
$$

$$= \left(0.5/\eta_{prop}\right) \times k_1 \times \left(\frac{1}{3}\right)^{3/4} 2W \sqrt{\frac{2W}{\rho S_W}} \times \left(\frac{C_{DPmin}}{\left(e\pi AR\right)^3}\right)^{1/4}$$

$$= \frac{4}{3\eta_{prop}} \times W \sqrt{\frac{2W}{\rho S_W}} \times \left(\frac{3C_{DPmin}}{\left(e\pi AR\right)^3}\right)^{1/4}$$

(10.39)

10.7.2.3 *Turboprop Velocity at V_{Dmin}* From parabolic drag polar relations, Equation 10.11 gives

$$V_{Dmin} = \sqrt{\frac{2W}{\rho S_W}} \left(\frac{k}{C_{DPmin}}\right)^{1/4} = \sqrt{\frac{2W}{\rho S_W}} \left(\frac{1}{e\pi ARC_{DPmin}}\right)^{1/4} \quad \text{occurs at } \left(L/D\right)_{max}$$

Equation 10.20 gives

$$V_{Dmin} = \left(\frac{1}{3e\pi ARC_{DPmin}}\right)^{1/4} \sqrt{\frac{2W}{\rho S_W}} = \left(\frac{1}{3}\right)^{1/4} V_{Pmin}$$

(10.40)

10.7.2.4 *Condition for V_{Dmin}* Equation 10.32 gives $(T_P)_{reqd} = DV.V_{Dmin} = (T_P)_{reqd}/D$ gives the minimum value. Figure 10.17a shows that this occurs by drawing a tangent from the origin to the $(T_P)_{reqd}$ curve.

10.7.2.5 *Summary of Expressions Specific to Propeller-Driven Aircraft*
Table 10.5 shows a summary of the equations derived in the sections above, which are specific to propeller-driven aircraft.

■ **TABLE 10.5**
Summary of equations specific to propeller-driven aircraft

Property	Equation	Equation no.
Power required, $(T_P)_{reqd}$	$(T_P)_{reqd} = DV = W\sqrt{\dfrac{2W \times C_D^2}{\left(\rho \times S_W \times C_L^3\right)}}$	10.33
	$(ESHP)_{reqd} = 0.5 \times \rho \times V^3 \times S_W \times \left(C_{DPmin}\right)/\eta_{prop}$ $+ k \times 0.5 \times \rho \times V^3 \times S_W \times C_L^2/\eta_{prop}$	10.36
Minimum power $(ESHP)_{min\_reqd}$ and velocity at minimum power V_{Pmin}	$V_{Pmin} = \left(\dfrac{1}{3e\pi ARC_{DPmin}}\right)^{1/4} \sqrt{\dfrac{2W}{\rho S_W}}$	10.37
	$(ESHP)_{min\_reqd} = \dfrac{4}{3\eta_{prop}} \times W \sqrt{\dfrac{2W}{\rho S_W}} \times \left(\dfrac{3C_{DPmin}}{\left(e\pi AR\right)^3}\right)^{1/4}$	10.39
Turboprop velocity at V_{Dmin}	$V_{Dmin} = 3^{1/4} \times V_{Pmin}$ (occurs at $(L/D)_{max}$)	10.40
Condition for V_{Dmin}	$V_{Dmin} = \left(T_P\right)_{reqd}/D$	

The advantages of parabolic drag polar are evident as it offers close-form solutions to bring out aircraft properties quickly. At the conceptual aircraft design phase, it gives good insight into aircraft capability to allow freeze of the design. As more information becomes available through tests and CFD analyses, more accurate actual drag polar is estimated. Actual drag polar requires laborious numerical calculations, as will be compared with parabolic drag polar in Section 10.10.1. The difference between them for low-speed aircraft is negligible.

10.8   Classwork Examples of Parabolic Drag Polar

Sections 9.18, 9.19 and 9.20 gave geometric details of the Bizjet, AJT and TPT respectively. This section gives the other pertinent details for these aircraft as coursework examples. The conceptual designs of the Bizjet and the AJT are carried out in [1]. Chapter 9 worked out the drag (see summary of drag equations of the Bizjet, AJT and TPT given in Section 10.8.4) and Chapter 16 generates the matched installed engine performances for the three aircraft.

10.8.1   Bizjet Market Specifications

The Bizjet C_{DPmin} is evaluated at Mach 0.65 (Section 9.18.1). The LRC (over 1000 nm mission range cruising at higher altitudes) is carried out at Mach 0.7.

The baseline aircraft specifications are as follows (geometry details are given in Section 9.18.1):

- Payload – 10 passengers + baggage = 1100 lb.
- Range = 2000 nm.
- HSC Mach = 0.74, LRC Mach = 0.65–0.7 (higher altitudes).
- Initial climb rate = 16 m/s.
- Initial cruise altitude > 40,000 ft.
- Takeoff field length = 1200 m ($\approx$ 4000 ft).
- Landing distance from 50 ft = 1200 m.
- MTOM = 9400 kg ($\approx$ 20,720 lb).
- OEM = 5800 kg ($\approx$ 12,780 lb).
- Fuel mass = 2500 kg ($\approx$ 5510 lb).

The Bizjet variant configurations and their CAD drawings are shown in Figure 10.12. For the two variants, all component dimensions except the fuselage length are kept invariant compared with the baseline aircraft.

Short variant: fuselage length = 12.71 m (41.71 ft), MTOM_{small} = 7800 kg (17,190 lb).
Long variant: fuselage length = 17.74 m (58.2 ft), MTOM_{long} = 10,900 kg (24,020 lb).

10.8.2   Turboprop Trainer Specifications

Figure 9.18 gives the three-view diagram of the TPT. Given below is the defence specification for which the design was developed [1]. Section 9.20 gives the TPT geometric dimensions and estimates the drag. Drag polar in the form of Equation 10.5 is given as: $C_D = 0.0226 + 0.052 C_L^{\,2}$ (C_{DPmin} = 0.0226, no graph plotted).

■ **FIGURE 10.12**
Short and long Bizjet variants

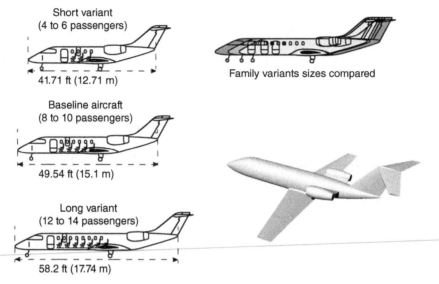

Short variant
(4 to 6 passengers)

41.71 ft (12.71 m)

Family variants sizes compared

Baseline aircraft
(8 to 10 passengers)

49.54 ft (15.1 m)

Long variant
(12 to 14 passengers)

58.2 ft (17.74 m)

The important TPT details are as follows:

- Practice altitude = 25,000 ft ($\rho = 0.00106$ slug/ft^3 and $\mu = 0.3216 \times 10^{-6}$ lb/ft^2).
- Maximum speed = 280 kt.
- Sustained speed = 240 kt.
- Initial climb rate = 40 m/s.
- Initial cruise altitude = 9 km.
- Takeoff distance = 500 m.
- Landing distance = 550 m.
- Manoeuvre: +8 to − 4g (full aerobatics).
- Roll rate: 75°/s at 250 kt.
- Turn performance: 4g at sea-level (at mean NTC weight).
- Range = 1000 nm at 7 km altitude.
- MTOM = 6600 lb (3000 kg).
- NTC-TOM = 5000 lb (2500 kg).
- OEW = 4400 lb (2000 kg).
- Typical mid-training aircraft mass = 5280 lb (2400 kg).
- Crew weight = 400 lb ($\approx$180 kg).
- Fuel weight = 1000 lb ($\approx$450 kg).
- Maximum speed = 280 kt.
- Sustained speed = 240 kt (405.2 ft/s).

Engine specifications are:

- Takeoff static power at ISA sea-level = 1075 SHP.
- Dry weight = 603 kg (1330 lb).
- Fan diameter = 0.56 m (22 in.).
- Length = 1.956 m (77 in.).
- Maximum depth = 1.04 m (3.4 ft).
- Maximum width = 0.75 m (2.46 ft).

10.8.3   Advanced Jet Trainer Specifications

Figure 9.17 gives the three-view diagram of the AJT (of the BAe Hawk trainer class) and its variant CAS aircraft. Section 9.19 gives the important external dimensions of the AJT and CAS aircraft. Given below is the specification for which the design was developed ($C_{DPmin} = 0.0212$):

- Payload = 1800 kg.
- Range = 1200 km at 9 km altitude.
- MTOM = 6500 kg (15,210 lb).
- NTCM = 4800 kg (10,800 lb).
- OEM = 3700 kg.
- Fuel mass = 1100 kg.
- HSC Mach = 0.75, LRC Mach = 0.7.
- Initial climb rate = 40 m/s.
- Initial cruise altitude = 9 km.
- Takeoff distance = 1100 m.
- Landing distance = 1100 m.
- Wing area, $S_W = 17 \, \text{m}^2 (183 \, \text{ft}^2)$.
- Wing loading, $W/S_W = 388.7 \, \text{kg/m}^2 (710.5 \, \text{lb}/\text{ft}^2)$.

The CAS role aircraft is the only variant of the AJT aircraft, and its configuration is achieved by splitting the AJT front fuselage, then replacing the tandem seat arrangement with a single seat cockpit. The length could be kept the same, as the nose cone needs to house more powerful acquisition radar. The front-loading of radar and single pilot is placed in such a way that the centre of gravity (CG) location is kept undisturbed. The CAS variant specifications are as follows:

- Payload = 2500 kg.
- Radius of action = 500 km.
- MTOM = 7500 kg (16,535 lb).
- Clean MTOM = 5000 kg (11,023 lb).
- OEM = 3700 kg.
- Fuel mass = 1300 kg.
- HSC Mach = 0.8, LRC Mach = 0.7.
- Initial climb rate = 50 m/s.
- Initial cruise altitude = 9 km.
- Takeoff distance = 1400 m.
- Landing distance = 1200 m.
- Wing area, $S_W = 17 \, \text{m}^2 (183 \, \text{ft}^2)$.
- Wing loading, $W/S_W = 441.2 \, \text{kg/m}^2 (90.36 \, \text{lb/ft}^2)$.
- Parasite drag increment due to armament, $\Delta C_{DPmin} = (5 \times 0.25 \times 0.8)/183 = 0.0055$.

Armaments and fuel could be traded for range. Drop tanks could be used for ferry range.

The drag level of the clean AJT and CAS aircraft may be considered to be about the same. There would be an increase of drag on account of the weapon load. For the CAS aircraft, there is a wide variety, but to give a general perception, the typical drag coefficient increment for armament load is $C_{D\pi} = 0.25$ (includes interference effect) each for five hard points as weapon carrier. Weapon drag is based on maximum cross-sectional area (say 0.8 ft²) of the weapons.

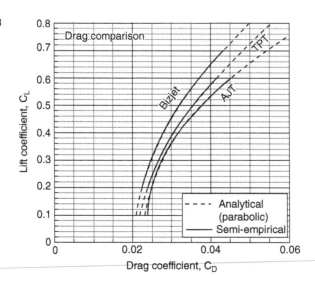

■ **FIGURE 10.13**
Comparison of parabolic and semi-empirical drag polar profiles

10.8.4   Comparison of Drag Polars

The following equations are the parabolic drag polar of the aircraft used as classwork examples, and are compared to the semi-empirically derived drag polars in Figure 10.13.

Bizjet: Oswald's efficiency factor, *e*, is determined in Section 10.6.2, Figure 10.10. Drag may be approximated as parabolic drag polar:

$$C_D = 0.0205 + 0.0447 C_L{}^2 \quad \text{with} \quad S_W = 323 \text{ ft}^2 \qquad \textbf{(10.41)}$$

which gives Oswald's efficiency factor, $e = 0.95$.

AJT: The AJT drag may be approximated as parabolic drag polar as given below:

$$C_D = 0.0212 + 0.068 C_L{}^2 \quad \text{with} \quad S_W = 183 \text{ ft}^2 \qquad \textbf{(10.42)}$$

which gives Oswald's efficiency factor, $e = 0.85$.

Turboprop (TPT): Maximum TPT speed is below Mach 0.5 (incompressible flow). Here, parabolic drag is close enough to actual drag polar:

$$C_D = 0.0226 + 0.052 C_L{}^2 \quad \text{with} \quad S_W = 200 \text{ ft}^2 \qquad \textbf{(10.43)}$$

which gives Oswald's efficiency factor, $e = 0.875$.

10.9   Bizjet Actual Drag Polar

Figure 10.14a is the drag polar of the Bizjet obtained by semi-empirical means (Chapter 9). Figure 10.14b shows the same drag polar plotted in a different manner, with the X-axis as the square of lift coefficient $(C_L{}^2)$ and the Y-axis as the drag coefficient, compared with straight lines which represent the parabolic drag polar.

The comparison indicates that at low and high C_L the semi-empirical drag polar shows non-linearity, with increased drag. This is because at low C_L (i.e. at high speeds) the compressibility effect produces local shock waves as wave drag, C_{Dw}. At high C_L (i.e. at low

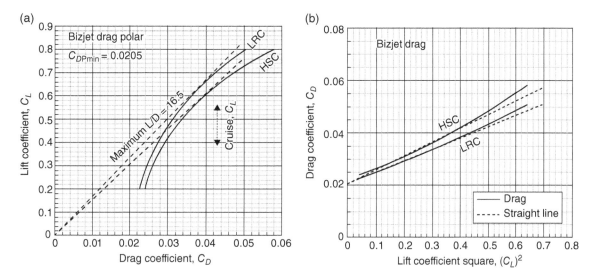

■ **FIGURE 10.14**   Bizjet semi-empirical drag polar. (a) Drag polar of Figure 9.3. (b) C_L^2 versus C_D

■ **TABLE 10.6**
Bizjet parabolic drag polar (lift/drag) and (Mach × lift/drag) ratios (weight = 18,000 lb at 41,000 ft altitude, ρ = 0.00055 lb/ft³, a = 968.08 ft/s)

Mach	Velocity (ft/s)	kt	C_L	C_D	L/D	M(L/D)	Induced drag	Zero-lift drag	Total drag
0.4	387.23	221.29	1.352	0.102	13.23	5.293	1087.36	273.04	1360.41
0.45	435.64	257.96	1.068	0.0715	14.94	6.724	851.15	345.57	1204.72
0.5	484.04	286.62	0.865	0.0539	16.04	8.018	695.91	426.63	1122.54
0.55	532.444	315.28	0.715	0.0433	16.49	10.071	575.14	516.22	1091.36
0.57	551.806	326.74	0.666	0.0403	16.52	10.413	535.48	554.45	1081.93
0.6	580.848	343.94	0.601	0.0366	16.4	10.839	483.27	614.35	1097.62
0.65	629.252	372.6	0.512	0.0322	15.89	10.328	411.78	721.0	1132.79
0.7	677.656	401.27	0.441	0.0292	15.11	10.577	355.06	836.2	1191.25
0.74	716.379	424.19	0.395	0.0275	14.37	10.637	317.71	934.49	1252.20
0.8	774.464	458.59	0.338	0.0256	13.2	10.557	271.84	1092.17	1364.01

speed), separation at the aft end adds to the viscous-dependent profile drag, ΔC_{Dp}. While C_{Dw} can be accurately computed, estimation of ΔC_{Dp} is not easy. These are relatively small effects, but industry cannot afford to have large error in its estimation.

10.9.1   Comparing Actual with Parabolic Drag Polar

Table 10.6 is worked out for Bizjet parabolic drag polar. The actual drag polar using semi-empirical relations are worked out in Chapter 9 (Table 10.7). Figure 10.15 plots the two and makes the comparison. It has a close fit for Mach 0.6–0.75; this is the speed range within which most of the Bizjet missions operate. Above and below this range it starts to deviate, and actual drag polar is more representative. The Bizjet C_{DPmin} is evaluated at Mach 0.65 (Section 9.18.1). The LRC (over 1000 nm mission range cruising at higher altitudes) is carried out at Mach 0.7.

■ **TABLE 10.7**
Bizjet actual drag polar (lift/drag) and (Mach × lift/drag) ratios (weight = 18,000 lb at 41,000 ft altitude, $\rho = 0.00055\,\text{lb/ft}^3$)

Mach	Velocity (ft/s)	kt	C_L	C_D	L/D	M(L/D)
0.5	484.04	286.62	0.865	0.058	14.91	7.456
0.55	532.444	315.28	0.715	0.045	15.88	8.737
0.6	580.848	343.94	0.600	0.0378	15.89	9.534
0.65	629.252	372.6	0.512	0.031	16.51	10.731
0.7	677.656	401.27	0.441	0.0288	15.32	10.726
0.74	716.379	424.19	0.395	0.0273	14.46	10.703
0.8	774.464	458.59	0.338	0.026	12.0	10.396

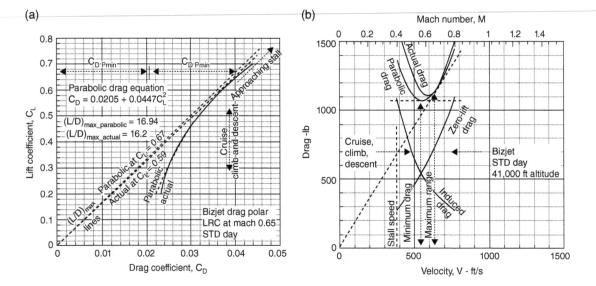

■ **FIGURE 10.15**   Bizjet actual and parabolic drag polar

10.9.2   (Lift/Drag) and (Mach × Lift/Drag) Ratios

From Table 10.6 the plots for (L/D) and (ML/D) for Bizjet at 41,000 ft are shown in Figure 10.16. This graph establishes the LRC speed schedule. This can be justified in a wider operational spectrum in Figure 13.2 giving the range factor (RF) and specific range (SR), dealt with in detail in Chapter 13.

Figure 10.16 shows that Mach 0.65 (at maximum ML/D) is selected for LRC. It is stressed here that these speed schedules are determined early in Phase I by analysing the close-form solution to get quick results to discuss with potential customers – the Bizjet speed schedule is not arbitrarily specified. In Phase II, refined performance analysis is carried out, guaranteeing the aircraft capability, and finally, after verifying with flight tests, an accurate performance flight manual is prepared.

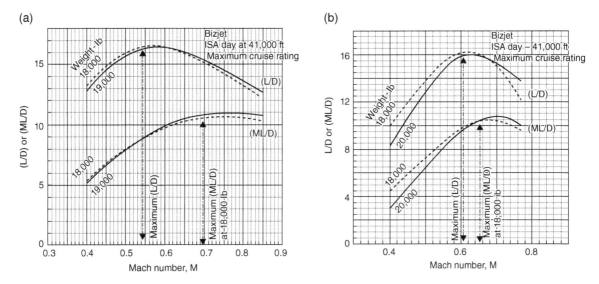

■ **FIGURE 10.16** (*L/D*) and (*ML/D*) plots to select LRC and HSC speed schedules for the Bizjet. (a) Parabolic drag polar. (b) Actual drag polar

10.9.3 Velocity at Minimum (D/V)

Equation 10.11 gives $V_{(D/V)min} = \sqrt{\dfrac{2W}{\rho S_w}\left(\dfrac{k}{C_{DPmin}}\right)^{0.25}}$. On substituting the values,

$$V_{(D/V)min} = \sqrt{\frac{2\times18000}{0.00055\times323}\left(\frac{0.0447}{0.0205}\right)^{0.25}} = 450.16\times1.215 = 546.5\,\text{ft/s}$$

This is equal to Mach 0.56, about the same as in Figure 10.15b. Figure 10.15b shows that actual drag polar value has $V_{(D/V)min}$ = Mach 0.61, slightly different from Figure 10.15a. (Chapter 12 shows that maximum range is obtained at *ML/D* at Mach 0.65.)

10.9.4 (Lift/Drag)$_{max}$, $C_{L@(L/D)max}$ and V_{Dmin}

Equation 10.18 can be used to evaluate $(C_L/C_D)_{max}$ for the approximated parabolic drag polar as follows ($k=0.0447$):

$$\left(\frac{C_L}{C_D}\right)_{max} = \frac{1}{2}\sqrt{\frac{1}{kC_{DPmin}}} = \frac{1}{2}\sqrt{\frac{1}{0.0447\times0.0205}} = 16.5$$

Figure 10.15a gives actual drag polar $\left(\dfrac{C_L}{C_D}\right)_{max} = 16.5$. Equation 10.16a gives

$$2kC_L = \frac{C_D}{C_L}, \quad \text{i.e.} \quad C_{L@(L/D)max} = \frac{1}{2\times0.0447\times16.5} = 1/1.475 = 0.678$$

Figure 10.15a compares the parabolic drag polar properties with actual drag polar, which gives $C_{L@(L/D)max} = 0.64$.

■ **TABLE 10.8**
Thrust required for the TPT aircraft at 5280 lb

Mach	V (ft/s)	C_L	C_D	C_L^3/C_D^2	L/D	ML/D	Induced drag (lb)	Zero drag (lb)	Total drag (lb)	SHP
0.185	188.0	1.41	0.126	176.6	11.2	2.07	387.13	84.63	471.76	190.0
0.2	203.2	1.206	0.098	181.8	12.28	2.455	331.23	98.911	430.15	186.96
0.232	235.25	0.9	0.0647	174.05	13.9	3.22	247.12	132.6	371.1	191.07
0.3	304.8	0.537	0.0376	101.33	14.28	4.284	147.22	222.55	361.1	241.07
0.32	325.12	0.471	0.034	81.75	13.8	4.416	121.4	253.2	382.6	266.072
0.35	355.6	0.394	0.031	65.0	12.84	4.495	108.16	302.92	411.07	312.67
0.4	406.4	0.302	0.027	36.73	11.04	4.414	82.81	395.64	478.45	415.91
0.45	457.2	0.238	0.0256	20.73	10.326	4.2	65.43	500.74	566.17	553.68
0.5	508.0	0.193	0.0245	11.95	7.867	3.93	53.0	618.2	671.2	721.323
0.55	558.8	0.16	0.0239	7.09	6.668	3.67	43.8	748.02	791.8	946.43
0.6	601.6	0.134	0.0235	4.35	5.7	3.42	36.8	890.2	927.0	1208.75
0.65	660.4	0.114	0.0233	2.75	4.89	3.2	31.36	1044.75	1076.1	1520.1
0.7	711.2	0.0985	0.0231	1.79	4.26	2.98	27.04	1211.66	1238.7	1884.0

Then Equation 10.21a can be used to evaluate Bizjet D_{min}, for the approximated parabolic drag polar for a 18,000 lb aircraft flying 41,000 ft altitude, ISA day, as follows (see Figure 10.15b):

$$D_{min} = 2W\sqrt{kC_{DPmin}} = (2 \times 18000)\sqrt{(0.0447 \times 0.0205)} = 1062 \text{ lb}$$

10.9.4.1   *Discussion* Figure 10.16 makes it clear that there is some difference between parabolic and actual drag polars. In industry the difference is sufficiently large not to be accepted in operational usage, yet the parabolic drag polar gives quick results for trend analyses. Industry benefits from exploring parabolic drag polar analyses, and uses the more accurate actual drag polar values for engineering analyses for commercial usage.

10.9.5   Turboprop Trainer (TPT) Example – Parabolic Drag Polar

Equation 10.43 gives the parabolic drag polar of the TPT: $C_D = 0.0226 + 0.052C_L^2$ with $S_W = 200 \text{ ft}^2$. TPT operating altitude is taken as 25,000 ft ($\rho = 0.00106$ for an ISA day) at 5280 lb (2400 kg). Take $\eta_{prop} = 0.85$. Table 10.8 gives the plotting details (see Figure 10.17).

10.9.6   TPT (Lift/Drag)$_{max}$, $C_{L@(L/D)max}$ and V_{Dmin}

Equation 10.18 can be applied to both turbofan and turboprop aircraft. It is applied to evaluate TPT $(C_L/C_D)_{max}$ as follows ($k = 0.052$):

$$\left(\frac{C_L}{C_D}\right)_{max} = \frac{1}{2}\sqrt{\frac{1}{kC_{DPmin}}} = \frac{1}{2}\sqrt{\frac{1}{0.052 \times 0.0226}} = 0.5 \times \sqrt{851} = 14.6 \ \left(\text{see Figure 10.17b}\right)$$

or

$$C_{L@(L/D)max} = \frac{1}{2 \times 0.052 \times 14.6} = 0.66$$

Velocity at 25,000 ft,

$$V_{(L/D)max} = \sqrt{\frac{(5280/200)}{(0.5 \times 0.00106 \times 0.66)}} = 275 \text{ ft/s or } 162.8 \text{ kt}$$

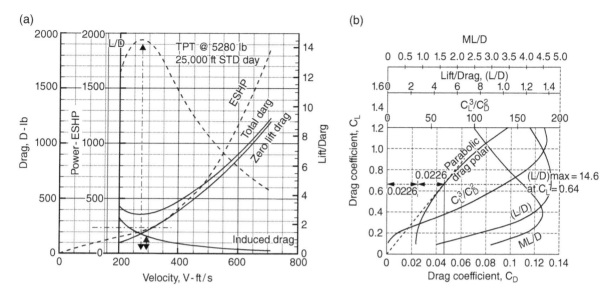

■ **FIGURE 10.17**   TPT aircraft drag. (a) Drag polar break-down. (b) Drag polar

The same value can be verified by using Equation 10.11 to evaluate TPT V_{Dmin}, as follows:

$$V_{Dmin} = \sqrt{\frac{2W}{\rho S_W}}\left(\frac{k}{C_{DPmin}}\right)^{1/4} = \sqrt{\frac{2\times 5280}{0.00106\times 200}}\times\left(\frac{0.052}{0.0226}\right)^{1/4}$$
$$= \sqrt{49\,811.3}\times(2.3)^{1/4} = 223.2\times 1.23 = 274.5\text{ ft/s}$$

It can also be obtained using Equation 10.40:

$$V_{Dmin} = 3^{1/4}\times V_{Pmin} = 3^{1/4}\times 209 = 275\text{ ft/s }(163\text{ kt})$$

A 5280 lb TPT aircraft flying at 25,000 ft altitude on an ISA day gives D_{min} as follows (for parabolic drag polar see Figure 10.17a):

$$D_{min} = 2W\sqrt{kC_{DPmin}} = (2\times 5280)\sqrt{(0.052\times 0.0226)} = 10560\times\sqrt{(0.0011752)} = 362\text{ lb}$$

$V_{Dmin} = (T_P)_{reqd}/D$ gives the minimum value. Figure 10.17a shows that this occurs by drawing a tangent from the origin to the $(T_P)_{reqd}$ curve. It gives $(T_P)_{reqd} = 222$ ESHP.

10.9.7   TPT $(ESHP)_{min\_reqd}$ and V_{Pmin}

1. *Minimum power $(ESHP)_{min\_reqd}$ and velocity at minimum power V_{Pmin}* Equation 10.37 gives the velocity for minimum power, V_{Pmin} as:

$$V_{Pmin} = \left(\frac{1}{3e\pi ARC_{DPmin}}\right)^{1/4}\sqrt{\frac{2W}{\rho S_W}} = \left(\frac{0.052}{3\times 0.0226}\right)^{1/4}\sqrt{\frac{2\times 5280}{0.00106\times 200}}$$
$$= (0.767)^{1/4}\sqrt{49811.3} = 0.936\times 223.2 = 209\text{ ft/s }(124\text{ kt})$$

2. *Minimum power* $(ESHP)_{min\_reqd}$

Equation 10.39 gives $(ESHP)_{min\_reqd}$ (note: $1/(e\pi AR) = 0.052$ and 1 SHP = 550 ft.lb):

$$(ESHP)_{min\_reqd} = \frac{4}{3\eta_{prop}} \times W \sqrt{\frac{2W}{\rho S_W}} \times \left(\frac{3C_{DPmin}}{(e\pi AR)^3}\right)^{1/4}$$

$$= \frac{4}{3 \times 0.85 \times 550} \times 5280 \sqrt{\frac{2 \times 5280}{0.00106 \times 200}} \times \left[3 \times 0.0226 \times (0.052)^3\right]^{1/4}$$

$$= 15.05 \times \sqrt{(49811.3)} \times 0.056 = 187 \text{ SHP}$$

3. *Turboprop* C_L *at maximum lift/drag ratio* $(L/D)_{max}$

$$C_L \text{ at } V_{Dmin} = \frac{2W}{\rho V_{Dmin}^2 S_W} = \frac{2 \times 5280}{0.00106 \times 275^2 \times 200} = \frac{10560}{16032.5} = 0.66$$

10.9.8   Summary for TPT

The following are the calculated values for the TPT at 25,000 ft and 5280 lb on an ISA day:

- $V_{stall} = 188$ ft/s (111.4 kt)
- $(ESHP)_{Vstall} = 190$ SHP
- $V_{Pmin} = 209$ ft/s (124 kt)
- $(ESHP)_{VPmin} = 187$ SHP
- $V_{Dmin} = 275$ ft/s (163 kt)
- $(ESHP)_{VDmin} = 222$ SHP
- C_L at $V_{Dmin} = 0.66$
- $\left(\dfrac{C_L}{C_D}\right)_{max} = 14.6$

10.10   Aircraft and Engine Grid

In a steady level flight such as in cruise, the aircraft drag equals the total thrust offered by the engine or engines. In other words, in equilibrium cruise, thrust $(T) = $ drag (D).

Aircraft designers must ensure that adequate thrust is available $(T_{av}$ or simply $T)$ to meet the thrust required $(T_{req}$, which is the drag $D)$. If T_{av} is more than T_{req}, that is, $(T - D) > 0$, then this excess thrust can be used in many possible ways, for example (i) accelerating until the drag increase equates to T_{av}; (ii) climbing, when the excess thrust overcomes the component of aircraft weight; or (iii) manoeuvring, which requires additional thrust to overcome the drag increase.

Thrust available (T) is an engine characteristic and is throttle-dependent, as well as depending on operating speed and altitude. The aircraft drag (D) depends on aircraft weight, speed and altitude. Therefore separate graphs are required to estimate $(T - D)$ at the speed, altitude, throttle setting and aircraft weights. Figure 10.18a shows one such graph for the Bizjet at 20,000 lb flying at 20,000 ft altitude at the maximum climb rating. Engine installed thrust (Figure 8.12) gives T, and Bizjet drag polar gives D.

A convenient way to show T_{av} and T_{req} in one graph for each engine rating is to construct an aircraft and engine grid in a non-dimensional form, as shown in Figure 10.18b for the

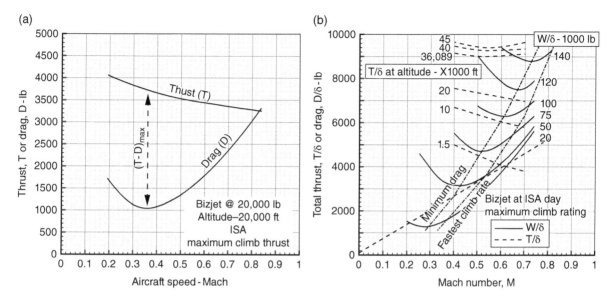

■ **FIGURE 10.18**   **Maximum climb thrust (Bizjet actual drag polar). (a) Thrust available and thrust required. (b) Thrust available and thrust required (standardized to sea-level)**

maximum climb rating. Section 8.9 has already used the engine grid in non-dimensional form, and the same graphs are used here, superimposed with the aircraft grid, together becoming the "aircraft and engine grid".

The aircraft and engine grid plays an important part in evaluating aircraft performance, in climb, cruise and descent. These kinds of graphs are used in the next few chapters. To apply to a wide range of non-standard conditions, the graphs are presented for sea-level standard conditions using the density ratio $\sigma = \rho/\rho_0$, pressure ratio $\delta = p/p_0$ and temperature ratio $\theta = T/T_0$.

10.10.1   Aircraft and Engine Grid (Jet Aircraft)

The main parameters to generate the aircraft and engine grid are lift (L) and drag (D).

$$C_L = \text{lift coefficient} = \text{lift}/qS_W = L/\left(0.5\rho V^2 S_W\right) \tag{10.44}$$

$$C_D = \text{drag coefficient} = \text{drag}/qS_W = D/\left(0.5\rho V^2 S_W\right) \tag{10.45}$$

In equilibrium flight, $T=D$. The following can be written for equilibrium flight in terms of coefficients:

$$C_D = \text{thrust}/qS_W = T/\left(0.5\rho V^2 S_W\right) \tag{10.46}$$

We need to use the definitions as follows:

- Density ratio = σ = (density at altitude ρ)/(density at sea-level ρ_0)
- Pressure ratio = δ = (pressure at altitude p)/(pressure at sea-level p_0)
- Temperature ratio = θ = (temperature at altitude T)/(temperature at sea-level T_0)
- From thermodynamics, $\delta = \sigma\theta$
- The ratio of speed of sound $a/a_0 = \sqrt{(T/T_0)}$, or $a = a_0\sqrt{\theta}$
- Mach number, $M = V/a$ where V is the aircraft velocity.

In equilibrium flight, the following can be written in terms of non-dimensional ratios (note: $\rho_0 = 0.002378$ slug/ft^3, $a_0 = 1116.4$ ft/s in Imperial system):

$$L = 0.5\rho V^2 S_w C_L = 0.5(\rho_0 \delta)(Ma_0)^2 S_w C_L = 1481 C_L M^2 S_w \delta$$
$$\text{or}\quad L/\delta = W/\delta = 1481 C_L M^2 S_w \tag{10.47}$$

$$D = 0.5\rho V^2 S_w C_D = 0.5(\rho_0 \delta)(Ma_0)^2 S_w C_D = 1481 C_D M^2 S_w \delta$$
$$\text{or}\quad D/\delta = T/\delta = 1481 C_D M^2 S_w a \tag{10.48}$$

With these relationships, the aircraft and engine grid can be generated for the classwork examples. Aircraft details are given in Section 10.8, and installed engine performance details are given in Section 8.6.

Section 8.9 computed the engine grid for the Bizjet aircraft, and only the maximum cruise rating grid is shown in Figure 8.19. When the aircraft grid is superimposed on the engine grid, then it becomes the aircraft and engine grid at maximum cruise rating. The aircraft and engine grid at maximum cruise rating is shown in Figure 10.19 for both the actual and parabolic Bizjet drag polar.

The aircraft and engine grid for the TPT using parabolic drag polar is shown in Figure 10.20a. Readers should generate the aircraft and engine grid for the AJT in the same way.

10.10.2   Classwork Example – Bizjet Aircraft and Engine Grid

Section 8.6.1 presents the installed thrust and fuel flow available for the Bizjet aircraft at maximum climb rating and maximum cruise rating. The engine grid for the maximum cruise rating is plotted in Figure 8.19. It will be used to generate the aircraft and engine grid for the Bizjet. Table 10.9 is the aircraft grid which is superimposed over Figure 8.19.

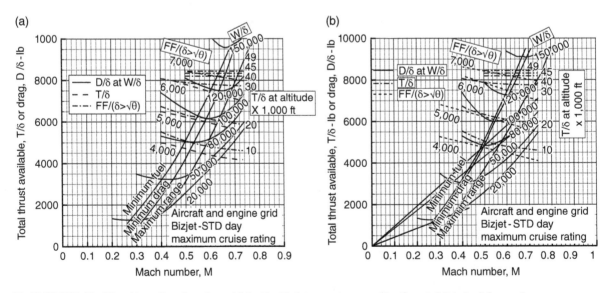

■ **FIGURE 10.19**   Aircraft and engine grid for the Bizjet – maximum cruise thrust. (a) Actual drag polar. (b) Parabolic drag polar

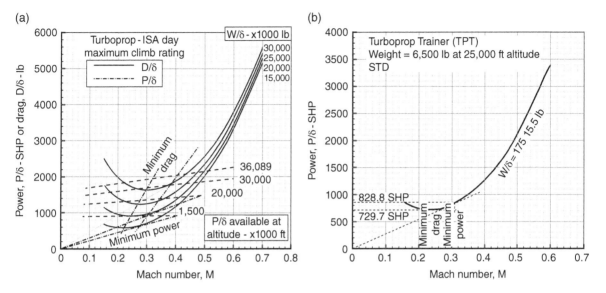

■ FIGURE 10.20   (a) Aircraft and engine grid (parabolic drag polar) for the TPT. (b) Classwork example

■ TABLE 10.9

Thrust required (actual drag polar) for the Bizjet aircraft ($C_L = 0.0418/M^2$, $S_W = 323\,\text{ft}^2$)

Mach	C_L	C_D	D/δ	Mach	C_L	C_D	D/δ
$W/\delta = 20{,}000\,\text{lb}$				$W/\delta = 50{,}000\,\text{lb}$			
0.2	1.045	0.082	1561.0				
0.3	0.46	0.03	1291.6	0.3	1.161	0.086	3702.5
0.4	0.26	0.0234	1791.0	0.4	0.653	0.0414	3168.7
0.5	0.167	0.0212	2535.3	0.5	0.418	0.0278	3324.6
0.6	0.116	0.021	3616.4	0.6	0.29	0.0233	4012.5
0.7	0.0853	0.0218	5101.9	0.7	0.213	0.023	5391.2
0.74	0.0763	0.0224	5869	0.74	0.191	0.024	6286.8
$W/\delta = 80{,}000\,\text{lb}$				$W/\delta = 100{,}000\,\text{lb}$			
0.4	1.045	0.086	6582.3	0.4	1.3	0.098	7500.7
0.5	0.669	0.0448	5357.7	0.5	0.84	0.055	6577.5
0.6	0.464	0.03	5166.3	0.6	0.58	0.0362	6234.0
0.7	0.341	0.026	6094.3	0.7	0.43	0.0285	6680.3
0.74	0.305	0.206	6810	0.74	0.382	0.0282	7387
$W/\delta = 120{,}000\,\text{lb}$				$W/\delta = 140{,}000\,\text{lb}$			
0.5	1.0	0.069	8251.8				
0.6	0.7	0.045	7715	0.6	0.813	0.054	9300
0.7	0.51	0.035	7735	0.7	0.597	0.039	9095
0.74	0.458	0.0318	8330	0.74	0.534	0.0354	9273

The thrust required grid can be generated in the following manner. Six aircraft weights at $W/\delta = 20{,}000 - 150{,}000$ lb are worked out in Imperial units. For a Bizjet the wing area, $S_W = 323$ ft^2. Use Equations 10.10 and 10.12.

Equation 10.47 gives:

$$W/\delta = 1481 C_L M^2 S_W = 1481 \times 323 C_L M^2 = 478{,}363 C_L M^2 \tag{10.49}$$

or

$$C_L = 0.209 \times 10^{-5} W/\left(\delta M^2\right) \tag{10.50}$$

Equation 10.48 gives

$$\text{Drag} = D/\delta = 1481 C_D M^2 S_W = 1481 \times 323 C_D M^2 = 478{,}363 C_D M^2 \tag{10.51}$$

Table 10.9 is plotted along with Tables 8.3 and 8.4 in Figure 10.19a (engine grid at maximum cruise rating). Figure 10.19a,b compares the aircraft and engine grid based on actual drag polar and on parabolic drag, respectively. Figure 10.19 superimposes Figure 8.19 showing the lines of $(T/\delta)_{avail}$ and $FF/(\delta\sqrt{\theta})$. The advantages of such plots are clear as they cover the full envelope of aircraft weights and altitude capabilities with the matched engine performances.

In Figure 10.19a, the dashed lines in the diagram represent thrust available at maximum cruise rating normalized for altitude as (T/δ) for altitudes from 10,000 to 45,000 ft. The full lines represent thrust required (D/δ, see Section 12.3) at altitudes for 20,000 lb weight of the Bizjet.

Figure 10.19a is extensively used in subsequent chapters. The pattern of the graph changes if the X-axis represents V_{EAS} instead of Mach number to represent aircraft speed. Note that at high altitudes the V_{TAS} is a lot higher for the prescribed V_{EAS}, that is, faster ground speed. Readers may plot and study the changes with X-axis for V_{EAS} and V_{TAS}. The graph still gives the same information in a different pattern.

Flying below minimum drag speed is not desirable as drag increases even when speed reduces. Pilots are trained to be aware of this phenomenon. During takeoff and landing, the aircraft is below minimum drag speed. Takeoff is at full thrust and the aircraft keeps accelerating towards speeds above the minimum drag speed. At landing, it is critical that the pilot should maintain the prescribed approach speed by adjusting the throttle and/or descent rate. In proper execution, this assists in setting the aircraft on the ground gently as drag increases.

Aircraft operation may require going the longest distance (range) or staying aloft for the longest time (endurance). These are dealt with in detail in Chapters 12 and 13. Operation for the longest range is carried out at minimum drag, that is, at $(L/D)_{max}$. Operation for the longest endurance is carried out at minimum fuel consumption.

From the conditions obtained in Figure 10.15b for the minimum drag, Figure 10.19b shows the lines of minimum drag and minimum fuel consumption for each W/δ. The minimum drag line is the locus of the lowest point of W/δ as the point of minimum drag V_{Dmin} at $(L/D)_{max}$. The minimum fuel line is the locus of points where $FF/(\delta\sqrt{\theta})$ (see Figure 8.19) is tangent to W/δ lines. At such low speeds, the line of minimum fuel is in the unstable part of aircraft velocity where lowering of velocity increases drag – an undesirable adverse situation. Since the lines of minimum fuel consumption and minimum drag are close to each other, endurance is carried out at velocities of minimum drag without much loss of endurance time.

If the plots are made for each altitude without normalizing to sea-level conditions, they will still show same general characteristics.

10.10.3 Aircraft and Engine Grid (Turboprop Trainer)

The equations derived in Section 10.10.1 can be used incorporating the propeller efficiency as given in Equation 10.34. In equilibrium flight, Equations 10.35 to 10.36 can be written in terms of non-dimensional ratios. (Note: $\rho_0 = 0.002378$ slug/ft^3, $a_0 = 1116.4$ ft/s in Imperial system.)

Equation 10.34 gives $(ESHP)_{reqd} = (T_P)_{reqd}/\eta_{prop}$. In steady flight,

$$L/\delta = W/\delta = 1481C_L M^2 S_W$$

(at quasi − level, quasi − steady flight, $L = W$),

or

$$C_L = (W/\delta)/(1481M^2 S_W)$$

Equation 10.51 gives thrust required

$$D/\delta = 1481C_D M^2 S_W$$

and

$$(ESHP)_{reqd}/\delta = TV/\eta_{prop}\delta = 1481C_D VM^2 S_W/\delta\eta_{prop} \qquad (10.52)$$

This follows the worked example of the Bizjet in Sections 10.10.1 and 10.10.2. TPT parabolic drag polar is $C_D = 0.0226 + 0.052C_L^2$, with $S_W = 200ft^2$. TPT operating altitude is taken as 25,000ft ($\rho = 0.00106$) at 6500lb (2955 kg). Take $\eta_{prop} = 0.85$.

For the TPT for the above condition,

$$C_L = (W/\delta)/(1481M^2 S_W) = 0.3376 \times 10^{-5} W/(\delta M^2)$$

$$(ESHP)_{reqd} = TV/\eta_{prop}\delta = 1481C_D VM^2 S_W/\eta_{prop}$$

$$= 296,200 \times C_D VM^2/0.85 = 348,470 \times C_D VM^2$$

Turboprop aircraft and engine grid generation follows the same routine, with SHP as the additional parameter to be evaluated to bring in the inter-phasing effects of an integrated propeller. To apply for a wide range of non-standard conditions, the graphs are presented standardized to sea-level conditions using the density ratios $\sigma = \rho/\rho_0$, $\delta = p/p_0$ and $\theta = T/T_0$. While Figure 10.20a shows all the W/δ lines for 15,000, 20,000, 25,000 and 30,000lb, the computation procedure is given in Table 10.10 only for 15,000lb. It gives SHP at 25,000ft altitude = 270.0 SHP.

$$V_{Pmin} = 235.25ft/s\,(1310.4\ kt)\ \text{at}\ C_L = 1.098$$

■ **TABLE 10.10**
Thrust required for the TPT aircraft (25,000ft altitude, $\delta = 0.371$, $\rho = 0.001066$ slug/ft³)

Mach	Velocity (ft/s)	C_L	C_D	Thrust $(T)_{reqd}$ (lb)	T/δ (lb)	Pr/δ SHP	Pr/$(\delta \times \eta_{prop})$
($W/\delta = 15,000$lb)							
0.15	152.4	2.251	0.307	754.966	2034.45	563.72	663.1973
0.2	203.2	1.266	0.1128	493.48	1321.81	491.3	577.9945
0.25	254.0	0.810	0.0598	408.69	1101.32	508.6	598.3505
0.3	304.8	0.5627	0.0407	401.075	1080.8	598.95	704.6455
0.4	406.4	0.3165	0.0286	500.851	1341.68	997.27	1173.256
0.5	508.0	0.2026	0.0253	691.986	1864.74	1722.31	2026.244
0.6	601.6	0.1407	0.0241	941.601	2558.95	2836.2	3336.697
0.7	711.2	0.1034	0.0236	1265.17	3401.34	4408.5	5186.47
($W/\delta = 17,000$lb) checking classwork example (see Section 10.7.4) at minimum drag							
0.232	235.71	1.0986	0.0906	533.5	1437.66	616.12	724.85

At V_{Pmin} the minimum power $(ESHP)_{min\_reqd} = 270.66$ SHP, which results in $Pr/\delta = 7210.33 SHP$.

$$V_{Dmin} = 301.06 \text{ ft/s, having } C_L \text{ at } V_{Dmin} = 0.64$$

Figure 10.20a gives the TPT aircraft and engine grid (parabolic drag polar) and Figure 10.20b plots the classwork example in Table 10.10.

There is a small difference between what is given in Section 10.9.7 and what is obtained from Figure 10.20. Reading from graphs is not accurate enough.

References

1. Miele, A., *Flight Mechanics: Theory of Flight Paths*, vol. 1. Addison-Wesley, Menlo Park, CA, 1962.

2. Mair, W.A., and Birdshall, D.L., *Airplane Performance*. Cambridge University Press, Cambridge, 1992.

3. Vinh, N.X., *Flight Mechanics of High-Performance Aircraft*. Cambridge University Press, Cambridge, 1993.

4. Phillipone, A., *Flight Performance of Fixed and Rotary Wing Aircraft*. Butterworth-Heinemann, Oxford, 2006.

5. Ruijgork, G.J.J., *Elements of Airplane Performance*. Delft University Press, Delft, 1996.

6. Perkins, C.D., and Hage, R.E., *Airplane Performance, Stability and Control*. John Wiley & Sons, Inc., Chichester, 1949.

7. Lan, C.T.E., and Roskam, J., *Airplane Aerodynamics and Performance*. DARcorporation, Lawrence, 1981.

8. Phillip, W.F., *Mechanics of Flight*. John Wiley & Sons, Inc., Chichester, 2004.

9. Saarlass, M., *Aircraft Performance*. John Wiley & Sons, Inc., Chichester, 2007.

10. Bernard, R.H., and Philpot, D.R., *Aircraft Flight*. Prentice-Hall, Harlow, 2010.

11. McCormick, B.W., *Aerodynamics, Aeronautics and Flight Mechanics*. John Wiley & Sons, Inc., Chichester, 1995.

12. Hale, F.J., *Aircraft Performance, Selection and Design*. John Wiley & Sons, Inc., Chichester, 1984.

13. Hull, D.G., *Fundamentals of Airplane Flight Mechanics*. Springer, Berlin, 2007.

14. Ojha, S.K., *Flight Performance of Airplane*. AIAA, Reston, VA.

15. Eshelby, M.E., *Aircraft Performance Theory and Performance*. AIAA, Reston, VA, 2000.

16. Aselin, M., *An Introduction to Aircraft Performance*. AIAA, Reston, VA, 1997.

17. Ashley, H., *Engineering Analyses of Flight Vehicles*. Addison-Wesley, Menlo Park, CA, 1974.

18. Anderson, J.D., *Aircraft Performance and Design*. McGraw-Hill, New York, 1999.

19. Shevell, R.S., *Fundamentals of Flight*. Prentice-Hall, New Jersey, 1989.

20. Talay, T.A., Introduction to the Aerodynamics of Flight. NASA SP-367. NASA, Langley, 1975.

21. Smetana, F.O., *Flight Vehicle Performance and Aerodynamic Control*. AIAA, Reston, VA, 1994.

22. Roskam, J., *Airplane Design*, DARcorporation, Lawrence, 2007.

23. Pamadi, B.N., *Performance Stability, Dynamics and Control of Aircraft*. AIAA, Reston, VA, 2015.

CHAPTER 11

Takeoff and Landing

11.1 Overview

Aircraft have to take off from an airfield to carry out the mission role, and eventually land at the destination. Takeoff and landing takes place on the airfield surface, and so are termed field performance. This chapter deals comprehensively with all aspects of the takeoff and landing performances of conventional aircraft. Second-segment climb and missed approach climb performance are part of field performance, but more appropriately treated in Chapter 12 dealing with climb and descent [1–3].

Takeoff and landing are critical segments of any mission. To ensure safety, airworthiness authorities have stringent stipulations that must be complied with. This chapter deals with the pertinent regulations associated with takeoff and landing. Operational aspects of field performance are also part of airworthiness requirements. Performance engineers engaged in preparing user operational manuals must be familiar with the FAA defined terminologies. It is for this reason this chapter introduces the operational terms associated with field performance. However, the theoretical derivations and worked examples use the conventional terms typically used in industry.

There are three main types of takeoffs and landings as listed below:

Conventional takeoff and landing   (CTOL): dealt with in detail in this chapter.
Short takeoff and landing   STOL): not considered in this book.
Vertical takeoff and landing   (VTOL): not considered in this book.

Since this book only deals with CTOL, the term "conventional" is dropped. Aircraft sizing and engine matching exercises also estimate takeoff and landing, but the procedure is not meant for certification purposes and hence is dealt with in a different manner as shown in Chapter 16.

Military aircraft performance uses the same equations but has different airworthiness requirements, and the worked examples in this book use simplistic analyses.

Readers are to compute takeoff and landing performances on classwork examples of a Bizjet for civil aircraft, and an advanced jet trainer (AJT) for military aircraft. Since there is extensive computational work involved, it is suggested that spreadsheets may be used for repeating computations. Computation for a turboprop trainer aircraft (TPT) field performance follows the same procedure as for the AJT, and is left to the readers to carry out. A problem set is given in Appendix E.

Theory and Practice of Aircraft Performance, First Edition. Ajoy Kumar Kundu, Mark A. Price and David Riordan.
© 2016 John Wiley & Sons, Ltd. Published 2016 by John Wiley & Sons, Ltd.

11.2   Introduction

Aircraft takeoff and landing offer the most demanding workload for the pilot other than in combat. Most of the accidents and incidents occur in these segments of a mission. There is a loose saying in operational circles that most accidents at takeoff are associated with engineering failure, and most at landing are due to human errors. Engineers take measures to add safety during takeoff, and performance engineers provide a proper pilots' manual so that human errors can be avoided through training. Airworthiness authorities have stringent design stipulations to ensure safety during these phases of operation. For engineers, takeoff demands stringent attention to ensure compliance with airworthiness requirements. Engineers may incorporate more than what is stipulated by the authorities, for example, use of thrust reversers, and so on.

During takeoff, the aircraft accelerates from a stationary condition to liftoff velocity and continues to climb. It requires adequate airfield length and unobstructed frontage to climb to complete the takeoff routine. The takeoff field length (TOFL) has three parts as follows: (i) the ground run; (ii) transition during rotation to takeoff; and (iii) flare to clear obstacle height (35 ft for civil aircraft and 50 ft military aircraft) as stipulated by airworthiness requirements. The takeoff process continues with climb up to 1500 ft in definite segments.

If there is an engine (critical) failure during takeoff, airworthiness authorities require aircraft to continue with the climb. Takeoff requirements for civil and military design differ to an extent, although the theory behind both of them is the same.

In a way, landing can be seen as similar to takeoff in reverse. The aircraft descends to approach height to land. It has four parts as follows: (i) glide down from 50 ft; (ii) flare to touch down; (iii) ground run at idle thrust until nose wheel touches down; and (iv) application of retarding device to stop. Landing also requires compliance with airworthiness requirements to ensure safety. Pilot skill is in demand during landing. In strong gusty conditions or a cross wind, landing could become dangerous and the pilot may have to divert to land at an alternative airfield.

Field performance document presentation for engineers differs from the pilots' manual. Engineers use graphs plotted in a manner not suitable for pilot usage. Both types are shown in Section 11.8.

11.3   Airfield Definitions

The takeoff procedure is an involved affair to ensure safety in various situations, for example, engine failure, wet runway, and so on. The general definitions of airfield lengths are given by Federal Aviation Regulations (FAR) (14CFR) 25 Part 1.1 in a formal manner, which are involved and are beyond the scope of this book.

Here we use the terms used in a typical design bureau. The field length definitions given in this section are meant for information to newcomers to the subject. Fresh graduates will have to deal with terms as soon as they join the industry. Aircraft performance engineers may have to work at the various levels of analyses as described in Section 1.6. As mentioned in Section 1.5, FAR (14CFR) documentations are written in an elaborate formal manner for clarity, and require cross-referencing between various parts on the same subject. For this reason the FAR regulations are not copied here but explained below, quoting the FAR (14CFR) part numbers for readers to consult. Those who are engaged in preparing operators' documents must use the formal FAR (14CFR) definitions and terminologies.

Readers may encounter some of the following terms in relation to the distances available for use in meeting an aeroplane's takeoff run (TOR), takeoff distance (TOD), accelerate-stop distance (ASD) and landing distance requirements. All of these are seen as *runway safety areas* (RSAs).

The runway is defined as the full strength pavement surface from threshold to threshold, but not the overrun areas at both ends. Runway thresholds are markings across the runways that denote the beginning and end of the designated space for landing and takeoff under non-emergency conditions. The RSA is the cleared, smoothed and graded area around the paved runway. It is kept free from any obstacles that might impede flight or ground roll of aircraft.

Some airfields are associated with a defined *stopway* (SWY) and/or *clearway* (CWY) for safe operations. The terminologies used here are explained in the later sections.

11.3.1   Stopway (SWY) and Clearway (CWY)

The associated symbols A, B, C and D in Figure 11.1 are defined in Section 11.6.2.

Stopway (SWY) A stopway is a safety provision extending beyond the end of the full strength pavement runway to allow aircraft a longer distance to stop under brake application, in case of a failed takeoff. The SWY is a hard surface of partial strength pavement, and its width has to be at least equal to the runway width or more. A SWY is not used for landing, but may also serving as a blast pad, to prevent erosion from gas turbine engine exhaust blasts.

■ **FIGURE 11.1**
Stopway and clearway

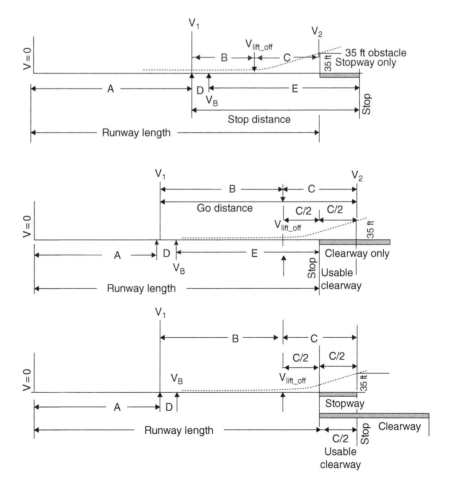

■ **FIGURE 11.2**
Airfield length
definitions. Top:
TORA and TODA.
Bottom: Takeoff field
length

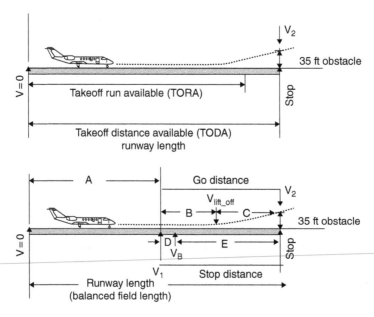

Clearway (CWY) The clearway is a non-paved area, clear of obstacles, placed at a runway's extremity, past the SWY, in order to ease takeoff. If there is not enough prepared runway length, then the CWY is provided at the extension of the runway to clear 35 ft height for aircraft that cannot stop within the SWY. The CWY is under the control of airport authorities and must be at least 500 ft wide. It can have a 1.25% positive gradient.

Most modern airfields are provided with CWY and/or SWY.

11.3.2   Available Airfield Definitions

Aircraft are to operate in a declared designated airfield by the appropriate authorities, and the operators must know the available length at the time of operation. In the rare event of an emergency, the available distance may be slightly shorter than the planned length. The FAA defines the following four declared airfield lengths, independent of aircraft capabilities (Figure 11.2 shows the generalized runway definitions):

Takeoff run available (TORA): the declared runway length available and suitable for satisfying takeoff ground run requirements. No aircraft is allowed to exceed TORA during TOR.

Takeoff distance available (TODA): this distance comprises the TORA plus the length of any remaining runway or if any CWY is available beyond the far end of the TORA.

Accelerate-stop distance available (ASDA): this is the runway plus any SWY length declared available and suitable for the acceleration and deceleration of an aircraft that must abort its takeoff.

Landing distance available (LDA): the runway length that is declared available and suitable for satisfying aircraft landing distance requirements.

11.3.3   Actual Field Length Definitions

The aircraft capabilities for takeoff and landing are the actual distances, and the operators must ensure that the aircraft is prepared to operate within the available declared airfield lengths described above. These are the capabilities as given in the operators' manual supplied by the aircraft manufacturers.

11.3.3.1   Takeoff Run (TOR)   This is the distance covered from brake release to a point equidistant between the point at which $V_{lift\text{-}off}$ is reached and the point at which the aeroplane is 35 ft above the takeoff surface, as determined under FAR (14CFR) 25.111 for a dry runway. It includes a CWY. If the TOD does not include a CWY, the TOR is equal to the TOD. It also has two definitions for a dry runway as follows (this is covered in FAR (14CFR) 25.113 Subpart B).

1. TOR_{1eng_failed} = one engine (critical) inoperative. This has to be the balanced field length (BFL – see Section 11.6.2).
2. TOR_{all_eng} = all engines operative.

FAR (14CFR) has a separate distance specification for each situation of TOR for all engines operative ($TOR_{all\_eng\_dry}$), one engine inoperative ($TOR_{1eng\_failed\_dry}$), a wet runway ($TOR_{all\_eng\_wet}$), and one engine inoperative on a wet runway ($TOR_{1eng\_failed\_wet}$). For most of the situations with one engine inoperative, TOR is higher than the all-engine TOR, but not necessarily. Therefore, for a mission, the maximum TOR for the situation is taken as the operational usage. This book computes these distances, termed as takeoff field length (TOFL). This offers simplicity, yet uses the relevant equations to obtain distances for any specific situation.

Dry runway For any given aircraft weight and ambient conditions, the TOR on a dry runway is the greater of the following values:

$TOR_{1eng\_failed\_dry}$ is measured from the start of takeoff to a point equidistant from lift-off to reaching 35 ft altitude. This adds C/2 to B in Figure 11.2, giving additional distance to stop in the clearway.

$TOR_{all\_eng\_dry}$ is 115% of the distance covered from the start of takeoff to a point equidistant from lift-off to reaching 35 ft altitude.

Wet runway For any given aircraft weight and ambient condition, the TOR on a wet runway is the greater of the following values:

$TOR_{1eng\_failed\_wet}$ is measured from the start of takeoff to a point reaching 15 ft altitude, ensuring than V_2 is reached by 35 ft altitude.

$TOR_{all\_eng\_wet}$ is 115% of the distance covered from the start of takeoff to a point equidistant from lift-off to reaching 35 ft altitude.

11.3.3.2   Takeoff Distance (TOD)   This is the distance covered from brake release to the point the aircraft reached 35 ft height. It does not include CWY. It has two definitions for a dry runway as follows. It is covered in FAR (14CFR) 25.113 Subpart B.

1. TOD_{1eng_failed} = one engine (critical) inoperative. This has to be the balanced field length (BFL – see Section 11.6.2).
2. TOD_{all_eng} = all engines operative.

This distance comprises the TOR plus the length of any remaining runway or CWY beyond the far end of the TOR. For runways without clearway, the takeoff run is equal to the takeoff distance. In most situations the one-engine inoperative TOD is higher than the all-engine TOD, but not necessarily. Therefore, for a mission, the maximum possible TOD for the situation is taken as the operational usage.

Dry runway For any given aircraft weight and ambient condition, the TOD on a dry runway is the greater of the following values:

$TOD_{leng\_failed\_dry}$ is measured the start of takeoff to a point equidistant from lift-off to reaching 35 ft altitude.

$TOD_{all\_eng\_dry}$ is 115% of the distance covered from the start of takeoff to a point equidistant from lift-off to reaching 35 ft altitude.

Wet runway For any given aircraft weight and ambient condition, the TOD on a wet runway is the greater of the following values:

$TOD_{leng\_failed\_wet}$ is measured from the start of takeoff to a point where 15 ft altitude is reached, ensuring that V_2 is reached at 35 ft altitude.

$TOD_{all\_eng\_wet}$ is 115% of the distance covered from the start of takeoff to a point equidistant from lift-off to reaching 35 ft altitude.

11.3.3.3   *Accelerate-Stop Distance (ASD)*

This is the distance covered from the point of brake application in the case of one engine being inoperative, plus the distance covered in 3 seconds when the engine fails, to cover pilot recognition time and the times for brakes to act, spoiler deployment, and so on. Upon engine failure the thrust decay is gradual, and within this reaction time before brake application there is a small speed gain. This is covered in FAR (14CFR) 25.109 Subpart B.

Engineers prefer to replace TOD as the aircraft takeoff field length (TOFL) at the maximum takeoff mass (MTOM), as will be used in this book. The rated TOFL at MTOM is determined by the BFL (see Section 11.6.2) in the event of an engine failure. The BFL must comply with FAR regulations.

11.4   Generalized Takeoff Equations of Motion

Takeoff progresses in an unsteady state as the aircraft accelerates. The generalized forces and the ground run in the takeoff procedure are shown in Figure 11.3, where the following considerations are depicted:

1. The runway has a mean inclination angle δ to the horizon. Typically, δ is small. To prepare the airfield, topography with a large inclination or having large undulations is bulldozed to an acceptable level, but a small inclination can be tolerated.
2. The component of aircraft weight along the velocity vector is $(W\sin\delta \approx W\delta)$, negative in uphill inclination and *vice versa*.
3. There is a wind component V_w along the aircraft velocity vector. A head wind has a negative $(-V_w)$ value and *vice versa*.
4. Aircraft lift is L normal to the velocity vector, and drag is D opposite to the velocity vector.
5. The wheel-rolling coefficient of friction of the tyres with the runway is μ. Therefore the friction force $\mu(W-L)$ is proportional to the aircraft weight acting opposite the aircraft velocity vector.

■ **FIGURE 11.3**
Takeoff. (a) Takeoff forces. (b) Takeoff ground run

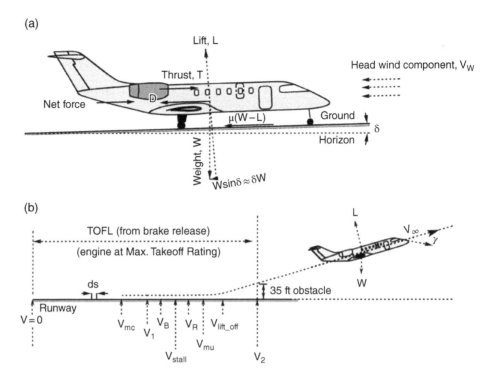

6. Thrust T is the total thrust with all engines operating, aligned to the fuselage axis. If T is in vectored position, then its components are resolved along and perpendicular to the velocity vector.

The following equation can be written in terms of the forces defined thus far. Let the acceleration of the aircraft be denoted by a. The force balance equation resolved along the aircraft velocity vector is given as:

Net force on the aircraft, $(W/g)a = \Sigma$ (applied forces along aircraft velocity vector)

V and a are instantaneous velocity and acceleration of the aircraft on the ground encountering friction. Then at any instant:

$$(W/g)a = T - D - \mu(W - L) - W\delta \tag{11.1}$$

or

$$a = (g/W)(T - D - \mu(W - L) - W\delta)\,(\text{all engines operating}) \tag{11.2}$$

In terms of coefficients and rearranging (all engines operating):

$$\begin{aligned} a &= (g/W)\big[(T - \mu W) - (D - \mu L) - W\delta\big] \\ &= (g/W)\big[(T - \mu W) - (C_D - \mu C_L)Sq - W\delta\big] \end{aligned} \tag{11.3}$$

where S is the wing area and q is the dynamic head.

11.4.1   Ground Run Distance

Let dS_G be the elemental ground run distance covered by the aircraft in infinitesimal time dt. If the instantaneous velocity of the aircraft at that point is V, then

$$dS_G = Vdt = V(dV/dV)dt = (V/a)dV$$

where $dV/dt = a$ (instantaneous acceleration). Takeoff field length (TOFL) is the integrated performance of dS_G expressed as:

$$\text{TOFL} = \int dS_G = \int_0^{V_2}(V/a)dV \qquad\qquad \textbf{(11.4a)}$$

In the generalized situation with a head wind component of velocity $(-V_W)$, Equation 11.4a can be written as follows:

$$\text{TOFL} = \int dS_G = \int_{V_W}^{V_2}\left[(V-V_W)/a\right]dV \qquad\qquad \textbf{(11.4b)}$$

(Wind close to ground has strong shear effects. If the wind velocity is measured at control tower height, then it has to be corrected to aircraft height with a suitable factor less than 1.0. This is not dealt with in detail in this book.)

Substituting the expression of acceleration a from Equation 11.3 into Equation 11.4b, the following can be written.

$$\text{TOFL} = \int dS_G = \int_{V_W}^{V_2}\left(\frac{W}{g}\right)\left(\frac{(V-V_W)dV}{(T-\mu W)-(C_D-\mu C_L)Sq-W\delta}\right)dV \qquad\qquad \textbf{(11.5a)}$$

To keep the computation simple, this book considers takeoff at level airfield $(\delta=0)$ and in still air $(V_W=0)$. Then Equation 11.5a can be simplified to Equation 11.5b.

$$\text{TOFL} = \int dS_G = \int_0^{V_2}\left(\frac{W}{g}\right)\left(\frac{VdV}{(T-\mu W)-(C_D-\mu C_L)Sq}\right)dV \qquad\qquad \textbf{(11.5b)}$$

Aircraft acceleration and engine thrust vary with aircraft speed. The terms of Equation 11.5b are plotted in Figure 11.4, which shows the unsteady nature of the forces resulting in varying levels of aircraft acceleration during the takeoff process.

Analytical integration of Equation 11.5b is difficult, but can be simplified. Integration of an approximated equation is good for analyses in making comparative studies, but suffers in accuracy. To improve accuracy, Equation 11.5b is discretized (Figure 11.4b) to small steps of velocity intervals within which variables are treated as constant at their average condition. The computation process is shown in subsequent sections.

Returning to Equation 11.4a $(\delta=0$ and $V_W=0)$, let $\bar{a}$ be the average aircraft acceleration in the velocity interval from initial velocity V_i to final velocity V_f. Then the TOFL can be discretized as:

$$S_G = \int dS_G \int_0^{V_2}(V/a)dV = \sum_{\Delta V_i}^{\Delta V_f}\Delta S_G \qquad\qquad \textbf{(11.6)}$$

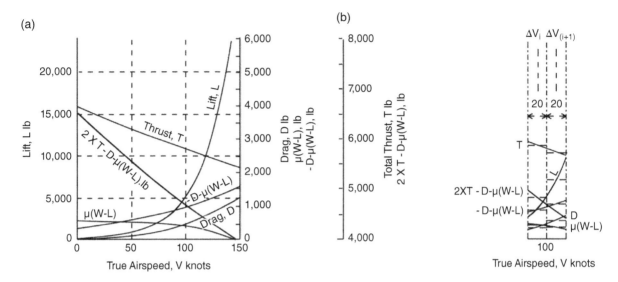

■ **FIGURE 11.4** **Bizjet example of forces during takeoff. (a) Forces in Equation 11.5b (Bizjet). (b) Small steps of ΔV of Figure 11.4a**

where

$$\Delta S_G = \frac{1}{a}\int_{V_i}^{V_f} V dV = \frac{V_f^2 - V_i^2}{2\bar{a}} = \frac{\left(V_f - V_i\right)V_{ave}}{\bar{a}} \qquad (11.7)$$

and where $\bar{a}$ = constant within the interval, noting that $(V_f + V_i)/2 = V_{ave}$ = average velocity, and

$$\bar{a} = \left(\frac{g}{W}\right)\left[\left(\bar{T} - \mu W\right) - \left(\bar{C}_D - \mu \bar{C}_L\right)Sq\right] = (g)\left[\left(\bar{T}/W - \mu\right) - \left(\bar{D}/W - \mu \bar{L}/W\right)\right] \qquad (11.8)$$

Equation 11.7 becomes,

$$\Delta S_G = \frac{W\left(V_f - V_i\right)V_{ave}}{g\left[\left(\bar{T} - \mu W\right) - \left(\bar{C}_D - \mu \bar{C}_L\right)Sq\right]} \qquad (11.9)$$

$$S_G = \int dS_G = \sum_{\Delta V_i}^{\Delta V_f}\Delta S_{Gi} = \sum_{\Delta V_i}^{\Delta V_f}\frac{W\left(V_f - V_i\right)V_{ave}}{g\left[\left(\bar{T} - \mu W\right) - \left(\bar{C}_D - \mu \bar{C}_L\right)Sq\right]} \qquad (11.10)$$

A simplified approach can be taken by taking the average acceleration $\bar{a}$ for aircraft at $0.707V_{LO}$ (average of V_{LO}^2 from zero, that is at $\sqrt{(V_{LO}^2/2)}$ of the following equation). Then

$$S_{G\_LO} = \int dS_G = \frac{1}{\bar{a}}\int_0^{V_{LO}} V dV = \frac{V_{LO}^2}{2\bar{a}} = \frac{WV_{LO}^2}{2g\left[\left(T - \mu W\right) - \left(C_D - \mu C_L\right)Sq\right]_{0.707V_{LO}}} \qquad (11.11a)$$

Naturally, Equation 11.10 is more accurate than Equation 11.11a, but the latter is faster to compute. The difference in the Bizjet example is shown in Section 11.7.

To simplify, the ground run S_{GR} up to V_R is sometimes computed as will be shown. Then the Equation 11.11a can be written as

$$S_{GR} = \int dS_G = \frac{1}{\bar{a}}\int_0^{V_R} V dV = \frac{V_R^2}{2\bar{a}} = \frac{W V_{LO}^2}{2g\left[(T-\mu W)-(C_D-\mu C_L)Sq\right]_{0.707VR}} \quad \textbf{(11.11b)}$$

11.4.2   Time Taken for the Ground Run S_G

At any instant, acceleration, $a = dV/dt$ or $dt = dV/a$. Substituting with Equation 11.3:

$$dt = dV/a = \frac{W dV}{g\left[(T-\mu W)-(C_D-\mu C_L)Sq\right]} \quad \textbf{(11.12)}$$

On integrating,

$$t_{SG} = \int_0^{V_{LO}} \frac{W dV}{g\left[(T-\mu W)-(C_D-\mu C_L)Sq\right]} \quad \textbf{(11.13a)}$$

This is discretized to

$$t_{SG} = \sum_{\Delta V_i}^{\Delta V_f} \Delta t_{SGi} \quad \textbf{(11.13b)}$$

Similar simplified computation for t_{SG} can be computed as $\Delta t_{SG} = (\Delta V/\bar{a})$, and on integrating at $0.707V_{LO}$ or 0.707_{VR} average conditions:

$$t_{SG} = (V_{LO}/\bar{a}) \quad \textbf{(11.13c)}$$

11.4.3   Flare Distance and Time Taken from V_R to V_2

Continuing with the next two phases, we compute distance covered and time taken from (i) rotation to lift-off and (ii) flaring to clearing 35 ft height (representing a possible obstacle – FAR requirement), with the aircraft continuing to climb. The time taken for this short phase needs to be substantiated through flight tests.

Flare distance covered and time taken during takeoff can be analytically determined from the velocity diagram as shown in Figure 11.5. However, a simplified method is presented here with little error, as the time taken to flare is short.

There are two methods to solve flare distance and time taken for both takeoff and landing as follows. Both are approximations, and the difference between the two methods is negligible; typical time taken during flare is about 3–6 seconds.

1. *Average speed/time method:* Here both ends of flare velocities are known and the average velocity for the small duration can be easily established. This method is adopted for takeoff and is quite accurate for the short transition during flaring of aircraft towards climb in Section 11.5.

■ FIGURE 11.5
Flare distance
(see Section 11.6.1
for speed schedules)

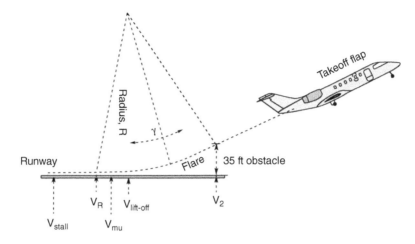

2. *Flare path (circular arc) method:* Here the flare path is approximated to be circular from or up to a point that could be below the obstacle height. This is a more laborious method and is shown for landing flare phase in Section 11.11.

Here the advantage of using Equation 11.13b is evident as the flare procedure starts from V_R. However, using Equation 11.13a will give about the same result. The time difference from V_R to V_{LO} is of the order of 2 seconds.

11.4.4   Ground Effect

During a TOR, the restrictive boundary of the ground affects airflow of the aircraft. The ground offers a cushion effect by restricting the downwash from wing trailing-edge vortices in an equivalent increase in wing aspect ratio, resulting in an increase in C_L and a decrease in induced drag C_{Di}.

Reference [4] gives the relationships of the induced drag changes as follows.

$$\left(C_{Di\_withGE}\right) / \left(C_{Di\_withoutGE}\right) = \left(16h / b\right)^2 / \left[1 + \left(16h / b\right)^2\right] \tag{11.14}$$

where h is the wing height from the ground and b is the wing span.

Typical values of h/b for a low wing are around 0.07–0.12. A high wing is about double that. The ground effect reduces TOFL to an extent. However, in this book, no credit is taken from this benefit from ground effects, adding to the margin of safety.

11.5   Friction – Wheel Rolling and Braking Friction Coefficients

During ground roll at TOR, the aircraft has to encounter drag associated with its contact with ground known as "rolling friction" or "ground friction". Rolling friction on a ground run during takeoff is treated as drag consuming engine power.

In this book we take the following typical values for rolling friction coefficient, μ:

Type 3 runway (concrete pavement)=0.02–0.025 (recommend 0.025 for classroom work)
Type 2 runway (tarmac)=0.025–0.04 (recommend 0.03 for classroom work)
Type 1 runway (unprepared)=0.04–0.3.

Civil aircraft examples in this book use Type 3 airfield, and military aircraft examples use Type 2 and Type 3 airfield. Club usage small aircraft can use Type 1 airfield.
Depending on the surface type, some values of μ are as follows (not considered in this book):

Hard turf=0.04
Grass field=0.04–0.1 (recommend 0.05 for maintained airfield)
Soft ground=0.1–0.3
Wet ice=0.01–0.02
Snow (less than 1 ft)=0.05–0.1.

Wheel braking friction coefficient, μ_b, (Figure 11.6 and Table 11.1) would be much higher depending on the runway surface conditions, for example, dry, wet, slushy, snowy or ice-covered. Locked wheels will skid and will wear out tyres to possible blow-out. Most high-performance

■ **FIGURE 11.6**
Braking friction coefficient

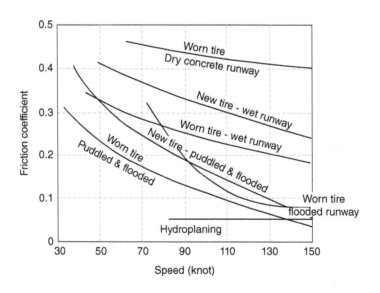

■ **TABLE 11.1**
Average wheel braking coefficient, μ_b (may interpolate in between values)

Aircraft speed (mph)	20	40	60	80	100
Dry concrete runway	0.8	0.7	0.6	0.5	0.4
Wet concrete runway	0.56	0.44	0.35	0.28	0.23
Iced runway	0.1–0.2	0.1–0.2	0.1–0.2	0.1–0.2	0.1–0.2
Light snow	≈0.2	≈0.2	≈0.2	≈0.2	≈0.2

aircraft touching down above 80 kt have anti-skid devices when μ_b could be as high as 0.7. Slipping wheels are not considered during the conceptual study phase. Tyre tread selection should be compatible with runway surface conditions, to avoid hydroplaning, and so on. Take braking friction coefficient, $\mu_b = 0.4$–0.5 as a typical value in this book.

11.6 Civil Transport Aircraft Takeoff

(FAR (14CFR) 25 requires takeoff substantiation for both dry and wet runways. This book only considers dry runway conditions.)

To ensure safety, the takeoff procedure must satisfy airworthiness requirements so that the aircraft will be able to climb to safety or stop safely in case of one critical engine failing. FAR (14CFR) 1.1 defines the "critical engine" whose failure would most adversely affect the aircraft takeoff procedure. Evidently, at failure this is the most outboard engine of multi-engine aircraft, which would generate maximum yaw moment requiring rudder action (as well aileron adjustment) to balance the thrust asymmetry to keep the aircraft flying straight in a yawed attitude. A failed engine loses considerable thrust (half in the case of a twin-engined aircraft) as well creates drag on account of rudder deflection and the yawed state of the aircraft. This severely affects the takeoff procedure, and there are rigorous design and operational requirements enforced by the airworthiness authorities.

Designing for critical engine failure also covers the safety aspects of the failure of non-critical engines for aircraft with more than two engines.

To ensure safety, certification authorities demand mandatory requirements to cater for taking off with one engine inoperative to clear a 35 ft height representing an obstacle, and continue with the required climb gradient (Figure 11.7). The TOFL with one engine inoperative is computed by considering BFL (see Section 11.6.2) when the stopping distance after engine failure at the decision speed, V_1, is the same as the distance taken to clear the obstacle at MTOM.

At takeoff, the ground run is initiated with the aircraft under maximum thrust (takeoff rating) to accelerate, gaining speed until a suitable safe speed is reached, when the pilot initiates rotation of the aircraft by gently pulling back the control stick/wheel (elevator going up) for a lift-off. In case of an inadvertent situation of critical engine failure, provision is made to cater for safety so that the aircraft can continue with takeoff and return immediately to land.

11.6.1 Civil Aircraft Takeoff Segments

The takeoff process with one engine inoperative (FAR (14CFR) 25.107 – Subpart B) can be divided into four segments (Figure 11.7) of climb procedure as follows:

First segment: From 35 ft until undercarriage is retracted.

Second segment: Climb from 35 ft to 400 ft. Engine power is kept at maximum takeoff rating. Drag reduction on account of undercarriage retraction is substantial.

Third segment: In this segment the aircraft stays at 400 ft altitude and accelerates to higher speeds ($V_2 > 10$ kt) to increase the stall speed margin. The aircraft is at clean configuration.

Fourth segment: This is the final segment under takeoff procedure. With one engine inoperative, the aircraft must land immediately at a designated airfield. With all engines operating the aircraft continues to accelerate to en-route climb at a specified speed schedule. The engine is throttled back to maximum continuous rating.

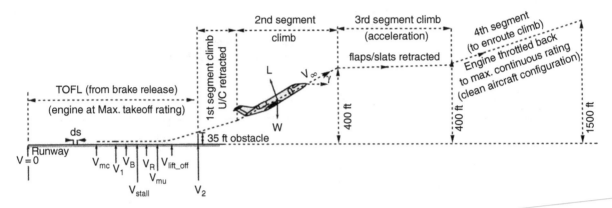

■ **FIGURE 11.7**   Takeoff, first, second, third and fourth segments of climb

■ **TABLE 11.2**
Civil aircraft takeoff segment requirements and status [FAR (14CFR) 25.121 Subpart B]

	First segment	Second segment	Third segment	Fourth segment
Altitude (ft)	35 ft to u/c retract	Reach 400 ft	Level at 400 ft	Reach 1500 ft
Speed (reference)	$V_{lift\text{-}off}$	V_2	$V_2 + 10$ kt	$V_2 + 10$ kt
Engine rating	Maximum takeoff	Maximum takeoff	Maximum takeoff	Maximum cont.
Undercarriage	*Retracting*	Retracted	Retracted	Retracted
Flaps/slats	Extended	Extended	*Retracting*	Retracted
Maximum climb gradient (%)				
Two-engine	0.0	2.4	0	1.2
Three-engine	0.3	2.7	0	1.5
Four-engine	0.5	3.0	0	1.7

After the fourth segment climb, the aircraft proceeds with all-engine en-route climb at a prescribed speed schedule, with the engine further throttled back to maximum climb rating. Table 11.2 gives the aircraft configuration and power settings in the climb segments.

The takeoff climb segments have FAR requirements, but the en-route climb capability is a customer requirement and not a FAR requirement.

The equations for climb are dealt with in Chapter 12, with worked examples. Figure 11.5 also gives the various speed schedules during the takeoff run. Since this is what the pilot reads on the flight-deck instrument, it has to be the calibrated speeds, V_{CAS}. At sea-level on an ISA day, $V_{CAS} = V_{EAS}$, being at low speed. These speed schedules are explained below.

V_{stall}: This is the aircraft stall speed covered in FAR (14CFR) 25.103. The formal definition of V_{stall} is as follows:

$$\text{Reference stall speed } V_{SR} \geq \frac{V_{CLmax}}{\sqrt{n}}, \quad \text{where } n \text{ is the load factor} \qquad (11.15)$$

The load factor occurs during rotation of the aircraft and depends on the rate of rotation. However, to keep treatment simple, this book considers $n = 1$ and uses the term V_{stall}.

V_{mc}: This is the minimum control speed covered in FAR (14CFR) 25.149 Subpart B. V_{mc} is the minimum speed when the aircraft can be aerodynamically controlled in case of a critical engine failed. During the control, the aircraft is to recover to fly straight, not deviating more than 30 ft from the centreline of the runway. Below this speed the rudder is not effective to control the asymmetry created by the critical engine failure. Aircraft designers must size the control surfaces to make $V_{mc} < V_1$ (decision speed – see below). (The FAR has finer definition of V_{mc} into two categories of V_{mcg} and V_{mca}, standing for minimum control speed on the ground and minimum control speed in the air, respectively. Generally, design for V_{mcg} covers the requirements for V_{mca}.)

V_1: This is the decision speed covered in FAR (14CFR) 25.107 Subpart B. This is the maximum speed below which critical engine failure would result in the aircraft not being able to satisfy takeoff within the specified field length, but will be able to stop with brake application. If an engine fails above V_1 speed, the aircraft should continue with takeoff operation. V_1 should be higher than V_{mc}, otherwise at the loss of one engine the aircraft cannot be controlled if continuing with the take-off procedure. Aircraft stall speed for the configuration has $V_{stall} > V_1$.

V_R: This is the rotation speed covered in FAR (14CFR) 25.107 Subpart B. As the aircraft gains speed, reaching V_R, the pilot gently pulls back the control-stick to rotate the aircraft, raising the nose up at increasing angle of incidence to generate lift for a takeoff. It should be $\geq 1.05 V_{mc}$ and higher than V_1. Once this is done, reaching V_{LO} and V_2 occurs as a fallout of the action. V_R should be more than V_{stall}. Flight test determines V_R in such a way that at the end of rotation the aircraft reaches V_{mu} (minimum unstick speed).

V_{mu}: This is a minimum unstick speed covered in FAR (14CFR) 25.107 Subpart B. V_{mu} is the speed above which the aircraft should be able to leave the ground as a result of the pilot rotating the aircraft. This should be slightly above V_R. In fact, V_{mu} decides V_R. If the pilot makes an early rotation, then V_{mu} may not be sufficient for a lift-off and the aircraft will tail-drag until it gains sufficient speed for the lift-off.

V_{LO}: This is the speed at which the aircraft lifts off the ground, covered in FAR (14CFR) 25.107 Subpart B. V_{LO} is closely associated with V_R. In the case of one engine inoperative, it should have $V_{LO} \geq 1.05 V_{mu}$ (one engine inoperative) and $V_{LO} \geq 1.08 V_{mu}$ (all engines operative).

V_2: This is the takeoff climb speed at 35 ft height, covered in FAR (14CFR) 25.107 Subpart B. The first segment climb starts from clearing the 35 ft height. V_2 is also known as the first segment climb speed. It is closely associated with V_R. FAR requires that $V_2 = 1.2 V_{stall}$ (can be higher) and $V_2 = 1.13 V_{mc}$.

V_{EF}: This is the engine failure speed covered in FAR (14CFR) 25.107 Subpart B. There is a recognition time of 1–3 seconds required for the pilot to take action.

V_B: This is the brake application speed at one engine failure, covered in FAR (14CFR) 25.109 Subpart B. The brake is applied after an engine has failed and $V_B > V_1$.

It may be noted that only V_{stall} is deterministic from aircraft weight, and C_{Lmax} and FAR require that $V_2 \geq 1.2 V_{stall}$. All other speed schedules are interrelated with the regulation requirements. At the conceptual design stage, the ratios are taken from the statistics of past designs, and as the project progresses these are refined initially from wind tunnel tests and finally from flight tests. Table 11.3 gives the important speed schedule ratios. (Climb is dealt with in Chapter 12.)

V_R is selected such that V_2 is reached at 35 ft altitude. Much depends on how the pilot executes rotation; FAR gives the details of requirements between maximum rate and normal rates of rotations. From the FAR requirements it can be seen that $V_2 > V_{LO} > V_{mu} > V_R > V_{stall} > V_1 > V_{mc}$ (see Table 11.4) .

■ **TABLE 11.3**
Airworthiness regulations at takeoff

| | Civil aircraft design | | Military aircraft design |
	FAR23	FAR25	MIL-C5011A
Take-off climb speed at 35 ft height V_2	$\geq 1.2V_{stall}$	$\geq 1.2V_{stall}$	$\geq 1.2V_{stall}$
Lift-off velocity, V_{LO}	$\geq 1.1V_{stall}$	$\geq 1.1V_{stall}$	$\geq 1.1V_{stall}$
Obstacle height at V_2 (ft)	50	35	50
TOFL (BFL)			
One engine failed	As computed	As computed	As computed
All engines operating	As computed	1.15×as computed	As computed
Rolling coefficient, µ	Not specified	Not specified	0.025

■ **TABLE 11.4**
Civil aircraft takeoff speed schedule (FAR requirements)

| Speed schedule | All engines operating | | | One engine inoperative | | |
	Two-engine	Three-engine	Four-engine	Two-engine	Three-engine	Four-engine
V_2/V_{stall}	≥ 1.20	≥ 1.20	≥ 1.20	≥ 1.20	≥ 1.20	≥ 1.20
V_{LO}/V_{mu} (approximately)	≥ 1.10	≥ 1.10	≥ 1.10	≥ 1.05	≥ 1.05	≥ 1.05
V_R/V_{mc} (approximately)	≥ 1.05	≥ 1.05	≥ 1.0	≥ 1.05	≥ 1.05	≥ 1.05

■ **FIGURE 11.8**
Takeoff velocity ratios at lift-off

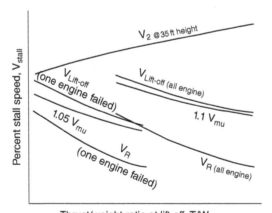

The higher the thrust loading (T/W), the higher the aircraft acceleration. For small changes, V_R/V_{stall} and V_{LO}/V_{stall} may be linearly decreased with increase in T/W. The decision speed V_1 with an engine failed is established through iterations as shown in Section 11.8. In a family of derivate aircraft, the smaller variant can have V_1 approach close to V_R. Figure 11.8 shows the relative magnitudes of speed schedules in terms of V_{stall} and $(T/W)_{LO}$. The FAR speed schedule at takeoff can be an involved one to ensure safety.

Some engines at takeoff rating can have an augmented power rating (APR) which can generate 5% higher thrust than the maximum takeoff thrust for short periods of time. These types of engines are not considered in this book, but that amount of extra thrust when an engine fails offers some enhanced safety margin.

11.6.2   Balanced Field Length (BFL) – Civil Aircraft

The rated TOFL at MTOM is determined by the BFL in the event of an engine failure. The BFL must comply with FAR regulations. Normal takeoff with all engines operating will take considerably lower field length than the rated TOFL, but FAR requires 15% additional field length over the estimated value. Designers must provide the decision speed V_1 to pilots below which, if an engine fails, then the takeoff has to be aborted for safety reasons. Figure 11.9 shows the segments involved in computing BFL. It can be seen that taking off with one engine inoperative (failed) above the decision speed V_1 has three segments to clear 35 ft height, as follows:

Segment A: Distance covered by all engines operating ground run until one engine fails at the decision speed V_1.

Segment B: Distance covered by one engine inoperative acceleration from V_1 to V_{LO}.

Segment C: Continue with the flare distance from lift-off speed V_{LO} to clear 35 ft obstacle height, reaching aircraft speed V_2.

For stopping at decision speed V_1, there are two segments (which replace segments B and C), as follows.

Segment D: Distance covered during the reaction time for the pilot to take braking action after an engine fail. (Typically, 3 seconds is used as the time for pilot recognition, brakes to act, spoiler deployment, and so on. At engine failure the thrust decay is gradual, and within this reaction time before brake application there is a small speed gain shown in the diagram.) Typically $V_B \geq V_1$, but can be taken as $V_B \approx V_1$.

Segment E: Distance to stop from V_B to V_0 (maximum brake effort).

Balanced field length is established when:

$$\text{Segments}\left(B+C\right)=\text{Segments}\left(D+E\right) \tag{11.16}$$

■ FIGURE 11.9
Segments of balanced field length

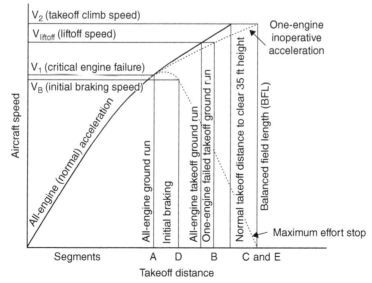

Normal takeoff with all engines operating (TOFL$_{all\_eng}$) will take considerably lower field length than the case of one engine failed case (TOFL$_{1eng\_failed}$). Therefore, TOFL must be at least equal to the greater of the two as shown below:

$$\text{TOFL} = \text{higher of BFL and } 1.15 \times \text{TOFL}_{all\_eng} \tag{11.17}$$

The equation for retardation (negative acceleration) after brake application can be taken from Equation 11.8. During brake application, thrust is kept as zero.

$$\bar{a} = \left(\frac{g}{W}\right)\left[(-\mu_B W) - \left(\bar{C}_D - \mu\bar{C}_L\right)Sq\right] = (g)\left[(-\mu_B) - \left(\bar{D}/W - \mu\bar{L}/W\right)\right] \tag{11.18}$$

11.6.2.1   Unbalanced Field Length (UBFL) – Civil Aircraft   At low aircraft weight the decision speed V_1 at one engine inoperative can become lower than V_{mc}. V_1 should not be below V_{mc} because it results in unbalanced field length (UBFL), as segments (B+C) < segments (D+E). In this case, TOFL is taken as segments (A+D+E). In another situation, if V_2 < 110% of V_{mc}, it results in UBFL when TOFL is taken as segments (A+B+C). UBFL is not computed in this book.

11.6.3   Flare to 35 ft Height (Average Speed Method)

Equations 11.10 and 11.13a give the distance covered and time taken respectively from brake release to lift-off. To continue with computation of the next phase is to compute distance covered and time taken from lift-off to clearing 35 ft height (FAR requirement) as the aircraft flares out towards climb. The time taken for this short phase needs to be ascertained through flight test.

During takeoff, FAR requires $V_2/V_{stall} > 1.2$. Typically around 5% change occurs in aircraft velocity through acceleration at the most critical condition; it amounts to about 5–7 kt speed increase.

Typically, the flare time t_{flare} is about 2–6 seconds depending on aircraft configuration, thrust level and ambient conditions. In extreme situations it can take more than 6 seconds. The Bizjet example in this book takes t_{flare} = 3 seconds to flare out and reach 35 ft height, and 4 seconds for larger jet transport aircraft.

More precisely, $V_{flare} = (V_{LO} + V_{@35ft})/2$. Then the distance covered during flare:

$$S_{flare} = \left(t_{flare}\right) \times \left(V_{flare}\right) \tag{11.19}$$

Given below are typical takeoff speeds, V_2 (with high-lift device deployed), for some of the larger jet transports currently in operation.

- Boeing 737: 150 mph
- Boeing 757: 160 mph
- Boeing 747: 180 mph
- Airbus 320: 170 mph
- Airbus 340: 180 mph
- Concorde: 225 mph

11.7   Worked Example – Bizjet

The first task is to prepare a graph like that of Figure 11.4 to obtain the varying force parameters during the ground run. Table 11.5 may be used for Bizjet aircraft takeoff estimation at sea-level on an ISA day, to prepare a spreadsheet for repeated computations. Aircraft lift develops from zero to the level of MTOW as speed reaches lift-off speed.

Relevant Bizjet data include:

- Ramp maximum mass = 20,720 lb (9420 kg).
- Wing area, S_W = 323 ft² (30 m²).
- MTOM = 20,680 lb (9400 kg) – 49 lb taxi fuel consumed.
- Landing weight = 15,800 lb (7182 kg).
- Rolling friction coefficient on paved runway, μ = 0.025.

Use a graph like that of Figure 11.4 and add undercarriage C_{D_UC} = 0.0124. C_L and C_D are taken from Figure 9.2 and adjusted for flap setting and one engine inoperative configuration (see Chapter 9). Intermediate C_L and C_D are extrapolated. Equation 11.7 gives the ground distance covered, $S_G = V_{ave} \times (V_f - V_i)/\bar{a}$.

$$\text{Dynamic head, } q = \tfrac{1}{2}\rho V^2 = 0.5 \times 0.002378 V^2 = 0.001189 V^2$$

Equations used are:

$$C_L = \frac{2 \times \text{MTOM}}{\rho S_W V^2} = \frac{20680}{0.001189 \times 323 \times V^2} = \frac{53847.6}{V^2}, \qquad V_{stall} = \sqrt{\frac{2 \times \text{MTOM}}{\rho S_W C_L}}$$

Table 11.5 gives the Bizjet aircraft data generated thus far. To make the best use of available data, all computations are done in the FPS system. The results can be converted to the SI system subsequently.

■ **TABLE 11.5**
Bizjet performance parameters (takeoff/landing – W/S_W = 64 lb/ft²)

Flap setting (°)	0	8	20	Landing[a]
C_{Lmax}	1.55	1.67	1.9	2.2
C_{DPmin}	0.0234	0.0234	0.0234	0.0234
ΔC_{Dflap}	0	0.013	0.032	0.06
$\Delta C_{D\_U/C}$	0.0124	0.0124	0.012	0.012
$\Delta C_{D\_one\_eng}$ (fuselage mount)	0.003	0.003	0.003	0.003
Rolling friction coefficient, μ	0.03	0.03	0.03	0.03
Braking friction coefficient, μ_B	0.45	0.45	0.45	0.45
V_{stall} at 20,680 lb (ft/s)	186.42	179.01	167.83	136.16
V_{stall} (kt)	110.38	106.00	99.38	80.66
V_R (ft/s) (1.11V_{stall})	205.08	196.91	184.61	149.77
V_R (kt)	121.5	116.60	109.32	88.7
V_2 (ft/s)	223.70	214.82	201.39	163.40
V_2 (kt) (1.2V_{stall})	132.46	127.20	119.25	96.75
T/W – all engine	0.34	0.34	0.34	0
T/W – single engine	0.17	0.17	0.17	0
V_{LO} (ft/s)	214.38	205.87	193.00	156.58
V_{LO} (kt) (1.15V_{stall})	126.94	121.90	114.28	92.72

[a] Landing flap is at 40°, engines are kept at idle and V_{stall} is computed at aircraft landing weight of 15,800 lb (Section 11.11 uses these data).

11.7.1   All-Engine Takeoff

Table 11.6 gives the values for a Bizjet all-engine TOFL with 20° flap. Due to the extended undercarriage and flap deflection, Bizjet aircraft lift to drag ratio degrades to a typical value of approximately 10.

11.7.2   Flare from V_R to V_2

Take the flare time as 3 seconds for the Bizjet. Take the average $V_{Rto\_V2} = (184.61 + 201.39)/2 = 193$ ft/s.

$$S_{Gflare} = 3 \times 193 = 579 \, \text{ft}$$

Computed all-engine operating field length $= 2182 + 579 = 2761 \, \text{ft}$

Time taken, $t_{all\_eng} \approx 23.71 + 3 \approx 27$ seconds

To ensure safety, the FAR requires the all-engine operating TOFL to have a 15% higher margin than the computed value, and it may exceed the value obtained for one-engine inoperative BFL. In that case the higher value of the two is used.

FAR all-engine operating field length at 20° flap, $FL_{all\_eng} = 1.15 \times 2761 = 3175 \, \text{ft}$.

Corresponding time taken, $t_{all\_eng} = 1.15 \times 27 \approx 31$ seconds.

Comparing with the simplified method to compute FL_{all_eng}, take average acceleration at $0.707 V_{LO} = 0.707 \times V_{LO} = 0.707 \times 192.91 = 136.387$ ft/s. Then $\bar{a} = 8.16$ ft/s².

$$S_{G\_all\_eng} = \frac{V_{LO}^2}{2\bar{a}} = \left(192.91^2\right)/\left(2 \times 8.16\right) = 2278 \, \text{ft/s}$$

This compares with 2182 ft in Table 11.7. It is ≈1% less than the detailed estimation. It is strongly recommended that the readers undertake the rigorous method of computation.

■ **TABLE 11.6**

Forces during Bizjet all-engine ground run with 20° flap

						V_{stall}	V_R	V_{LO}
V (kt)	0	20	40	60	80	99.38	109.3	114.28
V (mph)	0	23.02	46.04	69.06	92.1	114.4	125.8	131.54
V (ft/s)	0	33.76	67.52	101.28	135.04	167.8	184.5	192.91
q	0	1.355	5.421	12.196	21.682	33.5	40.5	44.25
$S_w q$	0	437.71	1750.85	3939.4	7003.4	10,807.6	13,077.6	14,292.2
C_L	0	0.2	0.4	0.6	0.8	1	1.2	1.446
C_D	0	0.02	0.04	0.06	0.08	0.1	0.12	0.1446
L	0	87.543	700.34	2363.65	5602.7	10,807.6	15,693.0	20,666.6
D	0	8.754	70.03	236.37	560.27	1080.8	1569.3	2066.7
$2T$	6630	6460	6280	6120	5960	5800	5720	5640
$\mu(W-L)$	517	514.81	499.5	457.91	376.93	246.81	124.7	0.3357
$[-D-\mu(W-L)]$	−517	−523.6	−569.53	−694.27	−937.2	−1327.6	−1694.0	−2067.0
$[2T-D-\mu(W-L)]$	6113	5936.4	5710.5	5425.73	5022.8	4472.4	4026.0	3573.0

■ **TABLE 11.7**
All-engine ground run calculation to V_{LO} (continuing from Table 11.6) at 20° flap, $W/g = 642.236$, $\mu W = 517$ (average values between the two speeds for ΔV)

V (kt)	0	20	40	60	80	99.38	109.3	114.23
ΔV (kt)	0	20	20	20	20	19.38	9.94	14.85
ΔV (ft/s)	0	33.76	33.76	33.76	33.76	32.71	16.8	25.0668
V_{ave} (ft/s)	0	16.88	50.64	84.4	118.16	151.40	176.2	180.29
q		0.34	3.05	8.47	16.60	27.25	36.9	38.646
$S_W q$		109.43	984.85	2735.71	5361.98	8802.73	11,915.6	12,482.81
$2T_{ave}$		6545	6370	6200	6040	5880	5760	5570
C_{Lave}		0.1	0.3	0.5	0.7	0.9	1.1	1.223
C_{Dave}		0.01	0.03	0.05	0.07	0.09	0.11	0.1223
$(C_D - \mu C_L)S_W q$		0.8207	22.2	102.6	281.504	594.2	983.0	1145.0
$2T_{ave} - \mu W$		6028	5853	5683	5523	5363	5243	5053
a (ft/s²)		9.385	9.079	8.689	8.161	7.425	6.63	6.085
ΔS_G (ft)	0	60.723	188.3	327.924	488.78	667.00	445.6	742.68
S_G (ft)	0	60.72	249.03	576.95	1065.73	1732.7	2182.0	2475.41
Δt (s)	0	3.597	3.72	3.9	4.15	4.41	2.53	1.40
t (s)	0	3.60	7.32	11.22	15.37	19.78	22.31	23.71

11.7.3 Balanced Field Takeoff – One Engine Inoperative

Section 11.6.2 gives the BFL takeoff to ensure that the aircraft can stop within the airfield in case an engine fails. It requires a decision speed V_1 below which the aircraft stops, and above which the aircraft continues with the takeoff operation with power available to maintain the FAR stipulated minimum climb gradient up to the second segment, and then return to land immediately. At first, V_1 is estimated, and subsequently BFL for the takeoff is established.

Table 11.5 gives the Bizjet aircraft data generated thus far. To make the best use of available data, all computations are done in the FPS system. The results can be converted to the SI system subsequently.

11.7.3.1 Segment A – All Engines Operating Up to the Decision Speed V_1
To determine decision speed V_1, estimate five speeds, for example 85, 90, 95, 100 and 105 kt. The all-engine operating case up to these speeds can be taken from Table 11.7.

11.7.3.2 Segment B – One Engine Inoperative Acceleration from V_1 to Lift-off Speed, V_R
This phase is a transient one; the aircraft leaves the ground to become airborne. The lift and drag characteristics change rapidly. For an aircraft in a high drag configuration with one engine inoperative, drag estimation is quite difficult. Therefore, a simplified approach is taken to compute distance covered and time taken from V_1 to lift-off speed, V_{LO}. The simplification gives a reasonable result and conveys the physics involved.

In such a dirty configuration the lift to drag ratio will be low, of the order of around 9.5. On the ground below rotation speed, the lift is low and so is drag is low; the simplification takes the same value of $(L/D) \approx 9.5$. Therefore an average value of $(C_D/C_L) \approx 0.105$ is taken. Typical average $C_L \approx 0.8$ is used, but the weight on the wheels is lower on account of some lift being generated on the wing. Because there is one engine inoperative, power is halved ($T/W = 0.17$). The simplified method is used as the difference will be small.

The velocity that would give average acceleration is $V_{0.7} = 0.7 \times (V_R - V_1) + V_1$

■ **TABLE 11.8**
Bizjet one-engine ground distance V_1 to V_{LO} (Segment B) at 20°

	85	90	95	100	105	109.32
V_1 (kt) (estimate)	85	90	95	100	105	109.32
V_1 (ft/s)	143.6	152.0	160.44	168.88	177.32	184.62
V_{stall} (ft/s)	167.83	167.83	167.83	167.83	167.83	167.83
V_{LO} (ft/s)	193	193	193	193	193	193
V_{ave} (ft/s)	168.3	172.5	176.72	181.0	185.16	188.81
ΔV (ft/s)	49.45	41.00	32.56	24.12	15.676	8.38
$V_{0.707}$ (ft/s)	178.16	180.7	183.23	185.76	188.3	190.5
$\bar{a}_{ave}$ (ft/s²) (one engine)	4.23	4.2	4.16	4.13	4.09	4.06
$\bar{a}_{ave}$ (ft/s²) (all engines)	7.98	7.7	7.55	7.4	7.0	6.5
$\bar{a}_{ave}$ (ft/s²)	6.10	5.95	5.86	5.76	5.55	5.28
S_G (ft) (Segment B)[a]	1162.94	1048.0	892.22	706.77	501.21	292.98
Segment (B+C)[b]	1557.33	1442.39	1286.62	1101.17	895.61	687.38
Δt (s)	6.91	6.08	5.05	3.90	2.71	1.55

[a] Can be refined by taking more step intervals of ΔV of 10 kt, that is, at steps of 90, 95, 100, 105, 110 and 114.28 kt (as in Tables 11.6 and 11.7), instead of taking average velocity as shown in the table. The refined value gives 1160 ft instead of 1048 ft when the decision speed slightly changes – iteration is required. Then segments (B+C)=1554 ft. The readers may do it the hard way.
[b] Segment C is given in Table 11.9.

Taking Equation 11.8,

$$\bar{a} = (g)\left[\left(\bar{T}/W - \mu\right) - \left(\bar{D}/W - \mu\bar{L}/W\right)\right]_{0.7V}$$

$$= (g)\left[\left(\bar{T}/W - \mu\right) - \left(\frac{0.5\rho S_w V^2 \bar{C}_L}{W}\right)\left(\frac{\bar{C}_D}{C_L} - \mu\right)\right]_{0.7V}$$

$$= 32.2 \times \left[\left(0.17 - 0.025\right) - \left(\frac{0.5 \times 0.002378 \times 323 \times V_{0.7}^2}{20{,}680}\right)\left(0.105 - 0.025\right)\right] \quad \textbf{(11.20)}$$

$$= 32.2 \times \left(0.1675 - 0.000951 \times V_{0.7}^2 / 831.2\right)$$

$$= 32.2 \times \left(0.1675 - 0.00000114 \times V_{0.7}^2\right)$$

Braking takes place after an engine fails. With the loss of thrust on account of engine failure, the aircraft retards to a lower level of acceleration. Therefore, for the interval of computation, the average value has to be taken.

$$V_{ave} = 0.5\left(a_{all\_engine} + a_{one\_engine\_failed}\right)$$

Equation 11.7 gives the ground distance covered, $S_G = V_{ave} \times (V_f - V_i)/\bar{a}$. Table 11.8 computes the ground distance covered from V_1 to V_R for the two flap settings.

11.7.3.3 Segment C– Flaring Distance with One Engine Inoperative from V_R to V_2
This is the flaring distance to reach V_2 from V_R. From statistics, time taken to flare is 2 seconds. Table 11.9 computes the ground distance covered from V_{LO} to V_2 with one engine inoperative. In this segment the aircraft is airborne; hence there is no rolling friction. Taking the average velocity between V_2 and V_R gives the distance covered during flare.

■ **TABLE 11.9**
Bizjet one-engine ground distance V_R to V_2 (Segment C)

V_{LO} at $1.15V_{stall}$	193.00
V_2 at $1.2V_{stall}$	201.39
V_{ave}	197.20
S_{Gflair} (3 s) (ft)	591.6

11.7.3.4 Segment D– Distance Covered during Pilot Recognition and Braking

The next step is to compute the stopping distance with maximum application of brakes. Flap settings are of little consequence. We assume it takes 1 second for the pilot to recognise the need and apply the brakes, and 2 seconds for the brakes to act. Table 11.10 shows the distance covered from V_1 to V_B during Segment D.

11.7.3.5 Segment E – Braking Distance from V_B to Zero Velocity

The reaction time to apply the brake after the decision speed, V_1, is 3 seconds. The aircraft continues to accelerate a little in the 3 seconds, but speed returns to V_B – this is ignored. Table 11.11 computes the ground distance covered from V_B to stop, Segment E. Again, the flap settings are of little consequence.

Assume the aircraft is in full brake with $\mu_B = 0.4$, all engines shut down and average $C_L = 0.5$. Using Equation 11.3, the average acceleration based on $0.707V_B$ ($\approx 0.707\,V_1$) reduces to:

$$\bar{a} = 32.2 \times \left[(-0.4) - (C_L q / 64)(0.1 - 0.4) \right] = 32.2 \times \left[-0.4 + (0.15q / 64) \right]$$
$$= 32.2 \times \left[-0.4 + (q / 426.7) \right] = 0.075q - 12.88$$

Select the distance when segments (B+C) equal segments (D+E) as shown in Figure 11.10a. With 20° flap and decision speed V_1 is 90 kt, the distance is 1420 ft. Then TOFL becomes

■ **TABLE 11.10**
Bizjet failure recognition distance (Segment D)

V_1 (kt) (estimate)	85	90	95	100	105
V_1 (ft/s)	143.55	152.00	160.45	168.88	177.32
Distance in 3 s at V_1, S_{G_B} (ft)	430.64	456.0	481.31	506.64	531.97

■ **TABLE 11.11**
Bizjet stop distance (Segment E)

$V_1 = V_B$ (kt)	85	90	95	100	105
V_1 (ft/s)	143.548	151.992	160.436	168.88	177.324
V_{ave} (ft/s)	71.774	75.996	80.218	84.44	88.662
ΔV (ft/s)	143.55	152.00	160.44	168.88	177.33
$V_{0.707}$ (ft/s)	100.48	106.39	112.30	118.22	124.13
q	12.00528	13.4592	14.99621	16.6163	18.31947
$\bar{a}_{ave}$ (ft/s²) (retard)	−11.9796	−11.8706	−11.7553	−11.6338	−11.506
Segment E	860.05	973.06	1094.81	1225.76	1366.40
Segment (D+E)	1290.69	1429.04	1576.12	1732.40	1898.38

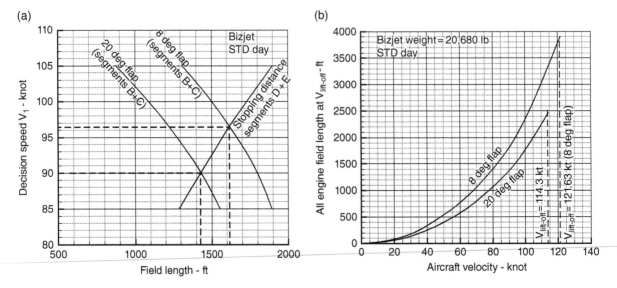

■ **FIGURE 11.10** Field length. (a) Balanced field length. (b) All-engine field length

■ **TABLE 11.12**
Bizjet decision speed (Figure 11.8)

	BFL	All engine	BFL	All engine
Flap (°)	8	(Figure 11.10b)	20	(Figure 11.10b)
Decision speed, V_1 (kt)	96.5		90	
Rotation speed, V_R (kt)	116.6		109.32	
Lift-off speed, V_{LO} (kt)	121.9		114.28	
V_2 (kt)	127.9		119.72	
Distance (ft) at (B+C)=(D+E) (ft)	1620		1440	
All-engine ground run S_G up to V_1 (ft)		≈1600		≈1500
TOFL (ft)	3220	Not computed	2940	2761
All-engine increase by 15%				3175

1355+1400=2755 ft. With 8° flap the decision speed V_1 is 96.5 kt, with the distance covered 1620 ft. This is summarized in Table 11.12. Both satisfy the specified TOFL requirement of 4400 ft.

The higher value of the field length is retained for the all-engine operating field length=3175 ft. The TOFL requirement is 4400 ft. The recommended takeoff procedure is with 20° flap. At lower takeoff weights and/or having a longer runway length (4400 ft), the pilot may choose 8° flap, which gives a better second segment climb gradient.

11.7.3.6 Discussion An increase of flap setting improves the BFL capability at the expense of loss of climb gradient (which will be shown in the next chapter). With one engine inoperative, the percentage loss of thrust for a two-engine aircraft is the highest (50% lost). At one engine failed the aircraft acceleration suffers and the ground run taken from V_1 to lift-off is higher. Table 11.13 summarizes the takeoff performance and the associated speed schedules for the two flap settings, while Figure 11.11 shows the summary graphically for the 8° flap setting. This is an example of the procedure. The ratio of speed schedules can be made to vary for pilot ease, as long as it satisfies FAR requirements.

■ **TABLE 11.13**
Bizjet takeoff field length summary (Figure 11.11)

Flap setting	8°		20°	
	kt	ft/s	kt	ft/s
V_{stall} at 20,680 lb	106	179.	99.38	167.83
$V_2 = 1.2 V_{stall}$ at 35 ft altitude[a]	127.2	214.82	119.26	201.39
V_1 decision speed	96.5	163.74	90.00	151.92
V_{LO} at $1.15 V_{stall}$	121.90	205.8	114.28	193.00
$V_R = 1.05 V_{stall}$	116.60	196.92	109.32	184.61
V_{mu} at $1.02 V_R$ (lower than V_{LO})	119.00	200.85	111.50	188.30
V_{mc} at $0.94 V_1$[b]	91.2	153.9	84.4	142.8
BFL (ft)	3220		2940	
All-engine takeoff time (s)	≈31			

[a] If required, V_2 can be higher than $1.2 V_{stall}$.

[b] Aircraft control surfaces (mainly rudder) are sized to get V_{mc} below V_1 so the pilot has the speed margin. Here, it is arbitrarily chosen as 75 kt – needs to be computed (a design task beyond the scope of this book). This is because the lowest V_1 is when the aircraft is lightest (say at the landing weight of 15,800 lb) and at the highest flap deployment (for an inadvertent case, it can be at 40° flap, otherwise 20° flap). This point is often overlooked in aircraft design coursework.

■ **FIGURE 11.11**
Takeoff speed schedule with example distances and speeds

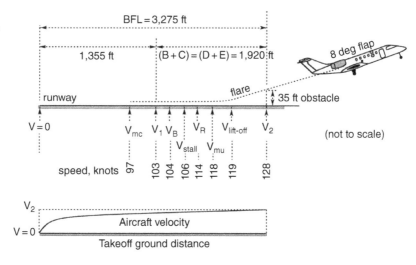

Higher flap settings would give more time between decision speed V_1 and the rotation speed V_R. However, it is not a problem if V_1 is close to V_R so long as there is sufficient pilot reaction time available if one engine fails; if not, then in a very short time the rotation speed V_R is reached and the aircraft takes off (typically, BFL could be considerably less than the available airfield length). Also V_{mu} is close to V_R, hence there is little chance for tail dragging. If the engine fails early enough, then the pilot has sufficient time to recognize it and act to abort takeoff.

With more than two engines, the decision speed V_1 is further away from the rotation speed V_R. The pilot must remain alert as the aircraft speed approaches the decision speed V_1 and must react quickly if an engine fails.

The readers should compute for other weights to prepare graphs for an ISA day and ISA + 20 °C day, as shown in Figure 11.12.

11.8   Takeoff Presentation

Section 1.15 indicated that aircraft performance is documented in different formats for specific usages. Takeoff presentation in graphical form differs from what aircraft performance engineers use and from what pilots have in their user manual. Engineers like to have specific takeoff plots for those parameters of topical analyses for each atmospheric day. Engineers make use of separate graphs for aircraft with different gradient and/or wind conditions. One of the standard plots shows BFL as a function of aircraft up to MTOM for various altitudes for each ambient condition, as shown in Figure 11.12. To meet the FAR climb gradient with one engine inoperative, aircraft need to unflap to 15° from 25° to operate from above sea-level altitude.

On the other hand, pilots find it convenient to have different types of takeoff charts with pertinent data in one page plotted in a *"web chart"*, with parameters arranged in a *"chase-around"* fashion to obtain the TOFL for the MTOM, as shown in Figure 11.13. Performance engineers prepare the web chart for the pilots.

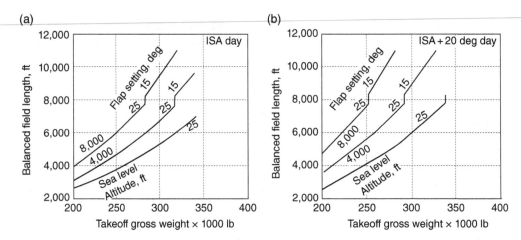

■ **FIGURE 11.12**   Balanced field lengths that engineers use (not of the Bizjet example, but of a similar class). (a) For an ISA day. (b) For an ISA + 20 °C day

■ **FIGURE 11.13**
Balanced field
length web chart
such as pilots use

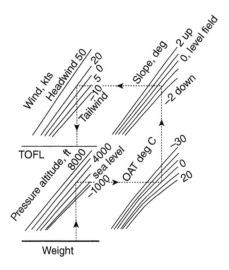

EXAMPLE 11.1

Using Figure 11.13, to obtain the TOFL for the MTOW at a 4000 ft altitude airfield on an ISA day with a runway slope 1° down and a headwind component of 20 mph, start with the MTOW, run vertically up to 4000 ft altitude runway curve then move right horizontally to the ambient condition of an ISA day. Next proceed vertically upward to

the curve for the 1° slope down runway, and then go left horizontally to the 20 mph headwind curve, and finally read vertically down to the TOFL.

Nowadays, the pilot manual is embedded in computers and chase-around graphs have become redundant. Accurate results can be obtained by a few clicks of buttons.

11.8.1 Weight, Altitude and Temperature Limits

Evidently, the aircraft takeoff performance is affected by its weight, runway altitude (elevation) and ambient temperature. BFL increases substantially with increase of airfield altitude and temperature. If the available runway length is not sufficient, then aircraft operation is not possible; there is an operational restriction on takeoff as a safety measure. The aircraft has to sacrifice loads by offloading payload and/or fuel to take off. This is what is known as weight, altitude and temperature limits, in short, WAT limits. These are not dealt with in this book, but readers can work them out if engine performance data for the hot day/elevated airfield are available – the computational methodology is the same. The readers are to estimate the maximum aircraft weight allowed to take off from the available airfield length, and the thrust or power available for the ambient conditions.

Airlines operators have regulatory takeoff weight (RTOW) charts for each aircraft type and for each airfield (runway) of operation. These charts give the limiting takeoff weights for the topographical situation of the runway at all possible weather conditions, and the associated takeoff speed schedules. These charts are embedded in microprocessor-based display units both on the flight deck and in the operations office. Pilots use this information for a safe takeoff. These charts are not available in the public domain. Instructors/readers may request local airlines to offer a hard copy as an example to study.

11.9 Military Aircraft Takeoff

For single-engine military aircraft, there is no question of BFL. Military aircraft TOFL must satisfy the *critical field length* (CFL). The CFL also has an associated decision speed V_1, which is determined by whether the pilot can stop within the distance available, otherwise the pilot ejects.

Figure 11.14 gives the speed schedules of single-engine military type aircraft. To obtain the CFL for a single-engine aircraft, the decision speed is at the critical time just before the pilot initiates rotation at V_R. Then the decision speed V_1 is worked out from the V_R by giving at least 2 seconds past V_1 as the recognition time. Engine failure occurring before this V_1 leads the pilot to stop the aircraft by applying full brakes and deploying any other retarding facilities available, for example brake-parachute, and so on. If there is not enough runway available, then a new decision speed V_1, considerably lower than the V_R, has to be worked out in a similar manner as done for the Bizjet. Engine failure occurring after V_1 means the pilot will have little option other than to eject (if there is not enough runway/SWY available to stop). For multi-engine aircraft, determining CFL follows the same procedure as in the civil aircraft computation, but complying with Milspecs.

MIL-C5011 requires a rolling coefficient $\mu = 0.025$ and minimum braking coefficient $\mu_B = 0.3$ (modern brakes on anti-skid wheels have higher μ_B). Training aircraft will have a good annual utilization operated by relatively inexperienced pilots (with about 200 hours of flying

■ **FIGURE 11.14**
AJT takeoff

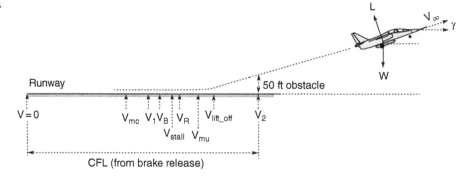

Runway   50 ft obstacle

V = 0 V_{mc} $V_1 V_B$ V_R V_{lift_off} V_2

V_{stall} V_{mu}

CFL (from brake release)

■ **TABLE 11.14**
Aerodynamic data for AJT

	Cruise	Takeoff	Takeoff	Landing[a]
Flap setting (°)	0	8	20	40
C_{Lmax} (from experiment)	1.54	1.67	1.9	2.2
V_{stall} (ft/s)	177.02	170.0	159.37	133.0
V_{stall} (kt)	104.82	100.7	94.37	78.75
V_R (ft/s)	194.72	187.00	175.31	146.29
V_R (kt) $(1.1V_{stall})$	115.30	110.72	103.80	86.62
V_{LO} (ft/s)	214.38	195.49	183.27	152.94
V_{LO} (kt) $(1.15V_{stall})$	126.94	115.75	108.52	90.56
V_2 (ft/s)	223.70	204.0	191.24	159.58
V_2 (kt) $(1.2V_{stall})$	132.46	120.78	113.24	92.50

[a] Landing at 8466 lb.

experience) carrying out a very large number of takeoffs and landings, on relatively shorter runways. AJT brakes are generally robust in design, giving a brake coefficient $\mu_B = 0.5$ for student pilot usage, much higher than the minimum Milspec requirement.

    To keep the workload simple, both the Bizjet and the AJT have the same class of aerofoil and the same types of high-lift devices. Therefore, there is a strong similarity in the wing aerodynamic characteristics. Table 11.14 gives the AJT data pertinent to takeoff performance at normal training configuration (NTC). In this book, the following CFL requirements for the single-engine AJT are worked out.

1. To meet the TOD of 3600 ft (1100 m) to clear 15 m.
2. To meet the initial rate of climb (unaccelerated) of 10,000 ft/min (50 m/s).

11.10   Checking Takeoff Field Length (*AJT*)

The first task is to prepare a graph like that of Figure 11.4 to obtain the varying force parameters during the ground run.

11.10.1   AJT Aircraft and Aerodynamic Data

■ Ramp maximum mass = 10,582 lb.
■ Wing area, $S_w = 17\,m^2$ (183 ft²), $W/S_w = 57.4\,lb/ft^2$.

- MTOM = 10,500 lb (82 lb of taxi fuel consumed).
- Landing weight = 8466 lb, W/S_w = 42.26 lb/ft², keeping reserve fuel of 440 lb (200 kg).
- Full flap extended C_{Lmax} = 2.2.
- Rolling friction coefficient on paved runway, μ = 0.025.
- Minimum brake coefficient, $μ_B$ = 0.3.

Use Figure 9.12 and add undercarriage C_{D_UC} = 0.0124 (Section 9.14.3). Intermediate C_L and C_D are extrapolated. Equation 11.7 gives the ground distance covered, $S_G = V_{ave} \times (V_f - V_i)/\bar{a}$. Other equations used are:

$$q = \frac{1}{2}\rho V^2 = 0.5 \times 0.002378 V^2 = 0.001189 V^2$$

$$C_L = \frac{2 \times \text{MTOM}}{\rho S_w V^2} = \frac{10500}{0.001189 \times 183 \times V^2} = \frac{48256.6}{V^2}$$

$$V_{stall} = \sqrt{\frac{2 \times \text{MTOM}}{\rho S_w C_L}}$$

Equation 11.8 gives average acceleration as

$$\bar{a} = g\left[(T/W - \mu) - (C_L Sq/W)(C_D/C_L - \mu)\right]$$

Using the values given in Table 11.15,

$$\bar{a} = 32.2 \times \left[(0.55 - 0.025) - (C_L q/57.4)(0.11 - 0.025)\right]$$
$$= 32.2 \times \left[0.525 - (C_L q/765.33)\right]$$

■ **TABLE 11.15**
AJT takeoff distance

Flap setting (°)	0	8	20	Landing[a]
C_{DPmin}	0.0212	0.0212	0.0212	0.0212
C_{Lmax}	1.5	1.67	1.85	2.2
ΔC_{Dflap}	0	0.012	0.03	0.3
$\Delta C_{D\_U/C}$	0.0222	0.022	0.0212	0.021
Rolling friction coefficient, μ	0.025	0.025	0.025	0.025
Braking friction coefficient, $μ_B$	0.3	0.3	0.3	0.3
V_{stall} at 10,500 lb (ft/s)	179.4	170	161.5	133
V_R (ft/s) at V_{stall}/1.1	163.1	154.5	146.82	121
V_{LO} (ft/s) at 1.12 V_{stall}	201	190.4	181	149
V_2 (ft/s) at 1.2 V_{stall}	215.3	204	194	159.6
T/W	0.55	0.55	0.55	0
W/S_w (lb/ft²)	57.4	57.4	57.4	46.26
C_D/C_L at ground run	0.1	0.1	1	0.12

[a] Takeoff at 8° and 20° flap at takeoff engine power. Landing at 35–40° flap, engines at idle engine power and V_{stall} at aircraft landing weight of 8466 lb.

11.10.2   Takeoff with 8° Flap

Refer to Figure 11.11 showing a typical takeoff profile. First a normal takeoff length is estimated, followed by the case of a failed engine CFL. If the engine is operating then the AJT continues with the take off to climb. To keep computation simple, the distance covered from zero to lift-off speed V_{LO} is computed, instead of up to V_R.

11.10.2.1   Distance Covered from Zero to Lift-Off Speed V_{LO}
Using the same equation for distance covered up to the decision speed V_1 with the change to lift-off speed, $V_{LO}=190.4$ ft/s. Aircraft velocity at $0.707V_{LO}=133.3$ ft/s, $V_{ave}=(190.4+0)/2=95.2$ ft/s.

$$q=\left(\text{at }0.7V_1\right)=0.5\times0.002378\times\left(0.7V_{LO}\right)^2=0.001189\times133.2^2=21.1$$

Up to V_{LO}, the average $C_L=0.5$ (still at low incidence). Then:

$$\bar{a}=32.2\times\left(0.525-C_L q/765.33\right)=32.2\times\left[0.525-\left(0.5\times21.1\right)/765.33\right]$$
$$=32.2\times0.5114=16.5\text{ ft}/\text{s}^2$$

Using Equation 11.7, the distance covered until lift-off speed is reached:

$$S_{G\_LO}=V_{ave}\times\left(V_1-V_0\right)/\bar{a}\text{ ft}$$
$$=95.2\times\left(190.4-0\right)/16.5\approx1099\text{ ft}$$

(A detailed computation will be some 8% higher (say around ≈1200 ft) as can be seen in Bizjet case. It is suggested that the readers may compute using the detailed method as shown for the Bizjet.)

11.10.2.2   Distance Covered During Flare from V_{LO} to V_2
The flaring distance is required to reach V_2 clearing a 50 ft obstacle height from V_{LO}. The time taken to flare is 2 seconds with 8° flap, and $V_2=204$ ft/s. $V_{ave}=(190.4+204)/2=197.2$ ft/s.
In 2 seconds the aircraft would cover $S_{G\_LOV2}=2\times197.2\approx394$ ft.

11.10.2.3   Total Takeoff Distance
The takeoff length is thus $S_{GLO}+S_{GLOV2}=(1099+394)=1493$ ft (say 1500 ft); the NTC configuration is lighter than a fully loaded AJT with armament (6800 kg ≈15,000 lb) for which a TOFL specification of 3500 ft is more than enough. At higher AJT weights, a higher flap setting of 20° is possible.

11.10.2.4   Determination of CFL
Provision must be made if the engine fails at the decision speed, V_1, when braking has to be applied to stop the aircraft. The designers must make sure that the operating runway length is adequate to stop the aircraft to establish the CFL.
The stopping distance has two segments. The methodology is the same as carried out for Bizjet.

Distance covered from V_1 to braking speed V_B, S_{GIB}.
Braking distance from V_B ($\approx V_1$) to zero velocity, S_{GB0}.

The distance covered from zero to the decision speed V_1, S_{GI} is computed next. The decision speed V_1 has to be lower than V_R (154.5 ft/s). Initially guess $V_1=130$ ft/s (typically takes about 2 seconds before V_R).
Aircraft velocity at $0.707V_1=91.9$ ft/s

$$q\left(\text{at }0.707V_1\right)=0.5\times0.002378\times\left(0.707V_1\right)^2=0.001189\times91.9^2=10.044$$

Up to V_1, the average $C_L = 0.5$ (still at low incidence). Then

$$\bar{a} = 32.2 \times (0.525 - C_L q / 765.33) = 32.2 \times \left[0.525 - (0.5 \times 10.044) / 765.33 \right]$$
$$= 32.2 \times 0.4934 = 15.88 \ \text{ft} / \text{s}^2$$

Using Equation 11.7, the distance covered until lift-off speed is reached,

$$S_{G1} = V_{ave} \times (V_1 - V_0) / \bar{a} \ \text{ft} = 65 \times (130 - 0) / 15.88 = 532 \ \text{ft}$$

For the distance covered from V_1 to braking speed V_B, assume 3 seconds for pilot recognition and taking action. Therefore, the distance covered from V_1 to V_B, $S_{G1B} = 3 \times 130 = 390 \, \text{ft}$.

For the braking distance from V_B ($\approx V_1$) to zero velocity, the flap settings are of little consequence. The most critical moment of brake failure is at the decision speed V_1 when the aircraft is still on the ground. With full brake coefficient, $\mu = 0.5$ and average $C_L = 0.5$, $V_1 = 130 \, \text{ft/s}$ and $0.707 \, V_1 = 91.9 \, \text{ft/s}$.

$$q(\text{at } 0.7 \ V_1) = 0.5 \times 0.002378 \times (0.707 \ V_1)^2 = 0.001189 \times 91.9^2 = 10.044$$

The equation for average acceleration is:

$$\bar{a} = (g) \left[(\bar{T} / W - \mu) - (q C_L S_W / W)(\bar{D} / \bar{L} - \mu) \right]_{0.7V}$$

$$\bar{a} = 32.2 \times \left[-0.5 - (4.923 / 57.4)(0.1 - 0.5) \right]$$
$$= 32.2 \times \left[-0.5 + (1.97 / 57.4) \right] = -32.2 \times 0.466 = -15 \ \text{ft} / \text{s}^2$$

$$S_{G0} = V_{ave} \times (V_1 - V_0) / \bar{a} \ \text{ft} = 65 \times (130 - 0) / 15 \approx 563 \ \text{ft}$$

Therefore the minimum runway length (CFL) for takeoff should be $= S_{G1} + S_{G1B} + S_{G0}$.

TOFL $= 532 + 390 + 563 \approx 1485 \, \text{ft}$. This is within the specified requirement of 3600 ft.

The takeoff length is thus 1500 ft, about the same as the CFL of 1485 ft computed above. The required TOFL is 3500 ft, sufficiently long to cater for the full weight 6800 kg ($\approx 15,000 \, \text{lb}$) for armament training (if required, with 20° flap setting). Provision for SWY is useful for training exercises.

11.11 Civil Transport Aircraft Landing

To ensure safety, like takeoff, the landing procedure must satisfy airworthiness requirements. The landing takes place within the RSA – see Section 11.4. Pilot user manuals may be prepared with these terms in mind.

FAR specifies for two kinds of landings: (i) manual landing and (ii) automatic landing. This book deals only with the manual landing procedures.

11.11.1 Airfield Definitions

The book uses the preferred terms used by some engineers in the design office, but those who are engaged in preparing user documents must use the formal FAR (14CFR) definitions and terminologies as given below. There are two kinds of landing: (i) manual landing and (ii) automatic landing. This book deals with manual landing only.

410   Chapter 11 ■ Takeoff and Landing

■ **TABLE 11.16**
Airworthiness regulations at landing

	Civil aircraft design		Military aircraft design
	FAR23	**FAR25**	**MIL-C5011A**
Approach velocity, V_{app}	$\geq 1.3 V_{stall}$	$\geq 1.3 V_{stall}$	$\geq 1.2 V_{stall}$
Touchdown velocity, V_{TD}	$\geq 1.15 V_{stall}$	$\geq 1.15 V_{stall}$	$\geq 1.1 V_{stall}$
Landing field length, LFL (from 50 ft height)	Distant to stop, S_G	$1.667 S_G$	Distant to stop, S_G
Braking coefficient, μ_B	Not specified	Not specified	Minimum of 0.30

Actual landing distance (ALD): the runway length required to land from 50 ft altitude at the runway threshold to full stop. Ambient conditions, such as whether the runway is wet or dry, will affect the required runway length.

Required landing distance (RLD): This is an additional safety measure to allow flight dispatch to assist operators to ensure the minimum distance required for the forecasted landing weight and ambient conditions at the destination airport, which must be less than the landing distance available (LDA). The RLD is based on ALD performance and is airfield-specific. Engineers prefer to call this the aircraft landing field length (LFL), as will be used in this book.

Airworthiness requirements for landing aircraft are given in Table 11.16. All regulations require a clearing threshold height of 50 ft, as a safety measure. The approach segment starts from 50 ft height. The landing speed schedules and landing procedure are given in Figure 11.15.

The landing configuration is with full flaps extended and the aircraft at landing weight. Drag coefficient, C_D, at landing configuration is higher than takeoff configuration (for the Bizjet, $C_D/C_L = 0.12$). The approach segment starts with the aircraft descending by gliding (at a gentle descent rate of ≈ 300 ft/min) to 50 ft threshold height, decelerating to approach velocity V_{app}. From 50 ft altitude, the aircraft flares out by the pilot gently pulling back the control stick/wheel, levelling up to the runway in a manner that the aircraft speed should not fall below $1.15 V_{stall@TD}$ (civil aircraft) as per the regulatory requirement. Until the aircraft touches down it is in airborne phase, and after touchdown it is in the ground phase. This flaring of the aircraft from glide path to level is approximated by a circular path which generates centripetal acceleration normal to the velocity vector. The aircraft angle of attack α_{app} stays relatively high. Touch-down aircraft attitude is at nose up with α_{TD} approaching the stall angle.

The typical civil aircraft descent rate at approach is anywhere between 12 and 22 ft/s. There are wide possibilities for arriving at the approach height depending on how steep the glide path is and how rapid is the flaring action. To clear awkward obstacles, the approach can be made at a steep angle and a rapid descent rate, as high as 22 ft/s requiring a rapid flaring action, or in a normal situation can be approached at a shallow angle requiring a shallow flare action. This book uses the typical descent rate of 12 ft/s. Even at fast descent, the glide angle γ is less than 10°. Typical normal acceleration at flare is taken as 1.2 g.

There is a floating period at flare to wash out the speed of the aircraft to sink to a gentle touchdown. At touchdown the aircraft speed is taken as $V_{TD} = 1.15 V_{stall@land}$; the nose wheel is still above the ground. After the main wheel touches down, the nose takes about 2–6 seconds to touch down. Typically, the brakes are applied 2 seconds after the nose touches down. This works out to be slightly less than the BFL at 0.95 × MTOM, but not necessarily.

■ **FIGURE 11.15**
**Typical landing
profile. (a) Landing
speed schedule.
(b) Landing
procedure**

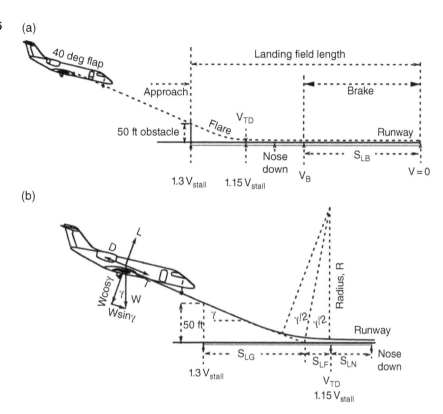

A typical example of a mid-size, mid-range class high subsonic narrow-body commercial jet aircraft (A320, B737 class) has the instrument landing system (ILS) approach at −3° and a flaring load of 1.15 g. It takes about 6 seconds to touch from the 50 ft threshold height, with a ground distance of 1200 ft covered. If floating longer, then the corresponding time taken to touch down could be as high as 12 seconds or more, consuming a greater runway distance. With a steep approach, the flaring load is around 1.4 g and may be associated with hard landings.

With all the possibilities, prediction of LFL can have some uncertainties. Therefore, to incorporate a safety margin, FAR 25 stipulates a landing runway length of 1.667 times the computed distance.

In the case of a baulked landing/missed approach at landing weight, the FAR requirements are given in Chapter 12. In general, baulked landing/missed approach is not a problem as the aircraft is lighter at the end of the mission and all engines are operating at takeoff rating.

Computation of LFL uses similar equations as for computing the TOFL. The values of rolling coefficient, μ, vary from main wheel touchdown to when the nose wheel touches down. After all-wheel touchdown, the brakes are applied. A considerable amount of heat is generated at full braking and may pose a fire hazard. If a brake parachute/thrust reverser is deployed, then the drag of the retarding mechanism is to be accounted for in the deceleration. Similarly, with a thrust reverser, the negative thrust needs to be taken into account as a decelerating force. With full flaps extended and spoilers activated, the aircraft drag is substantially higher than at takeoff.

The approach has three segments: (i) a steady straight glide path from 50 ft height; (ii) flaring in a nearly circular arc to level out for touchdown; and (iii) the braking phase.

11.11.2   Landing Performance Equations

Landing analyses have similarities with takeoff analyses; the difference is in the regulations and speed schedules. Landing can be seen as negative acceleration, and so some of the derivation of takeoff will be modified for landing analyses. Figure 11.15 gives a typical landing profile. The segments of the landing process are as follows:

1. Aircraft gliding down at $V_{app} = 1.3V_{stall@land}$ in a steady path from 50 ft. The distance covered is S_{LG}.
2. Pilots initiates rotation to flare (approximated as a circular arc) prior to touch down, with a high angle of attack at $1.15V_{stall@land}$. The aircraft experiences higher g force due to centripetal acceleration. The Bizjet example takes the ideal situation with no floating and touches down at the end of flare. Typically it takes 2–4 seconds to wash out the speed to gradually settle to touchdown. After main wheels touch down, the pilot eases the control stick/wheel as the nose wheel settles onto the ground. This takes about another 2 seconds. The distance covered is S_{LF}.
3. The braking phase to full stop. The distance covered is S_{LB}. Although a thrust reverser (TR) is not a FAR requirement, it is normally considered in the derivation, but the worked example does not have TR.

11.11.2.1   Ground Distance During Glide S_{LG}   Normally, at steady approach the aircraft is kept throttled back to idle, producing no thrust. Figure 11.15b gives the generalized force diagram. It shows thrust, T, assisted by component of weight, $W\sin\gamma$, in descent angle γ. In equilibrium, lift L is balanced by the weight component, $W\cos\gamma$. Equating for equilibrium conditions:

$$L = W\cos\gamma \approx W \quad (\text{for small }\gamma)$$

$$\text{and} \quad D = T + W\sin\gamma$$

or

$$\sin\gamma = (D-T)/W = D/W - T/W \approx \gamma \quad (\text{for small }\gamma)$$

or

$$\gamma = C_D/C_L - T/W \tag{11.21}$$

Figure 11.15b gives

$$S_{LG} = 50/\tan\gamma, \quad \text{for small }\gamma, \quad S_{LG} = 50/\gamma \tag{11.22}$$

Substituting Equation 11.21 into Equation 11.22 gives

$$S_{LG} = \frac{50}{\left(\dfrac{C_D}{C_L} - \dfrac{T}{W}\right)} \tag{11.23}$$

11.11.2.2   Ground Distance from End of Flare to Touchdown S_{LF}   Flare is approximated to a circular path of radius R. From Figure 11.15b,

$$S_{LF} = R\tan\gamma = R\gamma/2 \quad (\text{for small }\gamma) \tag{11.24}$$

Tangential velocity in a circular path is:

$$V = R(d\gamma / dt), \quad \text{or} \quad V / R = (d\gamma / dt) \tag{11.25}$$

As per Newton's law, centripetal force

$$N = m(V^2 / R) = ma_n = a_n(W / g) \tag{11.26}$$

Using Equation 11.25, normal acceleration

$$a_n = N/g = (V^2 / R) = V(V / R) = V(d\gamma / dt) \tag{11.27}$$

To compute g-load during flare, the centripetal force on the circular path of flare increases lift from L to $L' = (L + \Delta L)$ with corresponding coefficients C_L to $C_L' = (C_L + \Delta C_L)$. Note that:

$$L = W\cos\gamma \approx W = qS_wC_L \quad (\text{for small } \gamma)$$

and

$$L' = qS_wC_L' (= L + \Delta L) \tag{11.28}$$

The ratio of

$$(C_L / C_L') = (W / L') \quad \text{or} \quad L' = (WC_L')/C_L \tag{11.29}$$

Normal force to flight path:

$$N = L' - L = \Delta L = a_n(W / g)$$
$$= L' - W\cos\gamma \approx L' - W \quad (\text{for small } \gamma)$$

Substituting from Equations 11.26 and 11.29, it becomes

$$N = (WC_L')/C_L - W = W(C_L'/C_L - 1) = ma_n = a_n(W / g) \tag{11.30}$$

The resultant force equilibrium gives:

$$L + \Delta L = W + a \times W / g = W(1 + a / g)$$

Load factor, n, is defined as:

$$n = (1 + a_n / g) = L / W + \Delta L / W = 1 + \Delta L / W = (C_L'/C_L) \tag{11.31}$$

where a_n is derived in Equation 11.26 as $a_n = g(n-1)$. This gives $nW = W + \Delta L = L'$. On substitution into Equation 11.30, it becomes

$$N = W(n-1) = W[(1 + a_n / g) - 1] = (Wa_n)/g = W(V^2 / R)/g \tag{11.32}$$

Therefore

$$R = V^2 / [g(n-1)] \tag{11.33}$$

or

$$V^2 = R[g(n-1)] \tag{11.34}$$

Equation 11.28 gives $L' = qS_w C_L' = 0.5\rho V^2 S_w C_L'$
or

$$V^2 = (L'/S_w)(2/\rho C_L')$$

or

$$V^2 = (nW/S_w)(2/\sigma\rho_0 C_L')$$

where $\sigma = \rho/\rho_0$ the density ratio, and from above $L' = nW$. Substituting Equation 11.34, it gives

$$R[g(n-1)] = (nW/S_w)(2/\sigma\rho_0 C_L')$$

or
$$R = (2nW)/[gS_w(n-1)(\sigma\rho_0 C_L')] \qquad (11.35)$$

Substituting Equations 11.21 and 11.35 in Equation 11.24,

$$S_{LF} = R\gamma/2 = (\gamma nW)/[gS_w(n-1)(\sigma\rho_0 C_L')] = \frac{(W/S_w)\left(\dfrac{C_D}{C_L}-\dfrac{T}{W}\right)}{\sigma\rho_0 g(n-1)C_L} \qquad (11.36)$$

Velocity at touchdown $= 1.15V_{stall@land}$.
Ground distance from touchdown to nose wheel on ground,

$$S_{LN} = 2 \times 1.15 V_{stall@land} \qquad (11.37)$$

11.11.2.3   *Ground Distance from V_B to Stop, S_{LB}*   To keep the derivation generalized, the negative thrust of a thrust reverser is included. However, in the Bizjet example there is no thrust reverser. (Note: average condition is at $0.707V_{TD}$.)

$$S_{LB} = V_{ave} \times (V_{TD}-0)/\bar{a} \qquad (11.38)$$

where

$$\bar{a} = \left(\frac{g}{W}\right)\left[(-\mu_B W - T_{TR}) - (\bar{C}_D - \mu_B \bar{C}_L)Sq\right]$$
$$= (g)\left[(-\mu_B - T_{TR}/W) - (\bar{D}/W - \mu_B \bar{L}/W)\right]_{0.707VTD} \qquad (11.39)$$

11.11.3   Landing Field Length for the Bizjet

In the worked example of the Bizjet, the flap setting is 40°, giving $C_{Lmax} = 2.2$. The landing weight of the Bizjet is 15,800 lb (wing loading $= 48.92$ lb/ft²), and at full flap extended, $C_{Lmax\_land} = 2.2$. Therefore:

$$V_{stall@land} = \sqrt{\frac{2\times 15,800}{0.002378\times 323\times C_L}} = \sqrt{\frac{41,141}{2.2}} = 137\ \text{ft/s}$$

$$V_{appr} = 1.3V_{stall@land} = 178\ \text{ft/s}$$
$$V_{TD} = 1.15V_{stall@land} = 158\ \text{ft/s}$$
$$V_{0.707} = 0.707\times 158 = 111.7\ \text{ft/s}$$

1. *In glide.* Equation 11.23 gives:

$$S_{LG} = \frac{50}{\left(\dfrac{C_D}{C_L} - \dfrac{T}{W}\right)} = 50/0.12 = 417 \text{ ft}$$

2. *During flare.* Equation 11.36 gives:

$$S_{LF} = \frac{(W/S_w)\left(\dfrac{C_D}{C_L} - \dfrac{T}{W}\right)}{\sigma \rho_0 g (n-1) C_L}$$

At flare, $C_{Lave} = 1.3$ as the average of $C_{Lmax} = 2.2$ and $C_{LTD} = 0.5$.

$$S_{LF} = \frac{48.92 \times (0.1 - 0)}{0.002378 \times 32.2 \times (1.2 - 1) \times 1.3} = 4.892/0.02 \approx 246 \text{ ft}$$

3. *Reaction time.* Ideally the Bizjet touches down at the end of flare and there is no floating. Nose wheel touchdown and reaction time of 2 seconds takes $2 \times 158 = 316$ ft.

$$S_{LF} = 246 + 316 = 562 \text{ ft}$$

4. *In braking.* Equation 11.38 gives $S_{LB} = V_{ave} \times (V_{TD} - 0)/\bar{a}$. From Equation 11.39, without the thrust reverser:

$$\bar{a} = \left(\frac{g}{W}\right)\left[(-\mu_B W) - (\bar{C}_D - \mu_B \bar{C}_L) Sq\right] = (g)\left[(-\mu_B) - (\bar{D}/W - \mu_B \bar{L}/W)\right]$$

The equation for average acceleration is:

$$\bar{a} = (g)\left[(-\mu_B) - (q C_L S_w /W)(\bar{D}/\bar{L} - \mu)\right]_{0.707V}.$$

For an aircraft in full brake with $\mu_B = 0.4$, all engines shut down, average $C_L = 0.5$ $[(C_L/C_D)_{ave} = 0.12]$.
The average acceleration based on $0.7V_{TD} = 0.7 \times 158 = 111.6$ ft/s, the value of $q = 0.5 \times 0.002378 \times 111.6^2 = 14.544$, so then the above equation reduces to:

$$\bar{a} = 32.2 \times \left[(-0.4) - (C_L q/48.92)(0.12 - 0.4)\right] = 32.2 \times \left[-0.4 + (0.14 q/48.92)\right]$$
$$= 32.2 \times \left[-0.4 + (q/350)\right] = -32.2 \times 0.359 = -11.5 \text{ ft/s}^2 \text{ (retardation)}$$

Then $S_{LB} = V_{ave} \times (V_{TD} - 0)/\bar{a} + 316 + 316 = (79 \times 158)/11.5 = 1085$ ft

$$\text{Total landing field length (LFL)} = 1.667 \times (417 + 246 + 316 + 1085)$$
$$= 1.667 \times 2064 = 3440 \text{ ft}$$

This is within the requirement of 4400 ft. Compare this with TOFL = 3300 ft (20° flap) – about the same.

Had there been thrust reversers, spoilers and better brakes, the LFL could be reduced substantially; in this case they are not required, saving considerable cost. Advanced wheels and brakes have $\mu_B = 0.5$ or even a little more.

11.11.4   Landing Field Length for the AJT

Keeping reserve fuel of 440 lb (200 kg), the landing weight of the AJT is 8466 lb, (wing loading = 42.26 lb/ft^2) and at full flap extended $C_{Lmax} = 2.2$. Therefore:

$$V_{stall\,@\,land} = \sqrt{\frac{2 \times 8466}{0.002378 \times 183 \times C_L}} = \sqrt{\frac{38908}{2.2}} = 133 \text{ ft / s}$$

$$V_{app} = 1.3 V_{stall\,@\,land} = 172.9 \text{ ft / s}$$

$$V_{TD} = 1.1 V_{stall\,@\,land} = 146 \text{ ft / s}$$

$$V_{0.707} \text{ at touchdown} = 0.707 \times 146 = 103.2 \text{ ft / s}$$

1. *In glide.* Equation 11.23 gives

$$S_{LG} = \frac{50}{\left(\dfrac{C_D}{C_L} - \dfrac{T}{W}\right)} = 50 / 0.12 = 417 \text{ ft}$$

2. *During flare.* At flare, $C_{Lave} = 1.6$ as the average of $C_{Lmax} = 2.2$ and $C_{LTD} = 1.3$. Equation 11.36 gives:

$$S_{LF} = \frac{(W / S_W)\left(\dfrac{C_D}{C_L} - \dfrac{T}{W}\right)}{\sigma \rho_0 g (n-1) C_L}$$

$$= \frac{42.26 \times (0.12 - 0)}{0.002378 \times 32.2 \times (1.2 - 1) \times 1.6} = 5.07 / 0.0246 \approx 206 \text{ ft}$$

3. *Reaction time.* Ideally the AJT touches down at the end of flare and there is no floating. However, there are 2 seconds of reaction time before full brakes can be applied. Distance covered in 2 seconds = $2 \times 146 = 292$ ft.

4. *In braking.* Equation 11.38 gives:

$$S_{LB} = V_{ave} \times (V_{TD} - 0) / \bar{a}$$

where from Equation 11.39,

$$\bar{a} = \left(\frac{g}{W}\right)\left[(-\mu_B W) - \left(\bar{C}_D - \mu_B \bar{C}_L\right) Sq\right] = (g)\left[(-\mu_B) - \left(\bar{D}/W - \mu_B \bar{L}/W\right)\right]$$

The equation for average acceleration is:

$$\bar{a} = (g)\left[(-\mu_B) - \left(q C_L S_W / W\right)\left(\bar{D}/\bar{L} - \mu\right)\right]_{0.7V}$$

For the aircraft in full brake with $\mu_B = 0.4$ (Milspecs requires >0.3), all engines shut down, the average $C_L = 0.5$. The average retardation based on $0.707\,V_{TD} = 0.707 \times 146 = 103.2$ ft/s, the value of $q = 0.5 \times 0.002378 \times 103.2^2 = 12.42$, then the above equation reduces to:

$$\bar{a} = 32.2 \times \left[\left(-0.4 \right) - \left(C_L q / 42.26 \right) \left(0.12 - 0.4 \right) \right] = 32.2 \times \left[-0.4 + \left(0.14 q / 42.26 \right) \right]$$

$$= 32.2 \times \left[-0.4 + q / 302 \right] = -32.2 \times 0.359 = -11.6 \text{ ft} / \text{s}^2 \left(\text{retardation} \right)$$

Then

$$S_{LB} = V_{ave} \times \left(V_{TD} - 0 \right) / \bar{a} = \left(73 \times 146 \right) / 11.6 = 919 \text{ ft}$$

Considering all the above, then:

$$\text{Total landing field length} \left(\text{LFL} \right) = \left(417 + 206 + 292 + 919 \right) = 1834 \text{ ft}.$$

Military aircraft landing does not need the factor of 1.667. The LFL value is within the requirement of 3600 ft. It is of the order of the CFL of 1500 ft at 8° flap.

In general, military aircraft do not install thrust reversers, but instead use brake-parachutes, and combined with spoilers and better brakes, the LFL can be made substantially shorter. The AJT does not have brake parachute. Advanced wheels and brakes have $\mu_B = 0.5$, or even a little higher value.

11.12   Landing Presentation

There are external parameters that influence landing performance, such as deviations from the standard ambient conditions and runway conditions. Like takeoff, the presentation of landing in graphical form differs from what aircraft performance engineers use and what pilots have in their user manual. Typical landing performance for the pilot's manual is shown in Figure 11.16

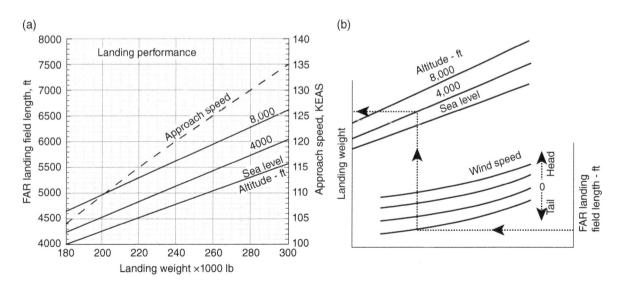

■ **FIGURE 11.16**   **Landing field length presentation (not of Bizjet, but a typical example). (a) Engineer's graph. (b) Pilot's manual**

in a "chase-around" manner in *web chart* form. Figure 11.16a is meant for a high subsonic jet transport aircraft. Figure 11.16b is the typical web chart form. Nowadays this information is embedded in microprocessor-based flight deck display instruments.

11.13   Approach Climb and Landing Climb

In case of a missed approach the aircraft should be able to climb away above 50 ft height to make the next attempt to land. The FAR (14CFR) 25.121 subpart B requirements for the approach climb gradient are 2.1% for twin-engine aircraft, 2.4% for three-engine aircraft, and 2.7% for four-engine aircraft with one engine inoperative and in landing configuration.

In case of baulked or called-off landing with all engines operating, the aircraft should be able to climb away from below 50 ft height to make the next attempt to land. The FAR (14CFR) 25.119 subpart B requirements for the landing climb gradient are 3.2%.

In both these cases, the aircraft is considerably lighter at the end of mission on account of consuming onboard fuel, less the reserve.

11.14   Fuel Jettisoning

In the case of one engine failure at takeoff when the aircraft is heavy, close to MTOM, fuel jettisoning arrangements must be installed for aircraft safety at the cost of environmental issues. This has to cater for 15 minutes of fuel to be left for the aircraft to complete the circuit to land back at the departing airfield. The airworthiness requirements are the same as for takeoff and landing procedures.

References

1. Lan, C.T.E., and Roskam, J., *Airplane Aerodynamic and Performance*. DARcorporation, Lawrence, 1981.

2. Ruijgork, G.J.J., *Elements of Airplane Performance*. Delft University Press, Delft, 1996.

3. Ojha, S.K., *Flight Performance of Airplane*. AIAA, Reston, VA, 1995.

4. McCormick, B.W., *Aerodynamics, Aeronautics and Flight Mechanics*. John Wiley & Sons, Inc., Chichester, 1995.

CHAPTER 12

Climb and Descent Performance

12.1   Overview

The last chapter dealt with takeoff and landing performances of aircraft. The next segment of a mission after takeoff to 35 ft is to climb away from the first segment takeoff procedure to cruise altitude. At the end of cruise, the aircraft descends to land. Climb and descent segments are carried out in the vertical plane of the aircraft. Since the equations of motions for climb and descent are similar, they are considered together in this chapter. Here we deal comprehensively with all aspects of the climb and descent segments of a mission, including both the point and integrated climb and descent performances. The values for distance covered, time taken and fuel consumed are required to obtain aircraft payload range. A methodology to establish a commercial aircraft climb schedule is shown.

Aircraft climb can be carried out in many ways depending on the flight planning. Civil and military climb schedules differ on account of differences in mission requirements. In civil applications, the aircraft climbs in a specified speed schedule for piloting ease, economy, and to clear any obstacle in the flight path. Military climb schedules demand faster climb rates and are more frequently required to clear an obstacle in low-level operations. Physically, descent can be seen as the opposite to climb. Civil aircraft descent is with zero or partial thrust setting and is restricted by the cabin pressurization schedule. Military aircraft descent is dictated by the mission requirements. There are few airworthiness requirements for en-route climb from 1500 ft altitudes and descent performances, such as for the case of the drift-down procedure at engine failure. The drift-down procedure is dealt with in Chapter 14, having similarity with drift-down in cruise.

Aircraft sizing and engine matching exercises also carry out initial climb analyses, but such analyses are not meant for certification purposes and hence are done in a different manner, as shown in Chapter 16.

Readers are to compute climb and descent performances in detail on classwork examples of a Bizjet for civil aircraft, and an AJT for military aircraft. Since extensive computational work is involved, it is suggested that spreadsheets may be used for repeating computations. The readers may also compute TPT climb and descent performance. It may be noted that values read from small graphs are not accurate enough (see Section 1.12). The readers may prepare large graphs to improve accuracy. A problem set for this chapter is given in Appendix E.

Theory and Practice of Aircraft Performance, First Edition. Ajoy Kumar Kundu, Mark A. Price and David Riordan.
© 2016 John Wiley & Sons, Ltd. Published 2016 by John Wiley & Sons, Ltd.

12.2   Introduction

Low density at high altitudes generates low dynamic head q for the same true airspeed compared with lower altitude cruise [1–4]. Low q offers low aircraft drag, demanding low thrust to maintain level flight. Considerable savings in fuel burn can be achieved by cruising at high altitude (at the prescribed flight level, FL). This is one of the ways to maximize financial returns for commercial aircraft operation. The operation strategy is to adopt a suitable climb schedule to reach the desirable/prescribed altitude. To ensure safety, the climb path of commercial transport aircraft operation takes place under traffic control to reach the prescribed cruise altitude. A military aircraft climb schedule depends on the mission demand, which differs between types of operational role. Engine power setting is at the maximum climb rating for civil aircraft, and maximum continuous rating for military operation.

There are several options to execute aircraft climb, as follows:

1. *Maximum gradient climb*: this is to clear obstacles in the climb flight path.
2. *Minimum time to climb or at maximum rate of climb*: this is to reach altitude as quickly as possible.
3. *Minimum fuel burn climb*: this is meant for best economy for the climb.
4. *Constant V_{EAS} (V_{CAS}) climb*: this is adopted for piloting ease and fuel economy as the normal practice. Typically, it is kept close to the maximum rate of climb. The constant V_{EAS} (V_{CAS}) climb speed schedule depends on the design of the aircraft (and/or operator's preference). Larger high subsonic commercial jet aircraft have two climb speeds; (i) normal climb schedule and (ii) high-speed climb schedule. Typically, turboprop aircraft have one climb speed schedule (normal climb schedule).

This chapter shows how to establish the climb schedule.

Having considerably higher thrust to weight ratios (T/W), military aircraft have the additional capability to climb above the altitude where not much excess thrust is available, but have high speed when its kinetic energy is converted to a gain in potential energy through zoom climb.

Climb altitude is decided by range considerations. The range in cruise segment should be at least equal to the range covered during climb and descent. The higher the cruise altitude, the more distance will be covered to climb to that altitude and to descend from there. Naturally, for lower range aircraft there is no need to climb to higher cruise altitudes. For example, a B737 from Belfast to Glasgow could cruise around 30,000 ft, while going to London it has the opportunity to cruise above 30,000 ft. (A typical transatlantic climb-cruise can reach 39,000 ft at the end of cruise.)

The physics of descent is easier compared with climb, as it is assisted by gravitational force. During descent the aircraft has the advantage of using its potential energy of cruise altitude; in principle does not require thrust application. However, its computational effort should not be underestimated on account of the cabin pressurization schedule. Descent proceeds in a specific speed schedule with engines at partial throttle setting, well below cruise rating, adjusting with the descent gradient to suit the cabin pressurization schedule. Fuel burn during descent is low. There are several options to execute aircraft descent, the important ones being given below:

1. *Minimum fuel burn descent*: this is meant for best economy.
2. *Maximum distance covered during descent*: this is to maximize range during descent.

3. *Maximum permissible descent rate as the cabin pressurization schedule permits*: design restriction.
4. *Constant equivalent airspeed (EAS) descent*: this is for piloting ease and adopted as the normal practice.

All these topics are considered in this chapter.

12.2.1 Cabin Pressurization

It is topical to introduce the role of cabin pressurization in assessing aircraft climb and descent performance. Ambient pressure reduces non-linearly with an increase in altitude, as can be seen in Figure 2.2. At the tropopause it is 472.68 lb/ft^2 (3.28 lb/in^2). For average human physiology, big jets maintain cabin pressure around 8000 ft altitude (ambient pressure of 1571.88 lb/ft^2 (10.92 lb/in^2), resulting in a differential pressure of 7.63 lb/in^2. Smaller Bizjets fly at higher altitudes to stay separate from big-jet traffic, as high as 50,000 ft (242.21 lb/ft^2, i.e. 1.68 lb/in^2) when the aircraft structural design needs to be able to maintain the inside cabin pressure at 10,000 ft altitude (1455.33 lb/ft^2, i.e. 10.11 lb/in^2, a differential pressure of 8.42 lb/in^2). For big jets a maximum of 8.9 lb/ft^2 (for Bizjets 9.4 lb/in.^{2}) of differential pressure is used in practice. Cabin pressurization is in use for certain types of military aircraft with large fuselage volume. Combat aircraft have different arrangements [1] with higher altitude provision in the flight deck – pilots wear pressure suits and helmets with masks supplying oxygen.

The differential pressure between outside and inside the cabin is substantial and it cycles through every sortie flown. Stress and fatigue considerations of fuselage structural design are constrained by weight considerations. Aircraft cabins are built as sealed pressure vessels (very low leakage), kept airtight, and are provided with a complex environmental control system (ECS) which has a pressure control valve to automatically regulate cabin pressure (and air-conditioning). The aircraft crew can select the desired cabin pressure within the design limits. Unless there is a catastrophic failure, the sealed cabin can hold pressure to give enough time for pilots to descend fast enough in case there is malfunction in pressurization. If cabin pressure exceeds the limit, then automatic safety valves can relieve the cabin pressure to bring it down within the limit.

The classwork example of Bizjet has cabin pressure selectable from sea-level to 10,000 ft altitude, limiting the differential pressure to within 9.4 lb/ft^2. During climb the equivalent Bizjet cabin pressure rate of change is designed to 2600 ft/min. Bizjet type aircraft engine power is adjusted to have cabin pressurization at the climb rate to a maximum of about 1800 ft/min to stay within the ECS system capability, sufficient for a sealed cabin that starts at sea-level pressure; as a result the en-route climb rate is not required to be restricted. Section 12.7 works out the Bizjet climb performance. However, during descent, the cabin needs to gain pressure as it descends to lower altitudes. Average human physical tolerance is taken at 300 ft/min descent, and the ECS system pressurization capability meets that.

12.2.2 Aircraft Ceiling

As an aircraft gains altitude, thrust decays, reducing the excess thrust margin over drag, and thereby the rate of climb capability degrades to a point when the engine cannot supply the excess thrust required to climb. There are three kinds of aircraft ceiling as follows (see Section 12.7.1):

1. Absolute ceiling
2. Service ceiling
3. Operating ceiling.

■ **TABLE 12.1**
FAR second segment climb gradient at missed approach

Number of engines	2	3	4
Minimum climb gradient at approach segment	2.1	2.4	2.7
Minimum climb gradient at landing segment	3.2	3.2	3.2

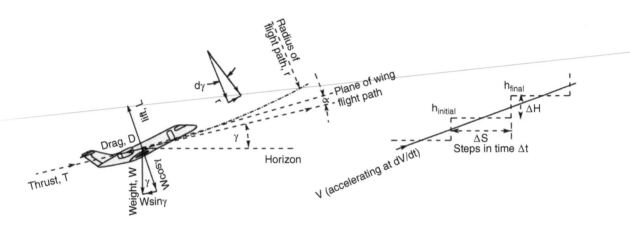

■ **FIGURE 12.1**   Generalized force diagram in pitch (vertical) plane

FAR requirements for takeoff procedures up to 1500 ft altitude climb are given in Table 11.2. In addition, the aircraft manufacturers need to substantiate the operators' all-engine operating initial en-route climb rate. For the Bizjet it is 2600 ft per minute.

At the end of mission the aircraft is light, and in the case of a missed approach/baulked landing the aircraft is required to climb away with higher flap settings at a steeper gradient, as given in Table 12.1.

12.3   Climb Performance

Aircraft climb is possible when the available engine thrust is in excess of aircraft drag, that is, the excess thrust $(T-D)$. The excess thrust is converted into the potential energy of height gain. To compute the climb performance, the first task is to generate an aircraft and engine grid (see Section 12.5.1) for the thrust required and thrust available at various speeds, altitudes and aircraft weights.

Figure 12.1 gives the generalized force diagram for the aircraft climb path in wind axes in pitch plane (vertical plane – the plane of symmetry). Here the thrust vector is aligned to the velocity vector. Aircraft velocity is V climbing at an angle γ in an angle of attack α. To compute the integrated climb performance (Section 12.5), the climb trajectory is discretized into small steps of altitude (ΔH) within which the variables are considered invariant.

12.3.1   Climb Performance Equations of Motion

Given below is the pair of equations 4.5a and 4.5b, as derived formally in Section 4.5.2. From Figure 12.1 the force equilibrium gives:

$$\sum \text{forces}_{\text{flight path}} = 0, \quad \text{i.e.,} \quad m\dot{V} = (T) - \left(C_D/2\rho V^2 S_W\right) - mg\sin\gamma$$

$$\sum \text{forces}_{\text{perpendicular}} = 0, \quad \text{i.e.,} \quad V(d\gamma/dt) = (C_L/2m) \times \rho V^2 S_W - g\cos\gamma$$

In climb, the flight path is almost straight when $(d\gamma/dt) \approx 0$. This may not be the case with high-performance combat aircraft. For civil aircraft, the above two equations can be written as follows.

$$m\dot{V} = (T) - \left(C_D/2\rho V^2 S_W\right) - mg\sin\gamma = T - D - W\sin\gamma \tag{12.1}$$

$$\left(C_L/2m\right)\rho V^2 S_W = mg\cos\gamma \quad \text{or} \quad L = W\cos\gamma \tag{12.2}$$

Note that, at climb, lift L is lower than aircraft weight, W. The residual component of the weight is balanced by the excess thrust (Equation 12.1). Without this excess thrust of $(T - D)$ the aircraft is incapable of climbing. Combat aircraft or aerobatic aircraft climb angle can be high. Typically, commercial aircraft climb angle is less than $15°$ ($\cos 15 = 0.96$ and $\sin 15 = 0.23$). Two situations can arise as follows:

Quasi-steady-state climb: This is when the acceleration term, $\dot{V}$, and the rate of change in climb angle term $\dot{\gamma}$ in climb, cruise and descent segments are small and can be neglected. That is:

$$\dot{V} \approx 0, \quad d\gamma/dt = \dot{\gamma} \approx 0$$

Quasi-level flight: This is when the climb angle γ is small (typically $<20°$) but the acceleration term, $\dot{V}$ is not:

$$\sin\gamma \approx \gamma \text{ radians}, \quad \cos\gamma \approx 1, \quad \gamma^2 \ll 1$$

For transport category aircraft in shallow climb ($<15°$), the equations can be simplified to the following.

$$dV/dt = (T/m) - \left(C_D q\right)/m \tag{12.3}$$

$$0 = \left(C_L q S_W\right) - mg \tag{12.4}$$

12.3.2   Accelerated Climb

En-route climb is performed in an accelerated climb. The equation for accelerated climb can be derived as follows. Note that h is the true altitude and on an ISA day is the same as the pressure altitude. [For a non-standard day, the following correction may be applied: True altitude = Pressure altitude $\times (T_{non\_STD}/T_{STD})$.]

$$\text{Rate of climb, } R/C = dh/dt = V\sin\gamma \tag{12.5a}$$

or

$$\text{sine angle of climb gradient } \sin\gamma = (dh/dt)/V \tag{12.5b}$$

Climb gradient, γ, can be expressed in term of $\sin\gamma$.

Equation 12.1 can be written as $(T-D) = mg\sin\gamma + (m)dV/dt$. This gives the gradient,

$$\sin\gamma = \left[T - D - (W/g)dV/dt\right]/W = \left[(T-D)/W\right] - \left[(1/g) \times dV/dt\right] \tag{12.6}$$

Write acceleration term, $\dot{V} = dV/dt = (dV/dh) \times (dh/dt)$, then substituting into Equation 12.6:

$$\text{Rate of climb, } R/C_{accl} = dh/dt = V\sin\gamma$$
$$= V(T-D)/W - (V/g) \times (dV/dh) \times (dh/dt) \tag{12.7}$$

By transposing and collecting dh/dt,

$$R/C_{accl} = dh/dt = \frac{V\left[(T-D)/W\right]}{1 + (V/g)(dV/dh)} \tag{12.8}$$

Making use of Figure 10.18 for the aircraft and engine grid for climb rating, the above can be written as follows:

$$R/C_{accl} = dh/dt = \frac{V\left[(T/\delta - D/\delta)/(W/\delta)\right]}{1 + (V/g)(dV/dh)} \tag{12.9}$$

In general, the maximum $(T/\delta - D/\delta)$ is not at the minimum D/δ, but at a slightly higher speed. The readers may find it difficult to establish the maximum $(T/\delta - D/\delta)$ from Figure 12.2. It is suggested that the readers generate their own data from industry-supplied engine performance figures and a high-resolution drag polar of the aircraft under study to obtain the maximum $(T/\delta - D/\delta)$, using graphs to determine the longest vertical distance between T/δ and D/δ for the altitude.

■ **FIGURE 12.2**
Aircraft and engine grid for Bizjet (maximum climb thrust)

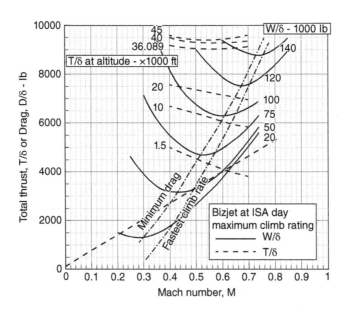

The rate of climb is a point performance and is valid at any altitude. The term (V/g) (dV/dh) is a dimensionless term. It penalizes unaccelerated rate of climb depending on how fast the aircraft is accelerating during climb. Part of the propulsive energy is consumed for speed gain instead of altitude gain. Military aircraft make accelerated climb in the operational arena when the $(V/g)(dV/dh)$ term reduces the rate of climb.

On the other hand, a civil aircraft has no demand for high accelerated climb; instead it makes an en-route climb to cruise altitude at a *quasi-steady-state* climb by holding the aircraft climb speed at constant EAS/Mach number. Constant indication of speed eases the pilot workload. At the flight deck, the pneumatic airspeed indicator shows V_E, and the electronic airspeed indicator shows V_{CAS}. At higher speeds the compressibility correction ΔV_C needs to be applied. Recall Section 3.2, where $V_{EAS} = V_{CAS} - \Delta V_C = V_{TAS} \sqrt{\sigma} = Ma\sqrt{\sigma}$, noting that V_C and V_E are abbreviations of V_{CAS} and V_{EAS}, respectively. At higher subsonic speeds, there is a small difference between V_{EAS} and V_{CAS}, typically less than 7% (for Bizjet less than 4%). Instrument and other errors are hardware-specific and are small (not applied in this book for performance analyses). Industry will obtain these hardware-specific errors and apply them during calculations.

Section 3.2 also mentioned that engineers prefer to work with V_{EAS}, while pilots observe V_{CAS}. The best rate of climb is nearly flat (see Section 12.5.1), that is, the difference between best rates of climb at V_E and at V_C is negligible. It is for this reason that in this book, $V_{EAS} \approx V_{CAS}$ at climb. At a given altitude and true speed V, aircraft equivalent airspeed V_{EAS} can be determined by $V_{EAS} = V_{TAS} \sqrt{\sigma} = Ma\sqrt{\sigma}$. It will be shown that best climb rates are close to a constant calibrated airspeed, V_{CAS} ($\approx V_{EAS}$), and as a consequence true airspeed V (V_{TAS}) increases with altitude until it reaches the limiting climb Mach number, when it is held constant to continue with climb (see Section 12.5.1 and Table 12.6).

Constant V_{EAS} ($\approx V_{CAS}$) is in *quasi-steady-state* climb and has small acceleration as TAS is increasing at constant EAS; the contribution of the $(V/g)(dV/dh)$ term is low. The magnitude of the acceleration term reduces with altitude gain. The following sections work out the relationship between $(V/g)(dV/dh)$ and the constant speed climb schedules above and below the tropopause.

12.3.3 Constant EAS Climb

The term $(V/g)(dV/dh)$ can be worked out in terms of constant EAS as follows.

$$\frac{V}{g}\left(\frac{dV}{dh}\right) = \frac{V_{EAS}V_{EAS}}{g\sqrt{\sigma}}\left(\frac{d(1/\sigma)}{dh}\right) = -\frac{V_{EAS}^2}{2g\sigma^2}\left(\frac{d\sigma}{dh}\right) = -\frac{M^2 a^2}{2g\sigma}\left(\frac{d\sigma}{dh}\right) \qquad \textbf{(12.10)}$$

12.3.3.1 *Troposphere*

$$\left[\text{For air, } \gamma = 1.4, \text{R} = 287 \text{ J/kg K}, \text{g} = 9.81 \text{ m/s}^2.\right]$$

In the SI system, in the troposphere Equation 3.1b gives $T = (288.16 - 0.0065\,h)$ and

$$\sigma = \rho/\rho_0 = \left(T/T_0\right)^{\left[(g/0.0065R)-1\right]} = \left(T/T_0\right)^{\left[(9.81/(0.0065\times287))-1\right]} = \left(T/T_0\right)^{4.255}$$

$$\left(d\sigma/dh\right) = 4.225 \times \left(T/T_0\right)^{3.255} \times \frac{d\left(T/T_0\right)}{dh}$$

$$= 4.225 \times \left(T/T_0\right)^{3.255} \times \frac{1}{T_0}\left(\frac{dT}{dh}\right) \text{ where } \left(\frac{dT}{dh}\right) = -0.0065$$

Substituting the values $-\dfrac{M^2 a^2}{2g\sigma}\left(\dfrac{d\sigma}{dh}\right)$ of Equation 12.10,

$$\frac{V}{g}\left(\frac{dV}{dh}\right) = -\frac{M^2 a^2}{2g\sigma}\left(\frac{d\sigma}{dh}\right) = -\frac{M^2 a^2}{2g\left(T/T_0\right)^{4.255}} \times 4.225 \times \left(T/T_0\right)^{3.255} \times \frac{(-0.0065)}{T_0}$$

or

$$\frac{V}{g}\left(\frac{dV}{dh}\right) = \frac{M^2 a^2}{2g\left(T/T_0\right)} \times 4.255 \times \frac{(0.0065)}{T_0} = \frac{M^2 \gamma R T}{2g\left(\dfrac{T}{288.07}\right)} \times 4.255 \times \frac{(0.0065)}{288.07}$$

or

$$\frac{V}{g}\left(\frac{dV}{dh}\right) = \frac{M^2 \times 1.4 \times 287}{\left(\dfrac{2 \times 9.81}{288.07}\right)} \times 4.255 \times \frac{(0.0065)}{288.07} = \frac{V}{g}\left(\frac{dV}{dh}\right) = 0.566 M^2 \quad \textbf{(12.11)}$$

12.3.3.2   *Above Tropopause*

$\left[\text{For air, } \gamma = 1.4, \text{R} = 287\,\text{J/kg K}, g = 9.81\,\text{m/s}^2, a = 295.07\,\text{m/s}.\right]$

From 11 to 20 km the ambient temperature is held constant at 216.65 K. Equation 2.8 g gives the density relation in this regime as $\rho\left(\text{kg/m}^3\right) = \rho_1 e^{-\left[\frac{g_0}{RT}\right](h-h_1)}$ where the base pressure ρ_1 is at 11 km. That gives:

$$\sigma = \rho/\rho_0 = \left(\rho/\rho_1\right)\left(\rho_1/\rho_0\right) = \left(\rho/\rho_1\right)\left(0.36392/1.225\right)$$

$$= 0.297 \times \left(\rho/\rho_1\right) = 0.297 \times e^{-\left[\frac{g_0}{RT}\right](h-h_1)}$$

or

$$\sigma = 0.297 \times e^{-\left[\frac{32.2}{287 \times 216.65}\right](h-11000)} = 0.297 \times e^{-[0.0001578](h-11000)}$$

$$= 0.297 \times e^{(1.7355 - 0.0001578 h)}$$

Then

$$\frac{d\sigma}{dh} = 0.297 \times e^{(1.7355 - 0.0001578 h)}\,[-0.0001578]$$

Then Equation 12.10 can be written as follows:

$$\frac{V}{g}\left(\frac{dV}{dh}\right) = -\frac{M^2 a^2}{2g\sigma}\left(\frac{d\sigma}{dh}\right)$$

$$= -\frac{M^2\,295.07^2\,(-0.0001578)}{2 \times 9.81} = 0.7 M^2 \qquad \textbf{(12.12)}$$

12.3.4   Constant Mach Climb

The term $(V/g)(dV/dh)$ can be worked out in terms of constant Mach number climb as follows:

$$\frac{V}{g}\left(\frac{dV}{dh}\right) = \frac{MaM}{g}\left(\frac{da}{dh}\right) \quad \left(M \text{ being constant taken out of differentiation}\right)$$

or

$$\frac{V}{g}\left(\frac{dV}{dh}\right) = \frac{aM^2\sqrt{\gamma R}}{g}\left(\frac{d\sqrt{T}}{dh}\right) = \frac{\gamma RM^2\sqrt{T}}{g}\left(\frac{d\sqrt{T}}{dh}\right)$$

or

$$\frac{V}{g}\left(\frac{dV}{dh}\right) = \frac{\gamma RM^2\sqrt{T}}{2g\sqrt{T}}\left(\frac{dT}{dh}\right) \tag{12.13}$$

Below tropopause:

[For air, $\gamma = 1.4$, $R = 287\,\text{J/kg K}$, $g = 9.81\,\text{m/s}^2$.]

From Equation 2.7, atmospheric temperature, T, can be expressed in terms of altitude, h, as follows:

$$T = \left(288 - 0.0065h\right)$$

where h is in metres. Substituting the value in Equation 12.13,

$$\frac{V}{g}\left(\frac{dV}{dh}\right) = \frac{1.4 \times 287 \times M^2}{2 \times 9.81}(-0.0065)$$

or

$$\frac{V}{g}\left(\frac{dV}{dh}\right) = \frac{401.8 \times M^2}{19.62}(-0.0065) = -0.133M^2 \tag{12.14}$$

Above tropopause:

[For air, $\gamma = 1.4$, $R = 287\,\text{J/kg K}$, $g = 9.81\,\text{m/s}^2$, $a = 295.07\,\text{m/s}$.]

In a similar manner, the relations above the tropopause can be obtained. Above the tropopause up to 25 km the atmospheric temperature remains constant at 216.65 K, therefore speed of sound remains invariant, that is, $(da/dh) = 0$, making $(V/g)(dV/dh) = 0$.

Table 12.2 summarizes the relationship between $(V/g)(dV/dh)$ and the constant speed climb schedules as derived in the above.

To prepare the pilot manual, the airspeed need to be in V_{CAS} and the conversion from V_{EAS} requires incorporating the relevant errors associated with the airspeed indicator (see Section 3.2). Except for the compressibility correction (see Figure 3.4), incorporation of other errors is not tackled in this book. Nowadays, ready data are available in the *electronic flight information system* (EFIS) memory.

■ **TABLE 12.2**
(*V*/g)(*dV*/*dh*) value (dimensionless quantity) at constant climb speeds

	Below tropopause	Above tropopause
At constant EAS	$0.566\,M^2$	$0.7\,M^2$
At constant Mach number	$-0.133\,M^2$	0 (Mach held constant)

12.3.5 Unaccelerated Climb

It is mentioned earlier that the term $(V/g)(dV/dh)$ penalizes unaccelerated rate of climb, depending on how fast the aircraft is accelerating during climb.

With the loss of one engine at second-segment climb, an accelerated climb would penalize the rate of climb. Therefore, second-segment climb with one engine inoperative is done at unaccelerated climb speed, at a speed a little above V_2 on account of undercarriage retraction. The unaccelerated climb equation is obtained by dropping the acceleration term in Equation 12.9, yielding the following equations.

$$T - D = W\sin\gamma; \quad \text{rewriting,} \quad \sin\gamma = (T - D)/W$$

$$\text{Unaccelerated rate of climb, } R/C = dh/dt = V\sin\gamma = V \times (T - D)/W \qquad \textbf{(12.15)}$$

It is the same by dropping the term from Equation 12.9:

$$R/C = dh/dt = V\big[(T - D)/W\big]$$

12.4   Other Ways to Climb (Point Performance) – Civil Aircraft

Section 12.2 lists four different ways to climb. The last one has been dealt in the previous section and this section deals with the first three. A summary of Sections 12.3 and 12.4 is given at the end of this section.

12.4.1 Maximum Rate of Climb and Maximum Climb Gradient

Equations 12.1 and 12.9 give

$$R/C_{accl} = dh/dt = \frac{V\big[(T - D)/W\big]}{1 + (V/g)(dV/dh)} = V\sin\gamma \qquad \textbf{(12.16a)}$$

To determine a maximum rate of climb and maximum climb gradient, the above equation needs to be maximized. Evidently, the acceleration term $(V/g)(dV/dh)$ is penalizing R/C, in other words, unaccelerated climb has a better rate of climb, that is $R/C_{unaccl} > R/C_{accl}$.

$$R/C_{unaccl} = dh/dt = V\big[(T - D)/W\big] \qquad \textbf{(12.16b)}$$

Therefore, maximum R/C is when Equation 12.16a is maximized when $(V/g)(dV/dh) = 0$, that is,

$$\big(R/C_{unaccl}\big)_{max} = dh/dt = \big\{V\big[(T - D)/W\big]\big\}_{max} = \big(V\sin\gamma\big)_{max} \qquad \textbf{(12.17)}$$

Note that the maximum climb gradient

$$\left(\sin\gamma\right)_{max} = \left(T-D\right)_{max} / W \tag{12.18}$$

These are not the same, as it will be shown that the steepest climb has a higher gradient than the maximum climb gradient, that is, $(\sin\gamma)_{steepest\_grad} > (\sin\gamma)_{max\_R/C}$, which gives $(R/C_{unaccl})_{max} > (R/C_{unaccl})_{steepest\_grad}$. A hodograph plot is a good way to see the difference. Bizjet hodograph plots are given in Section 12.6. Henceforth the subscript (*unaccl*) is dropped for maximizing exercises, unless it is attached for clarity.

Graphs of thrust available (T_{avail}) from the engine and thrust required (T_{reqd}), that is, drag (*D*) of the aircraft are required to obtain the climb performance. Use of graphs is required to obtain $\{V[(T-D)/W]\}_{max}$ and $(T-D)_{max}$ as shown in Problems 1 and 2 in the problem set for this chapter given in Appendix E.

The analytical derivation of $(R/C_{unaccl})_{max}$ and $(\sin\gamma)_{steepest\_grad}$ is not easy as both *T* and *D* vary with speed. The following section derives the analytical expressions by making some approximations.

12.4.1.1   Velocity ($V_{(R/C)max}$) at Maximum Rate of Climb

Differentiating Equation 12.16b with respect to V and setting it equal to zero gives the condition for maximum rate of climb. During climb the aircraft weight change due to fuel burn is small, and therefore the aircraft weight may be kept constant without incurring much error. At a constant throttle setting, engine thrust (*T*) variation is slow to change with altitude and can be treated as constant with tolerable error. On the other hand, drag changes with the square of the velocity. (The AK4 climb performance in Problem 2 of this chapter in Appendix E shows that thrust varies by less than 0.5% per second at sea-level, and becomes less as it gains altitude. Note that to gain accuracy, the interval of velocity increment needs to be small.)

Also in *quasi-steady* climb, $\dot{V} = 0$ and $\dot{\gamma} \approx 0$. Then

Differentiating equation 12.16b with respect V and setting equal to zero gives the maximum rate of climb as follows (dropping out subscript 'unaccelerated', being in quasi-steady climb).

$$\frac{d\left(R/C\right)}{dV} = \frac{d}{dV}\left[\frac{VT}{W}\right] - \frac{d}{dV}\left[\frac{VD}{W}\right] = 0$$

gives the maximum. With weight remaining constant, the equation as $V_{(R/C)max}$ becomes maximum is

$$\frac{d\left(R/C\right)}{dV} = \frac{1}{W}\left[T + \frac{VdT}{dV}\right] - \frac{1}{W}\left[D + \frac{VdD}{dV}\right] = 0$$

Climb is carried out at constant throttle. Therefore in the limiting sense of differentiation, thrust, $T \approx$ constant. Since $W \neq 0$, the above equation becomes

$$\frac{d\left(R/C\right)}{dV} = T - D - \frac{VdD}{dV} = 0 \tag{12.19}$$

In terms of coefficients:

$$T - 0.5\rho V^2 S_W C_D - \frac{Vd\left(0.5\rho V^2 S_W C_D\right)}{dV} = 0$$

or

$$2T - \rho V^2 S_W C_D - \frac{Vd\left(\rho V^2 S_W C_D\right)}{dV} = 0$$

$$2T - \rho V^2 S_W \left(C_{DPmin} + kC_L^2 \right) - \frac{V d\left(\rho V^2 S_W \left(C_{DPmin} + kC_L^2 \right) \right)}{dV} = 0$$

$$2T - \rho V^2 S_W C_{DPmin} - \rho V^2 S_W k \left(2W/\rho V^2 S_W \right)^2$$

$$-2\rho V^2 S_W C_{DPmin} + \rho S_W kVk \left[\frac{d\left(2W/\rho V^2 S_W \right)^2}{dV} \right] = 0$$

$$2T - 3\rho V^2 S_W C_{DPmin} - \frac{4kW^2}{\rho V^2 S_W} - \frac{4kVW^2/(\rho S_W)d\left(1/V^2 \right)^2}{dV} = 0$$

$$2T - 3\rho V^2 S_W C_{DPmin} - \frac{4kW^2}{\rho V^2 S_W} - \frac{4kVW^2}{(\rho S_W)\left(-2/V^3 \right)} = 0$$

$$2T - 3\rho V^2 S_W C_{DPmin} - \frac{4kW^2}{\rho V^2 S_W} + \frac{8kW^2}{\rho V^2 S_W} = 0$$

$$2T - 3\rho V^2 S_W C_{DPmin} + \frac{4kW^2}{\rho V^2 S_W} = 0$$

Transposing and rearranging

$$3\rho^2 V^4 S_W^2 C_{DPmin} - 4kW^2 - 2\rho S_W V^2 T = 0$$

or

$$3\rho^2 S_W^2 C_{DPmin} \left(V^2 \right)^2 - 2\rho S_W TV^2 - 4kW^2 = 0 \tag{12.20}$$

This is a quadratic whose roots can be easily found. The root is the velocity V^2, which gives the maximum rate of climb $(V_{(R/C)max})^2$. Keeping its positive real root,

$$V_{(R/C)max}^2 = \frac{\left[2\rho S_W T + \sqrt{\left(4\rho^2 S_W^2 T^2 + 48k\rho^2 S_W^2 W^2 C_{DPmin} \right)} \right]}{6\rho^2 S_W^2 C_{DPmin}}$$

or

$$V_{(R/C)max}^2 = (2\rho S_W T) \frac{\left[1 + \sqrt{\frac{4\rho^2 S_W^2 T^2}{(2\rho S_W T)^2} + \frac{48k\rho^2 S_W^2 W^2 C_{DPmin}}{(2\rho S_W T)^2}} \right]}{6\rho^2 S_W^2 C_{DPmin}}$$

or

$$V_{(R/C)max}^2 = \frac{T/S_W}{3\rho C_{DPmin}} \left[1 + \sqrt{1 + 12kC_{DPmin} \left(W/T \right)^2} \right] \tag{12.21}$$

Taking the root of the above equation and keeping the positive root,

$$V_{(R/C)max} = \left[\left(\frac{T/S_W}{3\rho C_{DPmin}} \right) \left\{ 1 + \sqrt{\left\{ 1 + 12kC_{DPmin} \left(W/T \right)^2 \right\}} \right\} \right]^{\frac{1}{2}}$$

(12.22a)

or

$$V_{(R/C)max} = \left[\left\{ \frac{(T/W)(W/S_W)}{3\rho C_{DPmin}} \right\} \left\{ 1 + \sqrt{\left\{ 1 + 12kC_{DPmin} \left(W/T \right)^2 \right\}} \right\} \right]^{\frac{1}{2}}$$

(12.22b)

To find L/D at maximum rate of climb ($V_{(R/C)max}$), use Equation 10.18:

$$\left(\frac{C_L}{C_D} \right)_{max} = \frac{1}{2} \sqrt{\frac{e\pi AR}{C_{DPmin}}} = \frac{1}{2} \sqrt{\frac{1}{kC_{DPmin}}}$$

or

$$\left(\frac{C_D}{C_L} \right)_{min}^2 = 4kC_{DPmin}$$

Substituting in Equation 12.22b it becomes

$$V_{(R/C)max} = \left[\left\{ \frac{(T/W)(W/S_W)}{3\rho C_{DPmin}} \right\} \left\{ 1 + \sqrt{\left\{ 1 + 3\left(\frac{C_D}{C_L} \right)_{min}^2 \left(W/T \right)^2 \right\}} \right\} \right]^{\frac{1}{2}}$$

(12.23)

The climb gradient at the maximum rate of climb, $(\sin\gamma)_{max\_R/C}$ can be obtained by using Equation 12.18.

$$\sin\gamma = (T - D)/W = T/W - D/W$$

It was mentioned earlier that during climb the aircraft weight change due to fuel burn is small, and also the thrust variation at V_{CAS} climb is small. Therefore the term T/W may be considered invariant. Also, load factor $n \approx 1$ during climb, therefore $W \approx L$. The above equation can now be written as

$$\sin\gamma = \frac{(T-D)}{W} = \frac{T}{W} - \frac{D}{L} = \frac{T}{W} - \frac{C_{DPmin} + kC_L^2}{C_L} = \frac{T}{W} - \frac{C_{DPmin}}{C_L} - kC_L$$

(12.24)

At maximum climb gradient,

$$\left(\sin\gamma\right)_{max\_R/C} = \frac{T}{W} - \frac{\rho \left(V_{(R/C)max}\right)^2 C_{DPmin}}{(2W/S_W)} - \frac{2k\left(W/S_W\right)}{\rho \left(V_{(R/C)max}\right)^2}$$

(12.25)

Substituting $(V_{(R/C)max})^2$ from Equation 12.21 in Equation 12.25, it becomes

$$(\sin\gamma)_{max\_R/C} = \frac{T}{W} - \rho\frac{\left(T/S_W\right)}{3\rho C_{DPmin}}\left[1+\sqrt{\left\{1+3\left(\frac{C_D}{C_L}\right)_{min}^2\left(\frac{W}{T}\right)^2\right\}}\right]\frac{C_{DPmin}}{(2W/S_W)}$$

$$-2k(W/S_W)/\rho\,\frac{\left(T/S_W\right)}{3\rho C_{DPmin}}\left[1+\sqrt{\left\{1+3\left(\frac{C_D}{C_L}\right)_{min}^2\left(\frac{W}{T}\right)^2\right\}}\right]$$

or

$$(\sin\gamma)_{max\_R/C} = \frac{T}{W} - \frac{\left(T/W\right)}{6}\left[1+\sqrt{\left\{1+12kC_{DPmin}\left(\frac{W}{T}\right)^2\right\}}\right]$$

$$-6kC_{DPmin}\left(\frac{T}{W}\right)\bigg/\left[1+\sqrt{\left\{1+12kC_{DPmin}\left(\frac{W}{T}\right)^2\right\}}\right]$$

or

$$(\sin\gamma)_{max\_R/C} = \frac{T}{W} - \frac{\left(T/W\right)}{6}\left[1+\sqrt{\left\{1+12kC_{DPmin}\left(\frac{W}{T}\right)^2\right\}}\right]$$

$$-\left(3/2\right)\left(\frac{C_D}{C_L}\right)_{min}^2\left(\frac{T}{W}\right)\bigg/\left[1+\sqrt{\left\{1+12kC_{DPmin}\left(\frac{W}{T}\right)^2\right\}}\right] \tag{12.26}$$

Typically, en-route climb gradient for civil aircraft is less than 15°; the approximation of $\sin\gamma \approx \gamma$ (radian) gives less than 3% difference between the two.

12.4.2 Steepest Climb

The steepest climb has a higher gradient than the maximum climb gradient, that is, $(\sin\gamma)_{steepest\_grad} > (\sin\gamma)_{max\_R/C}$. As in Section 12.4.1, this section derives the expressions for gradient $(\sin\gamma)_{steepest\_grad}$ and velocity $(V_{steepest\_grad})$ and the climb rate $(R/C_{steepest\_grad})$ at the steepest climb.

At constant thrust setting, the gradient at steepest climb $(\sin\gamma_{steepest\_grad})$ occurs at unaccelerated climb and can be obtained by using Equation 12.18,

$$\sin\gamma = (T-D)/W = T/W - D/W$$

During climb the aircraft weight change due to fuel burn is small, and also the thrust variation at V_{CAS} climb is small. Therefore the term T/W may be considered as invariant. Also the load factor $n \approx 1$ during climb, therefore $W \approx L$ (as $\cos\gamma \approx 1$). The above equation can now be written as

$$(\sin\gamma)_{steepest\_grad} = \frac{T}{W} - \left(\frac{D}{L}\right)_{min} = \frac{T}{W} - \frac{1}{\left(\frac{C_L}{C_D}\right)_{max}} \tag{12.27}$$

Using Equation 10.18 gives:

$$\left(\frac{C_L}{C_D}\right)_{max} = \frac{1}{2}\sqrt{\frac{1}{kC_{DPmin}}}$$

Substituting in the previous equation, it becomes

$$\left(\sin\gamma\right)_{steepest\_grad} = \frac{T}{W} - \frac{2}{\sqrt{\dfrac{1}{kC_{DPmin}}}}$$

(12.28)

For military aircraft at large climb angles ($\cos\gamma \ne 1$) the lift component is used.

Equation 10.18 gives the expression for velocity at $\left(\dfrac{C_L}{C_D}\right)_{max}$ in this case at steepest gradient. Therefore

$$V_{steepest\_grad} = \sqrt{\left(\frac{2W}{\rho S_W}\right)\left(\frac{k}{C_{DPmin}}\right)^{0.25}}$$

(12.29)

To find the steepest climb rate using Equations 12.28 and 12.29,

$$\left(R/C\right)_{steepest\_grad} = \left(V_{steepest\_grad}\right)\left(\sin\gamma_{steepest\_grad}\right)$$

Therefore the product of Equations 12.28 and 12.29 gives

$$\left(R/C\right)_{steepest\_grad} = \sqrt{\left(\frac{2W}{\rho S_W}\right)\left(\frac{k}{C_{DPmin}}\right)^{0.25}} \times \left\{\frac{T}{W} - \frac{2}{\sqrt{\dfrac{1}{kC_{DPmin}}}}\right\}$$

(12.30)

12.4.3   Economic Climb at Constant EAS

To save costs, commercial aircraft operators like to use a climb speed schedule at the minimum fuel burn to reach cruise altitude. Typically, the minimum fuel burn rate occurs at minimum drag for the climb speed. However, if it takes too long to climb then this does not guarantee use of the minimum amount of fuel during the climb. It is not easy to determine the best economy. However, Figure 12.2 shows that the maximum rate of climb and minimum drag climb speeds are close. It can be seen that the rate of climb at a given altitude is flat around its maximum value. Therefore it is advantageous to climb at a higher speed without losing much on the rate of climb. It covers greater distance, contributing to the range, yet not losing the good rate of climb capability. This gives a better economy for burning less fuel for the range covered during climb. This procedure requires continuous small adjustments of V_{EAS} in climb. For pilot ease, a constant V_{EAS} climb schedule is worked out around this small variation in speed adjustment.

It can be seen that at constant V_{EAS} climb schedule the dynamic head q remains constant, but V_{TAS} and hence the Mach number increases and can become supersonic. It is for this reason that there is a crossover altitude ($\approx$30,000 ft) when the climb schedule changes from constant V_{EAS} to a constant Mach number ($\approx$0.78 Mach) to stay within the design limitations (Example 3.1).

12.4.4   Discussion on Climb Performance

Sections 12.3 and 12.4 packed in considerable information depicting numerous possibilities for conducting climb performance analysis. The advantage of using close-form analysis is clearly evident as the climb characteristics can be quickly established (at the cost of some loss in accuracy) without the labour-intensive computational effort. This will be shown in the next section. (Unfortunately, aircraft drag and engine performance graphs are not sufficient since their results are not acceptable in industry for FAR substantiation/pilot manual.)

Pilots fly by observing flight deck instruments. In climb, the ambient pressure, temperature and density varies, and therefore the indicated airspeed varies. Pilots will find it convenient if any prescribed schedule is maintained at a constant value, that is, at a calibrated airspeed V_C, preferably kept at constant indication so that the pilots will not have to attend the gauge too frequently. Engineers work out a convenient speed schedule to climb; in lower altitudes at *constant EAS*, and when a Mach number limit is reached for the prescribed climb speed, at *constant Mach number*. Figures 12.3 and 12.4 in Section 12.5.2 establish the climb speed schedules close to the maximum rate of climb. Civil operators prefer the most economic climb schedule, unless obstacles demand a steeper climb schedule. As a good compromise is established through analyses, typically there are two schedules, for example: (i) *long-range climb* at the best economy yet keeping constant V_C until it reaches a Mach limit, when constant Mach is held during climb (see Section 12.5.1); and (ii) *high-speed climb* as a fast climb yet keeping constant V_C until it reaches the Mach limit. Hodograph plots (see Figures 12.5 and 12.6) compare explicitly the differences between various ways to climb.

Table 12.3 gives the summary of equations involved in climb performance. These equations are also valid for propeller-driven aircraft, bearing in mind that ESHP (equivalent shaft horse power) $= (TV)/\eta_{prop}$. In Imperial units, $T = (550 \times ESHP \times \eta_{prop})/V$. Also note that most of the currently flying turboprop aircraft operate below the tropopause (36,089 ft altitude) and below Mach 0.5, that is, in incompressible flow.

■ **FIGURE 12.3**
Bizjet rate of climb

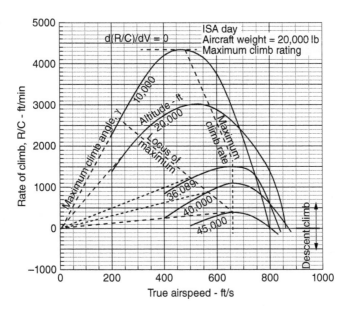

■ FIGURE 12.4
Bizjet economic
climb speed schedule

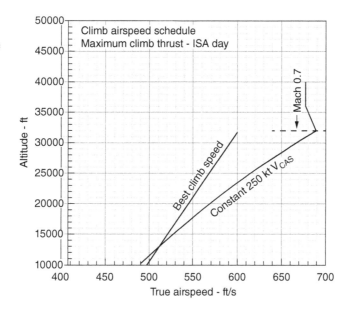

■ FIGURE 12.5
Bizjet hodograph
plot for two aircraft
weights

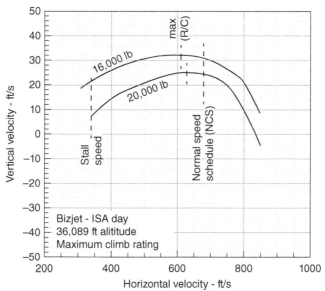

12.5   Classwork Example – Climb Performance (Bizjet)

12.5.1   Takeoff Segments Climb Performance (Bizjet)

To compute Bizjet takeoff segment climb, use the Bizjet drag polar to obtain drag, $C_{DPmin} = 0.0205$. Aircraft weight = 20,600 lb (80 lb fuel consumed) and kept invariant for simplicity. Section 12.9 substantiates the customer requirement of an all-engines operating en-route initial climb rate of 2600 ft/min to 1500 ft altitude. All-engines climb is accelerated climb. The cabin pressurization system should cope with the rate of climb (this is not a FAR

■ **FIGURE 12.6**
**Bizjet hodograph
plot at aircraft
weight of 20,000 lb**

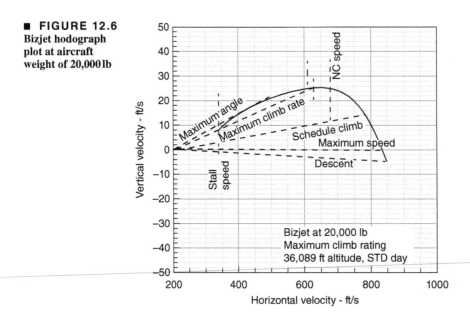

■ **TABLE 12.3**
Climb performance equations

Equation number	Item	Equation
12.11–12.14	Constant speed climbs	
		Below tropopause *Above tropopause*
	At constant EAS	$0.566\,M^2$ $0.7\,M^2$
	At constant Mach number	$-0.133\,M^2$ 0 (Mach held constant)
12.15	Unaccelerated R/C	$R/C = dh/dt = V \times (T - D)/W = V\sin\gamma$
12.16a	Accelerated R/C	$R/C_{accl} = dh/dt = \dfrac{V\left[(T/\delta - D/\delta)/(W/\delta)\right]}{1+(V/g)(dV/dh)} = V\sin\gamma$
Maximum rate of climb		
12.23	Velocity	$V_{(R/C)max} = \left[\left\{\dfrac{(T/W)(W/S_w)}{3\rho C_{DPmin}}\right\}\left\{1+\sqrt{1+3\left(\dfrac{C_D}{C_L}\right)_{min}^2 (W/T)^2}\right\}\right]^{1/2}$
12.26	Gradient at maximum R/C	$(\sin\gamma)_{max\_R/C} = \dfrac{T}{W} - \dfrac{\left(T/W\right)}{6}\left[1+\sqrt{\left\{1+12kC_{DPmin}\left(\dfrac{W}{T}\right)^2\right\}}\right]$
		$\qquad -\left(\dfrac{3}{2}\right)\left(\dfrac{C_D}{C_L}\right)_{min}^2 \left(\dfrac{T}{W}\right)\bigg/\left[1+\sqrt{\left\{1+12kC_{DPmin}\left(\dfrac{W}{T}\right)^2\right\}}\right]$
12.17	Maximum R/C	$(R/C)_{max} = dh/dt = V\left[(T-D)/W\right]_{max} = \left(V_{(R/C)max}\right)\left(\sin\gamma\right)_{max\_R/C}$

■ **TABLE 12.3**
(*Continued*)

Steepest rate of climb		
12.28	Gradient at steepest climb	$(\sin\gamma)_{steepest\_grad} = \dfrac{T}{W} - \dfrac{2}{\sqrt{\dfrac{1}{kC_{DPmin}}}}$
12.29	Velocity at steepest climb	$V_{steepest\_grad} = \sqrt{\left(\dfrac{2W}{\rho S_W}\right)}\left(\dfrac{k}{C_{DPmin}}\right)^{0.25}$
12.30	Steepest climb rate	$(R/C)_{steepest\_grad} = \sqrt{\left(\dfrac{2W}{\rho S_W}\right)}\left(\dfrac{k}{C_{DPmin}}\right)^{0.25} \times \left\{\dfrac{T}{W} - \dfrac{2}{\sqrt{\dfrac{1}{kC_{DPmin}}}}\right\}$

■ **TABLE 12.4**
Segment climb performance for Bizjet, 8° flap

	First segment	Second segment	Third segment	Fourth segment
One engine inoperative				
Final altitude (ft)	From 35	400	400	400–1500
C_{Lmax}	1.67	1.67	1.55	1.55
ρ_{ave} (lb/ft³)	0.002378	0.00235	0.0023	0.00228
$0.5\rho S_W$	0.384	0.3795	0.3797	0.3682
V_{stall} (kt)	106.4	106.4	110.8	112.6
V_{stall} (ft/s)	179.6	179.6	187.14	190
V_2 (ft/s)	216	216	234.6 (V_2 + 10 kt)	237.5 ($\geq 1.25V_{stall}$)
qS_W	17,930	17,720	20,898	20,769
C_L	1.15	1.163	0.986	0.992
C_{Dclean}	0.075	0.078	0.07	0.072
$\Delta C_{D\_one\_eng}$	0.003	0.003	0.003	0.003
ΔC_{Dflap}	0.013	0.013	–	–
$\Delta C_{D\_u/c}$	0.0124	–	–	–
C_D	0.1034	0.0823	0.073	0.075
Drag (lb)	1854	1458	1466	1420
$(Thrust)_{avail}$ (lb)	2800	2800	2800	2700
All-engine operating case (no asymmetric aircraft drag)				
C_D	0.1004	0.0793	0.07	0.072
Drag (lb)	1800	1405	1406	1363
$(Thrust)_{avail}$ (lb)	5600	5600	5600	5400

requirement). The one engine inoperative climb rate is unaccelerated. Stalling speed V_{stall} with one engine inoperative is taken as at all-engines operating.

Consult Table 11.2, which details the civil aircraft takeoff segment requirements [JAR (14CFR) 25.121 Subpart B]. Table 12.4 gives the segment climb performance with a flap setting of 8°, for reference.

12.5.1.1 Sample Calculation to Obtain Gradient (One Engine Failed – Unaccelerated Climb)

Equation 12.17 gives the relation for the climb gradient (one engine inoperative). In the fourth segment the engine is at maximum climb rating. See Section 12.9 for all engine rate of climb substantiations.

FAR first segment climb gradient = (2800 – 1854)/20,600 = 0.046, climb angle $\gamma = 2.6°$, meets positive gradient.

FAR second segment climb gradient = (2800 – 1458)/20,600 = 0.065, climb angle $\gamma = 3.7°$ exceeds 2.4%.

FAR third segment climb gradient = 0 (level flying).

FAR fourth segment climb gradient = (2700 – 1420)/20,600 = 0.062, climb angle $\gamma = 3.55°$, exceeds 1.1%.

En-route climb can commence with all engines operating, otherwise the aircraft lands as soon as possible. With all engines operating, the distance covered, time taken and fuel burnt to reach 1500 ft are computed below. These parameters are added to the mission performance in Chapter 14.

12.5.1.2 All Engines Operating Takeoff Segment Climb Performance to 1500 ft Altitude

FAR first segment climb gradient = (5600 – 1800)/20,600 = 0.1845, climb angle $\gamma = 10.6°$.

FAR second segment climb gradient = (5600 – 1405)/20,600 = 0.2036, climb angle $\gamma = 11.75°$.

FAR third segment climb gradient = 0 (level flying).

FAR fourth segment climb gradient = (5400 – 1363)/20,600 = 0.196, climb angle $\gamma = 11.3°$.

The aircraft attitude is slightly higher (see Figure 12.1) after adding the angle between the fuselage axis and the flight path.

For safety reasons, passengers are required to remain seated until a comfortable aircraft attitude is reached at higher altitude.

12.5.1.3 To Compute Distance Covered, Time Taken and Fuel Burned

Since this is a small fraction of the total mission, a simplified approach is taken, as worked out below.

First and second segment climb to 400 ft altitude:
(Take the average value of velocity = 216 ft/s. Take average climb gradient = 11°.)

Time taken: 400/(216 sin 11) = 400/41.2 = 9.7 seconds ≈ 10 seconds.
Distance covered: 9.7 × (216 cos 11) = 9.7 × 212 = 2057 ft.
Fuel burned (fuel flow = 2 × 1772 lb/h, see Section 8.6.1) = 9.6 lb ≈ 10 lb.

Third segment climb – level flying at 400 ft altitude (accelerate from 216 to 234.6 ft/s):
The case with one engine operating has low acceleration for the flaps to retract while accelerating. In the all-engine operating case the aircraft accelerates fast, hence a flap retraction may be arbitrarily taken after 8 seconds in this example (simplification). The aircraft reaches the velocity to make the fourth segment climb.

Time taken: 8 seconds
Distance covered: 8 × 225 = 1800 ft
Fuel burned: (fuel flow = 2 × 1772 lb/h, see Section 8.6.1) = 7.9 lb ≈ 8 lb.

Fourth segment climb from 400 to 1500 ft altitude:
(Take the average value of velocity = 235 ft/s. Take average climb gradient = 11°.)

Time taken: 1100/(235 sin 11) = 1100/44.82 = 24.5 seconds ≈ 25 seconds.
Distance covered: 24.5 × (235 cos 11) = 24.5 × 230.7 = 5652 ft.
Fuel burned: (fuel flow = 2 × 1745 lb/h, see Section 8.6.1) = 23.75 lb ≈ 24 lb.

Takeoff ground run to 35 ft altitude (from statistics):

Time taken: 35 seconds.
Distance covered: 4400 ft.
Fuel burned: 30 lb.

Total for the takeoff procedure (brake release to 1500 ft altitude):

Total distance = 4400 + 2057 + 1800 + 5652 = 13,909 ft ≈ 14,000 ft.
Total time = 35 + 10 + 8 + 25 = 78 seconds ≈ 80 seconds.
Total fuel burned = 30 + 10 + 8 + 24 = 72 lb ≈ 80 lb.
For the case with one engine inoperative, the numbers will be higher.
The readers may compute for 20° flaps. Take $\Delta C_{Dflap\_20deg} = 0.032$.

12.5.2 En-Route Climb Performance (Bizjet)

En-route climb commences at the end of the fourth segment of the takeoff procedure with all engines operating, further throttled back to the maximum climb rating. In fact, the prescribed normal climb speed (NCS) requires less than the maximum climb rating.

The aircraft rate of climb is the point performance at that instant of time of assessment. Substantiation of the initial rate of climb is a customer requirement and not an airworthiness requirement. The Bizjet customer specification requirement is 2600 ft/min.

Older flight deck airspeed indicators exhibit V_{CAS}, and the modern EFIS displays both V_{CAS} and V_{EAS}. The difference between V_{CAS} and V_{EAS} at the specified climb speed is small. This book uses V_{CAS} for the climb performance computation, facilitating the older aircraft display. The rate of climb at the speed schedule has negligible difference between them, being flat on the top of the graph.

For pilot ease, the climb procedure is carried out at constant specified V_{CAS} (V_{EAS}). Above crossover altitude (see Section 3.3) the schedule is switched to constant Mach climb. Generally, two types of constant climb speeds are prescribed, for example: (i) HSC as a time-saving procedure yet gaining in ground distance; and (ii) NCS which offers a compromise between best climb rate and minimum fuel burn (economy).

The aircraft and engine climb performance grid, given below in Figure 12.2, is extensively used in computation of the worked example of the Bizjet aircraft climb performance. The best climb rate of aircraft is when $(T/\delta - D/\delta)$ is the highest in unaccelerated climb. It shows that the maximum rate of climb and minimum drag climb speeds are close.

From Equation 12.9, quasi-steady-state rate of climb is given by

$$R/C_{accl} = \frac{V_\infty\left[(T-D)/W\right]}{1+(V/g)(dV/dh)}$$

To save space, only the 20,000 ft altitude case is shown in Table 12.5 and plotted in Figure 12.3 (some rounding discrepancies exist in all graphs). Computations for other altitudes and weights follow the same procedure.

There are two important parameters that can be shown in the graph (Figure 12.3): the locus of the maximum climb rate and the locus of maximum climb angle γ. Equation 12.5b gives $\sin\gamma = (dh/dt)/V$, which is maximum at a tangent to the rate of climb curve in Figure 12.3. It can be computed using Equation 12.28. The best climb is an unaccelerated climb as computed in Table 12.6. It gives the aircraft speed for the altitude that gives the maximum rate of climb. Figure 12.3 gives the aircraft speed at the best climb and at the steepest climb angle γ.

Table 12.6 gives the point climb performance of the Bizjet at an initial climb weight of 20,000 lb (use Figure 12.9b to obtain the climb fuel consumed) which gives the aircraft speed $V_{(R/C)max}$ at the best climb rate, in a quasi-steady (near unaccelerated) climb. Figure 12.2 shows that the best rate of climb and climb at minimum drag (minimum fuel burn) are close. A convenient constant V_{CAS} can be established as a compromise to have good economy and a good rate of climb speed schedule, and for the pilot's ease.

12.5.3 Bizjet Climb Schedule

The economic Bizjet climb schedule has constant speed climb at 250 kt V_{CAS} until it reaches 0.7 Mach at 31,000 ft altitude, and thereafter maintains climb at constant Mach 0.7. Table 12.7 gives an example to show that below the crossover altitude (31,000 ft) the difference between V_{CAS} and V_{EAS} is small. Above the crossover altitude the climb schedule is at constant Mach 0.7. Figure 12.4 gives the economic climb speed at 250 kt V_{CAS} until it reaches Mach 0.7 at 31,000 ft altitude. (The readers may use these values for most of the high subsonic jet transport aircraft, unless another speed schedule is prescribed.)

12.6 Hodograph Plot

A hodograph plot is a good way for engineers to analyse aircraft climb performance at a particular weight and altitude. A hodograph plot is nothing but a convenient replotting of Table 12.8 into Figures 12.5 and 12.6, with the X-axis as the aircraft horizontal velocity and the Y-axis as the vertical velocity during climb, where horizontal velocity,

$$V_{Hor} = (V_{TAS})\cos\gamma$$

and vertical velocity,

$$V_{Ver} = (V_{TAS})\sin\gamma = R/C \quad \text{in ft/s}$$

$$\tan\gamma = \frac{R/C \text{ in ft/s}}{\text{true airspeed in ft/s}}$$

Then

$$(V_{TAS})^2 = (V_{Hor})^2 + (V_{Ver})^2 = (V_{TAS})^2 \cos^2\gamma + (V_{TAS})^2 \sin^2\gamma$$

Therefore

$$V_{TAS} = \sqrt{(V_{Hor})^2 + (V_{Ver})^2} \tag{12.31}$$

TABLE 12.5
Rate of climb performance versus true airspeed at mid-cruise weight of 20,000 lb

V_{TAS} (ft/s)	Mach (ft/s)	C_L (ft/s)	C_{Di} (ft/s)	ΔC_{DP} (ft/s)	C_D (ft/s)	Drag (lb)	$2 \times T$ (lb)	$(2T-D)/\delta$ (lb)	$(V/g)(dV/dh)$ (ft/s²)	R/C_{acel} (ft/s)	R/C_{acel} (ft/min)
200	0.193	2.06	0.181	0.009	0.2104	1712.65	4000	4201.6	0.0208	22.41	1344.502
300	0.289	0.92	0.0357	0.007	0.0632	1158.12	3900	5036.53	0.0469	39.3	2357.38
400	0.386	0.52	0.0113	0.0006	0.0324	1055.15	3700	4858.3	0.0833	48.832	2929.913
500	0.4822	0.33	0.0046	0	0.025	1278.54	3620	4301.0	0.1302	51.797	3107.81
600	0.5786	0.229	0.0022	0.0001	0.0228	1672.77	3600	3540.1	0.1875	48.692	2921.52
700	0.675	0.1685	0.0012	0.0004	0.0221	2204.25	3440	2269.9	0.2552	34.46	2067.624
800	0.772	0.129	0.00070	0.00055	0.0218	2833.6	3360	966.97	0.3333	15.794	947.632
860	0.8294	0.1116	0.00053	0.00055	0.02158	3247.85	3300	95.80	0.3852	2.243	1.62

At altitude 20,000 ft, $W/\delta = 36,735$ lb, $a = 1036.95$ ft/s, $(V/g)(dV/dh) = 0.56M^2$, $C_{DPmin} = 0.0205$.

TABLE 12.6
Best rate of unaccelerated climb at initial weight of 20,000 lb (use Equation 12.30 and Figure 12.2)

Altitude (ft)	W (lb)	Thrust (lb)	W/S_W	$2T/W$	A^a	B^a	C^a	$B \times C$	$V_{(R/C)max}$ (ft/s)
10,000	19,900	2000	61.6	0.201	12.384	115,065.2	2.13	244,985.3	494.96
20,000	19,750	1750	61.5	0.1757	10.836	139,836.4	2.1664	302,851.8	550.32
30,000	19,640	1320	60.81	0.1344	8.17	151,023.2	2.271	339,818.6	582.94
32,000	19,600	1280	60.68	0.1306	7.93	157,162.3	2.186	381,855.7	617.95

^{a} Where $A = (T/W) \times (W/S_W)$, $B = \dfrac{\dfrac{T}{W}\left(\dfrac{W}{S_W}\right)}{3\rho C_{DPmin}}$ and $C = 1 + \sqrt{1 + 3\left[\left(\dfrac{C_D}{C_L}\right)^2_{min}\right]\left(\dfrac{W}{T}\right)^2}$.

■ **TABLE 12.7**
Bizjet climb speed schedule at constant 250 V_{CAS} and Mach 0.7

Altitude (ft)	ρ	ΔVc (kt)	V_{EAS}	σ	√σ	V_{TAS} (kt)	V_{TAS} (ft/s)	a (ft/s)	Mach
1500	0.00227	1	249	0.955	0.9772	254.8	430.1	1115.1	0.386
10,000	0.00175	2	248	0.736	0.8578	289.1	488.22	1077.38	0.453
20,000	0.00126	5	245	0.53	0.7279	336.58	568.41	1036.95	0.548
30,000	0.00088	9	241	0.3701	0.6083	396.17	669.05	986.0	0.678
32,000	0.00082	10	240	0.3448	0.58722	408.71	690.22	981.65	0.703
35,000	0.00073	13	237	0.307	0.55406	427.75	722.39	972.88	0.7425
40,000	0.00058	16	234	0.2439	0.49387	473.81	800.18	968.08	0.8266

■ **TABLE 12.8**
Hodograph plot of rate of climb performance (W = 20,000 lb, altitude = 36,089 ft)

V_{TAS} (ft/s)	R/C (ft/min)	R/C (ft/s)	γ (rad)	V_{Ver} (ft/s)	V_{Hor} (ft/s)
343.43	500	8.33	0.0243	8.333	343.33
400	860	14.33	0.0358	14.33	399.74
450	1200	20	0.0444	19.99	449.56
510	1220	20.33	0.0399	20.33	509.595
550	1350	22.5	0.04091	22.494	549.54
600	1480	24.67	0.0411	24.66	599.493
650	1500	25	0.03846	24.99	649.52
700	1420	23.67	0.03381	23.66	699.6
750	1180	19.67	0.02622	19.66	749.74
800	600	1	0.0125	9.999	799.94
850	−300	−5	−0.00588	−4.999	849.985

■ **TABLE 12.9**
Summary of hodograph plot

Hodograph (point)	Rate of climb (ft/min)	V_{TAS} (ft/s)	tan γ	Climb angle (γ-deg)	V_{Hor} (ft/s)	V_{Ver} (ft/s)
Stall	500	343.43	0.024	1.37	343.3	8.33
Max γ	1210	475	0.0425	2.43	474.57	20.17
Max R/C	1500	640	0.0384	2.2	649.5	25
Climb schedule	1480	677.75	0.0364	2.085	677.3	24.67
Descent	−300	840	−0.006	−0.344	839.9	−5
Maximum level speed	0	820	0	0	730	0

$$\gamma = \tan^{-1} \frac{(R/C)}{V_{TAS}} \tag{12.32}$$

A summary of the hodograph plot is given in Table 12.9.

It is interesting to note that while at lower altitudes, the best climb speed is close to the Bizjet NCS schedule, it gradually moves away at higher altitudes. The hodograph plots in

Figures 12.5 and 12.6 show that the rate of climb is flat around $V_{(R/C)max}$. The vertical velocity (rate of climb) at NCS at 36,086 ft is about 2.5% less than at $V_{(R/C)max}$, yet gives about 7% higher ground speed. The advantage of hodographs for engineering analyses is demonstrated here. It is suggested that readers may compute hodographs for other weights and altitudes

12.6.1   Aircraft Ceiling

There are three kinds of aircraft ceiling, as mentioned in Section 12.2.

1. *Absolute ceiling:* When the engine is unable to produce enough thrust at high altitudes, then the rate of climb reduces to zero. It is hard to establish the point performance as the time taken to climb increases, while fuel burn decreases the aircraft weight significantly. However, it can be extrapolated from the rate of climb values at lower altitudes, as shown in Figure 12.8. The last 500 ft climb takes a long time and the aircraft weight keeps changing in the meantime, making it difficult to achieve a zero rate of climb. The absolute ceiling can be only at one velocity, and if it is not hitting buffet (see Section 5.2.1) then it can also be seen as the *aerodynamic ceiling.*

2. *Service ceiling:* From the difficulty of attaining the absolute ceiling, the industry conveniently uses a climb rate of 100 ft/min as the service ceiling. It has little operational value, since to reach 100 ft/min for jet-propelled aircraft also takes some time. This is used for proving purposes and to compare aircraft capabilities.

3. *Operating ceiling:* From the operational point of view the operating ceiling for piston-engine aircraft is taken as a climb rate of 100 ft/min, and for jet propulsion aircraft, 500 ft/min. However, since it is the choice of industry, some take a climb rate of 300 ft/min as the operating ceiling, as shown in Figure 12.8.

The operational ceiling has to be computed. The Bizjet has an operating ceiling at the altitude where its rate of climb reaches 300 ft/min. Following the same methodology as carried out in the previous subsection, the rate of climb at 44,000, 46,000 and 48,000 ft altitudes for various aircraft weights are calculated to determine what aircraft weight reaches the ceiling capability. The results are plotted in Figure 12.8. At the end of climb, the aircraft weight will be around 19,500 lb and the ceiling will be in excess of 44,000 ft altitude, and at the end of cruise when the aircraft becomes lighter, the operating ceiling will be in excess of 47,000 ft.

12.7   Worked Example – Bizjet

12.7.1   Bizjet Climb Rate at Normal Climb Speed Schedule

The initial en-route climb weight may be taken as 20,600 lb (80 kg fuel consumed during takeoff procedure up to fourth segment climb) starting from 1500 ft altitude ($\rho = 0.00227$ slug/ft^3, $\sigma = 0.971$, $a = 1110$ ft/s) at the end of third segment climb. During en-route climb, all engines are throttled back to the maximum climb rating and the aircraft is in clean configuration. As indicated, economic climb suits commercial operations best.

We need to establish the rate of climb of Bizjet up to the ceiling of aircraft capability. Initial climb weight may be taken as 20,600 lb starting from 1500 ft altitude at the end of third segment climb. The first task in computing integrated climb performance is to establish the rate

■ **TABLE 12.10**
Rate of climb performance versus altitude

Altitude (ft)	δ	Mach	V_{TAS} (ft/s)	W/δ (lb)	(T/δ – D/δ)	(V/g)(dV/dh)	R/C_{accl} (ft/min)
1500	0.971	0.386	432.1	20,597	2840	0.083	3300
5000	0.832	0.42	461	24,038	3340	0.099	3500
10,000	0.688	0.457	492	29,070	3960	0.117	3600
15,000	0.564	0.509	538.2	35,460	4040	0.145	3300
20,000	0.49	0.56	580	40,418	3470	0.176	2540
31,000	0.284	0.71 ≈ 0.7	693.3	73,800	2890	0.274	1280
31,000	0.284	0.71 ≈ 0.7	693.3	73,800	2850	−0.065	1720
36,089	0.2233	0.7	677.6	89,566	2200	0	1000
40,000	0.185	0.7	677.6	108,108	1870	0	700
45,000	0.1455	0.7	677.6	137,457	1100	0	325
46,000	0.142	0.7	677.6	140,845	500	0	144

of climb for varying aircraft weights and altitudes. Use Equation 12.16a and the aircraft/engine grid from Figure 12.2 to compute the accelerated rate of climb:

$$R/C_{accl} = dh/dt = \frac{V\left[(T/\delta - D/\delta)/(W/\delta)\right]}{1+(V/g)(dV/dh)}$$

The Bizjet rate of climb is computed in Table 12.9.

Section 12.4 has established that the best normal Bizjet climb schedule has constant speed climb at 250 kt V_{CAS} until it reaches Mach 0.7 at 31,000 ft altitude, and thereafter maintains climb at constant Mach 0.7.

12.7.2 Rate of Climb Performance versus Altitude

Table 12.10 shows the rate of climb versus altitude, assuming weight $W = 20,000$ lb, constant $V_{CAS} = 250$ kt, and the acceleration term $(V/g)(dV/dh) = 0.56 M^2$ below 31,000 ft and $-0.133 M^2$ above 31,000 ft into the stratosphere.

The rate of climb performances are also plotted (Figure 12.7) as a function of aircraft true speed versus rate of climb at constant altitude, as computed in Table 12.9.

12.7.3 Bizjet Ceiling

The Bizjet ceiling capability is computed and shown in Table 12.11. The complete calculation is plotted in Figure 12.8. Table 12.12 gives a summary of the Bizjet operating ceiling capability.

Military aircraft have higher speeds and higher thrust; their ceiling capability is determined in a slightly different way. It uses an energy method, as discussed in Section 12.9.

12.8 Integrated Climb Performance – Computational Methodology

Integration of the climb equations may not be easy. One way to tackle this is to integrate in small steps in which the aerodynamic variables can be treated as constant to get the trajectory analysis.

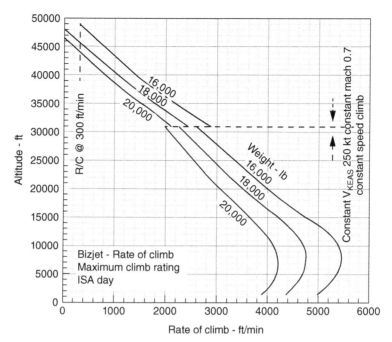

W (lb)	Mach	V_{TAS} (ft/s)	C_L	C_D	Drag, D (lb)	Thrust, T (lb)	$(T-D)/\delta$ (lb)	R/C (ft/s)	R/C (ft/min)
17,000	0.7	677.67	0.472687	0.03	1078.9	720	2354.0	14.4	863.7
17,500	0.7	677.67	0.48659	0.0309	1112.3	720	2143.0	12.7	763.8
18,000	0.7	677.67	0.500492	0.0318	1143.6	720	1932.0	12.2	669.5
18,500	0.7	677.67	0.514395	0.0323	1161.6	720	1814.7	10.2	612.8
19,000	0.7	677.67	0.528297	0.0328	1179.6	720	1697.5	9.3	557.3
19,500	0.7	677.67	0.5422	0.0338	1215.6	720	1463.1	7.8	468.0
20,000	0.7	677.67	0.556102	0.0348	1251.5	720	1228.6	6.4	383.2
22,000	0.7	677.67	0.611713	0.0388	1395.4	720	290.9	1.4	82.5

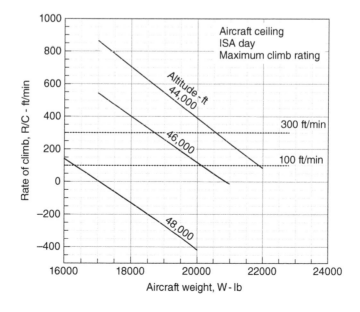

■ **TABLE 12.12**
Summary of Bizjet operating ceiling capability

Operating ceiling capability at 300 ft/min			
Altitude (ft)	44,000	46,000	48,000
Aircraft weight (lb)	20,500	18,200	≳15,000
Service ceiling capability at 100 ft/min			
Altitude (ft)	44,000	46,000	48,000
Aircraft weight (lb)	21,900	20,100	16,200
Absolute ceiling capability at 0 ft/min			
Altitude (ft)	44,000	46,000	48,000
Aircraft weight (lb)	22,500	21,000	17,000

The climb performance parameters vary with altitude. En-route climb performance up to cruise altitude is computed at discrete steps of altitude (for manual computation, say in steps of 5000 ft altitude – see Figure 12.1), within which all parameters are considered invariant and taken as an average value within the step altitudes. The engineering approach is to compute the integrated distance covered, time taken and fuel consumed to reach the cruise altitude in steps of small altitudes and then summed up. The procedure is shown below. (At a constant climb angle γ, the incremental $d\gamma = 0$. Incremental height gain is ΔH in time Δt.)

The infinitesimal time to climb is expressed as $dt = dh/(R/C_{accl})$. The integrated performance within the small steps of altitude can be written as

$$\Delta t = t_{final} - t_{initial} = \frac{\left(h_{final} - h_{initial}\right)}{\left(R/C_{accl}\right)_{ave}} = \Delta H / \left(R/C_{accl}\right) \tag{12.33}$$

The distance covered during climb can be expressed as

$$\Delta s = \Delta t \times V_{ave} = \Delta t \times V \cos\gamma \tag{12.34}$$

where V is the average aircraft speed within the step altitude.

Fuel consumed during climb can be expressed as

$$\Delta fuel = \text{average fuel flow rate} \times \Delta t \tag{12.35}$$

To summarize, the total time taken, distance covered and fuel consumed during climb is obtained by summing up the values in the small steps as follows.

Time to climb,	$time_{climb} = \Sigma \Delta t$	
Distance covered during climb,	$R_{climb} = \Sigma \Delta s$	(12.36)
Fuel consumed,	$fuel_{climb} = \Sigma \Delta fuel$	

12.8.1   Worked Example – Initial En-Route Rate of Climb (Bizjet)

En-route climb to destination is with all engines operating, otherwise the aircraft lands at the first opportunity with one engine failed. The aircraft initiates en-route climb from the fourth segment takeoff V_2 speed reaching 1500 ft altitude, to an accelerated climb to reach 250 kt V_{CAS} (=428.26 ft/s, Mach 0.386) and thereafter continues at a constant 250 kt quasi-steady-state climb until it reaches Mach 0.7 at a crossover altitude of about 31,000 ft, from which the Mach number is held constant until it reaches the cruise altitude.

Operators specify the initial en-route climb rate capability, which is not a airworthiness requirement. It is used to obtain the payload-range capability, which is to be substantiated. The Bizjet requirement for the initial en-route climb rate is 2600 ft/min. The aircraft weight at 1500 ft altitude is taken as 20,600 lb, taking into account the fuel burned to reach 1500 ft altitude.

The aircraft lift coefficient

$$C_L = M/qS_W = 20600 / \left(0.5 \times 0.00228 \times 428.26^2 \times 323\right) = 0.304$$

The clean aircraft drag coefficient from Figure 9.3 at $C_L = 0.304$ gives $C_{Dclean} = 0.0241$. Clean aircraft drag,

$$D = 0.0241 \times \left(0.5 \times 0.0023 \times 428.26^2 \times 323\right) = 0.0241 \times 68126.4 = 1641 \text{ lb.}$$

Available all-engine installed thrust at the maximum climb rating from Figure 8.12 at Mach 0.386 is $T = 2 \times 2300 = 4600$ lb. For the Bizjet example, Table 12.5 gives

$$\frac{V}{g}\left(\frac{dV}{dh}\right) = 0.566 \ M^2 = 0.566 \times 0.386^2 = 0.0834.$$

Hence,

$$R/C_{accl} = \frac{\left\{\left[428.26 \times \left(4600 - 1641\right)\right] / 20560\right\}}{\left[1 + 0.0834\right]} = 56.8 \text{ ft/s} = 3414 \text{ ft/min} \left(17.33 \text{ m/s}\right)$$

The capability exceeds the market requirement of 2600 ft/min (13.2 m/s).

12.8.2 Integrated Climb Performance (Bizjet)

This subsection computes the integrated climb performance. For operational convenience, the Bizjet initial cruise altitude starts at 43,000 ft with a mid-cruise step climb to 45,000 ft up to end cruise. Cruise can be performed at a gradual shallow climb. The aircraft and engine grid generated in Figure 12.2 is used to compute integrated climb performance in this section.

Integrated climb performance can be computed at 5000 ft intervals within which the parameters are treated as constant at its mid value. Table 12.13 tabulates the integrated climb performance for an ISA day, using Figure 12.2 to obtain $(T-D)$ values. Altitude corrections for a non-standard day have been discussed earlier. Figure 12.9 plots the Bizjet integrated climb performance, showing climb time, distance covered and fuel consumed to the altitude.

12.8.3 Turboprop Trainer Aircraft (TPT)

Readers are encouraged to compute all the climb graphs for the TPT aircraft as done for the Bizjet in the previous sections. The TPT aircraft and engine grid is given in Figure 10.20.

12.9 Specific Excess Power (SEP) – High-Energy Climb

The advent of jet propulsion expanded the flight envelope well beyond the capabilities of propeller-driven aircraft. The enhanced performance in speed and altitude gain opens the scope to exploit the use of aircraft energy, for example, to gain additional height. Use of energy

Integrated climb performance (accelerated climb), with initial weight = 20,600 lb (reducing in climb as fuel is consumed)

Altitude, h ($\times 1000$ ft)	Δh (ft)	W_{ave} (lb)	R/C (ft/min)	V (ft/s)	Δt (min)	$\Sigma \Delta t$ (min)	Fuel ($2 \times$ FF, lb)	$\Sigma \Delta FF$ (lb)	ΔS (nm)	$\Sigma \Delta S$ (nm)
0–5	5000	20,560	3300	480	1.52	1.52	83.5	83.3	7.20	7.20
5–10	5000	20,480	3500	470	1.43	2.95	72.8	156	6.64	13.84
10–15	5000	20,400	3480	512	1.44	4.39	70.6	226	7.28	21.12
15–20	5000	20,300	2820	550	1.77	6.16	79	305	9.61	30.73
20–25	5000	20,200	2200	610	2.27	8.43	87	392	13.67	44.40
25–31	6000	20,100	1620	664	3.7	12.13	120	512	24.26	68.66
31–36	5089	20,000	1380	682	3.67	15.85	104	616	24.71	93.37
36–40	3911	19,900	800	678	4.88	20.74	117	733	32.67	126.04
40–45	5000	19,700	520	678	9.61	30.35	192	925	64.33	190.37
45–46	1000	19,600	300	678	3.30	33.65	38.5	963	22	212.4

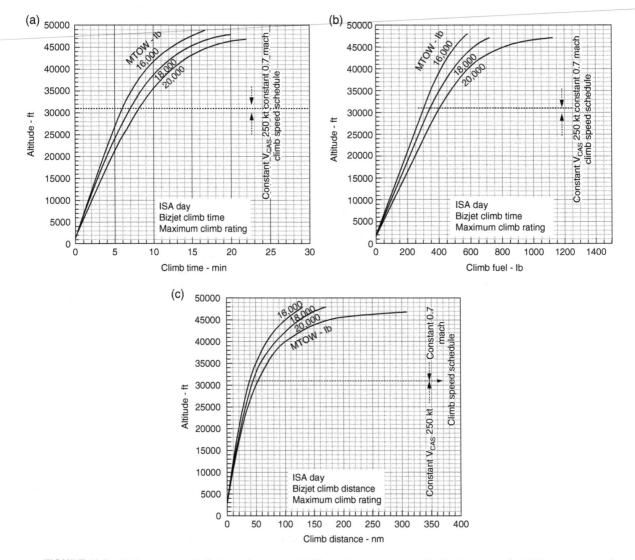

■ FIGURE 12.9   Bizjet integrated climb performance. (a) Time taken in minutes. (b) Fuel consumed. (c) Distance covered

methods in mechanics is not new, but applying it to aircraft performance analysis enabled the prediction of capability in a new way. Aircraft total energy is the sum of its kinetic and potential energies. Aircraft total energy dissipates through aircraft drag.

$$\text{Total energy}, E = \text{potential energy}, PE + \text{kinetic energy, KE}$$

where

$$PE = \int_0^h mgdh \text{ and } KE = \tfrac{1}{2}\, mV^2$$

or

$$E = \int_0^h Wdh + \tfrac{1}{2}(W/g)V^2 \;=\; Wh + \tfrac{1}{2}(W/g)V^2 \tag{12.37}$$

where aircraft weight $W = mg$ and V is the aircraft true airspeed.

Defining *specific energy* (SE), where h_e is total energy per unit aircraft weight, also seen as *energy height*:

$$SE = E/mg = h_e = h + \tfrac{1}{2}(V^2/g), \text{ units in feet or metres} \tag{12.38}$$

It shows that $h_e > h$ by the term $V^2/2g$. In other words, the aircraft can continue to climb by converting its kinetic energy to potential energy until it washes away its speed to the point where it is unable to sustain climb. It is not the altitude of the aircraft, but indicates how much height the aircraft can reach. This is the *energy height*. When KE is converted fully into PE, then it has reached its maximum height. Specific energy h_e is independent of any aircraft, and a generic plot can be made as shown in Section 12.9.2.

Defining excess power,

$$EP = V(T - D) \tag{12.39}$$

This represents excess energy, not total energy. Then specific excess power,

$$SEP = EP/mg = V(T - D)/mg \quad \left(\text{ft/s or m/s}\right)$$

Specific excess power, P_s, is defined as the rate of energy change (work done per unit aircraft weight). Differentiating Equation 12.39,

$$SEP = P_s = \frac{dh_e}{dt} = \frac{dh}{dt} + \frac{\left(\tfrac{1}{2}dV^2/g\right)}{dt} = \frac{dh}{dt} + \frac{(V/g)(dV)}{dt} \tag{12.40}$$

This can also be derived from the fundamentals. Using Equation 12.8:

$$R/C_{accl} = dh/dt = \frac{V\left[(T-D)/W\right]}{1 + (V/g)(dV/dh)}$$

or

$$(dh/dt)\left[1 + (V/g)(dV/dh)\right] = (dh/dt) + (V/g)(dV/dh)(dh/dt) = V\left[(T-D)/W\right]$$

or

$$\frac{dh}{dt}+\left(V/g\right)\frac{dV}{dt}=V\left[(T-D)/W\right]=\text{SEP}=P_s=\frac{dh_e}{dt} \tag{12.41}$$

which is the same as Equation 12.40.

12.9.1   Specific Excess Power Characteristics

This subsection examines the characteristics of SEP with variation of (i) normal acceleration, (ii) gross weight, (iii) drag, (iv) thrust, and (v) altitude for subsonic aircraft. These effects are plotted in Figure 12.10.

Increased normal acceleration: When force appears as an increment in lift, ΔL, in the pitch plane, it would overcome the weight, W, to an increased altitude (Figure 5.1), initiated by rotation of the aircraft associated with normal acceleration, n. Equation 5.2 shows that this can be expressed as an increase in lift as ΔL, causing an increase in induced drag as ΔD_i. Substituting the incremental drag in Equation 12.41, the following is obtained.

$$\text{SEP}_{accl}=P_{s\_accl}=\frac{dh_e}{dt}=V\left[(T-D-\Delta D_i)/W\right] \tag{12.42}$$

Increased gross weight: The effect of an increase in aircraft weight by ΔW can be included in Equation 12.41 as follows:

$$\text{SEP}_{\Delta W}=P_{s\_\Delta W}=\frac{dh_e}{dt}=V\left[(T-D)/(W+\Delta W)\right] \tag{12.43}$$

Increased drag: The effect of an increase in aircraft drag by ΔD (parasite drag) can be included in Equation 12.41 as follows:

$$\text{SEP}_{\Delta D}=P_{s\_\Delta D}=\frac{dh_e}{dt}=V\left[(T-D-\Delta D)/W\right] \tag{12.44}$$

Increased thrust: The effect of an increase in aircraft thrust by ΔT can be included in Equation 12.41 as follows:

$$\text{SEP}_{\Delta T}=P_{s\_\Delta T}=\frac{dh_e}{dt}=V\left[(T+\Delta T-D)/W\right] \tag{12.45}$$

Increased altitude: The effect of an increase in aircraft altitude can be included in Equation 12.41 as follows:

$$\text{SEP}_{\Delta T}=P_{s\_\Delta T}=\frac{dh_e}{dt}=V\left[(T-\Delta T-D-\Delta D)/W\right] \tag{12.46}$$

12.9.2   Worked Example of SEP Characteristics (Bizjet)

One characteristic of the effect of a thrust increase may be seen as a decrease in drag, as shown in Figure 12.10. Readers may compute the effect of a change in altitude. Table 12.14 shows the computation of SEP.

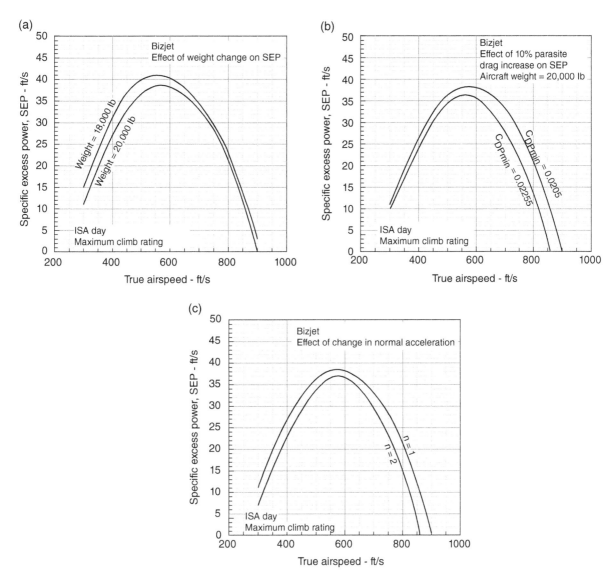

■ **FIGURE 12.10** Specific excess power explained (Bizjet). (a) Effect of weight change. (b) Effect of drag change. (c) Effect of normal acceleration change

■ **TABLE 12.14**
Bizjet specific excess power (Bizjet weight $W = 20{,}000$ lb at $30{,}000$ ft altitude, ISA day)

V (ft/s)	Mach	C_L	C_D	Drag, D	$2 \times T$ (lb)	SEP, P_s (ft/s)
200	0.201045	3.475843	–	–	–	–
300	0.301568	1.544819	0.16	2071.44	2800	10.9284
400	0.402091	0.868961	0.062	1426.992	2760	26.66016
500	0.502614	0.556135	0.035	1258.688	2720	36.53281
600	0.603136	0.386205	0.0268	1387.865	2660	38.16406
700	0.703659	0.283742	0.0238	1677.579	2620	32.98475
800	0.804182	0.21724	0.0225	2071.44	2600	21.1424
900	0.904704	0.171647	0.023	2679.926	2660	≈ 0

Figures 12.11 and 12.12 give the SEP characteristics of the Bizjet. The *SEP* graph (Figure 12.11) shows the points of maximum climb rate and climb angle. Figure 12.12 shows the lines of constant h_e, the stall line, the Bizjet speed schedule and the maximum level speed.

The complete picture of Bizjet climb and energy height performance can be generated by cross-plotting the computational results thus far in Figure 12.12. Equation 12.38 gives

■ **FIGURE 12.11**
Bizjet specific excess power

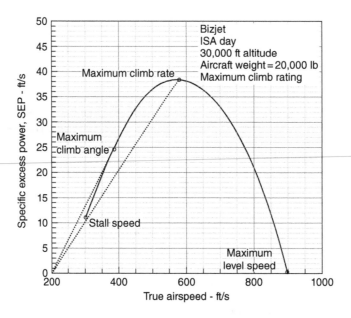

■ **FIGURE 12.12**
Bizjet energy height and specific excess power (aircraft weight = 20,000 lb)

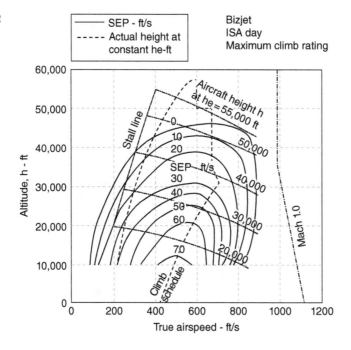

■ **TABLE 12.15**
Lines of constant $h_e = 40,000$ ft

V_{TAS} (ft/s)	$\frac{1}{2}(V^2)$ (g)	H (ft)
100	155.28	39,844.72
200	621.12	39,378.88
300	1397.52	38,602.48
400	2484.5	37,515.53
500	3882.0	36,118.01
600	5590.0	34,409.94
700	7608.7	32,391.3
800	9937.9	30,062.11
900	12,577.6	27,422.36

energy height $h_e = h + \frac{1}{2}(V_T^2/g)$ and lines are computed by holding h_e constant and finding h for each velocity. Table 12.15 gives an example for $h_e = 40,000$ ft. Note that Equation 12.38 is independent of aircraft. Lines of constant SEP are generated by cross-plotting P_s as computed in Table 12.15 and shown in Figure 12.12.

It can be seen that at Bizjet maximum energy height, h_e is 55,000 ft at 844 ft/s TAS (500 K_{TAS}, Mach 0.87), beyond the aircraft design limit. Note that the aircraft climb schedule is running close to best energy height and SEP for the altitude. The area to the left side of stall speed is outside the flight envelope of the Bizjet.

Transport aircraft have no requirement to attain energy height. Combat aircraft have a strategic interest to gain the maximum possible height by zoom climb, that is, gain energy height. The AJT example is worked out below.

12.9.3 Example of AJT

SEP variation for the AJT is shown in Figure 12.13. Although the AJT has a higher thrust loading (T_{SLS}/W) than the Bizjet, it decays faster with altitude. At 30,000 ft altitude, the SEP of the AJT is less than for the Bizjet at a higher speed.

To compute the fuel requirement, the climb capability is given in Figure 12.14. It is suggested that the readers may compute and generate a graph like Figure 12.12 for the AJT.

12.9.4 Supersonic Aircraft

Supersonic aircraft are slightly different as they are capable of much higher speeds and have a higher level of excess thrust.

A supersonic aircraft flight envelope is worked out in a similar manner as for the Bizjet and is replotted as Figure 12.15. The nature of the graph depends considerably on the engine performance characteristics and aircraft drag characteristics.

At transonic speeds the sudden drag rise (Figure 2.39) can be in excess of the ability of an engine to deliver thrust. In that case, a supersonic aircraft will not be able to accelerate at level flight, or be able to climb to higher altitudes. In that case, the aircraft could gain speed by a shallow dive to cross the drag hump and enter into the supersonic regime, when excess thrust is available on account of drag reduction.

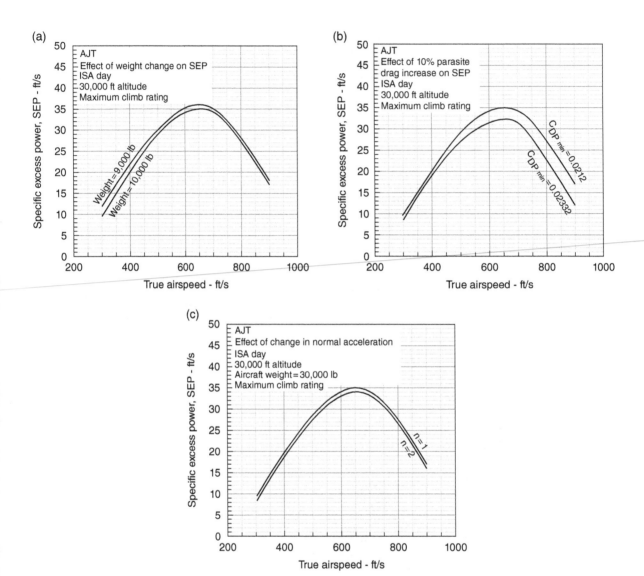

■ **FIGURE 12.13** AJT specific excess power. (a) Effect of weight change. (b) Effect of drag change. (c) Effect of normal acceleration change

12.10 Descent Performance

During descent the aircraft drag must exceed thrust, while gravitational pull assists the procedure. In the climb Equation 12.8, when drag exceeds applied thrust, descent occurs – the sign changes. The physics of descent is easier compared with climb, but its computational effort should not be underestimated. There are restrictions arising from the cabin pressurization schedule. Climb can be carried out at its maximum rate, but descent is restricted to what human tolerance allows. The descent segment is a very important part of a mission and demands detailed consideration.

■ **FIGURE 12.14**
AJT climb performance

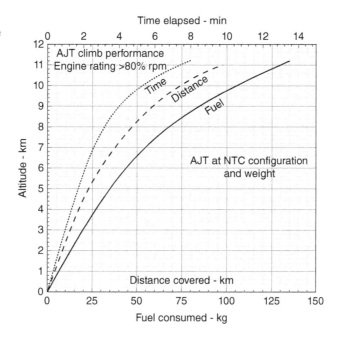

■ **FIGURE 12.15**
Example of supersonic aircraft specific excess power

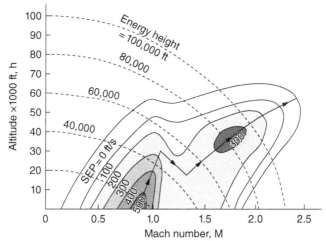

There are no FAR requirements for the descent procedure except at engine failure when the drift-down procedure is adopted (Chapter 14). Descent rate is limited by the cabin pressurization schedule for passenger comfort (ECS capability). FAR requirements are enforced during approach and landing. For the Bizjets at high altitude, inside cabin pressure is maintained at around 10,000 ft altitude. Depending on the structural design, the differential pressure between inside and outside is maintained around 9.4 lb/in.^{2}.

Engine rating during descent follows two possibilities.

1. Zero (almost) thrust at idle rpm at about 40–45% of the maximum rated power/thrust. This is the gliding descent of the aircraft.

2. Controlled descent under partial thrust carried out anywhere between 45–60% of the maximum rating.

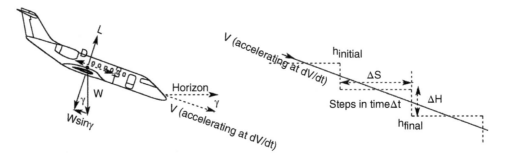

In the second case, the descent velocity schedule caters to the ECS capability.

From the force equilibrium as shown in Figure 12.16 (compare with Figure 12.1 for climb), the pair of equations 4.5a and 4.5b as derived formally in Section 4.5.2 can be rewritten as follows, bearing in mind that during descent, drag D is higher than thrust T.

$$\Sigma \text{forces}_{\text{flight path}} = 0, \text{i.e., } m\dot{V} = (C_D/2)\rho V^2 S_w - (T) - mg\sin\gamma \tag{12.47}$$

$$\Sigma \text{forces}_{\text{perpendicular}} = 0, \text{ i.e., } V(d\gamma/dt) = (C_L/2)\rho V^2 S_w - mg\cos\gamma \tag{12.48}$$

where descent angle

$$\gamma = \sin^{-1}\left[(\text{rate of descent})/V_{TAS}\right] \tag{12.49}$$

Equation 12.47 becomes $m\dot{V} = m dV / dt = D - T - mg\sin\gamma$

or

$$\sin\gamma = \left[(D-T)/W\right] - \left[(1/g)dV/dt\right] \tag{12.50}$$

But

$$\frac{dV}{dt} = \frac{dV}{dh} \times \frac{dh}{dt}$$

Using Equation 12.50, the accelerated rate of descent can be written as

$$R/D_{accl} = \frac{dh}{dt} = V\sin\gamma = V\left[(D-T)/W\right] - \left[(V/g)\frac{dV}{dt}\right] \tag{12.51}$$

or

$$R/D_{accl} = \frac{dh}{dt} = \frac{V\left[(D-T)/W\right]}{1+(V/g)(dV/dh)} \tag{12.52}$$

If the flight path is at constant descent angle, then $(d\gamma/dt) \approx 0$, and Equation 12.48 can be written as follows.

$$m\dot{V} = (C_D/2)\rho V^2 S_w - T - mg\sin\gamma = D - T - W\sin\gamma \tag{12.53}$$

$$(C_L/2)\rho V^2 S_w = mg\cos\gamma \text{ or } L = W\cos\gamma \tag{12.54}$$

At unaccelerated descent, the acceleration term $(V/g)(dV/dh)=0$. Then, using Equation 12.52, in unaccelerated descent:

$$\text{Rate of descent } = R/D_{unaccl} = \frac{dh}{dt} = \frac{V(D-T)}{W} = V\sin\gamma \qquad (12.55)$$

and

$$\text{Descent gradient } \sin\gamma_{unaccl} = \frac{(D-T)}{W} \approx \frac{D\left[(1-T/D)\right]}{L} \qquad (12.56)$$

12.10.1   Glide

Unlike climb, gravity assists descent, and hence can be performed almost without any thrust (engine kept at idle rating and considered to be producing zero thrust, i.e. $T\approx 0$). When descent is achieved at about zero thrust (idle engine rating), it is called *gliding* descent. Unaccelerated descent is achieved at glide by adjusting the glide angle.

At unaccelerated descent at zero thrust,

$$R/D_{glide} = \frac{dh}{dt} = \frac{VD}{W} \qquad (12.57\text{a})$$

$$\sin\gamma_{glide} = D/W \qquad (12.57\text{b})$$

This gives

$$D = W\sin\gamma \qquad (12.58)$$

and the force diagram (Figure 12.15) gives

$$L = W\cos\gamma \qquad (12.59)$$

At unaccelerated flight (zero thrust), this gives

$$L/D = (\cos\gamma)/(\sin\gamma) = \cot\gamma \qquad (12.60\text{a})$$

or

$$D/L = \tan\gamma \qquad (12.60\text{b})$$

Maximum L/D gives minimum descent angle, γ, which is shallow

$$1/(L/D)_{max} = (\tan\gamma)_{min} \qquad (12.60\text{c})$$

At a shallow angle, $L \approx W$, which makes Equation 12.57a as follows (i.e. $\sin\gamma \approx \gamma$ radian and $\cos\gamma \approx 1$):

$$R/D_{unaccl\_T=0} = \frac{dh}{dt} = \frac{VD}{L} \qquad (12.61\text{a})$$

$$\sin\gamma_{unaccl\_T=0} = (D/L) \qquad (12.61\text{b})$$

(at shallow angles, $\cos\gamma \approx 1$, then $L \approx W$, but $\sin\gamma$ is not approximated to zero since drag, D, is never zero).

To obtain the maximum range, the aircraft should ideally make its descent at the minimum rate by maintaining the best lift to drag (*L/D*) ratio, which also ensures descent well below the cabin pressurization limits. These adjustments will entail varying speed at each altitude (below tropopause, 36,089 ft). To ease pilot load, descent is made at a constant Mach number, and when the V_{CAS} limit is reached, it adopts a constant V_{CAS} descent, as done for climb. However, passenger comfort and aircraft structural considerations require a controlled descent with a maximum rate limited to a certain value depending on the design. Controlled descent is carried out at part throttle setting.

12.10.2 Descent Properties

Some special situations may be considered as follows.

For unaccelerated descent, Equation 12.55 gives

$$R/D_{unaccl} = \frac{dh}{dt} = \frac{V(D-T)}{W} \tag{12.62}$$

At higher altitudes, the prescribed speed schedule for descent is at constant Mach, hence above the tropopause V_{TAS} is constant and descent is kept in unaccelerated flight. At a shallow angle ($L \approx W$),

$$R/D_{unaccl} = \frac{dh}{dt} = \frac{V(D-T)}{L} = \frac{VD(1-T/D)}{L} \tag{12.63}$$

Equation 12.56 indicates that if $\dfrac{D[(1-T/D)]}{L}$ can be maintained at a constant value, then the descent angle γ can be kept invariant. Typically, a controlled descent will be carried out in part throttle well below aircraft drag. The dominant contribution comes from (*L/D*) which can be of the order of 14–20 depending on the civil aircraft design. By part throttle adjustment, a constant rate of descent can be achieved by the aircraft flying close to the best *L/D*.

At zero thrust, Equation 12.57a indicates that when descent is carried out at maximum *L/D*, then the glide angle γ is same and is independent of weight. When weight *W* increases, then the aircraft needs to fly at higher speed to maintain the minimum glide angle γ (both *L* and *D* increase). The rate of descent increases as the descent speed increases.

Evidently, a drag increase (say, flap/speed brake/undercarriage deployment) will increase rate of descent. In case of emergency, rapid descent may be required.

High-performance combat aircraft have a different type of pressurization system and are capable of a rapid descent, even in a vertical path. The AJT is assumed to be capable of descending in an unrestricted manner.

As for climb, the other parameters of interest during descent are: range covered ($R_{descent}$), fuel consumed ($Fuel_{descent}$) and time taken ($Time_{descent}$). It is convenient to establish first the descent velocity schedule (Figure 12.17a) and the point performances of rate of descent (Figure 12.17b) down to sea-level (valid for all weights – the difference between the weights is ignored). While redoing the calculations by the readers, there could be small differences in the result.

12.10.3 Selection of Descent Speed

Selection of descent speed schedule is different from a climb segment. Unlike climb, descent is gravity-assisted and the engine runs at part throttle/idle rating, keeping engine thrust always less than aircraft drag (i.e. *D>T*). While very rapid descent is possible with gravitational pull, it

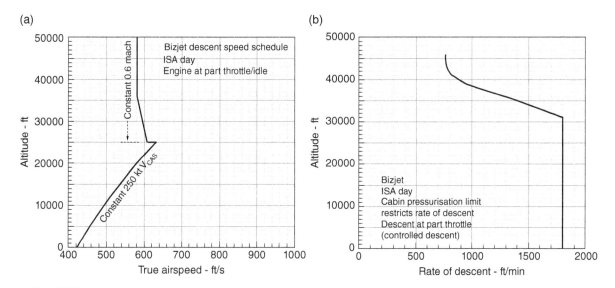

■ **FIGURE 12.17**   Bizjet descent speed schedule. (a) Descent velocity schedule. (b) Rate of descent

needs to be restricted to a maximum average human tolerance level and what the ECS system can support in terms of pressurization.

During civil aircraft descent, atmospheric pressure increases, and so the cabin needs to be pressurized to maintain the differential pressure and to adjust the human body to the ground level pressure for when the aircraft lands. The average human tolerance to pressurization is equivalent to 300 ft/min altitude change. The sealed cabin starts with a low pressure level equivalent to 10,000 ft altitude, with a differential pressure of 9.4 lb/in.² (Bizjet class, lower for big jets). While descent is performed within the limits of human comfort level, in emergency rapid descent it becomes necessary to compensate for loss of pressure and recover the oxygen level.

The Bizjet descent rate is restricted to a maximum of 1800 ft/min at any time (for higher performance, at lower altitudes it can be raised up to 2500 ft/min). The ECS system pressurizes at 300 ft/min rate of descent. Descent speed schedule is to continue at Mach 0.6 from cruise altitude until reaching the approach height, when it changes to constant $V_{EAS} = 250$ kt until the end. For higher performance (big jets), it can be raised to Mach 0.7 and $V_{EAS} = 300$ kt. The longest range can be achieved at minimum rate of descent (maximum L/D). Evidently, these will require throttle-dependent descent in order to stay within the various limits.

Therefore, the main design object is to find a descent speed schedule that will offer a good L/D ratio for efficiency (gain maximum range during descent) within the limits of ECS capability, that is, pressurization at a rate of 300 ft/min. Through flight tests it is found that Bizjet descent is best at a constant Mach of 0.6 down to 25,000 ft altitude, and thereafter maintains a constant V_{CAS} of 250 kt until landing. The worked example will justify the selected schedule. Figure 12.17 gives the Bizjet descent schedule.

12.11   Worked Example – Descent Performance (Bizjet)

There are three requirements to be substantiated for the Bizjet descent performance. The three requirements are given below for the *two-engine* aircraft. The first two are the FAR requirements as given in Table 12.1 to meet the second-segment climb gradient at missed approach.

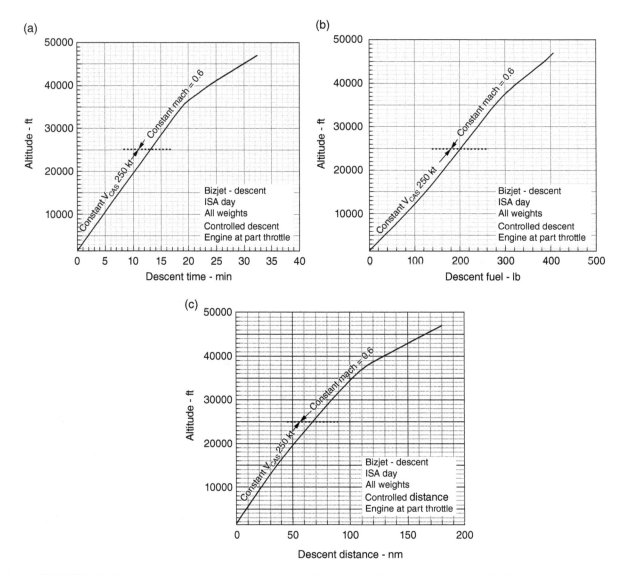

■ **FIGURE 12.18** Integrated descent performances for the Bizjet. (a) Time taken in minutes. (b) Fuel consumed. (c) Distance covered

1. To satisfy the FAR minimum climb gradient of 2.1% at approach segment.
2. To satisfy the FAR minimum climb gradient of 3.2% at landing segment.
3. To keep the maximum rate of descent within the limit of 1800 ft/min.

Aircraft weight is considerably lighter at the end of mission on account of fuel burn.

12.11.1 Limitation of Maximum Descent Rate

The related governing equations are explained in Section 12.10.3. Two of the difficulties in computing descent performance are to have the part-throttle engine performance and the ECS pressurization capabilities, which dictate the rate of descent and in turn stipulate the descent

■ **TABLE 12.16**
Rate of descent at 36,089 ft altitude

Altitude (×1000 ft)	Δh (ft)	r/d (ft/s)	V_{TAS} (ft/s)	Δt (min)	$\Sigma\Delta t$ (min)	Time taken[a] (min)	Fuel (lb/min)	$\Sigma\Delta FF$ (lb/min)	Fuel consumed[a] (lb/min)	ΔS (nm)	$\Sigma\Delta S$ (nm)	Distance covered[a] (nm)
47–45	2000	13	580	2.54	2.54	32.52	21.38	21.38	407.04	14.7	14.7	180.27
45–40	5000	16.67	580	6.41	8.97	29.96	59.83	82.2	372.69	36.76	51.46	165.57
40–36	3911	16.67	580	3.91	12.89	23.55	40.41	121.61	326.39	22.42	73.88	128.84
36–30	6089	26.67	590	3.18	16.69	19.64	46.93	168.54	282.88	22.18	96.06	106.42
30–25	5000	30	600	3.33	19.67	15.83	36.11	204.66	243.06	16.46	112.52	84.24
25–20	5000	30	600	2.78	22.25	13.05	38.89	243.55	204.17	16.46	129.98	67.78
20–15	5000	30	560	2.78	25.02	10.28	39.81	283.36	160.84	15.36	144.34	51.32
15–10	5000	30	520	2.78	27.8	7.5	43.52	326.88	121.61	14.26	158.6	35.96
10–5	5000	30	480	2.78	30.58	4.72	46.3	373.18	81.2	13.14	171.74	21.7
5–1.5	3500	30	445	1.944	32.52	1.944	46.35	407.53	21.37	8.53	180.27	8.53
1500	0	30	440	0	–	0	0	0	–	–	–	0

[a] These columns give the total amount, for example, total time taken, fuel consumed and distance covered to descent from the corresponding altitude.

■ **TABLE 12.17**
Hodograph plot at 36,089 ft altitude (16,000 lb)

V_{TAS} (ft/s)	R/D (ft/s)	$\tan\gamma$ (rad)	$V_{vertical}$ (ft/s)	$V_{horizontal}$ (ft/s)
500	31.87	0.06374	−31.85	498.98
580	30	0.051724	−29.98	579.22
650	35.28	0.054277	−35.26	649.04
700	42.1	0.060143	−42.07	698.73
800	53.02	0.066275	−52.98	798.24

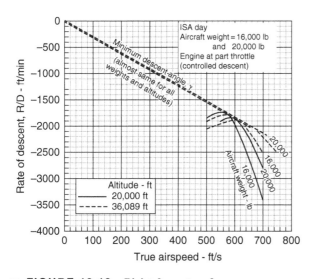

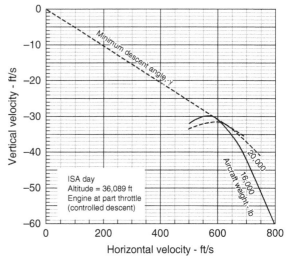

■ **FIGURE 12.19** Bizjet descent performances

■ **FIGURE 12.20**
Bizjet descent hodograph
with climb hodograph
plot

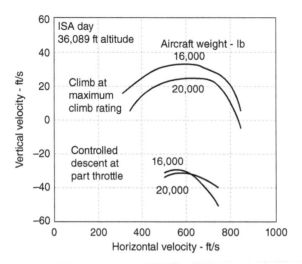

velocity schedule. These are not supplied in the book. Classroom instructors are to assist to establish these two graphs. In the absence of any information, Figure 12.17 may be used.

In industry, the exact installed engine performance at each part-throttle condition is computed from the engine deck supplied by the manufacturer. ECS design engineers remain in touch with aircraft designers and work in conjunction to develop the system specification. The ECS manufacturer supplies the cabin pressurization capability from which aircraft designers work out the velocity schedule.

It is convenient to establish first the Bizjet aircraft rate of descent corresponding to the restricted pressurization schedule, that is, 1800 ft/min (30 ft/s) below 31,000 ft altitude at constant 250 V_{CAS}, and as permitted above 31,000 ft at constant Mach 0. The Bizjet drag polar gives the maximum (L/D) ratio around 16.

Integrated descent performance is computed in the same way as done for climb, in steps of approximately 5000 ft altitude (or as convenient) in which the variables are kept invariant. No computation work is shown here

Figure 12.18 plots the fuel consumed, time taken and distance covered during descent from ceiling altitude to sea-level (Table 12.16). While redoing the integrated descent performances by the readers, there could be small differences in the result.

Table 12.17 tabulates the vertical and horizontal descent velocities of the Bizjet for 16,000 lb weight at 36,089 ft. Figure 12.19 is the hodograph plot for 16,000 and 20,000 lb aircraft weight at 20,000 and 36,089 ft. Figure 12.20 is the hodograph plot showing both the climb and descent performances, with the Y-axis indicating climb or descent.

References

1. Lan, C.T.E., and Roskam, J., *Airplane Aerodynamic and Performance*. DARcorporation, Lawrence, 1981.

2. Kundu, A.K., *Aircraft Design*. Cambridge University Press, Cambridge, 2010.

3. Ruijgork, G.J.J., *Elements of Airplane Performance*. Delft University Press, Delft, 1996.

4. Ojha, S.K., *Flight Performance of Airplane*. AIAA, Reston, US, 1995.

CHAPTER 13

Cruise Performance and Endurance

13.1 Overview

After estimating takeoff, climb, descent and landing, the last of the segments to complete the mission profile is the cruise segment in the middle. This is the longest segment of the mission. Operationally, it may appear to be the least stressful to pilot; the cruise segment continues in a quasi-steady state. There is no airworthiness requirement for the cruise segment (except at engine failure), but designers need to substantiate the customer requirement of meeting the maximum initial cruise speed.

Chapter 10 compared parabolic drag polar with computed actual drag polar to demonstrate that there is a difference between the two. Being more accurate, industries use computed actual drag polar verified in wind-tunnel tests for high subsonic aircraft application. The difference reduces for low-speed aircraft such as the TPT (turboprop trainer) flying below 0.5 Mach (no wave drag); use of parabolic drag polar is common in industrial practice. In civil applications, where economics and/or time constraints dictate, it is important to find an operational schedule at better accuracy to suit the requirements. For long ranges, substantial weight change takes place on account of fuel burnt, requiring fine-tuning of speed–altitude schedules as discussed in this chapter. For this reason, commercial transport aircraft have aircraft performance monitoring (APM) systems to fly at the most economic schedule.

Low drag at higher altitudes gives better fuel economy. In general, the higher the altitude, the better it is for cruising. Typically, a commercial aircraft cruise segment should be at least as long as the distances covered during climb and descent together. A short mission range need not climb very high, while longer ranges will benefit from higher altitude flights. The cruise segment schedule has several options, primarily depending on the distance the aircraft has to cover. Commercial aircraft cruise at a prescribed flight level (FL; see Section 3.4.1), while Bizjets cruise at a higher FL to keep the traffic separated. In the neighbourhood of terminals, the flight plan must comply with ATC regulations and remain under their surveillance. (Since the loss of flight MH370, the case for real-time monitoring during cruise is now under consideration.) For engineering computational purposes, this book assumes the cruise segment flight path is a straight line, which in reality may not be the case on account of unforeseen (weather) or imposed restrictions (territorial clearance) on the straight-line path.

To compute the cruise segment, the *specific range* of the aircraft needs to be estimated to assist in determining the fuel consumed during cruise. Once the fuel consumed, time and

Theory and Practice of Aircraft Performance, First Edition. Ajoy Kumar Kundu, Mark A. Price and David Riordan.
© 2016 John Wiley & Sons, Ltd. Published 2016 by John Wiley & Sons, Ltd.

distance taken to cover the cruise segment are determined, the information can be summed to obtain the payload–range capability of the aircraft. The mission profile to compute payload–range is covered in the next chapter.

Military aircraft intend to return to the base from where they took off, and so do not have a comparable cruise segment as in the case of civil operation, unless it is a ferry mission. Long-distance bombing missions may require flying not necessarily in a straight line and may require in-flight refuelling from tanker aircraft – quite different procedures from commercial flying. In a combat role, a military aircraft has to reach operational theatres that can have an involved flight profile with a schedule to fit the operational need. *Mission radius* or *combat radius* is a more suitable term than range for military aircraft. The combat mission is also discussed in the next chapter.

While the cruise segment is concerned with distance covered, there are situations that demand *endurance* to stay in flight for a long time, mostly maintaining altitude, for example, for surface observation (maritime, traffic control, crime watch, geological survey, etc.), or just holding on close to the destination waiting for clearance to land. Using the same equilibrium steady-state equations of motion, the calculation for endurance slightly differs from that for cruise, the unit being time instead of distance.

This chapter deals with the cruise segment and endurance in detail, mainly in still air and on an ISA day. High subsonic speed jet aircraft conduct cruise in the stratosphere and holding flight in the troposphere. Turboprops cruise in the troposphere. In still air, aircraft true airspeed is the same as ground speed covered for the mission range. Computation for a non-standard day and head/tail wind uses the same equations using corresponding atmospheric data and their influence on aircraft and engine performance data [1–3]. There is a small difference in formulation of the cruise segment between jet power and propeller aircraft. All these are briefly discussed in this chapter.

The readers may compute specific range and cruise time, distance and fuel consumed for the Bizjet and AJT aircraft. A problem set for this chapter is given in Appendix E.

13.2   Introduction

Technology levels of aircraft and engines determine the extent of cruise performance capabilities. High *L/D* (aircraft design parameter) and low specific fuel consumption (sfc) (engine design parameter) contribute to the overall fuel efficiency. A measure of fuel efficiency is the distance the aircraft covers for a unit mass of fuel burn; it gives the fuel-mileage. Since the type of fuel is supplied, the engine equations in this book do not incorporate fuel heat value for thrust generation. Standard fuel is used, and hence the role of fuel heat value is not considered here (it is more appropriate in engine design books).

Fuel prices fluctuate in cycles, usually increasing. The fuel cost accounted for a quarter to a third of aircraft operational costs at year 2000 prices. A global escalation of fuel prices since then has increased the share to nearly double that in 2000. This correspondingly raised the aircraft DOC (direct operating cost) by about 30–50%, to the point where growth in airline business seriously started to suffer. Airline operation is sensitive to airfare prices, and so growth is not sustainable with a rise in airfares.

Aircraft operate under the ATC regulations, flying within prescribed corridors especially close to terminal areas. An aircraft levels off at the end of climb, reaching the prescribed cruise altitude to initiate the cruise segment. The bulk of the fuel is consumed during cruise, and hence the initial cruise weight of the aircraft is considerably higher than the end cruise weight. Long-range Bizjet fuel consumption as a percentage of aircraft weight can be twice than that of smaller Bizjets flying less than half the range in comparison. Flight planning should take advantage of aircraft getting lighter, with the reduction of induced drag. Aircraft can then cruise at a better fuel economy (efficiency) by flying at higher altitude. There are several options for commercial aircraft cruise, as discussed in Section 13.8.

Note that drag changes as an aircraft gets lighter and/or changes speed and/or altitude. To maintain cruise at constant *L/D* (best value), pilots may need to adjust thrust (throttle) as the aircraft weight keeps changing on account of fuel burn. For operational ease, typically the pilot manual prescribes mainly two speeds: (i) long-range cruise (LRC) for good economy at speeds below M_{crit}; and (ii) high-speed cruise (HSC) to save time at M_{crit}. The difference between the two speed schedules varies between Mach 0.04 and 0.1, depending on the design and mission requirements. Designers must ensure that the aircraft is capable of flying at HSC at the initial cruise weight. Aircraft have maximum speed limits above the maximum cruise speed, but in civil usage this only serves as a safety margin. The speed schedule is determined by the *specific range* of the aircraft (see Section 13.5).

13.2.1   Definitions

The following definitions are used extensively in this and the following chapters.

Cruise range: This is the flight segment of a mission profile to cover distance from the end of climb to the beginning of descent, that is (seen as range for the segment), the distance between departure and arrival (destination) airports. One of the aims is to maximize fuel economy for the segment, or if required, minimize the time for a given range.

Mission range: This is the straight-line ground distance between two stations, say, from originating (departure) airport to destination (arrival) airport. Airline route-planning selects the pair, which becomes the sortie obligation of the aircraft. Engineers must ensure this can be accomplished safely and complying with all prevailing regulations.

Specific range (SR): This is the distance covered in the cruise segment per unit mass (weight) of fuel burn. In FPS the unit is nm/lb, and in SI it is km/kg.

Endurance: This is the time the aircraft can stay in flight irrespective of range considerations, for example, for observation, surveillance, and so on. The aim is to maximize the duration so that the task undertaken can be completed in fewer sorties.

Holding: In emergency situations, an aircraft may need to wait near the destination airport until it gets clearance to land. Certification authorities have laid down specific types of mandatory requirements to carry sufficient additional fuel to meet the emergency holding situation. However, operators can demand more than the minimum specified.

Direct operating cost (DOC): This is the cost of operation of a mission sortie, and is dealt with more thoroughly in Chapter 15. The unit is US dollars ($).

Seat–mile cost: This is the DOC per passenger per unit distance (mile) of the range and is commonly used for comparison instead the total cost of operation of a mission sortie. In loose terms, *seat–mile cost* is also seen as DOC.

Block time, block distance and block fuel: These terms are used in computing DOC and considered in Chapter 16. The terms represent time taken, distance covered and fuel burnt from engine start to engine stop from originating to destination airport. That accounts for time and fuel consumed even when the aircraft is on the ground with engines running, say during taxi, wait, and so on.

Quasi-level flight in cruise: This is slightly different from what is defined Chapter 12. This is when the climb angle term γ is small and the acceleration term $\dot{V}$ is very small and can be ignored.

Maximum operating speed, V_{MO}: This is the design limit of speed at level flying.

Never-exceed speed, V_D or V_{NE}: This is the maximum allowable speed, say in dive, which is higher than V_{MO}.

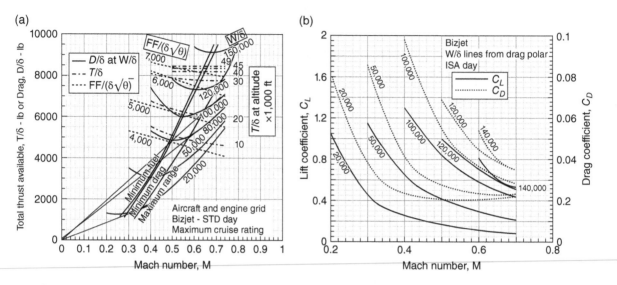

■ **FIGURE 13.1** Aircraft and engine grid for maximum cruise rating, Bizjet. (a) The grid (parabolic drag polar). (b) Lift and drag coefficients for W/δ

13.3 Equations of Motion for the Cruise Segment

From Figure 4.4, the force equilibrium gives the following pair of Equations 4.6 (θ replaced by γ) as derived in Section 4.5.2.

$$\Sigma \text{forces}_{\text{flight path}} = 0, \quad \text{i.e.} \quad m\dot{V} = T - \left(C_D/2\right)\rho V^2 S_W - mg\sin\gamma$$

$$\Sigma \text{forces}_{\text{perpendicular}} = 0, \quad \text{i.e.} \quad V\left(d\gamma/dt\right) = \left(C_L/2\right)\rho V^2 S_W - mg\cos\gamma$$

In steady-state cruise, the flight path is considered straight and level when $\gamma = 0$, making $d\gamma/dt = 0$. (There could be very small acceleration and/or very shallow *quasi-level climb-cruise*, $\gamma \approx 0$; these are discussed subsequently.) Then the two equations above reduce to:

$$T = \left(C_D/2\right)\rho V^2 S_W = D \quad \text{or} \quad T = D \qquad (13.1)$$

and

$$\left(C_L/2\right)\rho V^2 S_W = mg \quad \text{or} \quad L = W \qquad (13.2)$$

Equation 13.1 gives the thrust required (T_{reqd}) to equate with the drag generated by the aircraft. The engine must be capable of providing the required thrust. The thrust available (T_{avail}) graphs are to be used in conjunction to obtain the thrust level. Figure 13.1a gives the aircraft and engine grid at maximum cruise rating for the Bizjet aircraft. Figure 13.1b gives the lift and drag coefficients taken from actual Bizjet drag polar (Figure 9.3) for the speed, altitude and weight.

13.4 Cruise Equations

The Breguet range equation (Equation 13.11) was originally derived for propeller-driven (piston engine powered) aircraft using propeller parameters (reaction-type jet propulsion was not invented at that time). Therefore following the sequential development, the derivation of equations for propeller-driven aircraft (applies to both piston and turboprop engines) is presented first, followed by the equations meant for jet-propelled aircraft.

Let W_i = aircraft initial cruise weight (at end of climb) and W_f = aircraft final cruise weight (at beginning of descent). Then the fuel burnt during the cruise segment,

$$Fuel_{cruise} = W_i - W_f \tag{13.3}$$

Since the propeller is a separate component from the engine, the thrust depends on the choice of propeller fitted to the engine. Different types of aircraft design (especially in the same class) can have the same engine, but the choice of propeller could differ. For example, the RAF Tucano trainer aircraft has Garrett TPE331-12B engines (1100 SHP) with the four-bladed Hartzell propeller, while the same engine (civil version) is installed on Jetstream aircraft matched with another type of four-bladed propeller (e.g. Dowty propeller or McCauley propeller), offering slightly different thrust characteristics sufficient to meet the required performance. The choice depends on the cost frame versus the performance of the aircraft. Therefore, it is convenient to make aircraft equations in terms of engine power with propeller characteristics embedded to obtain thrust. The cruise equations for propeller-driven and jet-propelled aircraft bear similarity, as can be seen in the respective Breguet range equations.

13.4.1 Propeller-Driven Aircraft Cruise Equations

Thrust obtained from the propeller is the propulsive force for the aircraft. The equations derived in this section are meant for turboprops, and power is expressed as ESHP (equivalent shaft horse power), which includes the residual thrust from the exhaust of gas turbines (see Equation 7.20). For piston engines, ESHP is replaced by HP (horse power) and it does not have residual thrust from the engine.

Equation 7.19 gives thrust from propeller and can be expressed as

$$\text{ESHP} = \left(T_P\right)_{avail} / \eta_{prop}$$

where η_{prop} is the propeller efficiency (characteristic) and $(T_P)_{avail}$ is the thrust power available from the turboprop engine. It is power and not thrust force.

Power required by the aircraft is expressed as

$$\text{drag}\,(D) \times \text{aircraft velocity}\,(V) = DV = \left(T_P\right)_{reqd}$$

The fuel flow rate (FF) can then be expressed as

$$\text{FF} = \text{psfc} \times \left(\text{ESHP}\right)_{reqd} = \text{psfc} \times \left(T_P\right)_{reqd} / \eta_{prop} = \text{psfc} \times DV / \eta_{prop} \tag{13.4}$$

where psfc is the power specific fuel consumption for propeller-driven aircraft, with FPS units of fuel consumed in pounds per hour per unit of ESHP produced, that is, lb/ESHP/h. Note that aircraft velocity,

$$V = \sqrt{\frac{\left(W/S_W\right)}{\left(0.5 \times \rho \times C_L\right)}}$$

In steady level flight (typically in cruise), $L = W$.

$$\text{FF} = \text{psfc} \times \left(\text{ESHP}\right)_{reqd} = \left(\text{psfc} \times (TV)\right) / \left(550 \times \eta_{prop}\right) \qquad \text{units in lb/h} \tag{13.5a}$$

In equilibrium flight, $T=D=(WC_D/C_L)$. Upon substituting into Equation 13.2:

$$\text{FF} = \frac{V \times \text{psfc} \times C_D \times W}{500 \times \eta_{\text{prop}} C_L} = \frac{\text{psfc} \times C_D \times W}{500 \times \eta_{\text{prop}} C_L} \sqrt{\frac{(W/S_w)}{0.5 \times \rho \times C_L}} \qquad (13.5b)$$

The range equation uses the term V/FF, and therefore Equation 13.5b can be expressed as:

$$V/\text{FF} = 500 \times \eta_{\text{prop}}/(\text{psfc} \times D) = 500 \times \eta_{\text{prop}}/\left[\text{psfc} \times (T_P)_{reqd}\right] = \frac{500 \times \eta_{\text{prop}} C_L}{\text{psfc} \times C_D W} \quad (13.6)$$

At any instant, the rate of aircraft weight change $(-dW)=$ rate of fuel burnt (consumed). The negative sign indicates the aircraft weight reducing; in terms of fuel consumed, the negative sign is removed. In an infinitesimal time dt, the infinitesimal amount of fuel burnt, $dW=\text{FF} \times dt$

or

$$dt = dW/\text{FF} \qquad (13.7)$$

and elemental range

$$ds = V \times dt = V \times dW/\text{FF} \qquad (13.8)$$

Substituting Equation 13.6 into Equation 13.8,

$$ds = \frac{550 \times \eta_{\text{prop}} C_L \times dW}{\text{psfc} \times C_D \times W} \quad \text{ft} \qquad (13.9)$$

On integrating Equation 13.9 between any two weights, the distance covered can be obtained for propeller-driven aircraft as follows. Typically, the integration limits are initial cruise weight W_i and final cruise segment weight W_f.

$$\text{Range, } R = \int_{W_i}^{W_f} ds = \int_{W_i}^{W_f} \left(\frac{550 \times \eta_{\text{prop}} C_L}{\text{psfc} \times C_D}\right)\left(\frac{dW}{W}\right) \quad \text{ft}$$

$$= \int_{W_i}^{W_f} \left(\frac{375 \times \eta_{\text{prop}}}{\text{psfc}}\right)\left(\frac{dW}{W\sqrt{\left(\dfrac{0.5}{\rho}\right)\left(\dfrac{W}{S_w}\right)\dfrac{C_D^2}{C_L^3}}}\right) \quad \text{miles} \tag{13.10}$$

There are several options available to integrate Equation 13.10, as discussed in Section 13.5. Here, a simple approximated case is presented, such as was used in earlier times. Typically, propeller aircraft range was not as long as today's long-range aircraft; about 2000 nm was considered a good range. If cruising at constant altitude (ρ constant), constant speed (η_{prop} constant for the propeller) and the mid-cruise condition is considered, then wing loading W/S_w, C_L and C_D can be approximated as constant. In that case, most of the parameters in Equation 13.10 can be taken out. Then for the approximation considered above, Equation 13.10 can be written as

$$\text{Range, } R = \int_{W_i}^{W_f} ds = \left(\frac{375 \times \eta_{\text{prop}} C_L}{\text{psfc} \times C_D}\right)\int_{W_i}^{W_f} \frac{dW}{W} = \frac{375 \times \eta_{\text{prop}}}{\text{psfc}\sqrt{\left(\dfrac{0.5}{\rho}\right)\left(\dfrac{WC_D^2}{S_w C_L^3}\right)}} \int_{W_i}^{W_f} \frac{dW}{W} \quad (13.11)$$

This equation is the original form of the Breguet range formula.

Today's subsonic commercial aircraft are predominantly of jet propulsion type. Other cruise options are dealt with in the next section considering jet engines. Readers should be able to derive the related expressions for propeller-driven aircraft along the lines of what is done for jet-propulsion aircraft.

13.4.2   Jet Engine Aircraft Cruise Equations

Range equations for jet propulsion aircraft are simpler for not having the η term. The Breguet equation (Equation 13.11) is modified for turbofan engines, this time dealing directly with thrust (T) and not power.

Starting similarly to the last section, at any instant, the rate of aircraft weight change, dW = rate of fuel burnt (consumed). In an infinitesimal time dt, the infinitesimal weight change,

$$dW = \text{sfc} \times \text{thrust}\left(T\right) \times dt$$

or

$$dt = dW/\left(\text{sfc} \times T\right) \tag{13.12}$$

At quasi-level cruise, $T = D$ and $L = W$. Multiply both the numerator and denominator of Equation 13.12 by weight, W, and then equate $T = D$ and $W = L$.

$$dt = \frac{1}{\text{sfc}}\left(\frac{W}{T}\right)\left(\frac{dW}{W}\right) = \frac{1}{\text{sfc}}\left(\frac{L}{D}\right)\left(\frac{dW}{W}\right) \tag{13.13}$$

$$\text{Elemental range, } ds = V \times dt = \frac{V}{\text{sfc}}\left(\frac{L}{D}\right)\left(\frac{dW}{W}\right) = \text{RF} \times \left(\frac{dW}{W}\right) \tag{13.14a}$$

where range factor is defined as

$$\text{RF} = \frac{V}{\text{sfc}}\left(\frac{L}{D}\right) \tag{13.14b}$$

In terms of Mach number, velocity in Equation 13.14b becomes (note $V = a_0 M \sqrt{\theta_{\text{amb}}}$ where $M = V/a$).

$$ds = V \times dt = \frac{a_0 \sqrt{\theta_{\text{amb}}}}{\text{sfc}} M \left(\frac{L}{D}\right)\left(\frac{dW}{W}\right) = \text{RF}\left(\frac{dW}{W}\right) \tag{13.15}$$

The term $\frac{V}{\text{sfc}}\left(\frac{L}{D}\right)$ in Equation 13.14b is known as the range factor (RF). In terms of Mach number:

$$\text{RF} = \frac{V}{\text{sfc}}\left(\frac{L}{D}\right) = \frac{a_0 \sqrt{\theta_{\text{amb}}}}{\text{sfc}} M \left(\frac{L}{D}\right) \tag{13.16}$$

Here the term V/sfc is concerned with engine design; the lower the sfc (more efficient engine), the higher the RF contributing to the range. The second term (L/D) is the aerodynamic design parameter of the aircraft. A higher value of (L/D) increases the range. In summary, for the same aircraft weight, superior aerodynamics and better engine design increases the aircraft range. The goal of aircraft designers is to increase aircraft (L/D), and that of engine designers is to have lower sfc.

In the generalized expression, as derived in Equation 13.14b, aircraft designers aim to increase (VL/D) as best as possible to maximize the range capability. The aim is not just to maximize (L/D), but to maximize (VL/D), to obtain the best range. Expressing this in terms of

Mach number it becomes (ML/D). To have the best engine-aircraft gain, it is best to maximize (ML)/(sfc D), that is, the RF.

On integrating Equation 13.14b, the distance covered in the cruise segment (denoted by R_{cruise}) can be obtained. Then

$$R_{cruise} = \int ds = \int_{W_f}^{W_i} \frac{V}{\text{sfc}} \left(\frac{L}{D}\right) \left(\frac{dW}{W}\right) \tag{13.17a}$$

$$= \int_{W_f}^{W_i} \frac{1}{\text{sfc}} \left(\sqrt{\frac{2W}{\rho S_W C_L}}\right) \left(\frac{C_L}{C_D}\right) \left(\frac{dW}{W}\right) = \int_{W_f}^{W_i} \frac{1}{\text{sfc}} \left(\sqrt{\frac{2W C_L}{\rho S_W C_D^2}}\right) \left(\frac{dW}{W}\right) \tag{13.17b}$$

These are the two generalized forms of the Breguet range equation for jet propulsion aircraft.

The range covered during cruise (R_{cruise}) is the integration of Equation 13.17b from the initial cruise weight to the final cruise weight. There are various possibilities for the cruise segment of the mission profile, but some numerical examples would first prove useful for explanation. Section 13.8 gives the possible options.

Equation 13.17b gives good insight into what can maximize range, in order to stay ahead of the competition. It says:

1. Make the aircraft as light as possible without sacrificing safety. Material selection and structural efficiency are key. Integrate with lighter bought-out equipment.
2. Superior aerodynamics give lower drag, that is, maximize (L/D).
3. Choose the better aerofoil for good lift, keeping moment low (minimize trim drag).
4. Keep the aircraft centre of gravity at the position that will least penalize with trim drag.
5. Cruise as fast as possible within the M_{crit} (maximum ML/D).
6. Match with the best available engine with the lowest sfc.

However, these do not address the cost implications. In the end, the DOC will dictate market appeal, and the designers will have to compromise performance with cost. This is the essence of good civil aircraft design – easily said, but not so easy to achieve.

13.5   Specific Range

Specific range (SR) is a convenient way to present cruise fuel mileage. Specific range is defined as the distance covered (cruise segment range) per unit weight (or mass) of fuel burnt, that is, distance covered in the cruise segment at the expense of fuel consumed, depleting the aircraft weight. The units used in this book are nm/lb or km/kg (1 nm/lb = 4.083 km/kg). In still atmospheric wind, *nautical air miles* (nam) = *nautical miles* (nm) on the ground. (In a head wind, nam < nm and *vice versa*.)

This book deals only with still atmospheric wind, therefore *nautical air miles* (nam) and *nautical miles* (nm) are synonymous, but this section uses nam, in line with industry usage.

By definition, aircraft velocity, V (kn) = nam/h.

$$\text{sfc} = \text{fuel flow lb per h/thrust}$$

$$\text{Fuel flow rate, FF} = T \times \text{sfc} \left(\text{lb/h}\right)$$

$$\text{SR} = V/(T \times \text{sfc}) = \frac{V}{\text{sfc}} \left(\frac{1}{D}\right) \text{ nam/lb in equilibrium cruise flight } (T = D) \tag{13.18}$$

Equation 13.14b gives

$$RF = \frac{V}{sfc}\left(\frac{L}{D}\right)$$

SR and RF are related as

$$RF = L \times SR \quad (\text{convert nam to ft/s}) \tag{13.19a}$$

Equation 10.51 gives

$$L/\delta = W/\delta = 1481 C_L M^2 S_W \text{ in imperial units.}$$

Equation 10.52 gives

$$D/\delta = T/\delta = 1481 C_D M^2 S_W \text{ in imperial units.}$$

So

$$RF/\delta = \frac{V}{sfc \times \delta}\left(\frac{C_L}{C_D}\right) = \frac{V}{sfc \times \delta}\left(\frac{1481 C_L M^2 S_W}{1481 C_D M^2 S_W}\right) \tag{13.19b}$$

Figure 13.2 gives the specific range of the Bizjet at mid-cruise weight of 18,000 lb at 41,000 ft altitude, where the choice for cruise Mach number for the LRC is made around 0.65–0.7 at the maximum RF (at best specific range).

13.6   Worked Example (Bizjet)

Bizjet MTOW (maximum takeoff weight) = 20,680 lb, S_W = 323 ft², parabolic drag polar, $C_D = 0.0206 + 0.0447 C_L^2$.

13.6.1   Aircraft and Engine Grid at Cruise Rating

The aircraft and engine grid at maximum cruise rating summarizes the aircraft cruise capability and is an important one for computational purposes. The Bizjet grid is plotted in Figure 13.1a (re-plot of Figure 10.19b). In the figure, the dashed line represents thrust available at maximum cruise rating normalized for altitudes from 10,000 to 49,000 ft. The full lines represent thrust required (W/δ) for 20,000 lb weight of the Bizjet. Figure 13.1b gives the lift and drag coefficients taken from actual Bizjet drag polar for the speed, altitude and weight.

13.6.2   Specific Range Using Actual Drag Polar

Using Equation 13.18, the specific range is computed for cruise-climb. Table 13.1 gives the specific range for the Bizjet example for two different weights worked out for 45,000 ft altitude for an ISA day, and the data are plotted in Figure 13.2 (data for other weights not tabulated). While redoing the specific range computations by the readers, there could be small differences in the result. The readers should also examine the difference between the results when compared with the parabolic drag polar values.

Figure 13.2 indicates that the maximum SR at the crests are relatively flat, in general, covering 30–50 kt. It also indicates that SR improves and velocity at SR_{max} reduces as aircraft weight goes down. As aircraft weight reduces with fuel burn, the SR_{max} can be maintained by

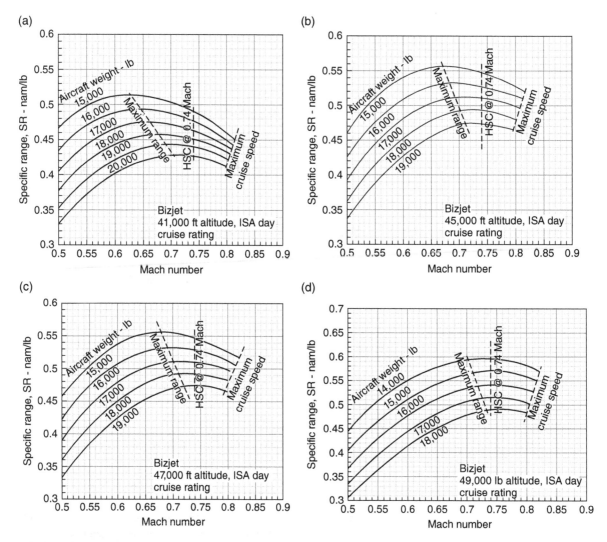

■ **FIGURE 13.2** Specific range of the Bizjet (parabolic drag polar). (a) SR at 41,000 ft altitude. (b) SR at 45,000 ft altitude. (c) SR at 47,000 ft altitude. (d) SR at 49,000 ft altitude

adjusting the speed. It also requires the aircraft to fly at quasi-level flight in cruise-climb. Section 13.7 discusses the possible ways the pilot can choose to cruise. At a constant aircraft weight, SR improves with altitude gain up to a certain height and then starts to degrade at still higher altitudes because the engine is unable to supply adequate thrust. Also note that the aircraft velocity at SR_{max} changes as weight and altitude changes. This is because at higher altitudes the rarefied air and higher demand for engine bleed for the environment control system (ECS) makes engine power suffer. From the SR values the distance covered is worked out, which in turn gives the time taken for the distance.

It was mentioned earlier that the aim is not just to maximize (L/D) but to maximize (VL/D). Expressing this in terms of Mach number it becomes (ML/D). To have the best

■ **TABLE 13.1**

Specific range and range factor (parabolic drag polar) for a Bizjet at an altitude of 45,000 ft (ISA day) at two different weights

Mach	V (ft/s)	C_L	C_D	Drag (lb)	sfc	FF (lb/h)	SR (nam/lb)	RF
Weight = 19,000 lb								
0.5	484.04	1.092	0.07376	1283.91	0.668	857.65	0.3344	10,723.2
0.55	532.44	0.902	0.05688	1197.94	0.68	814.60	0.3873	12,418.91
0.6	580.85	0.758	0.04619	1157.63	0.698	808.03	0.4259	13,658.12
0.65	629.25	0.646	0.03915	1151.60	0.712	819.94	0.4547	14,581.36
0.7	677.66	0.557	0.03437	1172.37	0.73	855.83	0.4691	15,044.44
0.74	716.38	0.498	0.03160	1204.83	0.742	893.98	0.4748	15,225.39
0.78	755.10	0.4486	0.02949	1249.30	0.756	944.48	0.4737	15,190.38
0.8	774.46	0.4264	0.028627	1275.60	0.762	972.01	0.4721	15,138.61
Weight = 18,000 lb								
0.5	484.04	1.0341	0.0683	1188.89	0.668	794.18	0.3611	10,970.75
0.55	532.44	0.8547	0.0532	1119.41	0.68	761.20	0.4144	12,590.65
0.6	580.85	0.7182	0.0436	1091.64	0.698	761.97	0.4516	13,721.4
0.65	629.25	0.6119	0.03724	1095.37	0.712	779.90	0.4780	14,522.98
0.7	677.66	0.5276	0.03294	1123.89	0.73	820.44	0.4894	14,867.42
0.74	716.38	0.4721	0.03046	1161.45	0.742	861.79	0.4925	14,962.8
0.8	774.46	0.40396	0.0278	1238.48	0.765	947.44	0.4843	14,713.73

engine-aircraft gain it is best to maximize $(ML)/(\text{sfc } D)$ for LRC. The best range at 41,000 ft altitude for mid-cruise weight $\approx 19,000$ lb is $\geq$ Mach 0.65, but with increased cruise altitude the best range is at Mach 0.7.

Observing Figure 13.2, the following can be summarized:

1. As aircraft weight reduces with fuel burn, the SR improves and the corresponding best speed reduces.
2. At a given weight, a higher cruise altitude gives better SR.

Therefore, the best fuel economy is obtained by climbing while cruising, as will be shown in Section 14.4.2.

13.6.3   Specific Range and Range Factor

A good way to examine cruise capability is to superimpose the lines of constant RF and the lines of nam/δ on the aircraft/engine grid for cruise altitudes in the stratosphere, as shown in Figure 13.3 for 41,000 ft altitude. The readers are to prepare data for 45,000 ft as done in Table 13.1 for aircraft weights of 19,000 and 18,000 lb. Such data for other weights and altitudes are not shown here.

The cruise capability can be determined by consolidating all the information in a single graph as in Figure 13.3a, which offers the best cruise speed at the point of highest RF (a point in the middle). Figure 13.3b is the uncluttered version of Figure 13.3a, with lines of minimum drag and minimum fuel flow. The graphs consolidate all the cruise performance data thus far computed to obtain the best LRC speed at 41,000 ft altitude. It shows that the best LRC speed is Mach 0.65. Such graphs are constructed for several altitudes to get the complete picture.

■ **FIGURE 13.3**
Generalized
performance graph
at 41,000 ft altitude
for the Bizjet
example. (a) Showing
lines of constant RF.
(b) Showing lines of
maximum range and
minimum drag

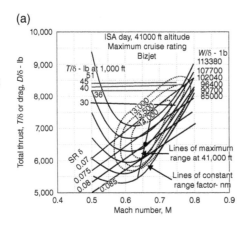

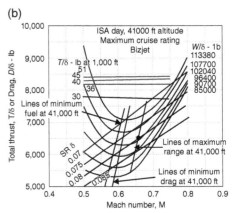

■ **TABLE 13.2**
Maximum range factor – altitude 41,000 ft and Mach 0.65 (LRC)

W (lb)	V (kt)	V (ft/s)	C_L	C_D	Drag (lb)	sfc	FF (lb/h)	SR	RF
20,000	372.78	629.25	0.569	0.036	1266.16	0.695	879.98	0.4236	14,302
19,000	372.78	629.25	0.540	0.034	1195.81	0.695	831.09	0.4485	14,386
18,400	372.78	629.25	0.523	0.0326	1146.57	0.695	796.87	0.4678	14,530
18,000	372.78	629.25	0.512	0.032	1125.47	0.695	782.20	0.4766	14,480
17,000	372.78	629.25	0.483	0.0314	1104.37	0.695	767.54	0.4857	13,937

The best LRC speed increases as altitude increases (e.g. at 45,000 ft the best LRC is at Mach 0.7). These generalized graphs are useful for preliminary design study and range estimation.

The readers should plot similar graphs for other altitudes. While it is easier to plot such graphs using parabolic drag polar, it is important that accurate data are used – proper engine performance data (best is to obtain the engine manufacturer supplied data) and actual drag polar. The readers may not be able to exactly duplicate the graphs presented here, but should get close results. It is better that readers attempt from scratch a design project similar to some existing aircraft so that at the end of the study a comparison can be made. These detailed manual computations are not easy, and the graphical solution can differ to an extent from solutions obtained from close-form equations. Use of computer programs improves accuracy, but it is recommended that the classroom exercise is done manually, using spreadsheets.

Table 13.2 gives the data for 41,000 ft altitude. The RF at various weights and altitudes is plotted in Figure 13.4. It is a composite plot with the X-axis representing aircraft weight – the top line shows weight variation where the RF_{max} is 14,480 at 41,000 ft altitude and 18,000 lb. The readers may do the same at 45,000 ft to find that the best LRC is at Mach 0.7.

Table 13.3 extracts the best weight–altitude relationship with corresponding RF and SF. Table 13.3 is extracted from Figure 13.4, which gives the optimum weight for best range at each altitude.

To extract more refined details, Table 13.4 is computed at several constant values of W/δ. Only $W/\delta = 107,700$ lb and $W/\delta = 102,040$ lb are shown for 41,000 ft altitude. Extracting from constant W/δ gives the plot in Figure 13.5, which gives the best value of $W/\delta = 104,308$ lb ($W = 18,400$ lb). The end of Table 13.1 indicates that a quasi-level cruise climb gives the longest range at LRC. Section 13.7 deals with various cruise options.

■ **TABLE 13.3**
Specific range and range factor – optimum values

Altitude (ft)	δ	Aircraft cruise weight (lb)	Optimum RF	SR (nm/lb)
41,000	0.1764	18,400	14,530	0.4678
45,000	0.1455	17,500	14,415	0.4880
49,000	0.1322	16,000	14,010	0.5188
51,000	0.1087	15,800	13,922	0.5220

■ **FIGURE 13.4**
Range factor at various altitudes

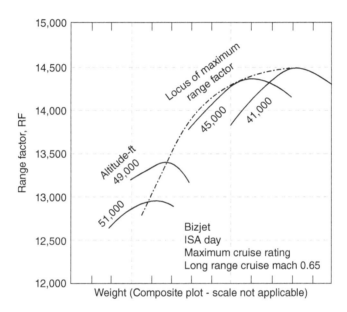

By cross-plotting Figure 13.2 (specific range of the Bizjet), the optimum LRC cruise-climb altitude can be established as shown in Figure 13.6 (it requires some fine-tuning which remains to be done). Figure 13.6 suggests that the Bizjet initial cruise weight at around 19,880 lb starts at around 45,000 ft and ends at around 16,310 lb at close to 48,000 ft altitude.

13.6.3.1   Summary   From Figure 13.2, the following observations are made:

1. As the aircraft becomes lighter with fuel burn, the SR improves, giving better range along with a reduction of speed at SR_{max}.
2. At a given aircraft weight, the higher the cruise altitude, the better is the SR, along with a speed increase at the SR_{max}.
3. The SR graphs are relatively flat around the maximum (SR_{max}) value. Therefore, considerable time can be saved by flying slightly faster without dropping below the maximum value. Typically, aircraft speed at 0.99 SR_{max} is taken as LRC. Equation 13.16 in Section 13.4.2 infers that the best range is obtained by maximizing (VL/D), that is, the RF, as shown in Figure 10.19.

■ TABLE 13.4

Specific range and range factor at 41,000 ft altitude (maximum total cruise thrust = 1500 lb)

Mach	Velocity (ft/s)	nm	C_L	C_D	Drag (lb)	D/δ (lb)	sfc (lb/lb/h)	2 × fuel (lb/h)	SR (nm/lb)	RF (nm)	L/D	M(L/D)
W/δ = 107,700 lb (weight 19,000 lb)												
0.5	484.04	286.78	0.913	0.07	1456.8	8258.42	0.65	984.8	0.2912	9713.4	13.04	6.52
0.55	532.44	315.46	0.754	0.05	1246.49	7066.3	0.665	862.07	0.3659	12,204.4	15.24	8.38
0.6	580.85	344.14	0.634	0.04	1183.74	6710.6	0.68	837.14	0.4111	13,710.4	16.05	9.63
0.65	629.25	372.82	0.54	0.0335	1174.7	6660.0	0.695	864.33	0.4313	14,385.7	16.12	10.47[a]
0.7	677.66	401.49	0.466	0.031	1264.5	7168.3	0.71	933.7	0.43	14,341.3	15.03	10.52[a]
0.74	716.38	424.44	0.417	0.03	1367.55	7752.5	0.72	1024.02	0.4145	13,823.6	13.89	10.28
0.78	755.1	447.38	0.375	0.03	1509.26	8555.9	0.735	1153.7	0.3878	12,933.3	13.59	9.82
0.8	774.46	458.85	0.357	0.029	1545.03	8758.7	0.75	1205.12	0.3808	12,698.7	13.30	9.84
W/δ = 102,040 lb (weight 18,000 lb)												
0.5	484.04	286.78	0.865	0.064	1331.92	7551.0	0.65	900.4	0.3185	10,063.8	13.51	6.76
0.55	532.44	315.46	0.715	0.046	1153.32	6538.0	0.665	797.63	0.395	12,496.2	15.61	8.58
0.6	580.85	344.14	0.601	0.037	1114.82	6319.8	0.68	788.4	0.436	13,791.9	16.15	9.69
0.65	629.25	372.82	0.512	0.0322	1143.1	6480.0	0.695	813.49	0.4583	14,480.3	15.9	10.34[a]
0.7	677.66	401.49	0.441	0.03	1215.54	6890.8	0.71	897.56	0.447	14,133.6	14.81	10.37[a]
0.74	716.38	424.44	0.395	0.029	1321.96	7494.1	0.72	989.89	0.429	13,547.7	13.62	10.08
0.8	774.46	458.85	0.338	0.028	1491.75	8456.6	0.75	1163.6	0.394	12,460.0	13.07	9.65

[a] Best M(L/D) is at LRC at around Mach 0.65 to 0.7, at a weight of 18,000 lb at 41,000 ft altitude.

■ **FIGURE 13.5**
Range factor versus
W/δ **at 41,000 ft**
altitude (aircraft
weight is 18,400 lb)

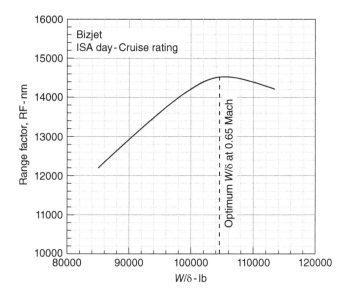

■ **FIGURE 13.6**
Optimum weight–
altitude relationships
for LRC cruise-climb
at Mach 0.65–0.7

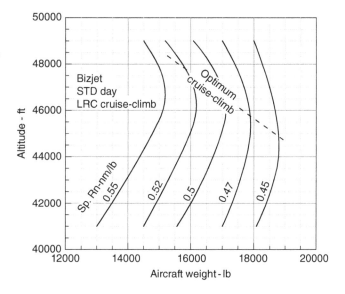

It suggests that the aircraft will give better range if flown at high altitude and to continue in a cruise-climb holding the best speed to get the better range. There are several options for cruise-climb dealt with subsequently in Section 13.8. Airlines make their flight plan to cruise at a schedule that will give the best economy, as considered in Chapter 17.

13.6.3.2 ***Comment*** There are many ways to plot SR and RF graphs for analyses. Figures 13.2–13.6 show some of the well-used types. These graphs give substantial information on aircraft cruise capability and are important ones that readers should carry out for their own project. Tables 13.1 and 13.4 are essential to construct graphs of the type shown in Figures 13.4–13.6. These are not easy to plot and require considerable amounts of cross-plotting, for example, lines of constant RF, and so on. These are obtained by using separate plots to generate the lines of constant parameter. The closed lines of constant RF will require smoothing. These are sensitive to C_D and sfc values. These are studied in detail during the detailed design in Phase II. At the end, flight tests generate accurate values for the preparation of the pilot's flight manual (nowadays embedded in flight-deck computers).

13.7   Endurance Equations

Endurance is the time spent in the cruise segment and is expressed as follows.

$$\text{Endurance, } E = \int_{t_1}^{t_2} dt \quad \text{units in time} \tag{13.20}$$

Elemental fuel dW (which is the reduction in aircraft weight) consumed in elemental time dt is expressed by Equation 13.12 as $dt = dW/FF$, where FF is the fuel flow rate. Substituting in Equation 13.20, it becomes:

$$E = \int_{t_1}^{t_2} dt \;=\; \int_{W_1}^{W_2} \frac{dW}{FF} \tag{13.21}$$

Define endurance factor, EF, as the time to consume unit fuel (units in h/lb or h/kg). EF is easily obtained by inverting fuel flow rate FF (units in lb/h or kg/h).

Computation of endurance has similarity with range equations, where plots for specific range, SR, are used. In the case of EF, plots for h/lb are to be prepared to obtain the best value to maximize endurance, which occurs at minimum fuel burn (Figure 13.7 is an example at 45,000 ft altitude). The EF values are obtained by inverting the lb/h fuel flow values given in Table 13.4. Figure 13.1a shows that the lowest fuel burn is at the unstable part of aircraft velocity, that is, where lowering of velocity increases drag. Therefore, best endurance is carried out close to the minimum drag point, that is, at the lowest point of the W/δ graph.

■ **FIGURE 13.7**
Bizjet endurance

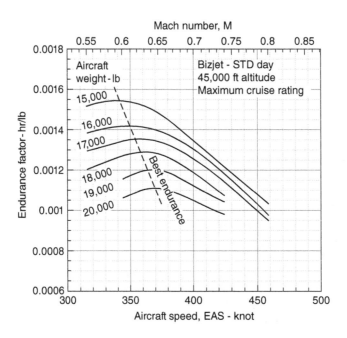

Turboprop and turbofan powered aircraft have slightly different forms on account of the former having a propeller interfaced with the gas turbine to obtain thrust. First we look at the expressions for turboprop aircraft, followed by turbofan aircraft derivations.

13.7.1   Propeller-Driven (Turboprop) Aircraft

Fuel flow rate (FF) is given by Equation 13.4 as:

$$\text{FF} = \text{psfc} \times \text{ESHP}_{\text{reqd}} = \text{psfc} \times (TV)/\left(550 \times \eta_{\text{prop}}\right) \quad \text{units in lb/h} \qquad (13.22)$$

where V is in ft/s and psfc is the specific fuel consumption for propeller-driven aircraft, with FPS units of lb/HP, that is, lb/ESHP/h.

$$V/\text{FF} = \frac{V}{\text{psfc} \times (TV)/\left(550 \times \eta_{\text{prop}}\right)} = \left(550 \times \eta_{\text{prop}}\right)/\left(\text{psfc} \times T\right) \qquad (13.23)$$

In steady level flight (typically in endurance), $T = D$ and $L = W$, which gives in coefficient form,

$$D = \frac{DW}{L} = \frac{C_D W}{C_L} \qquad (13.24)$$

Substituting Equations 13.22 and 13.24 (coefficient form) in Equation 13.23, it becomes,

$$V/\text{FF} = \left(550 \times \eta_{\text{prop}}\right)/\left(\text{psfc} \times T\right) = \frac{550 \times C_L \times \eta_{\text{prop}}}{\text{psfc} \times C_D W}$$

Note that aircraft velocity, $V = \sqrt{\dfrac{(W/S_W)}{(0.5 \times \rho \times C_L)}}$ and on substituting in the above expression and rearranging:

$$\sqrt{\frac{(W/S_W)}{(0.5 \times \rho \times C_L)}} = \frac{\text{FF} \times C_L \times \eta_{\text{prop}}}{\text{psfc} \times C_D W}$$

or

$$\text{FF} = \left(\frac{\text{psfc} \times C_D \times W}{550 \times C_L \times \eta_{\text{prop}}}\right) \sqrt{\frac{(W/S_W)}{(0.5 \times \rho \times C_L)}} \quad \text{per lb/h}$$

$$= \left(\frac{\text{psfc} \times W}{550 \times \eta_{\text{prop}}}\right) \sqrt{\frac{2 \times W \times C_D^{\,2}}{\rho \times S_W \times C_L^{\,3}}} \quad \text{units per lb/h} \qquad (13.25)$$

Wing area S_W is constant, the aircraft is made to fly at constant $\dfrac{C_L^{\,3}}{C_D^{\,2}}$ in steady cruise, and η_{prop} and psfc are kept constant. These four terms are taken out of the integral sign. Then after substituting Equation 13.25 into Equation 13.21 the endurance equation is expressed as

$$E = \int_{W_1}^{W_2} \frac{dW}{\text{FF}} = \int_{W_1}^{W_2} \left(\frac{550 \times \eta_{\text{prop}}}{\text{psfc}}\right) \frac{dW}{W \sqrt{\dfrac{2W}{\rho S_W}\left(\dfrac{C_D^{\,2}}{C_L^{\,3}}\right)}} \quad \text{units in h}$$

$$\qquad (13.26)$$

$$= \left(\frac{550 \times \eta_{\text{prop}}}{\text{psfc}}\right) \sqrt{\frac{\rho S_W \left(\dfrac{C_L^{\,3}}{C_D^{\,2}}\right)}{2}} \int_{W_1}^{W_2} \frac{dW}{W \sqrt{W}}$$

Upon integration, Equation 13.26 is reduced to

$$E = \left(\frac{550 \times \eta_{prop}}{psfc}\right)\sqrt{\frac{\rho S_W\left(\frac{C_L{}^3}{C_D{}^2}\right)}{2}}\left(\frac{2}{\sqrt{W_2}} - \frac{2}{\sqrt{W_1}}\right) \quad \text{units in h} \quad \textbf{(13.27)}$$

Evidently the best endurance is when $\frac{C_L{}^3}{C_D{}^2}$ is at its maximum.

Equation 10.26 gives $\left(\frac{C_L{}^3}{C_D{}^2}\right)_{max} = \left(\frac{3\sqrt{3}}{16k}\right)\sqrt{\frac{1}{kC_{DPmin}}}$ for parabolic drag polar (see Problem 14.2 in Appendix E).

13.7.2   Turbofan Powered Aircraft

Starting with Equation 13.20,

$$E = \int_{W_1}^{W_2} \frac{dW}{FF} \quad \text{units in time}$$

Fuel flow rate (FF) is given by Equation 13.4 as:

$$FF = tsfc \times T_{reqd}$$

where tsfc is the specific fuel consumption for turbofan aircraft with FPS units of lb/h per unit of thrust, T, that is, lb/lb/h. In steady level flight (typically in endurance) thrust equals drag, that is, $T=D$ and $L=W$. Then Equation 13.24 gives, in coefficient form,

$$T = \frac{DW}{L} = \frac{C_D W}{C_L}$$

By substituting the equations it becomes

$$V/FF = \frac{C_L}{tsfc \times C_D W} \quad \textbf{(13.28)}$$

Aircraft velocity

$$V = \sqrt{\frac{(W/S_W)}{(0.5 \times \rho \times C_L)}}$$

so substituting in the above expression and rearranging

$$\sqrt{\frac{(W/S_W)}{(0.5 \times \rho \times C_L)}} = \frac{FF \times C_L}{tsfc \times C_D W}$$

or

$$FF = \left(\frac{tsfc \times C_D \times W}{C_L}\right)\sqrt{\frac{(W/S_W)}{(0.5 \times \rho \times C_L)}} = (tsfc \times W)\sqrt{\frac{2(W \times C_D{}^2)}{(\rho \times S_W \times C_L{}^3)}} \quad \textbf{(13.29)}$$

Substituting Equation 13.29 into Equation 13.21 and noting that at steady level flight $L=W$,

$$E = \int_{W_1}^{W_2} \frac{dW}{tfsc \times W \times C_D}\sqrt{\frac{\rho S_W C_L{}^3}{2W}} = \int_{W_1}^{W_2} \frac{dW}{tfsc \times W \times C_D}\sqrt{\frac{LC_L{}^2}{W}} = \int_{W_1}^{W_2} \frac{C_L dW}{tfsc \times W \times C_D} \quad \textbf{(13.30)}$$

The wing area S_w is constant and at cruise the aircraft is made to fly at constant $\dfrac{C_L}{C_D}$ and constant *tsfc*. These terms are taken out of the integral sign.

$$E = \left(\frac{C_L}{\text{tsfc} \times C_D}\right) \int_{W_1}^{W_2} \frac{dW}{W} = \left(\frac{C_L}{\text{tsfc} \times C_D}\right) \ln\left(\frac{W_1}{W_2}\right) \tag{13.31}$$

Evidently the best endurance is when $\dfrac{C_L}{C_D}$ is at its maximum. For parabolic drag polar, Equation 10.18 gives

$$\left(\frac{C_L}{C_D}\right)_{max} = \frac{\sqrt{\dfrac{C_{DPmin}}{k}}}{2C_{DPmin}} = \frac{1}{2}\sqrt{\frac{e\pi\text{AR}}{C_{DPmin}}} = \frac{1}{2kC_L}$$

A worked example is given in the next chapter.

13.8 Options for Cruise Segment (Turbofan Only)

The cruise segment is the longest segment of the aircraft mission profile. Fuel burning in the cruise segment makes the aircraft lighter and can burn 15–40% of MTOW, depending on the distance covered. Aircraft-induced drag C_{Di} is weight dependent, and there is considerable weight change in the cruise segment. In the Bizjet example, cruising at $C_L = 0.5$–0.6, the drag change in the cruise segment is about 5%. The question is, can the drag reduction due to fuel burn be exploited for range gain?

There are two main ways to determine LRC with operational significance: (i) at constant altitude, and (ii) with varying cruise as seen in climb-cruise. These can be performed in the stratosphere above 36,056 ft altitude where the speed of sound is constant. If thrust is maintained at a rated thrust (at those altitudes, thrust does not vary much with small speed changes and also not much for small changes in altitude – see Figure 8.13), then there will be a speed gain cruising at constant altitude. Otherwise, there will be a corresponding altitude gain if speed is maintained constant. Given below are the five main possibilities for LRC.

At constant altitude (in stratosphere only):

1. *Constant altitude and L/D, but decrease in both thrust and speed*: Aircraft weight decreases on account of fuel depletion. This requires gradual reduction of thrust. However, if the best range at LRC is to fly at $(L/D)_{max}$, then the aircraft speed (Mach) needs to be reduced in a way that the RF change stays in the line of maximum range (both W/δ and T/δ change – see Figure 13.3 for a typical aircraft and engine grid for 41,000 ft. The worked example requires a similar graph at 45,000 ft not computed here; instead SR graphs from Figure 13.2 are used).

2. *Constant altitude and speed, but decrease thrust only*: Aircraft weight decreases on account of fuel depletion. This requires corresponding lowering of thrust to maintain constant speed. In this case, the RF will gradually degrade.

3. *Constant altitude and thrust, but increase speed*: Engine thrust at constant rating (throttle) that will gradually increase speed (Mach) on account of weight loss with fuel depletion. In this case also, the RF will gradually degrade.

At varying altitude (climb-cruise):

4. *Constant speed and constant W/δ*: Aircraft weight decreases on account of fuel depletion. To keep constant speed, as well as making a shallow climb to keep W/δ constant, thrust adjustment is required to maintain the RF unchanged. Examining Equation 10.50 indicates that maintaining W/δ constant keeps C_L constant, that is, L/D constant.

5. *Constant speed and constant thrust*: At constant thrust rating (throttle) and speed in the stratosphere, there will be a corresponding gain in altitude on account of fuel depletion. In this case, the RF gradually degrades.

Figure 13.8 depicts the above five options. Worked examples are given in Chapter 14. The above options are summarized in tabular form in Figure 13.9.

As can now be seen, cruise scheduling can be an involved affair for the route planners/pilots to follow. The readers are recommended to study this in depth and work out examples as shown in the next chapter.

Equation 13.17b gives the cruise equation in generalized form as follows:

$$R_{cruise} = \int ds = \int_{W_f}^{W_i} \frac{V}{sfc}\left(\frac{L}{D}\right)\left(\frac{dW}{W}\right) = \int_{W_f}^{W_i} \frac{1}{sfc}\left(\sqrt{\frac{2WC_L}{\rho S_w C_D^2}}\right)\left(\frac{dW}{W}\right)$$

The following discussion will facilitate deriving the various relations for the options available for cruise.

Equation 9.2 gives drag, *D*, in terms of coefficients as follows:

$$C_D = C_{Dpmin} + \Delta C_{Dp} + C_L^2/\pi AR + C_{Dw}$$

■ **FIGURE 13.8**
Cruise schedule options

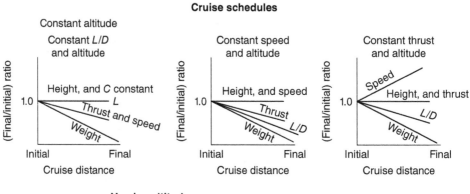

Cruise schedules

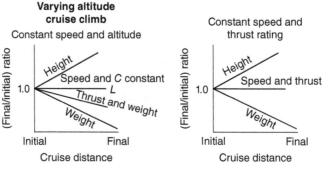

■ **FIGURE 13.9**
Summary of options
for cruise segment

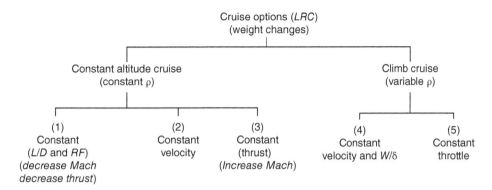

C_{Dpmin} is constant. Cruising at constant velocity (true airspeed) and constant altitude keeps C_{Dw} invariant. Finally, lift-dependent ΔC_{Dp} changes very little as it is a very small percentage of the overall aircraft C_D. Therefore the above drag equation can be written as

$$C_D = C_{Dconst\_comp} + C_L^2/\pi AR$$

This is almost same as expressing parabolic drag polar of the form

$$C_D = C_{Dpmin} + C_L^2/e\pi AR = C_{Dpmin} + kC_L^2 \quad \text{where } k = 1/e\pi AR$$

Figure 10.15 shows the difference between actual and parabolic drag at cruise. The worked examples in this and the next chapters are based on parabolic drag polar. Parabolic drag eases close-form solutions to allow rapid approximations of aircraft characteristics for subsequent engineering computations.

For operational ease, typically the pilot manual prescribes two speeds: (i) LRC for good economy at speeds below M_{crit}, and (ii) HSC at M_{crit} to save time. The difference between the two speed schedules varies between Mach 0.04 and 0.08 depending on the design and mission requirements.

The five main options given in Figure 13.9 are derived below. Note that at steady cruise the aircraft is in equilibrium flight. Chapter 14 gives the classwork example based on these equations.

1. *Constant altitude and L/D, but decrease in both thrust and speed*:

   Recall the Breguet equation (Equation 13.17). At constant altitude, atmospheric density ρ remains constant. As aircraft becomes lighter with fuel burn, its C_L also reduces slowly, hence also the drag. To hold constant C_L, the aircraft speed needs to be increased with throttle application in a manner that C_D also increases, to keep L/D ratio nearly constant. From the definition of the lift coefficient, C_L, the aircraft velocity V can be expressed as: $W = L$ and $V = \sqrt{\dfrac{2W}{\rho S_w C_L}}$; substituting in Equation 13.17a the cruise range R_{cruise} can be written as

   $$R_{cruise} = \int_{W_f}^{W_i} \frac{1}{\text{sfc}} \left(\sqrt{\frac{2W}{\rho S_w C_L}} \right) \left(\frac{L}{D} \right) \left(\frac{dW}{W} \right)$$

   by taking ρ, S_w and sfc outside the integral sign. C_L and (L/D) are also constants and can be taken out of the integral sign. The above equation can then be expressed as follows:

$$R_{cruise} = \int_{W_f}^{W_i} \frac{1}{sfc}\left(\sqrt{\frac{2W}{\rho S_w C_L}}\right)\left(\frac{L}{D}\right)\left(\frac{dW}{W}\right) = \frac{1}{sfc}\left(\sqrt{\frac{2}{\rho S_w C_L}}\right)\left(\frac{C_L}{C_D}\right)\int_{W_f}^{W_i}\left(\frac{dW}{\sqrt{W}}\right)$$

or

$$R_{cruise} = \frac{1}{sfc}\left(\sqrt{\frac{2C_L}{\rho S_w C_D^2}}\right)\int_{W_f}^{W_i}\left(\frac{dW}{\sqrt{W}}\right) = \frac{2}{sfc}\left(\sqrt{\frac{2C_L}{\rho S_w C_D^2}}\right)\left(\sqrt{W_i}-\sqrt{W_f}\right) \quad \textbf{(13.32a)}$$

In another form,

$$R_{cruise} = 2\sqrt{\frac{2C_L}{\rho S_w}}\left(\frac{\sqrt{W_i}-\sqrt{W_f}}{sfc \times C_D}\right) = \frac{2}{sfc \times C_D}\sqrt{\frac{2C_L W_i}{\rho S_w}}\left(1-\sqrt{\frac{W_f}{W_i}}\right)$$

$$= \frac{2C_L}{sfc \times C_D}\sqrt{\frac{2W_i}{\rho S_w C_L}}\left(1-\sqrt{\frac{W_f}{W_i}}\right) = \frac{2C_L V_i}{sfc \times C_D}\left(1-\sqrt{\frac{W_f}{W_i}}\right) \quad \textbf{(13.32b)}$$

It is clear from Equation 13.32a that the higher the cruise altitude, the longer the range, as a result of having lower density. In Equation 13.32b, the benefit of range increase at higher cruise altitude is embedded in the initial cruise velocity V_i. Both Equations 13.32a and 13.32b are slightly different in format but yield the same result.

2. *Constant altitude and speed, but decrease thrust only*:

Aircraft weight reducing at a constant altitude reduces W/δ, that is, C_L reduces. Therefore, engine thrust (throttle) has to be reduced to maintain constant velocity. In this case the RF changes. The Breguet equation (Equation 13.17b) gives

$$R_{cruise} = \int ds = \int_{W_f}^{W_i}\frac{V}{sfc}\left(\frac{L}{D}\right)\left(\frac{dW}{W}\right) = \int_{W_f}^{W_i}\frac{1}{sfc}\left(\sqrt{\frac{2WC_L}{\rho S_w C_D^2}}\right)\left(\frac{dW}{W}\right)$$

At level flight, $D = 0.5\rho V^2 S_w(C_{DPmin}+kC_L^2)$ and $L=W$.
Then (L/D) can be written as:

$$\frac{L}{D} = \frac{2W}{\rho V^2 S_w\left(C_{DPmin}+kC_L^2\right)}$$

Substituting this into the general form of the Breguet range equation,

$$R_{cruise} = \int ds = \int_{W_f}^{W_i}\frac{V}{sfc}\left(\frac{2W}{\rho S_w V^2\left(C_{DPmin}+kC_L^2\right)}\right)\left(\frac{dW}{W}\right)$$

The constants within the integral sign are V, sfc, S_w and ρ and can be taken out of the integral sign. That makes the above equation

$$R_{cruise} = \frac{2}{\rho V sfc S_w}\int_{W_f}^{W_i}\left(\frac{1}{\left(C_{DPmin}+kC_L^2\right)}\right)dW \quad \textbf{(13.33)}$$

To convert C_L into W, multiply and divide Equation 13.33 by $\left(\frac{\rho V^2 S_w}{2\sqrt{k}}\right)^2$

$$R_{cruise} = \frac{2}{\rho V \text{sfc} S_W} \int_{W_f}^{W_i} \left(\frac{\left(\frac{\rho V^2 S_W}{2\sqrt{k}} \right)^2 dW}{\left(\frac{\rho V^2 S_W}{2\sqrt{k}} \right)^2 \left(\sqrt{C_{DPmin}} \right)^2 + W^2} \right)$$

$$= \frac{V \rho V^2 S_W}{2k \text{sfc}} \int_{W_f}^{W_i} \left(\frac{dW}{\left(\frac{\rho V^2 S_W}{2\sqrt{k}} \right)^2 \left(\sqrt{C_{DPmin}} \right)^2 + W^2} \right)$$

This in the form of integration of $\int \frac{dW}{a^2 + W^2} = \frac{1}{a} \tan^{-1} \left(\frac{W}{a} \right)$ where $a = \left(\frac{\rho V^2 S_W}{2} \right) \sqrt{\frac{C_{DPmin}}{k}}$.

Then the above equation is integrated as

$$R_{cruise} = \left(\frac{\rho V^3 S_W}{2k \text{sfc}} \right) \left[\left(\frac{2}{\rho V^2 S_W \sqrt{\frac{C_{DPmin}}{k}}} \right) \tan^{-1} \left(\frac{W}{\left(\frac{\rho V^2 S_W}{2} \right) \sqrt{\frac{C_{DPmin}}{k}}} \right) \right]_{W_f}^{W_i}$$

Simplifying,

$$R_{cruise} = \left(\frac{V}{\text{sfc}\sqrt{kC_{DPmin}}} \right) \left[\tan^{-1} \left(\frac{2W\sqrt{k/C_{DPmin}}}{\rho V^2 S_W} \right) \right]_{W_f}^{W_i} \tag{13.34}$$

3. *Constant altitude and thrust, but increase speed*:

As aircraft weight reduces with fuel burn, C_L reduces and therefore drag D reduces. If engine thrust (throttle) has to be kept constant, then velocity would gradually increase. This is unlike option 2, holding Mach constant in a quasi-steady climb. As usual, sfc, S_W, ρ and C_{DPmin} are constant. At cruise, $W=L$ and thrust, T=constant. Also, $V = \sqrt{\frac{2W}{\rho S_W C_L}}$. The range equation becomes

$$R_{cruise} = \int_{W_f}^{W_i} \frac{V}{\text{sfc}} \left(\frac{L}{D} \right) \left(\frac{dW}{W} \right) = \int_{W_f}^{W_i} \frac{V}{\text{sfc}} \left(\frac{W}{T} \right) \left(\frac{dW}{W} \right) \quad \text{taking the constants out of integrals,}$$

$$= \frac{1}{\text{sfc} \times T} \int_{W_f}^{W_i} \sqrt{\frac{2W}{\rho S_W C_L}} dW = \frac{2}{\text{sfc} \times T \sqrt{\rho \times S_W}} \int_{W_f}^{W_i} \sqrt{\frac{W}{C_L}} dW \tag{13.35}$$

This is normally used for short ranges. The easiest way is to compute in steps of weight change, within which the variables are treated as constant.

4. *Constant speed and constant W/δ – cruise-climb*:

To keep constant speed, as well as make a shallow climb to keep W/δ constant, thrust adjustment is required to maintain the RF unchanged. Examining Equation 10.50

indicates that maintaining W/δ constant keeps C_L constant, that is, L/D constant. This is known as *cruise-climb* (typically the gradient is very shallow and can be treated as *quasi-level climb*). The gradient of the cruise-climb depends on the choice of initial and final cruise altitudes. (Operational flight planning can have any convenient rate of cruise-climb, close enough.) Equation 13.17a can be written as

$$R_{cruise} = \int_{W_f}^{W_i} \frac{V}{\text{sfc}}\left(\frac{L}{D}\right)\left(\frac{dW}{W}\right) = \frac{V}{\text{sfc}}\left(\frac{L}{D}\right)\ln\left(\frac{W_i}{W_f}\right) = \frac{a_0\sqrt{\theta_{amb}}}{\text{sfc}}\left(\frac{ML}{D}\right)\ln\left(\frac{W_i}{W_f}\right)$$
$$= RF\ln\left(\frac{W_i}{W_f}\right) \tag{13.36a}$$

The other form is

$$R_{cruise} = \frac{V}{\text{sfc}}\left(\frac{L}{D}\right)\ln\left(\frac{W_i}{W_f}\right) = \frac{1}{\text{sfc}}\left(\sqrt{\frac{2WC_L}{\rho S_W C_D^2}}\right)\ln\left(\frac{W_i}{W_f}\right) \tag{13.36b}$$

The terms W_i and W_f relate to fuel consumed during cruise and the term sfc stems from the matched engine characteristics. The three equations are different forms of the Breguet equation cruising at constant L/D and constant V_{TAS}. The value of ln $(W_i/W_f) = k_{1\_range}$ varies from 0.2 to 0.5 – the longer the range, the higher the value.

5. *Constant speed and constant thrust – cruise-climb*:
   At constant thrust rating (throttle) and speed in stratosphere there will be a corresponding gain in altitude on account of weight loss with fuel depletion. In this case, the RF gradually degrades. As aircraft drag reduces with fuel burn, then constant throttle will increase speed. Therefore to keep constant Mach number, the aircraft needs to make a quasi-level cruise-climb at a rate higher than the previous case on account of the thrust rating being kept constant. Therefore at constant Mach number the aircraft speed V (true airspeed) remains constant in the stratosphere. At cruise, $W=L$ and $T=D$ $= \frac{WC_D}{C_L}$. Note that $V = \sqrt{\frac{2W}{\rho S_W C_L}}$. The generalized Breguet equation can be written as follows:

$$R_{cruise} = \int_{W_f}^{W_i} \frac{V}{\text{sfc}}\left(\frac{L}{D}\right)\left(\frac{dW}{W}\right) = \int_{W_f}^{W_i} \frac{V}{\text{sfc}}\left(\frac{W}{T}\right)\left(\frac{dW}{W}\right)$$

where V, sfc and T are kept constant, so

$$R_{cruise} = \frac{V}{\text{sfc} \times T}\left(W_i - W_f\right) \tag{13.37}$$

For LRC (say around and over 2000 nm), the aircraft weight difference from initial to final cruise is considerable. There is a benefit if cruise is carried out at a higher altitude when the aircraft becomes lighter. This can be done either in stepped altitude changes or by making a gradual shallow climb matching with the gradual lightening of the aircraft.

Sometimes, a mission may demand HSC to save time. The following applies to HSC at constant altitude and speed (stratosphere only). The same Equation 13.33 for LRC is modified to include wave drag C_{DW} on account of reaching M_{crit}.

$$R_{cruise} = \frac{2}{\rho \times V \times \text{sfc} \times S_W} \int_{W_f}^{W_i} \left(\frac{1}{\left(C_{DPmin} + kC_L^{\,2} + C_{DW} \right)} \right) dW$$

$$= \frac{2}{\rho \times V \times \text{sfc} \times S_W} \int_{W_f}^{W_i} \left(\frac{1}{\left((C_{DPmin} + C_{DW}) + kC_L^{\,2} \right)} \right) dW \tag{13.38a}$$

To convert C_L into W, multiply and divide Equation 13.34 by $\left(\dfrac{\rho V^2 S_W}{2\sqrt{k}} \right)^2$:

$$R_{cruise} = \frac{2}{\rho \times V \times \text{sfc} \times S_W} \int_{W_f}^{W_i} \left(\frac{\left(\dfrac{\rho V^2 S_W}{2\sqrt{k}} \right)^2}{\left(\dfrac{\rho V^2 S_W}{2\sqrt{k}} \right)^2 \left((C_{DPmin} + C_{DW}) + kC_L^{\,2} \right)} \right) dW$$

$$= \frac{\rho V^3 S_W}{2 \times k \times \text{sfc}} \int_{W_f}^{W_i} \left(\frac{dW}{\left(\dfrac{\rho V^2 S_W}{2\sqrt{k}} \right)^2 \left((C_{DPmin} + C_{DW}) + W^2 \right)} \right)$$

This is the form of integration of $\displaystyle\int \frac{dW}{a^2 + W^2} = \frac{1}{a}\tan^{-1}\left(\frac{W}{a} \right)$, where $a = \left(\dfrac{\rho V^2 S_W}{2\sqrt{k}} \right) \sqrt{(C_{DPmin} + C_{DW})}$.

$$R_{cruise} = \frac{\rho V^3 S_W}{2 \times k \times \text{sfc}} \times \frac{1}{\left(\dfrac{\rho V^2 S_W}{2\sqrt{k}} \right) \sqrt{(C_{DPmin} + C_{DW})}} \tan^{-1}\left(\frac{W}{\left(\dfrac{\rho V^2 S_W}{2\sqrt{k}} \right) \sqrt{(C_{DPmin} + C_{DW})}} \right)_{W_f}^{W_i}$$

$$R_{cruise} = \frac{V}{\text{sfc} \times \sqrt{k(C_{DPmin} + C_{DW})}} \tan^{-1}\left(\frac{2W}{\left(\dfrac{\rho V^2 S_W}{\sqrt{k}} \right) \sqrt{(C_{DPmin} + C_{DW})}} \right)_{W_f}^{W_i} \tag{13.38b}$$

13.9   Initial Maximum Cruise Speed (Bizjet)

Civil aircraft maximum speed is executed during HSC in a steady level flight when the available thrust equals the aircraft drag. At the conceptual design phase (Phase I), the aircraft sizing and engine matching exercise promised the capability to meet the customer requirement of initial maximum cruise speed. In the next phase of the project (detailed design phase), with actual drag polar available, this capability needs to be guaranteed so that there will be no disappointment at the substantiation through flight tests.

The first task is to compute drag at the maximum cruise speed and then check whether the available thrust (at maximum cruise rating) is sufficient to achieve the required speed. In some cases, the available maximum cruise thrust is more than what is required; in that case the engine is adjusted to a slightly lower level. The LRC schedule is meant to maximize range and operates at a lower speed to avoid compressibility drag.

For high-speed initial cruise, the aircraft is at Mach 0.74 (716.4 ft/s) at 41,000 ft altitude ($\rho = 0.00055$ slug/ft^3). Fuel burned to climb is computed to be 700 lb. Aircraft weight at initial cruise is 20,000 lb. At Mach 0.74 the aircraft lift coefficient

$$C_L = \text{MTOM} / qS_w = 20,000 / \left(0.5 \times 0.00055 \times 716.4^2 \times 323\right)$$
$$= 20,000 / 45,627 = 0.438$$

Clean aircraft drag coefficient from Figure 9.2 at $C_L = 0.438$ gives $C_{Dclean} = 0.031$.
Then clean aircraft drag

$$D = 0.031 \times \left(0.5 \times 0.00055 \times 716.4^2 \times 323\right) = 0.0308 \times 45,627 = 1414 \text{ lb}$$

Available all engine installed thrust at maximum cruise rating at the speed and altitude from Figure 13.3 at Mach 0.74 is $T = 2 \times 790 = 1580$ lb (adequate). The capability satisfies the market requirement of Mach 0.74 at HSC.

13.10 Worked Example of AJT – Military Aircraft

The important engine data for the AJT (Adour class) are given in this section so the readers can work out the aircraft and engine grid, as for Bizjet.

13.10.1 To Compute the AJT Fuel Requirement

For military aircraft, short combat times at maximum engine rating, mostly at low altitudes, account for a good part of the fuel consumed. However, the range to the target area would dictate the fuel required. Long-distance ferry flight to the combat arena will require additional fuel carried by drop tanks. Just before combat, the drop tanks (by that time they are empty) could be jettisoned to gain aircraft performance capability. The CAS variant of the AJT has this type of mission profile. For military missions, *radius of action* is the most appropriate term, rather than range.

To compute fuel requirement, specific range graphs for the AJT at NTC are required. To compute the varied engine demand of a training mission profile, use sfc and Figure 8.15 to establish the trust for the throttle settings to obtain fuel flow rate. The graph is valid for 75% rpm to 100% ratings. Specific range for the AJT is given in Figure 13.10.

A training mission has varied engine demands, and it returns to its own base covering no net range. Mission fuel burn is computed sector by sector, as shown in Table 14.3.

13.10.2 To Check Maximum Speed

The aircraft at HSC is at Mach 0.85 (845.5 ft/s) at 30,000 ft altitude ($\rho = 0.00088$ slug/ft^3). Fuel burned to climb is computed to be 582 lb. Aircraft weight at that altitude is 10,000 lb. At Mach 0.85 the aircraft lift coefficient

$$C_L = \text{MTOM} / qS_w = 10,000 / \left(0.5 \times 0.00088 \times 845.5^2 \times 183\right)$$
$$= 10,000 / 57,561.4 = 0.174$$

■ **FIGURE 13.10**
AJT specific range

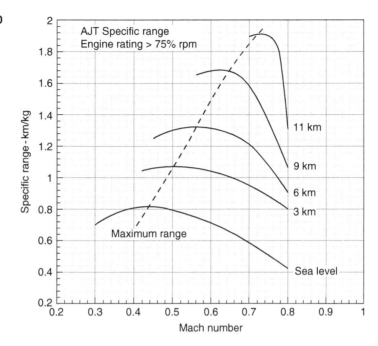

Clean aircraft drag coefficient from Figure 9.16 at C_L=0.174 gives C_{Dclean}=0.025 (high speed). Then clean aircraft drag

$$D = 0.025 \times \left(0.5 \times 0.00088 \times 858.5^2 \times 183\right) = 0.025 \times 57,561.4 = 1440 \text{ lb}$$

Available engine installed thrust at maximum cruise rating (90% of the maximum rating) is taken from Figure 13.3 at Mach 0.85 and 30,000 ft altitude as T=0.9×2000=1800 lb. Therefore the AJT satisfies the customer requirement of Mach 0.85 at HSC.

References

1. Lan, C.T.E., and Roskam, J., *Airplane Aerodynamics and Performance*, DARcorporation, Lawrence, 1981.

2. Ruijgork, G.J.J., *Elements of Airplane Performance*. Delft University Press, Delft, 1996

3. Ojha, S.K., *Flight Performance of Airplane*. AIAA, Reston, VA.

CHAPTER 14

Aircraft Mission Profile

14.1 Overview

The main objective of this chapter is to establish mission capability: payload-range for civil aircraft, and combat/training action capability for military aircraft [1–4]. The mission performance is an integrated aircraft performance by adding up the results obtained in each segment of the mission as shown in Chapters 11–13. This chapter is devoted to classroom examples of civil Bizjet and military AJT mission profiles. For civil aircraft it will be sufficient to estimate the integrated performance of the mission payload capability. Military aircraft segments are composed of different combinations of the segments as shown in Table 14.3 (taken from Table 10.1).

Chapter 10 derived the fundamentals of aircraft performance, laying the foundations for estimating performances for all the segments of a mission profile, for example, takeoff, landing, climb, descent and cruise. Chapter 13 derived the equations required for cruise segment computation, with Bizjet and AJT examples. Manoeuvring is also part of aircraft performance, and being transient in nature it is dealt with separately in Chapter 15.

Low drag at higher altitudes offers better fuel economy. In general, the higher the cruise altitude, the better is the fuel economy. Typically, commercial aircraft cruise segments should be at least as long as the distances covered during climb and descent together. A short mission range need not climb very high, while longer ranges will benefit from higher altitude flights. The cruise segment schedule has several options, primarily depending on the distance the cruise segment needs to cover. Commercial aircraft cruise at prescribed flight levels (FLs; see Section 3.3.1), while Bizjets cruise at higher FLs to keep traffic separated. In the neighbourhood of terminals, a flight plan must comply with ATC regulations and remain under their surveillance. For engineering computational purposes, this book performs cruise segment computations in still air and on an ISA day. In this case, aircraft true airspeed is the same as ground speed covered for the mission range. Computations for non-standard day and head/tail wind use the same equations with corresponding atmospheric data. Mission procedures at engine failure is discussed.

The readers are to compute specific range and cruise time, distance and fuel consumed for the Bizjet and AJT aircraft. A problem set for this chapter is given in Appendix E.

Theory and Practice of Aircraft Performance, First Edition. Ajoy Kumar Kundu, Mark A. Price and David Riordan.
© 2016 John Wiley & Sons, Ltd. Published 2016 by John Wiley & Sons, Ltd.

14.2   Introduction

The mission profile capability is a customer requirement that needs to be substantiated. There is no airworthiness requirement for the mission itself other than when an engine fails anywhere in the mission, emergency diversion, and when it comes close to ground segment operation.

The definitions in Sections 10.3 and 10.4 are expanded here to obtain the integrated performance of an aircraft for the intended mission. The specific missions are as follows:

1. *Payload-range*: Primarily meant for the commercial aircraft mission to join city pairs for the number of passengers or cargo-load or any combination of the two. Depending on the city-pair (its distance, geographical and atmospheric characteristics), the specified payload can vary but does not guarantee that the sortie will have full payload. Aircraft capability is identified with a specific payload-range, which is analysed. Military transport missions and ferry missions (may not have any payload) have similarity in analyses with commercial flight but are not treated here.

2. *Military combat mission*: These depend on the type of combat envisaged, and are analysed in a different manner using the same equations. They may be varied and can be quite complicated. In this chapter only an advanced military training profile (for airmanship) is analysed to give an idea of military type mission analysis. For military missions, *radius of action* is a more appropriate term than range.

3. *Other types of mission*: There many other types of mission, for example, fire-fighting, medical evacuation, surveillance, and so on (as well as endurance missions mentioned in the next paragraph). All these missions can be segmented to the fundamental types as shown in this chapter.

4. *Endurance*: Other missions are performed by both civil and military aircraft. Typical examples are observation of road traffic, crime watch, coastguard, anti-submarine observation, reconnaissance, and so on. The main objective is to hang on as long as possible to save cost and time without having to return to base for refuelling. Here speed is not important, and can be an impediment. Military anti-submarine missions sometimes stay aloft for about a day, if required, with in-flight refuelling. They carry additional crew to change over and have proper resting arrangements with sleeping facilities.

A *holding* situation is not a mission but arises out of an emergency. If the destination airport is not in a position to accept an arriving aircraft to land for whatever reason, the pilot may be ordered to hold in close proximity and return to land as soon as the destination airport is ready to accept. If the destination airport is not in a position to accept the aircraft within a safe time, then they instruct the pilot to fly to an alternative airport (diversion), which is within the plan before the mission begins (for commercial missions as a mandatory safety measure). Commercial aircraft must carry additional reserve fuel; a minimum amount is outlined by certification agencies.

The emergency procedure due to engine failure in the cruise segment requires an aircraft to land at the earliest opportunity. The pilot has to decide which airport the aircraft needs to head to, such as an alternative airport in close proximity. The decision to return to the originating airport or continue towards the destination airport depends on the position of the aircraft in the cruise segment.

In this book, the reserve fuel requirements for holding and/or diversion are outlined to give sufficient information to carry out industrial standard computational work. Operators work in details of requirements for the specific city-pair in question (see Section 14.3.1).

14.3   Payload-Range Capability

Civil aircraft must be able to meet the payload-range capability as specified by the market (customer) requirement. The mission range and fuel consumed during the mission are given by the following three equations, as given in Equations 10.1–10.3:

$$Mission\ fuel\ consumed\ (block\ fuel) = \Sigma\ fuel_{segments}$$
$$= \text{taxi}\left(fuel_{taxi}\right) + \text{takeoff}\left(fuel_{TO}\right) + \text{climb fuel}\left(fuel_{climb}\right) + \text{cruise fuel}\left(fuel_{cruise}\right) \quad \textbf{(14.1)}$$
$$+ \text{descent fuel}\left(fuel_{descent}\right) + \text{land}\left(fuel_{land}\right) + \text{return taxi}\left(fuel_{return}\right)$$

$$Mission\ distance\ covered\ (range) = \Sigma distance_{segments}$$
$$= \text{climb}\left(distance_{climb}\right) + \text{cruise}\left(distance_{cruise}\right) + \text{descent}\left(distance_{descent}\right) \quad \textbf{(14.2)}$$

There is no credit for distances covered during taxi, takeoff, land and return taxi.

$$Mission\ time\ taken\ (block\ time) = \Sigma time_{segments}$$
$$= \text{taxi}\left(time_{taxi}\right) + \text{takeoff}\left(time_{TO}\right) + \text{climb}\left(time_{climb}\right) + \text{cruise}\left(time_{cruise}\right) \quad \textbf{(14.3)}$$
$$+ \text{descent}\left(time_{descent}\right) + \text{land}\left(time_{land}\right) + \text{return taxi}\left(time_{return}\right)$$

Fuel is consumed during taxiing, takeoff and landing without any range contribution, and this fuel is to be added to the mission fuel and the total known as *block fuel*. The time taken from the start of the engine at the beginning of the mission to stop of the engine at the end of the mission is known as *block time*, in which a small part of time (e.g. taxi) does not contribute to a gain in range. The additional fuel burn and time consumed without contributing to range are taken from the statistics of operation for the class of aircraft.

The method to compute fuel consumption, distance covered, and time taken during climb and descent is discussed in Chapter 12. In this section, cruise fuel ($fuel_{cruise}$) and time taken during cruise are computed. Typical mission profiles of Bizjet are given in Figure 14.1.

14.3.1   Reserve Fuel

Safety considerations for airline operations are of paramount importance. To meet the emergency situation of the destination airport if it is not in a position to accept the arriving aircraft, airworthiness authorities stipulate mandatory requirements to ensure safety. There is a small difference in requirements between the two main airworthiness authorities, the Federal Aviation Administration (FAA) and the European Aviation Safety Agency (EASA). The former is outlined here.

The FAA has two categories of reserve requirements: (i) domestic reserve requirements to be applied within US airspace and (ii) international reserve requirements. There are differences between turbofan and turboprop aircraft. Here, only turbofan aircraft operation is outlined. Readers are advised to consult the details of regulations available in the public domain.

■ **FIGURE 14.1**
Bizjet aircraft mission profile – worked example, including step-climb option

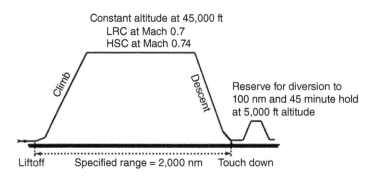

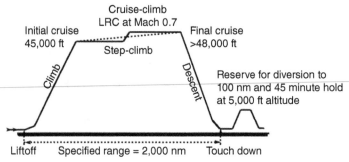

However, there is sufficient information given in this book to carry out industry-standard computation of reserve fuel requirements for turbofan aircraft. FAA regulations are updated periodically and are written in formal language and in detail. The amount of reserve fuel needs to take account of weather conditions. Given below is a simplified description to carry out computational work for an ISA day.

1. *Domestic reserve requirements* [FAR (14CFR) 121.639]: In addition to the block fuel required for the mission, the aircraft must carry reserve fuel to meet the following.

   (a) Must be able to fly to a specified alternate[1] airport and continue.

   (b) In addition, be able to fly another 45 minutes applicable only to transport airport.

2. *International reserve requirements* [FAR (14CFR) 121.645]: In addition to the block fuel required for the mission, the aircraft must carry reserve fuel to meet the following.

   (a) Must carry 10% of block fuel and

   (b) in addition, be able to fly to a specified alternate[1] airport and continue.

   (c) In addition, be able to hold another 30 minutes at 1500 ft at the alternate airport.

---

[1] An "alternate" airport is not necessarily the closest one but the closest one with requisite facilities to cater for any sort of emergency. For a regular city-pair, the alternative airport is specified in the flight planning. Note that the specified alternative airport depends on the location of the city-pair during mission operation, as the distance to the alternative airport varies. To set a standard, some industries take 200 nm for international operation and 100 nm for domestic operation to set the distance to the alternative airport.

Operators are required to carry contingency fuel to cater for any unforeseen situation, for example weather avoidance, traffic delays, accidents, and so on. The extent of contingency fuel depends on the sector of operation. There are also other reserve requirements for extended twin engine operations (ETOPS) over water. If desired, the operators can carry more reserve fuel than the minimum requirements as outlined above.

It can now be seen that the estimation of reserve fuel is quite involved. However, without losing the sight of the industry standard, the reserve fuel computation for the classwork example is outlined below in a simplified manner.

14.4   The Bizjet Payload-Range Capability

As mentioned earlier, the actual drag polar is more accurate as it is obtained through proven semi-empirical methods and refined by tests (wind-tunnel and flight). Operators require an accurate manual to maximize city-pair mission planning.

However, to establish the best combination of speed and altitude for the sector is not easy. Here the close-form analysis assists in rapidly providing answers for the best possibilities. This is then fine-tuned through flight tests, and then incorporated in the operator's manual; nowadays it is digitized and stored in several places, on the ground and in the aircraft. In the past flight engineers had to compute prior to any sortie and keep monitoring during the sortie flight. By today's standards, it seems a laborious and time-consuming process, now that these can be obtained almost instantaneously by the press of buttons.

There are several options for executing the cruise segment of a civil aircraft mission, as shown in Section 13.8. The most popular one is "*Cruise at constant Mach and constant altitude*", which can be conducted as LRC for economy and HSC to minimize time to reach destination. Long-range cruising at stepped altitudes (or step-climb) or at gradual cruise-climb proves economic. Other options are also computed for comparison.

Now we calculate the Bizjet reserve fuel needed according to domestic reserve requirements [FAR (14CFR) 121.639].

The Bizjet class aircraft must carry reserve fuel for an alternative airport at 100 nm and holding/diversion around the landing airfield for 45 minutes at 5000 ft altitude ($\rho = 0.00204$ lb / ft^3) and 0.35 Mach (384 ft/s).

At the end of the mission to the destination airport the aircraft weight is estimated at around 16,000 lb (it is evaluated in Section 14.4.1). Corresponding $C_L = 0.329$, resulting in $C_D = 0.0248$ (Figure 7.2). Equating thrust to drag $= 0.5 \times 0.00204 \times 323 \times 384^2 \times 0.0248 = 1205$ lb. Thrust per engine $= 603$ lb with sfc $= 0.6$ lb/lb/h. For 45 minutes (0.75 h) of hold, fuel consumed $= 0.75 \times 0.6 \times 1205 \approx 542$ lb.

In addition to the block fuel required for the mission, the reserve fuel is computed as:

1. Fuel for alternative airport at 100 nm (climb and descent included, with SR approximately $= 0.54$ nm/lb) $= 100 / 0.5 = 200$ lb.
2. Fuel to fly for another 45 minutes $= 542$ lb (≈ 250 kg).

Total reserve fuel $= 542 + 200 \approx 750$ lb (340 kg).

Bizjet aircraft data (see Table 3.6) are as follows:

- OEW (operating empty weight) $= 12,760$ lb (5800 kg)
- Payload $= 2420$ lb (1100 kg) at 90 kg per passenger $+ 100$ kg cargo
- OEW + payload $= 12,760 + 2420 \approx 15,180$ lb (6820 kg)

- Reserve fuel $= 750$ lb
- Onboard fuel $= 5500$ lb (2500 kg), includes reserve
- MTOW (maximum takeoff weight) $= 20{,}680$ lb (≈ 9400 kg) $=$ OEW + payload + fuel
- Final weight = OEW + payload + reserve fuel $= 12{,}760 + 2420 + 750 \approx 15{,}930$ lb (7241 kg) (return taxi taken from reserve fuel).

14.4.1   Long-Range Cruise (LRC) at Constant Altitude

A typical transport aircraft mission profile is shown Figure 14.1. Initial LRC cruise starts at 45,000 ft altitude and Mach 0.7 (677.8 ft/s). Initial cruise sfc $= 0.73$ lb/lb/h. Aircraft parabolic drag polar, $C_D = 0.0205 + 0.0447 C_L^2$, initial $L/D = 16.33$. Climb and descent fuel is taken from Figure 12.9.

- Climb to 45,000 ft $= 700$ lb to weight $20{,}680 - 100 - 700 = 19{,}880$ lb, distance $= 300$ nm.
- Descent from 45,000 ft $= 380$ lb to end cruise weight of $15{,}930 + 380 = 16{,}310$ lb, distance $= 170$ nm.
- Descent from 48,500 ft $= 420$ lb to end cruise weight of $15{,}930 + 420 = 16{,}350$ lb, distance $= 190$ nm.
- Approach fuel $= 70$ lb.

At the initial cruise weight of 19,880 lb:

$$C_L = 19{,}880 / \left(0.5 \times 0.00046 \times 677.6^2 \times 323\right) = 0.583$$

$$C_D = 0.0205 + 0.0447 \times 0.583^2 = 0.0205 + 0.0152 = 0.0357$$

Various LRC cruise options are now considered.

1. *Cruise at constant velocity (true airspeed) and constant altitude*:

   Aircraft weight reduction at a constant altitude reduces W/δ, that is, C_L reduces. Therefore engine thrust (throttle) has to be reduced to maintain constant velocity. In this case the sfc changes, hence the range factor (RF) changes. Equation 13.34 gives

$$R_{cruise} = \left(\frac{V}{\text{sfc}\sqrt{kC_{DPmin}}}\right)\left[\tan^{-1}\left(\frac{2W\sqrt{k/C_{DPmin}}}{\rho V^2 S_W}\right)\right]_{W_f}^{W_i}$$

and on substituting,

$$R_{cruise} = \left(\frac{3600 \times 677.6}{0.73\sqrt{0.0447 \times 0.0205}}\right)\left[\tan^{-1}\left(\frac{2 \times W\sqrt{0.0447/0.0205}}{0.00046 \times 677.6^2 \times 323}\right)\right]_{16,310}^{19,880}$$

$$= \left(\frac{3600 \times 677.6}{0.02215}\right)\left[\tan^{-1}\left(\frac{2 \times 1.4765 \times W}{0.00046 \times 677.6^2 \times 323}\right)\right]_{16,310}^{19,880}$$

$$= (3600 \times 30{,}591)\left[\tan^{-1}\left(\frac{2.953 \times W}{68{,}219.3}\right)\right]_{16,310}^{19,880}$$

$$= (3600 \times 30{,}591)\left[\tan^{-1}\left(\frac{2.953 \times 19{,}880}{68{,}219.3}\right) - \tan^{-1}\left(\frac{2.953 \times 16{,}310}{68{,}219.3}\right)\right]$$

Next convert into radians:

$$R_{cruise} = 3600 \times 30{,}591 \times \left(40.71/57.3 - 35.3/57.3\right)$$
$$= 3600 \times 30{,}591 \times 0.0944 = 3600 \times 2888 \text{ ft} = 1710 \text{ nm}$$

2. *Cruise at constant altitude and constant L/D ratio*:

   In this case also, aircraft weight reduction at a constant altitude reduces W/δ, that is, C_L reduces. Therefore, engine thrust (throttle) has to reduce velocity to maintain constant L/D. In this case the sfc changes, hence the RF changes.

   Initial cruise is at 45,000 ft altitude at Mach 0.7 (677.6 ft/s). The initial cruise weight of 19,880 lb gives

$$C_L = 19{,}880 / \left(0.5 \times 0.00046 \times 677.6^2 \times 323\right) = 0.583$$

$$C_D = 0.0205 + 0.0447 \times 0.583^2 = 0.0205 + 0.0152 = 0.0357$$
$$L/D = 16.33$$

Equation 13.32b gives $R_{cruise} = \dfrac{2C_L V_i}{\text{sfc} \times C_D}\left(1 - \sqrt{\dfrac{W_f}{W_i}}\right)$. On substitution

$$R_{cruise} = \frac{3600 \times 2 \times 16.33 \times 677.6}{0.73}\left(1 - \sqrt{\frac{16{,}310}{19{,}880}}\right)$$

$$= \frac{3600 \times 22{,}130}{0.73}\left(1 - \sqrt{0.8204}\right)$$

$$= 3600 \times 30{,}315 \times 0.0958 = 3600 \times 2904 \text{ ft} = 1720 \text{ nm}$$

3. *Cruise at constant altitude and constant throttle*:

Equation 13.33 gives $R_{cruise} = \dfrac{2}{\text{sfc} \times S_W \times \rho}\displaystyle\int_{W_f}^{W_i}\left(\dfrac{dW}{V\left(C_{DPmin} + kC_L^{\,2}\right)}\right)$.

Airspeed increases, so this is not adopted in operation except for short ranges (not computed here).

4. *Cruise at constant velocity and constant L/D ratio – cruise-climb* Initial cruise is at 45,000 ft altitude ($\rho = 0.00046$ slugs/ft^3, $\delta = 0.19344$) at Mach 0.7 (677.6 ft/s).

$$W/\delta = 19{,}880/0.19344 = 102{,}770 \text{ lb}$$

The initial cruise weight of 19,880 lb gives

$$C_L = 19{,}880/\left(0.5 \times 0.00046 \times 677.6^2 \times 323\right) = 0.583$$

$$C_D = 0.0205 + 0.0447 \times 0.583^2 = 0.0205 + 0.0152 = 0.0357$$

$$L/D = 16.33$$

Equation 13.36b gives $R_{cruise} = \dfrac{V}{\text{sfc}}\left(\dfrac{L}{D}\right)\ln\left(\dfrac{W_i}{W_f}\right)$

Then

$$R_{cruise} = \frac{3600 \times 677.6}{0.73} \times 16.33 \times \ln\left(\frac{19{,}880}{16{,}350}\right)$$
$$= 3600 \times 15{,}158 \times 0.1955 = 2964 \text{ ft} = 1755 \text{ nm}$$

Final altitude can be computed as the aircraft is flown at constant $W/\delta = 102,771$ at the initial condition. Final $W/\delta = 16,350/\delta = 102,770$ or $\delta = 0.159$ which gives $\rho = 0.0003785 \, \text{slugs/ft}^3$ at $\approx 48,500$ ft.

5. *Cruise at constant Mach number (true airspeed) and constant throttle – cruise-climb*

Equation 13.36b gives $R_{\text{cruise}} = \dfrac{1}{\text{sfc}} \left(\sqrt{\dfrac{2TC_L^2}{\rho S_W C_D^3}} \right) \ln \left(\dfrac{W_i}{W_f} \right)$

On substitution

$$R_{\text{cruise}} = \frac{1}{\text{sfc}} \left(\sqrt{\frac{2TC_L^2}{\rho S_W C_D^3}} \right) \ln \left(\frac{W_i}{W_f} \right) = \frac{3600}{0.745} \left(\sqrt{\frac{2 \times 1241 \times 0.583^2}{0.0005 \times 323 \times 0.0357^3}} \right) \ln \left(\frac{19,880}{16,350} \right)$$

$$= \frac{3600}{0.73} \left(\sqrt{\frac{843.6}{0.0000071}} \right) \ln(1.216) = \frac{3600 \times 10,900}{0.73} \times 0.1956$$

$$= 3600 \times 2920 \ \text{ft} = 1730 \text{nm}$$

14.4.1.1   Cruising in Step-Climb (Typically for Ranges ≥ 2000 nm)

In the stratosphere, a constant true airspeed keeps the Mach number constant. With fuel burn, weight decreases and hence drag reduces. Therefore, to maintain constant L/D (not necessarily at maximum value), a corresponding gain in altitude is allowed in a manner such that density reduction matches weight reduction, to maintain constant C_L. This is known as *cruise-climb* (typically the gradient is very shallow and can be treated as *quasi-level climb*). The gradient of the cruise-climb depends on the choice of initial and final cruise altitudes. For accuracy, computation needs to be done in small steps of altitude. To simplify, average conditions are taken as follows.

For pilot ease, cruise-climb can be simplified as a step-climb. Initial and final cruise weights are required to compute the cruise segment, fuel, distance and time. The cruise-climb is done at small steps of climb, within which the variables are treated as constant. The difference between the two is small and hence the cruise-climb is not computed here. The final altitude is not arbitrarily chosen but must match the aircraft L/D to the initial cruise value. With some iteration (shown below), a 3500 ft step climb starting from 45,000 ft will go up to 48,500 ft altitude at a cruise speed of 677.6 ft/s (Mach 0.7).

Estimate the descent fuel as 450 lb, and compute the end cruise weight as:

$$W_{\text{end\_cruise}} = \text{OEW} + \text{payload} + \text{descent fuel} + \text{reserve fuel} = 12,760 + 3270 + 450 + 750 = 16,310 \, \text{lb}$$

It is then iterated to correct the descent fuel in the final form, as given in Figure 12.18. The iterated value is taken as 450 lb.

For pilot ease, a cruise-climb can be simplified to step-climb with little sacrifice in range, as will be shown.

The mission range satisfies the requirement of 2000 nm by covering 2040 nm. Block time for the mission is 318 minutes (5.3 hours), and block fuel consumed is 4890 lb (≈ 2220 kg). On landing, the return taxi-in fuel of 20 lb is taken from the reserve fuel of 750 lb (45 minutes holding or diversion). Total minimum onboard fuel carried is therefore $(4890 - 20) + 750 = 5620$ lb. The fuel tank has a larger capacity than what is required for the design payload-range. Payload could be traded to increase range until the tanks fill up. Further reduction of payload would make the aircraft lighter, thereby increasing range – this is applied to ferry aircraft without any payload. The Bizjet range at full fuel and zero payload is 2300 nm (Figure 14.2).

■ **FIGURE 14.2**
Bizjet payload-range
capabilities

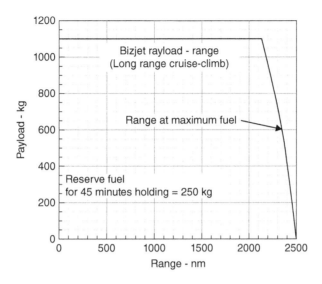

■ **TABLE 14.1**
Bizjet range at LRC (step-climb)

	Aircraft initial weight (lb)	Distance (nm)	Fuel (lb)	Time (min)	Aircraft final weight (lb)
Start and taxi out	20,720	0	40[a]	3[a]	20,680
Takeoff to 1500 ft[b]	20,680	2	80	2	20,600
Climb to 45,000 ft	20,600	220	700	20	19,880
Cruise at 45,000 ft (0.494 nm/lb) LRC at Mach 0.7	19,880	820	1660	122	18,220
Climb to 48,500 ft	18,220	60	270	6	17,950
Cruise at 48,500 ft (0.528 nm/lb) LRC at Mach 0.7	17,950	845	1600	127	16,350
Descent to 1500 ft	16,350	190	420	33	15,920
Approach and land	15,920	0	70[a]	5[a]	15,850
Landed weight	15,850	–	–	–	–
Taxi-in (from reserve)	–	0	20[a]	3[a]	–
Stage total	–	≈2100	4890	315	–

[a] From operational statistics.
[b] See Chapter 11.

14.4.1.2   Alternative Cruise Segment at Average Weight and Constant Average Altitude (Same Fuel Load)

The cruise altitude starts at 45,000 ft and ends at 48,500 ft (the design range requires a step cruise). The range, time and distance, as obtained in Table 14.1, can be evaluated in a rapid manner by taking an average altitude of $\approx 47,000$ ft ($\rho = 0.00041$ slug / ft^3).

Mid-cruise weight is worked out as $(19,880 + 16,310)/2 = 18,090$ lb $\approx 18,100$ lb. The aircraft is at LRC schedule operating at Mach 0.7 (677.6 ft/s, 401.4 kt). The specific range at average altitude of $\approx 47,000$ ft and 18,100 lb is 0.49 nm/lb. Figures 12.9 and 12.18 give $R_{climb} = 170$ nm, $fuel_{climb} = 750$ lb, $time_{climb} = 18$ minutes, $R_{descent} = 180$ nm, $fuel_{descent} = 400$ lb and $time_{descent} = 32$ minutes (in part-throttle gliding descent from 48,500 ft). From the initial and final cruise weights, the fuel burned during cruise is $(19,880 - 16,350) = 3530\,\text{lb}$. This gives cruise distance $= 3530 \times 0.49 = 1725$ nm and cruise time $= 4.4$ hours $= 246$ minutes.

$$\text{Mission distance covered } (\textit{\textbf{range}}) = \Sigma distance_{segments}$$
$$= 0 \ (\textit{taxi}) + 0 \ (\textit{takeoff}) + 170 \ (\textit{climb}) + 1619 \ (\textit{cruise})$$
$$+ 180 \ (\textit{descent}) + 0 \ (\textit{approach and land})$$
$$= 1969 \text{ nm.}$$

$$\text{Mission fuel consumed } (\textbf{block fuel}) = \Sigma fuel_{segments}$$
$$= 40 \ (\textit{taxi}) + 100 \ (\textit{takeoff}) + 750 \ (\textit{climb})$$
$$+ 3520 \ (\textit{cruise}) + 400 \ (\textit{descent})$$
$$+ 70 \ (\textit{approach}) + 20 \ (\textit{return taxi} - \text{from reserve})$$
$$= 4890 \text{ lb.}$$

$$\text{Mission time taken } (\textbf{block time}) = \Sigma time_{segments}$$
$$= 3 \ (\textit{taxi}) + 5 \ (\textit{takeoff}) + 18 \ (\textit{climb}) + 246 \ (\textit{cruise})$$
$$+ 32 \ (\textit{descent}) + 5 \ (\textit{approach}) + 3 \ (\textit{return taxi})$$
$$= 312 \text{ min.}$$

14.4.2 High-Speed Cruise (HSC) at Constant Altitude and Speed

HSC is at Mach 0.74 (716.4 ft/s) and 45,000 ft altitude. Engine power setting is at maximum cruise rating, and sfc $= 0.742\,\text{lb/lb/h}$.

Aircraft parabolic drag polar, $C_D = 0.0205 + 0.0447 C_L^2 + C_{DW}$

Equation 13.38b gives the cruise segment range equation (wave drag component C_{DW} included) as:

$$R_{cruise} = \frac{V}{\text{sfc} \times \sqrt{k\left(C_{DPmin} + C_{DW}\right)}} \tan^{-1}\left(\frac{2W}{\left(\dfrac{\rho V^2 S_W}{\sqrt{k}}\right)\sqrt{\left(C_{DPmin} + C_{DW}\right)}}\right)_{W_f}^{W_i}$$

On substitution it becomes

$$R_{cruise} = \frac{3600 \times 716.4}{0.742 \times \sqrt{0.0447 \times \left(0.0205 + 0.002\right)}}$$

$$\times \tan^{-1}\left(\frac{2W}{\left(\dfrac{0.00046 \times 716.4^2 \times 323}{\sqrt{0.0447}}\right) \times \sqrt{\left(0.0205 + 0.002\right)}}\right)_{16,310}^{19,880}$$

$$= \frac{3600 \times 716.4}{0.742 \times 0.03171} \times \tan^{-1}\left(\frac{2W}{\left(76,255.6\right) \times 0.71}\right)_{16,310}^{19,880} = 3600 \times 30,448 \times \tan^{-1}\left(\frac{2W}{54,142}\right)_{16,310}^{19,880}$$

$$= 3600 \times 30,448 \times \left(36.3/57.3 - 31.07/57.3\right) = 3600 \times 30,448 \times 0.0913 \text{ ft} = 1646 \text{ nm}$$

Total range $= 1646 + 370 = 2016$ nm.

Figure 14.3 compares ranges obtained by various cruise segment schedules.

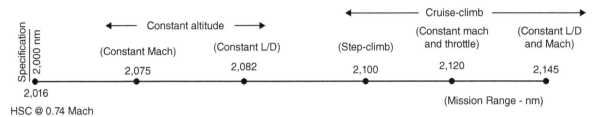

■ **FIGURE 14.3**   Bizjet cruise segment comparison of various options

■ **TABLE 14.2**
Bizjet range comparisons for the cruise segment

Method (LRC)	Constant altitude			Cruise-climb	
	Constant V	Constant L/D	Constant V and L/D	Step-climb	Constant V and T
Range (nm)	1710	1720	1755	1725	1730
Block fuel (lb)	3550	3550	3530	3530	3530
Block time (min)	251	≈253	257	255	253
Total range[a] (nm)	2075	2082	2145	2100	2120
Change with respect to maximum range capability of 2145 nm	−3.3%	−3%	0	−2.1%	−1.2%

[a] HSC total range = 2016 nm (−6%).

14.4.3   Discussion on Cruise Segment

As can be seen, there are many options for the operator to choose from. Table 14.2 gives a comparison of range capabilities of some of the options. Comparisons are made using parabolic drag (there will be a small difference if the actual drag polar is used, but the advantage of using the analytical equation is demonstrated).

With the same fuel load, as expected, the cruise-climb gives better range than constant-altitude cruise, taking advantage of drag reduction as a result of aircraft weight loss due to fuel consumed. A lighter aircraft can then gain in altitude, further lowering drag at the same cruise speed. Climbing continuously at a shallow gradient and maintaining a desirable L/D ratio gives the most efficient way to gain range. However, the pilot needs to adjust the thrust to maintain constant L/D, and therefore it is easier to fly by the gauge reading, keeping the throttle setting unchanged and maintaining the speed (Mach meter), trimming the aircraft to the climb gradient which will alter L/D but with insignificant loss of range. There is an even easier flying option, by flying level at the initial cruise and midway step-climbing to the final altitude, thereafter flying level, all the time with the pilot keeping speed (Mach meter) and altitude (altimeter) constant, with a loss of about 2% in the range capability.

For short distances, constant altitude cruise is the most suitable; flying by keeping Mach constant proves convenient, with a loss of 3.3% range from the maximum capability. In this example this equates to 70 nm, which may be considered significant.

HSC is used for need-based operation at a loss of 6% in range capability, to gain about quarter of an hour in time. The operator may also exercise other possibilities not shown here, such as HSC at cruise-climb. The basic methods are well covered, and readers may work out other possibilities.

The specification requirement for the range is 2000 nm, for which HSC substantiates the capability, and LRC has a longer range capability, in this case by a maximum of ≈ 130 nm extra.

Operators have the option to trade fuel with passengers to gain range. Figure 14.2 shows that the ferry range without any payload is 2500 nm. The aircraft could have gone further but the full capacity restricted the range gain by another ≈ 300 nm. With zero payload, it is left with 600 lb less than the maximum weight capability.

14.5   Endurance (Bizjet)

Section 10.10.2 reasoned that the minimum fuel operation is in the unstable part of aircraft velocity and is undesirable. Since the lines of minimum fuel consumption and minimum drag are close to each other, endurance is carried out at velocities of minimum drag without much loss of endurance time.

An example of endurance at 45,000 ft is worked out for comparison (see Figure 13.7). Initial cruise weight $W_1 = 19,780$ lb, and final cruise weight, $W_2 = 16,210$ lb. $(L/D)_{max} = 16.54$ occurs at Mach 0.63 at mid-endurance weight $\approx 18,000$ lb. Figure 13.7 gives best endurance as 0.0013 h/lb. Fuel consumed is $(19,780 - 16,210) = 3570$ lb.

Then endurance is $3570 \times 0.0013 = 4.641$ h.

It is checked again using Equation 13.31:

$$\text{Endurance, } E = \left(\frac{C_L}{\text{tsfc} \times C_D} \right) \int_{W_1}^{W_2} \frac{dW}{W} = \left(\frac{C_L}{\text{tsfc} \times C_D} \right) \ln \left(\frac{W_1}{W_2} \right)$$

Using Figure 8.13, it gives sfc $= 0.71$ lb/lb/h with installation losses.
Therefore,

$$E = \left(\frac{16.54}{0.71} \right) \ln \left(\frac{19,880}{16,310} \right) = 23.3 \times 0.199 = 4.64 \text{ hour}$$

which compares well with 4.641 hours as computed above. The difference between LRC Mach of 0.7 and endurance Mach of 0.63 is 0.07. From Table 14.1, the cruise segment time is about 4.3 hr, that is, the LRC is at an economic fuel burn, yet flying faster.

14.6   Effect of Wind on Aircraft Mission Performance

Although this book considers mainly still air conditions on an ISA day, the effect of a tail wind and head wind are worth studying. Here the component of velocity aligned to flight velocity as head or tail wind is considered. A strong tail wind assists aircraft covering the same cruise-segment distance in less time and with less fuel burn. Chapter 13 derived that cruising at constant W/δ is the optimum for fuel economy – it implies a cruise-climb procedure for the segment. If the gain on account of a tail wind offsets the loss on account of flying at altitudes other than keeping constant W/δ, then the pilot may decide to alter the altitude schedule. To benefit from a tail wind requires the aircraft to cruise at altitudes where the best tail-wind conditions exist.

A tail or head wind will affect the ground speed of an aircraft. Here V_{TAS_wind} is the true airspeed relative to wind, with units in *nautical air miles* (nam), in comparison with the ground speed which is denoted in *nautical miles* (nm).

It follows that:

- with a tail wind, $V_{TAS\_wind} > V_{TAS\_still\_air}$, nautical air miles (nam) > nautical miles (nm);
- with a head wind, $V_{TAS\_wind} < V_{TAS\_still\_air}$, nautical air miles (nam) < nautical miles (nm);
- in still air, nautical air miles (nam) = nautical miles (nm).

Let V_{wind} = wind velocity. Then $V_{TAS\_wind} = (V_{TAS\_still\_air} \pm V_{wind})$ where +ve is for a tail wind and −ve is for a head wind.

Then the distance covered in the cruise segment having head/tail wind will be different from the still air distance. In still air:

$$\text{Time, } t_{still\_air} = \left(\text{distance},D\right) / V_{TAS\_still\_air} \tag{14.4}$$

Flying with a wind:

$$t_{wind} = D / V_{TAS\_wind} = D / \left(V_{TAS\_still\_air} \pm V_{wind}\right) \tag{14.5}$$

The ratio between the two times

$$\left(t_{still\_air}\right) / \left(t_{wind}\right) = \left(V_{TAS\_still\_air}\right) / \left(V_{TAS\_still\_air} \pm V_{wind}\right) \qquad \leq 1 \text{ for tail wind}$$
$$\geq 1 \text{ for head wind}$$

Therefore, for the same amount of fuel burned, a tail wind will take an aircraft to greater distances compared with still air, and *vice versa*. The flight speed for the best range with a head wind will be higher than the flight speed with no wind, and *vice versa* for the tail wind case.

The pilot's manual includes a follow-through carpet plot for the pilot to select the altitude and compute time and distance for the airspeeds. An aircraft equipped with a microprocessor-based electronic flight instrument system (EFIS) has the manual embedded in it.

14.7 Engine Inoperative Situation at Climb and Cruise – Drift-Down Procedure

In the rare situation of an engine failure in any segment of a mission profile, the FAA stipulates the drift-down procedure to land at a designated airfield. Engine failure at takeoff was dealt with in Chapter 12. Since the drift-down procedures on account of engine failure are similar during climb and cruise, both are considered in this section. Engine failure during descent is less critical, since the aircraft is already on the path to land at part throttle. Only the outline of the procedure is given. Details are given in FAR (14CFR)/JAR 25.123. Computation of the procedure uses the same equations, so no example is given here.

Aircraft are designed to perform level flight with one engine failed. Four-engine aircraft are designed to fly with two engines inoperative. To compensate for the loss of an engine, all other operating engines are set to a higher throttle setting to increase thrust. The engine rating is increased from maximum climb thrust or cruise thrust to maximum continuous rating. The terrain of the regulatory corridor of the mission flight path must be known before flight dispatch.

Note that after engine failure there will be a drag increase on account of the failed engine and the additional trim required to fly in the flight path.

14.7.1 Engine Inoperative Situation at Climb

The FAA requires that in the event of an engine failure in the climb segment, the aircraft must land as soon as possible, in this case invariably returning to the originating airport, unless it is not available for some reason in the meantime.

The drift-down procedure is initiated as soon as an engine failure occurs. The engine throttle setting of the residual operating engines is increased to maximum continuous rating from maximum climb rating. The aircraft is decelerated to a speed to get the maximum lift/drag ratio (the EFIS panel shows a green dot). This is because if the terrain (say a mountainous region) does not allow descent, then the aircraft needs to climb at the highest climb gradient (1.1% for two-engine aircraft and 1.6% for four-engine aircraft) to reach a suitable height to initiate the drift-down procedure, clearing the highest obstacle by at least 1000 ft as vertical clearance. Thereafter, during drift-down a vertical separation of at least 1000 ft is maintained to clear any obstacle (see Figure 14.4).

14.7.2   Engine Inoperative Situation at Cruise (Figure 14.5)

The drift-down procedure is initiated as soon as an engine failure is recognized (see Figure 14.5). The engine throttle setting of the residual operating engines is increased to maximum continuous rating from cruise rating. The aircraft is decelerated to a speed to get the maximum lift/drag ratio (the EFIS panel shows a green dot). The aircraft is lighter as a considerable amount of fuel is consumed during the climb to cruise altitude.

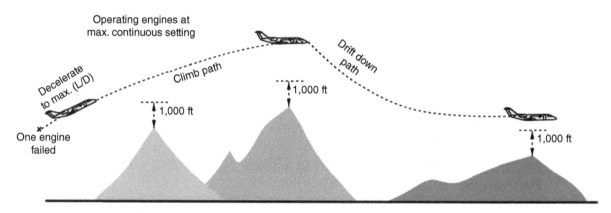

■ **FIGURE 14.4**   Drift-down after engine failure during climb

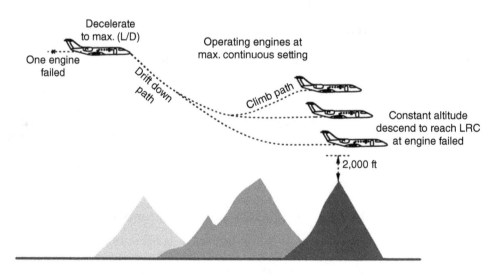

■ **FIGURE 14.5**
Drift-down after engine failure during cruise

The FAA requires that in the event of an engine failure in the cruise segment, the aircraft must land as soon as possible. In the absence of any nearby suitable airfield to land, the question is: to which airport should the aircraft head – the originating airport (departing airport) or the destination airport (arrival airport)?

14.7.3 Point of No-Return and Equal Time Point

The answer to the above question is based on where the aircraft is relation to the midpoint of the flight. In fact, the decision is based on the following two considerations.

1. The point of no-return (PNR): The greatest distance the aeroplane can fly after engine failure and stay on course towards reaching the destination airport, beyond which there will not be enough fuel left to return to the originating airport. This is the point when aircraft cannot return, hence known as PNR. In still air, the PNR is the midpoint.
2. The equal time point (ETP): The point of greatest distance from where the aircraft with an engine failure can either return to the departure airport or continue towards the destination airport. In still air the ETP is the midpoint.

In still air, ETP = midpoint; beyond the ETP the aircraft enters into the PNR domain. In a head or tail wind, the ETP and PNR are not at the midpoint. In a head wind, the PNR has to be well past the midpoint, since on return the head wind becomes a tail wind allowing greater distance (and *vice versa*).

14.7.4 Engine Data

The engine rating is increased from cruise thrust to maximum continuous rating. At higher altitudes, even engines operating at the maximum continuous rating may not be able to maintain level flight, and the aircraft loses altitude. In that case the aircraft needs to come down to a suitable lower altitude to maintain height. Aircraft then need to fly at a speed maintaining maximum $(L/D)_{eng\_inoper}$ as the most efficient and safe flight.

A generalized aircraft and engine grid for one engine inoperative can be prepared. This is not done in this book, but readers can generate this for their project, using the same procedure.

14.7.5 Drift-Down in Cruise

After loss of an engine at high cruise altitude, to minimize the loss of range a *drift-down* procedure is adopted by maintaining maximum $(L/D)_{eng\_inoper}$.

The pilot has the following three options for the *drift-down* procedure (Figure 14.5).

1. Maintain the speed at maximum $(L/D)_{eng\_inoper}$. With fuel burn the aircraft becomes lighter and the aircraft can enter cruise-climb if required to clear any obstacle.
2. Maintain a constant altitude. With fuel burn the aircraft becomes lighter and the aircraft increases speed at a slow acceleration towards the $(LRC)_{eng\_inoper}$ speed. This altitude should be able to clear any obstacle (say a mountain terrain) along the route.
3. If there is no terrain obstacle, then to reach the $(LRC)_{eng\_inoper}$ speed the pilot may descend to lower altitudes.

The FAA (Part 121) stipulates gradient requirements for one and two engines inoperative in separate specifications.

In the case of failure of the in-flight cabin environmental control system (pressurization/oxygen system/air-conditioning), the aircraft is required to descend below 10,000 ft altitude in a schedule not dealt with in this book. The readers may refer to FAR (14CFR) 121.329 for details.

14.8   Military Missions

Military aircraft certification standards are different from those for civil aircraft. There is one customer requirement, and the design standards vary from country to country based on their defence requirements. The safety issues have a similarity in their reasoning but differ in the required specifications. Readers are advised to obtain the regulatory military specifications from their respective Ministry/Department of Defence – these are generally available in the public domain. In this book, the substantiation procedure is the same as for the civil aircraft case covered in detail in the previous section.

Fuel load and management depends on the type of mission profile, which can be varied (see Section 10.4). Figure 10.5 gives a typical training profile to gain airmanship and navigational skills in an advanced aircraft. Armament training practice closely follows the combat mission profile discussed below.

Weapons load, target range and expected adversary defence capability will decide the combat mission profile. Two typical combat missions are shown in Figure 14.6. Air defence would require continuous intelligence feedback.

Study of the combat mission requires complex analysis by specialists. Defence organizations conduct these studies and understandably keep them confidential in nature. Game theory, twin dome combat simulations, and so on, are some of the tools for such analysis. Actual combat may yet prove quite different, as not all factors are known about adversary tactics and capabilities. Detailed study is beyond the scope of this book and possibly of most academies.

14.8.1   Military Training Mission Profile – Advanced Jet Trainer (AJT)

Figure 10.5 gives a typical AJT training profile to gain airmanship and navigational skills in an advanced aircraft. Block fuel and block time can be computed from adding fuel and time consumed in each segment. However, the distance covered is meaningless as the aircraft returns to base. Armament training practice closely follows the combat mission profile. The training profile segment breakdown is given in Table 10.1. Readers will require this kind of information for their project.

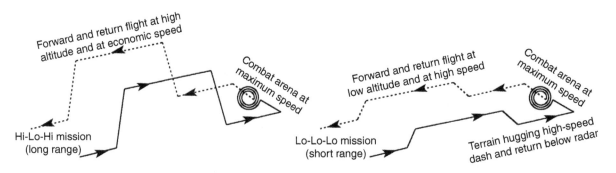

■ **FIGURE 14.6**   **Typical combat mission profiles**

■ **TABLE 14.3**
AJT mission fuel and time consumed – detailed segments

	Fuel burnt (kg)	Time (min)	Engine rating (% rpm)
Taxi and takeoff	60	6	60% (idle)
Takeoff and climb to 6 km altitude	125	5	TO at 100% then at 95%
Four turns	50	4	1 min at 95% + 3 min at 60%
Four stalls	60	5	1 min at 95% + 4 min at 60%
Climb from 5 to 8 km altitude	50	3	95%
Four turn spins	25	3	60%
Climb from 5 to 8 km altitude	50	3	95%
Four turn spins	25	3	60%
Climb from 5 to 6 km altitude	15	1	95%
Aerobatics practice	70	6	95%
Descent and practice force landing	95	8	2 min at 95% + 6 min at 60%
Three circuits for landing practice	110	10	Average 80%
Approach, land return taxi	40	4	60%
Trainee pilot allowance	30	2	95%
Total mission fuel	805	59	
		(≈ 60)	
Diversion	200		
Residual fuel	105		
Total onboard fuel	1110 (conservative estimate) (internal fuel capacity = 1400 kg)		

In the absence of actual engine data, the following may be taken:

$$\text{At idle} \left(60\% \text{ rpm}\right) \approx 8 \text{ kg / minute}$$

$$\text{At } 75\% \text{ rpm} \approx 11 \text{ kg / minute}$$

$$\text{At } 95\% \text{ rpm} \approx 16.5 \text{ kg / minute}$$

Following Table 10.1, fuel and time consumed for the normal training profile of the AJT is given in Table 14.3.

Combat missions as shown in Figure 14.6 can also be dealt with in a similar manner.

14.9   Flight Planning by the Operators

Dispatch of aircraft for the operational trip (sortie) is the obligation of the operator using manufacturer-supplied aircraft design data. They mainly use aircraft manuals, and when analytical studies are required use the same equations as outlined in this book. Therefore this book does not enter into the specifics of flight planning, which is a subject on its own.

The main obligation of the operators' dispatch office is first to ensure safety during the trip. The aircraft should carry sufficient fuel to reach the destination under the prevailing weather conditions, as well as meet any inadvertent situation, for which the minimum reserve fuel amount is stipulated by airworthiness authorities. The operator should ensure that payload and fuel loading into the aircraft follow the defined sequence to operate within safety boundaries at all times within the operation. Of course, there are a host of other criteria to be satisfied before a dispatch is authorized. Readers who wish to study flight planning may refer to the appropriate literature.

Proper distribution of mass (weight) over the aircraft geometry is the key to establishing the aircraft CG. It is important to locate the wing, undercarriage, engine and empennage for aircraft stability and control. The convenient way is first to estimate each component weight separately and then position them to satisfy the CG location with respect to the overall geometry. Typical aircraft CG margins affecting aircraft operation are shown in Figure 3.19.

It is observed that passengers choose to take the window seats first. Figure 3.19 shows the window seating first and the aisle seating last. Note the boundaries of the front and aft limits. The cargo and fuel loading is done to a schedule with the locus of CG travel in lines. In the diagram the CG of the operating empty mass (OEM) is at the rear, indicating that the aircraft has aft-mounted engines. For wing-mounted engines the CG at OEM moves forward shifting the shape of the loading curve. The operators must ensure safety at all times during the loading of aircraft.

References

1. Lan, C.T.E., and Roskam, J., *Airplane Aerodynamic and Performance*. DARcorporation, Lawrence, 1981.

2. Ruijgork, G.J.J., *Elements of Airplane Performance*. Delft University Press, Delft, 1996.

3. Ojha, S.K., *Flight Performance of Airplane*. AIAA, Reston, VA, 1995.

4. Airbus Customer Service, *Getting grips with aircraft performance*, January 2002.

CHAPTER 15

Manoeuvre Performance

15.1   Overview

This chapter deals with aircraft turn and manoeuvre performance. Turn performance capability is a customer requirement that needs to be substantiated. There is no airworthiness requirement for turn performance, except that it must operate within the certified flight envelope. Only the point performance of turn is dealt with here. Combat manoeuvres are complex. This chapter introduces some basic fighter manoeuvre (BFM) tactics.

The readers are to compute the AJT turn performance as done in Section 15.3. A problem set for this chapter is given in Appendix E.

15.2   Introduction

Aircraft manoeuvre is a transient operation meant for changing the aircraft flight path in a safe manner. It has many possibilities, making its analysis specific to each type of manoeuvre, for example, turning, pull-up, aerobatics, and so on. Aircraft manoeuvres are executed through the use of aircraft controls and thrust adjustment. Chapter 6 dealt with the aircraft stability and control considerations in which inherent aircraft motion characteristics, for example Dutch roll, spinning, and so on, are discussed. Aircraft manoeuvres are associated with some 'g' pull (load factor, n – Chapter 5).

Hard and sudden application will make the aircraft enter into an instantaneous turn, while a gentle, steady application will take the aircraft into a sustained steady turn at lower g loads. This book deals only with steady sustained turns. The worked examples are for the AJT, where the requirements for turn are more stringent than for the Bizjet.

Along with the separate treatise of sustained turns in the horizontal and vertical planes, this book also deals with turning in a helical path involving both the horizontal and vertical planes and some aerobatic manoeuvres. Readers may also see references [1–3].

Theory and Practice of Aircraft Performance, First Edition. Ajoy Kumar Kundu, Mark A. Price and David Riordan.
© 2016 John Wiley & Sons, Ltd. Published 2016 by John Wiley & Sons, Ltd.

■ **FIGURE 15.1**   (a)   (b)
Aircraft turn:
(a) Turning in the
horizontal plane.
(b) Climbing turn in
horizontal and
vertical planes
(see also Figure 15.5,
climbing turn)

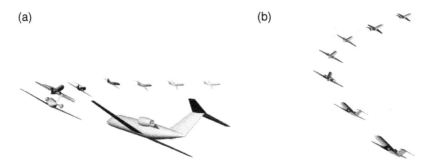

15.3   Aircraft Turn

Turning is a coupled motion; the aircraft changes direction with both yaw and roll. It is executed by using appropriate control authorities in a coordinated manner, with rudder for yaw and aileron for roll. Any imbalance of the coordinated control execution will have undesired effects, except in some special situation.

Turning in the horizontal plane (XY plane), that is, yawing, makes the outer wing move faster than the inner wing (Figure 15.1a). This will cause the outer wing to generate more lift and rise up in relation to the inner wing, resulting in a roll; that is, it has a coupled motion of yaw and roll. There is a tendency for aircraft to sideslip unless the aircraft is appropriately banked to make a coordinated turn; yaw with the right amount of roll prevents side-slip. Actually, an aircraft may have to keep its nose up to stay level. Section 15.3.5 tackles the case with nose up (Eulerian angle θ) for aircraft having a helical path either as a climbing turn or descending turn. The bank angle is the Eulerian angle Φ as shown in Figure 15.5.

The turn and side-slip indicator is a gyroscope-driven instrument as shown in Figure 15.2. In the diagram, a dial shows a black ball sealed in a curved, liquid-filled glass tube with two wires at its centre and a needle or aircraft symbol seen from aft end, in combination with aircraft attitudes. The ball measures gravity force and inertial force caused by the turn.

Aircraft manoeuvres may be as follows. A right turn is shown in Figures 15.1 and 15.2.

- *Straight level flying*: Aircraft wings level and ball in centre.
- *Coordinated turn*: In this case when the bank angle is correct, then the gravity and inertial forces are equal and the ball stays in the centre, but the right wing drops to show the turn (Figure 15.2, left). This is executed by coordinated application of right rudder and right aileron.
- *Skid turn*: When the bank is less, then inertia force is greater than gravity force, and the aircraft skids outward and the ball moves left, but the right bank angle is shown by the aircraft symbol (Figure 15.2, centre).
- *Slip turn*: When the bank is more, then inertia force is less than gravity force, and the aircraft slips inward and the ball moves right, but the right bank angle is shown by the aircraft symbol (Figure 15.2, right).

15.3.1   In Horizontal (Yaw) Plane – Sustained Coordinated Turn

Turning is a pilot-induced manoeuvre. A distinction is to be made between sustained turn in a steady state, and instantaneous turn which is an unsteady state. A steady, sustained turn is a coordinated turn where the load factor, *n,* remains constant and thrust is aligned with the velocity vector. Its analyses are relatively easier and may require iterative computation/graphical solutions.

■ **FIGURE 15.2**
Forces on aircraft
turn in the horizontal
plane. Left-hand
column: coordinated
turn. Centre column:
skidding out of turn.
Right-hand column:
slipping into turn

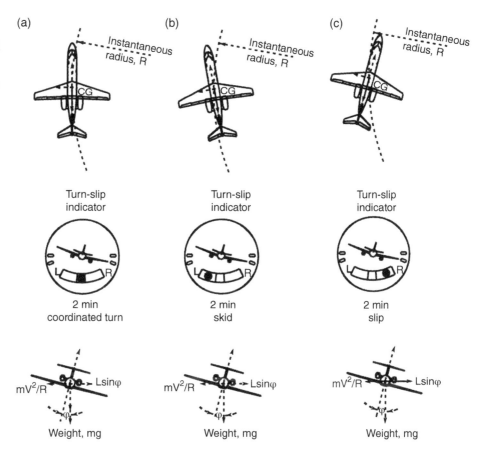

Instantaneous turn (sudden hard application of control) can generate higher load factors (*g*-load) and is not in steady state, and the thrust vector is generally not aligned with the velocity vector. If not properly handled, the aircraft may experience buffet – not a desirable situation. Treatment of instantaneous turn is beyond the scope of this book. Turning has to stay within the permissible load factor limit, n_{max} (structural consideration), or within the C_{Lstall} (aerodynamic consideration) limit. Operating within these limits, an aircraft is not designed to enter into buffet.

In a coordinated turn the Eulerian angles Φ and θ are the same in F_W and F_B. Forces in wind axes F_W are as shown in Figure 15.3.

15.3.1.1 ***Kinetics of Coordinated Turn in Steady (Equilibrium) Flight*** The computation for turn is a point performance evaluated at the instant. Integrated turn performance is carried out in small time steps in which the variables are treated as constant. The equation of motion at an instant of a steady turn has radius R of turn and aircraft velocity, V, tangential to the flight path. In elemental time Δt, the turning angle is $\Delta \psi$. Then the instantaneous angular velocity can be written as $(d\psi/dt)$ in radian/s. The instantaneous tangential velocity can be written as $V = R(d\psi/dt)$ in ft/s, or

$$\frac{d\psi}{dt} = \frac{V}{R} \tag{15.1}$$

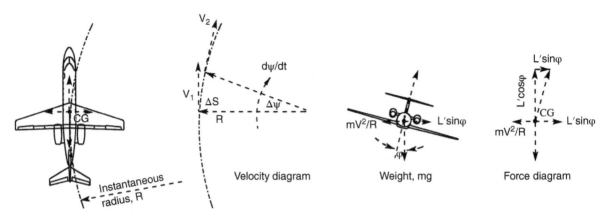

■ **FIGURE 15.3**   Forces on aircraft in a coordinated turn in the horizontal plane

The thrust line is taken along the flight path. Turn is in a bank angle of φ with L' as the lift force acting at the aircraft plane of symmetry normal to the instantaneous tangential velocity V (true airspeed). In the horizontal plane, thrust = drag. In the lateral plane,

$$\text{Weight, } W = mg = L'\cos\Phi$$

or

$$L' = \frac{W}{\cos\Phi} \tag{15.2}$$

where $L' - L = \Delta L = W/\cos\Phi - W = W(1/\cos\Phi - 1)$.

The ratio of the forces,

$$\frac{L'}{W} = \frac{1}{\cos\Phi} = n = \sec\Phi \quad \left(\text{aircraft load factor}\right) \tag{15.3}$$

where n = aircraft load factor acting in the plane of symmetry of the aircraft along L'.

[Note that for pull-up in the vertical (pitch) plane, L' and W are coplanar (Figure 15.3); therefore in Equation 15.3 the load factor n (seen as g-load) is seen as a multiple of acceleration. In the horizontal plane the load factor n is slightly different as L' and W are not coplanar.]

From the trigonometry identity, $\sin^2\Phi + \cos^2\Phi = 1$, this gives Equation 15.4 $\sin^2\Phi + 1/n^2 = 1$, or

$$\sin\Phi = \left(\frac{\sqrt{(n^2 - 1)}}{n}\right) \tag{15.4}$$

Perpendicular to the flight path towards the instantaneous centre of turn, force balance means that centrifugal force = centripetal force, that is:

$$\frac{mV^2}{R} = L'\sin\Phi \tag{15.5}$$

Therefore the centripetal acceleration acting radially (normal to flight path), $a_n = V^2/R$, and the tangential acceleration (acting along the flight path),

$$a_t = \frac{dV}{dt} \tag{15.6}$$

Combining Equations 15.2 and 15.5,

$$m = \frac{L'\cos\Phi}{g} = \frac{RL'\sin\Phi}{V^2}$$

or

$$\frac{V^2}{Rg} = \frac{\sin\Phi}{\cos\Phi} = \tan\Phi$$

Substituting Equation 15.1,

$$\tan\Phi = \frac{V(d\psi/dt)}{g} \tag{15.7}$$

or

$$\text{angular velocity, } \frac{d\psi}{dt} = \frac{g\tan\Phi}{V} \tag{15.8}$$

Equation 15.5 gives

$$R = \frac{mV^2}{L'\sin\Phi} \tag{15.9}$$

Using Equation 15.3 $n = L'/W = L'/mg$, Equation 15.9 gives radius of turn

$$R = \frac{mV^2}{L'\sin\Phi} = \frac{nV^2}{g\sin\Phi} = \frac{V^2}{g\sqrt{n^2-1}} \tag{15.10}$$

Again using Equation 15.2, $L' = mg/\cos\Phi$, the instantaneous radius of turn

$$R = \frac{V^2}{(g\tan\Phi)} \tag{15.11}$$

Equation 15.3 gives $n^2 = \sec^2\Phi$ or $n^2 - 1 = \tan^2\Phi$ or

$$\tan\Phi = \sqrt{(n^2-1)} \tag{15.12}$$

Equation 15.7 gives $\tan\Phi = V(d\psi/dt)/g$ or

$$\text{angular velocity } \frac{d\psi}{dt} = \frac{g\tan\Phi}{V} = \frac{g\sqrt{(n^2-1)}}{V} \tag{15.13}$$

To obtain the maximum angular velocity ($d\psi/dt$), the aircraft has to undergo highest possible load factor n and lowest possible aircraft velocity.

The maximum turn required to change aircraft direction is 180° either turning to left or right. Therefore time taken for 180° turn (π radian) is obtained as follows. Using Equation 15.7,

$$\text{Time}_\pi = \frac{\pi}{(d\psi/dt)} = \frac{\pi V}{g\tan\Phi} \tag{15.14}$$

When an aircraft performs a 180° coordinated turn in 1 minute, it is designated as a "*rate one*" turn; it is executing 3°/s. A "*rate two*" turn is when a 180° turn is carried out in half the time, that is, in 30 seconds.

15.3.1.2   *Load Factor, n*   Equation 15.9 gives $R = mV^2/L'\sin\Phi$. In equilibrium flight,

$$\text{thrust, } T = \text{drag, } D = 0.5\rho V^2 S_W\left(C_{DPmin} + kC_L'^2\right) \tag{15.15}$$

where C_L' is the lift coefficient while in a turn under load factor, n; where

$$L' = nmg = 0.5\rho V^2 S_W C_L'$$

or where

$$C_L' = 2nmg/\rho V^2 S_W \tag{15.16}$$

Substituting Equation 15.16 into Equation 15.15,

$$T = D = 0.5\rho V^2 S_W\left[C_{DPmin} + k\left(2nmg/\rho V^2 S_W\right)^2\right]$$

or

$$T = 0.5\rho V^2 S_W C_{DPmin} + \left[k\left(0.5\rho V^2 S_W\right)\left(2nW/\rho V^2 S_W\right)^2\right]$$

or

$$T = 0.5\rho V^2 S_W C_{DPmin} + k\left(2n^2 W^2/\rho V^2 S_W\right)$$

or

$$k\left(2n^2 W/\rho V^2 S_W\right) = T/W - 0.5\rho V^2\left(\frac{S_W}{W}\right)C_{DPmin} \tag{15.17}$$

Solve Equation 15.17 for load factor, n:

$$n^2 = \left[\frac{0.5\rho V^2}{k\left(\dfrac{W}{S_W}\right)}\left(\frac{T}{W} - 0.5\rho V^2\frac{C_{DPmin}}{W/S_W}\right)\right]$$

$$\text{so load factor, } n = \left[\frac{0.5\rho V^2}{k\left(\dfrac{W}{S_w}\right)} \left(\frac{T}{W} - 0.5\rho V^2 \frac{C_{DPmin}}{W/S_w} \right) \right]^{\frac{1}{2}} \tag{15.18}$$

15.3.1.3   *Angular Velocity*

$$\text{Angular velocity } \frac{d\psi}{dt} = \frac{g\sqrt{(n^2-1)}}{V} = \frac{g}{V} \sqrt{\left\{ \frac{0.5\rho V^2}{k\left(\dfrac{W}{S_w}\right)} \left(\frac{T}{W} - 0.5\rho V^2 \frac{C_{DPmin}}{W/S_w} \right) - 1 \right\}} \tag{15.19}$$

15.3.1.4   *Radius of Turn*   Substituting Equation 15.18 in Equation 15.10 gives

$$R = \frac{V^2}{g\sqrt{\dfrac{0.5\rho V^2}{k\left(\dfrac{W}{S_w}\right)}\left(\dfrac{T}{W} - 0.5\rho V^2 \dfrac{C_{DPmin}}{W/S_w} \right) - 1}}$$

Simplifying,

$$R = \frac{V^2}{g\sqrt{\dfrac{\rho(T/W)V^2}{2k\left(\dfrac{W}{S_w}\right)} - \left[\left(\dfrac{\rho^2}{(W/S_w)^2} \right)\left(\dfrac{C_{DPmin}}{4k} \right) \right]V^4 - 1}} \tag{15.20a}$$

Note that $\left(\dfrac{L}{D} \right)_{max} = \dfrac{1}{2\sqrt{kC_{DPmin}}}$ which gives

$$\frac{1}{4k} = C_{DPmin} \times \left(\frac{L}{D} \right)_{max}^2 \quad \text{or} \quad \frac{1}{k} = 4C_{DPmin} \times \left(\frac{L}{D} \right)_{max}^2 .$$

On substituting in the above equation, the following is obtained:

$$R = \frac{V^2}{g\sqrt{\dfrac{4\rho(T/W)\times C_{DPmin}\left(\dfrac{L}{D} \right)_{max}^2 V^2}{2\left(\dfrac{W}{S_w} \right)} - \left[\left(\dfrac{\rho^2}{(W/S_w)^2} \right)C_{DPmin}^2\left(\dfrac{L}{D} \right)_{max}^2 \right]V^4 - 1}} \tag{15.20b}$$

15.3.2   Maximum Conditions for Turn in Horizontal Plane

Differentiating Equation 15.18 with respect to V and setting it equal to zero gives the maximum conditions of V_{n_max} as follows.

$$2n\frac{dn}{dV}=0=\left[\frac{\rho V}{k\left(\dfrac{W}{S_W}\right)}\left(\frac{T}{W}\right)\right]-\left[\frac{\rho^2 V^3 C_{DPmin}}{k\left(\dfrac{W}{S_W}\right)^2}\right]=\left(\frac{\rho V}{k\dfrac{W}{S_W}}\right)\left[\left(\frac{T}{W}\right)-\frac{\rho V^2 C_{DPmin}}{\dfrac{W}{S_W}}\right]$$

This gives

$$\frac{\rho V_{n\_max}^2 C_{DPmin}}{\left(\dfrac{W}{S_W}\right)}=\left(\frac{T}{W}\right)$$

or

$$\left(V_{n\_max}\right)^2=\left(\frac{1}{\rho C_{DPmin}}\right)\left(\frac{T}{W}\right)\left(\frac{W}{S_W}\right)$$

or

$$V_{n\_max}=\sqrt{\left(\frac{1}{\rho C_{DPmin}}\right)\left(\frac{T}{W}\right)\left(\frac{W}{S_W}\right)}\qquad\text{(15.21)}$$

Substituting in Equation 15.18, the maximum load factor n can be obtained:

$$n_{max}=\left[\frac{0.5\rho\left(\dfrac{T}{W}\right)\left(\dfrac{W}{S_W}\right)}{k\rho C_{DPmin}\left(\dfrac{W}{S_W}\right)}\frac{T}{W}-\frac{0.5\rho\left(\dfrac{T}{W}\right)\left(\dfrac{W}{S_W}\right)}{\rho C_{DPmin}}\frac{C_{DPmin}}{W/S_W}\right]^{\frac{1}{2}}=\left[\frac{0.5\rho\left(\dfrac{T}{W}\right)^2}{k\rho C_{DPmin}}(1-0.5)\right]^{\frac{1}{2}}$$

or

$$n_{max}=0.5\left(\frac{T}{W}\right)\left[\frac{1}{kC_{DPmin}}\right]^{\frac{1}{2}}\quad\text{operating at }\left(\frac{T}{W}\right)_{max}\text{ as }T\text{ is the only variable.}\qquad\text{(15.22)}$$

Note that Equation 10.18 gives $\left(\dfrac{C_L}{C_D}\right)_{max}=\dfrac{1}{2}\sqrt{\dfrac{1}{kC_{DPmin}}}$, and substituting in Equation 15.22, then the equation for n_{max} can be written as

$$n_{max}=\left(\frac{T}{W}\right)\left(\frac{C_L}{C_D}\right)_{max}\qquad\text{(15.23)}$$

The maximum bank angle ϕ_{max} can be obtained from Equation 15.3, $\cos\phi=1/n$, and can be written as $\phi_{max}=\cos^{-1}(1/n_{max})$. Substituting n_{max} from Equation 15.22,

$$\phi_{max}=\cos^{-1}\left[\left(\frac{W}{T}\right)\left(\frac{C_L}{C_D}\right)_{max}\right]\qquad\text{(15.24)}$$

There is a quick way to obtain n_{max}. Noting

$$n = \frac{L'}{W} = \left(\frac{L'}{D}\right) \times \left(\frac{D}{W}\right) = \left(\frac{L'}{D}\right) \times \left(\frac{T}{W}\right)$$

then the maximum condition will be when

$$n_{max} = \left(\frac{L'}{D}\right)_{max} \times \left(\frac{T}{W}\right)_{max}$$

where $(L'/D)_{max}$ is obtained from AJT drag polar and $(T/W)_{max}$ from engine performance.

15.3.3   Minimum Radius of Turn in Horizontal Plane

In steady flight, $C_{Lmax} = \dfrac{2mg}{\rho V_{stall}^2 S_W}$ or $W = 0.5\rho V_{stall}^2 S_W C_{Lmax}$.

In a coordinated sustained turn, Equation 15.16 gives

$$C_{Lmax} = \frac{2nmg}{\rho V_{stall\_turn}^2 S_W}$$

or

$$nW = 0.5\rho V_{stall\_turn}^2 S_W C_{Lmax}$$

Equating the two,

$$V_{stall\_turn} = V_{stall}\sqrt{n} \tag{15.25}$$

For given aircraft settings, Equation 15.20a states that the radius of turn, R, is only dependent upon aircraft velocity. Evidently, the minimum radius of turn is at the limiting value of the aircraft speed, that is, at the stall speed.

Recall Equation 15.10, which gives radius of turn, $R = \dfrac{V^2}{g\sqrt{n^2-1}}$. Therefore, minimum radius of turn

$$R_{min} = \frac{V_{stall\_turn}^2}{g\sqrt{n^2-1}} = \frac{nV_{stall}^2}{g\sqrt{n^2-1}} \tag{15.26}$$

For a given aircraft weight, the minimum radius of turn R_{min} depends on the thrust level T, which decides the level of load factor until it reaches n_{max} or C_{Lmax}. The minimum radius of turn does not necessarily give the maximum load factor n_{max} (see Figure 15.6).

Setting Equation 15.20a to $dR/dV = 0$ to find the minimum radius of turn will require an iterative/graphical solution and is cumbersome.

15.3.4   Turning in Vertical (Pitch) Plane

A transitional pull-up manoeuvre (turning in a vertical plane) can be formulated in a similar manner. Figure 15.4 represents a typical turn in the vertical plane. The aircraft is in a curvilinear path in the vertical plane with an instantaneous radius of curvature R at an angle θ with the vertical. The aircraft is at steady turn at a constant velocity V tangential to the flight path.

■ **FIGURE 15.4**
**Turn in the vertical
plane**

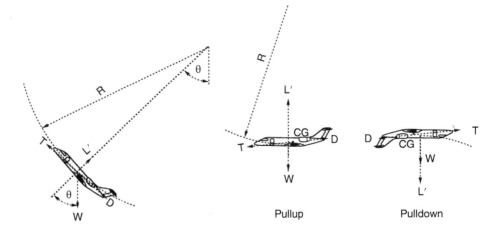

A pull-up manoeuvre in sustained load factor need not be in a high g-load. For club flying it should be uniformly less than $3\,g$ – it can be monitored in a g-meter fitted in the instrument panel. At the instant of point performance the equations of motion in equilibrium give:

horizontal components : $\Sigma F_H = 0 = T - D - W \sin\theta$ **(15.27a)**

vertical components : $\Sigma F_V = 0 = L' - W \cos\theta - \dfrac{mV^2}{R}$ **(15.27b)**

(the last term is centripetal force), and velocity, $V = R\dot\theta$.

A **pull-up** starts from horizontal flight ($\theta = 0$) when the point performance gives:

horizontal components : ΣF_{H_pullup} gives $T = D$ **(15.28a)**

vertical components : ΣF_{V_pullup} gives $L' = W + \dfrac{mV^2}{R}$ **(15.28b)**

and velocity $V = R(\mathrm{d}\theta/\mathrm{d}t)$.

Equation 15.3 gives

$$n_{pullup} = \frac{L'}{W} = 1 + \frac{V^2}{Rg}$$ **(15.29)**

By definition,

$$V_{pullup} = \sqrt{\frac{2L'}{\rho S_W C_L}} = \sqrt{\frac{2n_{pullup}\left(\dfrac{W}{S_W}\right)}{\rho C_L}}$$ **(15.30)**

and radius of turn

$$R = \frac{V^2}{g\left(n_{pullup} - 1\right)}$$ **(15.31a)**

Substituting for V from Equation 15.30,

$$R = \frac{2n_{pullup}\left(\dfrac{W}{S_W}\right)}{g\rho\left(n_{pullup}-1\right)C_L}$$

(15.31b)

Rate of pull-up can be derived from

$$\frac{d\theta}{dt} = \frac{V}{R}$$

(15.31c)

A ***pull-down*** starts in inverted horizontal flight ($\theta = 0$) when the point performance gives

horizontal components : $\Sigma F_{H\_pulldown}$ gives $T = D$ (15.32a)

vertical components : $\Sigma F_{V\_pulldown}$ gives $L' = \dfrac{mV^2}{R-W}$ (15.32b)

and velocity $V = R\dot\theta$.
    Equation 15.3 gives

$$n_{pulldown} = \frac{L'}{W} = \frac{V^2}{Rg} - 1$$

(15.33)

By definition,

$$V_{pulldown} = \sqrt{\frac{2L'}{\rho S_W C_L}} = \sqrt{\frac{2n_{pulldown}\left(\dfrac{W}{S_W}\right)}{\rho C_L}}$$

(15.34)

and from Equation 15.31a, radius of turn, $R = \dfrac{V^2}{g\left(n_{pulldown}+1\right)}$.

Substituting for V from Equation 15.30,

$$R = \frac{2n_{pulldown}\left(\dfrac{W}{S_W}\right)}{g\rho\left(n_{pulldown}+1\right)C_L}$$

(15.35)

Rate of pull-down can be derived from $\dfrac{d\theta}{dt} = \dfrac{V}{R}$.

15.3.5   In Pitch-Yaw Plane – Climbing Turn in Helical Path

A more generalized motion of climbing turn is a helical path in the pitch and yaw plane, with the aircraft turning in climb or descent (Figure 15.5).
    A climbing turn formulation of equations of motion is more complex. To keep it simpler, only a coordinated turn is treated. In this case there is no side-slip or skidding, hence $\beta = 0$. Refer to Section 4.5.4 on deriving equations of motion in the pitch-yaw plane (Eulerian angles, elevation angle θ and bank angle Φ). Forces involved in a helical turn (say climb) are as follows (in wind axes F_B).

■ **FIGURE 15.5**
Climbing turn
(see also
Figure 15.1b)

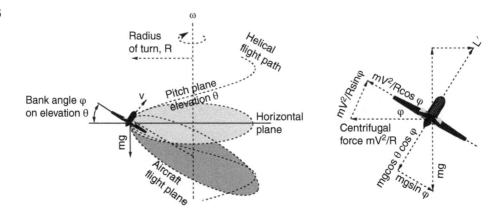

Along a flight path along the X-axis of F_B,

$$\text{thrust}\,(T) = \text{drag}\,(D) + mg\sin\theta \qquad (15.36)$$

Perpendicular to the flight path in the Y-axis,

$$\frac{mV^2\cos\Phi}{R} = L'\sin\Phi \qquad (15.37)$$

and in the Z-axis of F_B

$$mg\cos\theta\cos\Phi = L' + mV^2\sin\Phi \qquad (15.38)$$

The instantaneous radius of turn R can be obtained as follows. Combining Equations 15.37 and 15.38

$$\frac{mV^2\cos\Phi}{R} = \frac{V^2 L'\cos\Phi}{Rg} = L'\sin\Phi$$

$$R = \frac{V^2}{g\tan\Phi} \qquad (15.39)$$

For wind axes F_W, replace θ by θ_W and the rest are same in coordinated turn on account of $\beta = 0$, and bank angle is the same in both axes. If the aircraft is not in a coordinated turn, that is, it is in side-slip or skid, then $\beta \neq 0$ and $\theta \neq \theta_W$.

15.4 Classwork Example – AJT

A Bizjet turn is limited by airworthiness requirements related to passenger comfort. An AJT, as a military trainer, has higher performance requirements. For this reason, AJT examples are used in this section. Following are the aircraft data used in the worked tables below.

For a classwork example, AJT drag may be approximated as parabolic drag polar as given below (see Equation 9.38).

$$C_D = 0.0212 + 0.068 C_L^{\,2} \quad \text{with} \quad S_W = 183 \text{ ft}^2$$

(gives Oswald's efficiency factor, $e = 0.85$).

Equations 15.4 ($\sin\Phi$), 15.11 (radius of turn) and 15.14 (rate of turn) are independent of any aircraft; they give the capabilities associated with the load factor, n. Table 15.1 tabulates the capabilities. Table 15.2 gives the AJT capabilities for a coordinated turn at sea level on an ISA day. Sea-level installed thrust is given in the table. AJT weight at normal training configuration (NTC) is 9000 lb. (Requirement is 4 g at sea-level.)

Tables 15.1 and 15.2 are plotted in an integrated manner in Figure 15.6. It requires some cross-plotting to obtain the lines of constant radius of turn, R, in the finished graph. AJT turn performance capability is superimposed upon it.

This kind of graph is useful in obtaining the full aircraft turn performance up to its maximum capability. Of course, aircraft maximum turn capabilities can be analytically estimated by differentiating Equations 15.23 and 15.24 and setting them equal to zero. The AJT maximum load factor, n, is 5.6 at Mach 0.59 with a radius of turn of 3000 ft at 15°/s. At lower speeds it can turn at a higher rate with a smaller radius.

Use Equation 15.22 to analytically determine the maximum load factor, n, at sea-level.

$$n_{max} = 0.5\left(\frac{T}{W}\right)\left[\frac{1}{kC_{DPmin}}\right]^{\frac{1}{2}} \quad \text{operating at } \left(\frac{T}{W}\right)_{max} \text{ as } T \text{ is the only variable}$$

$$= 0.5 \times 0.4333 \times \left[\frac{1}{0.0706 \times 0.0212}\right]^{\frac{1}{2}}$$

$$= 0.21665 \times \sqrt{668.13} = 0.21665 \times 25.85 = 5.6$$

■ **TABLE 15.1**
Aircraft turn capabilities

n	n^2	$\sqrt{(n^2-1)}$	$\sin\Phi$	$\sin^{-1}\Phi$	Φ (°)	R (ft)	$d\psi/dt$	$d\psi/dt$(°)
2	4	1.732	0.866	1.0472	59.837	2011.6	0.1665	9.514
3	9	2.828	0.9428	1.231	70.337	1232.0	0.2719	15.54
4	16	3.873	0.9683	1.318	75.317	899.62	0.3723	21.274
5	25	4.899	0.9798	1.369	78.25	711.2	0.471	26.91
6	36	5.916	0.986	1.40	80.187	588.94	0.5687	32.5

■ **TABLE 15.2**
AJT coordinated turn capabilities on an ISA day

Mach	V (ft/s)	q (lb/ft²)	T (lb)	n^2	n	$\sqrt{(n^2-1)}$	R (ft)	$d\psi/dt$	$d\psi/dt$ (deg/s)	$\sin\Phi$	Φ (°)
0.2	223.3	59.29	3900	6.97	2.64	2.44	633.77	0.352	20.13	0.9255	67.558
0.3	334.95	133.4	3850	15.24	3.77	3.64	957.51	0.35	19.99	0.964	74.43
0.4	446.6	237.15	3800	21.88	4.68	4.57	1355.52	0.35	18.83	0.977	77.445
0.5	558.25	370.5	3800	28.05	5.3	5.2	1860.93	0.3	17.14	0.982	78.9
0.6	669.9	533.58	3800	29.58	5.44	5.35	2606.83	0.257	15.68	0.983	79.19
0.7	781.55	726.27	3800	22.88	4.78	4.68	4055.3	0.193	11.01	0.978	77.72
0.8	893.2	948.6	3800	3.69	1.92	1.64	15,116.8	0.059	3.38	0.854	58.45

■ **FIGURE 15.6**
AJT coordinated turn performances

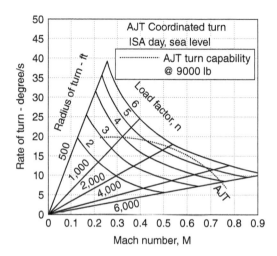

Velocity at n_{max} can be estimated by using Equation 15.21:

$$V_{n\_max} = \sqrt{\left(\frac{1}{\rho C_{DPmin}}\right)\left(\frac{T}{W}\right)\left(\frac{W}{S_W}\right)} = \sqrt{\left(\frac{1}{0.002378 \times 0.0212}\right)(0.4333)(49.2)}$$

$$= \sqrt{\left(\frac{21.32}{0.00005}\right)} = \sqrt{426367} = 653 \text{ ft/s} = 0.58 \text{ Mach} \approx 0.59 \text{ Mach as in Figure 15.6.}$$

15.5 Aerobatics Manoeuvre

Aerobatics are aircraft manoeuvres through changing flight profiles with combinations of turns in any plane in six degrees of freedom. This section briefly describes the basic aerobatic manoeuvres capabilities within the permissible flight envelope (*V-n* diagram). These are primarily functions of aircraft control in stable operation. Its analytical treatment is beyond the scope of this book, but is included as a discussion to show the transient performance in short durations lasting for seconds (freestyle competition gives 4 minutes to conduct multiple manoeuvres in a sequence; three-quarters of the time is spent in getting into position to initiate the manoeuvres). Fuel consumed in this short duration is the main parameter to estimate. Fuel flow rate is obtained from the sfc at the relevant rated engine power setting. The AJT training profile is an example (see Section 14.8.1).

Aerobatics is not an essential requirement for private pilots. Elementary aerobatics can be performed under visual flying rules. Aerobatics require skilled airmanship and can be dangerous in unskilled hands. Safety considerations are of the utmost importance, and preparation for aerobatic flying is essential. The following must be adhered to when undertaking any aerobatic flying.

1. The performing pilot must have the proven skills and be cleared by appropriate authorities.
2. The pilot has to be medically fit at the time of undertaking the sortie. There are club aircraft capable of reaching 6 g for recreational aerobatic flying – this is high enough to disorient the pilot, and can cause a temporary black-out.

3. The aircraft must be certified for the type of aerobatic manoeuvres, as not all aircraft are designed to perform aerobatics. Commercial transport aircraft are not allowed to undertake aerobatics; their structural load limit is $3.2\,g$, but the main reason is passenger safety.

4. All aerobatics must be performed at safe heights. The minimum height of any manoeuvre to full recovery must be at least $3000\,\text{ft}$. Special authorization is required for low-altitude aerobatics. There is no scope for so-called "dare-devil" flying to impress. Exhibition flying is conducted under strict controls.

5. The performing aircraft should be checked before flight. There should be no scope for control jamming. Accidents can happen if loose items enter the control circuit to make it inoperable. Loose objects can even injure the pilot. There should be an appropriate harness suited to aerobatic manoeuvres.

6. The aircraft CG position and weight distribution are to be kept as specified for aerobatics. A CG too far forward increases control forces, but too far rearward relaxes the stability margin, which could facilitate manoeuvres but approaches instability.

7. All pilots must be exposed to incipient spin stage and must be able to recognise it, know the procedure to avoid it, and if entered, recover from it.

8. The aerobatic operational zone should be a designated area free from the public and buildings.

External landmarks can help the pilot to conduct manoeuvres, especially to keep the wings level and stay in the appropriate flight plane. Straight roads, railtracks, streams, and so on, can help aircraft aligned in awkward attitudes (looping over a straight road can help to keep the aircraft in a vertical plane, as disorientation can easily make it leave the plane). There should be a defined margin between stall speed and minimum aerobatic speed, and buffeting should be avoided.

Military pilots must undergo aerobatic training as these are parts of combat manoeuvres. Advanced training to operational conversion completes extreme manoeuvres at very high g when pilot black-out is possible. Combat manoeuvres rely on instrument indications. A high-g suit is integral equipment for pilot safety. Military aircraft have a considerably higher thrust to weight ratio (T/W). If $(T/W) > 1$, then a vertical climb can be performed at any time.

Aerobatics are regular features in air shows. However, aerobatics are not mere eye-catching gymnastics. It has to be performed artistically, like a ballet in the sky. Pilots need to watch airspeed all the time. Military aerobatics are breath-taking; the MG29 "cobra" display stunned the world during the 1990s. In the 2013 Paris Air Show, the Sukhoi 35S stunned the public again with its eye-catching manoeuvres.

There are many types of aerobatic manoeuvres, still evolving as aircraft designs explore new boundaries. Given below are ten well-known, basic aerobatic manoeuvres; eight of them are illustrated and have some explanation, including the "cobra" manoeuvre (Figure 15.13). The chandelle is 180° coordinated climbing turn as discussed in Section 15.3.5. The chandelle and Lazy-8 may be considered as the most elementary aerobatics all aircraft can perform, and serve as useful practice to gain airmanship. Each aerobatic manoeuvre is associated with a specified entry airspeed; the magnitude depends on the design of the aircraft and its (T/W) ratio.

15.5.1   Lazy-8 in Horizontal Plane

This is an S-like (180° turns) flight trajectory in the horizontal plane and can be done softly or in hard coordinated turns. The throttle level is raised in turn and level depending on how hard the turn is. Figure 15.7 gives a typical Lazy-8 trajectory. A precision trajectory determines how artistically it can be presented by the pilot – it is a happy movement, waltzing in the sky.

■ **FIGURE 15.7**
Lazy-8 in the
horizontal plane

■ **FIGURE 15.8**
Slow roll

15.5.2   Chandelle

A chandelle is a 180° coordinated climbing turn as shown in Figure 15.1b, that is, on the way to making a 180° turn.

15.5.3   Slow Roll

This is a relatively difficult manoeuvre at the initial stages of experience. It is 360° rolling about the X-axis in a straight line in the horizontal plane (Figure 15.8). Keeping the aircraft nose at the proper attitude requires pilot airmanship. After gaining requisite airspeed, the roll is initiated with aileron application along with appropriate rudder and elevator to keep it coordinated and in a straight line. As bank reaches 90°, opposite rudder is progressively applied to keep the aircraft nose above the horizon, and similar but opposite application is required at the other side when coming out of it. When inverted the pilot has to push the stick forward to keep the nose above the horizon; at this stage the pilot experiences negative g-force, which some pilots are not comfortable with. If not executed in a precise manner, the aircraft can lose height. Prescribed entry airspeed and continuous throttle adjustment are required for its execution. It can now be appreciated why it is a relatively difficult manoeuvre; the pilot has to execute all controls including throttle in a continuous and alternating manner executed in harmony.

A *snap roll* is almost the same but performed very quickly. A higher entry speed is required to execute a fast snap roll; the excess kinetic energy holds the aircraft at desired attitudes.

15.5.4   Hesitation Roll

This is almost same as the slow roll described above. The difference is that there are prescribed positions (points) where the aircraft is held as if hesitating (Figure 15.9).

The minimum points are four bank positions at 0° (wings level), 90°, 180° (inverted), 270°, and return to level flying. The positions can be as many as 32 points. Typical display points are 8 at 45° intervals. A higher-powered aircraft proves advantageous in keeping the nose above the horizon for a longer time.

■ **FIGURE 15.9** Hesitation roll

■ **FIGURE 15.10**
Barrel roll about the
horizontal axis

15.5.5 Barrel Roll

A barrel roll is also about the X-axes, which are in general in the horizontal position, but need not be (Figure 15.10). This time the roll is opened up as if travelling along the inner surface of a barrel. A higher entry speed is required, for which the aircraft starts with a shallow dive to the requisite entry airspeed, then is pulled up with aileron and rudder application until it comes out at the other end.

15.5.6 Loop in Vertical Plane

A loop has simpler controls, applying elevator only and keeping the wings level. Pulling out at the bottom of the loop produces high g, possibly up to 4–6 g. However, a pull-up manoeuvre in sustained load factor need not be in a high g-load. For club flying, it should uniformly be less than 3 g; it can be monitored in a g-meter fitted in the instrument panel. High g-pulls may lead to aircraft stall with buffeting.

The loop is entered in a shallow dive to gain the requisite entry speed to a prescribed level at full throttle, followed by uniform stick pull (the pilot should monitor the g-meter if installed, otherwise develop a feel for the g-load) until it reaches the top at an inverted attitude. Then the pilot should gradually pull back throttle to almost idle when the aircraft is pointing down, and at the end of the loop gradually returning to normal throttle setting (Figure 15.11).

A lower entry speed increases the turn rate and reduces the turn radius. Increasing the load factor also does the same. The entry speed must provide sufficient energy along with available thrust to reach the top of the loop manoeuvre without stalling (buffeting). Coming down the loop is gravity-assisted with speed gain, and the recovery should maintain a uniform g-level.

■ **FIGURE 15.11**
Loop in the vertical plane (AJT)

■ **FIGURE 15.12**
Immelmann – roll off the top (AK4)

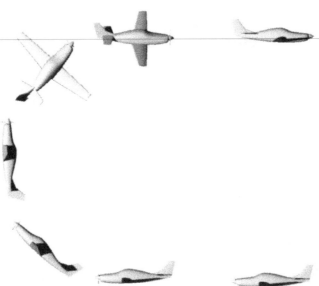

Keeping the aircraft precisely in the vertical plane requires skill to keep the wings level throughout. Without any reference, it is easy to lose precision. A loop above a straight road gives an excellent reference point to the pilot to stay in the vertical plane. Experienced pilots can manage well with cockpit instrument indications. If speed is not maintained properly, then buffet may be experienced near the top of the loop. In a wrongly executed loop, the aircraft may even inadvertently enter into an inverted spin, a dangerous situation from which to recover. It is hard to recognise inverted spin, and use of a normal spin recovery aggravates the situation. While the loop is simple to execute with single control application of elevator, it requires careful attention throughout.

15.5.7   Immelmann – Roll at the Top in the Vertical Plane

A roll at the top in the vertical plane is like a loop but less intensive, as the aircraft rolls to level attitude out at the top. It is also known as an Immelmann (Figure 15.12). High entry speed is required as in the case of the loop, and in addition aileron has to be applied to roll out at the top. Use of a little rudder may be required to maintain the in-plane flight profile.

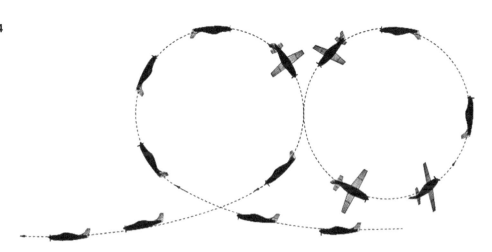

15.5.8 Stall Turn in Vertical Plane

This may be considered as a basic aerobatic manoeuvre. A stall turn is making a 180° turn at the top of a vertical climb at full throttle and as an aircraft loses speed approaching close to stall (Figure 15.13).

Full rudder application at the vertical position is made before the aircraft stalls to make the sharp 180° yaw turn. When the aircraft comes down vertically, throttle has to be brought to idle as gravitational pull assists the speed gain. At the end, the stick is pulled back to level flight along with the throttle opening up.

15.5.9 Cuban-Eight in Vertical Plane

The Cuban-Eight in the vertical plane is another relatively difficult manoeuvre. It is a combination of two loops, both rolling out at the end of the loop. The control applications follow the combination of operations as described earlier (Figure 15.14).

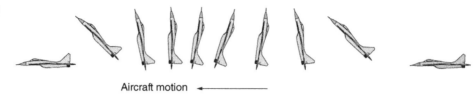

Aircraft motion ◄─────────────

15.5.10   Pugachev's Cobra Movement

This is a super manoeuvre undertaken by high-performance military aircraft with a high T/W ratio capable of high angle of attack flying. It first came to public attention in the 1989 Paris Air Show with a Sukhoi-27 twin jet combat aircraft flown by Soviet test pilot Viktor Pugachev. It was a breath-taking, dramatic spectacle never seen before by the public. It was an outcome of superior aerodynamic design of the aircraft with a vector-thrusting capability. Later, the MIG29 also performed similar manoeuvres. Initially Western experts debated its combat effectiveness. Subsequently, the aircraft capability, as demonstrated by the manoeuvre, proved to offer advantages in allowing a rapid change of attitude in the combat arena.

Figure 15.15 shows the sequence of aircraft positions in the vertical plane. At a prescribed airspeed the pilot initiates a rapid attitude change in the post-stall state well past 90°, and then the nose drops abruptly to recovery like a cobra strike. Superior aerodynamics with wing leading-edge vortex covering large portions of the wing, offering additional lift, a high T/W ratio and a powerful control authority contribute to such manoeuvres.

15.6   Combat Manoeuvre

A combat manoeuvre involves at least two aircraft, one friend (attacker in defensive role) and one foe (adversary), but engagement could be between two groups of aircraft. In this book only the basics are introduced to offer an overview of the complexity of performance, demanding extreme tasks to be carried out by the pilot in a life and death situation.

Combat technology is continually getting more complex and evolving for improvements. Adversary capability is kept secret; therefore an approach to engagement faces considerable unknown factors. With improved weapons and electronic support systems, the pilot workload has increased, demanding intensive training and practice.

One of the primary aims in aerial combat is for the attacker to get behind the adversary in a position, as fast as possible, to aim and fire. In the past, up to World War II, it was close range air combat manoeuvre (ACM – in loose terms, "*dog fights*"), with the attacker firing from behind in a straight line in short bursts of few minutes duration, to conserve bullets in short supply.

With advancement of missile systems, the zone to capture the target has widened, and the self-propelled missile with homing device allows launch of an attack from further behind, with enough gap to allow repositioning manoeuvres. Limitations in earlier missile capabilities made evasive action by the adversary easier, compared with the latest missile designs. Ground and airborne electronic support platforms allow beyond visual range (BVR) engagement, and proximity fuses in missiles make combat deadly.

15.6.1   Basic Fighter Manoeuvre

Basic fighter manoeuvre (BFM) is the tactical manoeuvre between attacker and adversary to gain an advantageous position to launch a weapon. Simultaneously, the adversary takes evasive actions to come out of the threat. The tussle to gain an advantageous position could be strenuous

■ **FIGURE 15.16**
Basic fighter manoeuvre (BFM) – terminologies

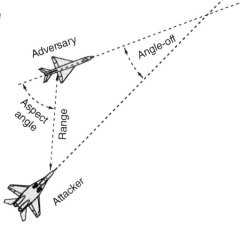

and lengthy. The most elementary BFM comprises a combination of aerobatic manoeuvres by both parties. It is a developed operational procedure with formal training methods. References [4, 5] covermodern air-combat fundamentals in some detail with formal terminology. Some of the standard manoeuvres are identified by their typical names.

The fundamental of any air combat is how fast one can attain the position to launch an attack first. Hard manoeuvres with extreme agility are the basis to attain superiority over the adversary. Modern combat aircraft design incorporating fly-by-wire (FBW) technology and vector thrust with a high T/W ratio to allow flying at high AOA and high angle of bank (AOB) can achieve the desired relative angle between the velocity vectors of attacker and adversary, called the *angle off* (AO). On the other hand, a low AOA is more suited to aim.

The following terminology concerning the positional advantage of the BFM is important to understand air combat (see Figure 15.16).

1. Angle off: This is the difference between the headings of the attacker and the defender.
2. Range: This is the distance between the two aircraft.
3. Aspect angle: This is the angle of the adversary in relation to the attacker aircraft.
4. Pursuit: This is the approach heading of the attacker to the adversary target. There are three fundamental positions: lead (approach aims ahead of the target), pure (approach aims at the target), and lag (approach aims behind the target). Lag pursuit has its advantages (see Figure 15.17).

The fundamental elements of air-combat tactics are surprise and manoeuvrability, supported by weapon types, teamwork and electronic support to detect the adversary within or beyond visual range. At all times the pilot has to stay alert to take evasive action or disengage when required. Pilots are given extensive training on positional geometry, attack geometry and weapons envelope.

BVR combat is the ultimate combat, and a successful engagement can result in an effective attack on multiple targets in rapid succession. It requires advanced electronic system integration with sensors (mainly radar) to acquire targets long before the adversary has warning. External platforms (e.g. AWACS) can offer detection over wider ranges and can assist the attacker to engage to intercept early. The radar cone can be swivelled in all directions to some extent to look up, down, right and left. Radar superiority is vital in BVR combat. Targets need to differentiate between friend or foe, which is done automatically.

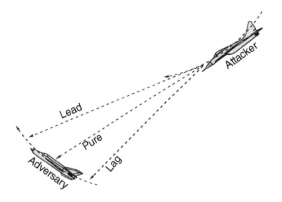

■ **FIGURE 15.18**
**Basic fighter
manoeuvre
(BFM) – an example
of out-of-plane
pursuit**

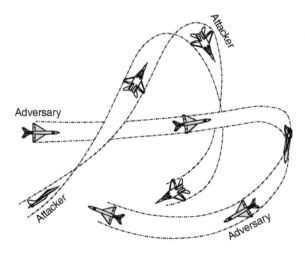

BFM as a tactical manoeuvre can be either offensive or defensive, and the techniques can be quite different. Offensive BFM can initiate from any direction: from the back or the front, from above or below, in-plane or out-of-plane pursuits with intricate manoeuvres. One example out of many possibilities is shown in Figure 15.18.

There are infinite possible combinations, and it is a whole subject of its own. Analytical methods are used to assess BFM techniques and to try them in practice before pilots are shown newer methods. Twin-dome simulation is an effective assessment method, but not many countries have twin-dome simulators.

15.7   Discussion on Turn

Transient though it may be, aircraft turning capability is the core of aircraft flying capability. Aircraft turn performance is control-dependent. Turning need not only be in a circular path. It can tighten inside or loosen outside the circular/helical trajectory in any plane at varying rates of turn. The computation for turn is a point performance evaluated at an instant. Integrated turn performance is carried out in small time steps for which the variables are treated constant. If the bank angle Φ is less than what is required for a coordinated turn, then the aircraft side-slips (skids) outward, and if it is higher then it will side-slip inward of the intended trajectory.

Energy manoeuvrability is part of aircraft performance. This gives an overall perspective of aircraft performance which is also dependent on aircraft design, its control and stability aspects mainly through superior aerodynamics and engine integration.

References

1. Lan, C.T.E., and Roskam, J., *Airplane Aerodynamic and Performance*. DARcorporation, Lawrence, 1981.

2. Ruijgork, G.J.J., *Elements of Airplane Performance*. Delft University Press, Delft, 1996.

3. Ojha, S.K., *Flight Performance of Airplane*. AIAA, Reston, VA, 1995.

4. Spick, M., *An Illustrated Guide to Modern Fighter Combat*. Prentice Hall Trade, 1987.

5. Lynch, S. (1999) *The Falcon 3.0 Manual Tactics Section – Introduction to the Geometry of Air Combat*, http://www.simhq.com/_air/air_038a.html (accessed 17 November 2015).

CHAPTER 16

Aircraft Sizing and Engine Matching

16.1 Overview

Aircraft sizing and engine matching is a necessary procedure for a new aircraft project, starting early at the conceptual design stage (Phase 1, see Section 1.7). This is to freeze the aircraft configuration to obtain the project go-ahead decision. The exercise is to ensure the best configuration to satisfy customer specifications and mandatory airworthiness requirements. The procedure does not require full aircraft performance analyses. Only the airworthiness requirements and customer specification requirements are checked out in a parametric search, as will be shown in this chapter. The aircraft sizing and engine matching exercise does not guarantee overall aircraft performance in detail, but makes certain that the new project is capable of meeting the necessary specifications and requirements. The exercise makes use of the aircraft performance analyses, and therefore has to wait until all the related equations are derived as in Chapters 10–15. If required, a family of derivative aircraft can be offered to cover a wider market at a lower development cost by retaining component commonality.

It is assumed here that preliminary aircraft geometry and its component weights are available for the analyses. Engine performance details are given in Chapter 8. Methods for aircraft drag estimation are given Chapter 9. From these building blocks, finally, the aircraft size can be fine-tuned to a satisfactory configuration offering a family of variant designs. None may be the optimum, but together they offer the best fit to satisfy many customers (i.e. operators) and to encompass a wide range of payload-range requirements, resulting in increased sales and profitability. Typically, the variant designs (at least three in the family) are offered with the baseline aircraft as the middle one, plus a bigger version and a smaller version retaining maximum component commonality within all the three designs.

This chapter proposes a methodology to conceive final configuration from a *"first-cut"* (i.e. preliminary) aircraft configuration derived primarily from statistical information, except for the fuselage (transport aircraft), which is deterministic. A designer's past experience is vital in producing the preliminary configuration; it has to be practised [1–3]. Initial weights are guessed from statistical data. Through an iterative process it is fine-tuned to an accurate final aircraft configuration. Typically, a good starting configuration can be converged in one iteration.

The two classic important parameters – wing-loading (W/S_w) and thrust-loading (T_{SLS}/W) are instrumental in the methodology for aircraft sizing and engine matching. This chapter

Theory and Practice of Aircraft Performance, First Edition. Ajoy Kumar Kundu, Mark A. Price and David Riordan.
© 2016 John Wiley & Sons, Ltd. Published 2016 by John Wiley & Sons, Ltd.

presents a formal methodology to obtain the sized W/S_w and T_{SLS}/W for a baseline aircraft. These two loadings alone provide sufficient information to conceive an aircraft configuration in a preferred aircraft size with a family of variant designs. Empennage size is governed by wing size and location of the centre of gravity (CG). This study is possibly the most important aspect in the development of an aircraft – finalizing the external geometry.

Because the preliminary configuration is based on past experience and statistics, an iterative procedure ensues to fine-tune the aircraft, converging on the correct size of the wing reference area for a family of variant aircraft designs and matched engines. Wallace [1] provides an excellent presentation on the subject.

This chapter is an important one. Readers are to compute the parameters that establish the criteria for aircraft sizing and engine matching. The final size is unlikely to be identical to the preliminary configuration; the use of spreadsheets facilitates the iterations. A problem set for this chapter is given in Appendix E.

16.2   Introduction

In a systematic manner, the conception of a new aircraft progresses from generating market specifications, followed by the preliminary candidate configurations that rely on statistical data of past designs in order to arrive at a baseline design, selected from several candidate configurations. In this chapter, the baseline design of an aircraft is formally sized with a matched engine (or engines) along with the family of variants to finalize the configuration (i.e. external geometry and takeoff gross weight). An example from each class of civil (i.e. the Bizjet) and military (i.e. the AJT) aircraft are used to substantiate the methodology.

As of year 2000 fuel prices, the aircraft cost contributes to the direct operating cost (DOC) three to four times the contribution made by the fuel cost. It is not cost-effective for aircraft manufacturers to offer custom-made new designs to each operator with varying payload-range requirements. Therefore, aircraft manufacturers offer aircraft in a family of variant designs. This approach maintains maximum component commonality within the family to reduce development costs, and is reflected in aircraft unit-cost savings. In turn, it eases the amortization of non-recurring development costs, as sales increase. It is therefore important for the aircraft sizing exercise to ensure that the variant designs are least penalized to maintain commonality of components. This is what is meant by producing satisfying, robust designs; these are not necessarily the optimum designs. Sophisticated multidisciplinary optimization is not easily amenable to ready use. The industries also use parametric searches to produce satisfying, robust designs.

To generate a family of variant civil aircraft designs, the tendency is to retain the wing and empennage almost unchanged while plugging and unplugging the fuselage to cope with varying payload capacities (see Section 16.5.1). Typically, the baseline aircraft remains as the middle design. The smaller aircraft results in a wing that is larger than necessary, providing better field performance (i.e. takeoff and landing); however, the cruise performance is slightly penalized. Conversely, larger aircraft have smaller wings that improve the cruise performance; the shortfall in takeoff is overcome by providing a higher thrust-to-weight ratio (T_{SLS}/W) and possibly with better high-lift devices, both of which incur additional costs. The baseline aircraft approach speed, V_{app}, initially is kept low enough so that the growth of V_{app} for the larger aircraft is kept within the specifications. Of late, high investment into advanced composite wing-manufacturing methods has put us in a better position to produce adjusted wing sizes for each variant (large aircraft) in a cost-effective manner, offering improved economics in the long run. However, for some time to come, metal wing construction is to continue with minimum change in wing size to maximize component commonality.

Matched engines are also in a family to meet the variation of thrust (or power) requirements for the aircraft variants. The sized engines are bought-out items supplied by engine manufacturers. Aircraft designers stay in constant communication with engine designers in order to arrive at the family of engines required. A thrust variation of up to ±30% from the baseline engine is typically sufficient for larger and smaller aircraft variants from the baseline. Engine designers can produce scalable variants from a proven core gas-generator module of the engine – these scalable projected engines are known loosely as *rubberized* engines. The thrust variation of a rubberized engine does not significantly affect the external dimensions of an engine (typically, the bare engine length and diameter change by only around ±2%). This book uses an unchanged nacelle external dimension for the family variants, although there is some difference in weight for the different engine thrusts. The generic methodology presented in this chapter is the basis for the sizing and matching practice.

16.3　Theory

The parameters required for aircraft sizing and engine matching derive from market studies and must satisfy user specifications and the certification agencies' requirements. In general, both civil and military aircraft use similar specifications, as given below, as the basic input for aircraft sizing. All performance analyses in this chapter are for an ISA day and all field performances are at sea-level.

1. Payload and range (fuel load): these determine the maximum takeoff weight (MTOW).
2. Takeoff field length (TOFL): this determines the engine power ratings and wing size.
3. Landing field length (LFL): this determines wing size (baulked landing included).
4. Initial maximum cruise speed and altitude capabilities determine wing and engine sizes.
5. Initial rate of climb establishes wing and engine sizes.

These five requirements must be satisfied simultaneously. The governing parameters to satisfy TOFL, initial climb, initial cruise and landing are wing-loading (W/S_w) and thrust-loading (T_{SLS}/W).

Additional specifications for military aircraft sizing are as follows:

6. Turn performance g-load (in horizontal plane).
7. Manoeuvre g-load (turning in any plane).
8. Roll rate (control sizing issue – not dealt with here).

Normally, thrust sizing for initial climb rates proves sufficient for satisfying the turn (g-load) requirements. It is assumed there is control authority available to execute the manoeuvres. A lower wing aspect ratio is considered for higher roll rates to reduce the wing-root bending moments.

As mentioned previously, an aircraft must simultaneously satisfy the TOFL, initial climb rate, initial maximum cruise speed-altitude capabilities and LFL requirements. Low wing-loading (i.e. a larger wing area) is required to sustain low speeds at lift-off and touchdown (for a pilot's ease), whereas high wing-loading (i.e. a low wing area) is suitable at cruise because high speeds generate the required lift on a smaller wing area. A larger wing area necessary for takeoff/landing results in excess wing for high-speed cruise. This may require suitable high-lift devices to keep the wing area smaller. The wing area is sized in conjunction with a matched engine for takeoff, climb, cruise and landing; landing is performed at the idle-engine rating. To obtain the minimum wing area matched with sized engine to satisfy all requirements is the core of the aircraft sizing and engine matching exercise.

In general, W/S varies with time as fuel is consumed, and T/W is throttle-dependent. Therefore, the reference design condition of the MTOW and T_{SLS} at sea-level ISA day is used for sizing considerations. This means that the MTOW, T_{SLS} and S_W are the only parameters considered for aircraft sizing and engine matching.

In general, wing-size variations are associated with changes in all other affecting parameters (e.g. AR, λ and wing sweep). However, at this stage, they are kept invariant – that is, the variation in wing size scales the wing span and chord, leaving the general planform unaffected (like zoom in/zoom out).

At this point, readers require knowledge of aircraft performance, and the important derivations of the equations used are provided in Chapters 10–15. Other proven semi-empirical relations are given in [2]. Although the methodology described herein is the same, industry practice is more detailed and involved, in order to maintain a high degree of accuracy.

Worked examples continue with Bizjet (Learjet 45 class) for civil aircraft and the AJT (BAe Hawk class) for military aircraft. Throughout this chapter, wing loading (W/S_W) in the SI system is in N/m² (or kg/m²) to align with the thrust (in N) in thrust loading (T_{SLS}/W) as a non-dimensional parameter. FPS units are also given.

16.3.1   Sizing for Takeoff Field Length – Two Engines

To keep analyses simple, only a two-engine aircraft is sized and engine matched. This is because sizing and engine matching requires aircraft drag polar, and in this book only the two-engine Bizjet drag polar is worked out. However, three- and four-engine aircraft use exactly the same procedure, with equations modified by the number of engines in question.

The TOFL is the field length required to clear a 35-ft (10-m) obstacle while maintaining a specified minimum climb gradient, γ, with one engine inoperative and flaps and undercarriage extended (Figure 16.1). The FAR requirements for a two-engine aircraft minimum climb gradient are given in Section 11.6.1.

For sizing, field-length calculations are at the sea-level ISA day (no wind) and a zero airfield gradient of paved runway. For further simplification, drag changes are ignored during the transition phase of lift-off to clear the obstacle (flaring takes about 2–4 seconds); in other words, the equations applied to $V_{lift\text{-}off}$ are extended to V_2.

■ **FIGURE 16.1**
Sizing for takeoff

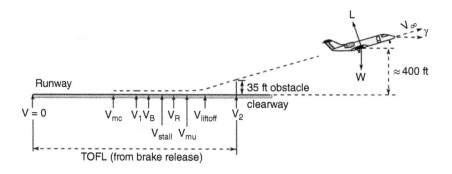

Chapter 11 addresses takeoff performance in detail. Section 11.4.1 gives

$$\text{TOFL} = \int dS_G = \int Vdt = \int V\frac{dV}{dV}dt = \int_0^{V_2}\frac{V}{a}dV = \sum_{\Delta V_i}^{\Delta V_f}\Delta S_G \quad \text{where } \frac{dV}{dt} = a \qquad \textbf{(16.1)}$$

where $dV/dt = a$ is the instantaneous acceleration and V is the instantaneous velocity. The aircraft on the ground encounters rolling friction (coefficient $\mu = 0.025$ for a paved, metalled runway). Average acceleration, $\bar{a}$, is taken at $0.7V_2$. FAR requires $V_2 = 1.2V_{stall}$. This gives $V_2^2 = [2 \times 1.44 \times (W/S_W)]/(\rho C_{Lstall})$. An aircraft stalls at C_{Lmax}. By replacing V_2, Equation 16.1 reduces to:

$$\text{TOFL} = (1/\bar{a})\int_0^{V_2}VdV = \frac{V_2^2}{2\bar{a}} = \frac{1.44W/S}{\rho C_{Lstall}\bar{a}} \qquad \textbf{(16.2)}$$

Writing in terms of wing loading, Equation 16.2 can be written as

$$(W/S_W) = (\text{TOFL} \times \rho \times \bar{a} \times C_{Lstall})/1.44 \qquad \textbf{(16.3)}$$

where average acceleration, $\bar{a} = F/m$ and applied force $F = (T-D) - \mu(W-L)$.

Note that until lift-off is achieved, $W > L$ and F is the average value at $0.7V_2$. Therefore,

$$\bar{a} = \frac{[(T-D)-\mu(W-L)]\text{g}}{W} = \text{g}\left(\frac{T}{W}\right)[1 - D/T - \mu W/T + \mu L/T]_{@0.7V_2} \qquad \textbf{(16.4)}$$

Substituting into Equation 16.3 it becomes

$$(W/S_W) = \left\{\text{TOFL} \times \rho \times \left[\text{g}\left(\frac{T}{W}\right)(1 - D/T - \mu W/T + \mu L/T)\right] \times C_{Lstall}\right\}\Big/1.44 \qquad \textbf{(16.5)}$$

In the FPS system ($\rho = 0.00238$ slugs and $g = 32.2\text{ft}/s^2$) it can be written as:

$$(W/S_W) = \left\{\text{TOFL} \times \left(\frac{T}{W}\right)(1 - D/T - \mu W/T + \mu L/T) \times C_{Lstall}\right\}\Big/18.85 \qquad \textbf{(16.6a)}$$

In the SI system it becomes ($\rho = 1.225\text{kg}/m^3$ and $g = 9.81\text{m}/s^2$):

$$(W/S_W) = 8.345 \times \left\{\text{TOFL} \times \left(\frac{T}{W}\right)(1 - D/T - \mu W/T + \mu L/T) \times C_{Lstall}\right\} \qquad \textbf{(16.6b)}$$

Here W/S_W is in N/m² to remain in alignment with the units of thrust in N.

Checking of the second-segment climb gradient occurs after aircraft drag estimation, which is explained in Chapter 12. If the climb gradient falls short of the requirement, then T_{SLS} must be increased. In general, the stringent TOFL requirements are likely to satisfy the second-segment climb gradient requirement.

16.3.1.1   *Civil Aircraft Design: Takeoff*

The contribution of the last three terms $(-D/T - \mu W/T + \mu L/T)$ in Equation 16.4 is minimal and can be omitted at this stage for the sizing calculation. In addition, for the one engine inoperative condition after the decision speed V_1, the acceleration slows, making the TOFL longer than in the case of all engines operative. Therefore, in the sizing computations to produce the specified TOFL, further simplification is possible by applying a semi-empirical correction factor primarily to compensate for loss of an engine. The correction factors are as follows [2]; all sizing calculations are performed at the MTOW and with T_{SLS}.

For two engines, use a factor of 0.5 (loss of thrust by a half). Then, Equation 16.6a in the FPS system reduces to:

$$\left(W \,/\, S_W\right) = \left\{\text{TOFL} \times \left(\frac{T}{W}\right) \times C_{Lstall}\right\} \Big/ 37.5 \qquad \textbf{(16.7a)}$$

For the SI system:

$$\left(W \,/\, S_W\right) = 4.173 \times \text{TOFL} \times \left(\frac{T}{W}\right) \times C_{Lstall} \qquad \textbf{(16.7b)}$$

For three engines, use a factor of 0.66 (loss of thrust by a third). Then, Equation 16.6a in the FPS system reduces to:

$$\left(W \,/\, S_W\right) = \left\{\text{TOFL} \times \left(\frac{T}{W}\right) \times C_{Lstall}\right\} \Big/ 28.5 \qquad \textbf{(16.8a)}$$

For the SI system:

$$\left(W \,/\, S_W\right) = 5.5 \times \text{TOFL} \times \left(\frac{T}{W}\right) \times C_{Lstall} \qquad \textbf{(16.8b)}$$

For four engines, use a factor of 0.75 (loss of thrust by a quarter). Then, Equation 16.6a in the FPS system reduces to:

$$\left(W \,/\, S_W\right) = \left\{\text{TOFL} \times \left(\frac{T}{W}\right) \times C_{Lstall}\right\} \Big/ 25.1 \qquad \textbf{(16.9a)}$$

For the SI system:

$$\left(W \,/\, S_W\right) = 6.25 \times \text{TOFL} \times \left(\frac{T}{W}\right) \times C_{Lstall} \qquad \textbf{(16.9b)}$$

16.3.1.2   *Military Aircraft Design: Takeoff*   Because military aircraft mostly have a single engine, there is no requirement for one engine being inoperative; ejection is the best solution if the aircraft cannot be landed safely. Therefore, Equation 16.6 can be directly applied (for a multi-engine design, the one-engine-inoperative case generally uses measures similar to the civil aircraft case). In the FPS system, this can be written as:

$$\left(W \,/\, S_W\right) = \left\{\text{TOFL} \times \left(\frac{T}{W}\right)\left(1 - D/T - \mu W / T + \mu L / T\right) \times C_{Lstall}\right\} \Big/ 18.85 \qquad \textbf{(16.10a)}$$

In the SI system, it becomes:

$$\left(W \,/\, S_W\right) = 8.345 \times \left\{\text{TOFL} \times \left(\frac{T}{W}\right)\left(1 - D/T - \mu W / T + \mu L / T\right) \times C_{Lstall}\right\} \qquad \textbf{(16.10b)}$$

Military aircraft have a thrust, T_{SLS}/W, that is substantially higher than civil aircraft, which makes $(D/T - \mu W / T + \mu L / T)$ even smaller. Therefore, for a single-engine aircraft, no correction is needed and the simplified equations are as follows:

In FPS units
$$(W/S_W) = \left\{ \text{TOFL} \times \left(\frac{T}{W}\right) \times C_{Lstall} \right\} \Big/ 18.85 \qquad \textbf{(16.11a)}$$

In SI units
$$(W/S_W) = 8.345 \times \left\{ \text{TOFL} \times \left(\frac{T}{W}\right) \times C_{Lstall} \right\} \qquad \textbf{(16.11b)}$$

16.3.2 Sizing for the Initial Rate of Climb (All Engines Operating)

The initial unaccelerated rate of climb is a user specification and not a FAR requirement; this is when the aircraft is heaviest in climb. In general, the FAR requirement for the one-engine-inoperative gradient provides sufficient margin to give a satisfactory all-engine initial climb rate. However, from the operational perspective, higher rates of climb are in demand when it is sized accordingly. Military aircraft (some with a single engine) requirements stipulate faster climb rates, and sizing for the initial climb rate is important. The methodology for aircraft sized to the initial climb rate is described in this section. The sizing exercise for climb is at an unaccelerated rate of climb. Figure 16.2 shows a typical climb trajectory.

For a steady-state climb, Equation 12.5a gives the expression for rate of climb, $RC = V \sin\gamma$. Steady-state force equilibrium gives $T = D + W \times \sin\gamma$, or $\sin\gamma = (T-D)/W$. This gives

$$RC = \frac{(T-D) \times V}{W} = \left(\frac{T}{W} - \frac{D}{W}\right) \times V \qquad \textbf{(16.12)}$$

Equation 16.12 is written as $\dfrac{T}{W} = \dfrac{RC}{V} + \dfrac{D}{W}$

or
$$\frac{T}{W} = \frac{RC}{V} + \frac{(C_D \times 0.5 \times \rho \times V^2 \times S_W)}{W} \qquad \textbf{(16.13)}$$

Equation 16.13 is based on climb thrust rating which is lower than T_{SLS}. Equation 16.13 needs to be written in terms of T_{SLS}. The ratio, $T_{SLS}/T = k_{cl}$, varies depending on the engine bypass ratio.

$$\frac{[T_{SLS}/W]}{k_{cl}} = \frac{RC}{V} + \frac{(C_D \times 0.5 \times \rho \times V^2 \times S_W)}{W} \qquad \textbf{(16.14)}$$

$$[T_{SLS}/W] = \left(k_{cl} \times \frac{RC}{V}\right) + \left(k_{cl} \times \frac{(C_D \times 0.5 \times \rho \times V^2 \times S_W)}{W}\right) \qquad \textbf{(16.15)}$$

■ **FIGURE 16.2**
Aircraft climb trajectory

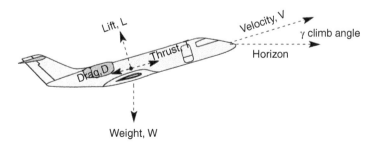

16.3.3   Sizing to Meet Initial Cruise

There are no regulations in force to meet the initial cruise speed; initial cruise capability is a user requirement. Therefore, both civil and military aircraft sizing for initial cruise use the same equations. At a steady-state level flight, thrust required (aeroplane drag, D) = thrust available (T_a), that is:

$$D = T_a = 0.5\rho V^2 C_D \times S_W \quad \text{where } T_a \text{ is thrust available} \quad (16.16)$$

Dividing both sides of the equation by the initial cruise weight, $W_{in\_cr} = k \times \text{MTOW}$ due to fuel burned to climb to the initial cruise altitude. The factor k lies between 0.95 and 0.98 (climb performance details are not available during the conceptual design phase), depending on the operating altitude for the class of aircraft, and it can be fine-tuned through iteration – in the coursework exercise, one round of iterations is sufficient. The factor cancels out in the following equation, but is required later. Henceforth, in this part of cruise sizing, W represents the MTOW, in line with takeoff sizing:

$$\frac{0.5\rho V^2 C_D \times S_W}{W} = \frac{T_a}{W} \quad (16.17)$$

The drag polar is now required to compute the relationship given in Equation 16.17. Use the C_D value to correspond to the initial cruise C_L (because they are non-dimensional, both the FPS and SI systems provide the same values). Initial cruise:

$$C_L = \frac{k \times \text{MTOW}}{0.5 \times \rho \times V^2 \times S_W} \quad (16.18)$$

The thrust-to-weight ratio sizing for initial cruise capability is expressed in terms of T_{SLS}. Equation 16.18 is based on the maximum cruise thrust rating, which is lower than the T_{SLS}. Equation 16.18 must be written in terms of T_{SLS}. The ratio, $T_{SLS} / T_a = k_{cr}$, varies depending on the engine BPR. The factor k_{cr} is computed from the engine data supplied. Then, Equation 16.18 can be rewritten as:

$$\frac{T_{SLS}}{W} = \frac{k_{cr} \times 0.5\rho V^2 \times C_D}{(W / S_W)} \quad (16.19)$$

Variation in wing size affects aircraft weight and drag. The question now is: how does the C_D change with changes in W and S_W? (T_a changes do not affect the drag because it is assumed that the physical size of an engine is not affected by small changes in thrust.) The solution method is to work with the wing only – first by scaling the wing for each case, and then by estimating the changes in weight and drag and iterating, which is an involved process.

This book simplifies the method by using the same drag polar for all wing loadings (W/S_W) with little loss of accuracy. As the wing size is scaled up or down (the AR invariant), it changes the parasite drag. The induced drag changes as the aircraft weight increases or decreases. However, the C_D increase with wing growth is divided by a larger wing, which keeps the C_D change minimal.

16.3.4   Sizing for Landing Distance

The most critical case is when an aircraft must land at its maximum landing weight of 0.98 MTOW. In an emergency, an aircraft lands at the same airport for an aborted takeoff procedure, assuming a 2% weight loss due to fuel burn in order to make the return circuit. Pilots prefer to approach as slowly as possible for ease of handling at landing.

For the Bizjet class of aircraft, the approach velocity, V_{app} (FAR requirement at $1.3\,V_{stall}$) is less than 125 kt to ensure that it is not constrained by the minimum control speed, V_c. Wing C_{Lstall} is at the landing flap and slat setting. For sizing purposes, an engine is set to the idle rating to produce zero thrust.

At approach:

$$\sqrt{V_{stall}} = \frac{(0.98W/S_W)}{0.5 \times \rho \times C_{Lstall}} \tag{16.20}$$

At landing, $V_{app} = 1.3V_{stall}$. Therefore,

$$V_{app} = 1.3 \times \sqrt{\left[\frac{(0.98W/S_W)}{0.5 \times \rho \times C_{Lstall}}\right]} = 1.793 \times \sqrt{\left[\frac{(W/S_W)}{\rho \times C_{Lstall}}\right]} \tag{16.21}$$

or

$$V_{app}{}^2 \times C_{Lstall} = 3.211 \times \frac{(W/S_W)}{\rho}$$

or

$$W/S_W = 0.311 \times \rho \times V_{app}{}^2 \times C_{Lstall} \tag{16.22}$$

16.4 Coursework Exercises: Civil Aircraft Design (Bizjet)

Both FPS and SI units are used in the examples. Sizing calculations require the generic engine data in order to obtain the k factors used. The Bizjet drag polar is provided in Figure 9.3.

16.4.1 Takeoff

The requirements are for TOFL of 4400 ft (1341 m) to clear a 35 ft height on an ISA day at sea-level. The maximum lift coefficient at takeoff (i.e. flaps down, to 20° and no slat) is $C_{Lstall(TO)} = 1.9$ (obtained from testing and CFD analysis). The result is computed in Table 16.1. Using Equation 16.7a, the expression reduces to:

$$W/S_W = 4400 \times 1.9 \times \left(\frac{T}{W}\right) \Big/ 37.5 = 222.9 \times \left(\frac{T}{W}\right)$$

■ **TABLE 16.1**
Bizjet takeoff sizing

W/S_W (FPS) (lb/ft²)	40	50	60	70	80
W/S_W (SI) (N/m²)	1916.9	2395.6	2874.3	3353.7	3832.77
T/W (non-dimensional)	0.18	0.225	0.27	0.315	0.36

■ **TABLE 16.2**
Bizjet climb sizing

W/S_W(lb/ft²)	40	50	60	70	80
W/S_W(N/m²)	1916.9	2395.6	2874.3	3353.7	3832.77
C_{Lclimb}	0.194	0.242	0.290	0.339	0.378
C_D (from drag polar)	0.024	0.0246	0.0256	0.0266	0.0282
T_{SLS}/W	0.34	0.31	0.286	0.272	0.265

Using Equation 16.7b, it becomes

$$W / S_W = 4.173 \times 1341 \times 1.9 \times \left(\frac{T}{W}\right) = 10{,}633.55 \times \left(\frac{T}{W}\right)$$

The industry must also examine other takeoff requirements (e.g. an unprepared runway) and hot and high ambient conditions.

16.4.2   Initial Climb

From the market requirements, an initial climb starts at 1400 ft altitude ($\rho = 0.00232$ slugs / ft³) at a speed $V_{EAS} = 250$ kt (Mach 0.38) $= 250 \times 1.68781 = 422$ ft/s (128.6 m/s) and the required rate of climb, $RC = 2600$ ft/min (792.5 m/min) $= 43.33$ ft/s (13.2 m/s). From the TFE731 class engine data, $T/T_{SLS} = 0.67 = k_{cl}$, that is, $T_{SLS}/T = 1.5$ (Figure 8.12). The undercarriage and high-lift devices are in a retracted position. The result is computed in Table 16.2.

Lift coefficient, $C_{Lclimb} = \dfrac{W}{\left(0.5 \times 0.00232 \times 422^2 S_W\right)} = 0.00484 \times \left(W / S_W\right)$

Using Equation 16.14,

$$\frac{\left[T_{SLS} / W\right]}{1.5} = \frac{43.33}{422} + \frac{\left(C_D \times 0.5 \times 0.00232 \times 422^2 \times S_W\right)}{W}$$

$$T_{SLS} / W = 0.154 + \left[310 \times C_D \times \left(S_W / W\right)\right]$$

16.4.3   Cruise

The requirements are that the initial cruise speed must meet high-speed cruise at Mach 0.74 and at 41,000 ft (flying higher than bigger jets in less congested traffic corridors). Using Equation 16.19, the result is computed in Table 16.3. (For the initial cruise aircraft weight, use $k = 0.972$ in.)
    In FPS at 41,000 ft:

$$\rho = 0.00056 \text{ slug / ft}^2 \text{ and } V^2 = \left(0.74 \times 968.076\right)^2 = 513{,}195 \text{ ft}^2 / \text{s}^2$$

In SI at 12,192 m:

$$\rho = 0.289 \text{ kg / m}^3, \ V^2 = \left(0.74 \times 295.07\right)^2 = 47{,}677.5 \text{ m}^2 / \text{s}^2$$

■ **TABLE 16.3**
Bizjet cruise sizing

W/S_W(lb/ft^2)	40	50	60	70	80
W/S_W(N/m^2)	1916.9	2395.6	2874.3	3353.7	3832.77
C_L	0.339	0.4064	0.474	0.542	0.619
C_D (from drag polar)	0.0255	0.0269	0.0295	0.033	0.0368
T_{SLS}/W at 41,000 ft	0.36	0.305	0.278	0.267	0.26

Equation 16.18 gives the initial cruise:

$$C_L = \frac{0.972 \ \text{MTOW}}{0.5 \times 0.289 \times 47 \ 677.5 \times S_W}$$
$$= 0.0001414 (W / S_W), \quad \text{where } W / S_W \text{ is in N / m}^2$$

Equation 16.19 gives:

$$\frac{T_{SLS}}{W} = \frac{k_{cr} \times 0.5 \rho V^2 \times C_D}{(W / S_W)}$$

(From Figure 8.13, the TFE731 class engine data has $T_{SLS} / T = k_{cr} = 4.5$.)

In FPS units: $\quad \dfrac{T_{SLS}}{W} = \dfrac{4.5 \times 0.5 \times 0.00056 \times 459,208.2 \times C_D}{(W / S_W)} = 565.73 \times \dfrac{C_D}{(W / S_W)}$

In SI units: $\quad \dfrac{T_{SLS}}{W} = \dfrac{4.5 \times 0.5 \times 0.289 \times 42,662.5 \times C_D}{(W / S_W)} = \dfrac{27,741.3 \times C_D}{(W / S_W)}$

16.4.4 Landing

From the market requirements, $V_{app} = 120 \text{kt} = 202.5 \ \text{ft/s}$ (61.72 m/s). Landing $C_{Lmax} = 2.1$ at a 40° flap setting (from testing and CFD analysis). For sizing purposes, the engine is set to the idle rating, producing zero thrust. Using Equation 16.22, the following is obtained.

In the FPS system, $W / S_W = 0.311 \times 0.002378 \times 2.1 \times (202.5)^2 = 63.8 \text{lb/ft}^2$. In the SI system, $W / S_W = 0.311 \times 1.225 \times 2.1 \times (61.72)^2 = 3052 \text{N/m}^2$. Because the thrust is zero (i.e. idle rating) at landing, W/S_W remains constant.

16.5 Sizing Analysis: Civil Aircraft (Bizjet)

The four sizing relationships for wing loading, W/S_W and thrust loading, $T_{SLSinst}/W$, are for (i) takeoff, (ii) approach speed for landing, (iii) initial cruise speed, and (iv) initial climb rate. These are plotted in Figure 16.3. The circled point in Figure 16.3 is the most suitable for satisfying all four requirements simultaneously. To ensure performance, there is a tendency to use a slightly higher thrust loading $T_{SLS-installed}/W$; in this case, the choice becomes $T_{SLSinst} / W = 0.32$ at a wing loading of $W/S_W = 63.75 \text{lb/ft}^2 (2885 \ \text{N/m}^2)$.

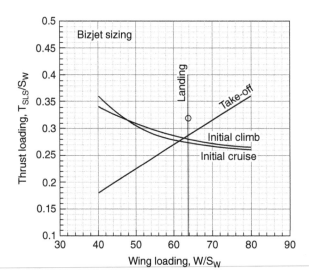

So far, the Bizjet aircraft data have been given. This chapter sizes the Bizjet to check the extent of differences in aircraft geometry, engine size and weights. Now is the time for the iterations required to fine-tune the given preliminary configuration. At 20,720 lb (9400 kg) MTOM, the wing planform area is 325 ft², close to the original area of 323 ft²; hence, no iteration is required. Otherwise, it is necessary to revisit the empennage sizing and revise the weight estimates. The $T_{SLS-installed}$ per engine then becomes $0.32 \times 20,720 / 2 = 3315$ lb. At a 7% installation loss at takeoff, this gives uninstalled $T_{SLS} = 3315 / 0.93 = 3560$ lb / engine ($T_{SLS} / W = 3560 \times 2 / 20,720 = 0.344$). This is very close to the TFE731-20 class of engine; therefore, the engine size and weight remain the same. For this reason, iteration is avoided; otherwise, it must be carried out to fine-tune the mass estimation.

The entire sizing exercise could have been conducted well in advance, even before a configuration was settled – if the chief designer's past experience could "guesstimate" a close drag polar and mass. Statistical data of past designs are useful in estimating aircraft close to an existing design. Generating a drag polar requires some experience with extraction from statistical data. Subsequently, the sizing exercise is fine-tuned to better accuracy.

In industry, more considerations are addressed at this stage – for example what type of variant design in the basic size can satisfy at least one larger and one smaller capacity (i.e. pay-load) size. Each design may have to be further varied for more refined variant designs.

16.5.1   Variants in the Family of Aircraft Design

The family concept of aircraft design is discussed in previous chapters and highlighted again at the beginning of this chapter. Maintaining high component commonality in a family is a definite way to reduce design and manufacturing costs – in other words, "design one and get two or more almost free". This encompasses a much larger market area and, hence, increased sales to generate resources for the manufacturer and nation. The amortization is distributed over larger numbers, thereby reducing aircraft costs.

Today, all manufacturers produce a family of derivative variants. The Airbus 320 series has four variants and by the end of March 2016 about 6,974 have been built (excludes A320neo family). The Boeing 737 family has six variants, offered for nearly four decades, and by the end of March 2016 about 8,966 have been built. It is obvious that in three decades, aircraft manufacturers have continuously updated later designs with newer technologies. The latest version of

■ **FIGURE 16.4**
Variant designs in a family of civil aircraft

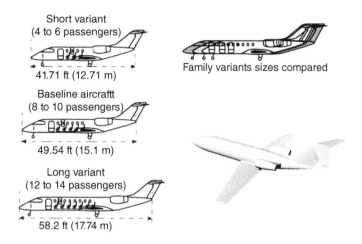

Short variant
(4 to 6 passengers)

41.71 ft (12.71 m)

Baseline aircraftt
(8 to 10 passengers)

49.54 ft (15.1 m)

Long variant
(12 to 14 passengers)

58.2 ft (17.74 m)

Family variants sizes compared

■ **TABLE 16.4**
Bizjet family variant summary ($S_W = 323\,\text{ft}^2$)

	Short variant	Baseline	Long variant
MTOM, kg (lb)	7800 (17,190)	9400 (20,680)	10,900 (23,580)
W/S_W(lb/ft^2)	53.22	64.0	73.0[a]
T_{SLS}/W	(0.32)	0.32	(0.35)
T_{SLS_inst}/per engine (lb)	2750	3315	4130
T_{SLS_uninst}/per engine (lb)	2950	3560	4435

[a]The long variant requires superior high-lift devices for landing.

the Boeing 737-900 has vastly improved technology compared with the late 1960s 737-100 model. The latest design has a different wing; the resources generated by large sales volumes encouraged investing in upgrades. In this case, a significant investment was made in a new wing, advanced cockpit/systems and better avionics, which has resulted in continuing strong sales in a fiercely competitive market.

The variant concept is market- and role-driven, keeping pace with technology advancements. Of course, derivatives in the family are not the optimum size (more so in civil aircraft design), but they are a satisfactory size that meets the need. The unit-cost reduction, as a result of component commonalities, must compromise with the non-optimum situation of a slight increase in fuel burn. Readers are referred to Figure 16.4, which highlights the aircraft unit-cost contribution to DOC as more than three to four times the cost of fuel, depending on payload-range capability. The worked examples in the next section offer an idea of three variants in a family of aircraft.

16.5.2   Example: Civil Aircraft

Figure 16.4 shows the final configuration of a family of variants; the baseline aircraft is in the middle. It proposes one smaller (i.e. 4–6 passengers) and one larger (i.e. 12–14 passengers) variant from the baseline design that carries 8–10 passengers by subtracting and adding fuselage plugs from the front and aft of the wing box. Table 16.4 gives a summary of the Bizjet variants.

The short variant engine is derated by about 17%, resulting in slightly lower engine weight, and the long variant engine is uprated by about 15%, resulting in slightly higher engine weight.

16.6 Classroom Exercise – Military Aircraft (AJT)

Both FPS and SI units are used in these examples. Figure 9.16 gives the AJT drag polar. The military aircraft example of the AJT operates at two takeoff weights: (i) normal training configuration (NTC– clean) at 4800 kg; and (ii) fully loaded for armament training at 6800 kg, that is, an increase of 41.7%. In this example the NTC is more critical to meet the specification of TOFL = 800m. The readers may work out both cases. The fully loaded aircraft needs to satisfy the longer field length requirement of 1800 m (< 6000ft), while the climb and cruise capabilities are taken as fallout of the design. After the armament practice run the payload is dropped and the landing weight is the same for both missions. It may be noted that the AJT is to have a close air support (CAS) variant. Only initial MTOW and wing area require sizing.

16.6.1 Takeoff

The requirements are that TOFL = 800 m (≈ 2600 ft) to clear 35 ft (10.7m) on an ISA day at sea-level and NTC. The maximum lift coefficient at TO (20° flaps down and no slat) is taken as $C_{Lmax\_TO} = C_{Lstall\_TO} = 1.85$. Military designs follows Milspecs and not FAR for airworthiness requirements. The result is computed in Table 16.5.

Using Equation 16.11a, the expression becomes

$$(W / S_W) = 2600 \times 1.85 \times \left(\frac{T}{W}\right) \Big/ 18.85 = 255.2 \times \left(\frac{T}{W}\right)$$

Using Equation 16.11b, it becomes

$$W / S_W = 8.345 \times 800 \times 1.85 \times \left(\frac{T}{W}\right) = 12350.6 \times \left(\frac{T}{W}\right)$$

16.6.2 Initial Climb

From market requirements, initial climb speed $V = 350$ kt = 590.7 ft /s, and the required rate of climb, $RC = 10,000$ ft / min = 164 ft /s (50m/s). From the Adour 861 class engine data T_{SLS} / Tratio, $k_{cl} = 1.06$. The result is computed in Table 16.6.

■ **TABLE 16.5**
AJT takeoff sizing

W/S_W (FPS) (lb/ft²)	40	50	60	70	80	90
W/S_W (SI) (N/m²)	1916.2	2395.6	2874.3	3353.7	3832.77	4311.5
T/W (non-dimensional)	0.157	0.2	0.235	0.274	0.313	0.353

■ **TABLE 16.6**
AJT climb sizing

W/S(lb/ft²)	40	50	60	70	80
W/S(N/m²)	1916.2	2395.6	2874.3	3353.7	3832.77
C_{Lclimb}	0.097	0.12	0.145	0.169	0.193
C_D	0.0222	0.0225	0.0258	0.026	0.0263
T_{SLS}/W	0.538	0.492	0.483	0.457	0.439

■ **TABLE 16.7**
AJT cruise sizing

W/S_w(lb/ft²)	50	60	70	80	100
W/S_w(N/m²)	2395.6	2874.3	3353.7	3832.8	4791
C_L	0.262	0.314	0.367	0.419	0.524
C_D	0.026	0.0292	0.0315	0.035	0.042
T_{SLS}/Wat 41,000 ft	0.346	0.324	0.30	0.29	0.279

Lift coefficient,

$$C_{Lclimb} = \frac{W}{0.5 \times 0.002378 \times 590.7^2 \times S_w} = 0.00241 \times (W / S_w)$$

Using Equation 16.15,

$$\frac{[T_{SLS} / W]}{1.06} = \frac{164}{590.7} + \frac{C_D \times 0.5 \times 0.002378 \times 590.7^2 \times S_w}{W}$$

$$\frac{T_{SLS}}{W} = 0.294 + \left[440 \times C_D \times (S_w / W) \right]$$

16.6.3 Cruise

From market requirements, initial cruise speed and altitude is Mach 0.75 and 36,000 ft (most training takes place below the tropopause). For the initial cruise aircraft weight take $k = 0.975$ in Equation 16.14. The result is computed in Table 16.7.

In FPS units at 36,000 ft, $\rho = 0.0007$ slug / ft² and $V^2 = (0.75 \times 968.07)^2 = 726.05^2 = 527,152.2$ ft² / s² .

In SI units, altitude $= 11,000$ m, $\rho = 0.364$ kg / m³, $V^2 = (0.75 \times 295.07)^2 = 221.3^2 = 48,974.8$ m² / s².

Equation 16.18 gives initial cruise

$$C_L = \frac{0.975 \times \text{MTOW}}{0.5 \times 0.364 \times 48974.8 \times S_w}$$

$$= 0.0001094 \times (W / S_w) \text{ where } W / S_w \text{ is in N / m}^2$$

Equation 16.19 gives

$$\frac{T_{SLS}}{W} = \frac{k_{cr} \times 0.5 \rho V^2 \times C_D}{(W / S_w)}$$

(take factor $1 / k_{cr} = T_{SLS} / T_a = 3.6$.)

In FPS units

$$\frac{T_{SLS}}{W} = \frac{3.6 \times 0.5 \times 0.0007 \times 527,152.2 \times C_D}{(W / S_w)} = \frac{664.2 \times C_D}{(W / S_w)}$$

In SI units

$$\frac{T_{SLS}}{W} = \frac{3.6 \times 0.5 \times 0.364 \times 48,974.8 \times C_D}{(W / S_w)} = \frac{32,088.2 \times C_D}{(W / S_w)}$$

Once again, make a table and plot.

16.6.4 Landing

From market requirements, $V_{app} = 110\text{kt} = 185.7\text{ft}/\text{s}$ (56.6 m/s). Landing $C_{Lstall} = 2.5$ at $40°$ double-slotted flap setting. Using Equation 16.22:

In FPS units

$$W/S_W = 0.311 \times 0.002378 \times 2.5 \times (185.7)^2 = 63.75 \text{ lb}/\text{ft}^2$$

In SI units

$$W/S_W = 0.311 \times 1.225 \times 2.5 \times (56.6)^2 = 2885 \text{ N}/\text{m}^2$$

Because at landing the thrust is taken as zero, W/S_W remains constant.

16.6.5 Sizing for Turn Requirement of 4 g at Sea-Level

In a way, turn sizing is relatively simpler, and here the parabolic drag polar is used to make use of the equations derived Chapter 14. AJT drag is approximated as parabolic drag polar as given below:

$$C_D = 0.0212 + 0.0706 C_L^2 \quad \text{with } S_W = 183 \text{ ft}^2$$

(gives Oswald's efficiency factor, $e = 0.85$).

The turn requirement is $n = 4$ on an ISA day at sea-level, but there is stipulation for speed at which it can be achieved. It gives the opportunity to initiate sizing analyses in many ways.

Consider that the requirement is the maximum capability when Equation 15.22 can be used:

$$n_{max} = 0.5 \left(\frac{T}{W}\right) \left[\frac{1}{k C_{DPmin}}\right]^{\frac{1}{2}}$$

operating at $(T/W)_{max}$, as T is the only variable. On substitution, this becomes:

$$4 = 0.5 \left(\frac{T}{W}\right) \left[\frac{1}{0.0706 \times 0.0212}\right]^{\frac{1}{2}} \quad \text{or } 8 = \left(\frac{T}{W}\right) \times \sqrt{668.12} \quad \text{or } \left(\frac{T}{W}\right) = 0.309$$

Engine setting at cruise is at maximum continuous rating at 95% of takeoff rating. Thrust decay from static to Mach 0.55 is $0.86 \times T_{SLS}$. At sea-level and around Mach 0.55 (see later), $k = 1.22$, giving

$$\left(\frac{T_{SLS}}{W}\right) = 0.309/(0.86 \times 0.95) = 1.22 \times 0.309 = 0.378 \approx 0.38$$

Note that if the sized value gives a higher value of (T/W), then the AJT will perform better turning with higher n_{max}. The velocity to obtain 4 g depends on the wing loading, and Equation 15.21 gives the required relationship as given below.

$$V_{n\_max} = \sqrt{\left(\frac{1}{\rho C_{DPmin}}\right)\left(\frac{T}{W}\right)\left(\frac{W}{S_W}\right)}$$

Substituting the AJT values for $(T/W) = 0.309$, it becomes:

$$V_{n\_max} = \sqrt{\left(\frac{0.309}{0.002378 \times 0.0212}\right)\left(\frac{W}{S_W}\right)} = \sqrt{6129.3 \times \left(\frac{W}{S_W}\right)} = 78 \times \left(W / S_W\right)$$

This equation gives the values as shown in Table 16.8.

To generate the relation between (W/S_W) and (T/W), Equation 15.18 is used as follows:

$$n^2 = 16 = \frac{0.5\rho V^2}{k\left(\dfrac{W}{S_W}\right)}\left(\frac{T}{W} - 0.5\rho V^2 \frac{C_{DPmin}}{W / S_W}\right)$$

$$= \frac{0.5 \times 0.002378 \times V^2}{0.0706 \times \left(\dfrac{W}{S_W}\right)}\left(\frac{T}{W} - 0.5 \times 0.002378 \times V^2 \frac{0.0212}{W / S_W}\right)$$

or

$$16 = \frac{0.001684 \times V^2}{\left(\dfrac{W}{S_W}\right)}\left(\frac{T}{W} - \frac{0.0000252 \times V^2}{W / S_W}\right) \tag{16.23}$$

From Table 16.8 take three convenient speeds, for example Mach 0.5, 0.55 and 0.6, and work out as follows:

At Mach 0.5 ($V = 558.25\text{ft} / \text{s}$), Equation 16.23 reduces to (see Table 16.9):

$$16 = \frac{0.001684 \times V^2}{\left(\dfrac{W}{S_W}\right)}\left(\frac{T}{W} - \frac{0.0000252 \times V^2}{W / S_W}\right) = \frac{5248}{\left(\dfrac{W}{S_W}\right)}\left(\frac{T}{W} - \frac{7.87}{W / S_W}\right)$$

$$0.003 \times \left(\frac{W}{S_W}\right) = \frac{T}{W} - \frac{7.87}{W / S_W}$$

$$T / W = 0.003 \times \left(\frac{W}{S_W}\right) + \frac{7.87}{W / S_W}$$

■ **TABLE 16.8**
AJT turn sizing

W/S_W(lb/ft^2)	50	60	70	80	100
W/S_W(N/m^2)	2395.6	2874.3	3353.7	3832.77	4791
V_{n_max}(ft/s)	553	606.43	655	700	782.9
Mach	0.495	0.543	0.587	0.627	0.7

■ **TABLE 16.9**
AJT turn sizing ($n = 4$ at Mach 0.5)

W/S_w (lb/ft²)	50	60	70	80	90
W/S_w (N/m²)	2395.6	2874.3	3353.7	3832.77	4312
$0.003 \times (W/S_w)$	0.15	0.18	0.21	0.24	0.27
$7.87/(W/S_w)$	0.157	0.131	0.1124	0.0984	0.0874
T/W ($n = 4$ at Mach 0.5)	0.307	0.311	0.322	0.338	0.36
$T_{SLS}/W = 1.22 \times T/W$	0.375	0.38	0.393	0.41	0.436

■ **TABLE 16.10**
AJT turn sizing (Mach 0.55)

W/S_w (lb/ft²)	50	60	70	80	90
W/S_w (N/m²)	2395.6	2874.3	3353.7	3832.77	4312
$0.00252 \times (W/S_w)$	0.126	0.1512	0.17	0.202	0.227
$9.53/(W/S_w)$	0.191	0.1588	0.136	0.119	0.106
T/W ($n = 4$ at Mach 0.55)	0.317	0.311	0.306	0.321	0.333
$T_{SLS}/W = 1.22 \times T/W$	0.387	0.38	0.373	0.392	0.406

At Mach 0.55 ($V = 614.075$ ft/s), Equation 16.23 becomes (see Table 16.10):

$$16 = \frac{0.001684 \times V^2}{\left(\dfrac{W}{S_w}\right)}\left(\frac{T}{W} - \frac{0.0000252 \times V^2}{W/S_w}\right) = \frac{6350}{\left(\dfrac{W}{S_w}\right)}\left(\frac{T}{W} - \frac{9.53}{W/S_w}\right)$$

$$0.00252 \times \left(\frac{W}{S_w}\right) = \frac{T}{W} - \frac{9.53}{W/S_w}$$

$$T/W = 0.00252 \times \left(\frac{W}{S_w}\right) + \frac{9.53}{W/S_w}$$

At Mach 0.6 ($V = 669.9$ ft/s), Equation 16.23 becomes (see Table 16.11):

$$16 = \frac{0.001684 \times V^2}{\left(\dfrac{W}{S_w}\right)}\left(\frac{T}{W} - \frac{0.0000252 \times V^2}{W/S_w}\right) = \frac{7557.12}{\left(\dfrac{W}{S_w}\right)}\left(\frac{T}{W} - \frac{11.333}{W/S_w}\right)$$

$$0.00212 \times \left(\frac{W}{S_w}\right) = \frac{T}{W} - \frac{11.333}{W/S_w}$$

$$T/W = 0.00212 \times \left(\frac{W}{S_w}\right) + \frac{11.333}{W/S_w}$$

■ **TABLE 16.11**
AJT turn sizing (Mach 0.6)

W/S_W (lb/ft²)	50	60	70	80	90
W/S_W (N/m²)	2395.6	2874.3	3353.7	3832.77	4312
$0.00212 \times (W/S_W)$	0.106	0.1272	0.1484	0.1696	0.191
$11.333/(W/S_W)$	0.227	0.1889	0.162	0.1417	0.126
T/W ($n = 4$ at Mach 0.6)	0.332	0.316	0.31	0.312	0.317
$T_{SLS}/W = 1.22 \times T/W$	0.405	0.386	0.378	0.38	0.387

■ **FIGURE 16.5**
Military aircraft sizing – AJT

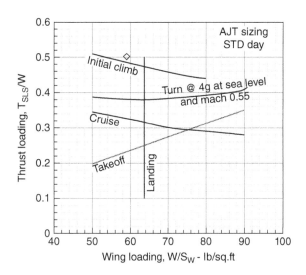

Tables 16.9–16.11 give three lines with close values of (T/W). Only the line for Mach 0.55 is plotted in Figure 16.5 as the middle one of the three.

16.7   Sizing Analysis – Military Aircraft

The methodology for military aircraft is the same as in the case of civil aircraft sizing and engine matching. The five sizing relationships between wing loading, W/S_W and thrust loading, T_{SLS}/W, need to meet capability requirements for (i) takeoff, (ii) approach speed for landing, (iii) initial cruise speed, (iv) initial climb rate, and (v) turn rate. These are plotted in Figure 16.5.

Military aircraft sizing poses an interesting situation. A variant in a combat role, for example in the CAS role, has to carry more armament load externally, increasing the drag. The overall geometry does not change much, except that the front fuselage is now redesigned for one pilot, saving about 100 kg (the weight of seat, escape system, etc., are replaced by radar and combat avionics). The aircraft still has the same engine, tweaked to an uprated thrust level.

Therefore a conservative sizing for the AJT should benefit the CAS variant. Figure 16.5 shows that the sizing point is at slightly lower wing loading at $W/S_W = 59 \text{lb}/\text{ft}^2$ to benefit CAS performance. Thrust loading is taken as $T_{SLS}/W = 0.5$. The circled point in Figure 16.5 satisfies all requirements simultaneously. A slightly higher value of T_{SLS}/W would benefit the takeoff performance with a full practice armament load.

Chapter 8 calculated the mass of the preliminary configuration of the AJT aircraft as: MTOM = 4800 kg (10,582 lb) at NTC, which gives the matched engine thrust $T_{SLS} = 0.5 \times 10,582 \approx$ 5300 lb (23,583 N). Checking with the sized wing loading $W/S_W = 59 \text{lb/ft}^2$, the wing area comes out 185 ft² (17.2 m²), only about 1% different from the preliminary wing area, hence kept unchanged. The matched engine thrust gives a lower value compared with the statistical estimate of 5860 lb, which is good. Once again, iteration is avoided.

16.7.1   Single Seat Variants

Military aircraft are no exception in offering variant designs depending on their mission role, in addition to the typical payload-range variation. The F16 and F18 have had modifications since they first appeared, with an increasing envelope of combat capabilities. The F18 has increased in size. The BAe Hawk100 jet trainer now has a single-seat close support combat derivative, the Hawk200.

The CAS role aircraft is the only variant of the AJT (Figure 16.6). The details on how it is achieved, with associated design changes, are described below.

16.7.1.1   Configuration   Configuration of the CAS variant is achieved by splitting the AJT front fuselage, then replacing the tandem seat arrangement with a single seat cockpit. The length could be kept the same, as the nose cone needs to house more powerful acquisition radar. The front-loading of radar and single pilot is placed in such a way that the CG location is kept undisturbed. Wing area = 17 m² (183 ft²).

16.7.1.2   Weights   A summary of mass changes is given in Table 16.12. Armaments and fuel could be traded for range. Drop tanks could be used for ferry range.

16.7.1.3   Thrust   The CAS variant would require an engine variant with 30% higher thrust. This is possible without a change in external dimensions, but would incur an increase of 60 kg in engine mass.

The CAS turbofan thrust (with small bypass) = $1.3 \times 5300 \approx 6900 \text{lb}$ (30,700 N). Thrust loading at MTOM becomes $T_{SLS}/W = 0.417$ (a satisfactory value).

16.7.1.4   Drag   The drag levels of the clean AJT and CAS aircraft may be considered to be about the same. There would be an increase in drag on account of the weapon load. For the CAS aircraft, a typical drag coefficient increment for armament load is $C_{D\pi} = 0.25$ (includes interference effect) each for five hard points as weapon carrier. Weapon drag is based on maximum cross-sectional area (say 0.8 ft²) of the weapons.

Parasite drag increment, $\Delta C_{Dpmin} = (5 \times 0.25 \times 0.8)/183 = 0.0055$, where $S_W = 183 \text{ft}^2$.

■ **FIGURE 16.6**
Variant designs in the family of military aircraft

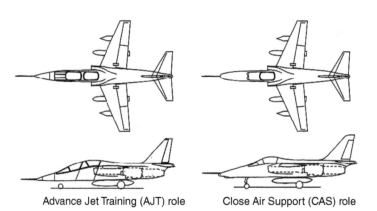

Advance Jet Training (AJT) role   Close Air Support (CAS) role

■ **TABLE 16.12**
AJT/CAS sized mass

	AJT (kg)	CAS (kg)	Remark
OEM	3700	3700	Remove one pilot, ejection seat, and so on (260 kg) and include radar, combat avionics (100 kg). There is an increase of 60 kg in engine mass
Fuel	1100	1300	Internal capacity 2390 kg (max)
Clean aircraft MTOM	4800 (10,582 lb)	5000 (11,023 lb)	
Wing loading[a], W/S_W	282 kg/m²(57.8 lb/ft²)	294 kg/m²(60.23 lb/ft²)	
Armament mass	1800	2500	
MTOM, kg (lb)	6600 (14,550 lb)	7500 (16,535 lb)	
Wing area, S_W	17 m (183 ft²)	17 m (183 ft²)	
Wing loading, W/S_W	388.7 kg/m²(79.5 lb/ft²)	441.2 kg/m²(90.36 lb/ft²)	

[a] Sized wing loading for AJT at NTC came close to this.

16.8   Aircraft Sizing Studies and Sensitivity Analyses

Aircraft sizing studies and sensitivity analyses are not exactly the same thing, as outlined below. Using both the sizing and sensitivity study exercises, aircraft designers are able to find and optimize a satisfying design to meet certification agency and customer requirements simultaneously. The designers and the users will have better insight to make tailored choices.

16.8.1   Civil Aircraft Sizing Studies

The aircraft sizing exercise freezes an aircraft to a unique geometry with little scope to make variations. In this kind of study, aircraft performance requirements (FAA) and specifications (customer) are given, such as (i) TOFL and climb gradients, (ii) initial en-route climb, (iii) initial HSC capability and (iv) LFL and climb gradients at baulked landing. These are simultaneously satisfied with a matched engine, and a unique solution emerges. The search for a compromise to offer a family of variants will depend on the objectives, for example, the degree that components can be retained.

It is a broad-based analysis and could be extensive. Basically, it is an optimization process to find the best value solution with opportunities to make compromises, if required. The easiest method is to take one variable at a time in a *parametric* search, to find how it affects other parameters. Normally, the variations are made in small steps to keep changes in the affected parameters within the range of acceptable error. In civil aircraft design, the study should continue to examine the objective function, such as the DOC. Typically the results are shown in carpet plots (Figure 16.7) showing up to as many as five variables [4]. This carpet plot shows four variables in one graph with mutual interdependency. It shows how aircraft DOC is affected by aircraft drag change, fuel price and aircraft cost for an Airbus 320 class aircraft.

Apart from the parametric method of searching, there are other theoretical methods available. The simpler ones are the Simplex and Gradient methods, which are used in industry, but the parametric method is popular.

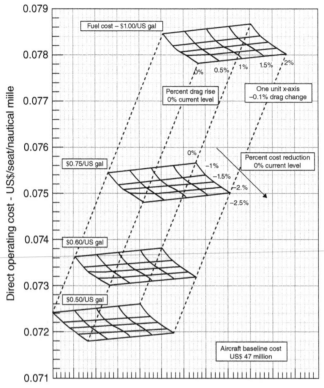

■ FIGURE 16.7
**Typical parametric
study of a mid-sized
aircraft (A320 class)
displayed as a carpet
plot showing DOC
variations with aircraft
drag and fuel costs**

16.8.2   Military Aircraft Sizing Studies

Sensitivity studies can be made to determine how individual component geometries (for example, wing area, aspect ratio, thickness to chord ratio, sweep, and so on) can affect aircraft capabilities, as shown in Table 16.13 for the AJT. This example sensitivity study shows what happens with small changes in wing reference area, S_w, aspect ratio, AR, aerofoil thickness to chord ratio, t/c, and wing quarter chord sweep, $\Lambda_{1/4}$.

A more refined analysis could be made with a detailed sensitivity study on various design parameters, such as other geometric details, materials, structural layout, and so on, to address cost versus performance issues to arrive at a "satisfying" design. This may require some local optimization with full awareness that the global optimization is not sacrificed. While a broad-based multidisciplinary optimization is the ultimate goal, dealing with large numbers of parameters in a sophisticated algorithm may not prove easy. It is still researched intensively within academic circles, but industry tends to use multidisciplinary optimization (MDO) in a conservative manner, if required in a parametric search, tackling one variable at a time. Industry cannot afford to take risks with any unproven algorithm just because it is elegant and bearing promise. Industry takes a more holistic approach to minimize costs without sacrificing safety, but may compromise performance, if it pays.

16.9   Discussion

The aircraft sizing exercise can keep open the possibilities for future aircraft growth. Even a radically new design could salvage components from older designs to maintain parts commonality.

■ **TABLE 16.13**
AJT sensitivity study

Sized geometry	Perturbed geometry
$S_W = 17\text{m}^2$	$\Delta S_W = 1.0\text{m}^2$
	$\Delta\text{Mach} = \pm 0.004$
	$\Delta V = \pm 4.35\text{mph}\,(3.78\text{kt})$
	$\Delta V_{touch\_down} = \pm 3.12\text{mph}\,(2.71\text{kt})$
	$\Delta M = \pm 65\text{kg}\,(143.3\text{lb})$
$AR = 5.0$	$\Delta AR = \pm 0.5$
	$\Delta M = \pm 40\text{kg}\,(88.2\text{lb})$
	$\Delta n\,(\text{accln}) = \pm 0.18$
$t/c = 10\%$	$\Delta t/c = \pm 2\%$
	$\Delta\text{Mach} = \pm 0.023$
	$\Delta V = \pm 16.6\text{mph}\,(13.6\text{kt})$
	$\Delta M = \pm 45\text{kg}\,(99.2\text{lb})$
$\Lambda_{\frac{1}{4}} = 20°$	$\Delta\Lambda_{\frac{1}{4}} = \pm 10°$
	$\Delta\text{Mach} = \pm 0.01$
	$\Delta V = \pm 6.25\text{mph}\,(5.43\text{kt})$

This culminating section presents some discussion on the classroom examples and their performance options. The sizing exercise gives a simultaneous solution to satisfy the airworthiness and market requirements. This is an activity done in the conceptual design phase when not much aircraft data are available. The sizing exercise requires specific aircraft performance evaluation in a concise manner using the relevant equations derived thus far. Sizing studies are management issues that are reviewed along with the potential customers to decide to give a "go-ahead" or not. In the sizing and engine matching exercise, the wing loading (W/S_w) and thrust loading (T/W) are the dictating parameters in equations for takeoff, second segment climb, en-route climb and maximum speed capability. The first two are FAR requirements and the last two are customer requirements. Information on detailed engine performance is not required during the sizing exercise; projected *rubberized* engine data are used during the sizing and engine matching studies. Payload-range capability of the proposed design needs to be demonstrated as a customer requirement. Subsequently, with the details of engine performance and aircraft data, relevant aircraft performance analyses are carried out more accurately to satisfy airworthiness and market requirements.

Once a go-ahead is obtained, then a full-blown detailed definition study ensues as Phase II activities, with large financial commitments. Refinement continues until accurate data are obtained from flight tests and other refined analyses. Most of the important aircraft performance equations are in this book in reasonable detail, supported by worked examples.

Figure 16.3 (Bizjet) shows the lines of constraint for the various requirements. The sizing point to satisfy all requirements will show different margins for each capability. Typically, initial en-route climb rate is the most critical to sizing. Therefore, the takeoff and maximum speed capabilities have considerable margins, which is good as the aircraft can do better than what is required.

From the statistics, experience shows that aircraft mass grows with time. This occurs primarily on account of modifications arising from minor design changes and with changing requirements, sometimes even before the first delivery is made. If the new requirements demand large numbers of changes, then a civil aircraft design may appear as a new variant, but military aircraft hold on a little longer before a new variant emerges. It is therefore prudent for the designers to keep some margin, especially with some reserve thrust capability, that is, to keep the thrust loading, (T/W), slightly higher to begin with. Re-engining with an uprated version is expensive.

It can be seen that field performance would require a bigger wing planform area (S_w) than at cruise. It is advisable to keep the wing area as small as possible (i.e. high wing loading) by incorporating superior high-lift capabilities, which is not only heavy but also expensive. Designers must seek a compromise to minimize operating costs. No iterations were needed for the designs worked out in this book.

Section 13.3 derived all the necessary relations required to estimate the aircraft mission payload-range capability for the aircraft cruise segment, and the block time and block fuel required for the mission. In summary, both close-form analytical methods for trend analyses using parabolic drag polar, and numerical analyses using actual drag polar, are analysed and the results compared. The question now is which method is to be applied, when each one has its own reason for application.

The first thing that needs to be pointed out is that actual drag polar is more accurate since it is obtained through proven semi-empirical methods and refined by tests (wind-tunnel and flight). Operators require an accurate manual to maximize city-pair mission planning. To establish the best possibility, for example to decide the speed and altitude for the sector, is not easy. Here the close-form analysis assists with quickly providing answers for the best possibilities. This is then fine-tuned through flight tests and then incorporated into the operator's manual; nowadays it is digitized and stored in several places, on the ground and on the aircraft.

The sizing point in Figure 16.3 gives a wing loading, $W/S_w = 64$ lb/ft^2 and a thrust loading, $T/W = 0.32$. Note that there is little margin given for the landing requirement. The maximum landing mass for this design is at 95% of MTOM. If for any reason the aircraft operating empty weight (OEW) increases, then it would be better if the sizing point for the W/S_w is taken a little lower than 64 lb/ft^2, say at 62 lb/ft^2. A quick iteration would resolve the problem. But this choice is not exercised, to keep the wing area as small as possible. Instead the aircraft could be allowed to approach to land at a slightly higher speed, as LFL is generally smaller than TOFL. This is easily achievable as the commonality of undercarriage for all variants would start with the design for the heaviest variant, and its bulky components are shaved to lighten for the smaller weights. The middle variant is taken as the baseline version. Its undercarriage can be made to accept MTOM growth as a result of OEW growth, instead of making the wing larger.

It should be borne in mind that it is recommended that civil aircraft should come in a family of variants to cover wider market demand to maximize sales. Truly speaking, none of the three variants are optimized, although the baseline is carefully sized in the middle to accept one larger and one smaller variant. Even when the development cost is front-loaded, the cost of development of the variant aircraft is kept low by sharing components. The low cost is then translated to lowering aircraft prices, which can absorb the operating costs of the slightly non-optimized designs.

It is interesting to examine the design philosophy of the Boeing 737 family and the Airbus 320 family of aircraft variants in the same market arena. Together, more than 8000 of these aircraft have been sold on the world market. This is a no small achievement in engineering. The cost of these aircraft is about $50 million a piece (2005). For airlines with deregulated fare structures, making a profit involves complex dynamics of design and operation. The cost and operational scenario changes from time to time, for example, rise in fuel costs, terrorist threats, and so on.

As early as the 1960s, Boeing saw the potential of keeping component commonality in offering new designs. The B707 was one of the earliest commercial jet transport aircraft carrying passengers. It was followed by the shorter B720. Strictly speaking the B707 fuselage relied on the KC135 tanker design of the 1950s. From the four-engine B707 came the three-engine B727, and then the two-engine B737, both retaining considerable fuselage commonality. This was one of the earliest attempts to utilize the benefits of maintaining component commonality. Subsequently, the B737 started to emerge in different sized variants, maximizing component commonality. The original B737-100 was the baseline design for all other variants that came later, up to the B737-900. It posed certain constraints, especially on the undercarriage length. On the other hand, the A320 serving as the baseline design was in the middle of the family; its growth variant is the A321 and shrunk variants are the A319 and A318. Figure 18.8 gives a good study of how the OEW is affected by showing examples of different sized families of variants. A baseline aircraft starting at the middle of the family would be better optimized and hence in principle would offer a better opportunity to lower production costs of the family.

16.9.1   The AJT

Military aircraft serve only one customer: the armed forces of the nation in question. Front-line combat aircraft incorporate the newest technologies at the cutting edge to stay ahead of potential adversaries. Development costs are high, and only a few countries can afford to produce advanced designs. International political scenarios indicate a strong demand for combat aircraft, even for developing nations who must purchase from abroad. This makes the military aircraft design philosophy different from civil aircraft design. Here, the designers/scientists have a strong voice, unlike in civil design where the users dictate terms. Selling combat aircraft to restricted foreign countries is one way to recover some finance.

Once the combat aircraft performance is well understood over its years of operation, consequent modifications follow capability improvements. Eventually a new design replaces the older design – there is a generation gap between the designs. Military modifications for the derivative design are substantial. Derivative designs primarily come from revised combat capabilities with newer types of armament, along with all-round performance gains. There is also a need for modifications, seen as variants rather than derivatives, to sell to foreign customers. These are quite different to civil aircraft variants, which are simultaneously produced for some time, serving different customers, with some operating all the variants.

Advanced trainer aircraft designs have variants serving as combat aircraft in close air support. AJTs are less critical in design philosophy in comparison with front-line combat air-craft, but bear some similarity. Typically, AJTs will have one variant in the CAS role produced simultaneously. There are fewer restrictions to exporting these kinds of aircraft.

The military infrastructure layout influences aircraft design, and here the life cycle cost (LCC) is the prime economic consideration. For military trainer aircraft, it is better to have a training base close to the armament practice arena, saving time. A dedicated training base may not have as long a runway as major civil runways. These aspects are reflected in the user speci-fications needed to start a conceptual study. The training mission includes aerobatics and flying with onboard instruments for navigation. Therefore, the training base should be far away from airline corridors.

The AJT sizing point in Figure 16.5 gives a wing loading, $W/S_W = 59$ lb/ft^2 and a thrust loading, $T/W = 0.5$. It may be noted that there is a large margin, especially for the landing requirement. The AJT can achieve maximum level speed over Mach 0.88, but this is not demanded as a requirement. Mission weight for the AJT varies substantially; the NTC is at 4800 kg and for armament practice it is loaded to 6600 kg. The margin in the sizing graph covers some increase in loadings (the specification taken in the book is for the NTC only). There is a big demand for higher power for the CAS variant. The choice of having an uprated engine or to have an afterburner depends on the choice of engine and the mission profile.

Competition for military aircraft sales is not as critical compared with the civil aviation sector. The national demand would support a national capacity for producing a tailor-made design with manageable economics. But the trainer aircraft market can have competition, unfortunately sometimes influenced by other factors that may fail to bring out a national product even if the nation is capable of doing so. The Brazilian design Tucano was re-engined and underwent extensive modifications by Short Brothers plc of Belfast for the Royal Air Force (RAF), UK. The BAe Hawk (UK) underwent major modifications in the US for domestic use.

References

1. Wallace, R.E., *Parametric and Optimisation Techniques for Airplane Design Synthesis*. AGARD-LS-56, 1972.

2. Loftin, L.K. Jr, *Subsonic Aircraft: Evolution and Matching of Size to Performance*. NASA RP 1060, 1980.

3. Kundu, A.K., *Aircraft Design*. Cambridge University Press, New York, 2010.

4. Jenkinson, L.R., Simpson, P., and Rhodes, D., *Civil Jet Aircraft Design*. Arnold Publishers, London, 1999.

CHAPTER 17

Operating Costs

17.1 Overview

The civil aviation industry expects a return on investment, with cash flowing back to allow growth in a self-sustaining manner, with or without some government assistance. Sustenance and growth of civil aviation depends upon profitability. In a free market economy, industry and operators face severe competition to survive, forcing them to operate under considerable pressure to manage manufacture and operation in a lean manner. Economic considerations are the main drivers for commercial aircraft operation, and in some ways also for military aircraft. This chapter primarily deals with commercial aircraft operation, and discusses the pertinent aspects of cost implications related to military aircraft.

Aircraft direct operating cost (DOC) is the most important parameter that concerns airline operators. The DOC will depend on how many passengers the aircraft carries for what range, and the unit is expressed in $/seat-nm. There are standard rules, for example from the Association of European Airlines (AEA) [1] and the Aircraft Transport Association (ATA), for the DOC comparison. Each industry/airline has their own DOC ground rules suiting their administrative infrastructure, which results in different values compared with what is obtained from standard methods. Typically, aircraft performance engineers do the DOC estimation.

For mid-size commercial aircraft operation, the ownership cost contribution to the DOC could be as high as a third to a half of the total DOC (at year 2000 fuel prices). It is for this reason that the industry was driven to reduce the cost of manufacture by as much as 25%. Clearly, the design and manufacturing philosophy play a significant role in facilitating manufacturing cost reduction. To understand the cost implications, performance engineers could benefit from a brief overview on the various aspects of costing to appreciate the competition and perform route planning accordingly. When the fuel price is low, then drag can be sacrificed to reduce manufacturing costs; on the other hand, as the fuel price increases, the role of drag starts to dominate and aircraft costs could go up. The DOC estimation at the initial project phase is continuously revised as the project progresses with more detailed cost information.

During the last two decades, the aerospace industry is increasingly addressing factors such as cost, performance, delivery schedule and quality to satisfy "customer driven" requirements of affordability by reducing aircraft acquisition costs. Steps to address these factors are to synchronize and integrate design with manufacturing and processes planning as a business strategy that will lower the cost of production, at the same time ensuring reliability and maintainability at lower operational costs. Therefore there is a need for more rigorous cost assessment at each stage of design in order to cater for the objective of making a more effective

Theory and Practice of Aircraft Performance, First Edition. Ajoy Kumar Kundu, Mark A. Price and David Riordan.
© 2016 John Wiley & Sons, Ltd. Published 2016 by John Wiley & Sons, Ltd.

value-added "customer driven" product. At this time, data from the emerging geopolitical situation, national economic infrastructure, fluctuating fuel prices, emerging technological considerations, for example on sustainable developments, anti-terrorism design features, passenger health issues, and so on, are difficult to predict.

Visibility of costing forces long-range planning, helps with a better understanding of system architecture of the design for trade-off studies to explore alternative designs, and helps to explore the scope for sustainability and eco-friendliness of the product line. The product passes through well-defined stages during its lifetime, for example conception, design, manufacture, certification, operation, maintenance/modification, and finally disposal at the end of the life cycle. Cost information for a past product should be sufficiently comprehensive and available during the conceptual stages of a new project. Differential evaluation of product costs and technology, offering reliability and maintainability, along with risk analysis, are important considerations in cost management. Cost details also assist the preliminary planning for procurement and partnership sourcing through an efficient bid process. The final outcome would be to ensure acquisition of aircraft and its components with an objective to balancing the tradeoff between cost and performance, which will eventually lead to ensuring affordability and sustainability for the operators over the product life cycle. Cost analysis stresses the importance in playing a more rigorous role, as an integrated tool embedded in the multidisciplinary systems architecture of aircraft design, to help arrive at a "best value", specifically aimed at manufacturing and operational needs.

On the other hand, military aircraft are driven by defence requirements as the primary objective. Their export potential is a byproduct, and restricted to friendly nations with some risk of disclosure of technical confidentiality. There is some difference between the ground rules for costing of military and civil aircraft manufacture and operation. Because of the need to stay ahead of adversary capability, military aircraft designs have to explore newer technologies, which are expensive and require laborious proving to ensure safety and effectiveness. Many military projects have had to be abandoned even after prototypes were flown, for example the TSR2 (UK), Northrop F20 (USA), and so on; the reasons could be different, but the common factor would always be their cost-effectiveness. The product must demonstrate best value for money. The readers are encouraged to look into aircraft project histories. This chapter considers primarily the civil aircraft cost considerations, with passing mention of military aircraft costing.

The readers are to estimate Bizjet DOC. All relevant information to estimate aircraft DOC is given here. A problem set on this chapter is given in Appendix E.

17.2   Introduction

Aircraft DOC is dependent on aircraft price (see Figure 17.2), and it is to be predicted at the conceptual design stage well ahead of when the aircraft is built, even though detailed information is not available. The post conceptual design study phase leads into the project definition phase, followed by the detailed design phase when the manufacturing activities build up to produce the finished aircraft. At later stages of the project when more accurate cost data is available, then the DOC estimates become much more accurate. Figure 17.1 shows the levels of cost model architecture to serve various groups at different programme milestones.

Industry needs to recover all the investment at the point of selling around 400 aircraft, and preferably fewer. Typically, about 4–6% of the aircraft price is meant to recover the project cost, known as amortization of investment made. Offering aircraft in a family concept covers a wider market at considerable lower investment, so the cost of amortization will drop down closer to 2–4%. Smaller aircraft should break even at around 200 aircraft sales. In today's practice, civil manufactures sell *pre-production* aircraft used for flight testing to recover costs. Military aircraft

■ **FIGURE** 17.1
Various levels of cost prediction methodologies at various phases of a project

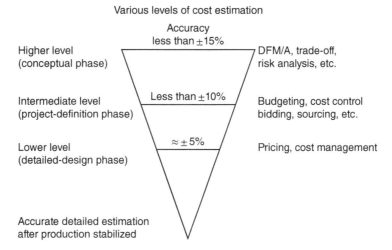

Various levels of cost estimation

Accuracy
less than ±15%

Higher level
(conceptual phase)

DFM/A, trade-off,
risk analysis, etc.

Less than ±10%

Intermediate level
(project-definition phase)

Budgeting, cost control
bidding, sourcing, etc.

≈ ± 5%

Lower level
(detailed-design phase)

Pricing, cost management

Accurate detailed estimation
after production stabilized

incorporate unproven new technologies and invest in *technology demonstrator* aircraft (at a reduced scale) to prove the concept and subsequently substantiate the design by flight testing on pre-production aircraft; some of them could be retained by the manufacturer for future tests.

The general definition of aircraft price is as follows: the cost includes amortization of research design, development, manufacture and cost (RDDMC), but does not include spares and support costs.

$$\text{Aircraft price} = \text{aircraft cost} + \text{profit} = \text{aircraft acquisition cost}$$

In this book, the aircraft price and cost are taken synonymously. In fact, the aircraft price is also known as the aircraft acquisition cost. The profit margin is a variable quantity and depends on what the market can bear. This book does not deal with the aircraft pricing method. In general, profit from a new aircraft sale is rather low. Most of the profits come from sale of affordable spares and maintenance support. Operators will have to depend on the manufacturer as long the aircraft are in operation, say for two to three decades. Manufacturers are in a healthy position for several decades if their products sell in large numbers.

There are many cost methodologies; the appropriate ones are kept "commercial in confidence" by the industry. Reference [2] suggests a rapid costing methodology reflecting an industrial approach.

17.3   Aircraft Cost and Operational Cost

Figure 17.2 gives typical high subsonic civil aircraft costs at the year 2000 price levels in millions of US dollars. It reflects the basic (lowest) cost of the aircraft. This graph is generated from some accurate industrial data which are kept "commercial in confidence".

In general, exact aircraft cost data are not readily available, and the overall accuracy of the graph is not substantiated. The aircraft price varies for each sale depending on the terms, conditions, support and economic climate involved. The values in the figure are crude but offer a feel for the newly initiated on the expected cost of the class of aircraft. Statistics could be used to obtain the relationship between maximum takeoff weight (MTOW) versus number of passengers. The basic price of a mid-range 150-passenger class, high subsonic turbofan aircraft is taken as US$ 47 million (year 2000 level).

■ **FIGURE 17.2**
Aircraft cost factors
[2]. Reproduced with
permission of
Cambridge
University Press

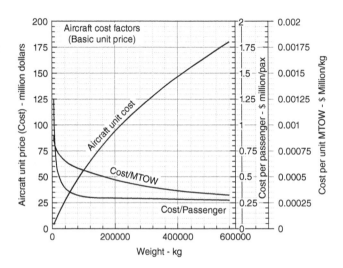

Weight - kg

■ **TABLE 17.1**

Typical cost fractions for a civil aircraft (two engines) at the shop floor level

	Cost fraction (%)
1. Aircraft empty shell structures[a]	
Wing shell structure	10–12
Fuselage shell structure	6–8
Empennage shell structure	1–2
Two nacelle shell structure[b]	2–3
Miscellaneous structures	0–1
Sub-total	20–25
2. Bought-out vendor items	
Two turbofan dry bare engines[b]	18–22
Avionics and electrical system	8–10
Mechanical systems[c]	6–10
Miscellaneous[d]	4–6
Sub-total	40–50
3. Final assembly to finish (labour-intensive: includes component sub-assembling, final assembling, equipping/ installing, wiring, plumbing, furnishing, finishing, testing)	25–30

[a] Individual component sub-assembly cost fraction.
[b] Smaller aircraft engine cost fraction is higher (up to 25%).
[c] Includes control linkages, servos, undercarriage, and so on.
[d] Cables, tubing furnishing, and so on.

Aircraft MTOW reflects the range capability, which varies from type to type. Therefore, strictly speaking, cost factors should be based on the manufacturer's empty weight (MEW). In general, bigger aircraft have a longer range.

Typical cost fractions (with respect to aircraft cost) of various groups of civil aircraft components are given in Table 17.1. It serves as preliminary information for two-engine aircraft (four-engine costs are slightly higher). It is better to obtain actual data from industry, if possible.

■ **TABLE 17.2**
Typical cost fractions of combat aircraft (two engines) at the shop floor level

	Cost fraction (%)
1. Aircraft empty shell structures[a]	
Wing shell structure	6–7
Fuselage shell structure	5–7
Empennage shell structure	≈1
Two nacelle shell structure[b]	Part of fuselage
Miscellaneous structures	0–1
Sub-total	12–15
2. Bought-out vendor items	
Two turbofan dry bare engines[b]	25–30
Mechanical systems[c]	5–8
Miscellaneous[d]	1–2
Sub-total	30–40
3. Avionics and electrical system	30–35
4. Final assembly to finish (labour-intensive: component sub-assembling, final assembling, equipping/installing, wiring, plumbing, furnishing, finishing, testing)	12–15

[a] Individual component sub-assembly costs fraction.
[b] Single engine at lower cost fraction.
[c] Includes control linkages, servos, undercarriage, and so on.
[d] Cables, tubing furnishing, and so on.

Combat aircraft cost fractions are different. Here, the empty shell structure is smaller but houses very sophisticated avionics black boxes serving the very complex tasks of combat and survivability. Typical cost fractions of various groups of combat aircraft components are given in Table 17.2. In the table, the avionics cost fraction is separated. It serves as preliminary information for two-engine aircraft. It is better to obtain actual data from industry, if possible.

In the US, military aircraft costing uses AMPR (aeronautical manufacturer's planning report) weight, also known as DCPR (defence contractor's planning report) weight, for the manufacturer to bid. The AMPR weight represents the weight of the empty aircraft shell structure without any bought-out vendor items, such as engines, undercarriages, avionics packages, and so on.

17.3.1 Manufacturing Cost

The total manufacturing cost of the finished product is taken as the sum of the items given in Table 17.3. The cost of manufacture is not the selling price.

Given below are the 11 specific parameters, in two groups, identified as the design and manufacture sensitive cost drivers. Group 1 consists of eight cost drivers, which relate to in-house data within the organization. The Group 2 cost drivers are not concerned with in-house capability issues, and so are not within the scope of this treatise.

Group 1 (concerned with in-house issues)
1. *Size*: Size is considered as the main parameter in establishing the base cost. Size and weight are correlated.

■ **TABLE 17.3**
Components of manufacturing cost

1. Cost of material (raw and finished product)
2. Cost of parts manufacture
3. Cost of parts assembly to finish the product
4. Cost of support (e.g. rework/concessions/quality)
5. Amortization of non-recurring costs
6. Miscellaneous costs (other direct costs, contingencies)

■ **FIGURE 17.3**
**Cost versus tolerance
[2]. Reproduced with
permission of
Cambridge
University Press**

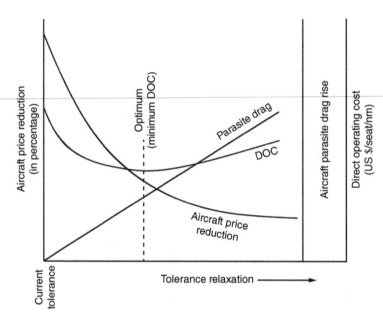

2. *Material:* Weight data for parts will give a more accurate cost of material than applying a size factor K_{size}, which may be used when weight details are not available. There
   are two kinds of material considered, based on industrial terminology: (i) raw material
   and (ii) finished material. The latter comprise the subcontracted items.

3. *Geometry:* Double curvature at the nacelle surface requires stretch-formed sheet metal
   or a complex mould for composites in shaping the mouldlines. Straight longitudinal and
   circumferential joints would ease auto-riveting. In short, there are four "Cs" associated
   with geometric cost drivers: Circularity, Concentricity, Cylindricity and Commonality.

4. *Technical specification:* These standards form the finish and maintainability of the aircraft, for example, the surface smoothness requirements (manufacturing tolerance at the
   surface), safety issues (fire detection), interchangeability criteria, pollution standards,
   and so on. Figure 17.3 shows the cost versus tolerance relationship taken from [2].

5. *Structural design concept:* Component design concepts contribute to the cost drivers
   and are taken as non-recurring costs (NRCs) amortized over the production run (typically 400 units). The aim is to have a structure with a low part count involving low
   production man-hours. Manufacturing considerations are integral to structural design
   as a part of the design for manufacture/assembly (DFM/A) obligation.

6. *Manufacturing philosophy:* This is closely linked to the structural design concept as described above. It has two components of cost drivers: (i) the NRC of tool and jig design, and (ii) the recurring cost during production (parts manufacture and assembly). An expensive tool set-up to cater for a rapid learning process and a faster assembly time with lower rejection rates (concessions and rework) makes budgetary provision front-loaded, but considerable savings can be obtained.

7. *Functionality:* This concerns the enhancement required compared with the baseline nacelle design, for example, anti-icing, thrust reversing, treatment of environmental issues of pollution (noise and emissions), position of engine accessories, bypass duct type, and so on. A factor of "complexity" may be used to describe the level of sophistication incorporated into the functionality.

8. *Man-hour rates, overheads, and so on:* Man-hour rates and overheads are held constant, and therefore the scope of applicability becomes redundant in this study.

   Group 2 (do not relate to in-house issues, therefore not considered further)

9. *Role:* Basically it describes the difference between military and commercial aircraft design.

10. *Scope and condition of supply:* This concerns the packaging quality of the aircraft supplied to the customer. This is not a design/manufacture issue.

11. *Programme schedule:* This is an external cost driver not discussed here.

Aircraft component manufacture case studies and operating cost reduction benefits are discussed in [3, 4].

17.3.2 Operating Cost

This cost arises during operation after the aircraft is sold – it is seen as the *operating cost* [5]. This is the concern of airline operators. Revenue earned from passenger airfares covers the full expenditure of airline operators including the aircraft price and support costs. The DOC gives a measure of cost involved with the aircraft mission. Standards for DOC ground rules exist – in the US they are proposed by the ATA (1967), and in Europe by the AEA (1989). Aircraft designers must be aware of operational needs to make sure that their design serves the expectations of the operators. In fact, the manufacturers and the operators are in constant dialogue to make the current and future products fine-tuned to the optimum profitability. The big airline operators have permanent representatives located at the manufacturers to enable all-round support and dialogue on all aspects of the product line. Civil aircraft operating cost is of two types, as follows.

1. *Direct operating cost (DOC):* These are the operational costs directly involved with the mission flown. Each operator will have their own ground rules depending on the country, pay scales, management policies, fuel price, and so on. The standard ground rules are used for comparison of a similar class of product manufactured by different companies. In Europe, the AEA ground rules are accepted as the basis for comparison. This gives a good indication of aircraft capability. A cheaper aircraft may not prove profitable on the long run, if its operating costs are high.

2. *Indirect operating cost (IOC):* The IOC breakdown in the US is slightly different from European standards. The IOC components are given below.

The airline operators have other costs that involve training, evaluation, logistic support, special equipmens and ground-based resource management that are not directly related to the aircraft design and mission sector operation. This is independent of aircraft type. Together they form the

■ **TABLE 17.4**
Life cycle cost (civil aircraft)

Aircraft-related elements	Passenger-related elements	Cargo-related elements
Property depreciation[a]	Handling and insurance	Handling and insurance
Property amortization[a]	Baggage handling	Administration/office
Property maintenance[a]	Emergencies	Sales and support
Ground support	Administration/office	Fees/commissions
Administration costs	Sales and support	Publicity
Ground handling/control	Publicity	
Training	Fees/commissions	

[a] Ground property (e.g. hangars, equipment, etc.).

■ **TABLE 17.5**
Life cycle cost (military aircraft)

Research/development	Production	In-service	Disposal
Engineering	Parts manufacture	Operation	Storage
Ground testing	Assembly	Maintenance	Recycling
Technology demonstrator	Tooling	Ground support	Scrapping
Prototype flight test	Deliveries	Training	
Technical support		Post-design services	
Publication		Administration	

total cost to the operator, termed the life cycle cost (LCC). Unlike DOC, there is no standard for LCC proposed by any established associations for commercial aircraft operation. Currently, each organization has its own ground rules to compute LCC. Together with the DOC, they give the total operating cost (TOC).

Military operating costs involve a different book-keeping method. Unlike military aircraft, the impact of "*other costs*" on the LCC (Table 17.4) in commercial aircraft application may be considered to be of a lower order – the DOC covers the bulk of the cost involved. This book covers the DOC only. The breakdown of LCC components is given below.

The military uses LCC rather than DOC for the ownership of aircraft in service. In loose terms, this is the cost involved for the entire fleet from cradle to grave including disposal. Military operation has no cash flowing back – there are no paying customers like passengers and cargo handlers. Tax-payers bear the full costs of military design and operation. Military aircraft operating cost ground rules are based on total support by the manufacturer for the entire operating life-span, which can be extended with fresh contracts. It is for this reason that military aircraft operation deals with the LCC, although there are various levels of cost breakdown including aircraft-related and sortie-related costs. Given below is an outline of the breakdown by categorizing the elements affecting the military aircraft LCC model (Table 17.5).

Of late, the customer-driven civil market desires LCC estimation. Academics and researchers are suggesting various LCC models. The principles of such an approach are directed at cost management and cost control, offering advice in assigning responsibilities, effectiveness, and other administrative measures at the conceptual stages in an integrated product and process development (IPPD) environment.

17.4   Aircraft Direct Operating Cost (DOC)

Each airline generates its own in-house DOC computations. There are variations in the man-hour rates and schedules. Although the ground rules for DOC vary from company to company, standardization made by the AEA (1989 – short–medium range) has been accepted as the basis for comparison. ATA rules are used in the US.

A NASA report [6] gives American Airlines-generated airline economics in detail. The NASA document (1978) proposed an analytical model associated with advanced technologies in aircraft design, but it is yet to be accepted as a standard method for comparison. The AEA ground rules appear to have taken into account all the pertinent points and have become the standard for the operational and manufacturing industries for benchmarking/comparison.

Commercial aircraft operation is singularly dependent on revenue earned from fare-paying passengers/cargo. Operating sector "*passenger load factor*" (LF) is defined as the ratio of occupied seats to available seats. Typically, for aircraft of medium size and above, aircraft operational cost breaks even at around a third full (varies from airline to airline – fuel cost at year 2000 levels and having regular airfare) – that is, a LF of around 0.33 (with inflation and rises in oil price, the LF is around 0.5). Naturally, the empty seats could be filled up with cut-price fares, contributing to the revenue earned.

At this point, it is pertinent to introduce the definition of the dictating parameter of "seat-mile cost", representing the unit of the aircraft DOC that determines the airfare to meet the operational cost and sustain a profit. The DOC is the total cost of operation for the mission sector.

$$\text{Seat-mile cost} = \frac{\text{direct operating cost} \left(\text{DOC} \right)}{\text{number of passengers} \times \text{range in nm}} \tag{17.1}$$

Units are therefore in cents per seat-nautical mile.

Evidently, the higher the denominator, the lower would be the seat-mile cost. Therefore the longer it flies and/or the more it carries, the seat-mile cost would reduce accordingly. Up to the 1960s the passenger fare was kept fixed under regulations. The 1970s onwards saw deregulated fare structures – an airline can now choose its own airfare and vary as the market demands.

Table 17.6 gives the breakdown of DOC components under two headings: fixed cost and trip cost. The ownership cost element depends on aircraft acquisition cost.

The NRC (design and development) of a project and cost of aircraft manufacture contribute to "ownership cost", while costs of fuel, landing fees and maintenance contribute to "trip cost elements". Once the aircraft is purchased, the ownership cost runs even when crews are hired but no flight operation is carried out. Crew cost added to ownership cost becomes the "fixed cost" elements. Crew cost added to trip cost is the "running cost" of the trip (mission

■ **TABLE 17.6**
DOC components

Fixed cost elements	Trip cost elements
1. Ownership cost	3. Fuel charges
(a) Depreciation	4. Maintenance (airframe and engine)
(b) Interest on loan	5. Navigational and landing charges
(c) Insurance premium	
2. Crew salary and cost	

■ **FIGURE 17.4**
DOC breakdown (at
0.75 cents/US gallon)
[7]. Reproduced
from the American
Institute of
Aeronautics and
Astronautics, Inc.

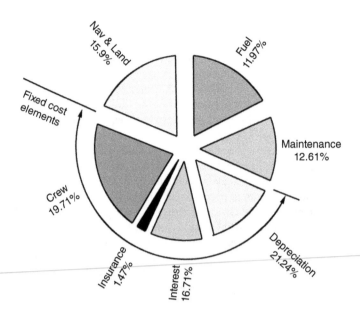

sortie). Aircraft price-dependent DOC contributions are depreciation, interest, insurance and maintenance (airframe and engine). Crew salary/cost and navigational/landing charges are aircraft weight-dependent, which is second-order aircraft price-dependent, but here it is taken based on man-hour rates.

The ownership cost contribution to DOC could be as high as a third to a half of the total DOC (at year 2000 fuel prices). It is for this reason that industry was driven to reduce costs of manufacture by as much as 25%. Clearly, the design philosophy plays a significant role in facilitating such cost reduction. One of the considerations was to relax some quality issues [2], sacrificing aerodynamic and structural considerations without sacrificing safety. However, when fuel prices rise, then considerations for such drivers would be affected. For example, fuel prices rose very high in 2008. Big increases would require drastic measures on many fronts. These are important considerations during the conceptual design stage. Research and development efforts look further into the future, sometimes through crystal-ball gazing. There is a diminishing return on investment for aerodynamic gains. The fuel price fluctuates severely, and we would require a stable situation to make big efforts to invest in reducing drag. Parallel efforts to use cheaper alternative biofuels are underway. A demand for turboprop operation is a reality.

Figure 17.4 shows the DOC components of a mid-sized aircraft. For an average mid-range sector, mid-sized aircraft, the cost contribution to aircraft DOC is three to four times higher than the contribution made by fuel costs (year 2000 prices).

Typically, an airline would like to break even for the sector DOC by about a third full, that is, at a LF of 0.33 (33%). Of late, on account of fuel price increases, the figure has gone up. Revenue earned from passengers carried above the break-even LF comes as profit. While some flights could run at 100% LF, the yearly average for a high-demand route could be much lower. Passenger accommodation could be in different classes or tiers of fare structure, or in one class as decided by the airline. Even within the same class, airfares can vary depending on the type of offer made. From airline to airline the break-even LF varies. With deregulated airfares, the ticket price could vary by the hour depending on passenger demand. The standard fare is the ceiling fare of the class, and offers better privileges.

17.4.1   Formulation to Estimate DOC

Given below is the DOC formulation, based on the AEA ground rules [1]. The formulae computes component DOC per block hour. To obtain trip cost, the DOC per block hour is multiplied by the block time. Aircraft performance calculates the block hour and block time for the mission range. The next section works out the DOC values using the Bizjet example obtained so far.

Normally, the DOC is computed for a fleet of aircraft. The AEA suggests an aircraft fleet of 10, having 14 years of life-span with a residual value of 10% of the total investment made. These values can be changed, as will be shown in the next section. Fuel price, insurance rates, salaries, and shop-floor man-hour rates vary with time. Engine maintenance cost depends on the type of engine. Here, only the turbofan type is treated. For ground rules for other kinds of power plant, the readers are advised to look into [1].

Aircraft price:

Total investment = (aircraft + engine price) × (1 + spares allowance fraction)

Readers need to be careful to obtain the standard study price of the manufacturer. The AEA uses total investment, which includes the aircraft delivery price, cost of spares, any change in order and other contractual financial obligations. In the example, the aircraft and engine price is taken as the total investment per aircraft.

Outstanding Capital = total capital cost × (1 − purchase downpayment fraction)

Utilization – per block hour per annum (hours/year)

$$\text{Utilisation, } U = \left(\frac{3750}{t+0.5}\right) \times t$$

where t = block time for the mission

Fixed cost elements:

$$\text{Depreciation} = \frac{0.9 \times \text{total investment}}{14 \times \text{utilization}}$$

$$\text{Loan interest repayment} = \frac{0.053 \times \text{total capital cost}}{\text{utilization}}$$

$$\text{Insurance premium} = 0.005 \times \left[\frac{\text{aircraft cost}}{\text{utilization}}\right]$$

Crew salary and cost:

Flight crew – AEA has taken $493 per block hour for two-crew operation.

Cabin crew – AEA has taken $81 per block hour for each cabin crew.

Trip cost elements:

$$\text{Landing fees} = \frac{7.8 \times \text{MTOW in tonnes}}{t}$$

where t = block time for the mission.

$$\text{Navigational charges} = \left(\frac{0.5 \times \text{range in km}}{t}\right) \times \sqrt{\frac{\text{MTOW in tonnes}}{50}}$$

$$\text{Ground handling charges} = \frac{100 \times \text{payload in tonnes}}{t}$$

The landing/navigational/ground handling charges are MTOW-dependent, and ground handling charges are payload-dependent. In practice, crew salary is also MTOW-dependent, but the AEA keep this invariant.

Airframe maintenance, material and labour

Airframe labour

$$= \left(0.09 \times W_{airframe} + 6.7 - \frac{350}{\left(W_{airframe} + 75 \right)} \right) \times \left(\frac{0.8 + 0.68 \times \left(t - 0.25 \right)}{t} \right) \times R$$

where $W_{airframe}$ is the MEW less engine weight in tonnes, R is the man-hour rate of $63 per hour at 1989 levels, and t is the block time for the mission.

$$\text{Airframe material cost} = \left(\frac{4.2 + 2.2 \times \left(t - 0.25 \right)}{t} \right) \times C_{airframe}$$

where $C_{airframe}$ is the price of the aircraft less engine price in millions of dollars.

Engine maintenance, material and labour

$$\text{Engine labour} = 0.21 \times R \times C_1 \times C_3 \times \left(1 + T \right)^{0.4}$$

where R is the man-hour rate of $63 per hour at 1989 levels; T is the sea-level static thrust in tonnes; $C_1 = 1.27 - 0.2 \times \text{BPR}^{0.2}$ (where BPR is the bypass ratio); and $C_3 = 0.032 \times n_c + K$ (n_c is the number of compressor stages), and K = 0.50 for one shaft, 0.57 for two shafts, and 0.64 for three shafts).

$$\text{Engine material cost} = 2.56 \times \left(1 + T \right)^{0.8} \times C_1 \times \left(C_2 + C_3 \right)$$

where T is the sea-level static thrust in tonnes; C_1 and C_3 are as previously; and $C_3 = 0.4 \times (\text{OAPR}/20)^{1.3} + 0.4$ (OAPR is the overall pressure ratio).

Direct engine maintenance cost $\left(\text{labour} + \text{material} \right)$

$$= Ne \times \left(\text{engine labour cost} + \text{material cost} \right) \times \frac{\left(t + 1.3 \right)}{\left(t - 0.25 \right)}$$

where Ne is the number of engines.

$$\text{Fuel charges} = \frac{\text{block fuel} \times \text{fuel cost}}{\text{block time}}$$

Then, DOC per hour = (fixed charges + trip charges)$_{\text{per\_hour}}$, and DOC per trip = $t \times (\text{DOC})_{\text{per\_hour}}$.

$$\text{DOC per aircraft mile} = \frac{\text{DOC} \times 100 \times \text{block time}}{\text{range}}$$

$$\text{DOC per passenger mile per nautical mile} = \frac{\text{DOC} \times 100 \times \text{block time}}{\text{range} \times \text{no. of passengers}}$$

17.4.2 Worked Example of DOC – Bizjet

Based on the formulation given above, this section works out the DOC of the Bizjet in question. Instead of working on a fleet, only one aircraft is worked out here. The information needed as input for the DOC calculation is given below. The man-hour rates are given in [1]. All costs figures are in US dollars and rounded up to the next highest figure. Table 17.7 gives the aircraft and engine details.

Aircraft price = $7 million

Engine price = $1 million

Total aircraft acquisition cost = $8 million (this is taken as total investment per aircraft – price includes spares, etc.).

The DOC is computed for a single aircraft to obtain the trip cost instead of per hour cost. A life span of 14 years is taken, with a residual value of 10% of the total investment.

Utilization – per block hour per annum (hours/year):

$$U = \left(\frac{3750}{5.38 + 0.5} \right) \times 5.38 = 637.75 \times 5.38 = 3431 \text{ hours per year}$$

where t = block time for the mission = 5.38 hours.
Fixed cost elements:

$$\text{Depreciation} = \frac{0.9 \times 8 \times 10^6}{14 \times 3431} \times 5.38 = \$807 \text{ per trip.}$$

$$\text{Loan interest repayment} = \frac{0.053 \times 8 \times 10^6}{3431} \times 5.38 = \$665 \text{ per trip.}$$

$$\text{Insurance premium} = 0.005 \times \left[\frac{8 \times 10^6}{3431} \times 5.38 \right] = \$63 \text{ per trip.}$$

■ **TABLE 17.7**
Bizjet data for DOC estimation

Aircraft details	Turbofan details (two engines)	Conversion factors
MTOW: 9400 kg	T_{SLS}/engine: 17.23 kN	1 nm = 1.852 km
OEW: 5800 kg	Dry weight: 379 kg	US gallon = 6.78 lb
MEW: 5519 kg	Bypass ratio: 3.2	1 lb = 0.4535 kg
Payload: 1100 kg[a]	Number of compressor stages: 10[b]	1 ft = 0.3048 m
Range: 2000 nm	Overall compressor ratio: 14	1 kg fuel = 0.3245 gal
Block time: 5.38 h	Number of shafts: 2	
Block fuel: 2233 kg	Fuel cost: $0.75 per US gallon	

[a] Ten passengers.
[b] It has one high-pressure compressor, four-stage low-pressure compressor and one fan.

Crew salary and cost:

> Flight crew = $493 \times 5.38 = \$2652$ per trip for two crew.
> Cabin crew = 0 as there is no cabin crew.

Trip cost elements:

$$\text{Landing fees} = \frac{7.8 \times 9.4}{5.38} \times 5.38 = \$74 \text{ per trip.}$$

$$\text{Navigational charges} = \left(\frac{0.5 \times 2000 \times 1.852}{5.38} \right) \times \sqrt{\frac{9.4}{50}} \times 5.38 = \$803 \text{ per trip.}$$

$$\text{Ground handling charges} = \left(\frac{100 \times 1.1}{5.38} \right) \times 5.38 = \$110 \text{ per trip.}$$

Airframe maintenance, material and labour

$$\text{Airframe labour} = \left(0.09 \times 5.52 \times 1.02 + 6.7 - \frac{350}{5.52 \times 1.02 + 75} \right)$$
$$\times \left(\frac{0.8 + 0.68 \times (5.38 - 0.25)}{5.38} \right) \times 63$$
$$= (1.853) \times 1.448 \times 63 = \$169 \text{ per trip.}$$

$$\text{Airframe material cost} = \left(\frac{4.2 + 2.2 \times (5.38 - 0.25)}{5.38} \right) \times 7 \times 5.38 = \$109 \text{ per trip}$$

where $C_{airframe} = \$7$ million.

Total airframe maintenance (material + labour) = $\$910 + \$109 = \$1019$ per trip.

Engine maintenance, material and labour

$$\text{Engine labour} = 0.21 \times 63 \times 1.018 \times 0.89 \times (1 + 1.72)^{0.4} = \$17.88 \text{ per hour}$$

where $T = 17.23 \text{ kN} = 1.72$ tonnes; $C_1 = 1.27 - 0.2 \times 3.2^{0.2} = 1.018$;   and
$C_3 = 0.032 \times 10 + 0.57 = 0.89$.

$$\text{Engine material cost} = 2.56 \times (1 + 1.72)^{0.8} \times 1.018 \times (0.652 + 0.89)$$
$$= 5.7 \times 1.542 = \$8.79 \text{ per hour}$$

where $C_2 = 0.4 \times (14/20)^{1.3} + 0.4 = 0.652$, and C_1 and C_3 are as above.

Direct engine maintenance cost $(\text{labour} + \text{material})$
$$= 2 \times (17.88 + 8.79) \times \frac{(5.38 + 1.3)}{(5.38 - 0.25)} = \$308 \text{ per trip.}$$

$$\text{Fuel charges} = \frac{2233 \times 0.3245 \times 0.75}{5.38} \times 5.58 = \$544 \text{ per trip.}$$

■ **TABLE 17.8**
Bizjet summary of DOC per trip (all figures in US dollars)

Fixed cost elements	
Depreciation	807
Interest on loan	665
Insurance premium	63
Total ownership cost	1535
Flight crew	2652
Cabin crew	0
Total fixed cost elements	4187
Trip cost elements	
Fuel charges	544
Navigational charges	803
Landing charges	74
Ground handling charges	110
Maintenance (airframe)	1019
Maintenance (engine)	308
Total trip cost elements	2858
Total direct operating cost	$7045 per trip

The baseline Bizjet DOC is summarized in Table 17.8. DOC per hour = 7045/5.38 = $1309.5 per hour.

$$\text{DOC per aircraft nautical mile} = \frac{7045}{2000} = \$3.52/\text{nm}/\text{passenger}.$$

$$\text{DOC per passenger mile per nautical mile} = \frac{7045}{2000 \times 10} = \$0.352/\text{nm}/\text{passenger}.$$

Appendix C gives DOC details for a mid-size high subsonic transport aircraft.

17.4.2.1   Operating Costs of the Variants in the Family

Larger variant:

- MTOM = 10,800 kg and 14 passengers.
- Aircraft price = $9 million.
- Ownership cost element = $1727.
- Crew cost = $3047.
- Fuel cost = $625.
- Maintenance/operational charges = $2650.
- Total operational cost = $8049.

Then, direct operating cost (DOC) per hour = 8049/5.38 = $1496.4 per hour.

$$\text{DOC per aircraft nautical mile} = \frac{8049}{2000} = \$4.025 \text{ per nautical mile per trip}.$$

$$\text{DOC per passenger mile per nautical mile} = \frac{8049}{2000 \times 14} = \$0.2875/\text{nm}/\text{passenger}.$$

Smaller variant:

- MTOM = 7600 kg and 6 passengers.
- Aircraft price = $6 million.
- Ownership cost element = $1220.
- Crew cost = $2201.
- Fuel cost = $449.
- Maintenance/operational charges = $1910.
- Total operational cost = $5780.

Then, direct operating cost (DOC) per hour = 5780/5.38 = $1075 per hour.

$$\text{DOC per aircraft nautical mile} = \frac{5780}{2000} = \$2.898 \text{ per nautical mile per trip.}$$

$$\text{DOC per passenger mile per nautical mile} = \frac{5796}{2000 \times 6} = \$0.482 \,/\, nm \,/\, passenger.$$

17.5   Aircraft Performance Management (APM)

Commercial aircraft design is sensitive to the DOC, with the fuel-burn rate during operation (mainly climb and cruise) contributing nearly a third of the DOC. Just 1% of fuel saved in every mission could account for several million dollars in costs over the life-span of a small fleet of mid-sized aircraft. These become more important for larger aircraft operations, and increasingly important with increasing fuel costs.

Therefore it is important to reduce DOC by reducing fuel burn during operation without sacrificing the mission goal. In the past, for pilot ease, the older aircraft operational schedules were set to constant equivalent airspeed (EAS) operations, as considered in Chapter 12 for climb and Chapter 13 for long-range cruise/high-speed cruise. Modern aircraft operarion with microprocessor-based fly-by-wire/full authority digital engine control (FBW/FADEC) allows fine-tuning with synchronized tweaking of the engines to operate at lowest fuel burn around the prescribed speed schedules. This is seen as a part of the APM. The APM is nothing new but has steadily gained importance, facilitated by microprocessor-based aircraft systems architecture to improve all-round performance gains, safety, and real-time flight-tracking during operation.

These data can be used in a number of ways as listed below.

1. *Safety:* Early warning of a potentially fatal failure event.
2. *Fleet maintenance:* Fewer mission aborts, fewer grounded aircraft, simplified logistics for fleet deployment.
3. *Aircraft maintenance:* Aircraft structure/system maintenance management to cut down costs by avoiding unscheduled inspection, maintenance and trouble-shooting by recognizing incipient failure stages and taking action in the appropriate time in anticipation.
4. *Operational maintenance:* Improved flight safety, mission reliability and effectiveness.
5. *Performance:* Improved aircraft performance and reduced fuel consumption.

6. *Maintain status deck:* Gathering aircraft data in order to determine the actual performance level of each aeroplane (old and new) of the fleet with respect to the manufacturer's book level.

7. *Flight tracking (under study for implementation):* Real time-tracking and data streaming.

Today's aeronautical industry is involved with generation and acquisition of large amounts of data in all airline departments, in particular, in airline flight operations. The crucial issue is to identify the type of data needed and for what purpose. From organization to organization the aircraft management terminologies differ based on which aspects are considered in the monitoring process. Some of the terminologies have even been patented, signifying their importance. At the core, it means detecting all-round anomalies and taking corrective actions in time. Basically, it is a cost-saving measure that includes the following, as classified in this book.

Aircraft performance management (APM)

- Engine management (FADEC-based) to obtain least fuel consumption for the planned mission.
- Aircraft to operate in least trim drag (FBW-based).

Aircraft health monitoring (AHM)

Aircraft structure and system maintenance management to cut down costs by recognising incipient failure stages and taking action in anticipation. It is meant for timely and streamlined identification and diagnosis of issues. Performing repairs when the damage is minor increases the aircraft mean time before failure (MTBF) and decreases the mean time to repair (MTTR). It reduces the down-time; an aircraft that is not utilized is not returning on the investment.

The AHM not only monitors the aircraft but also the engines, providing significant overall fuel and emission performance measures. In earlier times, maintenance monitoring was carried out as technology permitted, for example by oil inspection, vibration checks, and so on, but nowadays microprocessor-based technology can have embedded internal sensors that can offer good insight to aircraft/engine health. These sensors send their data in real time to the aircraft engine monitoring system running with specialized software. It detects whether there are anomalies that require corrective actions.

The AHM also facilitates the spare parts flow schedule and stock inventory in a manner that does not block cash. It monitors performance measures for individual aeroplanes, enabling operators to improve overall average fleet performance.

In military operations, monitoring aircraft structural integrity is an important factor to determine how their fatigue life is being consumed. It generally involves the process of recording aircraft *g*-levels in three axes (sometimes also including data from the engine thermal cycling), downloading them into a database with the fatigue model of the aircraft, and then calculating the remaining Fatigue Index (FI) of the airframe.

Proposed flight track monitoring

The case of a state-of-the-art aircraft, the MH370 Boeing 777-200, disappearing, raises the question of how it can vanish. It brings back the debate about the need for global flight-tracking and data streaming that can act as back-up if the emergency locator transmitters fail to emit signals. The aviation authorities are expected to announce new global flight-tracking standards soon. A task force set up by the International Civil Aviation Organization (ICAO) is due to publish global flight-tracking standards in 2016.

An on/off switch is offered to the pilot to make a choice of whether to activate APM or not.

17.5.1   Methodology

The cash operating cost (COC), also known as mission trip cost (T_c), consists of the following elements:

1. Fuel cost (F_c – depends on the price at the time).
2. Crew cost (when airline fleet is in operation).
3. Maintenance (airframe and engine).
4. Navigational and landing charges (aircraft weight-dependent).

The higher the aircraft operational speed, the lower is the trip time (block-time); the saved time can offer higher levels of utilization of the aircraft, that is, more sorties can be accommodated per year, earning more revenue. However, an aircraft flying faster or slower than the speed which offers the maximum specific range (SR) will increase costs of fuel burn, as fewer nautical miles are covered per unit of fuel burn (Figure 13.2).

Also, with higher yearly utilization, the crew cost per flight hour goes down as they are available to make more sorties. When an aircraft is sitting on the ground, crew costs are not attributed to the costs operating that same aircraft. Their salaries are then accounted as a fixed cost element (see Table 17.6).

Maintenance costs goes up with faster speeds on account of greater wear and tear.

In summary, trip cost is dependent on block-time. An increase in aircraft cruise speed above the speed at maximum SR will decrease the block-time, resulting in increase in fuel and maintenance costs but a decrease in crew cost. Flying below the speed at maximum SR will increase the block-time, resulting in increased fuel and crew costs, but a decrease in maintenance costs. Figure 17.5 gives a representative graph depicting the relationship as given above. Landing/navigation (aircraft weight-dependent) costs are not shown, being relatively flat (small percentage) for the aircraft type, although higher utilization will incur additional costs.

Section 13.6.2 showed that to optimize aircraft performance, it is best to maximise (ML)/(sfc D) for LRC. The APM continually adjusts the aircraft flight parameters to fly around this point. The top line in Figure 17.5 is the total trip cost, which has a minimum at a slightly higher speed than the point of maximum SR. It is closer to maximum (ML)/(sfc D).

This book deals only with that part of the APM involving mission trip cost, primarily in fuel saving as a topic of aircraft performance. Saving fuel consumption in a mission sortie is based on (i) flying aircraft with minimum trim drag, which can be easily achieved with FBW aircraft that continuously adjust the aircraft CG position by fuel transfer to keep

■ **FIGURE 17.5**
Trip cost elements

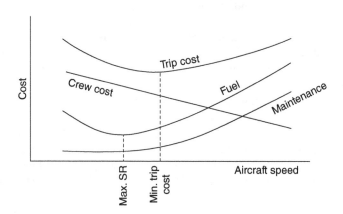

trimmed, and (ii) continuous thrust adjustment to offer minimum fuel burn with FADEC-controlled engines.

It is considered that the aircraft is flying at minimum trim drag and, to keep the explanation simple, only the fuel cost is taken as a significant variable as given in the following expression.

$$T_c = f(V)$$

that is, trip cost is a function of aircraft velocity, V. The objective of the APM is to minimise F_c by flying at the optimum V. Differentiating and setting equal to zero, then solving for V, gives the optimum. The relation $T_c = f(V)$ is not easily integrated, but the optimum V can be established through parametric stuzzdies by lengthy manual computation using the same equations derived in this book. Nowadays, with an FBW/FADEC-controlled onboard computer with the aircraft manufacturer's installed software incorporating all the relevant variables, an aircraft can continuously fly at optimum V to minimize trip costs.

17.5.2   Discussion – the Broader Issues

With the advent of modern microprocessor-based technologies, each industry is stretching the scope of APM, in a broader meaning of the acronym. It can cover a wide range of aircraft fleet operations to maintain a "status deck". It includes a procedure devoted to gathering aircraft data in order to determine the actual performance level of each aeroplane of the fleet with respect to the manufacturer's book level. The performance data given in the flight crew operating manual reflects this book value. It represents a fleet average for a brand new airframe and engines. This level is established in advance of production. Normal scatter for brand new aircraft leads to individual performances above and below the book value.

An APM program uses information from engine condition monitoring and air data computers to provide fleet operators with fuel burn information. If the fuel burn is higher than expected and the engines are operating as expected, then it would imply the engines are pulling an aircraft with a higher than normal aerodynamic drag. This would be caused typically by misrigged/out-of-trim control surfaces or an airframe with poor surface contouring. The APM calculates the SR during the cruise segment. Airlines compare the SR across the fleet to identify poorly performing aircraft and pull them in for overnight maintenance inspections or attaching rigging boards to adjust control surface rest positions, if necessary.

The performance levels are measured in their variations over time. Resulting trends can be made available to the operators' various departments, which perform corrective actions to keep the airline fleet in operation-ready condition.

References

1. Association of European Airlines, *Short-Medium Range Aircraft – AEA Requirements*. Association of European Airlines Publication, 1989.
2. Kundu, A.K., *Aircraft Design*. Cambridge University Press, Cambridge, 2010.
3. Kundu, A., Curran, R., Crosby, S., *et al.*, *Rapid Cost Modelling at the Conceptual Stage of Aircraft Design*. AIAA Aircraft Technology, Integration and Operations Forum, Los Angeles, CA, 2002.
4. Kundu, A.K., Crosby, S., Curran, R., *et al.,* Aircraft component manufacture case studies and operating cost reduction benefit. AIAA Conference ATIO, Denver, CO, November 2003.
5. Apgar, H., Design to life-cycle cost in aerospace. *IAA/AHS/ASEE Part G, Journal of Aero. Engineering,* Aerospace Design Conference Feb. 16–19, 1993.
6. NASA, *A New Method for Estimating Current and Future Transport Aircraft Operating Cost*. NASA CR 145190, March 1978.
7. Kundu, A.K., Watterson, J., *et al.* Parametric optimisation of manufacturing tolerances at the aircraft surface. *Journal of Aircraft*, **39**, 2, 271–279, 2002.

CHAPTER 18

Miscellaneous Considerations

18.1 Overview

Although the main task of the aircraft performance analyses is now complete, there are a few more topics of interest, varied in nature, which should be exposed to broader understanding on the subject. This chapter briefly overviews the history of US certification, now manifested as the FAA, and briefly outlines flight testing, describing the influence of wind, ground effects during field performance and effects of ice accretion on the wing. Detailed study of these topics is beyond the scope of this book.

18.2 Introduction

Described below are the few topics chosen here to broaden the horizon of readers on areas associated with aircraft performance.

1. *History of the FAA*

   A newcomer will find a maze of documents pertaining to airworthiness on the top of what the FAA issues. This is because there is an evolutionary past. The FAA, after its formation, continued with older issues by previous authorities. A brief history should help the newcomer with the structure of *type-certification*.

2. *Flight testing*

   To satisfy the mandatory requirements of the FAA, manufacturers' check out the design and guarantee customer specification. It is essential that the aircraft go through a lengthy series of flight tests for substantiation. Aircraft performance engineers play a significant role in analysing the flight test data to reduce it in prescribed format for documentation and issuance. The scope of flight tests is outlined here.

3. *Contribution of the ground effect on takeoff*

   This book does not deal in detail with the ground effects on aircraft performance when flying close to the ground, but some elementary exposure is given in this chapter.

---

4. *Flying in adverse environments*

This book primarily deals with zero wind conditions, with passing mention of head/tail wind effects. Head/tail winds also influence field performance at takeoff and landing. With cross-winds there are control issues, adding to the drag to keep the aircraft aligned to the airfield. Industry makes an effort to estimate their effects for the pilot manual. This chapter outlines those truly adverse environments, for example foggy weather, icing conditions, turbulent weather, wind shear, jet-stream and thunderstorms.

18.3   History of the FAA

Within a decade of the first flight by the Wright brothers, the potential of aircraft application in both civil and military sectors showed unprecedented promise of growth, as the free-market economy rushed in for quick gains through new discoveries to stay ahead of the competition. Proliferation in designs started to show compromises, and there were avoidable accidents/ incidents. Time and financial constraints forced industries, in certain situation, to rush into production without systematic substantiation of design to guarantee safety.

The history of regulation in the US began within that first decade. Its intention was to improve safety in the new aviation industry. The Federal Aviation Administration (FAA) can trace its formation to the 1958 Federal Aviation Act, although predecessor agencies have dated back to 1926, when the US Government first stepped in. The Air Commerce Act brought in a formal method to maintain a minimum standard within which aircraft could be flown safely. The Act was codified in the US Department of Commerce Bureau of Air Commerce Aeronautics Bulletin No.7, *Airworthiness Requirements for Aircraft.* In time, along with rapid progress, it was revised in October 1934 as Aeronautics Bulletin No.7A in more detail. Special requirements for commercial transport operation were detailed in Aeronautics Bulletin No.7-J.

As aircraft operation in the civil sector grew very rapidly, the US Government felt the need for a dedicated department, and in 1930 the Civil Aeronautics Authority was formed. The new set-up allowed more authoritative regulation, as Civil Air Regulation (CAR) Part 4 replaced the Aeronautics Bulletins issued by the earlier set-up. The CAR was in two parts: one for larger transport aircraft as Part 4T, and one for smaller aircraft as Part 4. The regulations were systematically kept updated, and in 1938 new regulations were issued: one for small aircraft as CAR Part 4A, and one for large aircraft as CAR Part 4B. What is small and what is large are defined as follows (from these FAR23 and FAR25 evolved):

CAR Part 3 for aircraft MTOW $\leq$ 12,500 lb;

CAR Part 4b for aircraft MTOW $>$ 12,500 lb;

CAR Part 8 for restricted category aircraft.

These regulations came in manual form as Civil Aeronautics Manuals; CAM3 for Part 3, CAM4b for Part 4b and CAM8 for Part 8. These contain interpretation of regulations, methods of tests, and so on.

The two World Wars were clear drivers for the explosive growth of the aircraft industry, followed by civil aircraft expansion on an unprecedented scale. The government needed to extend their administrative arm to regulate aircraft design. In 1958 the Federal Aviation Administration (FAA) replaced the Civil Aeronautics Agency, but maintained the CARs. With progress, in 1965 the CARs were replaced by new regulations issued by the FAA, as follows:

FAR 23 replacing CAR Part 3;

FAR 25 replacing CAR Part 4b.

In the period 1938 and 1965 there were other minor amendments. While old designs continued with the CARs, from 1965 new aircraft had to substantiate their designs under the FARs. Once the substantiation formalities are met, the certification authority issues *Type Certification* for the aircraft type, allowing them to be operated in the marketplace. Every modification to the aircraft requires approval from the certification authority.

Like the CAMs for the CARs, outlining detailed explanations and procedures for type certification, the FAA issued documents as follows:

Engineering Flight Test Guide (FAA Order 8110.7) in three categories of aircraft such as (i) Normal Category, (ii) Utility Category and (iii) Aerobatic Category starting from 1970; and
Engineering Flight Test Guide (FAA Order 8110.8) for Transport Category.

Because administrative responsibility meant that circulation of these documents was restricted, they were modified and issued as Advisory Circulars as follows. Advisory Circulars are not laws, unlike FARs.

Advisory Circular No: 23-8A for *Flight Test Guide for Certification of Part 3 Airplanes*, replacing *Engineering Flight Test Guide* (FAA Order 8110.7); and
Advisory Circular No: 25-7 for Transport Category Airplanes replacing *Engineering Flight Test Guide* (FAA Order 8110.8).

With the advent of helicopters and other special purpose aircraft, for example, agriculture spray aircraft, and so on, and new environment-related regulations, each area has their respective FARs. In fact, every country has their certification agencies. Governments in Western countries have developed and published thorough and systematic standards to enhance safety. These regulations are available in the public domain (see relevant websites). In civil applications, the main regulations are FAR and Joint Airworthiness Regulation (JAR) (the newly formed designation is European Aviation Safety Agency – EASA); both are quite close. The author prefers to work with the established FAR at this point. FAR documentation for certification has branched into many specialist categories. FAA publications include Airworthiness Directives (ADs), Orders and Notices, Advisory Circulars (ACs) and the all-important Federal Aviation Regulations (FARs). Aircraft designers and engineers deal with the FAA Aircraft Certification Offices (ACOs) and the Aircraft Evaluation Group (AEG), whose mission is to ensure that aircraft manufacturers offer adequate instructions for continued airworthiness, and do so prior to certification.

Table 18.1 gives FAR Categories of Airworthiness standards.

■ **TABLE 18.1**
FAR Categories of Airworthiness standards

Applications	General aviation	Normal	Transport
Aircraft	FAR Part 23	FAR Part 23	FAR Part 25
Engine	FAR Part 33	FAR Part 33	FAR Part 33
Propeller	FAR Part 35	FAR Part 35	FAR Part 35
Noise	FAR Part 36	FAR Part 36	FAR Part 36
General operations	FAR Part 91	FAR Part 91	FAR Part 91
Agriculture	FAR Part 137		
Large commercial transport			FAR Part 121

■ **TABLE 18.2**
Aircraft category definition

Aircraft features	General aviation	Normal	Transport
MTOW (lb)	Less than 12,500	Less than 12,500	Over 12,500
Number of engines	Zero or more	More than one	More than one
Type of engine	All types	Propeller only	All types
Flight crew	One	Two	Two
Cabin crew	None	None up to 19 passengers	None up to 19 passengers
Maximum number of occupants	≤9	≤23	Unrestricted
Maximum operating altitude	25,000 ft	25,000 ft	Unrestricted

FAR criteria for certification include many specialized categories and basic applicability. A partial listing is shown in Table 18.2 under the headings of General Aviation, Normal Category and Transport Category. Aerobatics type aircraft come under General Aviation. (New categories are evolving for the home-built/light aircraft types of small aircraft.)

18.3.1   Code of Federal Regulations

Section 1.5 mentioned that the US Government has 50 titles of Code of Federal Regulations (CFRs) published in the Federal Register. The FARs are rules prescribed by the FAA governing all aviation activities in the US under title 14 of the CFRs, but title 48 of the CFRs cover "Federal Acquisitions Regulations". To avoid confusion, of late the FAA began to refer to aerospace-specific regulations by the term "14 CFR part XX" instead of FAR.

In the UK, the Civil Aviation Authority (CAA), established in 1972, oversees every aspect of aviation in the UK. It had predecessor agencies dating many years prior to its formation. Recently some regulatory responsibilities in the UK have been passed to the EASA, which became operational in 1972. Based in Cologne, Germany, it is independent under European law, with an independent Board of Appeal. The EASA is currently developing close working relationships with its worldwide counterparts.

In military applications, the standards are Milspecs (US) and Defence Standard 970 (earlier AvP 970 – UK); they do differ in places.

18.3.2   The Role of Regulation

Aeroplane design and operations are reliant on regulatory standards and controls, as well as aeronautical science and the physics of flight. This reliance on regulations is mandatory in order to obtain or retain regulatory agency approval and certification. It is important to maintain the experience-proven safety standards, represented by mandated regulations, which transcend all other considerations in aviation.

The role of regulation in the design and operation of civil aircraft needs to be acknowledged by all interested parties. Regulation has been founded on empiricism as much as on analytical methods. Many regulations have been written into the law as a result of accidents and/or incidents, many of these being tragic. The regulatory authorities have taken an empirical approach to verifying the integrity of complex aircraft systems, in order to avoid the uncertainty of sole reliance on analytical solutions which may prove to be unreliable in actual practice. The air regulations generally pertain to aviation safety, including exhaust and noise emissions.

All civil aircraft must hold a *Type Certificate,* or its equivalent, issued by one or more of the responsible regulatory authorities where the aircraft is to be manufactured and/or entered

■ **TABLE 18.3**
Important regulatory standards from the FAA and EASA

FAR	EASA	Application
Part 23	CS-23	Normal, utility, acrobatic and commuter category airplanes
Part 25	CS-25	Transport category airplanes
Part 27	CS-27	Normal category rotorcraft
Part 29	CS-29	Transport category rotorcraft
Part 33	CS-E	Aircraft engines
Part 34	CS-34	Fuel venting and engine exhaust emissions
Part 35	CS-P	Propellers
Part 36	CS-36	Aircraft noise standards

into operation. The aircraft must also possess an Airworthiness Certificate in its operating venue. The exceptions are aircraft permitted to operate with an experimental certificate in restricted airspace to minimize danger to other aircraft and to those below. These aircraft are generally under development and strict scrutiny of the responsible regulatory authority.

Topics of particular concern regarding aircraft performance requirements that must be complied with include the following:

Stalling speed

Takeoff speeds, path, distance, takeoff run and flight path

Accelerate stop distance

Climb with all engines operating and one engine inoperable

Landing and baulked landing.

These requirements must be met under applicable atmospheric conditions, relative humidity and particular flight conditions, using available propulsive power or thrust, minus installation losses and power absorbed by the accessories such as fuel and oil pumps, alternators or generators, fuel control units, environmental control units, tachometers and services, for the appropriate aeroplane configuration.

Special tests required by the responsible regulatory authority will also have their influence, as completion and passage of these tests is mandatory. Testing to FAA, CAA and/or EASA regulations includes simulated bird strikes to engine inlets and windshields, operation under actual lightning conditions, one engine cut out during the most critical flight segment (usually second-segment climb for twin-engine aircraft), flyover testing of emanated noise levels during takeoff, fly-by and landing, cabin pressurization, engine heating, airframe vibration and flutter, as well as certain operating limitations, to name but a few. Table 18.3 gives a listing of important regulatory standards from the FAA and EASA.

An important advisory publication issued by the EASA is AMC-20, *General Means of Compliance for Airworthiness Products, Parts and Appliances.*

18.4 Flight Test

Aircraft capabilities have to be demonstrated through flight tests – most of them are mandatory requirements [1]. Flight test is a subject on its own, covering a large task carried out with interdisciplinary specialization, for example, aircraft performance and flight dynamics, dealing with a wide variety of instrumentation, avionics, computers, and so on. This chapter merely outlines the scope of the role required by the aircraft performance engineers. Stability and control tests are intertwined with aircraft performance.

Flight test obligations cover the following:

1. to substantiate requirements by the certification agencies to ensure safety;
2. to substantiate customer specification to guarantee performance;
3. to check the design as needed by the manufacturer (e.g. position errors of pitot-static tube, etc.);
4. to check modification/repair work carried out on an existing design;
5. to prepare operational manuals for aircrew and maintenance engineers; and
6. to prove new technology.

Flight testing is a protracted programme, in most industries conducted by a separate department in conjunction with the design bureau, primarily with the aerodynamics department. The design office has to prepare standards, schedules and checklists adhering to prescribed flight test manuals (FTMs) to give a streamlined procedure.

Test pilots are specially trained with considerable aircraft engineering background, at least up to the level of graduate engineers; some even have doctoral degrees. Pilot skill to maintain accurate and steady flying is essential. Stability and control response tests are carried out under the strictest precautions for safety – fatal accidents have occurred in the past.

Fight test engineers are also specially trained engineers fully conversant with all kinds of instruments. Calibration of test instruments is a prerequisite, and an elaborate procedure is followed to ensure credibility of data acquisition. Some instruments have a lag-time; pilots and test engineers work together to record steady-state stabilized data. Confidence in raw data is graded from poor to excellent in five classes. Recognizing and deleting poor or marginal data improves the test result for usage. The internal cabin volume of a larger transport category aircraft can be utilized to carry on-board recorders and processors to analyse real-time data for the flight-test engineers to decide acceptance or to repeat. Military aircraft use telemetry to record real-time data. The bulk of the test data are analysed by office-based engineers to prepare reports and manuals.

Typical time frames taken and flight test details for various designs are given in Table 18.4. It is assumed that tests continue uninterrupted without major design revision, and that the aircraft is not of a new class.

Aircraft performance figures depend on atmospheric conditions, which can categorically be said never to be the same or uniform. It is therefore important to reduce the recorded data to a standard condition, such as zero-wind on an ISA day – this helps comparisons to meet requirements in the specified format. The test volume is large, and data reduction time using computers is also large. Considerable improvement in data fidelity can be achieved if spurious data can be recognized and eliminated, and there are formal methods available for its execution. Skilled test pilots use precise techniques to minimize bad data.

It is possible to reduce testing time if more aircraft are deployed. Data acquisition and reduction to a useable form takes substantial amounts of time. To ease financial and time constraints, meticulous planning of the test schedule is to be done in consultation with the certification agencies and customers.

All flight tests are preceded by accurate ground-based calibration of test equipment onboard. Possibly one of the most important aircraft characteristics is to establish its drag polar. Engine manufacturer-supplied calibrated engines are required to be installed in the aircraft. The calibrated engines are ground-tested before being used at altitude. A series of tests is required, sometimes in two aircraft.

■ **TABLE 18.4**
Typical aircraft flight test details

Type of aircraft	Typical number of aircraft	Typical time taken (yr)	Typical number of sorties
Large commercial transport aircraft	≈4–6	≈2–3	≈2000
Medium commercial transport aircraft	≈3–5	≈2–3	≈1500
Business/executive aircraft	≈2–4	≈2	≈1200
Small aircraft (club flying type)	≈2–3	≈1–2	≈400–800
Military combat aircraft[a]	≈4–6 (2)	≈3–4 (2)	≈2500 (500)
Military training aircraft[a]	≈2–4 (1)	≈2–3 (1)	≈1500 (200)

[a] In brackets is the additional requirement for armament tests.

Stall and spin tests are essential. Aerodynamic analyses and wind tunnel tests offer some characteristics, but with a multitude of deviations in variables, a limitation in manufacturing technology can invalidate analyses, and in marginal situations the actual tests could prove dangerous. Section 6.7 describes spin tests in more detail.

Stability and control tests to establish aircraft handling qualities also require participation by aircraft performance engineers in establishing speed schedules, and so on. Establishing flutter speed, high speed at dive and trim run-out tests are critical and could also prove dangerous.

Military aircraft go through considerably more stringent flight tests as they invariably try to incorporate innovative technologies. In general, a technology demonstrator aircraft is used as the prototype, possibly in a scaled-down model as a proving platform that may undergo many modifications. This can be a lengthy programme until a satisfactory outcome emerges. Details in Table 18.4 do not include these kinds of flight tests.

Both civil and military aircraft must undertake all-weather trials, for which aircraft are taken to designated test centres that have polar or hot weather climates, or to high-altitude stations to prove takeoff capability. Penalties associated with environment control systems (engine air bleeds for cabin requirements, anti-icing protection, other mechanical off-takes, and so on, penalize engine power) have to be substantiated to ensure safety within the specified flight envelope.

In summary, the role of aircraft performance engineers in aircraft flight test programmes is substantial. This book covers the fundamental theories for what are required for the flight test data reduction. (Aircraft performance engineers play some role in aircraft simulator design to supplement the control laws given by stability/control engineers devoted to flight testing.)

18.5 Contribution of the Ground Effect on Takeoff

Chapter 11 dealt with field performance without considering ground effects. A conservative estimate is accepted by the certification agencies. This section is to make readers aware of aerodynamic effects on aircraft performance and subsequently explore them when required.

Ground proximity affects aircraft aerodynamics. As a restricting surface, the ground acts as a constraint to the development of wing tip vortices. As a result, there is an increase in C_L and reduction in C_D.

The wing tip vortices hit the ground, making their vertical velocity zero. A mathematical model was proposed as early as the 1920s using Prandtl's lifting-line theory. A mirror image of the wing vortices below the ground plane replaces the ground interaction with wing vortices, resulting in zero vertical velocity. The upward velocity of the mirrored vortices increase lift, and there is an associated reduction in induced drag at the same angle of incidence, α. The closer the wing to the ground, the greater the effect, and similarly reduces in a non-linear manner to zero as the gap between wing and ground increases to a height equal to the wing span (or depending on the design, half to twice the wing span). There is also reduction of parasite drag due to some reduction in effective velocity. Textbooks divide the zones as *in-ground effect* (IGE) and *out-of-ground effect* (OGE). A good analysis of ground effect is given in [2]. It also lists other publications available in the public domain.

However, the accuracy of the theoretical results suffers when flaps are deployed, as the flow-field over the wing gets complex. At high flap deflection, experiments show a reduction in lift. The IGE has other adverse effects on modern military combat aircraft that use a thrust-reverser for short landing; recirculating the reverse flow of engine exhaust gases can result in possible re-ingestion, affecting engine performance. Even vector-thrust for short takeoff can affect engine performance at this critical moment. Other stability and control problems may also appear.

Motor gliders with a large span and low-wing design staying close to the ground will benefit from ground effects at takeoff, but at landing it tends to float and takes time to settle down when lift-dumpers (an appropriate terminology for a spoiler or air-brake) are deployed.

Taking the benefits of the ground effect, the Russians successfully designed several versions of the *Ekranoplan* and even operated on the Black Sea. These designs never surfaced in the Western market, but subsequently Boeing studied a ground-effect craft (*Pelican*). It has not been built, but the potential still exists.

18.6   Flying in Adverse Environments

This section briefly discusses problems associated with flying in icing conditions, foggy weather, turbulent weather, wind shear, jet-stream and thunderstorms. Any of these are adverse environments for an aircraft to fly, and can be classified into two main groups: (i) adverse environment due to loss of visibility; and (ii) adverse environment due to aerodynamic and stability/control degradation.

18.6.1   Adverse Environment as Loss of Visibility

Loss of forward visibility and horizontal reference can affect pilot performance, and accidents happen when unable to see or judge flight obstacles and/or aircraft attitude. Lack of visibility is like flying on a dark cloudy night. Fortunately, degraded visibility does not affect aircraft performance, stability and control, other than affecting pilot skill. Flight deck instruments prove adequate under skilled hands. There are mainly three types of environmental conditions that affect pilot visibility, as described below.

1. *Atmospheric fog*: There are different kinds of reasons for fog formation and it can appear suddenly. Normally it occurs at lower altitudes; at higher altitudes the visibility problems arise from flying in cloud instead. Takeoff and landing become difficult, or an obstacle is not seen until it is too late. Lack of a horizon reference can make aircraft enter a spiral dive, which is not easy to detect at the initial stages by inexperienced pilots. Commercial aircraft pilots must have instrument-flying training. Aircraft fitted with forward-scanning radar is of considerable assistance. In any case, it is better to wait until fog has cleared, normally a relatively short time. In case a pilot confronts foggy conditions at landing, it should be undertaken using the instrument landing

■ **FIGURE 18.1**
BAe Regional Jet
family rain-repellent
system. Reproduced
with permission from
BAe Systems

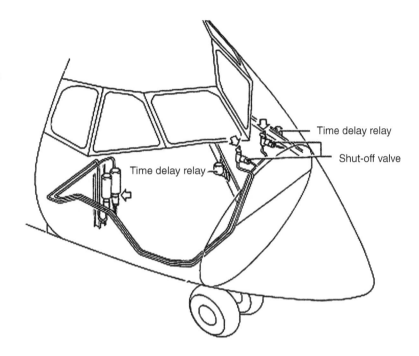

system (ILS) supported by the airport control tower. If possible, a pilot should loiter around close by where there is visibility until the destination airport fog condition improves. In the worst case, a pilot could opt for an alternative airport provided there is sufficient fuel.

2. *Flight-deck fog*: "Fogging up" of the flight deck is similar to what can happen in a car on a cold and wet day. Flight-deck fog has to be removed. The defogging system is like that of a car, with an embedded electrical wire in the windscreen. Figure 18.1 shows a generic layout with wipers, representing the BAe regional jet family in better detail.

3. *Heavy rain*: Heavy rain can also impair vision. The rain-removal system is also like that of a car, using windshield wipers. Rain-repellent chemicals also assist in rain removal. Figure 18.1 shows the regional jet family rain-repellent system.

18.6.2   Adverse Environment Due to Aerodynamic and Stability/Control Degradation

Adverse environments that affect aircraft performance, stability and control fall into two main types: icing hazards, and stability/control hazards.

18.6.2.1   Ice Formation   Ice formation on aircraft surfaces affects skin condition and considerably increases drag. In addition, ice formation on critical areas of lifting surfaces, intake and internal engine components can develop into a dangerous situation with a loss of lift and power. Ice accretion at the leading edge deforms the required aerofoil contour, disturbing the smooth streamline flow and causing premature stall. Ice accumulation at intake will degrade engine performance and can even damage the engine if large chunks are ingested.

It is difficult to notice visually clear ice forming at the early stages, and therefore engineers make provision to remove ice accretion as described below. Figure 3.32 shows typical icing envelopes.

It is a mandatory requirement to keep aircraft free from icing degradation. It can be achieved either by anti-icing, which stops ice from forming on critical areas, or by de-icing, which allows ice build-up to a point and then sheds it just before it becomes harmful. De-icing results in blowing chunks of ice away, and these should not hit or get ingested into an engine.

There are several methods to anti-icing and de-icing, and not all of them use pneumatics. The following methods are currently in practice [3].

1. *Use of hot air blown through ducts:* This is a pneumatic system, and the dominant system for larger civil aircraft. It is achieved by routing high-pressure hot air bled from the mid-compressor stage to the critical areas through perforated ducts (known as piccolo tubes). Pressure and temperature in the duct are about 25 psi and 200 °C. Designers must ensure that damage does not occur on account of overheating. Figure 18.2 shows a typical system.

2. *Use of boots:* Both anti-icing and de-icing can use boots. Rubber boots are wrapped (capped) around the critical areas (e.g. leading edges of lifting surfaces, propeller leading edges, intake lips, etc.) and heated either by electrical elements or by passing hot air. Electrically heated boots are lighter but can be relatively more expensive. Figure 18.3 shows a typical boot system.

3. *Use of electrical impulse:* This is not common but is quite an effective de-icing system. Ice is allowed to accumulate up to a point when vibration generated by an electrical impulse breaks the ice layer. It has low power consumption, but can be a heavy and expensive system.

4. *Use of chemicals:* This is also not a common anti-icing system. Glycol-based antifreeze is allowed to sweat through small holes in critical areas where the chemical is stored. It is limited by the amount of antifreeze carried onboard.

The piston engine carburettor and critical instruments must be kept heated for them to keep functioning without icing problems.

18.6.2.2   Stability/Control Degradation   Turbulent weather, wind shear, jet stream and thunderstorms are the main causes of problems with aircraft stability/control, and can degrade to dangerous conditions. These are described separately below. Weather-related disturbances primarily concern the control of the aircraft, bearing in mind that there is sufficient stability mar-

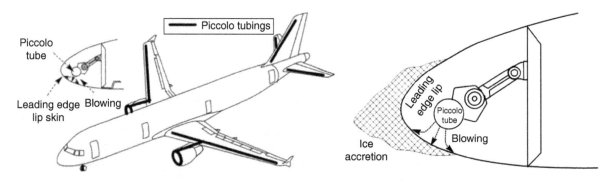

■ **FIGURE 18.2**   Civil aircraft anti-icing subsystem

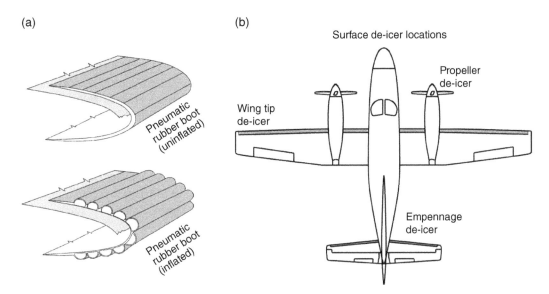

■ FIGURE 18.3   Anti-icing subsystem using boots (Goodrich). (a) pneumatically inflated rubber boot. (b) boot with electric element to heat

gin to control satisfactorily in any inadvertent situation. Meteorological offices routinely circulate weather forecasts, which all pilots must study and consider in their flight plan. Meteorology is an involved science beyond the scope of this book. Here, only the effects of adverse weather are touched upon. Flight planning must accommodate necessary provisions for bad weather.

1. *Turbulent weather:* Random, unsteady atmospheric wind flow that can change direction is seen as turbulent weather. There are several ways that turbulence develops, for example, ground contour/heat distribution developing convective currents. These can come as gusts of wind vibrating the aircraft, and sometimes can be severe enough to throw passengers from their seats if not secured with restraints such as seat-belts; injuries have happened. Modern transport aircraft are fitted with auto-stabilization systems that dampen the shaking of the aircraft to a considerable extent, but in strong turbulence the aircraft can suffer a very bumpy ride. All passengers must adhere to the cabin safety instructions to sit with seat-belts on for their own safety.

   The FAA has requirements for transport aircraft to withstand an upwards gust of 66 ft/s. Typically, weather reports can identify areas under turbulent conditions, and good weather radar can detect strong turbulences. Pilots should avoid flying through these conditions by making diversions; normally flight planning accommodates extra fuel for possible diversions.

   Strong downward gusts can cause an aircraft to lose considerable height. At low altitudes, say at approach, a strong down-wind can prove dangerous. Destination airfields offer local weather conditions, and pilots need to remain alert to control the aircraft in any unexpected condition.

   At high altitudes around the tropopause, turbulence can appear as clear air turbulence (CAT), which is not easy to detect and can be severe. Wind shear and jet streams are associated with CAT.

2. *Jet stream:* Around the tropopause there is a global pattern of periodic tube-like winds (relatively thin but very wide) flowing in long stretches from west to east in the mid-

latitudes of the northern hemisphere. When the average wind velocity exceeds 50 kt, it is called a jet stream; its core can have high velocities exceeding 100 kt. It is in an unsteady state, and its size and strength keep changing. There can be considerable fuel savings in the jet stream when flying eastwards, and *vice versa* in the southern hemisphere.

3. *Wind shear:* An aircraft enters into wind shear when it enters a zone where wind speed changes in magnitude and direction. The aircraft must be able to adjust to these changes; generally it is controlled through the disturbance in a relatively short time. A downdraft will cause the aircraft to sink rapidly, which can be observed using the flight-deck instruments. Aircraft sinking when flying close to the ground, say at approach to land, can pose a hazardous situation. Pilots need to remain alert at takeoff and landing phases to prevent the aircraft from hitting the ground.

4. *Thunderstorm/cloud:* Thunderstorms are violent outbursts in the atmosphere associated with cumulonimbus clouds (cumulonimbus clouds are nimbus clouds in the form of rising cumulus cloud). The violent nature can be observed by their swirling pattern moving mainly upwards. Gusts can easily exceed the design limits and can have catastrophic consequences if flown through. Fortunately, thunderstorm areas can be easily located and pilots must detour the area. A pilot may fly through a low-intensity thunderstorm after studying its nature, and precautionary measures must be taken for both passengers and crew. There are prescribed procedures to fly through permissible thunderstorms.

Lightning and hail are also associated with thunderstorms. Lightning is electrical discharge of the accumulated static electricity. The discharge is towards the area of maximum potential difference, branching out in different directions; this can be within the cloud or may strike the ground. The vast expanses of thunderstorm cloud can develop very high voltages and current that can be destructive, and it is another reason for pilots to detour the area. Aircraft are equipped to discharge static electricity build-up, and the shell insulates the interior. However, they are not designed to fly through severe thunderstorms.

Glider pilots seek updraft from thermals to gain altitude. They often take advantage of the upward flow of small cumulus clouds, which at the early stages may be seen like patches of cotton floating in interesting shapes. Pilots must be trained to recognize which ones are harmless.

18.7 Bird Strikes

Birds can create an adverse flying environment. The open span of grass field around the runway is an ideal feeding ground for birds – seeds, grains, insects, worms, and so on, abound. Some even make their permanent home there, and keep multiplying. They can come from outside in large flocks. Although the birds get used to aircraft noise, they may be spooked into flying into the aircraft flight path during takeoff and landing, when their ingestion into the engine is not uncommon. In many instances, single small bird ingestion by the engine may not be felt, but experienced pilots can tell by observing engine gauges and from the feel. On some occasions, the aircraft continues towards the destination airport and finds bird strike evidence after landing. When bird strikes happen at high speed, the impact force is much higher and can be felt by pilots.

Aircraft certification authorities have mandatory requirements to demonstrate safety by throwing fresh dead chickens into engines, known as the *chicken test*. Most bird strikes occur around airfields where birds rarely exceed that size. However, a larger bird strike is a different matter.

The 2009 Hudson River landing of an Airbus 320 after both engines failed on account of bird strike (Canada Geese?) is a well-known case of the kind of accident that can happen. At higher altitudes, vulture strikes have occurred many times. There are cases when older military aircraft with sensitive engines have been brought down by bird strikes as small as sparrows.

Birds around airports can be driven out, at least for a temporary period. Various attempts have been made to drive them out, with limited success. Constant surveillance, various scaring techniques, and environmental planning can allow flight operations to continue. Over long periods of time these may become an effective measure to some extent. However, stray birds away from the airfield are a random occurrence. Pilots must then land at the nearest available airport for safety reasons; the successful landing on the Hudson River is now incorporated as a pilot drill as a measure to be taken if landing at an airfield is not possible. Without a suitable stretch of water or an airfield, a pilot has to take action as he/she judges.

18.8 Military Aircraft Flying Hazards and Survivability

In additions to the hazards of flying as described above, military aircraft have to face additional hazards when flying at low altitudes, especially through terrain as a cover. There are hazards in hard manoeuvres to engage or disengage in combat at lower altitudes. As some measure of safety, military aircraft are fitted with *ground proximity warning systems*. Of late, modern civil transport aircraft above 30,000 lb MTOW are also fitted with such systems. These instruments sense the terrain and ground below to warn pilots in time to take action (typically 1 second). All types of aircraft, even small recreational aircraft flying through valleys for sight-seeing, must keep adequate separation distance if a manoeuvre, for example turning and climb-out, is required. A pilot's understanding of the surroundings, be it terrain or other aircraft, is essential even when terrain maps and ground proximity warning systems are integrated into the aircraft.

Jettisoning stores in certain manoeuvres can be hazardous if they hit any part of aircraft. There have been cases of light empty drop-tanks tumbling at jettison, hitting the empennage.

New combat aircraft capabilities are evaluated in simulated environment in twin platforms, the other platform being that of adversary capabilities to the extent that they can be ascertained in terms of combat scores [4]. These are based on mathematical modelling with measurable parameters, for example *susceptibility* (inability to avoid adversary interception), *vulnerability* (inability to withstand hit), and *killability* (inability to both). The general term *aircraft combat survivability* combines these parameters, and those working in the aircraft survivability disciplines attempt to maximize the survival of aircraft in wartime. These topics are beyond the scope of this book, but readers may refer to [4].

18.9 Relevant Civil Aircraft Statistics

Figure 10.3 gives the important MTOW–range–passenger number statistics to size civil aircraft [5, 6]. Figure 10.3 is summarized in Table 18.5.

As an example, for a mission profile with 300 passengers to 5000 nm range, it is expected to have maximum takeoff mass (MTOM) of around $750 \times 300 = 225,000$ kg (may compare with Airbus 300-300).

18.9.1 Maximum Takeoff Mass versus Operational Empty Mass

Figure 18.4 provides crucial information to establish the relationship between the MTOM and the operational empty mass, OEM. The ratio of (OEM/MTOM), known as the operational empty mass fraction (OEMF), can be found from the graph.

■ **TABLE 18.5**
MTOM per passenger versus range (mid-size aircraft and above)

Range (nm)	MTOM (kg) per passenger
1500	400
3500	600
6500	900
8000	1050

■ **FIGURE 18.4**
OEM versus MTOM

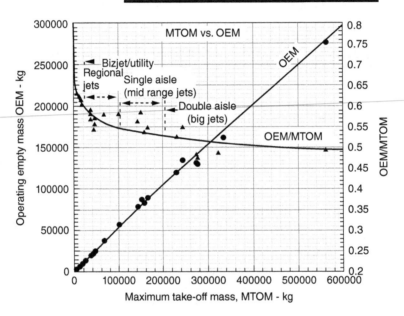

The graph indicates a predictable OEM growth with MTOM in a nearly linear manner. Observing more closely it can be seen that at the lower end, aircraft with less than 70 passengers (Bizjet, utility, RJ class) have a higher OEMF (around 0.6 – sharply decreasing), in the mid range (70–200 passenger class – single-aisle narrow body) OEMF is holding around 0.56, and at the higher end (above 200 passengers – double-aisle wide body) it levels out at around 0.483 (MTOM is slightly over twice the OEM). The decreasing trend of the weight fraction is on account of better structural efficiencies achieved with larger geometries, use of lighter materials, and more accurate design/manufacturing methods for the later designs.

18.9.2 MTOM versus Fuel Load, M_f

Figure 18.5 shows the relationship between fuel load, M_f, and MTOM for 20 turbofan aircraft. This graph gives the fuel fraction, $M_f/$MTOM. It may be examined in conjunction with Figures 10.3 and 18.4, showing a range increase with MTOM increase. Fuel mass increases with aircraft size, reflecting today's market demand for longer range. For long-range aircraft, the fuel load including the reserve is slightly less than half of the MTOM. At the low end, for the same passenger capacity there is wide variation, indicating aircraft having a selection of comfort levels and a choice of aerodynamic devices, as well as varied market demand – from short ranges of around 1400 nm to cross-country ranges of around 250 nm. At the higher end the selection narrows down, showing a linear trend.

■ **FIGURE 18.5**
Fuel load versus
MTOM (Ref. [7]).
Reproduced with
permission from
Dover Publications

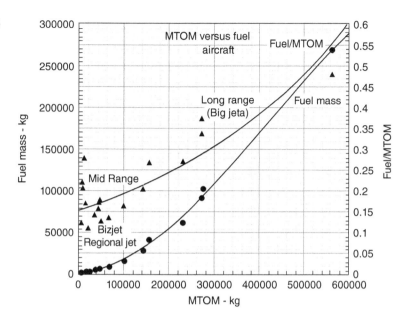

18.9.3   MTOM versus Wing Area, S_W

While fuselage size is derived from the design passenger capacity, the wing has to be sized to meet the performance constraints through a matched engine. Chapter 16 offers a formal presentation of the aircraft sizing and engine matching procedure. Figure 18.6 shows the relationship between wing planform reference area, S_W and wing loading (definition given in the next paragraph) versus MTOM. These graphs are a useful starting point (preliminary sizing) for a new aircraft design to be refined through sizing analysis.

Wing loading, W/S_w, is defined as the ratio of MTOM to the wing plan form reference area (W/S_W=MTOM/wing area, kg/m²; if expressed in terms of weight then the unit becomes N/m² or lb/ft²). This is a very significant sizing parameter and would play a dominant role in aircraft sizing, design and performance analyses.

The influence of wing loading is clear in Figure 18.6. The tendency is to have lower wing loading for smaller aircraft and higher wing loading for bigger aircraft operating at high subsonic speeds. High wing loading would require the assistance of better high-lift devices to operate at low speeds. Better high-lift devices are heaver and more expensive.

Growth of wing area with aircraft mass is necessary to sustain flight. A large wing planform area is required for better low-speed field performance, which is in excess of the cruise requirement. Therefore, wing sizing (Chapter 11) offers the minimum wing planform area to satisfy both the takeoff and cruise requirements simultaneously. Determination of wing loading is a result of the wing sizing exercise.

Smaller aircraft operate in smaller airfields, and to keep the weight and cost down, simpler types of high-lift devices are in use. This would result in lower wing loading (200–500 kg/m²), as can be seen in Figure 18.6a. Aircraft with ranges of more than 3000 nm could consider more efficient high-lift devices.

The trends for variants in the family of aircraft design can be examined. The Airbus 320 baseline aircraft was the middle in the family. The A320 family retained the wing to maintain component commonality, which reduced manufacturing costs substantially as no modifications have to be carried out for the variants. This resulted in large changes in wing loading – the smallest A318 has low wing loading with excellent field performance, and the

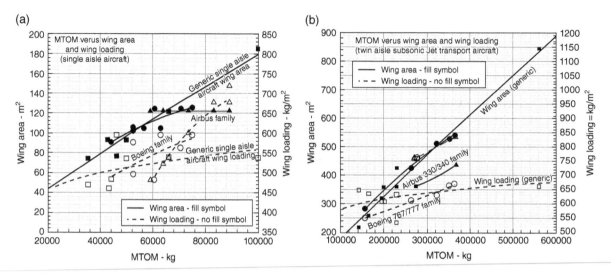

■ **FIGURE 18.6** Wing area, S_w, versus MTOM. (a) Mid-range single-aisle aircraft. (b) Large twin-aisle aircraft

largest in the family A321 has high wing loading that requires higher thrust loading to keep field performance from degrading below requirements. On the other hand, the Boeing 737 baseline aircraft started with the smallest in the family and was forced into wing growth, with growth in weight and cost. This keeps the changes in wing loading to a moderate level.

Larger aircraft have longer ranges; hence wing loading is higher to keep wing area low, thus lowering drag. For large twin-aisle subsonic jet aircraft, the picture is fairly similar to the mid-range sized single-aisle aircraft, but with higher wing loadings (500–900 kg/m²) to keep the wing size relatively small (countering the square-cube law as discussed in Section 2.17.1). It is necessary to have advanced high-lift devices and longer runways for large aircraft.

18.9.4 MTOM versus Engine Power

The relationship between engine size and MTOM is given in Figure 18.7. Turbofan engine size is expressed as sea-level static thrust (T_{SLS}) on an ISA day at takeoff ratings, when the engines produce maximum thrust. These graphs may only be used for preliminary sizing. Formal sizing and engine matching is done in Chapter 16.

Thrust loading, T/W, is defined as the ratio of total thrust (T_{SLS_tot}) of all the engines to the weight of the aircraft. Once again a clear relationship can be established through regression analysis. Mandatory airworthiness regulations require that multi-engine aircraft should able to climb at a specified gradient (see FAA requirements in Chapters 10 and 11) with one engine inoperative. For a twin-engine aircraft, failure of an engine amounts to loss of 50% of power, while for a four-engine aircraft it amounts to 25% loss of power. Therefore, T/W for two-engine aircraft would be higher than for a four-engine aircraft.

The constraints for engine matching would be that it should simultaneously satisfy sufficient takeoff thrust to meet: (i) field length specifications; (ii) initial climb requirements; and (iii) initial high-speed cruise requirements from the market specifications. Growth of engine thrust with aircraft mass is obvious to meet takeoff performance. The engine matching is dependent on wing size, number of engines and the type of high-lift device used. Propeller-driven aircraft are rated in power P in kW (HP or SHP), which in turn gives the thrust. Turboprop cases use power loading, P/W instead of T/W.

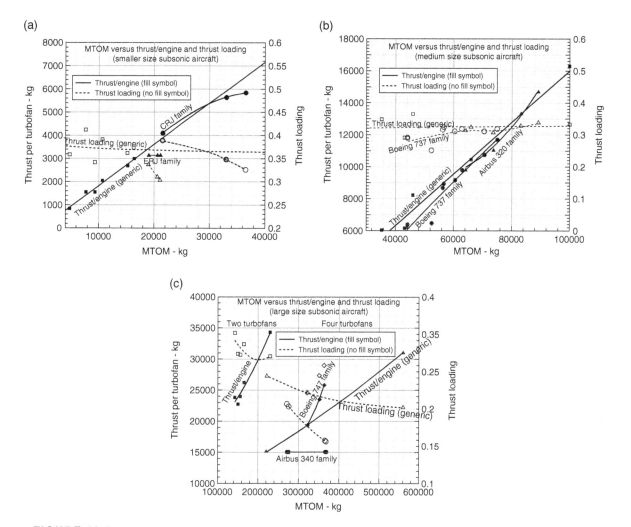

■ **FIGURE 18.7**   **Total sea-level static thrust versus MTOM. (a) Small aircraft. (b) Mid-size aircraft. (c) Large aircraft**

Smaller aircraft operate in smaller airfields and are generally configured with two engines and simpler flap types to keep costs down. Figure 18.7a shows thrust growth with size for small aircraft. Here thrust loading is held within 0.35–0.45. Figure 18.7b shows mid-range statistics, mostly for two engines. Mid-range aircraft operate in better and longer airfields than smaller aircraft; hence the thrust loading range is at a lower value between 0.3 and 0.37. Figure 18.7c shows long-range statistics; the aircraft have two or four engines – the three engine configuration is not currently in practice. Long-range aircraft with superior high-lift devices and long runways means that thrust loading could be kept between 0.22 and 0.33 (lower values are for four-engine aircraft). Trends in family variants in each of the three classes are also shown.

18.9.5   Empennage Area versus Wing Area

Once the wing area is established along with fuselage length and matched engine size, the empennage areas (horizontal tail, S_H and vertical tail, S_v) can be estimated from the static stability requirements. Chapter 2 deals in detail with empennage tail volume coefficients, yielding the figures to determine empennage areas.

Figure 18.8 gives growth for horizontal and vertical tail surface areas with MTOM. The variants in the families do not show changes in empennage areas, to maintain component commonality.

■ **FIGURE 18.8**
Empennage area versus MTOM. (a) Smaller aircraft. (b) Larger aircraft

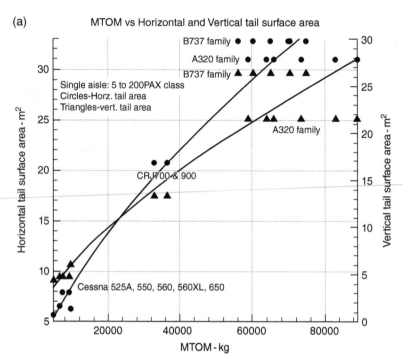

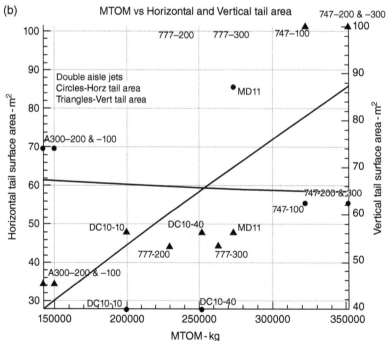

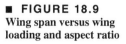

■ FIGURE 18.9
Wing span versus wing loading and aspect ratio

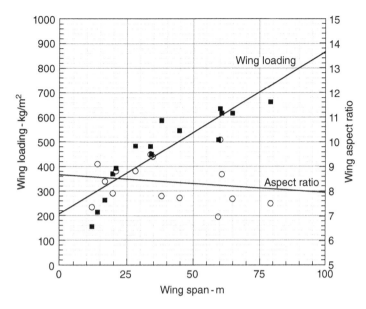

18.9.6   Wing Loading versus Aircraft Span

Figure 18.9 demonstrates that growth in wing-span is associated with growth in wing loading.

Along with steady improvement in new materials, miniaturization of equipment, and better fuel economy, wing-spans are increasing with the introduction of bigger aircraft (e.g. Airbus 380). Growth in size means the wing root thickness becomes big enough to swallow the whole fuselage depth, at which point a blended-wing body (BWB) configuration becomes a very attractive proposition for very large capacity aircraft. Although technically it is feasible, it awaits market readiness, especially from the ground handling point of view at the airports.

The aspect ratio shows scatter. In the same class of span, the aspect ratio could be increased with advancing technology, but kept restricted with increase in wing load. Current technology gives aspect ratios from 8 to 16.

18.10   Extended Twin-Engine Operation (ETOP)

In the past, twin-engine transport aircraft were not allowed to fly over water for safety reasons, in case an engine fails losing half the power, coupled with any other reason for the other engine not to supply adequate power. Any twin-engine commercial transport aircraft were restricted to operate within the land mass by a route where emergency airfields were available in case of emergency requirements to land. In this section, ETOP range is explained, and readers may attempt to compute range using the same procedure with the effects of an engine failure and drift-down procedure (Section 14.7).

With time, confidence has built up as engines have become more reliable, and failure at cruise power rating has become a rarity. In 1980, after discussions with operators and manufacturers, the FAA allowed flying over water up to a distance that ensures that, in case of emergency, the aircraft can land at a suitable airfield within 120 minutes under one engine powered up to maximum climb (i.e. continuous) rating. The operation is known as *extended twin-engine operation*. After the success of ETOPs under the 120 minutes restriction, the FAA subsequently raised the time limit to 180 minutes. This extension of time limit improved the fuel efficiency of operation. Since 1985, the acronym, ETOPS, has been defined as "extended twin operations" and has been limited to part 121 airplanes with only two engines. Current regulations have extended these applications to airplanes operating in both parts 121 and 135, and the acronym has now been refined to mean "extended operation".

18.11   Flight and Human Physiology

Defying gravity can be dangerous if the laws of nature are not fully understood [8, 9]. Apart from solving the laws of physics to make aircraft fly and gain altitude, there is the additional issue of understanding the nature of human physiology and its limitations. Inadequacy in the medical condition of a pilot is an actual threat to survival. Qualifying for medical airworthiness is mandatory for all pilots; its level of stringency differs for the class of aircraft to be operated, and combat pilots have the strictest requirements. Even for fit pilots, the influence of drugs/alcohol can affect medical airworthiness. Aviation physiology is a subject on its own, which is now a mature science, to understand medical human factors, their limitations and pilot response characteristics.

There are two main limitations based on (i) altitude effects and (ii) gravitational loads which can be different from the unit load (1 g) one experiences in day-to-day living. These two aspects are briefly described below.

1. *Effects of altitude:* Pressure and density of atmosphere decreases in a non-linear manner (see Section 2.3). Human habitation can exist above 10,000 ft in mountainous terrain, but by far the majority of normal habitation is below 3000 ft. It has been found that human beings can exist unassisted up to 14,000 ft. Aircraft flying above that altitude must have an environmental control system (ECS) to keep pressure at an acceptable limit, typically equivalent to that around 7000–8000 ft, depending on design (see Section 12.2). With adequate pressurization, the oxygen supply for human needs is simultaneously met. However, in inadvertent situations of cabin depressurization, then an emergency oxygen supply is necessary (see Section 3.9.3). The ECS takes engine airflow bleed at about 2–3% degradation of engine performance, affecting fuel economy (see Section 7.7). The aircraft makes a rapid descent to a safe altitude below 10,000 ft.

2. *Effects of g-load:* In manoeuvre (see Chapters 15), an aircraft experiences g-load. There two kinds of g-load: instantaneous and sustained. Instantaneous g-load is of short duration lasting few seconds, and humans can tolerate a higher load than under sustained g-load, which lasts longer (in space travel, initial booster g-load on astronauts can last in excess of 4 minutes). Military pilots with g-suits have experienced an instantaneous g-load of 12 g; extreme aerobatic aircraft are designed to withstand 9 g. Sustained g-load for military aircraft is 6 g. A typical fairground ride might deliver around 2–3 g. Aircraft pulling out from a dive/loop at a high g-load will prevent blood from reaching the brain, and a pilot can temporarily lose consciousness, which is of course dangerous. It is for this reason that military aircraft seats are more inclined to reduce the carotid distance, easing blood flow to the brain. Astronauts lie horizontal to withstand sustained g-load during launch.

   At negative g, a pilot can enter into weightlessness with a different kind of reaction, even enjoyment. At extremely high negative g, disorientation of the pilot can occur.

References

1. Kimberlin, R.D., *Flight Testing of Fixed-Wing Aircraft*. AIAA, Reston, VA, 2003.

2. Torenbeek, E., *Synthesis of Subsonic Airplane Design*. Delft University Press, Delft, 1982.

3. Roskam, J., *Airplane Design*. DARcorporation, Lawrence, 2007.

4. Ball, R.E., *The Fundamentals of Aircraft Combat Survivability Analysis and Design*, 2nd edn. AIAA, Reston, VA, 2003.

5. Adair, K., and McAneney, C., *Parametric Sensitivity Study of Fuselage Parasite Drag*. Final year project at Queen's University, Belfast, 2007.

6. McCanny, R., *Statistics and Trends in Commercial Transport Aircraft*. Final year project at Queen's University, Belfast, 2005.

7. Etkin, B., *Dynamics of Atmospheric Flight*. Dover Publications Inc., New York, 2005.

8. Naval Aerospace Medical Institute, *US Naval Flight Surgeons Manual*. The Bureau of Medicine and Surgery, Department of the Navy. US Government Office, Washington, DC, 1991.

9. Shannon, R.H., Operational aspects of forces on man during escape in the U.S. Air Force, 1 January 1968–31 December 1970. In *AGARD Linear Acceleration of Impact Type*, June 1971.

APPENDIX A

Conversions

Linear dimensions

1 in.	=	2.54 cm		1 cm	=	0.3937 in
1 ft	=	30.48 cm		1 cm	=	0.0328 ft
1 yd	=	0.9144 m		1 mile	=	5280 ft
1 mile	=	1.6093 km		1 m	=	1.0936 yd
1 nm	=	1.852 km		1 m	=	3.28084 ft
1 nm	=	1.1508 mile		1 km	=	0.54 nm
1 nm	=	6076 ft		1 ft	=	0.000146 nm
1 mile	=	0.869 nm		1 km	=	0.6214 mile

Area

1 in.^2	=	6.5416 cm^2		1 cm^2	=	0.155 in.^2
1 ft^2	=	929.03 cm^2		1 cm^2	=	0.00108 ft^2
1 ft^2	=	0.092903 mile^2		1 m^2	=	10.764 ft^2
1 yd^2	=	0.8361 mile^2		1 m^2	=	1.196 yd^2
1 mile^2	=	2.59 km^2		1 km^2	=	0.3861 mile^2

Volume

1 in.^3	=	16.387 cm^3		1 cm^3	=	0.061 in.^3
1 ft^3	=	28316.85 cm^3		1 cm^3	=	0.0000353 ft^3
1 yd^3	=	0.764555 cm^3		1 m^3	=	1.308 yd^3
1 ft^3	=	28.3171		1 l	=	0.0353 ft^3
1 US gallon	=	3.78541		1 l	=	0.2642 US gallon
1 UK gallon	=	4.5461		1 l	=	0.22 UK gallon
1 US pint	=	0.0004732 m^3		1 m^3	=	2113.376 US pint
1 UK pint	=	0.0005683 m^3		1 m^3	=	1759.754 UK pint
1 qt	=	946.353 cm^3		1 cm^3	=	0.001057 qt

Theory and Practice of Aircraft Performance, First Edition. Ajoy Kumar Kundu, Mark A. Price and David Riordan.
© 2016 John Wiley & Sons, Ltd. Published 2016 by John Wiley & Sons, Ltd.

Speed

1 ft/s	=	1.0973 km/h	1 km/h	=	0.9113 ft/s
1 ft/s	=	0.59232 kt	1 kt	=	6077.28 ft/h = 1.6881 mph
1 ft/s	=	0.000189394 mps	1 kt	=	5280 ft/h = 1.4667 ft/s
1 ft/min	=	0.00508 m/s	1 m/s	=	196.85 ft/min
1 mph	=	0.447 m/s	1 m/s	=	2.237 mph
1 mph	=	0.869 kt	1 kt	=	1.151 mph
1 kt	=	0.51444 m/s	1 m/s	=	1.944 kt
1 kt	=	1.853 km/h	1 kt	=	1.688 ft/s

Angle

1°	=	0.01716 rad	1 rad	=	57.296°

Mass

1 lb	=	0.454 kg	1 kg	=	2.2046 lb

Density

1 lb/in.^{3}	=	27.68 g/cm^3	1 g/cm^3	=	0.03613 lb/in.^{3}
1 lb/ft^3	=	16.1273 kg/m^3	1 kg/m^3	=	0.06243 lb/ft^3
1 lb	=	4.4482 N	1 N	=	0.2248 lb
1 lb	=	0.454 kg	1 kg	=	2.2046 lb
1 oz	=	28.35 g	1 g	=	0.3527 oz
1 oz	=	0.278 N	1 N	=	3.397 oz

Pressure

1 lb/in.^{2}	=	6894.76 Pa	1 Pa	=	0.000145 lb/in.^{2}
1 lb/ft^2	=	44.88 Pa	1 Pa	=	0.02089 lb/ft^2
1 lb/in.^{2}	=	703.07 kg/m^2	1 kg/m^2	=	0.0001422 lb/in.^{2}
1 lb/ft^2	=	4.8824 kg/m^2	1 kg/m^2	=	0.020482 lb/ft^2
1 atm	=	1013.25 millibar	1 millibar	=	0.000987 atm
1 bar	=	14.5 lb/in.^{2}	1 atm	=	14.7 lb/in.^{2}

Energy

1 lb-ft	=	1.356 J	1 J	=	0.7376 lb-ft
1 W-h	=	3600 J	1 J	=	0.000278 W-h
1 lb-ft	=	1.356 J	1 J	=	0.7376 lb-ft
1 W-h	=	0.00134 HP-h	1 HP-h	=	745.7 W-h

Power

1 hp (550 ft lb$_f$)	=	0.7457 kW	1 kW	=	1.341022 HP

Thermodynamics

$$R_{air} = 287.06 \, \text{J/kgK} = 53.35 \, \text{ft} \, lb_f / \text{lb}R$$

Constants

AVGAS		
1 US gallon	=	5.75 lb
1 ft^3	=	43 lb
AVTUR		
1 US gallon (JP4)	=	6.56 lb
1 ft^3	=	48.6 lb
AVTUR		
1 US gallon (JP5)	=	7.1 lb
1 ft^3	=	53 lb

APPENDIX B

International Standard Atmosphere Table

Altitude (m)	Pressure (N/m²)	Temperature (K)	Density (kg/m³)	Viscosity (N s/m²)	Sound speed (m/s)	$C_{f\_turbulent}$
0	101,327	288.15	1.225	0.00001789	340.3	0.00263
500	95,463	284.9	1.16727	0.00001773	338.37	0.00264
1000	89,876.7	281.65	1.11164	0.00001757	336.44	0.00266
1500	84,558	278.4	1.05807	0.00001741	334.49	0.00267
2000	79,497.2	275.15	1.00649	0.00001725	332.53	0.00269
2500	74,684.4	271.9	0.95686	0.00001709	330.56	0.00271
3000	70,110.4	268.65	0.90912	0.0001693	328.58	0.00273
3500	65,765.8	265.4	0.86323	0.00001677	326.59	0.00274
4000	61,641.9	262.15	0.81913	0.00001661	324.58	0.00276
4500	57,729.9	258.9	0.77678	0.00001644	322.56	0.00278
5000	54,021.5	255.65	0.73612	0.00001628	320.53	0.0028
5500	50,508.3	252.4	0.69711	0.00001611	318.49	0.00282
6000	47,182.5	249.15	0.6597	0.00001594	316.43	0.00284
6500	44,036.2	245.9	0.62385	0.00001577	314.36	0.00286
7000	41,062.1	242.65	0.5895	0.0000156	312.28	0.00288
7500	38,252.7	239.4	0.55663	0.00001543	310.18	0.0029
8000	35,601	236.15	0.52517	0.00001526	308.07	0.00292
8500	33,100.2	232.9	0.49509	0.00001509	305.94	0.00294
9000	30,743.6	229.65	0.46635	0.00001492	303.8	0.00297
9500	28,524.7	226.4	0.4389	0.00001474	301.64	0.00299
10,000	26,437.3	223.15	0.41271	0.00001457	299.47	0.00301
10,500	24,475.3	219.9	0.38773	0.00001439	297.28	0.00304

(Continued)

Theory and Practice of Aircraft Performance, First Edition. Ajoy Kumar Kundu, Mark A. Price and David Riordan.
© 2016 John Wiley & Sons, Ltd. Published 2016 by John Wiley & Sons, Ltd.

Altitude (m)	Pressure (N/m²)	Temperature (K)	Density (kg/m³)	Viscosity (N s/m²)	Sound speed (m/s)	$C_{f\_turbulent}$
11,000	**22,633**	**216.65**	**0.36392**	**0.00001421**	**295.07**	**0.00306**
tropopause						
11,500	20,916.8	216.65	0.33633	0.00001421	295.07	0.0031
12,000	19,331	216.65	0.31983	0.00001421	295.07	0.00314
12,500	17,865	216.65	0.28726	0.00001421	295.07	0.00318
13,000	16,511	216.65	0.26548	0.00001421	295.07	0.00323
13,500	15,259.2	216.65	0.24536	0.00001421	295.07	0.00327
14,000	14,102.3	216.65	0.22675	0.00001421	295.07	0.00331
14,500	13,033.2	216.65	0.20956	0.00001421	295.07	0.00336
15,000	12,045.1	216.65	0.19367	0.00001421	295.07	0.0034

Altitude (ft)	Pressure (lb/ft²)	Temperature (R)	Density (slug/ft³)	Viscosity 10^{-7} (lb s/ft²)	Sound speed (ft/s)	$C_{f\_turbulent}$
0	2116.22	518.67	0.00238	3.7372	1116.5	0.01449
1000	2040.85	515.1	0.0023	3.7172	1112.6	0.01459
2000	1967.68	511.54	0.00224	3.6971	1108.75	0.0147
3000	1896.64	507.97	0.00217	3.677	1104.88	0.0148
4000	1827.69	504.41	0.00211	3.657	1100.99	0.01491
5000	1760.79	500.84	0.00204	3.637	1097.09	0.1502
6000	1695.89	497.27	0.00198	3.616	1093.178	0.01513
7000	1632.93	493.71	0.00192	3.596	1089.25	0.01525
8000	1571.88	490.14	0.00186	3.575	1085.31	0.01536
9000	1512.7	486.57	0.00181	3.555	1081.35	0.01548
10,000	1455.33	483.01	0.00175	3.534	1077.38	0.0156
11,000	1399.73	479.44	0.0017	3.513	1073.4	0.01572
12,000	1345.87	475.88	0.00164	3.4927	1069.4	0.01585
13,000	1293.7	472.31	0.00159	3.4719	1065.39	0.01597
14,000	1243.18	468.74	0.00154	3.451	1061.36	0.0161
15,000	1194.27	465.18	0.00149	3.43	1057.31	0.01623
16,000	1146.92	461.11	0.00144	3.4089	1053.25	0.01637
17,000	1101.11	458.05	0.0014	3.388	1049.17	0.0165
18,000	1056.8	454.48	0.00135	3.3666	1045.08	0.01664
19,000	1013.93	450.91	0.0013	3.3453	1040.97	0.01678
20,000	1036.85	447.35	0.00126	3.324	1036.95	0.01693
21,000	932.433	443.78	0.00122	3.3025	1032.71	0.01707
22,000	893.72	440.21	0.00118	3.281	1028.55	0.01722
23,000	856.32	436.65	0.00114	3.26	1024.38	0.01738
24,000	820.19	433.08	0.0011	3.238	1020.18	0.01753
25,000	785.31	429.52	0.00106	3.216	1015.98	0.01769
26,000	751.64	425.95	0.00102	3.1941	1011.75	0.01785
27,000	719.15	422.38	0.00099	3.1722	1007.5	0.01802
28,000	687.81	418.82	0.00095	3.1502	1003.24	0.01819
29,000	657.58	415.25	0.00092	3.128	998.96	0.01836
30,000	628.43	411.69	0.00088	3.1059	994.66	0.01854
31,000	600.35	408.12	0.00085	3.0837	990.35	0.01872
32,000	573.28	404.55	0.00082	3.0614	986.01	0.0189
33,000	547.21	400.97	0.00079	3.0389	981.65	0.01909
34,000	522.12	397.42	0.00076	3.0164	977.28	0.0193
35,000	497.96	393.85	0.00073	2.9938	972.88	0.01948

Altitude (ft)	Pressure (lb/ft^2)	Temperature (R)	Density (slug/ft^3)	Viscosity 10^{-7} (lb s/ft^2)	Sound speed (ft/s)	$C_{f\_turbulent}$
36,089	**472.68**	**389.97**	**0.0007**	**2.969**	**968.08**	**0.0197**
tropopause						
37,000	452.43	389.97	0.00067	2.969	968.08	0.01999
38,000	431.2	389.97	0.00064	2.969	968.08	0.02032
39,000	410.97	389.97	0.00061	2.969	968.08	0.02065
40,000	391.68	389.97	0.00058	2.969	968.08	0.02099
41,000	373.3	389.97	0.00055	2.969	968.08	0.02134
42,000	355.78	389.97	0.00053	2.969	968.08	0.02169
43,000	339.09	389.97	0.0005	2.969	968.08	0.02205
44,000	323.08	389.97	0.00048	2.969	968.08	0.02243
45,000	308.01	389.97	0.00046	2.969	968.08	0.02281
46,000	299.56	389.97	0.00043	2.969	968.08	0.0232
47,000	279.78	389.97	0.00041	2.969	968.08	0.02359
48,000	266.65	389.97	0.00039	2.969	968.08	0.024
49,000	254.14	389.97	0.00037	2.969	968.08	0.02442
50,000	242.21	389.97	0.00036	2.969	968.08	0.02485

APPENDIX C

Fundamental Equations

Some elementary yet important equations are listed herein. Readers must be able to derive them and appreciate the physics of each term for intelligent application to aircraft design and performance envelope. The equations are not derived herein – readers may refer to any aerodynamic textbook for their derivation.

C.1   Kinetics

Newton's Law:

Applied force, $F = \text{mass} \times \text{acceleration} = \text{rate of change of momentum}$

$$\text{force} = \text{pressure} \times \text{area}$$

$$\text{work} = \text{force} \times \text{distance}$$

Therefore energy (i.e. rate of work) for the unit mass flow rate m is as follows:

$$\text{energy} = \text{force} \times \left(\frac{\text{distance}}{\text{time}}\right) = \text{pressure} \times \text{area} \times \text{velocity} = pAV$$

Steady-state conservation equations are as follows:

$$\text{Mass conservation:mass flow rate,}\ m = \rho AV = \text{constant} \tag{C1}$$

$$\text{Momentum conservation}: \text{dp} = -\rho V\, dV\ \left(\text{known as Euler's equation}\right) \tag{C2}$$

With viscous terms, it becomes the Navier–Stokes equation. However, friction forces offered by the aircraft body can be accounted for in the inviscid flow equation as a separate term:

$$\text{Energy conservation}:\ C_p T + \frac{1}{2}V^2 = \text{constant} \tag{C3}$$

Theory and Practice of Aircraft Performance, First Edition. Ajoy Kumar Kundu, Mark A. Price and David Riordan.
© 2016 John Wiley & Sons, Ltd. Published 2016 by John Wiley & Sons, Ltd.

From the thermodynamic relationships for a perfect gas between two stations, this gives

$$s - s_1 = C_p \ln(T/T_1) - R\ln(p/p_1) = C_v \ln(T/T_1) + R\ln(v/v_1) \tag{C4}$$

C.2   Thermodynamics

Using perfect gas laws, entropy change through shock is an adiabatic process, as given below.

$$(s_2 - s_1)/R = \ln(T_2/T_1)^{\gamma(\gamma-1)} - \ln(p_2/p_1)$$

$$= \ln\frac{(T_2/T_1)^{\frac{\gamma}{(\gamma-1)}}}{(p_2/p_1)} = \ln\frac{(p_2/p_1)^{\gamma/(\gamma-1)}/(\rho_2/\rho_1)^{\gamma/(\gamma-1)}}{(p_2/p_1)} \tag{C5}$$

or
$$(s_2 - s_1)/R = \ln\left[(p_2/p_1)^{\gamma/(\gamma-1)}\left(\frac{\rho_2}{\rho_1}\right)^{-\gamma/(\gamma-1)}\right] = \ln(p_{t1}/p_{t2})$$

When velocity is stagnated to zero (say in the hole of a pitot tube), then the following equations can be derived for an isentropic process. Subscript "t" represents a stagnation property, which is also known as the "total" condition. The equations represent point properties, valid at any point of a streamline (γ stands for the ratio of specific heat and M for Mach number).

$$\frac{T_t}{T} = \left(1 + \frac{\gamma-1}{2}M^2\right) \tag{C6}$$

$$\frac{\rho_t}{\rho} = \left(1 + \frac{\gamma-1}{2}M^2\right)^{\frac{\gamma}{\gamma-1}} \tag{C7}$$

$$\frac{p_t}{p} = \left(1 + \frac{\gamma-1}{2}M^2\right)^{\frac{\gamma}{\gamma-1}} \tag{C8}$$

$$\left(\frac{p_t}{p}\right) = \left(\frac{\rho_t}{\rho}\right)^\gamma = \left(\frac{T_t}{T}\right)^{\frac{\gamma}{\gamma-1}} \tag{C9}$$

The conservation equations yield many other significant equations. In any streamline of a flow process, the conservation laws exchange pressure energy with the kinetic energy. In other words, if the velocity at a point is increased, then the pressure at that point falls and *vice versa* (i.e. Bernoulli's and Euler's equations). Following are a few more important equations. At stagnation the total pressure, p_t, is given.

Bernoulli's equation for incompressible isentropic flow,

$$(p_t - p) = \rho V^2/2$$

or
$$p + \rho V^2/2 = \text{constant} = p_t \tag{C10}$$

Clearly, at any point, if the velocity is increased, then the pressure will fall to maintain conservation. This is the crux of lift generation: the upper surface has lower pressure than the lower surface. For compressible isentropic flow, Euler's (Bernoulli's) equation gives the following:

$$\left[\frac{(p_t - p)}{p} + 1\right]^{\frac{\gamma-1}{\gamma}} - 1 = \frac{(\gamma-1)V^2}{2a^2} \tag{C11}$$

There are other important relations using thermodynamic properties, as follows.

From the gas laws (combining Charles Law and Boyle's Law) the equation for state of gas for unit mass is

$$pv = RT \tag{C12}$$

where for air, $R = 287\,\text{J/kg\,K}$.

$$\gamma = C_p/C_v \text{ and } C_p - C_v = R, \text{ so } C_p/R = \gamma/(\gamma-1) \tag{C13}$$

where $C_p = 1004.68\,\text{m}^2/\text{s}^2\text{K}$; $C_v = 717.63\,\text{m}^2/\text{s}^2\text{K}$; gas constant, $R = 287.053\,\text{m}^2/\text{s}^2\text{K}$ (Universal Gas constant, $R = 8314.32\,\text{J/(K\,kmol)}$); mean molecular mass of air (28.96442 kg/kmol).

From the energy equation,

$$\text{Total temperature, } T_t = T + \frac{T(\gamma-1)V^2}{2RT\gamma} = T + \frac{V^2}{2C_p} \tag{C14}$$

$$\text{Mach number} = V/a, \text{where } a = \text{speed of sound}$$
$$\text{and } a^2 = \gamma RT = \gamma(p/\rho) = (dp/d\rho)_{\text{isentropic}} \tag{C15}$$

C.3 Aerodynamics

Some supersonic 2-D relations are given below, and Figure C.1 depicts normal and oblique shocks. Upstream flow properties are represented by subscript "1" and downstream by subscript "2". Process across shock is adiabatic, therefore the total energy remains invariant, that is, $T_{t1} = T_{t2}$.

C.3.1 Normal Shock

In this situation, relations between upstream and downstream are unique and do not depend on geometry (Figure C.1). If upstream conditions are known, the downstream properties can be computed from the following relations.
Mach number:

$$M_2^{\ 2} = \frac{M_1^2 + \frac{2}{(\gamma-1)}}{\frac{2\gamma}{(\gamma-1)}M_1^2 - 1} \tag{C16}$$

Density ratio:

$$\frac{\rho_2}{\rho_1} = \frac{M_1^2}{\frac{(\gamma-1)}{(\gamma+1)}M_1^2 + \left(\frac{2}{\gamma+1}\right)} = \frac{(\gamma+1)M_1^2}{(\gamma-1)M_1^2 + 2} \tag{C17}$$

Static pressure ratio:

$$\frac{p_2}{p_1} = \frac{2\gamma}{(\gamma+1)} M_1^2 - \frac{(\gamma-1)}{(\gamma+1)} \tag{C18}$$

Total pressure ratio:

$$\frac{p_{t2}}{p_{t1}} = \frac{\left[\dfrac{\left((\gamma+1)/2\right)M_1^2}{1+\left((\gamma-1)/2\right)M_1^2}\right]^{\frac{\gamma}{\gamma-1}}}{\left[\dfrac{2\gamma}{(\gamma+1)}M_1^2 - \dfrac{(\gamma-1)}{\gamma+1}\right]^{\frac{1}{\gamma-1}}} = \frac{\left[\dfrac{(\gamma+1)M_1^2}{2+(\gamma-1)M_1^2}\right]^{\frac{\gamma}{(\gamma-1)}}}{\left[\dfrac{2\gamma M_1^2 - (\gamma-1)}{(\gamma+1)}\right]^{\frac{1}{(\gamma-1)}}} \tag{C19}$$

or

$$\frac{p_{t2}}{p_1} = \left(\frac{p_{t2}/p_{t1}}{p_1/p_{t1}}\right) = \frac{\left[\dfrac{(\gamma+1)M_1^2}{2+(\gamma-1)M_1^2}\right]^{\frac{\gamma}{(\gamma-1)}}}{\left[\dfrac{2\gamma M_1^2 - (\gamma-1)}{(\gamma+1)}\right]^{\frac{1}{(\gamma-1)}}} \left(1+\frac{\gamma-1}{2}M^2\right)^{\frac{\gamma-1}{\gamma}}$$

$$= \left[\frac{(\gamma+1)}{2}M_1^2\right]^{\frac{\gamma}{(\gamma-1)}} \left[\frac{2\gamma M_1^2}{(\gamma+1)} - \frac{(\gamma-1)}{(\gamma+1)}\right]^{\frac{1}{(1-\gamma)}} \tag{C20}$$

Total temperature:

$$T_{t1} = T_{t2} \tag{C21}$$

Total temperature:

$$\frac{T_2}{T_1} = \frac{\left(\dfrac{2\gamma}{(\gamma-1)}M_1^2 - 1\right)\left(1+\dfrac{(\gamma-1)}{2}M_1^2\right)}{\dfrac{(\gamma+1)^2}{2(\gamma-1)}M_1^2} \tag{C22}$$

Once the Mach numbers are known, the static temperatures each side of normal shock can be computed out from the isentropic relation given in Equation C22.

C.3.2 Oblique Shock

In this situation (see Figure C.1), relations between upstream and downstream depend on the geometry of deflection angle θ and free upstream Mach number M_1 developing oblique shock angle, β. The higher the upstream Mach number M_1, the lower the oblique shock angle, β. For the same free upstream Mach number M_1, loss through oblique shock is less than normal shock. If upstream conditions are known, the downstream properties can be computed from the following relations. In this book, for given M_1 and θ, β is taken from reference tables (not given here).

$$\tan\beta = 2\cot\theta\left[\frac{M_1^2\sin^2\theta - 1}{M_1^2(\gamma-\cos 2\theta) + 2}\right] \tag{C23}$$

■ **FIGURE C.1**
Shock waves

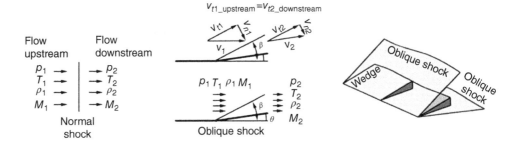

Geometry relation:

$$\tan \beta = V_{n1}/V_{t1}$$
$$\text{and } \tan(\beta - \theta) = V_{n2}/V_{t2}$$

(C24)

From continuity:

$$\frac{V_{n2}}{V_{n1}} = \frac{\rho_2}{\rho_1} = \frac{\tan\beta}{\tan(\beta - \theta)}$$

(C25)

Normal velocities V_{n1} and V_{n2} on oblique shock can be treated as normal shock. In that case, substituting $M_1 \sin \beta$ for M_1 in Equations C18–C20 we obtain relations between properties across oblique shock.

Static pressure ratio:

$$\frac{p_2}{p_1} = \frac{2\gamma}{(\gamma+1)} M_1^2 \sin^2 \beta - \frac{(\gamma-1)}{(\gamma+1)}$$

(C26)

Density ratio:

$$\frac{\rho_2}{\rho_1} = \frac{M_1^2 \sin^2 \beta}{\dfrac{(\gamma-1)}{(\gamma+1)} M_1^2 \sin^2 \beta + \left(\dfrac{2}{\gamma+1}\right)} = \frac{\tan\beta}{\tan(\beta - \theta)}$$

(C27)

Static temperature ratio:

$$\frac{T_2}{T_1} = \frac{\left(\dfrac{2\gamma}{(\gamma-1)} M_1^2 \sin^2 \beta - 1\right)\left(1 + \dfrac{(\gamma-1)}{2} M_1^2 \sin^2 \beta\right)}{\dfrac{(\gamma+1)^2}{2(\gamma-1)} M_1^2 \sin^2 \beta}$$

(C28)

Total temperature:

$$T_{t1} = T_{t2}$$

(C29)

Total pressure ratio:

$$\frac{p_{t2}}{p_{t1}} = \left[\frac{(\gamma+1)M_1^2 \sin^2 \beta}{2 + (\gamma-1)M_1^2 \sin^2 \beta}\right]^{\frac{\gamma}{(\gamma-1)}} \left[\frac{(\gamma+1)}{2\gamma M_1^2 \sin^2 \beta - (\gamma-1)}\right]^{\frac{1}{(\gamma-1)}}$$

(C30)

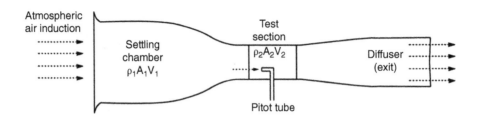

■ **FIGURE C.2**
Open circuit low speed wind tunnel

$$M_2^{\,2} \sin^2 (\beta - \theta) = \dfrac{M_1^{\,2} \sin^2 \beta + \dfrac{2}{(\gamma - 1)}}{\dfrac{2\gamma}{(\gamma - 1)} M_1^{\,2} \sin^2 \beta - 1} \qquad \textbf{(C31)}$$

C.3.2.1   *Incompressible Low-Speed Wind Tunnel (Open Circuit)*   Refer to Figure C.2 below. Let ρ, A and V be the density, cross-sectional area and velocity in the settling chamber with subscript 1, and in the test section with subscript 2, respectively. The contraction ratio of the wind tunnel is defined by A_1/A_2.

From conservation of mass flow, $m = \rho_1 A_1 V_1 = \rho_2 A_2 V_2$ (Equation C1). Being incompressible, $A_1 V_1 = A_2 V_2$. This gives velocity in the test section:

$$V_2 = (A_1 / A_2) / V_1 \qquad \textbf{(C32)}$$

Using Bernoulli's theorem (Equation C10),

$$p_1 + \rho V_1^{\,2}/2 = p_2 + \rho V_2^{\,2}/2 \quad (\rho_1 = \rho_2 = \rho, \text{ incompressible})$$

or $\qquad V_2^{\,2} = (p_1 - p_2)(\rho/2) + V_1^{\,2}/2 = 2(p_1 - p_2)/\rho + \left[(A_2/A_1)\right]^2 V_2^{\,2}$

Solving,

$$V_2 = \sqrt{\dfrac{2(p_1 - p_2)}{\rho\left[1 - \left(\dfrac{A_2}{A_1}\right)^2\right]}} \qquad \textbf{(C33)}$$

Airbus 320 Class Case Study

All computations carried out here follow the book instructions. The results are not from Airbus and they are not responsible for the figures given here. They are used only to substantiate the book methodology to get some confidence. Industry drag data are not available, but at the end, check whether the payload-range matches with the published data. We will investigate the case at long-range cruise (LRC) speed of Mach 0.75, and altitude of 36,089 ft.

D.1 Dimensions

See Figure D.1 for the configuration and dimensions of the Airbus 320.

- Fuselage length = 123.16 ft (scaled measurements differ slightly from the drawings).
- Fuselage width = 13.1 ft.
- Fuselage depth = 13 ft.
- Wing reference area (trapezoidal part only) = 1202.5 ft², add yehudi area = 118.8 ft².
- Span = 11.85 ft, MAC_{wing} = 11.64 ft, AR = 9.37, $\Lambda_{\frac{1}{4}}$ = 25°, C_R = 16.5 ft, λ = 0.3.
- H-tail reference area = 330.5 ft², $MAC_{H\text{-tail}}$ = 8.63 ft.
- V tail reference area = 235.6 ft², $MAC_{V\ tail}$ = 13.02 ft.
- Nacelle length = 17.28 ft, maximum diameter = 6.95 ft.
- Pylon = to measure from the drawing.

Reynolds number per foot is given by:

$$\text{Re}_{per\ foot} = (V\rho)/\mu = (aM\rho)/\mu$$

$$= \left[(0.75 \times 968.08)(0.00071) \right] / \left(0.7950 \times 373.718 \times 10^{-9} \right) = 1.734 \times 10^{6} \text{ per foot.}$$

Theory and Practice of Aircraft Performance, First Edition. Ajoy Kumar Kundu, Mark A. Price and David Riordan.
© 2016 John Wiley & Sons, Ltd. Published 2016 by John Wiley & Sons, Ltd.

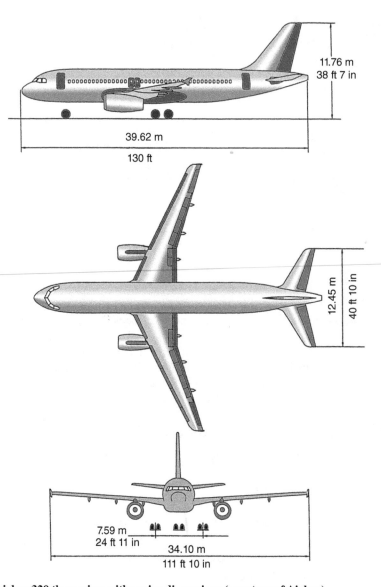

■ **FIGURE D.1**   **Airbus 320 three-view with major dimensions (courtesy of Airbus)**

D.2   Drag Computation

D.2.1   Fuselage

Table D.1 gives the basic average 2D flat plate for the fuselage, $C_{Ffbasic} = 0.00186$. Table D.2 summarizes the 3D and other shape effect corrections, ΔC_{Ff}, needed to estimate the total fuselage C_{Ff}.

The total fuselage $C_{Ff} = C_{Ffbasic} + \Delta C_{Ff} = 0.00186 + 0.0006875 = 0.002547$.

Flat plate equivalent f_f (see Section 9.21.3) $= C_{Ff} \times A_{wf} = 0.002547 \times 4333 = 11.03 \text{ ft}^2$.

■ **TABLE D.1**
Reynolds number and 2D basic skin friction C_{Fbasic}

Parameter	Reference area (ft²)	Wetted area (ft²)	Characteristic length (ft)	Reynolds number	2D C_{Fbasic}
Fuselage	n/a	4333	123.16	2.136×10^8	0.00186
Wing	1202.5	2130.94	11.64 (MAC$_w$)	2.02×10^7	0.00255
V-tail	235	477.05	13.02 (MAC$_{VT}$)	2.26×10^7	0.00251
H-tail	330.5	510.34	8.63 (MAC$_{HT}$)	1.5×10^7	0.00269
2 × nacelle	n/a	2 × 300	17.28	3×10^7	0.00238
2 × pylon	n/a	2 × 58.18	12 (MAC$_p$)	2.08×10^7	0.00254

■ **TABLE D.2**
Fuselage ΔC_{Ff} correction (3D and other shape effects)

Item	ΔC_{Ff}	% of $C_{Ffbasic}$
Wrapping	0.00000922	0.496
Supervelocity	0.0001	5.36
Pressure	0.0000168	0.9
Fuselage upsweep of 6°	0.000127	6.8
Fuselage closure angle of 9°	0	0
Nose fineness ratio	0.000163	8.7
Fuselage non-optimum shape	0.0000465	2.5
Cabin pressurization/leakage	0.000093	5
Passenger windows/doors	0.0001116	6
Belly fairing	0.000039	2.1
Environment control system (ECS) exhaust	−0.0000186	−1
Total ΔC_{Ff}	0.0006875	36.9

Add canopy drag, $f_c = 0.3\,\text{ft}^2$.
Total fuselage parasite drag in terms of $f_{f+c} = 11.33\,\text{ft}^2$.

D.2.2   Wing

Table D.1 gives the basic average 2D flat plate for the wing, $C_{Fwbasic} = 0.00257$, based on the MAC$_w$. Important geometric parameters are wing reference area (trapezoidal planform) = 1202.5 ft² and the gross wing planform area (including yehudi) = 1320.8 ft². Table D.3 summarizes the 3D and other shape effect corrections needed to estimate the total wing C_{Fw}.

Therefore, the total wing $C_{Fw} = C_{Fwbasic} + \Delta C_{Fw} = 0.00257 + 0.000889 = 0.00345$.
Flat plate equivalent, $f_w = C_{Fw} \times A_{ww} = 0.00345 \times 2130.94 = 7.35\,\text{ft}^2$.

D.2.3   Vertical Tail

Table D.1 gives the basic average 2D flat plate for the V-tail, $C_{FVTbasic} = 0.00251$ based on the MAC$_{VT}$. V-tail reference area = 235 ft². Table D.4 summarizes the 3D and other shape effects corrections (ΔC_{FVT}) needed to estimate the V-tail C_{FVT}.

Therefore V-tail $C_{FVT} = C_{FVTbasic} + \Delta C_{FVT} = 0.00251 + 0.000718 = 0.003228$.
Flat plate equivalent $f_w = C_{Fw} \times A_{ww} = 0.00345 \times 2130.94 = 7.35\,\text{ft}^2$.

■ **TABLE D.3**
Wing ΔC_{F_w} correction (3D and other shape effects).

Item	ΔC_{F_w}	% of $C_{Fwbasic}$
Supervelocity	0.000493	19.2
Pressure	0.000032	1.25
Interference (wing-body)	0.000104	4.08
Excrescence (flaps and slats)	0.000257	10
Total ΔC_{F_w}	0.000887	34.53

■ **TABLE D.4**
V-tail ΔC_{FVT} correction (3D and other shape effects)

Item	ΔC_{FVT}	% of $C_{FVTbasic}$
Supervelocity	0.000377	15
Pressure	0.000015	0.6
Interference (V tail-body)	0.0002	8
Excrescence (rudder gap)	0.0001255	5
Total ΔC_{FVT}	0.000718	28.6

D.2.4   Horizontal Tail

Table D.1 gives the basic average 2D flat plate for the H-tail, $C_{FHTbasic} = 0.00269$ based on the MAC_{HT}. The H-tail reference area $S_{HT} = 330.5$ ft². Table D.5 summarizes the 3D and other shape effects corrections (ΔC_{FHT}) needed to estimate the H-tail C_{FHT}.

Therefore H-tail $C_{FHT} = C_{FHTbasic} + \Delta C_{FHT} = 0.00269 + 0.000605 = 0.003295$.

Flat plate equivalent $f_{HT} = C_{FHT} \times A_{wHT} = 0.003295 \times 510.34 = 1.68$ ft².

D.2.5   Nacelle, C_{Fn}

Because a nacelle is a fuselage-like axisymmetric body, the procedure follows the method used for fuselage evaluation, but needs special attention on account of the throttle-dependent considerations.

Important geometric parameters are the nacelle length = 17.28 ft; maximum nacelle diameter = 6.95 ft; average diameter = 5.5 ft; nozzle exit plane diameter = 3.6 ft; maximum frontal area = 37.92 ft²; wetted area per nacelle $A_{wn} = 300$ ft².

Table D.1 gives the basic average 2D flat plate for the nacelle, $C_{Fnbasic} = 0.00238$ based on the nacelle length. Table D.6 summarizes the 3D and other shape effects corrections, ΔC_{Fn}, needed to estimate the total nacelle C_{Fn} for one nacelle.

For nacelles, a separate supervelocity effect is not considered as it is taken into account in the throttle-dependent intake drag. Also pressure drag is accounted for in throttle-dependent base drag.

■ **TABLE D.5**
H-tail ΔC_{FHT} correction (3D and other shape effects)

Item	ΔC_{FHT}	% of $C_{FHTbasic}$
Supervelocity	0.0004035	15
Pressure	0.0000101	0.3
Interference (H tail-body)	0.0000567	2.1
Excrescence (elevator gap)	0.0001345	5
Total ΔC_{FHT}	0.000605	22.4

■ **TABLE D.6**
Nacelle ΔC_{Fn} correction (3D and other shape effects).

Item	ΔC_{Fn}	% of $C_{Fnbasic}$
Wrapping (3D effect)	0.0000073	0.31
Excrescence (non-manufacture)	0.0005	20.7
Boat tail (aft end)	0.00027	11.7
Base drag (aft end)	0	0
Intake drag	0.001	41.9
Total ΔC_{Fn}	0.001777	74.11

D.2.6   Thrust Reverser Drag

The excrescence drag of the thrust reverser is included in Table D.7 as it does not emanate from manufacturing tolerances. The nacelle is placed well ahead of the wing, hence the nacelle-wing interference drag is minimized and is assumed to be zero.

Therefore nacelle $C_{Fn} = C_{Fnbasic} + \Delta C_{Fn} = 0.00238 + 0.001777 = 0.00416$.
Flat plate equivalent $f_n = C_{Fnt} \times A_{wn} = 0.00416 \times 300 = 1.25\,\text{ft}^2$ per nacelle.

D.2.7   Pylon

The pylon is a wing-like lifting surface and the procedure is identical to the wing parasite drag estimation. Table D.1 gives the basic average 2D flat plate for the pylon, $C_{Fpbasic} = 0.0025$ based on the MAC_p. The pylon reference area $= 28.8\,\text{ft}^2$ per pylon. Table D.7 summarizes the 3D and other shape effects corrections (ΔC_{Fp}) needed to estimate C_{Fp} (one pylon).

Therefore pylon $C_{Fp} = C_{Fpbasic} + \Delta C_{Fp} = 0.0025 + 0.00058 = 0.00312$.
Flat plate equivalent $f_p = C_{Fp} \times A_{wp} = 0.182\,\text{ft}^2$ per pylon.

D.2.8   Roughness Effect

The current production standard tolerance allocation would offer some excrescence drag. The industry standard takes 3% of the total component parasite drag, which includes the effect of surface degradation in use. The value is $f_{roughness} = 0.744\,\text{ft}^2$ given in Table D.8.

D.2.9   Trim Drag

Conventional aircraft will produce some trim drag in cruise and this varies slightly with fuel consumption. For a well-designed aircraft of this class, the trim drag may be taken as $f_{trim} = 0.1\,\text{ft}^2$.

■ **TABLE D.7**

Pylon ΔC_{Fp} correction (3D and other shape effects)

Item	ΔC_{Fp}	% of $C_{Fpbasic}$
Supervelocity	0.000274	10.78
Pressure	0.00001	0.395
Interference (pylon-wing)	0.0003	12
Excrescence	0	0
Total ΔC_{Fp}	0.000584	23

■ **TABLE D.8**

Aircraft parasite drag build-up summary and C_{Dpmin} estimation

	Wetted area, A_w (ft²)	Basic C_F	ΔC_F	Total C_F	f (ft²)	C_{Dpmin}
Fuselage & U/C fairing	4333	0.00186	0.00069	0.00255	11.03	0.00918
Canopy	–	–	–	–	0.3	0.00025
Wing	2130.94	0.00255	0.00089	0.00346	7.35	0.00615
V-tail	477.05	0.00251	0.00072	0.00323	1.54	0.00128
H-tail	510.34	0.00269	0.00061	0.0033	1.68	0.0014
2×nacelle	2×300	0.00238	0.00178	0.00415	2.5	0.00208
2×pylon	2×58.18	0.00254	0.000584	0.00312	0.362	0.0003
Rough (3%)	–	–	–	–	0.744	0.00062
Aerial	–	–	–	–	0.005	0.000004
Trim drag	–	–	–	–	0.1	0.00008
Total					25.611	0.0213

D.2.10   Aerial and Other Protrusions

For this class of aircraft, $f_{arial} = 0.005 \text{ft}^2$.

D.2.11   Air-conditioning

This is accounted for in fuselage drag as environmental control system (ECS) exhaust. This could provide a small amount of thrust.

D.2.12   Aircraft Parasite Drag Build-Up Summary and C_{Dpmin}

Table D.8 gives the aircraft parasite drag build-up summary in a tabular form.

D.2.13 ΔC_{Dp} Estimation

The ΔC_{Dp} table (Table D.9) is constructed, corresponding to the C_L values, as given below.

D.2.14   Induced Drag, C_{Di}

$$\text{Wing aspect ratio, AR} = \frac{\text{span}^2}{\text{gross wing area}} = (111.2)^2 / 1320 = 9.37$$

■ **TABLE D.9**
ΔC_{Dp} estimation

C_L	0.2	0.3	0.4	0.5	0.6
ΔC_{Dp}	0.00044	0	0.0004	0.0011	0.0019

■ **TABLE D.10**
Induced drag

C_L	0.2	0.3	0.4	0.5	0.6	0.7	0.8
C_{Di}	0.00136	0.00306	0.00544	0.0085	0.01224	0.0167	0.0218

■ **TABLE D.11**
Total aircraft drag coefficient, C_D

C_L	0.2	0.3	0.4	0.5	0.6
C_{Dpmin}		0.0213 from Table D.8			
ΔC_{Dp}	0.00038	0	0.0004	0.0011	0.0019
C_{Di}	0.00136	0.00306	0.00544	0.0085	0.01224
Total aircraft C_D	0.0231	0.02436	0.02714	0.0309	0.03544

$$\text{Induced drag, } C_{Di} = \frac{C_L^2}{\pi \text{AR}} = 0.034 C_L^2.$$

Table D.10 gives the C_{Di} corresponding to each C_L.

D.2.15 Total Aircraft Drag

Aircraft drag is given as $C_D = C_{Dpmin} + \Delta C_{Dp} + C_{Di} + [C_{Dw} = 0]$; the total aircraft drag is obtained by adding all the drag components in Table D.11. Note that the low and high values of C_L are beyond the flight envelope.

Table D.11 is drawn in Figure D.2.

D.2.16 Engine Rating

Uninstalled sea-level static thrust = 25,000 lb per engine.
Installed sea-level static thrust = 23,500 lb per engine.

D.2.17 Weights Breakdown

Design cruise speed, V_C = 350 KEAS (knots equivalent air speed).
Design dive speed, V_D = 403 KEAS.

■ **FIGURE D.2**
Aircraft drag polar
at LRC

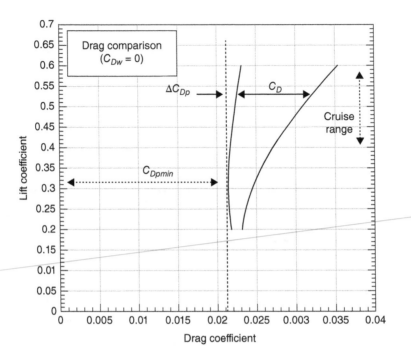

Design dive Mach number, $M_D = 0.88$.

Limit load factor $= 2.6$.

Ultimate load factor $= 3.9$.

Cabin differential pressure limit $= 7.88$ psi.

There could be some variation in the component weights given in Table D.12.

This gives a wing loading $= 162,310/1202.5 = 135$ lb/ft^2, and thrust loading $= 50,000/162,310 = 0.308$. The aircraft is sized to this with better high-lift devices.

D.2.18   Payload-Range

The payload-range is calculated assuming 150 passengers.

MTOM $= 162,000$ lb.

Onboard fuel mass $= 40,900$ lb.

Payload $= 200 \times 150 = 30,000$ lb.

Long-range cruise $=$ Mach 0.75, 36,089 ft (constant condition).

Initial cruise thrust per engine $= 4500$ lb.

Final cruise thrust per engine $= 3800$ lb

Average specific range $= 0.09$ nm/lb fuel.

The aircraft climbs at 250 KEAS reaching Mach 0.7 (Figure D.3).

Diversion fuel calculation:

Diversion distance $= 2000$ nm, cruising at Mach 0.675 and at 30,000 ft altitude.

Diversion fuel $= 2800$ lb, contingency fuel (5% of mission fuel) $= 1700$ lb.

■ **TABLE D.12**
Weights breakdown

Component	Weight (lb)
Wing	14,120
Flaps & slats	2435
Spoilers	380
Aileron	170
Winglet	265
Wing group total	17,370
Fuselage group	17,600 (Toernbeek's method)
H-tail group	1845
V-tail	1010
Undercarriage group	6425
Total structure weight	44,250
Power plant group (two)	15,220
Control systems group	2280
Fuel systems group	630
Hydraulics group	1215
Electrical systems group	1945
Avionics systems group	1250
APU	945
ECS group	1450
Furnishing	10,650
Miscellaneous	4055
MEW	83,890
Crew	1520
Operational items	5660
OEW	91,070
Payload (150×200)	30,000
Fuel (see range calculation)	41,240
MTOW	162,310

Summary of the mission segments is given in Table D.13.

■ **TABLE D.13**
Summary of the mission sector

Sector	Fuel consumed (lb)	Distance covered (nm)	Time elapsed (min)
Taxi-out	200	0	8
Takeoff	300	0	1
Climb	4355	177	30
Cruise	28,400	2560	357
Descent	370	105	20
Approach/land	380	0	3
Taxi-in	135	0	5
Total	34,140	2842	424

■ **FIGURE D.3**
Aircraft performance

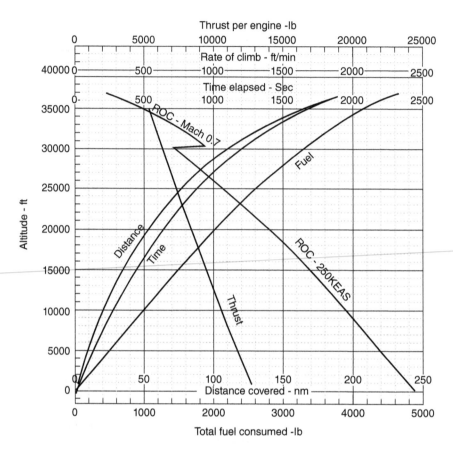

Holding fuel calculation:

    Holding time = 30 minutes at Mach 0.35 and at 5000 ft altitude.

    Holding fuel = 2600 lb.

    Total reserve fuel carried = 2800 + 1700 + 2600 = 7100 lb.

    Total onboard fuel carried = 7100 + 34,140 = 41,240 lb.

    A consolidated aircraft performance is given in Figure D.3.

D.2.19   Cost Calculations

Assuming 150 passengers, and prices in US dollars at year 2000 levels:

Yearly utilization:	497 trips per year.
Mission (trip) block time:	7.05 h.
Mission (trip) block distance	2842 nm.
Mission (trip) block fuel:	34,140 lb (6.68 lb/US gallons).
Fuel cost:	0.6 US$ per US gallon.
Airframe price:	$38 million.
Two engines price:	$9 million.
Aircraft price:	$47 million.

Operating costs per trip – AEA 89 ground-rules for medium jet transport aircraft:

Depreciation:	$6923
Interest:	$5370
Insurance:	$473
Flight crew:	$3482
Cabin crew:	$2854
Navigation:	$3194
Landing fees:	$573
Ground handling:	$1365
Airframe maintenance:	$2848
Engine maintenance:	$1208
Fuel cost:	$3066 (5110.8 US gallons)
Total DOC:	$31,356
DOC/block hour:	$4449
DOC/seat:	$209
DOC/seat/nm:	$0.0735/seat/nm.

The readers may compare with data available in the public domain.

A P P E N D I X   E

Problem Sets

Given below are two aircraft:

1. Model Belfast (B100) – a Fokker F100 class, narrow body, 110-passenger turbofan aircraft (FAR 25 certification); and
2. Model AK4 – a four-place utility aircraft (FAR 23 certification).

These aircraft serve as problem assignments for the readers to evaluate their full performances as covered in this book. Solutions for some problems are not supplied.

E.1   The Belfast (B100)

Readers may estimate aircraft performance of the B100 as given in the 3-view diagram below. Related problem sets are assigned chapter-wise. (We suggest copying the diagram, enlarging it to A3 size, and setting the scale of the drawing from the dimensions given below.)

E.1.1   Geometric and Performance Data

The geometric and performance parameters discussed herein were used in previous chapters. Figure E.1 illustrates the anatomy of the given aircraft. The B100 is a 5-abreast, 110-passenger aircraft.

E.1.1.1   Customer Specification

- Payload = 110 passengers
- Range = 1500 nm.
- Takeoff distance (BFL) = 5200 ft.
- Landing distance = 5000 ft.
- Maximum speed = Mach 0.81.

Theory and Practice of Aircraft Performance, First Edition. Ajoy Kumar Kundu, Mark A. Price and David Riordan.
© 2016 John Wiley & Sons, Ltd. Published 2016 by John Wiley & Sons, Ltd.

■ **FIGURE E.1**
Belfast 100 (B100)

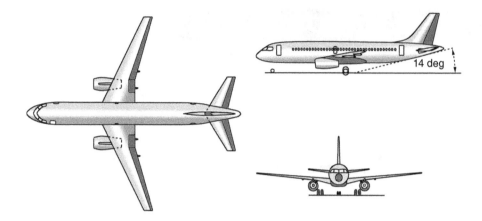

- Maximum initial cruise speed=455 kt (TAS).
- Initial en-route climb rate=3000 ft/s.
- Initial cruise altitude=37,000 ft.
- Maximum dive speed (structural limit)=Mach 0.85.
- LRC altitude=37,000 ft.
- LRC Mach=0.75.
- C_L at Mach 0.77=0.5.
- C_L at Mach 0.82=0.43.

E.1.1.2   Family Variants   Short variant B90, 90 passengers.
Long variant B130, 130 passengers.

E.1.1.3   Fuselage   Assume a constant circular cross-section. Fuselage length, L_f=33.5 m (110 ft), diameter=3.5 m (11.5 ft). Fuselage upsweep angle is 14°. Measure the fuselage closure angle and fineness ratio from the drawing.

E.1.1.4   Wing   Aerofoil: NACA 65-412, having 12% thickness to chord ratio for design C_L=0.4.

Planform reference area, S_W=88.26 m² (≈950 ft²), span=29 m (95 ft), aspect ratio (AR)=9.5.
Wing MAC=3.52 m (11.55 ft), dihedral=5°, twist=−1° (wash out), t/c=12%.
Root chord at centreline, C_R=5.0 m (16.4 ft), tip chord, C_T=1.3 m (4.26 ft), taper ratio=0.26.
Quarter chord wing sweep=23.5°.

E.1.1.5   Other Surfaces   V-tail (aerofoil 64-010): planform (reference) area, S_V=23 m² (248 ft²), AR=2.08.

H-tail (Tee tail, aerofoil 64-210 – installed with negative camber): planform (reference) area S_H=26.94 m² (290 ft²), AR=4.42, dihedral=5°.
Nacelle (each – two required): measure the nacelle length, maximum diameter and nacelle fineness ratio from the drawing.

E.1.2   The B100 Component Weights

Weights are given in kg, assuming 20% composite in the structure.

Wing (20% composite)	4400
Fuselage	6000
H-tail	600
V-tail	500
Pylon	500
Undercarriage	1600
Total structure	**13,600**
Engine and nacelle (two)	4000
Engine fuel system	500
Engine accessories	500
Total installed power plant	**5000**
Flight control	630
Instruments and autopilot	100
Navigation and communication	460
Electrical system	960
APU	160
Hydraulic system	540
Environmental control system	470
Anti/de-icing	40
Fire protection and toilet system	140
Oxygen system	100
Total systems	**3600**
Furnishing system	2000
Contingencies	300
MEW	**24,500**
Operator's items (consumables)	2500
OEW	**27,000 (≈59,400 lb)**
Payload (100 passengers)	9000 (19,800 lb)
Fuel	9500 (20,900 lb)
MTOM	**45,500 (≈100,000 lb)**

E.1.2.1   Weight Summary

MTOM = 100,000 lb (45,455 kg)
Wing = 9800 lb
Fuselage = 13,900 lb
H-tail = 1100 lb
V-tail = 850 lb
2 × pylon = 1150 lb
Undercarriage = 3600 lb

OEW = 59,500 lb (27,050 kg)
Installed engine unit = 10,500 lb
Systems = 7900 lb
Furnishing = 4700 lb
Contingency = 800 lb
Passenger = 200 lb each
Operator's items (consumables) = 5200 lb

E.1.2.2   Engine   Takeoff static thrust at ISA sea-level, T_{SLS}/engine = 17,000 lb (≈75,620 N), with BPR = 5.

sfc at T_{SLS} = 0.3452 lb/lb/h, sfc at T_{max_cruise} = 0.6272 lb/lb/h.
(Use Figures 8.5–8.7 – uninstalled values.)

E.1.2.3   Other Pertinent Data   Reserve fuel = 200 nm diversion + 30 minutes of holding + 5% sector fuel at the specified payload-range.

E.1.2.4   High-Lift Devices (Flaps and Slats)

	Clean	Takeoff (low setting)	Takeoff (mid setting)	Landing
C_{LMAX}	1.74	2.6	2.7	3.25

E.1.3   B100 Drag

The B100 drag polar equation $C_D = 0.0204 + 0.04 C_L^2$.
  One engine inoperative $\Delta C_D = 0.0023$.
  Control and asymmetry $\Delta C_D = 0.0015$.
  Undercarriage $\Delta C_{D\_UC} = 0.0025$.
  Flap drag ΔC_{D_flap} (flap and slat) $= 0.03$ (takeoff) and 0.068 (landing).

E.2   The AK4

Readers may estimate aircraft performance of the AK4, with retractable undercarriage, as given in Figure E.2 below. Related problem sets are assigned chapter-wise. (We suggest copying the diagram, enlarging it to A3 size, and setting the scale of the drawing from the dimensions given below.)

E.2.1   Geometric and Performance Data

The geometric and performance parameters of the AK4 are given below. Figure E.2 illustrates the anatomy of the given aircraft.

E.2.1.1   *Customer Specification*

- Payload = 4 occupants.
- Range = 1000 nm.
- Takeoff distance (BFL) = 1000 ft.
- Landing distance = 800 ft.
- Cruise altitude = 10,000 ft.
- Maximum cruise speed = 200 mph.
- Maximum dive speed (structural limit) = 280 mph.

E.2.1.2   *High-Lift Devices (Slotted Flaps)*

	Clean	Takeoff	Landing
C_{LMAX}	1.4	2.0	2.6

■ **FIGURE E.2**
The AK4

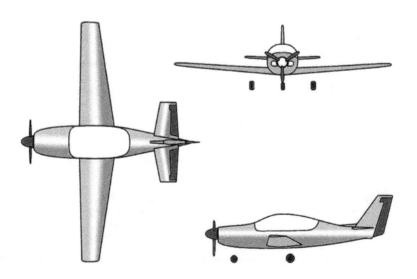

E.2.1.3 *Engine Data* Lycoming IO-390 performance graphs are given in Figure E.3.

E.2.1.4 *Propeller* Three-bladed propeller with AF (activity factor) = 100, integrated design C_{Li} = 0.5, diameter = 5.5 ft.
Thrust characteristics are given in Figure E.4.

E.2.1.5 *Fuselage* Fuselage length is 24 ft, maximum width is 50 in., and maximum height is 50 in. Fuselage upsweep angle is 14°. Measure the fuselage closure angle and fineness ratio from the drawing.

E.2.1.6 *Wing* Aerofoil: NACA 63_2-415, having 15% thickness to chord ratio for design C_L = 0.4.

Planform reference area, S_W = 140 ft², span = 36 ft, aspect ratio = 9.257.
Wing MAC = 3.89 ft, dihedral = 3°, twist = −1.5° (wash out).
Root chord at centreline, C_R = 5 ft, tip chord, C_T = 2.5 ft, taper ratio = 0.5.
Quarter chord wing sweep = 0°.

E.2.1.7 *Other Surfaces* V-tail (aerofoil 64-010): planform (reference) area, S_V = 23 m² (248 ft²), AR = 2.08.

H-tail (Tee tail, aerofoil 64-210 – installed with negative camber): planform (reference) area, S_H = 26.94 m² (290 ft²), AR = 4.42, dihedral = 5°.
Nacelle (each – two required): measure the nacelle length, maximum diameter and nacelle fineness ratio from the drawing.

E.2.1.8 *Engine* Consider performance on an ISA day. The engine is a single Lycoming IO-390 piston engine – four cylinders.
Sea-level – 210 HP at 2575 rpm, fuel flow 120 lb/h.
Maximum cruise – 154 HP at 2400 rpm, fuel flow 74 lb/h.
Normal cruise – 135 HP at 2200 rpm, fuel flow 60 lb/h.
Dry weight – 380 lb, oil – 8 qt.
Fuel – aviation grade 100 octane (AVGAS).
Propeller – three-bladed, 60 in. diameter, activity factor – 100.

E.2.2 The AK4 Component Weights

Wing	300 lb
Fuselage and canopy	250 lb
H-tail	70 lb
V-tail	50 lb
Undercarriage	140 lb
Installed engine unit	450 lb
Systems	160 lb
Furnishing	150 lb
Contingency	30 lb
Total	**1600 lb**
Payload (4 × 200 + 100)	900 lb
Fuel and oil	300 lb
MTOM	**2800 lb**
OEW	**1600 lb**

Reserve fuel = 200 nm diversion + 30 minutes of holding + 5% sector fuel at the specified payload-range.

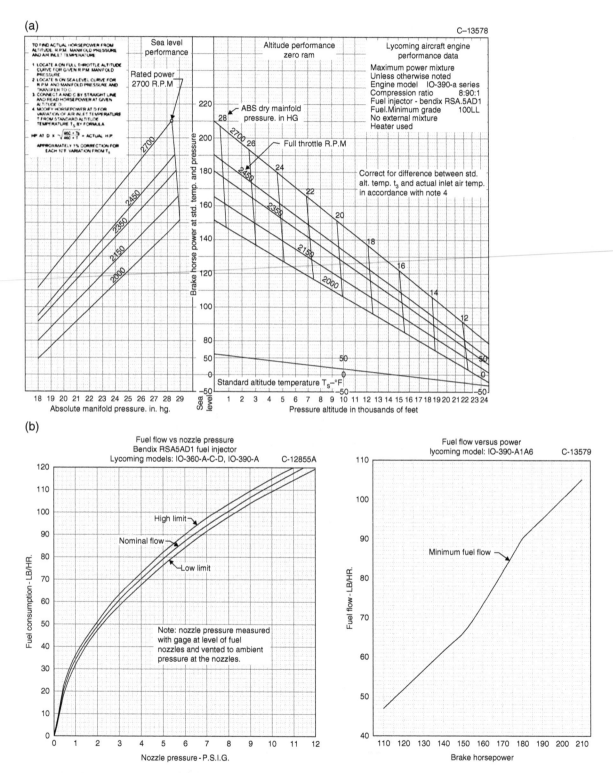

■ **FIGURE E.3** Lycoming IO-390 performance graphs (courtesy of Textron Lycoming Engines). (a) Lycoming IO-390 horsepower graphs. (b) Lycoming IO-390 fuel-flow graphs

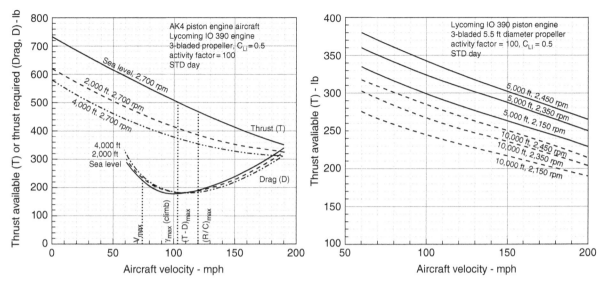

■ **FIGURE E.4** Propeller performance

E.2.3   Drag Coefficient at 5000 ft Altitude

Component	ΔC_{DPmin}
Fuselage	0.00697
Canopy	0.000714
Wing	0.00754
H-tail	0.00228
V-tail	0.00127
Spinner	0.000286
Intake	0.0012
Dorsal	0.000353
Sub-total	**0.0226**
Roughness at 3%	0.000678
Ariel + ECS + cxhaust	0.0007
Trim	0.0001
Subtotal	**0.00148**
Total C_{DPmin}	**0.024078 (≈0.0241)**
$\Delta C_{DPmin\_undercarriage}$	0.008
ΔC_{DPmin_flap}	0.008 (takeoff) 0.01 (landing)

Drag polar equation, $C_D = 0.0241 + 0.043 C_L^2$.

E.3 Problem Assignments

E.3.1 Chapter 1

A descriptive chapter with no problem assignments.

E.3.2 Chapter 2

Problem

2.1   Properties of an aerofoil (close to NACA 65-410) sectional with standard roughness are given as follows within the operating range of −6° to 6° (angles are in degrees: 1 rad = 57.29578°). Find the aerofoil centre of pressure, c.p. at 4° incidence, and its aerodynamic centre, a.c.

[*Ans: a.c.* = 0.261, *c.p.* = 0.361.]

2.2   Take a tapered wing with geometric details as given below (this is close to the isolated Bizjet wing).

Planform reference area, $S_W = 30\,m^2$ (323 ft²), span = 15 m (49.2 ft), aspect ratio = 7.5.

Wing MAC = 2.132 m (7 ft), dihedral = 0°, twist = 0°, t/c = 10%.

Root chord at centreline, $C_R = 2.86$ m (9.38 ft), tip chord, $C_T = 1.143$ m (3.75 ft).

Quarter chord wing sweep = 0°. Oswald's efficiency factor, e = 0.96.

Aerofoil: NACA 65-410 and design $C_L = 0.4$. Take the properties at the wing MAC to estimate 3D aerofoil

characteristics representing the wing. Aircraft weight = 18,000 lb.

Find the wing-lift-curve slope and the angle of incidence α of the wing flying at Mach 0.7 and 43,000 ft altitude with and without compressibility effects.

[*Ans:* $a_{incomp} = 0.105/°$, α = 5.8°; and $a_{comp} = 0.15/°$, α = 3.23°.]

2.3   What is the gravitational acceleration, g, at 128,000 ft geometric altitude (this is about the altitude from which Herr Baumgartner jumped)? What is the corresponding geopotential altitude? Using the hydrostatic equations, find the ISA day ambient temperature, pressure and density at that altitude. Hydrostatic equations may be applied to 47 km altitude. Take the lapse rate above 32 km as 2.766 °C/km.

[*Ans:* $h_p = 127{,}189$ ft, $T = 248$ K, $p = 319$ N/m², $\rho = 0.00448$ kg/m³.]

2.4(a)   The following table gives wind tunnel test results of C_l and C_d values for a 2D aerofoil (α = angle of incidence, degrees).

α	−2	0	2	4	6	8	10	12	14	16
C_l	0.11	0.34	0.56	0.76	0.96	1.12	1.26	1.32	1.3	1.25
C_d	0.0114	0.0104	0.01	0.0102	0.011	0.0122	0.0154	0.0176		

Evaluate graphically the (i) stall angle, (ii) maximum lift to drag ratio, (iii) α at $(L/D)_{max}$ and (iv) $dC_l/d\alpha$.

Referring to [23] in Chapter 2, identify this aerofoil. At what Reynolds number and condition is the test is done? What is the effect of Reynolds number on C_l? (Note that for a wing (3D case), the L/D ratio drops because of an increase in drag on account of wing tip effects.) Make sure that you evaluate $dC_l/d\alpha$, at lower angles of attack, α, where the line is straight.

(b)   From the 2D test results given in Appendix F, compare the NACA 4415 with the NACA 23015 aerofoil section. To design an *ab-initio* trainer aircraft, which one is better suited? When would the other kind be preferred?

2.5   If the pressure coefficient distribution over the upper and lower surfaces of the aerofoil is given along the chordline as (Courtesy of J. Anderson, [30] in Chapter 2):

Upper surface:

$C_{pu} = 1 - 300(x/c)^2$ from leading edge (LE) to 10% of chord length.

$C_{pu} = -2.23 + 2.23(x/c)$ from 10% chord to the trailing edge (TE).

This is because up to 10% it has laminar flow, and then it becomes turbulent.

Lower surface:

$C_{pl} = 1 - 0.95(x/c)$ from LE to TE.

Find the C_l per unit span of the aerofoil. (Hint: Use Equation 2.13.)

[Ans: $C_l = 1.43$.]

2.6   The mid-cruise Bizjet at 19,000 lb weight is cruising at 43,000 ft at Mach 0.7 on an ISA day. The Bizjet wing area = 323 ft², aspect ratio = 7.5, and Oswald's efficiency factor, $e = 0.98$. The wing has a NACA 65-410 aerofoil section. What angle of attack does it have? If the aircraft speed is increased to Mach 0.75, what will be the angle of attack this time? Make your comments. (To simplify, the influence of the fuselage on the wing may be ignored at this point. Approximate the aircraft as a flying wing. The wing is positioned at 3° incidence with respect to the fuselage axis.)

(*Hint*: From Appendix F, establish the $dC_l/d\alpha$ for the 2D NACA 65-410 aerofoil for the highest Re available in the graph. Next, correct the $dC_l/d\alpha$ for the 3D case using Equation 2.21. Next, get the C_L of the aircraft at the flight conditions and then obtain the angle of incidence, α. Use the atmospheric table given in Appendix B. *Comment:* The fuselage affects the lift, therefore will require a slightly higher α. The wing-fuselage angle is at about 3°. With an increase in speed, α reduces.)

2.7(a)   Find the boundary layer thickness, δ, and the skin friction D_f over one side of a rectangular flat plate of 4 cm chord and 10 cm span in an airflow of 50 m/s at zero angle of attack (at sea-level).

(b)   At what speed does transition takes place at 90% from the LE?

(c)   How long should the chord length be to have transition start at the TE in an airflow of 50 m/s? ($Re_{crit} = 500,000$ and $\mu = 1.789 \times 10^{-5}$ kg/ms.)

(d)   What happens if the LE is roughened up and airflow at 50 m/s is tripped at the LE?

(e)   Compare δ and D_f of the two types of flows (take ratios).

[Ans: (a) $\delta_{lam} = 0.562$ mm, $C_{F\_lam} = 0.00359$, $D_{F\_lam} = 0.022$ N, (b) $V = 20.83$ m/s, (c) $l = 146$ mm, (d) $\delta_{tur} = 1.39$ mm, $C_{F\_tur} = 0.00695$, $D_{F\_tur} = 0.0426$ N, $D_{F\_lam}/D_{F\_tur} = 1.936$.]

2.8(a)   For a tapered wing, derive the following relation for the chord length c at any distance y from the centreline:

$$c(y) = \left[\frac{2S_W}{(1+\lambda)b}\right]\left[1 - \frac{2(1-\lambda)}{b}y\right]$$

where λ = taper ratio, b = span and S_W = wing area.

(b)   Take a tapered wing with geometric details as given below (this is close to the isolated Bizjet wing).

Planform reference area, $S_W = 30$ m² (323 ft²), span = 15 m (49.2 ft), dihedral = 0°.

Twist = 0°, root chord at centreline, $C_R = 2.86$ m (9.38 ft), tip chord, $C_T = 1.143$ m (3.75 ft).

Quarter chord wing sweep = 0°.

(c)   Find the wing aspect ratio and its MAC and the mean geometric chord (MGC) and their positions from centreline.

(d)   If the tip chord is made equal to the root chord, $C_R = 2.86$ m, find the wing planform reference area, S_W, aspect ratio, MAC, MGC, and their positions from the centreline.

2.9   In the same way, find the following for the B100 as shown in Figure E.1. This aircraft performance estimation is to be carried out in all related chapters.

Find the wing aspect ratio and its MAC and MGC and their positions from the centreline.

(Hint: compute for a tapered wing, considering the yehudi as an extension to the main tapered wing. Answers can be checked out with the values given in Section E.1.)

[Ans: MAC = 3.512 m, $y_{MAC} = 6$ m, $y_{MGC} = 7.425$ m.]

2.10   A small aircraft with MTOW of 1400 lb has a rectangular wing area 100 ft² with a NACA 4415 aerofoil section. The wing aspect ratio is 6 and uses split flaps.

(a)   What are the span and chord of the wing? What is the Reynolds number at the stall speed? What is the stall speed (take wing C_{Lmax} as 85% of the aerofoil C_{lmax}, although lift curve slope changes)?

At takeoff with 20° flap setting ($\Delta C_L = 0.4$), what will be the stall speed?

(b)   There is a requirement to reduce stall speed by 5 mph. What percentage improvement in C_{Lmax} is required to achieve this? The new flap system has a weight penalty of 50 lb. What will be C_{Lmax} for the new system?

[Ans: (a) span = 25 ft, chord = 4 ft, $V_{stall} = 67.54$ mph, $V_{stall\_20flap} = 66.5$ mph, and (b) $C_{L\_stall\_20\_imp\_flap} = 1.55$.]

E.3.3   Chapter 3

3.1   A light aircraft with rectangular wings weighing 1000 kg is flying at 300 kmph at sea-level on a standard day (Figure E.5). Using a pitot tube, the maximum velocity is measured to be 110 m/s at the point over a wing section where it represents the average aerodynamic values of the wing (being a rectangular wing, it is the MAC).

■ **FIGURE E.5**
Wing section for Problem 3.1

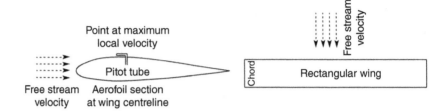

(a)   What is the local static pressure at that point where the maximum local velocity is recorded? Compare it with the ambient conditions. Find the difference between the two, Δp_{max}. Find the total pressure, p_t. Would p_t vary at different locations? Find C_p at the point of measurement.

(b)   If the average difference in pressure between the top and the bottom wing surfaces is a quarter of Δp_{max}, how big a rectangular wing area would you require to sustain flight? If the aspect ratio (span/chord) of the wing is 7, then find the dimensions of the wing.

[Ans: (a) $V_{max} = 110$ m/s, $p_2 = 98,167.72$ N/m², $C_p = -0.7423$; and (b) $S_w = 12.43$ m², span = 9.33 m, chord = 1.333 m.]

3.2(a)   How much does 1 mb represent in geometric height on an ISA day at sea-level?

The altimeter of an aircraft on the ground is set to 1020 mb (Kollsman window), representing the height of the ground to be zero. Later, a change of weather caused the altimeter Kollsman window to read 950 mb. What altitude would the altimeter show with the change?

(Since ambient pressure changes with the weather, all aircraft have the provision to reset the Kollsman window of the altimeter so that it reads zero on the ground, even if it is on a plateau. During flight, a change of weather (i.e. pressure) poses a problem. What should the pilot do?)

(b)   A low-speed tunnel has a contraction ratio of 5. During a certain run, the test section manometer shows a reading of 100 mm of water (Betz reading). What is the speed in the test section? If you have assumed that the speed in the settling chamber is low enough to be ignored, what would the velocity be then?

[Ans: (a) 1 mb = 100 N/m², $h = 582.4$ m; and (b) $V_{test\ section} = 40.8$ m/s.]

3.3   An aircraft is flying at 5 km altitude.

(a)   Given true airspeed (V_{TAS}) of 200 km/h, that is its ground speed. Find its V_{EAS}, V_{IAS}, V_{CAS} and Mach number for an ideal gauge with no errors. What will be the case if there are instrument errors of -3 kmph and a position error $+1.07$ kmph?

(b)   A pilot reads 200 km/h at 5 km altitude in his perfect cockpit airspeed indicator. What will be the ground speed? If the indicator has the above errors, then what will be the ground speed?

[Answer]

	Problem (a)		Problem (b)		
	Perfect instrument	**Real instrument**	**Perfect instrument**	**Real instrument**	**Remark**
V_{TAS} (kmph)	200	200	258.06	255.57	Range computation
V_{EAS} (kmph)	155.07	155.07	200	198.07	Engineers will use compressible $V_{EAS} \neq V_{CAS}$
V_{CAS} (kmph)	155.07	155.07	200	198.07	
V_{IAS} (kmph)	155.07	154	200	197	
V_i (kmph)	155.07	157	200	200	This is what pilot reads
Mach no.	0.1733	0.1733	0.224	0.221	Based on V_{TAS}

3.4 A pilot of an aircraft flying at 36,000 ft altitude reads a total pressure of 712 lb/ft² at an ambient temperature of 220.83 K. Find the flight Mach number, V_{TAS} and V_{EASi} in miles per hour.

[*Ans*: M = 0.784, V_{TAS} = 522.5 mph and V_{EASi} = 283.7 mph.]

3.5 Find the CG position of a Belfast 100 aircraft as given in Figure E.1. The geometric and component weight details are given along with the figure. (*Hint*: Find the *MAC* (Problem 2.8) and its position and proceed. The CG position should be typical for the class.) Note that the CG position of an aircraft is not directly computed in this book. In many industries, especially in smaller companies, aircraft performance engineers also have to estimate aircraft CG position.

E.3.4 Chapters 4 and 5

4-5.1 An aircraft is in steady level flight due north ($\psi_W = \theta_W = \phi_W = 0$) with no sideslip, at a speed of 100 m/s. The angle of attack $\alpha = 5°$. Calculate u_B, v_B and w_B.

[*Ans*: u_B = 99.6 m/s, v_B = 0 and w_B = −8.7 m/s.]

4-5.2 An aircraft is in steady level flight. The forward speed of the aircraft is 100 m/s (i.e. along body axes F_B) but the aircraft is being flown with 5° of yaw. Calculate U_w, V_w and W_w in the wind axis system.

[*Ans*: 99.62, −8.72, and 0 m/s.]

4-5.3 A navigational aid shows an aircraft in the vehicle frame F_V is at a steady flight speed of 100 m/s moving north and climbing straight with wings level at a vertical speed of 5 m/s. The pilot is flying the aircraft with Euler angles ($\phi = 6°$, $\theta = 4°$, $\psi = 5°$). Calculate the components of the aircraft velocity along each of the body axes.

[*Ans*: u_B = 99.6 m/s, v_B = −8.46 and w_B = 2.86 m/s.]

4-5.4 Complete the following table.

Vector components in initial frame	Euler angles ψ, θ, ϕ (°)	Vector components in final frame
(10, 13, 9)	(−10, 15, 45)	
(−33, 15, 27)	(60, 30, 45)	
(100, −5, 4)	(−18, 25, 36)	
	(22, −58, 10)	(230, 34, −15)
	(30, −30, 45)	(18, −4, −19)
	(20, 80, −45)	(−120, 45, 10)

4-5.5 An aircraft is in a steady climb such that its body axis OX_b is inclined at 15° to the horizon and its climb angle is 7°. The TAS is 100 m/s and there is no sideslip. Calculate α_e, U_e and W_e in the body axis system.

[*Ans*: 8°, 99.03 m/s, −13.92 m/s.]

4-5.6 An aircraft is equipped with airspeed sensors that indicate that in steady flight (u_B, v_B, w_B) = (95, −5, 5) m/s and (ψ, θ, ϕ) = (3, 4, 5) degrees, where θ is measured relative to the horizontal and ψ is measured relative to due north. Calculate the velocity components of the aircraft (U_E, V_E, W_E) where Ox_E is due north, Oy_E is due east and Oz_E is vertically downwards.

[*Ans*: 95.19, 3.05, and −1.91 m/s.]

4-5.7 An aircraft is performing a coordinated banked turn at a bank angle of 60°. Its turn rate about the Earth's vertical axis is 0.1047 rad/s. Assume that the aircraft axis is in the horizontal plane, and calculate the pitch rate and yaw rate about the aircraft y and z axes.

[*Ans*: 0.0907, and 0.0524 rad/s.]

4-5.8 An aircraft performs a coordinated banked turn at Euler pitch and roll angles of 6° and 50° respectively (the Euler angles are with respect to the Earth axis system). Rate sensors aligned to the aircraft x, y and z axes measure rates of −0.0073, 0.0532 and 0.0446 rad/s respectively. Calculate the rate of turn.

[*Ans*: 0.0698 rad/s.]

4-5.9 Qualitatively plot Euler angles for the following flight profiles (roll and turn with $\theta = 0$):

Loop (vertical plane)

Roll

Turn (horizontal plane).

4-5.10 Consider an aircraft performing a coordinated banked turn at 200 m/s and 40° of bank. Calculate the turn rate $\dot{\psi}$, the yaw rate r and the pitch rate q.

4-5.11 An aircraft performs a loop in 20 seconds at a uniform speed of 100 m/s. Calculate the normal load factor experienced by the pilot 5 m forward and 1 m above the centre of gravity.

4-5.12 Construct the manoeuvre envelope at 20,000 ft altitude on an ISA day for a Bizjet weighing 19,000 lb. Maximum dive (structure-limited) speed = 0.8 M.

[*Ans*: See Figure E.6.]

4-5.13 Construct the manoeuvre envelope on an ISA day for the B100 at MTOM as given at the beginning of this appendix. Maximum dive (structure limited) speed = 0.88 M.

■ **FIGURE E.6**
Answer to Problem 4-5.12

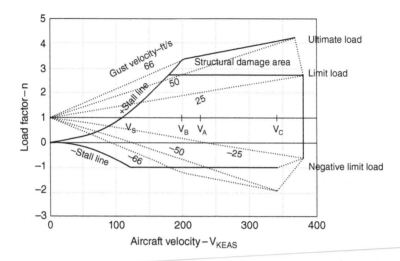

E.3.5   **Chapter 6**

This is a descriptive chapter; hence no problem set is assigned.

E.3.6   **Chapters 7 and 8**

7-8.1   A single turbojet powered test aircraft of mass 61,162 kg cruises at a steady level flight Mach number of 0.8 with lift to drag ratio of 15 at an altitude with atmospheric temperature and pressure of 230 K and 30.7 kPa, respectively. Engine intake mass flow rate is 295 kg/s of air and the exhaust mass flow rate is 300 kg/s. Exhaust nozzle exit area = 0.6 m² and exit pressure is 32 kPa. Calculate the net thrust, gross thrust, exit velocity, V_e, the propulsive efficiency, the fuel-to-airflow ratio, the sfc and specific thrust.

[*Ans:* $F_n = 40,000$ N, $F_g = 111,744$ N, $V_e = 369.8$ m/s, sfc = 0.125 gm/Ns, $\eta_p = 0.794$.]

7-8.2   A turbofan with a BPR = 4 is flying at 238 m/s. The following information is given:

The primary core flow exit velocity, $V_{ep} = 450$ m/s and flow rate $\dot{m}_p = 40$ kg/s.
The bypass flow exit velocity is equal to $V_{es} = 320$ m/s.

Find the thrust generated by the turbofan (ignore fuel flow rate contribution) and the propulsive efficiency, η_{pf}.

[*Ans:* $F / \dot{m}_p$ = 540 N/kg, $F_n = 21,600$ N, $V_{eq} = 346$ m/s, $\eta_{pf} = 0.81$.]

7-8.3   A typical BMW-RR 710 type turbofan having a short-duct nacelle has the following operating conditions

on a standard day. The primary nozzle exit plane of the nacelle has an area $A_e = 0.6$ m². (In a short-duct nacelle the primary flow does not mix with the bypassed secondary flow.)

■ Cruise at Mach 0.75 at 11 km altitude
■ Intake mass flow = 160 kg/s
■ BPR = 4
■ Nozzle exit plane velocity = 500 m/s
■ Nozzle exit plane is perfectly expanded ($p_\infty = p_e$)
■ Fan exit plane velocity = 280 m/s
■ Fan exit plane pressure = p_∞

Ignore intake losses and fuel flow rate. Find net thrust, gross thrust and ram drag.

[*Ans:* $F_{primary} = 8916$ N, $F_{secondary} = 7505$ N, $F_{net} = 16\ 421$ N, $F_{gross} = 51\ 840$ N.]

7-8.4   A mixed flow turbofan with BPR of 5 has the following data at takeoff at sea-level and static conditions. The nozzle is perfectly expanded.

■ Intake mass flow = 180 kg/s
■ Nozzle exit plane velocity = 600 m/s
■ Fan exit plane velocity = 300 m/s
■ Fuel flow rate = 2 kg/s
■ Primary flow air-bleed = 2 kg/s

Find net thrust, gross thrust and ram drag.

If you ignore fuel flow rate, what will be the net thrust? If you ignore air-bleed, what will be the net thrust? If there is 2% thrust loss due to intake recovery, what will be the net thrust?

[Ans: $F_{primary} = 18,000\,\text{N}$, $F_{secondary} = 45,000\,\text{N}$, $F_{net} = 63,000\,\text{N}$, $F_{net\_inst} = 61,740\,\text{N}$.]

7-8.5 Generate the installed engine performance of a medium-sized turbofan with bypass ratio (BPR) of 5 and $T_{SLS} = 17,000\,\text{lb/engine}$. The non-dimensional uninstalled performance of this class of engine is given in Figures 8.5–8.7.

7-8.6 Generate the installed thrust of a medium-sized turboprop engine (sea-level ISA day) with uninstalled SHP (shaft horse power) of 1075, using a four-bladed Hamilton standard propeller having activity factor of 180. The non-dimensional uninstalled performance of this class of engine is given in Figures 8.9 and 8.10.

7-8.7 Generate the installed thrust of a small piston engine (sea-level ISA day) having uninstalled HP of 100 using a three-bladed Hamilton standard propeller with activity factor of 100. The non-dimensional uninstalled performance of this class of engine is given in Figures 8.17 and 8.18.

E.3.7   Chapter 9

The problem set in this chapter involves primarily computational practices. Two types of aircraft have to be evaluated, one a high subsonic turbofan-powered 100 passenger capacity commercial transport aircraft, and one a low subsonic piston engine powered four-place utility aircraft, to encompass the wide range of subsonic aircraft types.

The results are important and should be worked out as these will be required in the rest of the chapters. Results computed by readers may not exactly match the answers given here, as they are deliberately kept approximate. A variation of ±10% is acceptable.

9.1 Estimate the aircraft component drag for the B100 aircraft at the LRC condition. Details of the aircraft are given at the beginning of this appendix.

[Ans: given in the table.]

Component	ΔC_{DPmin}
Fuselage (includes wind shield and undercarriage fairing)	0.0083
Wing (includes interference, control and flap)	0.006
V-tail (includes interference, control and flap)	0.00148

H-tail (includes interference, control and flap)	0.00156
Pylon and nacelle	0.00254
Excrescences	0.0006
Total C_{DPmin}	0.0208
Undercarriage (retractable)	$\Delta C_{D\_UC} = 0.0025$

9.2 Generate the drag polar for the B100 aircraft at the LRC and HSC conditions, given the values below. (Readers are to find suitable ΔC_{DP} and ΔC_{DW} to match the drag polar equation given to plot the graph at Mach 0.76 and Mach 0.82.)

C_L	0.2	0.3	0.4	0.5	0.6
ΔC_{DP}	0.005	0.002	0.0005	0.000	0.0001
$\Delta C_{DW@0.82Mach}$	0.0007	0.0013	0.0022	0.003	0.005

[Ans: The plot should come close to this equation: $C_D = 0.0205 + 0.035 C_L^2 + \Delta C_{DW}$.]

9.3 Estimate the aircraft component drag for the AK4 aircraft at 5000 ft altitude. Details of the aircraft are given at the beginning of this appendix.

[Ans: given in the table.]

Component	Wetted area (ft²)	f (ft²)	ΔC_{DPmin}
Fuselage	278	0.975	0.00697
Wing	220	1.056	0.00754
H-tail	70	0.3192	0.00228
V-tail	40	0.1775	0.00127
Dorsal	12	0.05	0.00353
Spinner	10	0.04	0.000286
Canopy		0.1	0.000714
Subtotal			0.0226
Excrescences (3%)			0.000678
Intake and exhaust			0.0012
Ariel, ECS, trim			0.0008
Total C_{DPmin}			0.024078 ≈ 0.0241
Undercarriage (retractable)			$\Delta C_{DPmin} = 0.0085$

9.4 Generate the drag polar for the AK4 aircraft at the cruise condition. Details of the aircraft are given at the beginning of this appendix.

[Ans: The plot should come close to this equation: $C_D = 0.0241 + 0.043 C_L^2$.]

E.3.8   Chapter 10

10.1   A light propeller-driven aircraft weighing 521.6 kg has a rectangular wing planform area of 9.236 m², AR=6, flying level at 160 kmph on an ISA day at sea-level.

(a)   What is its wing loading? Find the aircraft C_L at cruise. What will be the induced drag if $e=0.78$? Given $C_{DPmin}=0.04$, what will be the aircraft drag and L/D ratio?

(b)   If the aspect ratio is increased to 7 by increasing the span only, with associated changes in $e=0.8$, $C_{DPmin}=0.037$ and aircraft weight increased to 540 kg, what will be the new L/D? All other parameters are unchanged. (Why did these changes take place by changing the aspect ratio? Any comments?)

[*Ans:* (a) Wing loading = 554 N/m², $C_L=0.458$, $L/D=8.435$; and (b) $L/D=9.52$.]

10.2(a)   A turboprop aircraft is cruising at 250 kmph at 1 km altitude ($\rho=1.111$ kg/m³) at a weight of 2000 kg with a wing area of 30 m², aspect ratio of 7 and Oswald's efficiency, $e=0.8$. The wing is rectangular in shape. Given $C_{DPmin}=0.03$, what is the wing span and chord? At what lift and drag coefficients is the aircraft is flying? What is the maximum lift to drag ratio? What are the lift and drag coefficients at the maximum lift to drag ratio? At what speed will it give the maximum lift to drag ratio?

(b)   The aspect ratio is increased to 8 by increasing the span and area, keeping the chord unchanged so as to keep the same wing, adding wing tip only. The following changes take place: $e=0.82$, $C_{D0}=0.027$, and cruise weight increases to 2050 kg. What are the reasons for the changes? Find the new lift to drag ratio and compare it with the original design. Any comments?

[*Ans:* (a) $c=2.07$ m, $b=14.49$ m, $(L/D)_{max}=12.1$; and (b) $(C_L/C_D)_{max}=13.8$.]

10.3   Plot graphs like Figure 10.17a and find the lift to drag ratio for the aircraft given in Problem 10.2.

10.4   Plot the parabolic drag polar graph like Figure 10.17b for the aircraft given in Problem 10.2.

10.5   For the aircraft given in Problem 10.2:

(a)   Derive $C_{DPmin}=3C_{Di}$, and determine when it occurs.

(b)   Derive $C_{DPmin}=1/3C_{Di}$, and determine when it occurs.

[*Ans:* (a) $\left(\dfrac{C_L^3}{C_D^2}\right)_{max}=234.6$; and (b) $\left(\dfrac{C_L^3}{C_D^2}\right)_{max}=419.6$.]

The AK4 small utility aircraft drag polar equation at cruise is given as $C_D=0.0241+0.043C_L^2$. Find the values when the above two cases occur in Problem 10.5 part (a).

10.6   The AK4 small utility aircraft drag polar equation at cruise is given as $C_D=0.0241+0.043C_L^2$. What will be the power required for the AK4 cruising at 180 mph TAS at 5000 ft altitude ($\rho=0.0020482$ slugs) on an ISA day. Take weight at cruise as 2700 lb and propeller efficiency $\eta=0.82$. Then check with the Lycoming IO-390 engine performance graph given in Section E.2 (maximum cruise revolutions per minute rating = 2450 rpm).

[*Ans:* $P=148$ HP, $V_{max}\approx200$ mph at $P_{max}=192$ HP.]

10.7   The AK4 small utility aircraft drag polar equation at cruise is given as $C_D=0.0241+0.043C_L^2$. At this low speed, parabolic and actual drag polar are close enough. Plot L/D graphs like Figures 10.17a,b and 10.20a for two weights of 2800 and 2400 lb.

10.8   For the B100 high subsonic jet transport aircraft, the parabolic drag polar equation at LRC is given as $C_D=0.0205+0.035C_L^2$. Plot the actual drag polar (see Problem set 9) and the parabolic drag polar and compare. Plot the L/D and ML/D graphs for two weights of 98,000 and 95,000 lb. Discuss the graphs.

10.9   Generate the aircraft (D/δ) and engine (T/δ) grid for the B100 high subsonic jet transport aircraft, as done in Figure 10.19 for the Bizjet.

E.3.9   Chapter 11

11.1   Show that the simplified expression for all-engine takeoff field length (TOFL) can be approximated as:

$$\text{TOFL}=\frac{V_{V2}^2}{2\bar{a}}=\frac{WV_{LO}^2}{2g\left[(T-\mu W)-(C_D-\mu C_L)Sq\right]_{0.707V2}}$$

where $\bar{a}=g\left[(\bar{T}/W-\mu)-(\bar{D}/W-\mu\bar{L}/W)\right]_{0.707V2}$ (note the simplification is extended to V_2).

$$\bar{a}=\left(\frac{g}{W}\right)\left[(-\mu_B W)-(\bar{C}_D-\mu_B\bar{C}_L)Sq\right]$$
$$=g\left[(-\mu_B)-(\bar{D}/W-\mu_B\bar{L}/W)\right]$$

The equation for average acceleration is

$$\bar{a}=g\left[(-\mu_B)-(qC_L S_W/W)(\bar{D}/\bar{L}-\mu)\right]_{0.707V}$$

At $0.707\,V_{LO}$, further simplify the above relation by taking

$\bar{T} = 0.85T_{SLS}$ for civil turbofans;

$\bar{T} = 0.9T_{SLS}$ for military low BPR turbofans;

$\bar{T} = 2.6\,\text{HP}$ (or SHP) for propeller engines; and

$C_L = 0.8–1.0$ depending on the high-lift device used.

(Both the military and civil aircraft may use the same equation for a rapid TOFL estimation to give a feel, but it is not as accurate as the long-handed estimations in incremental steps as shown in Chapter 11.)

11.2   The Bizjet details are given in the book.

Compute the TOFL using the simplified method for the Bizjet, and then compare with the worked example evaluated by the long-hand method in the book. This gives the validity of the approximation, which is a good way to get a feel for the TOFL capability.

[*Ans*: TOFL = 3073 ft.]

11.3   The AK4 has the following specifications (see the details at the front): MTOW = 2800 lb, wing area = 140 ft², $C_{Lmax\_TO} = 2.0$, $C_{L@0.707V2} = 1.0$, $\mu = 0.025$. At takeoff, take the drag polar to include undercarriage and flap drag as $C_D = 0.0266 + 0.042C_L{}^2$.

Compute the TOFL using the simplified method and then do the long-hand process in steps, and compare.

[*Ans*: TOFL = 1400 ft.]

11.4   The effect of runway gradient can be incorporated into the takeoff approximation. If ϕ is the gradient, then the component of gravitational force affecting field length performance is $W\sin\varphi = W\varphi$ (radians).

Then the average acceleration $\bar{a} = g[(\bar{T}/W - \mu) - (\bar{D}/W - \mu\bar{L}/W \pm W\varphi)]_{0.707V2}$, where up gradient is negative and *vice versa*.

With 2° uphill gradient, what will be the TOFL for the AK4? Also, find the TOFL with a 5° uphill runway gradient. Compare the results.

[*Ans*: At 2° uphill, TOFL = 1405 ft, and at 5° uphill, TOFL = $1.15 \times 2798 = 3218$ ft.]

11.5(a)   Estimate the all-engine TOFL for the B100 high subsonic jet transport aircraft using the simplified method as given in Problem 11.1.

(b)   Estimate the one-engine inoperative BFL for the B100 high subsonic jet transport aircraft using the simplified method as given in Problem 11.1.

- B100 drag polar equation $C_D = 0.0204 + 0.04C_L{}^2$
- One engine inoperative $\Delta C_D = 0.0023$

- Control and asymmetry $\Delta C_D = 0.0015$
- Undercarriage $\Delta C_{D\_UC} = 0.025$
- Flap drag $\Delta C_{D\_flap} = 0.013$ (takeoff)

[*Ans*: (a) Around 4500 ft; and (b) BFL value given in the next problem.]

11.6(a)   Estimate the all-engine TOFL for the B100 high subsonic jet transport aircraft using the long step-by-step method as given in Chapter 11.

(b)   Estimate the one-engine inoperative BFL for the B100 high subsonic jet transport aircraft using the long step-by-step method as given in Chapter 11.

[*Ans*: BFL should be around 5200 ft.]

11.7   Estimate the landing distance of the AK4 at 0.95 MTOW (see the details earlier): wing area = 140 ft², $C_{Lmax\_land} = 2.6$, $C_{L@0.707Vapp} = 1.0$, $\mu = 0.025$, $\mu_B = 0.4$, flap drag $\Delta C_{D\_flap} = 0.04$ (land). At landing, drag polar to include undercarriage and flap drag $C_D = 0.0266 + 0.042C_L{}^2$.

Like the takeoff simplification in Problem 11.1, take the simplified expression for landing field length (LFL) as (this is further simplification of Equations 11.38 and 11.39):

$$S_{TD} = V_{ave} \times \left(V_{appr} - 0\right)/\bar{a}$$

where $\bar{a} = g\left[(-\mu_B) - \left(\bar{D}/W - \mu_B \bar{L}/W\right)\right]_{0.707V_{TD}}$

$$= g\left[(-\mu_B) - \left(qC_L S_w/W\right)\left(\bar{D}/L - \mu_B\right)\right]_{0.707V_{TD}}$$

[*Ans*: LFL = 1300 ft.]

11.8   Estimate the landing distance of the Bizjet at a landing weight of 15,800 lb (see Section 11.11.2): wing area = 323 ft², $C_{Lmax\_Land} = 2.2$, $C_{L@0.707Vapp} = 1.0$, $\mu = 0.025$, $\mu_B = 0.4$. At takeoff, take the drag polar to include undercarriage and flap drag, $C_D = 0.0205 + 0.0447C_L{}^2$.

Like the takeoff simplification in Problem 11.1, take the simplified expression for LFL as (this is further simplification of Equations 11.38 and 11.39):

$$S_{TD} = V_{ave} \times \left(V_{appr} - 0\right)/\bar{a}$$

where $\bar{a} = g\left[(-\mu_B) - \left(\bar{D}/W - \mu_B \bar{L}/W\right)\right]_{0.707V_{TD}}$

$$= g\left[(-\mu_B) - \left(qC_L S_w/W\right)\left(\bar{D}/L - \mu_B\right)\right]_{0.707V_{TD}}$$

[*Ans*: LFL ≈ 3450 ft.]

11.9   Estimate the landing distance of the B100 at a landing weight of 80,000 lb. Wing area = 950 ft², $C_{Lmax\_land} = 3.25$, $C_{L@0.707Vapp} = 1.0$, $\mu = 0.025$, $\mu_B = 0.4$. The B100 drag polar equation $C_D=0.0204+0.04C_L^2$. Undercarriage $\Delta C_{D_{UC}}=0.025$, flap drag $\Delta C_{D\_flap}=0.06$ (land).

[*Ans*: LFL should be = 4400 ft.]

E.3.10   Chapter 12

12.1   Plot the thrust required and thrust available graph for the AK4 at 2800 lb having drag polar ($C_D = 0.0241 + 0.043C_L^2$) at sea-level, 2000 ft and 4000 ft on an ISA day.

12.2   For the AK4 piston engine aircraft, compute the following at sea-level ISA day conditions. Use a Lycoming IO-390 engine operating at 2700 rpm (the graph from Problem 12.1 is required.)

(a)   Estimate the following:

(i) maximum rate of climb, $(R/C_{unaccl})_{max}$, (ii) velocity at maximum rate of climb $V_{(R/C)max}$, (iii) $(L/D)_{max}$ at maximum rate of climb, (iv) climb gradient at maximum rate of climb $(\sin\gamma)_{max\_R/C}$, (v) maximum climb gradient $(\sin\gamma)_{steepest\_grad}$, (vi) speed at maximum climb gradient $(V_{steepest\_grad})$ and (vii) rate of climb at maximum gradient $(R/C)_{steepest\_grad}$.

(b)   Plot the rate of climb and angle of climb for velocities from 60 to 190 mph and show the stall velocity.

[*Ans*: (i) $(R/C_{unaccl})_{max}=1121$ ft/m, (ii) $(V_{(R/C)max})=124$ mph, (iii) $(L/D)_{(R/C)max}=14.76$, (iv) $\gamma_{max\_R/C}=6°$, (v) $V_{steepest\_grad}=102$ mph, (vi) $\gamma_{max}=6.6°$, and (vii) $(R/C)_{steepest\_grad}=1121$ ft/m.]

12.3   Make a hodograph plot for the AK4 on an ISA day and 5000 ft altitude with a Lycoming piston engine set at 2450 rpm.

12.4   Construct the ISA day integrated climb performance for the AK4 showing (i) distance covered, (ii) fuel consumed, and (iii) time taken up to 14,000 ft altitude, as done in Figure 12.9 for the Bizjet.

12.5   For the B100 (drag polar equation is $C_D=0.0204+0.04C_L^2$), make the aircraft and engine grid at the maximum climb rating as in Figure 12.2 for the Bizjet aircraft. Make a hodograph plot for the B100 on an ISA day at 36,000 ft altitude.

12.6   Construct the ISA day integrated climb performance for the B100 (drag polar equation is $C_D=0.0204+0.04C_L^2$) showing (i) distance covered, (ii) fuel consumed, and (iii) time taken up to 40,000 ft altitude, as done in Figure 12.9 for the Bizjet.

[*Ans*: Climb 36,000 ft at design MTOW: (i) distance ≈ 120 nm, (ii) fuel ≈ 420 lb, (iii) time ≈ 18 minutes.]

12.7   A retractable engine motor-glider is in a glide path at 2000 ft altitude (ISA day and still air) in retracted engine configuration. The glider weight is 1200 lb and wing area is 150 ft².

Drag polar equation with engine retracted:
$C_D=0.012+0.022C_L^2$

Drag polar equation with engine extended and running:
$C_D=0.015+0.024C_L^2$.

At this low altitude, take the atmospheric properties at the mean altitudes to keep the computation simple ($\rho_{1000ft}=0.0023$ slugs/ft³ and $\rho_{500ft}=0.00234$ slugs/ft³.)

(i)   How far can the glider go with engines retracted, and at what speed and gradient?

(ii)   At 1000 ft, realizing that he has undershot, the pilot extended and powered up the engine to gain another 500 ft distance to be on the safe side. How much power would be needed to maintain the same speed as in the engine retracted case? Take propeller efficiency, $\eta_{prop}=0.7$.

(iii)   In case the pilot decides to go around at full power on account of baulked landing and climbing at 4% gradient at $1.2V_{stall}$ ($C_{Lstall}=1.5$, no flaps), what will be the engine rating? Take propeller efficiency, $\eta_{prop}=0.68$.

(iv)   Find the aspect ratio, root chord and tip chord if the glider has a tapered wing with tip/chord ratio of 0.4 and Oswald's efficiency, $e=0.95$ when the engine is retracted.

[*Ans*: (i) $R_{glide}=61,540$ ft, (ii) HP = 1.25, (iii) HP = 28.4 ≈ 30 (a matched and sized engine will have higher HP), and (iv) AR = 15.23, $C_R=4.48$ ft and $C_T=1.8$ ft.]

12.8   A single engine fighter aircraft is at weight $W=20,000$ lb with wing area $S_w=300$ ft². Find the specific excess power (SEP) and energy height, h_e, on an ISA day for the following two situations.

(i)   *Subsonic*: At 36,000 ft altitude ($\rho_{36,000ft}=0.0007$ slugs/ft³) at Mach 0.9 (drag polar, $C_D = 0.02+0.15C_L^2$). Thrust = 10,000 lb dry.

(ii)   *Supersonic*: At 45,000 ft altitude ($\rho_{45,000ft}=0.00046$ slugs/ft³) at Mach 1.8 (drag polar, $C_D=0.045+0.3C_L^2$). Thrust = 10,500 lb with afterburning.

[*Ans*: (i) $Ps=333.6$ ft/s, $h_e=47,787$ ft, (ii) $Ps=44.2$ ft/s, $h_e=92,148$ ft.]

E.3.11   Chapter 13

13.1(i)   Show that the equation for specific range, SR, and range factor, RF, can be expressed as given below.

$$SR = \frac{550\times\eta_p}{sfc_{hp}\times W}\left(\frac{L}{D}\right)$$ in Imperial units, and sfc is expressed in lb/HP/h.

$$RF = L\times SR$$

(ii) Find the RF and the SR of the AK4 piston engine aircraft weighing 2700 lb flying at 10,000 ft altitude ($\rho = 0.00175$ slugs/ft³) and 140 mph EAS (equivalent air speed) on an ISA day. Take propeller efficiency, $\eta_{prop} = 0.85$.

(iii) What the maximum speed an AK4 can achieve at 2450 rpm at 10,000 ft altitude?

(iv) Find the best SR by plotting the graph for the AK4 flying at 10,000 ft altitude.

[*Ans*: (ii) RF = 13,813 miles.]

13.2 What will be the endurance of the AK4 flying at 5000 ft altitude and 2750 lb weight?

13.3 Plot a SR graph as in Figure 13.2 for the B100 high subsonic jet transport aircraft flying at 36,000 ft altitude at 85,000 and 80,000 lb weights. Show the maximum range line, and lines of LRC and HSC.

[*Ans*: should be around ≈ 0.128 nm/lb at around 420 KTAS at 85,000 lb.]

13.4 Plot the generalized performance graph as in Figure 13.3 for the B100 high subsonic jet transport aircraft flying at 36,000 ft altitude, showing lines of constant RF.

13.5 What will be the endurance of the B100 flying at 36,000 ft altitude, and the initial weight at that altitude?

13.6 What is the maximum speed of the B100 at the initial cruise weight at 36,000 ft altitude?

13.7 What is the maximum speed of a 25,000 lb fighter aircraft flying at 45,000 ft altitude with the following technical information? Wing area = 300 ft², subsonic compressible drag polar, $C_D = 0.02 + 0.15C_L^2$, supersonic drag polar, $C_D = 0.045 + 0.3C_L^2$. Thrust values at 45,000 ft altitude are as follows (may interpolate):

Mach	0.8	1.0	1.2	1.4	1.6	1.8	2.0	2.2	2.4
Thrust (lb)	5500	6200	8000	9500	10,000	10,800	11,000	10,800	9800

E.3.12 Chapter 14

14.1(i) At what condition will the AK4 give maximum range? Compare the expression with the condition for maximum range for jet aircraft.

(ii) Find the maximum distance covered and time taken by the AK4 for the cruise segment at 10,000 ft (ISA day) having initial cruise weight, $W_1 = 2700$ lb and final cruise weight $W_2 = 2500$ lb, consuming 200 lbs of fuel. What is the cruise speed?

[*Ans*: (ii) $R_{prop} = 952$ miles, $t_{cruise} = 8.28$ hours.]

14.2(i) Show that the endurance equation for a piston engine aircraft as given in Equation 13.27 can be expressed in Imperial units as follows.

$$E = \left(\frac{550 \times \eta_{prop}}{sfc_{hp}}\right)\sqrt{\rho S_W \left(\frac{C_L^3}{C_D^2}\right)} \left(\frac{2}{\sqrt{W_2}} - \frac{2}{\sqrt{W_1}}\right)$$

$$= \left(\frac{37.92 \times \eta_{prop}}{sfc_{hp}}\right)\left(\frac{C_L^{3/2}}{C_D}\right)\sqrt{\frac{\sigma \times S_W}{W_1}}\left[\sqrt{\frac{W_1}{W_2}} - 1\right]$$

(ii) Compute the endurance E and distance covered by the AK4 for the cruise segment at 10,000 ft ($\rho = 0.00175$ slugs/ft³) and 120 mph consuming 200 lbs (initial cruise weight = 2700 lb and final cruise weight = 2500 lb). Take propeller efficiency, $\eta_{prop} = 0.85$.

(iii) What will be the maximum endurance and distance covered by the AK4 for the cruise segment at 10,000 ft (initial cruise weight = 2700 lb and final cruise weight = 2500 lb). Take propeller efficiency, $\eta_{prop} = 0.85$. What is the aircraft speed?

(iv) Compare results for Problems 14.1 and 14.2.

[*Ans*: (ii) $E = 6.32$ hours, $R_{max\_E} = 758$ miles, and (iii) $E = 9.63$ hours, $R_{max\_E} = 835$ miles.]

14.3 Find the range of the B100 for the cruise segment at 36,000 ft (ISA day) flying at LRC of Mach 0.77 (climb and descent performances are estimated in the Chapter 12 problem set). Construct the payload-range graph as in Figure 14.2 for the Bizjet.

What will be the range for cruising at the HSC Mach 0.82?

[*Ans*: should be ≈ 1200 nm.]

14.4(i) Compute the endurance E of the B100 flying at 36,000 ft (ISA day).

(ii) Find the range cruising at constant altitude and constant L/D ratio as shown in Case 2 of Section 14.4.1.

(iii) Find the cruise range flying at constant velocity and constant L/D ratio (velocity of LRC in cruise-climb) as shown in Case 4 in Section 14.4.1.

(iv) Compare the cruise ranges obtained in Problems 14.3 and 14.4.

14.5 Open-ended assignment – readers may adjust the problem as required.

What is the maximum speed of a fighter aircraft at MTOW of 27,000 lb flying at 45,000 ft altitude with the following technical information? Wing area = 300 ft², subsonic com-pressible drag polar, $C_D = 0.02 + 0.15C_L^2$, supersonic drag polar, $C_D = 0.045 + 0.3C_L^2$. Thrust values at 45,000 ft altitude are as follows (may interpolate):

Mach	0.8	1.0	1.2	1.4	1.6	1.8	2.0	2.2	2.4
Thrust (lb)	5500	6200	8000	9500	10,000	10,800	11,000	10,800	9800

Takeoff thrust at static condition with after-burner = 29,000 lb and dry thrust = 17,500 lb.

Find the total fuel carried and time taken for the following mission profile. Keep diversion fuel for 200 nm and resid-ual tank fuel = 200 lb.

Template for fighter aircraft mission profile – detailed segments

	Fuel burnt (kg)	Time (min)	Engine rating = % rpm
Taxi			60% (idle)
Takeoff			With full afterburner
Climb to 45,000 ft altitude at 50,000 ft/min			At 95%
Cruise at Mach 1.4 for 400 miles			
Dash at Mach 2 for 2 minutes to engage (fire missiles)			With afterburner
Turn and high-speed return at Mach 2 for 5 minutes			With afterburner
Cruise at Mach 1.4 for 100 miles			
Slow down to Mach 0.8 for the rest			
Descent			
Approach, land, return taxi			

E.3.13   Chapter 15

15.1   Find the maximum load factor, n_{max}, in a sus-tained turn in the horizontal plane of the AK4 at 2600 lb weight flying at 5000 ft altitude ($\rho = 0.00204$ slugs/ft³) with engine set at 2450 rpm. What is the velocity, turn rate and bank angle φ at n_{max}? Also, determine the time taken for 180° turn.

[*Ans*: $V_{n\_max} = 144.7$ mph, $n_{max} = 1.84$ g, $\varphi_{max} = 57.3°$, $time_\pi = 13.33$ seconds.]

15.2   The AK4 has a limit of $n_{max} = 4$ and $C_{Lstall} = C_{Lmax} = 1.4$. Find the minimum radius of turn, R_{min}, at 5000 ft altitude ($\rho = 0.00204$ slugs/ft³) and 2600 lb weight. What is the velocity, turn rate and bank angle φ at $n_{max} = 4.0$? Check whether there is enough thrust available to reach the minimum radius of turn, R_{min}.

[*Ans*: $R_{min} = 417$ ft, turn rate = $31.34°$/s, $\varphi = 75.5°$.]

15.3   Find the pull-up rate and radius of turn, R, at sustained 2 g and 3 g manoeuvres in the vertical plane for the AK4 (2600 lb) at 5000 ft altitude ($\rho = 0.00204$ slugs/ft³), engine set at 2450 rpm in each case, with aircraft entry speeds of 180 and 200 mph. Compare and comment on the two cases with particular reference to a loop manoeuvre (Section 15.5.6). The AK4 has $C_{Lmax} = 1.4$.

[*Ans*: at $n = 3$, $d\theta/dt = 6.5°$/s and at $n = 4$, $d\theta/dt = 14.98°$/s.]

15.4   A commercial transport aircraft is restricted in g-load. The B100 aircraft flying at 5000 ft altitude ($\rho = 0.00204$ slugs/ft³) and Mach 0.4 is in 1.5 g, and at 1.2 g in a sustained turn in the horizontal plane.

(i)   What will be the aircraft velocity, turn rate, turn radius, and bank angle? Take it that the B100 is at 80,000 lb, wing area, $S_w = 1000$ ft² and the drag polar equa-tion $C_D = 0.0204 + 0.04C_L^2$. Also find the required applied total thrust.

(ii)   What will be the maximum turn rate, and the associated turn radius and bank angle, with power setting at maximum cruise thrust?

[*Ans*: At $n = 1.5$, $R = 4784$ ft, $\dot\psi = 5.56°$/s, $\varphi = 48.2°$, $T_{total} = 6834.8$ lb, $time_\pi = 32.4$ seconds. At $n = 1.2$, $R = 9015$ ft, $\dot\psi = 2.8°$/s, $\varphi = 33.6°$, $T_{total} = 5816.5$ lb, $time_\pi = 64.3$ seconds.]

15.5   Open-ended assignment – readers may adjust the problem as required.

Find the turn rate in the horizontal plane of a fighter air-craft at 45,000 ft altitude ($\rho = 0.00046$ slugs/ft³ and speed of sound = 968.08 ft/s) at Mach 0.8 and Mach 1.8 at the n_{max} of 9 g. What is the radius of turn, R, bank angle ϕ, and thrust required for the sustained turn rate?

What is the maximum speed of a fighter aircraft at MTOW of 27,000 lb flying at 45,000 ft altitude with the following technical information? Wing area = 300 ft², $C_{Lmax} = 1.3$ (clean) and 1.8 (high-lift extended), sub- sonic compressible drag polar, $C_D = 0.02 + 0.015C_L^2$, supersonic drag polar, $C_D = 0.045 + 0.03C_L^2$. Thrust values at 45,000 ft altitude are as follows (may interpolate):

Mach	0.8	1.0	1.2	1.4	1.6	1.8	2.0	2.2	2.4
Thrust (lb)	5500	6200	8000	9500	10,000	10.800	11.000	10.800	9800

E.3.14 Chapters 16–17

16-17.1 Open-ended problem – no answer given. The readers may adjust/add data as required.

For a marketing study, perform a sensitivity study for a two-seat club-trainer type aircraft, designated the AK2, to suggest a suitable fixed undercarriage configuration that fits the market requirement around the following specification. Make a carpet plot like Figure 16.7 with variable parameters of 90, 100 and 110 ft² wing area, S_w, weight variations of 1200, 1300 and 1400 lb, and takeoff field length of 700, 800 and 900 ft.

Specification: payload of 2 persons, range = 500 nm, take-off distance (BFL) = 800 ft, landing distance = 800 ft, cruise altitude = 10,000 ft, maximum cruise speed = 150 mph, maximum dive speed (structural limit) = 200 mph.

From the carpet plot, select a suitable configuration and size it with a matched engine.

16-17.2 How much would a club like to charge for hourly rates for flying the AK2? This type of aircraft does not have flight and cabin crew, and has no fare-paying payload. Therefore, a total operating cost per hour needs to be computed. In the absence of any standard ground rules like ATA/AEA, the readers may use a similar logic to arrive at a total operating cost per hour. Take 500 hours of utilization per annum and a labour rate of US$20 per hour (a total of 100 hours of maintenance per annum) and AVGAS price of $4 per US gallon. Aircraft price is $60,000 (5% depreciation per year up to 10 years, then flattens out to a residual value of $25,000 and interest rate of 10%) which includes a piston engine price of $10,000. Increase operating costs by 100% to cover overheads, taxes and club profit.

[*Ans*: US$100 per hour.]

16-17.3 Perform an aircraft sizing and engine matching exercise to arrive at a B100 type configuration, which may not be exactly the same as given above (which emerged taking into consideration a family of aircraft not given in this problem set).

16-17.4 Compute the direct operating cost (DOC) of the B100, as shown in Chapter 16 for the Bizjet (today's DOC will be roughly doubled). Show in a pie-chart each component percentage of the DOC.

APPENDIX F

Aerofoil Data

The data and figures in this Appendix are provided courtesy of I.R. Abbott and A.E. Von Doenhoff, *Theory of Wing Sections*.

Throughout, stations and ordinates are given in percentage of aerofoil chord.

NACA 4412

Upper surface		Lower surface	
Station	Ordinate	Station	Ordinate
0	0	0	0
1.25	2.44	1.25	−1.43
2.5	3.39	2.5	−1.95
5.0	4.73	5.0	−2.49
7.5	5.76	7.5	−2.74
10	6.59	10	−2.86
15	7.89	15	−2.88
20	8.80	20	−2.74
25	9.41	25	−2.50
30	9.76	30	−2.26
40	9.80	40	−1.80
50	9.19	50	−1.40
60	8.14	60	−1.00
70	6.69	70	−0.65
80	4.89	80	−0.39
90	2.71	90	−0.22
95	1.47	95	−0.16
100	(0.13)	100	(−0.13)
100	−	100	0

LE radius: 1.58.

Slope of radius through LE: 0.20.

Theory and Practice of Aircraft Performance, First Edition. Ajoy Kumar Kundu, Mark A. Price and David Riordan.
© 2016 John Wiley & Sons, Ltd. Published 2016 by John Wiley & Sons, Ltd.

NACA 4415

Upper surface		Lower surface	
Station	Ordinate	Station	Ordinate
0	–	0	0
1.25	3.07	1.25	−1.79
2.5	4.17	2.5	−2.48
5.0	5.74	5.0	−3.27
7.5	6.91	7.5	−3.71
10	7.84	10	−3.98
15	9.27	15	−4.18
20	10.25	20	−4.15
25	10.92	25	−3.98
30	11.25	30	−3.75
40	11.25	40	−3.25
50	10.53	50	−2.72
60	9.30	60	−2.14
70	7.63	70	−1.55
80	5.55	80	1.03
90	3.08	90	−0.57
95	1.67	95	−0.36
100	(0.16)	100	(−0.16)
100	–	100	0

LE radius: 2.48.
Slope of radius through LE: 0.20.

NACA 23012

Upper surface		Lower surface	
Station	Ordinate	Station	Ordinate
0	–	0	0
2.15	2.67	1.25	−1.23
2.5	3.61	2.5	−1.71
5.0	4.91	5.0	−2.26
7.5	5.80	7.5	−2.61
10	6.43	10	−2.92
15	7.19	15	−3.50
20	7.50	20	−3.97
25	7.60	25	−4.28
30	7.55	30	−4.46
40	7.41	40	−4.48
50	6.41	50	−4.17
60	5.47	60	−3.67
70	4.36	70	−3.00
80	3.08	80	−2.16
90	1.68	90	−1.23
95	0.92	95	−0.70
100	(0.13)	100	(−0.13)
100	–	100	0

LE radius: 1.58.
Slope of radius through LE: 0.305.

NACA 65-410

	Upper surface		Lower surface	
Station	Ordinate		Station	Ordinate
0	0		0	0
0.372	0.861		0.628	−0.661
0.607	1.061		0.893	−0.784
1.089	1.372		1.411	−0.944
2.318	1.935		2.682	−1.191
4.797	2.800		5.203	−1.536
7.289	3.487		7.711	−1.791
9.788	4.067		10.212	−1.999
14.798	5.006		15.202	−2.314
19.817	5.731		20.183	−2.547
24.843	6.290		25.157	−2.710
29.872	6.702		30.128	−2.814
34.903	6.983		35.097	−2.863
39.936	7.138		40.064	−2.854
44.968	7.153		45.032	−2.773
50.000	7.018		50.000	−2.606
55.029	6.720		54.971	−2.340
60.053	6.288		59.947	−2.004
65.073	5.741		64.927	−1.621
70.085	5.099		69.915	−1.211
75.090	4.372		74.910	−0.792
80.088	3.577		79.912	−0.393
85.076	2.729		84.924	−0.037
90.057	1.842		89.943	0.226
95.029	0.937		94.971	0.327
100.00	0		100.00	0

LE radius: 0.687.

Slope of radius through LE: 0.168.

NACA 64$_2$-415

	Upper surface		Lower surface	
Station	Ordinate		Station	Ordinate
0	0		0	0
0.299	1.291		0.701	−1.091
0.526	1.579		0.974	−1.299
0.996	2.038		1.504	−1.610
2.207	2.883		2.793	−2.139
4.673	4.121		5.327	−2.857
7.162	5.075		7.838	−3.379
9.662	5.864		10.338	−3.796
14.681	7.122		15.319	−4.430
19.714	8.066		20.286	−4.882
24.756	8.771		25.244	−5.191
29.803	9.260		30.197	−5.372
34.853	9.541		35.147	−5.421

(Continued)

Upper surface		Lower surface	
Station	**Ordinate**	**Station**	**Ordinate**
39.904	9.614	40.096	−5.330
44.954	9.414	45.046	−5.034
50.000	9.016	50.000	−4.604
55.040	8.456	54.960	−4.076
60.072	7.762	59.928	−3.478
65.096	6.954	64.904	−2.834
70.11	6.055	69.889	−2.167
75.115	5.084	74.885	−1.504
80.109	4.062	79.062	−0.878
85.092	3.020	84.908	−0.328
90.066	1.982	89.934	0.086
95.032	0.976	94.968	0.288
100.00	0	100.00	0

LE radius: 1.590.
Slope of radius through LE: 0.168.

NACE 64-210

Upper surface		Lower surface	
Station	**Ordinate**	**Station**	**Ordinate**
0	0	0	0
0.431	0.867	0.563	−0.767
0.676	1.056	0.827	−0.916
1.163	1.354	1.337	−1.140
2.401	1.881	2.599	−1.512
4.890	2.656	5.110	−2.024
7.387	3.248	7.613	−2.400
9.887	3.736	10.113	−2.702
14.894	4.514	15.106	−3.168
19.905	5.097	20.095	−3.505
24.919	5.533	25.081	−3.743
29.934	5.836	30.066	−3.892
34.951	6.010	35.049	−3.950
39.968	6.059	40.032	−3.917
44.985	5.938	45.015	−3.748
50.000	5.689	50.000	−3.483
55.014	5.333	54.987	−3.143
60.025	4.891	59.975	−2.749
65.033	4.375	64.967	−2.315
70.038	3.799	69.963	−1.855
75.040	3.171	74.96	−1.386
80.038	2.518	79.962	−0.926
85.033	1.849	84.968	−0.503
90.024	1.188	89.977	−0.154
95.012	0.564	94.988	0.068
100.00	0	100.00	0

LE radius: 0.720.
Slope of radius through LE: 0.084.

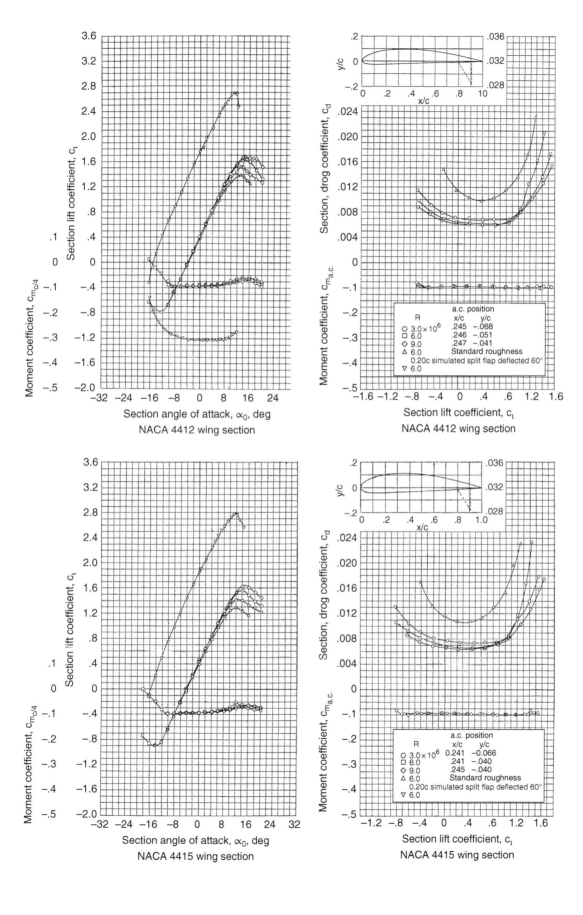

NACA 4412 wing section

NACA 4415 wing section

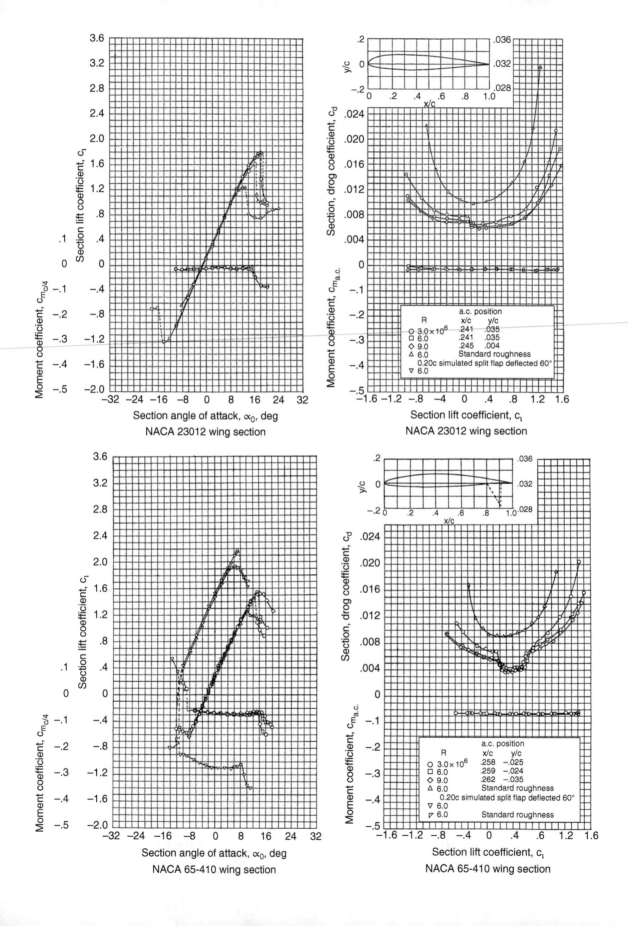

NACA 23012 wing section

NACA 23012 wing section

NACA 65-410 wing section

NACA 65-410 wing section

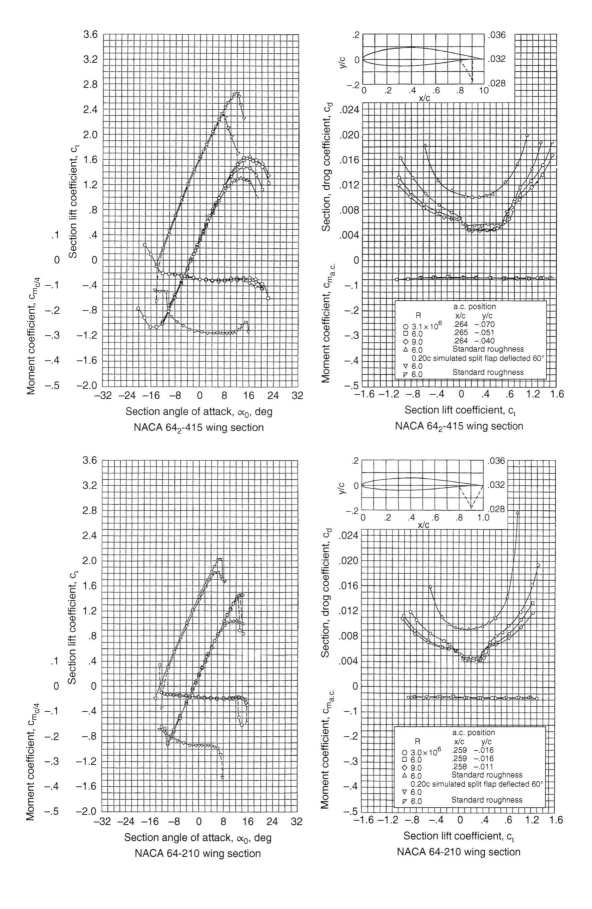

Section lift coefficient, c_l

Moment coefficient, $c_{m_{c/4}}$

Section angle of attack, α_0, deg

NACA 64_2-415 wing section

Section, drag coefficient, c_d

Moment coefficient, $c_{m_{a.c.}}$

Section lift coefficient, c_l

NACA 64_2-415 wing section

		a.c. position	
	R	x/c	y/c
○	3.1×10^6	.264	−.070
□	6.0	.265	−.051
◇	9.0	.264	−.040
△	6.0	Standard roughness	
	0.20c simulated split flap deflected 60°		
▽	6.0		
▽	6.0	Standard roughness	

Section lift coefficient, c_l

Moment coefficient, $c_{m_{c/4}}$

Section angle of attack, α_0, deg

NACA 64-210 wing section

Section, drag coefficient, c_d

Moment coefficient, $c_{m_{a.c.}}$

Section lift coefficient, c_l

NACA 64-210 wing section

		a.c. position	
	R	x/c	y/c
○	3.0×10^6	.259	−.016
□	6.0	.259	−.016
◇	9.0	.258	−.011
△	6.0	Standard roughness	
	0.20c simulated split flap deflected 60°		
▽	6.0		
▽	6.0	Standard roughness	

INDEX